Conformal Blocks, Generalized Theta Functions and the Verlinde Formula

In 1988, E. Verlinde gave a remarkable conjectural formula for the dimension of conformal blocks over a smooth curve arising from representations of affine Lie algebras. Verlinde's formula arose from physical considerations, but it attracted further attention from mathematicians when it was realized that the space of conformal blocks admits an interpretation as the space of generalized theta functions. A proof followed through the work of many mathematicians in the 1990s.

This book gives an authoritative treatment of all aspects of this theory. It presents a complete proof of the Verlinde formula and full details of the connection with generalized theta functions, including the construction of the relevant moduli spaces and stacks of G-bundles. Featuring numerous exercises of varying difficulty, guides to the wider literature and short appendices on essential concepts, it will be of interest to senior graduate students and researchers in geometry, representation theory and theoretical physics.

SHRAWAN KUMAR is John R. and Louise S. Parker Distinguished Professor of Mathematics at the University of North Carolina, Chapel Hill. He was an invited speaker at the 2010 International Congress of Mathematicians and was elected a Fellow of the American Mathematical Society in 2012. This is his third book.

NEW MATHEMATICAL MONOGRAPHS

All the titles listed below can be obtained from good booksellers or from Cambridge University Press. For a complete series listing visit www.cambridge.org/mathematics.

1. M. Cabanes and M. Enguehard *Representation Theory of Finite Reductive Groups*
2. J. B. Garnett and D. E. Marshall *Harmonic Measure*
3. P. Cohn *Free Ideal Rings and Localization in General Rings*
4. E. Bombieri and W. Gubler *Heights in Diophantine Geometry*
5. Y. J. Ionin and M. S. Shrikhande *Combinatorics of Symmetric Designs*
6. S. Berhanu, P. D. Cordaro and J. Hounie *An Introduction to Involutive Structures*
7. A. Shlapentokh *Hilbert's Tenth Problem*
8. G. Michler *Theory of Finite Simple Groups I*
9. A. Baker and G. Wüstholz *Logarithmic Forms and Diophantine Geometry*
10. P. Kronheimer and T. Mrowka *Monopoles and Three-Manifolds*
11. B. Bekka, P. de la Harpe and A. Valette *Kazhdan's Property (T)*
12. J. Neisendorfer *Algebraic Methods in Unstable Homotopy Theory*
13. M. Grandis *Directed Algebraic Topology*
14. G. Michler *Theory of Finite Simple Groups II*
15. R. Schertz *Complex Multiplication*
16. S. Bloch *Lectures on Algebraic Cycles (2nd Edition)*
17. B. Conrad, O. Gabber and G. Prasad *Pseudo-reductive Groups*
18. T. Downarowicz *Entropy in Dynamical Systems*
19. C. Simpson *Homotopy Theory of Higher Categories*
20. E. Fricain and J. Mashreghi *The Theory of H(b) Spaces I*
21. E. Fricain and J. Mashreghi *The Theory of H(b) Spaces II*
22. J. Goubault-Larrecq *Non-Hausdorff Topology and Domain Theory*
23. J. Śniatycki *Differential Geometry of Singular Spaces and Reduction of Symmetry*
24. E. Riehl *Categorical Homotopy Theory*
25. B. A. Munson and I. Volić *Cubical Homotopy Theory*
26. B. Conrad, O. Gabber and G. Prasad *Pseudo-reductive Groups (2nd Edition)*
27. J. Heinonen, P. Koskela, N. Shanmugalingam and J. T. Tyson *Sobolev Spaces on Metric Measure Spaces*
28. Y.-G. Oh *Symplectic Topology and Floer Homology I*
29. Y.-G. Oh *Symplectic Topology and Floer Homology II*
30. A. Bobrowski *Convergence of One-Parameter Operator Semigroups*
31. K. Costello and O. Gwilliam *Factorization Algebras in Quantum Field Theory I*
32. J.-H. Evertse and K. Győry *Discriminant Equations in Diophantine Number Theory*
33. G. Friedman *Singular Intersection Homology*
34. S. Schwede *Global Homotopy Theory*
35. M. Dickmann, N. Schwartz and M. Tressl *Spectral Spaces*
36. A. Baernstein II *Symmetrization in Analysis*
37. A. Defant, D. García, M. Maestre and P. Sevilla-Peris *Dirichlet Series and Holomorphic Functions in High Dimensions*
38. N. Th. Varopoulos *Potential Theory and Geometry on Lie Groups*
39. D. Arnal and B. Currey *Representations of Solvable Lie Groups*
40. M. A. Hill, M. J. Hopkins and D. C. Ravenel *Equivariant Stable Homotopy Theory and the Kervaire Invariant Problem*
41. K. Costello and O. Gwilliam *Factorization Algebras in Quantum Field Theory II*

Conformal Blocks, Generalized Theta Functions and the Verlinde Formula

SHRAWAN KUMAR
University of North Carolina, Chapel Hill

CAMBRIDGE
UNIVERSITY PRESS

University Printing House, Cambridge CB2 8BS, United Kingdom

One Liberty Plaza, 20th Floor, New York, NY 10006, USA

477 Williamstown Road, Port Melbourne, VIC 3207, Australia

314–321, 3rd Floor, Plot 3, Splendor Forum, Jasola District Centre,
New Delhi – 110025, India

103 Penang Road, #05–06/07, Visioncrest Commercial, Singapore 238467

Cambridge University Press is part of the University of Cambridge.

It furthers the University's mission by disseminating knowledge in the pursuit of education, learning, and research at the highest international levels of excellence.

www.cambridge.org
Information on this title: www.cambridge.org/9781316518168
DOI: 10.1017/9781108997003

First published 2022

A catalogue record for this publication is available from the British Library.

Library of Congress Cataloging-in-Publication Data
Names: Kumar, S. (Shrawan), 1953– author.
Title: Conformal blocks, generalized theta functions and
the Verlinde formula / Shrawan Kumar.
Description: Cambridge ; New York, NY : Cambridge University Press, 2022. |
Series: New mathematical monographs | Includes bibliographical references and index.
Identifiers: LCCN 2021029703 (print) | LCCN 2021029704 (ebook) |
ISBN 9781316518168 (hardback) | ISBN 9781108997003 (epub)
Subjects: LCSH: Lie algebras. | Affine algebraic groups. | Conformal invariants. |
Functions, Theta. | Fiber bundles (Mathematics) | Moduli theory. |
BISAC: MATHEMATICS / Topology
Classification: LCC QC20.7.L54 K86 2022 (print) | LCC QC20.7.L54 (ebook) |
DDC 512/.482–dc23
LC record available at https://lccn.loc.gov/2021029703
LC ebook record available at https://lccn.loc.gov/2021029704

ISBN 978-1-316-51816-8 Hardback

Contents

Preface

The main aim of this book is to give a self-contained proof of the Verlinde formula for the dimension of the space of conformal blocks and prove the connection between conformal blocks and generalized theta functions.

Let Σ be a smooth projective irreducible s-pointed ($s \geq 1$) curve of any genus $g \geq 0$ with marked points $\vec{p} = (p_1, \ldots, p_s)$ and let G be a simply-connected simple algebraic group with Lie algebra $\mathfrak{g}$. We fix a positive integer c called the *level* and let D_c be the set of dominant integral weights of $\mathfrak{g}$ of level at most c. We attach weights $\vec{\lambda} = (\lambda_1, \ldots, \lambda_s)$ (with each $\lambda_i \in D_c$) to the marked points $\vec{p}$, respectively. Associated to the triple $(\Sigma, \vec{p}, \vec{\lambda})$, there is the space $\mathscr{V}_\Sigma^\dagger(\vec{p}, \vec{\lambda})$ of *conformal blocks* (also called *space of vacua*), which is a finite-dimensional space given as the dual of $\mathfrak{g} \otimes \mathbb{C}[\Sigma \setminus \vec{p}]$-coinvariants of a tensor product of s copies of integrable highest-weight modules of level c with highest weights $\vec{\lambda}$ of the affine Kac–Moody Lie algebra $\hat{\mathfrak{g}}$ associated to $\mathfrak{g}$. This space is a basic object in rational conformal field theory arising from the Wess–Zumino–Witten model associated to G. Now, E. Verlinde gave a remarkable conjectural formula for the dimension of $\mathscr{V}_\Sigma^\dagger(\vec{p}, \vec{\lambda})$ in 1988. This conjecture was 'essentially' proved by a pioneering work of Tsuchiya–Ueno–Yamada, wherein they proved the *Factorization Theorem* and the *invariance of dimension* of the space of conformal blocks under deformations of the curve Σ, which allow one to calculate the dimension of the space of conformal blocks for a genus g curve from that of a genus $g-1$ curve. Thus, the problem gets reduced to a calculation on a genus 0 curve, i.e., on $\Sigma = \mathbb{P}^1$. The corresponding algebra for $\Sigma = \mathbb{P}^1$ is encoded in the fusion algebra associated to $\mathfrak{g}$ at level c, which gives rise to a proof of an explicit Verlinde dimension formula for the space $\mathscr{V}_\Sigma^\dagger(\vec{p}, \vec{\lambda})$.

Classical theta functions can be interpreted in geometric terms as global holomorphic sections of a certain determinant line bundle on the moduli space $\mathrm{Pic}^{g-1}(\Sigma)$ of line bundles of degree $g-1$ on Σ. This has a natural non-abelian generalization, where one replaces the line bundles on Σ by principal G-bundles on Σ to obtain the parabolic moduli space $M^G_{\text{par}, \vec{\tau}}(\Sigma)$ (or stack $\mathbf{Parbun}_G\left(\Sigma, \vec{P}\right)$) and certain determinant line bundles over them. Holomorphic sections of these determinant line bundles over these moduli spaces or stacks are called the generalized theta functions (generalizing the classical theta functions).

The Verlinde dimension formula attracted considerable further attention from mathematicians and physicists when it was realized that the space of conformal blocks admits an interpretation as the space of generalized theta functions. This interpretation was rigorously established independently in the 'non-parabolic' case by Beauville–Laszlo (for the special case $G = \mathrm{SL}_n$), Faltings and Kumar–Narasimhan–Ramanathan (for general G); and in the 'parabolic' case by Pauly (for the special case $G = \mathrm{SL}_n$), Laszlo–Sorger (for classical G and G_2 for the stack) and here in this book it is proved for general G.

The theory has undergone tremendous development in various directions and connections with diverse areas abound. The Verlinde formula and the ideas behind its proof have found numerous applications, e.g., in the theory of moduli spaces of vector bundles (and, more generally, principal G-bundles) on curves, the multiplicative eigenvalue problem, rank-level duality, moduli of curves (just to name a few). The works leading to the Verlinde dimension formula and connection between conformal blocks and generalized theta functions, as well as various applications, are scattered through the literature. For example, apart from the research papers, there is a Bourbaki talk and also lecture notes by C. Sorger, and a monograph by Ueno. But there is no single source containing various developments in and around the Verlinde formula explaining both of its aspects: the space of conformal blocks and the space of generalized theta functions, with details of proofs. This book attempts to fill this void in the literature.

As mentioned above, we give a self-contained proof of the Verlinde formula for the dimension of the space of conformal blocks (derived from the Factorization Theorem and the invariance of dimension of the space of conformal blocks under deformations of Σ, among others) and full details of the connection between conformal blocks and generalized theta functions. The proofs require techniques from algebraic geometry, geometric invariant theory, representation theory of affine Kac–Moody Lie algebras, topology and Lie algebra cohomology.

The main results covered in this book are: propagation of vacua; Factorization Theorem; flat projective connection on the sheaf of conformal blocks (thereby its local freeness); explicit Verlinde dimension formula for the space of conformal blocks; uniformization theorem for the moduli stack of quasi-parabolic G-bundles; identification of parabolic G-bundles over Σ with equivariant bundles on a certain Galois cover $\hat{\Sigma}$ of Σ; Harder–Narasimhan reduction of G-bundles; Narasimhan–Seshadri theorem for topological realization of polystable G-bundles over Σ; construction of the moduli *space* of parabolic semistable G-bundles over Σ; canonical identification of the space of conformal blocks with the space of generalized theta functions (over both parabolic moduli space and moduli stack); an explicit determination of the Picard group of the moduli space (as well as moduli stack) of G-bundles; and higher cohomology vanishing of the determinant line bundles on the moduli space. In addition, Chapter 1 is devoted to recalling the basic theory of affine Kac–Moody Lie algebras and their representations; and construction of the associated groups and their flag varieties to the extent we need them in the book. We have also added four appendices: one on the Dynkin index, which plays an important role in the identification of determinant line bundles on the moduli space; and the second and the third giving a crash (and hopefully quite palatable) course on $\mathbb{C}$-space (and $\mathbb{C}$-group) functors and stacks. The fourth appendix (due to S. Mukhopadhyay) gives a survey of rank-level duality.

This book should be useful for senior graduate students, postdocs and faculty members interested in the interaction between algebraic geometry, representation theory, topology and mathematical physics. Depending upon the interests of the audience, parts of the book are suitable for a one-year advanced graduate course. We have added numerous exercises of varying difficulty at the end of practically each section. We do require some knowledge of representation theory of (finite-dimensional) semisimple Lie algebras (roughly Chapters II and VI of Humphreys (1972)) and some algebraic geometry (roughly the first three chapters of Hartshorne (1977)).

I am indebted to M. S. Narasimhan, who introduced me to this beautiful garden. I am also grateful to my collaborators on the subject: A. Boysal, M. S. Narasimhan, and A. Ramanathan, and to P. Belkale and V. Balaji for numerous consultations. My special thanks are also due to A. Boysal, B. Conrad, C. Damiolini, N. Nitsure, S. Mukhopadhyay and X. Zhu, who carefully looked at parts of the book and pointed out some errors and made various comments to improve the exposition. I gave a semester-long course covering Chapters 1 through 4 at the University of Sydney during Fall 2015. It is my pleasure to thank the audience, especially Anthony Henderson, Gustav Lehrer, Alex Molev, Hoel Queffelec, Oded Yacobi and Ruibin Zhang, for their comments. I also gave a series of lectures (covering parts of the book) at Duke University (2006–07); the University of Georgia, Athens (May 2010); the Université Claude Bernard Lyon 1 (June 2017); and the Tata Institute of Fundamental Research, Mumbai (January 2018). The feedback from the audiences in these institutions was helpful. Finally, I acknowledge the continued support from NSF over several years during which the book was written. The typing of the book from my handwritten manuscript was done by M. P. Raghavendra Prasad from Sriranga Digital Technologies, Srirangapatna. I also thank Neeraj Kumar for his help in some formatting issues.

I dedicate this book to my wife Shyama and our children, Neeraj and Niketa.

Introduction

There follows a more detailed description of the contents of the book. For simplicity and uniformity in this Introduction, we let G be a connected simply-connected simple algebraic group over $\mathbb{C}$ with Lie algebra $\mathfrak{g}$ (though many of the results in the text are proved, more generally, for connected reductive groups) and let $\hat{\Sigma}$ be a connected smooth projective curve with faithful action of a finite group A and we set $\Sigma := \hat{\Sigma}/A$.

Chapter 1. *Section 1.1* lays out the basic notation to be used throughout the book. It also includes the Yoneda Lemma.

Section 1.2 introduces the basic theory of affine Kac–Moody Lie algebras $\hat{\mathfrak{g}}$. The main result here is the classification of integrable highest-weight $\hat{\mathfrak{g}}$-modules.

In *Section 1.3* we realize the loop group $G((t))$, its various subgroups and the infinite Grassmannian as ind-schemes. Specifically, consider the functors which assign, to any $\mathbb{C}$-algebra R, the groups (or set)

$$G(R((t))), G(R[[t]]), G(R[t^{-1}]), G(R((t)))/G(R[[t]]).$$

Then we show that they are representable functors represented respectively by ind-schemes $\bar{G}((t)), \bar{G}[[t]], \bar{G}[t^{-1}], \bar{X}_G$. In fact, we show that all these ind-schemes are reduced. The construction of $\bar{X}_G$ for $G = \mathrm{SL}_N$ proceeds via the so-called special lattice functor.

In *Section 1.4* we construct and discuss the central extensions of the loop group. This is essentially obtained by exponentiating the integrable highest-weight $\hat{\mathfrak{g}}$-modules $\mathscr{H}(\lambda_c)$, thereby realizing these representations as projective representations of the loop group $\bar{G}((t))$. We further show that the $\mathbb{G}_m$-central extension $\bar{\hat{G}}_{\lambda_c}$ thus obtained splits over $\bar{G}[[t]]$ and $\bar{G}[t^{-1}]$. In fact, the splitting is unique as shown in Section 8.2.

Chapter 2. Let Σ be a reduced projective curve with at worst only nodal singularity and let $\vec{p} = (p_1, \ldots, p_s)$ be a collection of distinct marked smooth points of Σ. We fix a central charge $c > 0$ and associate integrable highest-weight $\hat{\mathfrak{g}}$-modules with highest weights $\vec{\lambda} = (\lambda_1, \ldots, \lambda_s)$ (all with central charge c) to the points $\vec{p}$, respectively. To this data, there is associated the *space of vacua* $\mathscr{V}_\Sigma^\dagger(\vec{p}, \vec{\lambda})$ defined as the space of certain invariants in the dual of the tensor product $\mathscr{H}(\vec{\lambda}) := \mathscr{H}(\lambda_1) \otimes \cdots \otimes \mathscr{H}(\lambda_s)$. This space of vacua is a fundamental object for this book. It is shown that it is a finite-dimensional space.

In *Section 2.2* we prove *propagation of vacua*, which asserts that adding one extra point to $\vec{p}$ and associating $\mathscr{H}(0)$ to this extra point does not change the space of vacua.

In *Section 2.3* we give a manifestly finite-dimensional expression for the space of vacua on $\Sigma = \mathbb{P}^1$ in terms of the action of sl_2 passing through the highest root space of $\mathfrak{g}$.

Chapter 3. In *Section 3.1* we prove the basic *Factorization Theorem*, which explicitly relates the space of vacua on an s-pointed curve $(\Sigma, \vec{p})$ of genus g with a single node with that of the space of vacua on the normalization $\tilde{\Sigma}$ (which is of genus $g - 1$) marked with $s + 2$ points.

In *Section 3.2* we recall the definition of the *Sugawara elements* in the completion of the enveloping algebra of the affine Kac–Moody Lie algebra $\tilde{\mathfrak{g}}$ (the non-completed version of $\hat{\mathfrak{g}}$). These elements allow us to give the action of the Virasoro algebra on any smooth representation of $\tilde{\mathfrak{g}}$.

In *Section 3.3* we sheafify the construction of the space of vacua for a family $\mathfrak{F}_T$ of s-pointed curves with formal parameters at the marked points parameterized by a smooth variety T. We show that this sheaf $\mathscr{V}_{\mathfrak{F}_T}(\vec{p}, \vec{\lambda})$ is a coherent sheaf of $\mathscr{O}_T$-modules. We also sheafify the Virasoro algebra to allow its action on the sheafified version $\mathscr{H}(\vec{\lambda})_T$ of the tensor product $\mathscr{H}(\vec{\lambda})$ of integrable highest-weight $\hat{\mathfrak{g}}$-modules.

Then, in *Section 3.4* we show that the sheaf $\mathscr{V}_{\mathfrak{F}_T}(\vec{p}, \vec{\lambda})$ for a smooth family is locally free and admits a functorial flat projective connection. This connection generalizes the Knizhnik–Zamolodchikov connection for $\Sigma = \mathbb{P}^1$.

In *Section 3.5*, using the stack of stable s-pointed connected curves of fixed genus g and the local freeness of $\mathscr{V}_{\mathfrak{F}_T}(\vec{p}, \vec{\lambda})$ for a smooth family (proved in the previous section), we show that the dimension of the space of vacua is independent of the choice of the marked points $\vec{p}$ as well as the connected smooth curve Σ, as long as the genus of Σ is fixed and of course $\vec{\lambda}$ is fixed. Let us denote this dimension by $F_g(\vec{\lambda})$. Further, using the 'smoothing deformation' and the Factorization Theorem, we extend the above result to allow curves with nodes. This enables us to prove the following inductive formula to calculate the dimension $F_g(\vec{\lambda})$:

$$F_g(\vec{\lambda}) = \sum_{\mu} F_{g-1}(\vec{\lambda}, \mu^*, \mu),$$

where μ runs over dominant integral weights of $\hat{\mathfrak{g}}$ with central charge c. Thus, the problem to calculate $F_g(\vec{\lambda})$ for any g reduces to that for $g = 0$, i.e., $\Sigma = \mathbb{P}^1$ (though with $s + 2g$ marked points). Using a similar decomposition, the problem of calculating $F_0(\vec{\lambda})$ with n marked points reduces to that for three marked points on $\mathbb{P}^1$.

Chapter 4. As mentioned above, to determine $F_g(\vec{\lambda})$ we only need to determine $F_0(\vec{\mu})$ with three marked points. To be able to calculate $F_0(\vec{\mu})$, a general algebraic framework in the form of *fusion ring* $\mathbb{Z}[A]$ is introduced in *Section 4.1*. It is shown that the corresponding complexified algebra $\mathbb{C}[A]$ is a (finite-dimensional) reduced algebra.

In *Section 4.2* we consider a specific fusion ring $R_c(\mathfrak{g}) := \mathbb{Z}[D_c]$, called the *fusion ring of* $\mathfrak{g}$ *at level* c, with product structure constants coming from F_0 with three marked points. Simple algebraic manipulation in this ring allows us to give an explicit expression for $F_g(\vec{\lambda})$ once we are able to explicitly determine the set of characters S_{D_c} of $\mathbb{C}[D_c]$ (i.e., algebra homomorphisms to $\mathbb{C}$). This section is devoted to solve this problem by using the combinatorics of the affine Weyl group and its action on the dual $\mathfrak{h}^*$ of the Cartan subalgebra $\mathfrak{h}$ of $\mathfrak{g}$. One other important ingredient in determining S_{D_c} is the result that a certain linear map ξ_c from the representation ring $R(\mathfrak{g})$ of $\mathfrak{g}$ to the fusion ring $R_c(\mathfrak{g})$ at level c is a ring homomorphism. To prove that ξ_c is a ring homomorphism, we use the affine analogue of the Borel–Weil–Bott (BWB) theorem as well as a Lie algebra cohomology vanishing result of Teleman. (For the classical $\mathfrak{g}$ as well as $\mathfrak{g}$ of type G_2, there is a more direct proof that ξ_c is a ring homomorphism avoiding the Lie algebra cohomology vanishing result, as shown in Exercises 4.2.E.) Once we have explicitly determined S_{D_c} (as we have), one of the most important results of the book – the *Verlinde dimension formula* – follows easily by using simple representation theory for finite groups.

Chapter 5. Let $\mathfrak{S}$ be the category of quasi-compact separated schemes over $\mathbb{C}$ and let $\mathbf{Bun}_G(\Sigma)$ be the groupoid fibration over $\mathfrak{S}$ whose objects over $S \in \mathfrak{S}$ are G-bundles on $\Sigma \times S$ and morphisms are the G-bundle morphisms. Similarly, for an s-pointed curve $(\Sigma, \vec{p})$ together with a choice of standard parabolic subgroups $\vec{P} = (P_1, \ldots, P_s)$ attached to the marked points, we define the groupoid fibration $\mathbf{Parbun}_G(\Sigma, \vec{P})$ of quasi-parabolic G-bundles of type $\vec{P}$ over $(\Sigma, \vec{p})$. Then, as stated in *Section 5.1*, both these are smooth (algebraic) stacks.

In *Section 5.2* we define a 'tautological' G-bundle $\mathbf{U}$ over $\Sigma \times \bar{X}_G$, where (as earlier) $\bar{X}_G$ is the infinite Grassmannian. Consider the functor which assigns to any $\mathbb{C}$-algebra R the group $\mathrm{Mor}(\Sigma^* \times \operatorname{Spec} R, G)$, where $\Sigma^* := \Sigma \backslash \vec{p}$. Then we show that it is a representable functor represented by an ind-affine group variety denoted $\bar{\Gamma} = \bar{\Gamma}_{\vec{p}}$. Then we prove the *Uniformization Theorem* for both $\mathbf{Bun}_G(\Sigma)$ and $\mathbf{Parbun}_G(\Sigma, \vec{P})$. Specifically, they are realized as quotient stacks:

$$\mathbf{Bun}_G(\Sigma) \simeq \left[\bar{\Gamma} \backslash \bar{X}_G\right]$$

and

$$\mathbf{Parbun}_G(\Sigma, \vec{P}) \simeq \left[\bar{\Gamma} \backslash \left(\bar{X}_G \times \Pi_{i=1}^s (G/P_i)\right)\right],$$

where $\bar{\Gamma} = \bar{\Gamma}_\infty$ for a single point $\infty \in \Sigma$ different from any p_i in $\vec{p}$. An important ingredient in the proof of the above two uniformization theorems is a result due to Drinfeld–Simpson asserting that for a family of G-bundles over Σ parameterized by a scheme S, the pull-back of the family to some étale cover $\tilde{S}$ of S is trivial restricted to any affine open subset of Σ.

As an immediate consequence of the uniformization theorems specialized to $\operatorname{Spec} \mathbb{C}$, we get the following bijections:

$$\mathrm{Bun}_G(\Sigma) \simeq \Gamma \backslash X_G$$

and

$$\mathrm{Parbun}_G(\Sigma, \vec{P}) \simeq \Gamma \backslash \left(X_G \times \Pi_{i=1}^s (G/P_i)\right),$$

where Bun_G (resp. Parbun_G) denotes the set of isomorphism classes of G-bundles (resp. quasi-parabolic G-bundles) over Σ and $\Gamma := \bar{\Gamma}(\mathbb{C})$ and similarly $X_G := \bar{X}_G(\mathbb{C})$.

Chapter 6. In *Section 6.1* we define the *stability, semistability* and *polystability* of vector bundles over Σ and extend these notions to G-bundles over Σ. The semistabilty of a G-bundle is equivalent to the semistability of its adjoint bundle. More generally, we define the *parabolic stability* and *parabolic semistability* for parabolic G-bundles over an s-pointed curve $(\Sigma, \vec{p})$. We extend the notions of stability, semistability and polystability to *A-stability, A-semistability* and *A-polystability* in the case a finite group A acts faithfully on a smooth projective curve $\hat{\Sigma}$. The main aim of this section is to prove an equivalence between the groupoid fibration of A-equivariant G-bundles on $\hat{\Sigma}$ and quasi-parabolic G-bundles on $(\Sigma, \vec{p})$, where $\vec{p} \subset \Sigma$ denotes the set of all the ramification points under the action of A on $\hat{\Sigma}$. But first we need to define the *local type* of A-equivariant G-bundles, which is achieved by proving the following result.

Let A act on the formal disc $\mathbb{D} := \text{Spec}(\mathbb{C}[[t]])$. For a $\mathbb{C}$-algebra R, let $\mathbb{D}_R := \text{Spec}(R[[t]])$ be the formal disc over $\text{Spec}\, R$ and let $\mathscr{E}$ be an A-equivariant G-bundle over $\mathbb{D}_R$ (with the trivial action of A on R) such that it is trivial as a G-bundle. Then, there is a trivialization of the G-bundle $\mathscr{E}$ such that the action of A is a product action, i.e.,

$$\gamma \odot (x, g) = (\gamma x, \theta_\gamma(x(0))g), \text{ for } \gamma \in A, x \in \mathbb{D}_R \text{ and } g \in G,$$

for a morphism $\theta_\gamma \colon \text{Spec}\, R \to G$. For any $x^o \in \text{Spec}\, R$, we get a group homomorphism $\theta(x^o) \colon A \to G$ taking $a \mapsto \theta_\gamma(x^o)$. This homomorphism $\theta(x^o)$ is unique up to conjugation.

Now, given an A-equivariant G-bundle E over $\hat{\Sigma}$, for any ramification point $p_i \in \Sigma$, we take a point $\hat{p}_i$ in $\hat{\Sigma}$ over p_i. Applying the above result to the restriction of E to the formal disc in $\hat{\Sigma}$ around $\hat{p}_i$ and A replaced by the isotropy subgroup A_{p_i} of A at $\hat{p}_i$ (which is a cyclic group and, up to a conjugation, does not depend upon the choice of $\hat{p}_i$ over p_i), we get a homomorphism $A_{p_i} \to G$ (unique up to conjugation). This is, by definition, the *local type of E at p_i*. Let $\vec{p} = (p_1, \ldots, p_s)$ be the set of all the ramification points in Σ and let $\vec{\tau} := (\tau_1, \ldots, \tau_s)$ be the local type respectively at $\vec{p}$. Similar to the definition of the stack $\mathbf{Bun}_G(\Sigma)$, define the groupoid fibration $\mathbf{Bun}_G^{A,\vec{\tau}}(\hat{\Sigma})$ of A-equivariant G-bundles over $\hat{\Sigma}$ of local type $\vec{\tau}$, whose objects over any scheme S are A-equivariant G-bundles E_S over $\hat{\Sigma} \times S$ (A acting trivially on S) such that for any $t \in S$, $E_{S|\hat{\Sigma}\times t}$ is of local type $\vec{\tau}$. Then we prove that $\mathbf{Bun}_G^{A,\vec{\tau}}(\hat{\Sigma})$ is a stack and there is an isomorphism of stacks:

$$\mathbf{Bun}_G^{A,\vec{\tau}}(\hat{\Sigma}) \simeq \mathbf{Parbun}_G(\Sigma, \vec{P}),$$

where $\vec{P}$ corresponds to the Kempf parabolic subgroups attached to $\vec{\tau}$. In particular, $\mathbf{Bun}_G^{A,\vec{\tau}}(\hat{\Sigma})$ is also a smooth (algebraic) stack. Moreover, specializing the above isomorphism of stacks to $\text{Spec}\, \mathbb{C}$, we get a bijective correspondence between the set of isomorphism classes of A-equivariant G-bundles on $\hat{\Sigma}$ of local type $\vec{\tau}$ with the set of isomorphism classes of quasi-parabolic G-bundles over Σ of type $\vec{P}$:

$$\text{Bun}_G^{A,\vec{\tau}}(\hat{\Sigma}) \simeq \text{Parbun}_G(\Sigma, \vec{P}).$$

Under this correspondence, A-semistable (resp. A-stable) G-bundles over $\hat{\Sigma}$ correspond to the parabolic semistable (resp. stable) bundles over Σ with the parabolic weights given by $\vec{\tau}$.

In *Section 6.2* we discuss *Harder–Narasimhan (HR) reduction* (also called *canonical reduction*) of any G-bundle E over Σ. A P-subbundle E_P of E (for a standard parabolic subgroup P of G) is called a HN reduction if the associated $L \simeq P/U$-bundle $E_P(P/U)$ is a semistable L-bundle and for any nontrivial character λ of P which lies in the positive cone generated by the simple roots of $\mathfrak{g}$,

$$\deg(E_P \times^P \mathbb{C}_\lambda) > 0,$$

where L is a Levi subgroup of P and U is the unipotent radical of P. We prove that for any G-bundle E over Σ, the HN reduction E_P exists and is unique. As a consequence, it is shown that for an embedding $G \hookrightarrow G'$ of connected reductive groups and a G-bundle E over Σ, if $E(G')$ is semistable then so is E. Conversely, if E is semistable then so is $E(G')$ if G is not contained in any proper parabolic subgroup of G'. As another consequence (cf. Exercises 6.2.E), one gets that an A-equivariant G-bundle over $\hat{\Sigma}$ is A-semistable if and only if it is semistable.

Section 6.3 is devoted to constructing stable (more generally, polystable) G-bundles over Σ topologically from a homomorphism $\rho \colon \pi_1(\Sigma) \to K \subset G$ of the fundamental group, where K is a maximal compact subgroup of G. Specifically, define the corresponding holomorphic G-bundle over Σ by

$$E_\rho := \tilde{\Sigma} \times^{\pi_1(\Sigma)} G \to \Sigma,$$

where $\tilde{\Sigma}$ is the simply-connected cover of Σ. These bundles E_ρ are called *unitary bundles*. Then it is shown that E_ρ is semistable. Further, for two such homomorphisms ρ and ρ', the bundles E_ρ and $E_{\rho'}$ are isomorphic if and only if ρ and ρ' are conjugate. Moreover, E_ρ is stable if and only if ρ is irreducible in the sense that the image of ρ is not contained in any proper parabolic subgroup of G. The irreducibility of ρ is also shown to be equivalent to the corresponding adjoint representation $\operatorname{ad}\rho$ having no $\pi_1(\Sigma)$-invariants (assuming G to be semisimple). The proof requires, in particular, an identification of a certain group cohomology of $\pi_1(\Sigma)$ with the cohomology of a certain vector bundle over Σ. Because of the standard presentation of $\pi_1(\Sigma)$, the set of all homomorphisms from $\pi_1(\Sigma) \to K$ can be identified with $\beta^{-1}(e)$, where

$$\beta \colon K^{2g} \to K, \big((h_1,k_1),\ldots,(h_g,k_g)\big) \mapsto \Pi_{i=1}^{g}[h_i,k_i],$$

where g is the genus of Σ. For $\bar{\rho} \in \beta^{-1}(e)$, let ρ be the corresponding representation of $\pi_1(\Sigma)$. Then it is shown that ρ is irreducible if and only if the tangent map $(d\beta)_{\bar{\rho}}$ is of maximal rank. In particular,

$$M_g(K) := \{\bar{\rho} \in \beta^{-1}(e) : \rho \text{ is irreducible}\}$$

is a smooth manifold of dimension $(2g - 1)\dim K$ (for semisimple G). Moreover, it supports an $\mathbb{R}$-analytic family of holomorphic G-bundles over Σ such that its Kodaira–Spencer infinitesimal deformation map is surjective everywhere. We now come to the following celebrated result (generalization of the classical Narasimhan–Seshadri result to G-bundles).

Any stable G-bundle E over Σ (for $g \geq 2$) is realized as E_ρ for an irreducible representation $\rho\colon \pi_1(\Sigma) \to K$ (and conversely). In fact, the result is valid for any connected reductive G provided we assume that E is of degree 0. The result can easily be extended for any polystable G-bundle E. Conversely, for any (not necessarily irreducible) representation $\rho\colon \pi_1(\Sigma) \to K$, E_ρ is polystable.

Let us point out the main strategy behind its proof. Let $\mathscr{F} \to \Sigma \times T$ be a $\mathbb{C}$-analytic family of stable G-bundles over Σ. Then we prove that

$$T_u := \{t \in T : \mathscr{F}_t \simeq E_\rho \text{ for some unitary representation } \rho\}$$

is a closed subset of T. Moreover, for any $\mathbb{R}$-analytic family of holomorphic G-bundles over Σ,

$$T_o := \{t \in T : \mathscr{F}_t \simeq E_\rho \text{ for some unitary irreducible } \rho\}$$

is an open subset of T, which follows from the surjectivity of the Kodaira–Spencer infinitesimal deformation map (mentioned above). Further, there exists an irreducible representation $\rho_o\colon \pi_1(\Sigma) \to K$ (this is where we need $g \geq 2$). Finally, we construct a $\mathbb{C}$-analytic family $\mathscr{E}$ of stable holomorphic G-bundles over Σ parameterized by a connected open subset V of $\mathbb{C}$ containing $\{0, 1\}$ such that $\mathscr{E}_0 \simeq E$ and $\mathscr{E}_1 \simeq E_{\rho_o}$. Observe that, for this family, $T_u = T_o$ since the family consists of stable bundles and (as observed above) ρ is irreducible if and only if E_ρ is stable. Now, V being connected and $T_u = T_o$ being both open and closed and nonempty, $T_o = V$. This proves that $E \simeq E_\rho$ for some irreducible ρ.

We extend these results to the setting of A-equivariant G-bundles over $\hat{\Sigma}$. Specifically, let π_1 be the fundamental group of $\hat{\Sigma}$. Then, π_1 is a normal subgroup of a group π such that π acts on the simply-connected cover $\tilde{\hat{\Sigma}}$ of $\hat{\Sigma}$ (having fixed points in general) with $\tilde{\hat{\Sigma}}/\pi \simeq \Sigma$ and $\pi/\pi_1 \simeq A$. Given a representation $\hat{\rho}\colon \pi \to K$, we can construct (as above) the holomorphic G-bundle over $\hat{\Sigma}$:

$$\hat{E}_{\hat{\rho}} := \tilde{\hat{\Sigma}} \times^{\pi_1} G \to \hat{\Sigma}.$$

Since $\hat{\rho}$ is a representation of π, $\hat{E}_{\hat{\rho}}$ acquires the canonical structure of an A-equivariant G-bundle. These bundles $\hat{E}_{\hat{\rho}}$ are called *A-unitary*. Conversely, if $\hat{E}_{\hat{\rho}}$ (the definition of which only requires the homomorphism $\hat{\rho}_{|\pi_1}$) acquires the structure of an A-equivariant G-bundle, then $\hat{\rho}_{|\pi_1}$ extends to π. We extend various results proved for E_ρ to that for $\hat{E}_{\hat{\rho}}$. In particular, $\hat{E}_{\hat{\rho}}$ is A-stable if and only if $\hat{\rho}$ is irreducible. Moreover, for any homomorphism $\hat{\rho}: \pi \to K$, $\hat{E}_{\hat{\rho}}$ is A-semistable (in fact, A-polystable). Conversely, we have the following equivariant generalization of the Narasimhan–Seshadri theorem for any G:

Any A-polystable G-bundle $\hat{E}$ over $\hat{\Sigma}$ (when the genus $g \geq 2$ of Σ) is realized as $\hat{E}_{\hat{\rho}}$ for a representation $\hat{\rho}: \pi \to K$ (and conversely).

Let $G \to \mathrm{GL}_V$ be a representation with finite kernel. Then we show that an A-equivariant G-bundle $\hat{E}$ over $\hat{\Sigma}$ is A-unitary if and only if the corresponding vector bundle $\hat{\Sigma}(V)$ is A-unitary.

Chapter 7. Let us first recall the following result due to Grothendieck.

Let X be a projective scheme with a very ample line bundle $\mathscr{L}$ over X and let E be a coherent sheaf on X. Then, for any fixed polynomial $P(z) \in \mathbb{Q}[z]$, define a contravariant functor which associates to any noetherian scheme S, set of all $\mathscr{O}_{X\times S}$-module quotients $\mathscr{F}$ of $E \boxtimes \mathscr{O}_S$ such that $\mathscr{F}$ is flat over S and $\mathscr{F}_{|X\times t}$ has Hilbert polynomial $P(z)$ (with respect to $\mathscr{L}$) for any $t \in S$. Then, this functor is representable by a projective scheme $Q = Q(E, P)$ called the *quot scheme*. Moreover, there is a 'tautological' coherent sheaf $\mathscr{U}$ over $X \times Q$.

Take a pair of positive integers (r,d) such that $d > r(2g-1)$, where g is the genus of Σ. We specialize the above general result to $X = \Sigma, E = \mathscr{O}_\Sigma \otimes \mathbb{C}^N$ and $P(z) = N + rhz$, where $N := d + r(1-g)$ and h is the degree of a fixed very ample line bundle over Σ. Thus, we get the quot scheme $Q = Q(E, P(z))$ and the tautological coherent sheaf $\mathscr{U}$ over $\Sigma \times Q$. Moreover, GL_N acts canonically on Q making $\mathscr{U}$ a GL_N-equivariant sheaf (with the trivial action of GL_N on Σ). Define the subset

$$\begin{aligned} R^{ss} := \{q \in Q : \bar{q} \text{ is a semistable vector bundle over } \Sigma \text{ and} \\ \mathbb{C}^N = H^0(\Sigma, E) \to H^0(\Sigma, \bar{q}) \text{ is an isomorphism}\}, \end{aligned}$$

where $\bar{q}$ is the restriction $\mathscr{U}_{|\Sigma\times q}$. Then we prove that R^{ss} is a GL_N-stable irreducible smooth open subset of Q and $\mathscr{U}_{|\Sigma\times R^{ss}}$ is a rank-r vector bundle. Let $\mathscr{M}(r,d)$ (resp. $\mathscr{M}^s(r,d)$) be the functor of semistable (resp. stable) vector bundles over Σ of rank r and degree d. Then the main result of *Section 7.1* asserts that $\mathscr{M}(r,d)$ has a coarse moduli space $M(r,d) := R^{ss}//\mathrm{SL}_N$, which is an irreducible, normal projective variety with rational singularities of dimension $(g-1)r^2 + 1$ if $g \geq 2$. Moreover, the subfunctor $\mathscr{M}^s(r,d)$ has a coarse moduli space $M^s(r,d)$, which is an open subset of $M(r,d)$. The canonical map $\mathcal{M}(r,d)(\operatorname{Spec}\mathbb{C}) \to M(r,d)$ is surjective such that its fibers are precisely the equivalence classes of semistable vector bundles and its restriction $\mathcal{M}^s(r,d)(\operatorname{Spec}\mathbb{C}) \to M^s(r,d)$ is a bijection.

In *Section 7.2* we extend the above results for vector bundles to G-bundles and even more generally to A-equivariant G-bundles over $\hat{\Sigma}$. To this end, we fix an embedding $i: G \hookrightarrow \mathrm{SL}_r \subset \mathrm{GL}_r$ and realize G-bundles as GL_r-bundles via the embedding i (equivalently, rank-r vector bundles) together with a G-subbundle. To take into consideration the A-action, we fix an A-stable finite subset $\{y_1, \ldots, y_b\}$ of $\hat{\Sigma}$ and a positive integer d' such that the divisor $\vec{y} := d' \sum_j y_j$ has degree $\geq 2\hat{g}$, $\hat{g}$ being the genus of $\hat{\Sigma}$. Now, we consider the quot scheme as above:

$$Q = Q\big(E = \big(\mathcal{O}_{\hat{\Sigma}} \otimes \mathbb{C}^N\big) \otimes \mathcal{O}_{\hat{\Sigma}}(-\vec{y}), P(z)\big),$$

together with the tautological sheaf $\mathcal{U}$ over $\hat{\Sigma} \times Q$, where $N := r(d + 1 - \hat{g})$ and $P(z) = r(1 - \hat{g}) + rhz$ (h being the degree of a fixed very ample A-equivariant line bundle H over $\hat{\Sigma}$). Depending on the fixed local type $\vec{\tau}$ of A-equivariant G-bundles over $\hat{\Sigma}$, we fix a representation $\mathring{\tau}$ of A on $\mathbb{C}^N$. In fact, this representation of A on $\mathbb{C}^N$ is obtained from taking any A-semistable G-bundle F over $\hat{\Sigma}$ of topological type $\vec{\tau}$ and then taking the action of A on $\mathbb{C}^N \simeq H^0(\hat{\Sigma}, F(\vec{y}))$. (This action of A does not depend upon the choice of F.) This gives rise to a canonical action of A on Q making $\mathcal{U}$ an A-equivariant sheaf. Define the subset

$$\begin{aligned} R^{ss}_{\mathring{\tau}} :=& \{q \in Q^A : \bar{q} \text{ is an } A\text{-semistable vector bundle over } \hat{\Sigma} \text{ and} \\ & \mathbb{C}^N = H^0\big(\hat{\Sigma}, E(\vec{y})\big) \to H^0\big(\hat{\Sigma}, \bar{q}(\vec{y})\big) \text{ is an isomorphism}\}, \end{aligned}$$

where Q^A is the subscheme of A-invariants in Q and $\bar{q}$ is the restriction $\mathcal{U}_{|\hat{\Sigma}\times q}$. Let $\mathcal{U}^{ss}_{\mathring{\tau}}$ denote the restriction $\mathcal{U}_{|\hat{\Sigma}\times R^{ss}_{\mathring{\tau}}}$. Let $\mathfrak{G} := \mathrm{GL}_N^A$, the A-invariants under the conjugation action of A on GL_N induced from the representation $\mathring{\tau}$. Then, $R^{ss}_{\mathring{\tau}}$ is a $\mathfrak{G}$-stable open subset of Q^A and $\mathcal{U}^{ss}_{\mathring{\tau}}$ is an A-equivariant rank-r vector bundle with the action of $\mathfrak{G}$.

For any scheme S, let $\mathfrak{S}_S$ be the category consisting of morphisms $T \to S$ as objects and S-morphisms between them as morphisms. Define the contravariant functor to the category of sets (abbreviating the frame bundle of $\mathcal{U}^{ss}_{\mathring{\tau}}$ by $\mathcal{F}$):

$$\Gamma(i, \mathcal{F}) \colon \mathfrak{S}_{R^{ss}_{\mathring{\tau}}} \to \mathbf{Set}$$

by $\Gamma(i, \mathcal{F})(f: T \to R^{ss}_{\mathring{\tau}}) =$ the set of A-equivariant sections σ of $\mathcal{F}_f/G$, where $\mathcal{F}_f := \big(\mathrm{Id}_{\hat{\Sigma}} \times f\big)^*(\mathcal{F})$. Clearly, giving any such section σ is equivalent to giving an A-equivariant G-subbundle $\mathcal{F}_f(\sigma)$ of $\mathcal{F}_f$ over $\hat{\Sigma} \times T$. For an A-equivariant topological G-bundle τ over $\hat{\Sigma}$, define a subfunctor

$\Gamma^\tau(i,\mathscr{F})\colon \mathfrak{S}_{R^{ss}_{\mathring{\tau}}} \to$ **Set** of $\Gamma(i,\mathscr{F})$ by demanding that for any $f\colon T \to R^{ss}_{\mathring{\tau}}$, the G-subbundle $\mathscr{F}_f(\sigma)$ restricted to any $t \in T$ is topologically A-equivariant isomorphic with τ. Then, by a general result, the functor $\Gamma^\tau(i,\mathscr{F})$ is representable by a separated scheme of finite type $f_\tau\colon R^{ss}_\tau(G) \to R^{ss}_{\mathring{\tau}}$. Moreover, there exists a 'universal' A-equivariant G-bundle

$$\mathscr{U}^{ss}_\tau(G) \in \Gamma^\tau(i,\mathscr{F})(f_\tau) \text{ over } \hat{\Sigma} \times R^{ss}_\tau(G).$$

Now, the main result of this section asserts that the A-semistable G-bundles over $\hat{\Sigma}$ of topological type τ admit a coarse moduli space

$$M^G_\tau(\hat{\Sigma}) := R^{ss}_\tau(G)//\mathfrak{G}.$$

Moreover, it is an irreducible, normal variety with rational singularity; and nonempty and projective if the genus g of Σ is at least 2. We further prove that any element in $M^G_\tau(\hat{\Sigma})$ contains a unique A-polystable representative. Because of the correspondence between A-equivariant G-bundles over $\hat{\Sigma}$ and the parabolic G-bundles over Σ (as in Section 6.1), these results readily translate to the results about the moduli space of parabolic semistable G-bundles.

In the case $A = (1)$ so that $\hat{\Sigma} = \Sigma$, $M^G_\tau(\Sigma)$ is the (non-parabolic) moduli space of semistable G-bundles over Σ.

Chapter 8. Recall the definition of the ind-affine group variety $\bar{\Gamma}$ from above (summary of Chapter 5). Then, in *Section 8.1*, we prove that it is irreducible. The proof relies on showing that (under the analytic topology) Γ^{an} is path-connected, where $\Gamma := \bar{\Gamma}(\mathbb{C})$. As a corollary of this, we show that the infinite Grassmannian $\bar{X}_G$ is an irreducible ind-projective variety.

In *Section 8.2* we prove that the $\mathbb{G}_m$-central extension $\bar{\hat{G}}_{\lambda_c}$ described above in the summary of Section 1.4 splits uniquely for $\lambda_c = 0_c$ over $\bar{\Gamma} = \bar{\Gamma}_p$ for a single point $p \in \Sigma$.

Section 8.3: we prove that the space of vacua $\mathscr{V}^\dagger_\Sigma(\vec{p},\vec{\lambda})$ for any s-pointed smooth curve $(\Sigma,\vec{p})$ is canonically identified (up to scalar multiples) with the space of global sections of the moduli stack $\mathbf{Parbun}_G(\Sigma,\vec{P})$ with respect to a certain line bundle $\bar{\mathscr{L}}(\vec{\lambda})$, where the parabolic subgroups $\vec{P}$ and $\bar{\mathscr{L}}(\vec{\lambda})$ are given in terms of $\vec{\lambda}$ and the central charge c. The main ingredient in the proof is the analogue of the Borel–Weil theorem for affine Lie algebras and the propagation of vacua. We also explicitly determine the Picard group of the moduli stack $\mathbf{Parbun}_G(\Sigma,\vec{P})$.

Section 8.4: we first define the determinant and theta line bundles ($\mathrm{Det}(\mathscr{V})$ and $\Theta(\mathscr{V})$, respectively) of a family $\mathscr{V}$ of vector bundles over Σ parameterized by a noetherian scheme S. These are line bundles over S. Thus, for a family

$\mathscr{E}$ of G-bundles over Σ parameterized by a noetherian scheme S and a representation V of G, we have the corresponding line bundles $\mathrm{Det}(\mathscr{E}(V))$ and $\Theta(\mathscr{E}(V))$, where $\mathscr{E}(V)$ denotes the corresponding family of vector bundles induced by the G-module V. In fact, the theta bundle (for any G-module V) can even be defined on the moduli space $M^G(\Sigma)$ of semistable G-bundles over Σ (even though it does not support a family). We denote this line bundle over $M^G(\Sigma)$ by $\Theta(V)$. Then, $\Theta(V)$ pulls to $\Theta(\mathscr{E}(V))$ for any family $\mathscr{E}$ of semistable G-bundles over Σ.

For the 'tautological' G-bundle $\mathbf{U} = \mathbf{U}_G$ over $\Sigma \times \bar{X}_G$ (as in Section 5.2), we determine the corresponding theta bundle $\Theta(\mathbf{U}_G(V))$ for any G-module V in terms of the Dynkin index d_V of V. In fact, as shown in this section, identifying

$$\mathrm{Pic}(\bar{X}_G) \simeq \mathbb{Z},$$

$\Theta(\mathbf{U}_G(V)) = d_V$. Its proof relies on reducing the problem from G first to $G = \mathrm{SL}_N$ and then to SL_2 and then applying the Grothendieck–Riemann–Roch theorem.

The main result of this section asserts that there is a canonical identification (up to scalar multiples) between the space of global sections of a line bundle over $M^G(\Sigma)$ and the space of vacua. Specifically, we have the following (along with its equivariant generalization in the next section) second most important result of the book:

$$H^0(M^G(\Sigma), (\Theta(V))^{\otimes a}) \simeq \mathscr{V}^\dagger_\Sigma(p, 0_{d_V \cdot a}), \text{ for any } a \geq 0 \text{ and } G\text{-module } V.$$

In particular, $\dim\left(H^0(M^G(\Sigma), (\Theta(V))^{\otimes a})\right)$ is given by the Verlinde dimension formula (for one marked point and trivial weight 0 at any central charge $c > 0$). The most important technical result towards the proof of the above theorem asserts that for any line bundle $\mathscr{L}(0_d)$ over $\bar{X}_G$ (for any $d \geq 1$), any $\bar{\Gamma}$-invariant section of $\mathscr{L}(0_d)_{|\bar{X}_G^{ss}}$ extends to $\bar{X}_G$, where $\bar{X}_G^{ss}$ denotes the open subset consisting of those points which correspond to semistable G-bundles for the family $\mathbf{U}_G$.

Section 8.5: we extend the results from the last section to the moduli space $M^G_\tau(\hat{\Sigma})$ of A-semistable G-bundles over $\hat{\Sigma}$ of topological type τ. For any family $\mathscr{E}$ of quasi-parabolic G-bundles of type $\vec{P} := (P_1, \ldots, P_s)$ parameterized by a noetherian scheme S, a G-module V, an integer d and characters $\vec{\mu} = (\mu_1, \ldots, \mu_s)$ (where μ_j is a character of P_j), we define the *quasi-parabolic determinant line bundle* $\mathrm{Det}_{\mathrm{par}}(\mathscr{E}(V), d, \vec{\mu})$, which is a line bundle over S. We also have the *parabolic theta bundle* $\Theta_{\mathrm{par},G}(V, \tau, d)$ over $M^G_\tau(\hat{\Sigma})$ corresponding to any G-module V and positive integer d satisfying some integral weight condition with respect to the topological type τ. Then, for any family of A-semistable G-bundles over $\hat{\Sigma}$ of topological type τ parameterized by a noetherian scheme S, the induced map $S \to M^G_\tau(\hat{\Sigma})$ pulls the theta bundle $\Theta_{\mathrm{par},G}$ to the quasi-parabolic determinant bundle $\mathrm{Det}_{\mathrm{par}}$.

Finally, we prove the following equivariant generalization of the main result of the previous section identifying the space of global sections of line bundles over the equivariant moduli space with the space of vacua. Specifically, for any G-module V, any s-pointed smooth projective irreducible curve $(\Sigma, \vec{p})$ and any dominant integral weights $\vec{\lambda}$ at level c attached to $\vec{p}$ with some additional integrality condition, we have a canonical (up to scalar multiples) isomorphism:

$$H^0\left(M^G_\tau(\hat{\Sigma}), \Theta_{\mathrm{par},G}(V,\tau,c)\right) \simeq \mathscr{V}^\dagger_\Sigma(\vec{p}, d_V\vec{\lambda}),$$

where $\tau = \tau(\vec{\lambda})$ is given in terms of the weights $\vec{\lambda}$, d_V is the Dynkin index of V and the space of vacua is taken at the central charge cd_V (see Theorem 8.5.9 for precise details).

Chapter 9. The main aim of this chapter is to determine the Picard group of the moduli space $M^G(\Sigma)$ of semistable G-bundles over Σ explicitly. In *Section 9.1* we show that there is an injective map

$$\mathrm{Pic}(M^G(\Sigma)) \hookrightarrow \mathrm{Pic}_{\bar{\Gamma}}(\bar{X}_G) \simeq \mathbb{Z},$$

where $\bar{\Gamma}$ corresponds to deleting a single point p in Σ and the last identification is established in Section 8.3. In particular, $\mathrm{Pic}(M^G(\Sigma)) \simeq \mathbb{Z}$. We further prove that $M^G(\Sigma)$ is Gorenstein and identify its dualizing line bundle. Moreover, we prove the following vanishing theorem for any G-module V:

$$H^i(M^G(\Sigma), \Theta(V)) = 0, \quad \text{for all } i > 0.$$

Section 9.2: the moduli space $M^G(\Sigma)$ is identified as a weighted projective space for Σ an elliptic curve. This identification allows us to explicitly determine the Picard group of $M^G(\Sigma)$ as well as determine the space of global sections for any line bundle over $M^G(\Sigma)$ (for an elliptic curve Σ).

Section 9.3: by the main result of Section 9.1 (described above), we have an injection

$$\bar{f}^*\colon \mathrm{Pic}(M^G(\Sigma)) \to \mathrm{Pic}_{\bar{\Gamma}}(\bar{X}_G) \simeq \mathbb{Z}.$$

The main aim of this section is to determine the image of $\bar{f}^*$. Specifically, we show that the image is the abelian group generated by $\langle \Theta(V) \rangle$ as V ranges over G-modules. In fact, we show that the image is generated by the line bundle $\theta(V_o)$, where V_o is any fundamental representation of G with minimal Dynkin index (the list of such G-modules with minimal Dynkin index is given in Appendix A). This whole section is devoted to its proof. We first identify (under the first Chern class) $c\colon \operatorname{Pic}(M^G(\Sigma)) \simeq H^2(M^G(\Sigma),\mathbb{Z})$. We next show that, for any G-module V, $c(\Theta(V))$ does *not* depend upon the choice of the projective variety structure on Σ. Thus, it only depends upon the genus of Σ. We next show that this H^2 class does not depend upon the genus of Σ either once we prove it for $G = \mathrm{SL}_2$. Finally, we prove the last statement for $G = \mathrm{SL}_2$ by using the Grothendieck–Riemann–Roch theorem together with the Poincaré bundle. As a consequence of the main result of this section, we show that the moduli space $M^G(\Sigma)$ is *not* locally factorial for any G of type $B_\ell(\ell \geq 3); D_\ell(\ell \geq 4); G_2; F_4; E_6; E_7; E_8$. It is known to be locally factorial for G of type $A_\ell(\ell \geq 1)$ or $C_\ell(\ell \geq 2)$.

Appendix A. The Dynkin index of a G-module has crucially been used in Chapters 8 and 9 and hence we have included this appendix. For any Lie algebra homomorphism $f\colon \mathfrak{g}_1 \to \mathfrak{g}_2$ between simple Lie algebras, its *Dynkin index* d_f is defined by

$$\langle f(x), f(y)\rangle = d_f\langle x,y\rangle, \quad \text{for all } x,y \in \mathfrak{g}_1,$$

where $\langle\cdot,\cdot\rangle$ is the normalized invariant bilinear form on $\mathfrak{g}_i$ normalized so that $\langle\theta_i,\theta_i\rangle = 2$ for the highest root θ_i of $\mathfrak{g}_i$. For a G-module V, its Dynkin index d_V is defined to be the Dynkin index of the corresponding Lie algebra homomorphism $f_V\colon \mathfrak{g} \to sl(V)$. We give various expressions for d_V; in particular, we give an explicit expression for the Dynkin index of any irreducible G-module purely in terms of its highest weight. This allows us to determine the Dynkin index of any fundamental representation of any G. For example, the Dynkin index of the adjoint representation is shown to be two times the dual Coxeter number of G. For any G, we determine those nontrivial irreducible G-modules which have the minimum Dynkin index. In fact, it turns out that there is only one such G-module up to the diagram automorphism except in the case G of type B_3 (and it is always a fundamental representation). There is a homomorphism $\gamma_\theta\colon \mathrm{SL}_2 \to G$ such that its derivative passes through the highest root space $\mathfrak{g}_\theta$ of $\mathfrak{g}$. We show that γ_θ induces an isomorphism in the third singular cohomology $H^3(G,\mathbb{Z}) \to H^3(\mathrm{SL}_2,\mathbb{Z}) \simeq \mathbb{Z}$. Actually, we give two different proofs: one by using the topology of the corresponding loop groups and the other using Morse theory. Finally, we give a topological characterization of the Dynkin index. Specifically, we prove that for a Lie algebra homomorphism $f\colon \mathfrak{g}_1 \to \mathfrak{g}_2$ as above, its Dynkin index is given by the corresponding map in the third cohomology:

$$f^*\colon H^3(G_2,\mathbb{Z}) \simeq \mathbb{Z} \to H^3(G_1,\mathbb{Z}) \simeq \mathbb{Z}, \quad f^*(n) = d_f n,$$

where G_i are the connected, simply-connected algebraic groups with Lie algebra $\mathfrak{g}_i$. In particular, d_f is an integer (in fact, a non-negative integer).

Appendix B. Since we use the $\mathbb{C}$-space and $\mathbb{C}$-group functors extensively in this book, we have included this appendix to outline the general theory in the form needed for the book. A $\mathbb{C}$-space (resp. $\mathbb{C}$-group) functor $\mathscr{F}$ is a covariant functor from the category of $\mathbb{C}$-algebras to the category of sets (resp. groups) such that it is a sheaf for the fppf topology. We show that any covariant functor from the category of $\mathbb{C}$-algebras to the category of sets can be 'sheafified' to a $\mathbb{C}$-space functor. In the category of $\mathbb{C}$-space functors, we define the notion of fiber product; open subfunctors; open covering; tangent space; condition (E) and condition (E) finitely; condition (L) and condition (L) finitely; adjoint representation and, more generally, any representation of $\mathbb{C}$-group functors satisfying the condition (E); and the Lie algebra of a $\mathbb{C}$-group functor satisfying the condition (L). We discuss in detail the examples of $\mathbf{GL}_V$ and $\mathbf{PGL}_V$ for any $\mathbb{C}$-vector space V (not necessarily finite-dimensional) as $\mathbb{C}$-group functors.

Let Γ be an ind-group scheme and let X be an ind-finite type scheme with an action of Γ and let $\mathscr{V}$ be a Γ-equivariant vector bundle over X. Then we define the action of Γ and the action of its Lie algebra on the cohomology of X with coefficients in $\mathscr{V}$.

Appendix C. Since we use the *stacks* extensively in this book, similar to Appendix B, we have included this appendix to outline the general theory in the form needed for the book. We recall the definition of stacks as a groupoid fibration $\mathscr{T}$ over the category $\mathfrak{S}$ of schemes such that the isomorphisms are a sheaf for $\mathscr{T}$ and every fppf descent datum is effective. This generalizes the notion of space functors. One of the most important example of stacks (at least in the context of this book) is the quotient stack $[\Gamma\backslash X]$ obtained from an ind-group scheme Γ over $\mathbb{C}$ acting on an ind-scheme from the left. We define the fiber product of stacks; representable stacks; representable morphisms between stacks; Deligne–Mumford stacks; algebraic stacks; smooth stacks; dimension of algebraic stacks (which can be a negative integer); and union and disjoint union of stacks. We show that if an affine algebraic group Γ acts on an equi-dimensional scheme X of finite type over $\mathbb{C}$, then

$$\dim [\Gamma\backslash X] = \dim X - \dim \Gamma.$$

Moreover, $[\Gamma\backslash X]$ is a smooth stack in this case if and only if X is smooth. We define the notion of Γ-torsors over $\mathbb{C}$-space functors for a $\mathbb{C}$-group functor Γ. We show that when an ind-group Γ acts on an ind-scheme X from the left, then the canonical morphism of stacks $X \to [\Gamma\backslash X]$ is a Γ-torsor. We define the notion of quasi-coherent sheaves (in particular, vector bundles) over algebraic stacks and global sections of vector bundles over stacks. Let X and Γ be as in the preceding line such that $[\Gamma\backslash X]$ is an algebraic stack and let $\mathscr{V}$ be a Γ-equivariant vector bundle over X. Then it induces a vector bundle $\bar{\mathscr{V}}$ over the quotient stack $[\Gamma\backslash X]$. Moreover, the pull-back map induces an isomorphism

$$H^0([\Gamma\backslash X], \bar{\mathscr{V}}) \simeq H^0(X, \mathscr{V})^\Gamma.$$

Further, as shown in Exercises C.E,

$$\mathrm{Pic}([\Gamma\backslash X]) \simeq \mathrm{Pic}_\Gamma(X),$$

where $\mathrm{Pic}_\Gamma(X)$ denotes the group of isomorphism classes of Γ-equivariant line bundles over X.

Appendix D. This appendix is devoted to a brief survey of the *rank–level duality* including an application of the Verlinde formula to rank-level duality.

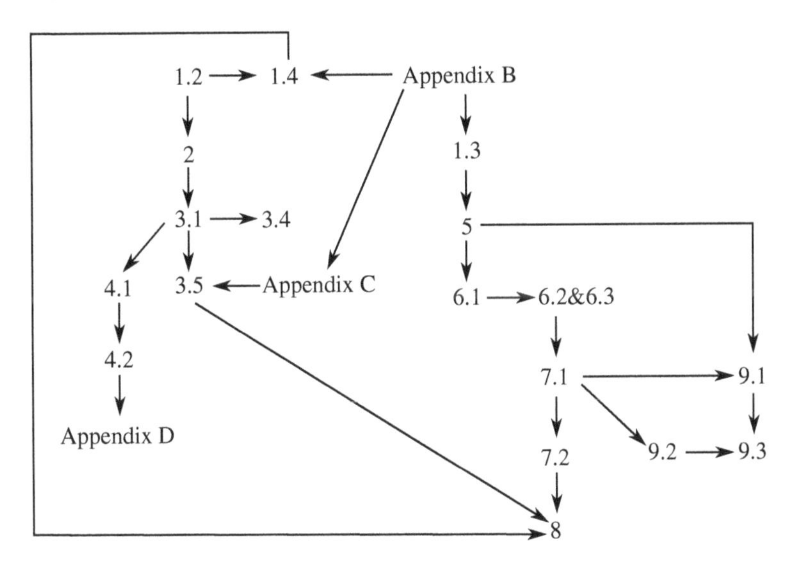

Figure 1 Flowchart of the book

1

An Introduction to Affine Lie Algebras and the Associated Groups

The aim of this chapter is first to set some basic notation and preliminaries (to be used throughout the book) and then recall the definition of affine Kac–Moody Lie algebras and their basic representation theory and to study the associated groups and their flag varieties.

In Section 1.1 we recall the basic notation and preliminaries centered around schemes, varieties, ind-schemes, ind-group schemes, representable functors, quasi-coherent sheaves and vector bundles over ind-schemes. We also recall the Yoneda Lemma (cf. Lemma 1.1.1). *The notation set here will implicitly be used throughout the book.*

Let $\mathfrak{g}$ be a finite-dimensional simple Lie algebra over $\mathbb{C}$ and let G be the connected, simply-connected complex algebraic group with Lie algebra $\mathfrak{g}$.

In Section 1.2 we recall the definition of the associated affine Kac–Moody Lie algebra $\tilde{\mathfrak{g}}$ and its completion $\hat{\mathfrak{g}}$ and their various subalgebras, including the standard Cartan $\hat{\mathfrak{h}}$, standard Borel $\hat{\mathfrak{b}}$ and standard maximal parabolic subalgebra $\hat{\mathfrak{p}}$. Our $\tilde{\mathfrak{g}}$ and $\hat{\mathfrak{g}}$ do *not* include the degree derivation. Then we define their Verma and generalized Verma modules and give an explicit construction of integrable highest-weight modules $\mathscr{H}(\lambda_c)$ (cf. Definition 1.2.6). Further, we show that this explicit construction exhausts all the integrable highest-weight modules of $\hat{\mathfrak{g}}$ and, moreover, these modules are irreducible (cf. Theorem 1.2.10). We also define the affine Weyl group and its action on the Cartan subalgebra of $\mathfrak{g}$ (by affine transformations).

In Section 1.3 we define the loop group $G((t))$ (without the central extension) associated to the Lie algebra $\hat{\mathfrak{g}}$ and its various subgroups, e.g., $G[[t]]$, $G[t^{-1}]$. We define the affine group scheme $\bar{G}[[t]]$ which is a non-noetherian scheme and ind-affine group schemes $\bar{G}((t))$ and $\bar{G}[t^{-1}]$ (cf. Definition 1.3.1). They respectively represent the functors $G(R[[t]]), G(R((t)))$ and $G(R[t^{-1}])$

from the category **Alg** of $\mathbb{C}$-algebras to the category of groups (cf. Lemma 1.3.2). In particular, $\bar{G}[[t]]$, $\bar{G}((t))$ and $\bar{G}[t^{-1}]$ have $\mathbb{C}$-points $G[[t]]$, $G((t))$ and $G[t^{-1}]$, respectively. Then we study the associated infinite Grassmannian $X_G = G((t))/G[[t]]$. Consider the functor $\mathscr{X}_G^o \colon R \in \mathbf{Alg} \rightsquigarrow G(R((t)))/G(R[[t]])$ and let its sheafification be denoted by $\mathscr{X}_G$. We first take $G = \mathrm{SL}_N$ and prove that $\mathscr{X}_{\mathrm{SL}_N}$ is represented by an ind-projective scheme $\bar{X}_{\mathrm{SL}_N}$ using the lattice construction (cf. Theorem 1.3.8). Moreover, $\bar{X}_{\mathrm{SL}_N}(\mathbb{C}) = X_{\mathrm{SL}_N}$. We further observe that the ind-group scheme $\overline{\mathrm{SL}}_N((t))$ acts on the ind-scheme $\bar{X}_{\mathrm{SL}_N}$ (cf. Definition 1.3.10). Then we prove that the product map $\overline{\mathrm{SL}}_N([t^{-1}])^- \times \overline{\mathrm{SL}}_N[[t]] \to \overline{\mathrm{SL}}_N((t))$ is an isomorphism onto an open subset of $\overline{\mathrm{SL}}_N((t))$, where $\overline{\mathrm{SL}}_N([t^{-1}])^-$ is the ind-scheme theoretic kernel of the evaluation homomorphism $\overline{\mathrm{SL}}_N([t^{-1}]) \to \mathrm{SL}_N$, $t^{-1} \mapsto 0$ (cf. Corollary 1.3.15). This last result is generalized for any connected reductive G in Lemma 1.3.16. This allows us to realize the infinite Grassmannian X_G as the $\mathbb{C}$-points of an ind-projective scheme $\bar{X}_G$ which represents the functor $\mathscr{X}_G$ (cf. Proposition 1.3.18). The projection $\pi \colon \bar{G}((t)) \to \bar{X}_G$ is a locally trivial principal $\bar{G}[[t]]$-bundle and $\bar{X}_G$ is an ind-projective scheme as proved in Corollary 1.3.19. This result is extended to $\bar{X}_G$ replaced by $\bar{G}((t))/\mathcal{P}$ for any parahoric subgroup $\mathcal{P} \subset \bar{G}[[t]]$ in Exercise 1.3.E.11.

We prove the following general result (cf. Theorem 1.3.22).

Theorem *Let $\mathscr{G}$ be an ind-affine group scheme filtered by (affine) finite type schemes over $\mathbb{C}$ and let $\mathscr{G}^{\mathrm{red}}$ be the associated reduced ind-affine group scheme. Assume that the canonical ind-group morphism $i \colon \mathscr{G}^{\mathrm{red}} \to \mathscr{G}$ induces an isomorphism of the associated Lie algebras. Then i is an isomorphism of ind-groups, i.e., $\mathscr{G}$ is a reduced ind-scheme.*

The basic idea of the proof involves considering the completion $\hat{\mathscr{G}}$ of $\mathscr{G}$ at the identity e, which is a formal group. Further, the formal groups in characteristic zero are determined by their Lie algebras. Moreover, the Lie algebras of $\mathscr{G}$ and $\hat{\mathscr{G}}$ are isomorphic. Thus, by assumption, we get that $\hat{\mathscr{G}}$ is isomorphic with the completion $\hat{\mathscr{G}}^{\mathrm{red}}$ of $\mathscr{G}^{\mathrm{red}}$ at e (and hence the completions of $\mathscr{G}$ and $\mathscr{G}^{\mathrm{red}}$ at any $\mathbb{C}$-point are isomorphic). From the isomorphism of the completions of $\mathscr{G}$ and $\mathscr{G}^{\mathrm{red}}$ at any $\mathbb{C}$-point, we conclude that $\mathscr{G}$ and $\mathscr{G}^{\mathrm{red}}$ themselves are isomorphic.

As a consequence of the above theorem, we get that the ind-affine group scheme $\bar{G}[t^{-1}]$ is reduced and hence so is $\bar{G}[t^{-1}]^-$ (cf. Theorem 1.3.23). In particular, the infinite Grassmannian $\bar{X}_G$ is a reduced ind-scheme. Moreover, $\bar{G}[[t]]$ is reduced. Thus, so is $\bar{G}((t))$ (cf. Remark 1.3.26(b)). It is shown that the ind-scheme $\bar{X}_G$ coincides with the ind-variety X_G^r defined via the representation theory (cf. Proposition 1.3.24).

We show that for any algebraic group H with a surjective algebraic group homomorphism $H \to \mathbb{C}^*$, $\bar{H}[t]$ is *not* reduced (cf. Example 1.3.25 and Remark 1.3.26(a)).

In Section 1.4 we study the central extension(s) of the ind-group scheme $\bar{G}((t))$. We define the adjoint representation of $G(R((t)))$ in Definition 1.4.2. The projective representation of $\mathfrak{g} \otimes \mathbb{C}((t))$ in any integrable highest-weight module $\mathscr{H}(\lambda_c)$ integrates to a projective representation of $G(R((t)))$ (cf. Proposition 1.4.3 and Theorem 1.4.4). The projective representation of the loop group $G(R((t)))$ in any $\mathscr{H}(\lambda_c) \otimes R$ gives rise to a central extension $\hat{\bar{G}}_{\lambda_c} \to \bar{G}((t))$, which is a $\mathbb{G}_m$-principal bundle, where $\hat{\bar{G}}_{\lambda_c}$ is a reduced ind-group scheme (cf. Definition 1.4.5 and Proposition 1.4.12). In particular, the projective representation of $G(R((t)))$ in $\mathscr{H}(\lambda_c) \otimes R$ lifts to an actual representation of $\hat{\bar{G}}_{\lambda_c}$ in $\mathscr{H}(\lambda_c)$ (cf. Corollary 1.4.7). Further, the central extension $\hat{\bar{G}}_{\lambda_c} \to \bar{G}((t))$ splits over $\bar{G}[[t]]$ as well as $\bar{G}[t^{-1}]^-$ (cf. Theorem 1.4.11).

1.1 Preliminaries and Notation

Unless otherwise explicitly stated, we take the base field to be the field of complex numbers $\mathbb{C}$. Though the bulk of the content of this book generalizes easily to any algebraically closed field of characteristic 0. The identity map of a set X is denoted by I_X , I_X or Id_X (or when no confusion is likely, by I , I or Id itself).

By *schemes* we mean quasi-compact (i.e., finite union of open affine subschemes) separated schemes over $\mathbb{C}$ but not necessarily of finite type over $\mathbb{C}$ (cf. (Mumford, 1988, §II.6, Definition 3), though quasi-compactness is not assumed here). Let $\mathfrak{S}$ be the category of schemes and morphisms between them. For a fixed scheme $S \in \mathfrak{S}$, let $\mathfrak{S}_S$ be the category of S-schemes whose objects are morphisms $f: T \to S$ (with target S) and the set of morphisms Mor (f, f') $(f': T' \to S)$ consists of morphisms $h: T \to T'$ making the following triangle commutative:

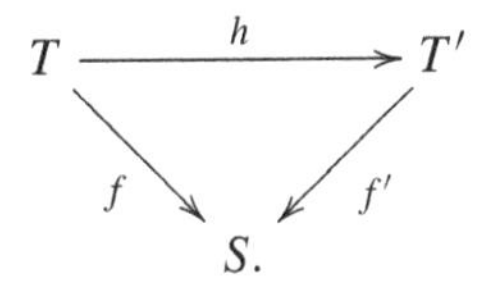

By a *variety* we mean a reduced scheme which is of finite type over $\mathbb{C}$. We do *not* require varieties to be irreducible. When we talk of *points of a variety or scheme* X, we always mean closed points, i.e., points in $X(\mathbb{C})$ (see below).

By an *ind-scheme* $X = (X_n)_{n\geq 0}$ we mean a collection of schemes X_n together with closed embeddings $i_n\colon X_n \hookrightarrow X_{n+1}$ for all $n \geq 0$. We thus think of X_n as a closed subscheme of X_{n+1}. Let $Y = (Y_n)_{n\geq 0}$ be another ind-scheme with closed embeddings $j_n\colon Y_n \hookrightarrow Y_{n+1}$. By a *morphism* $f\colon X \to Y$ we mean a sequence of non-negative integers $(m(0) \leq m(1) \leq m(2) \leq \cdots)$ and a collection of morphisms $f_n\colon X_n \to Y_{m(n)}$ (for all $n \geq 0$) such that the following diagram is commutative:

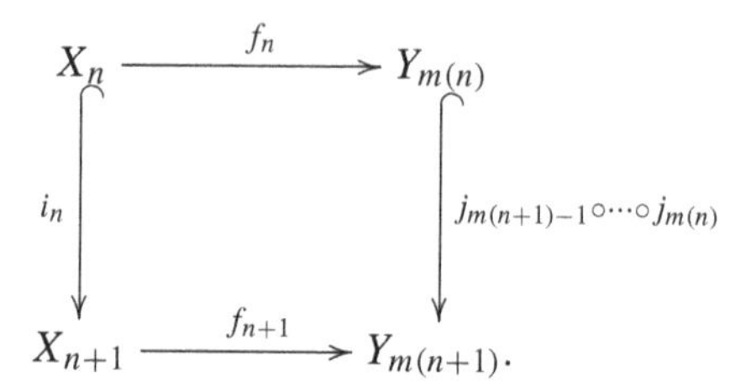

If $f'\colon X \to Y$ is another morphism with the underlying sequence $(m'(0) \leq m'(1) \leq m'(2) \leq \cdots)$, then we say that f and f' are *equivalent* if the following diagram is commutative for all $n \geq 0$ (assuming $m(n) \leq m'(n)$, otherwise we reverse the arrow in the following diagram to $Y_{m'(n)} \to Y_{m(n)}$):

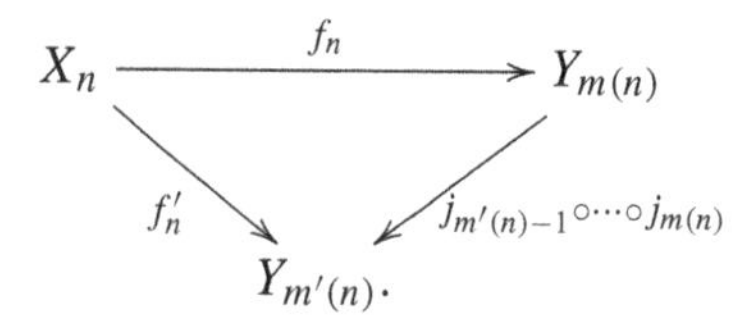

We do not distinguish between two equivalent morphisms. This allows us to talk about *isomorphisms of ind-schemes*.

Let $\bar{X} := \cup_{n\geq 0} X_n$ endowed with the direct limit Zariski topology, where X_n is identified as a closed subspace of X_{n+1} via i_n. Then, a morphism $f\colon X \to Y$ clearly gives rise to a continuous map $\bar{f}\colon \bar{X} \to \bar{Y}$ which only depends upon the equivalence class of f.

A scheme X can be thought of as an ind-scheme by taking $X_n = X$. We call an ind-scheme $X = (X_n)_{n\geq 0}$ of *ind-finite type* if each scheme X_n is of finite type over $\mathbb{C}$. If each X_n is a projective (resp. affine) scheme over $\mathbb{C}$, we call X an *ind-projective scheme* (resp. *ind-affine scheme*). If each X_n is a variety we call X an *ind-variety*. If each X_n is a projective (resp. affine) variety we call X an *ind-projective variety* (resp. *ind-affine variety*). An ind-scheme X is called *irreducible* if under the (direct limit) Zariski topology on $\bar{X}$, it is an irreducible space.

A morphism $f\colon X \to Y$ between ind-schemes is called a *closed embedding* (also called a *closed immersion*) if for each $n \geq 0$, $f_n\colon X_n \to Y_{m(n)}$ is a closed embedding, $\bar{f}(\bar{X})$ is closed in $\bar{Y}$ and $\bar{f}\colon \bar{X} \to \bar{f}(\bar{X})$ is a homeomorphism under the subspace topology on $\bar{f}(\bar{X})$. In this case we also say that X is a *closed ind-subscheme* of Y.

Let **Alg** be the category of commutative algebras over $\mathbb{C}$ with identity (which are not necessarily finitely generated) and all $\mathbb{C}$-algebra homomorphisms between them. Also, let **Set** be the category of sets. For any ind-scheme X, define the covariant functor

$$h_X : \mathbf{Alg} \to \mathbf{Set}, \quad R \rightsquigarrow \mathrm{Mor}(\mathrm{Spec}\, R, X),$$

where Mor is the set of all the morphisms. The functor h_X extends to a contravariant functor

$$\tilde{h}_X : \mathfrak{S} \to \mathbf{Set}, \quad Y \rightsquigarrow \mathrm{Mor}(Y, X).$$

Recall the Yoneda Lemma (cf. (Mumford, 1988, §II.6, Proposition 2) for schemes; its extension to ind-schemes is straightforward).

Lemma 1.1.1 *For any ind-schemes X, Y,*

$$\mathrm{Mor}(X, Y) \simeq \mathrm{Hom}(h_X, h_Y),$$

where Hom denotes the set of natural transformations. Hence, h is a fully faithful functor from the category of ind-schemes to the category of functors from **Alg** *to* **Set**.

By *R-points of an ind-scheme* we mean

$$X(R) := \mathrm{Mor}(\mathrm{Spec}\, R, X). \tag{1}$$

Then, $X(\mathbb{C})$ are the *closed points* of X.

Let **Var** be the category of ind-varieties and morphisms between them. Then, the functor

$$\mathbf{Var} \to \mathbf{Set}, \quad X \rightsquigarrow X(\mathbb{C}),$$

is a faithful functor, i.e., for $X, Y \in \mathbf{Var}$,

$$\mathrm{Mor}(X, Y) \to \mathrm{Maps}(X(\mathbb{C}), Y(\mathbb{C})) \text{ is injective} \tag{2}$$

(cf. (Mumford, 1988, §II.6, p. 162)).

We sometimes abuse the notation and denote ind-scheme X by $X(\mathbb{C})$.

By an *affine algebraic group* we mean an affine algebraic group of finite type over $\mathbb{C}$.

An ind-scheme $X = (X_n)_{n \geq 0}$ is called an *ind-group scheme* if it is equipped with morphisms

$$\mu : X \times X \to X, \; \tau : X \to X \text{ and } \epsilon : \mathrm{Spec}\, \mathbb{C} \to X$$

playing the role of multiplication, inverse and the identity element, respectively. Thus, they are required to satisfy the following three conditions:

(a) Associativity: $\mu \circ (\mu \times I_X) = \mu \circ (I_X \times \mu) \colon X^3 \to X$.

(b) Identity: The two morphisms $\mu \circ (I_X \times \epsilon)$ and $\mu \circ (\epsilon \times I_X) \colon \operatorname{Spec} \mathbb{C} \times X \to X$ coincide with I_X.

(c) Inverse: The morphism $\mu \circ (I_X, \tau) \colon X \to X$ coincides with the composite morphism $X \to \operatorname{Spec} \mathbb{C} \xrightarrow{\epsilon} X$.

In this book we only consider ind-affine group schemes, i.e., ind-group schemes $X = (X_n)_{n \geq 0}$ such that each X_n is an affine scheme. *So, by ind-group schemes, we will always mean ind-affine group schemes.*

For an ind-group scheme X and any $R \in \mathbf{Alg}$, $X(R)$ is clearly an abstract group given by the multiplication μ_R, inverse τ_R and the identity ϵ_R. If an ind-group scheme X is an ind-variety, then we call X an *ind-group variety*.

Let $X = (X_n)_{n \geq 0}$ be an ind-scheme. By a *quasi-coherent sheaf* $\mathcal{F}$ over X, we mean a collection of quasi-coherent sheaves $\mathcal{F}_n$ over X_n together with an isomorphism of $\mathcal{O}_{X_n}$-modules:

$$\theta_n \colon \mathcal{F}_n \simeq i_n^*(\mathcal{F}_{n+1}),$$

for all $n \geq 0$, where $i_n \colon X_n \to X_{n+1}$ is the closed embedding.

If each $\mathcal{F}_n$ is a locally free $\mathcal{O}_{X_n}$-module of rank r, then we call $\mathcal{F}$ a *rank-r vector bundle* over X. If $r = 1$, then, of course, $\mathcal{F}$ is called a *line bundle*.

For a quasi-coherent sheaf $\mathcal{F}$ over X, define

$$H^p(X, \mathcal{F}) = \varprojlim_n H^p(X_n, \mathcal{F}_n),$$

where the map $H^p(X_{n+1}, \mathcal{F}_{n+1}) \to H^p(X_n, \mathcal{F}_n)$ is defined as the composite

$$H^p(X_{n+1}, \mathcal{F}_{n+1}) \to H^p(X_{n+1}, i_{n*}\mathcal{F}_n) \simeq H^p(X_n, \mathcal{F}_n),$$

where the first map is obtained from the $\mathcal{O}_{X_{n+1}}$-module map $\mathcal{F}_{n+1} \to i_{n*}(\mathcal{F}_n)$ via the adjoint of the isomorphism θ_n (cf. (Hartshorne, 1977, Chap. II, §5)) and the second isomorphism is obtained from the closed embedding i_n (cf. (Hartshorne, 1977, Chap. III, Lemma 2.10)).

If an ind-group scheme Γ acts on ind-scheme X, then by a *Γ-equivariant vector bundle* $\mathcal{V}$ over X we mean a vector bundle $\mathcal{V}$ over X with an isomorphism of vector bundles $\phi \colon \mu^*(\mathcal{V}) \simeq \pi_X^*(\mathcal{V})$ over $\Gamma \times X$ satisfying the standard cocycle condition as in Mumford, Fogarty and Kirwan (2002, Definition 1.6), where $\pi_X \colon \Gamma \times X \to X$ is the projection and $\mu \colon \Gamma \times X \to X$ is the action map. For a Γ-equivariant vector bundle $\mathcal{V}$ over X, there is a natural action of $\Gamma(\mathbb{C})$ on $H^p(X, \mathcal{V})$ as follows (also see Definition B.22).

Take $\gamma\colon \operatorname{Spec}\mathbb{C} \to \Gamma$. This gives rise to a morphism $\mu_\gamma\colon X \to X$ by restricting μ to $\operatorname{Spec}\mathbb{C} \times X$ via γ and identifying $\operatorname{Spec}\mathbb{C} \times X$ with X. Thus, we get a canonical map

$$H^p(X, \mathcal{V}) \to H^p(X, \mu_\gamma^* \mathcal{V}) \simeq H^p(X, \mathcal{V}),$$

where the second isomorphism is obtained by restricting the isomorphism ϕ to $\operatorname{Spec}\mathbb{C} \times X \simeq X$. This is the required action of $\Gamma(\mathbb{C})$ on $H^p(X, \mathcal{V})$.

A covariant functor $\mathscr{F}$ from **Alg** $\to$ **Set** is called a *representable functor* if there exists an ind-scheme X such that there is a natural equivalence of functors between $\mathscr{F}$ and h_X. By Lemma 1.1.1, if such an X exists, then it is unique up to an isomorphism. Of course, we can extend this definition for any contravariant functor $\mathfrak{S} \to$ **Set**.

For any $S \in \mathfrak{S}$ and any ind-scheme $X \to S$, define the contravariant functor

$$\tilde{h}_{X/S}\colon \mathfrak{S}_S \to \textbf{Set}, \quad (Y \to S) \rightsquigarrow \operatorname{Mor}_S(Y, X).$$

Then, a contravariant functor $\mathfrak{F}\colon \mathfrak{S}_S \to$ **Set** is called *representable by an ind-scheme X over S* if there is a natural equivalence of functors between $\mathfrak{F}$ and $\tilde{h}_{X/S}$.

For a projective variety X and an affine algebraic group G, any $\mathbb{C}$-analytic G-bundle over X has a unique algebraic G-bundle structure and any analytic morphism between G-bundles is an algebraic morphism (cf. (Serre, 1958, §6.3)). The same is true for vector bundles.

1.2 Affine Lie Algebras

For a more exhaustive treatment of the theory, we refer to the standard text (Kac, 1990).

Let $\mathfrak{g}$ be a finite-dimensional simple Lie algebra over $\mathbb{C}$. Choose a Cartan subalgebra $\mathfrak{h}$ and a Borel subalgebra $\mathfrak{b} \supset \mathfrak{h}$. Let $\Delta^+ \subset \mathfrak{h}^*$ be the set of positive roots (i.e., the roots for the subalgebra $\mathfrak{b}$) and let $\Delta = \Delta^+ \sqcup \Delta^-$ be the set of all the roots of $\mathfrak{g}$, where $\Delta^- := -\Delta^+$. Let $\{\alpha_1, \ldots, \alpha_\ell\} \subset \Delta^+$ be the set of simple roots and let $\{\alpha_1^\vee, \ldots, \alpha_\ell^\vee\} \subset \mathfrak{h}$ be the set of corresponding simple coroots, where $\ell := \dim \mathfrak{h}$ is the rank of $\mathfrak{g}$. Let $\langle \cdot, \cdot \rangle$ be the invariant (symmetric, nondegenerate) bilinear form on $\mathfrak{g}$ normalized so that the induced form on the dual space $\mathfrak{h}^*$ satisfies $\langle \theta, \theta \rangle = 2$ for the highest root θ of $\mathfrak{g}$. *Unless otherwise* *stated, we will always take the invariant form on $\mathfrak{g}$ to be normalized as above.* For any $\alpha \in \Delta$, let $\mathfrak{g}_\alpha \subset \mathfrak{g}$ denote the root space corresponding to the root α.

Definition 1.2.1 Let $\mathfrak{g}$ be (as above) a finite-dimensional simple Lie algebra over $\mathbb{C}$ and let $\mathcal{A} := \mathbb{C}[t, t^{-1}]$, resp. $K = \mathbb{C}((t)) := \mathbb{C}[[t]][t^{-1}]$ be the algebra of Laurent polynomials, resp. the field of Laurent power series. Define the *affine Kac–Moody Lie algebra* (for short *affine Lie algebra*)

$$\tilde{\mathfrak{g}} := \left(\mathfrak{g} \otimes_{\mathbb{C}} \mathcal{A}\right) \oplus \mathbb{C}C, \tag{1}$$

under the bracket

$$[x[t^m] + zC, x'[t^{m'}] + z'C] = [x, x'][t^{m+m'}] + m\delta_{m,-m'} \langle x, x'\rangle C, \tag{2}$$

for $z, z' \in \mathbb{C}, m, m' \in \mathbb{Z}$ and $x, x' \in \mathfrak{g}$, where $x[P]$ denotes $x \otimes P$.

We will be particularly interested in the following 'completion' $\hat{\mathfrak{g}}$ of $\tilde{\mathfrak{g}}$ defined by

$$\hat{\mathfrak{g}} := \mathfrak{g} \otimes_{\mathbb{C}} K \oplus \mathbb{C}C, \tag{3}$$

under the bracket

$$[x[P] + zC, x'[P'] + z'C] = [x, x'][PP'] + \operatorname*{Res}_{t=0} \left((dP)P'\right)\langle x, x'\rangle C, \tag{4}$$

for $P, P' \in K$, $z, z' \in \mathbb{C}$ and $x, x' \in \mathfrak{g}$, where $\operatorname*{Res}_{t=0}$ denotes the coefficient of $t^{-1}dt$.

Clearly, $\tilde{\mathfrak{g}}$ is a Lie subalgebra of $\hat{\mathfrak{g}}$.

The Lie algebra $\hat{\mathfrak{g}}$ admits a derivation d defined by

$$d(x[P]) = x\left[t\left(\frac{dP}{dt}\right)\right], \; d(C) = 0, \; \text{ for } P \in K \text{ and } x \in \mathfrak{g}. \tag{5}$$

Clearly, d keeps $\tilde{\mathfrak{g}}$ stable. Thus, we have semidirect product Lie algebras $\mathbb{C}d \ltimes \hat{\mathfrak{g}}$ and $\mathbb{C}d \ltimes \tilde{\mathfrak{g}}$.

Define the (formal) *loop algebra*

$$\mathfrak{g}((t)) := \mathfrak{g} \otimes_{\mathbb{C}} K, \tag{6}$$

under the bracket

$$[x[P], x'[P']] = [x, x'][PP'], \text{ for } P, P' \in K, \text{ and } x, x' \in \mathfrak{g}. \tag{7}$$

Then, $\hat{\mathfrak{g}}$ can be viewed as a 1-dimensional central extension of $\mathfrak{g}((t))$:

$$0 \to \mathbb{C}C \to \hat{\mathfrak{g}} \xrightarrow{\pi} \mathfrak{g}((t)) \to 0, \tag{8}$$

where the Lie algebra homomorphism π is defined by $\pi(x[P]) = x[P]$, for $P \in K$ and $x \in \mathfrak{g}$, and $\pi(C) = 0$. As proved by Garland (1980, Theorem 3.14) and also independently by V. Chari (unpublished), the above is a universal central extension of $\mathfrak{g}((t))$ (see also Kac (1990, Exercises 3.14 and 7.8)). (For a geometric proof, see Kumar (1985, Corollary 1.9(c)).)

Definition 1.2.2 (Some subalgebras of $\hat{\mathfrak{g}}$) The Lie algebra $\mathfrak{g}$ is embedded in $\hat{\mathfrak{g}}$ as the subalgebra $\mathfrak{g} \otimes t^0$. Define the (standard) *Cartan subalgebra* of $\hat{\mathfrak{g}}$:

$$\hat{\mathfrak{h}} := \mathfrak{h} \otimes t^0 \oplus \mathbb{C}C, \tag{1}$$

the (standard) *Borel subalgebra*:

$$\hat{\mathfrak{b}} := \mathfrak{g} \otimes (t\mathbb{C}[[t]]) \oplus \mathfrak{b} \otimes t^0 \oplus \mathbb{C}C, \tag{2}$$

and the (standard) *maximal parabolic subalgebra*

$$\hat{\mathfrak{p}} := \mathfrak{g} \otimes \mathbb{C}[[t]] \oplus \mathbb{C}C. \tag{3}$$

Also, define the following subalgebras of $\hat{\mathfrak{g}}$:

$$\hat{\mathfrak{g}}_+ := \mathfrak{g} \otimes (t\mathbb{C}[[t]]),\ \hat{\mathfrak{g}}_- := \mathfrak{g} \otimes (t^{-1}\mathbb{C}[t^{-1}]),\ \hat{\mathfrak{l}} := \mathfrak{g} \otimes t^0 \oplus \mathbb{C}C. \tag{4}$$

Then, $\hat{\mathfrak{g}}_+$ is an ideal of $\hat{\mathfrak{p}}$ and we have the *Levi decomposition* (as vector spaces):

$$\hat{\mathfrak{p}} = \hat{\mathfrak{l}} \oplus \hat{\mathfrak{g}}_+. \tag{5}$$

Also, as vector spaces:

$$\hat{\mathfrak{g}} = \hat{\mathfrak{p}} \oplus \hat{\mathfrak{g}}_-. \tag{6}$$

We can similarly define $\tilde{\mathfrak{b}}, \tilde{\mathfrak{g}}_+, \tilde{\mathfrak{g}}_-, \tilde{\mathfrak{p}}$.

Finally, define the 3-dimensional subalgebra of $\hat{\mathfrak{g}}$:

$$\mathfrak{r} := \mathfrak{g}_\theta \otimes t^{-1} \oplus \mathfrak{g}_{-\theta} \otimes t \oplus \mathbb{C}(C - \theta^\vee), \tag{7}$$

where $\mathfrak{g}_\theta$ is the root space corresponding to the highest root θ and $\theta^\vee \in \mathfrak{h}$ is the coroot corresponding to θ.

Let $X = \left(\begin{smallmatrix} 0 & 1 \\ 0 & 0 \end{smallmatrix}\right)$, $Y = \left(\begin{smallmatrix} 0 & 0 \\ 1 & 0 \end{smallmatrix}\right)$, $H = \left(\begin{smallmatrix} 1 & 0 \\ 0 & -1 \end{smallmatrix}\right)$ be the standard basis of $s\ell_2$. Take any $x_\theta \in \mathfrak{g}_\theta$ and $y_\theta \in \mathfrak{g}_{-\theta}$ satisfying $\langle x_\theta, y_\theta \rangle = 1$.

The following lemma is trivial to verify using the commutation relations in $s\ell_2$.

Lemma 1.2.3 *The Lie algebra $\mathfrak{r}$ defined above is isomorphic with the Lie algebra $s\ell_2$ under an isomorphism $\gamma\colon s\ell_2 \to \mathfrak{r}$ taking $X \mapsto y_\theta \otimes t$, $Y \mapsto x_\theta \otimes t^{-1}$ and $H \mapsto C - \theta^\vee$.*

Definition 1.2.4 (a) Let $\mathfrak{s}$ be a Lie algebra and let V be an $\mathfrak{s}$-module. Then V is called a *locally finite* $\mathfrak{s}$-module if, for any $v \in V$, there exists a finite-dimensional $\mathfrak{s}$-submodule $V_v \subset V$ containing v.

In particular, a linear transformation $T \colon V \to V$ (for a vector space V) is called *locally finite* if, for any $v \in V$, there exists a finite-dimensional T-stable subspace V_v containing v. Similarly, T is called *locally nilpotent* if, for any $v \in V$, there exists $n_v \in \mathbb{Z}_{\geq 1}$ such that $T^{n_v}(v) = 0$.

(b) A representation V of $\hat{\mathfrak{g}}$ (or $\tilde{\mathfrak{g}}$) is called *integrable* if V is a locally finite $\mathfrak{g}$-module as well as a locally finite $\mathfrak{r}$-module.

Clearly, any submodule of an integrable module is integrable and so is any quotient.

(c) A representation V of $\hat{\mathfrak{g}}$ is called a *highest-weight module* if V contains a nonzero vector $v_+ \in V$ satisfying the following two properties:

(c_1) The line $\mathbb{C}v_+$ is stable under the action of $\hat{\mathfrak{b}}$.
(c_2) v_+ generates the $\hat{\mathfrak{g}}$-module V, i.e., the only $\hat{\mathfrak{g}}$-submodule of V containing v_+ is the whole of V.

For a Lie algebra $\mathfrak{s}$, let $U(\mathfrak{s})$ denote its *enveloping algebra.*

Any highest-weight $\hat{\mathfrak{g}}$-module V decomposes into homogeneous components:

$$V = \oplus_{d \in \mathbb{Z}_+} V_d, \quad \text{where } V_d := U_d\left(\mathfrak{g} \otimes \mathbb{C}[t^{-1}]\right) \cdot v_+, \ \mathbb{Z}_+ := \mathbb{Z}_{\geq 0},$$

$x[n]$ denotes $x[t^n]$ and $U_d\left(\mathfrak{g} \otimes \mathbb{C}[t^{-1}]\right)$ is the span of $x_1[n_1] \dots x_k[n_k] \in U(\hat{\mathfrak{g}})$ with $n_i \leq 0$ and $\sum_{i=1}^k n_i = -d$.

In exactly the same way we can define the highest-weight modules for the Lie algebra $\tilde{\mathfrak{g}}$, where we replace the Borel subalgebra $\hat{\mathfrak{b}}$ of $\hat{\mathfrak{g}}$ by the standard Borel subalgebra

$$\tilde{\mathfrak{b}} := \mathfrak{g} \otimes (t\mathbb{C}[t]) \oplus \mathfrak{b} \otimes t^0 \oplus \mathbb{C}C. \tag{1}$$

Since $\hat{\mathfrak{g}} = \hat{\mathfrak{p}} \oplus \hat{\mathfrak{g}}_-$ (cf. identity (6) of Definition 1.2.2) and $\tilde{\mathfrak{g}} = \tilde{\mathfrak{p}} \oplus \hat{\mathfrak{g}}_-$, it is easy to see that any highest-weight module of $\hat{\mathfrak{g}}$ is also a highest-weight module for $\tilde{\mathfrak{g}}$, where

$$\tilde{\mathfrak{p}} := (\mathfrak{g} \otimes \mathbb{C}[t]) \oplus \mathbb{C}C. \tag{2}$$

Any quotient module of a highest-weight module is clearly a highest-weight module.

(d) (Verma modules) For any $\hat{\lambda} \in \hat{\mathfrak{h}}^*$, define the *Verma module*

$$\hat{M}(\hat{\lambda}) := U(\hat{\mathfrak{g}}) \otimes_{U(\hat{\mathfrak{b}})} \mathbb{C}_{\hat{\lambda}}, \tag{3}$$

where $U(\hat{\mathfrak{b}})$ acts on $U(\hat{\mathfrak{g}})$ via right multiplication and $\mathbb{C}_{\hat{\lambda}}$ is the 1-dimensional $\hat{\mathfrak{b}}$-module so that the commutator $[\hat{\mathfrak{b}}, \hat{\mathfrak{b}}]$ of course acts trivially on $\mathbb{C}_{\hat{\lambda}}$ and $\hat{\mathfrak{h}}$ acts via the character $\hat{\lambda}$. (Observe that $\hat{\mathfrak{b}} = \hat{\mathfrak{h}} \oplus [\hat{\mathfrak{b}}, \hat{\mathfrak{b}}]$.) The action of $U(\hat{\mathfrak{g}})$ on $\hat{M}(\hat{\lambda})$ is via left multiplication on the first factor.

Clearly, $\hat{M}(\hat{\lambda})$ is a highest-weight $\hat{\mathfrak{g}}$-module. Further, any highest-weight $\hat{\mathfrak{g}}$-module is a quotient of $\hat{M}(\hat{\lambda})$ for some $\hat{\lambda} \in \hat{\mathfrak{h}}^*$.

In exactly the same way, for any $\hat{\lambda} \in \hat{\mathfrak{h}}^*$, we can define the Verma module $\tilde{M}(\hat{\lambda})$ of $\tilde{\mathfrak{g}}$. Then, the canonical map $i\colon \tilde{M}(\hat{\lambda}) \to \hat{M}(\hat{\lambda})$ (induced from the inclusion $\tilde{\mathfrak{g}} \hookrightarrow \hat{\mathfrak{g}}$) is an isomorphism. In particular, the $\tilde{\mathfrak{g}}$-module structure on $\tilde{M}(\lambda)$ extends to a $\hat{\mathfrak{g}}$-module structure.

Similarly, we define the *generalized Verma module* $\hat{M}(V,c)$ for any $\mathfrak{g}$-module V and any $c \in \mathbb{C}$ as follows:

$$\hat{M}(V,c) := U(\hat{\mathfrak{g}}) \otimes_{U(\hat{\mathfrak{p}})} I_c(V) = U(\tilde{\mathfrak{g}}) \otimes_{U(\tilde{\mathfrak{p}})} I_c(V), \tag{4}$$

where $\hat{\mathfrak{p}}$ (resp. $\tilde{\mathfrak{p}}$) is defined by identity (3) of Definition 1.2.2 (resp. identity (2) of Definition 1.2.4), $U(\hat{\mathfrak{p}})$ acts on $U(\hat{\mathfrak{g}})$ via right multiplication and $I_c(V)$ is the vector space V on which $\hat{\mathfrak{p}}$ acts via $(x[P] + zC) \cdot v = P(0)x \cdot v + zcv$, for $P \in \mathbb{C}[[t]]$, $x \in \mathfrak{g}$, $v \in V$, $z \in \mathbb{C}$. Here $P(0)$ denotes the constant term of P. To prove the second equality in (4), use identity (6) of Definition 1.2.2 and the analogous identity for $\tilde{\mathfrak{g}}$.

Let V be a highest-weight $\mathfrak{g}$-module generated by a highest-weight vector $v_+ \neq 0 \in V$ of weight $\lambda \in \mathfrak{h}^*$ (i.e., the line $\mathbb{C}v_+$ is stable under $\mathfrak{b}$, v_+ generates V as a $\mathfrak{g}$-module and the action of $\mathfrak{b}$ on v_+ is via the weight λ). Then, for any $c \in \mathbb{C}$, there is a unique $\hat{\mathfrak{g}}$-module map

$$\pi\colon \hat{M}(\lambda_c) \longrightarrow \hat{M}(V,c),$$

taking $1 \otimes 1_{\lambda_c} \mapsto 1 \otimes v_+$, where $\lambda_c \in \hat{\mathfrak{h}}^*$ is defined by $\lambda_{c|\mathfrak{h}} = \lambda$ and $\lambda_c(C) = c$. Since V is a highest-weight $\mathfrak{g}$-module (by assumption), $U(\mathfrak{g}) \cdot v_+ = I_c(V)$. Thus, π is surjective. In particular, in this case $\hat{M}(V,c)$ is a highest-weight $\hat{\mathfrak{g}}$-module.

Lemma 1.2.5 *For any locally finite $\mathfrak{g}$-module V and any $c \in \mathbb{C}$, $\hat{M}(V,c)$ is locally finite as a $\mathfrak{g}$-module.*

Proof Recall the decomposition $\hat{\mathfrak{g}} = \hat{\mathfrak{p}} \oplus \hat{\mathfrak{g}}_-$ from identity (6) of Definition 1.2.2. Then, by the Poincaré–Birkhoff–Witt (PBW) theorem,

$$U(\hat{\mathfrak{g}}) = U(\hat{\mathfrak{p}}) \otimes_{\mathbb{C}} U(\hat{\mathfrak{g}}_-)$$

as vector spaces, and hence the inclusion

$$\iota\colon U(\hat{\mathfrak{g}}_-) \otimes_{\mathbb{C}} I_c(V) \longrightarrow U(\hat{\mathfrak{g}}) \otimes_{U(\hat{\mathfrak{p}})} I_c(V)$$

is an isomorphism of vector spaces. We next claim that ι is an isomorphism of $\mathfrak{g}$-modules, where $\mathfrak{g}$ acts on $U(\hat{\mathfrak{g}}_-)$ via the adjoint action: $(\operatorname{ad} x)a = xa - ax$, for $x \in \mathfrak{g}$, $a \in U(\hat{\mathfrak{g}}_-)$, and $\mathfrak{g}$ acts on $U(\hat{\mathfrak{g}}_-) \otimes_{\mathbb{C}} I_c(V)$ via the standard tensor product action. (Of course, $\mathfrak{g}$ acts on the range of ι via its standard embedding $\mathfrak{g} \hookrightarrow \hat{\mathfrak{g}}$.) To prove the claim, for $x \in \mathfrak{g}$, $a \in U(\hat{\mathfrak{g}}_-)$ and $v \in I_c(V)$, we have

$$
\begin{aligned}
\iota(x \cdot (a \otimes v)) &= \iota\big((\operatorname{ad} x)a \otimes v\big) + \iota(a \otimes x \cdot v) \\
&= (\operatorname{ad} x)a \otimes v + a \otimes x \cdot v \\
&= (xa - ax) \otimes v + ax \otimes v \\
&= xa \otimes v \\
&= x \cdot \iota(a \otimes v).
\end{aligned}
$$

This proves that ι is a $\mathfrak{g}$-module isomorphism. Now, by assumption, the action of $\mathfrak{g}$ on $I_c(V)$ is locally finite and it is easy to see that the adjoint action of $\mathfrak{g}$ on $U(\hat{\mathfrak{g}}_-)$ is locally finite. This proves the lemma. □

Definition 1.2.6 Let $D \subset \mathfrak{h}^*$ be the set of *dominant integral weights* for $\mathfrak{g}$, i.e.,

$$D := \big\{\lambda \in \mathfrak{h}^* : \lambda(\alpha_i^\vee) \in \mathbb{Z}_+ \text{ for all the simple coroots } \alpha_i^\vee\big\}.$$

For any $\lambda \in D$, let $V(\lambda)$ be the finite-dimensional irreducible $\mathfrak{g}$-module with highest weight λ.

Define the set of dominant integral weights $\hat{D}$ for $\hat{\mathfrak{g}}$ as follows:

$$\hat{D} = \big\{\hat{\lambda} \in \hat{\mathfrak{h}}^* : \hat{\lambda}_{|\mathfrak{h}} \in D \quad \text{and} \quad \hat{\lambda}(C) - \hat{\lambda}(\theta^\vee) \in \mathbb{Z}_+\big\}.$$

We will denote $\hat{\lambda} \in \hat{\mathfrak{h}}^*$ by λ_c, where $\lambda := \hat{\lambda}_{|\mathfrak{h}}$ and $c = \hat{\lambda}(C)$.

For any $\hat{\lambda} = \lambda_c \in \hat{D}$, define the $\hat{\mathfrak{g}}$-module

$$\mathscr{H}(\lambda_c) := \frac{\hat{M}(V(\lambda), c)}{U(\hat{\mathfrak{g}}) \cdot \Big((x_\theta[t^{-1}])^{c-\lambda(\theta^\vee)+1} \otimes v_+\Big)},$$

where x_θ is a nonzero element of $\mathfrak{g}_\theta$ and v_+ is a nonzero vector in the unique line $\mathbb{C}v_+ \subset V(\lambda)$ stabilized by $\mathfrak{b}$.

We prove that $\mathscr{H}(\lambda_c)$ is $\hat{\mathfrak{g}}$-integrable, for which we need the following general result.

Lemma 1.2.7 (a) *Let $\mathfrak{s}$ be any Lie algebra and let $x \in \mathfrak{s}$. Define*

$$\mathfrak{s}_x := \{y \in \mathfrak{s} : (\operatorname{ad} x)^{n_y} y = 0, \quad \textit{for some } n_y \in \mathbb{N}\},$$

where $\mathbb{N} := \mathbb{Z}_{\geq 1}$. Then, $\mathfrak{s}_x$ is a Lie subalgebra of $\mathfrak{s}$.

(b) *For any representation* (V, π) *of* $\mathfrak{s}$ *and* $x \in \mathfrak{s}$, *define* $V_x = \{v \in V : \pi(x)^{n_v} v = 0$, *for some* $n_v \in \mathbb{N}\}$. *Then* V_x *is a* $\mathfrak{s}_x$*-submodule of* V.

(c) *Let* (V, π) *be a representation of* $\mathfrak{s}$ *such that* $\mathfrak{s}$ *is generated (as a Lie algebra) by the set* $F_V = \{x \in \mathfrak{s} : \operatorname{ad} x$ *acting on* $\mathfrak{s}$ *is locally finite and* $\pi(x)$ *is locally finite*$\}$. *Then*

(c_1) $\mathfrak{s}$ *is spanned over* $\mathbb{C}$ *by* F_V. *In particular, if* $\mathfrak{s}$ *is generated by the set* F *of its* ad *locally finite vectors, then* F *spans* $\mathfrak{s}$.

(c_2) *If* $\dim \mathfrak{s} < \infty$, *then any* $v \in V$ *lies in a finite-dimensional* $\mathfrak{s}$*-submodule of* V.

Proof (a) follows immediately from the Leibnitz formula (i.e., $\operatorname{ad} x$ is a derivation)

$$(\operatorname{ad} x)^n [y, z] = \sum_{j=0}^{n} \binom{n}{j} \left[(\operatorname{ad} x)^j \, y, (\operatorname{ad} x)^{n-j} \, z \right].$$

For a locally finite $T : V \to V$, we can define an automorphism $\exp T : V \to V$ in the usual manner:

$$\exp T = I + \sum_{n=1}^{\infty} \frac{T^n}{n!}. \tag{1}$$

Then,

$$\exp(kT) = (\exp T)^k, \qquad \text{for any } k \in \mathbb{Z}. \tag{2}$$

In an associative algebra R, we have the identity (for any $a, b \in R$ and $k \in \mathbb{N}$)

$$(\operatorname{ad} a)^k \, b = \sum_{r=0}^{k} (-1)^r \binom{k}{r} a^{k-r} b a^r, \tag{3}$$

where $\operatorname{ad} a : R \to R$ is defined by

$$(\operatorname{ad} a) b = ab - ba.$$

To obtain (3), apply the Binomial Theorem to $(L_a - R_a)^k$ for the two commuting operators L_a and R_a given respectively by $L_a b = ab$, $R_a b = ba$.

From (3) it is easy to see that for two linear maps $T, S : V \to V$ such that T is locally finite and $\{(\operatorname{ad} T)^n \, S, n \in \mathbb{N}\}$ spans a finite-dimensional subspace of $\operatorname{End} V$, we have

$$(\exp T) S \exp(-T) = \sum_{n \geq 0} \frac{(\operatorname{ad} T)^n}{n!} (S) \tag{4}$$

as operators on V, where $\operatorname{ad} T$ on the right-hand side is to be thought of as an operator on the associative algebra $\operatorname{End} V$ (of all the linear operators of V).

Similar to identity (3), considering the Binomial Theorem for the operator $L_x^n = (\operatorname{ad} x + R_x)^n$, we obtain in any associative algebra R and any elements $x, a \in R$,

$$x^n a = \sum_{j=0}^{n} \binom{n}{j} \big((\operatorname{ad} x)^j\, a\big) x^{n-j}.$$

Applying the above identity to v, the (b)-part follows.

We first show that for $a, x \in F_V$ and $t \in \mathbb{C}$, $(\exp(t \operatorname{ad} a))\, x \in F_V$: Since π is a Lie algebra representation, for any $y, z \in \mathfrak{s}$ and $n \in \mathbb{Z}_+$,

$$\pi\big((\operatorname{ad} y)^n\, z\big) = \big(\operatorname{ad} \pi(y)\big)^n\, \pi(z), \tag{5}$$

as elements of $\operatorname{End}(V)$. In particular, for $a, x \in F_V$,

$$\begin{aligned} \pi\big((\exp(\operatorname{ad} a))x\big) &= \big(\exp(\operatorname{ad} \pi(a))\big)\pi(x) \\ &= \exp(\pi a)\, \pi(x)\, \exp(-\pi a), \quad \text{by (4).} \end{aligned} \tag{6}$$

(Observe that, since $a \in F_V$, $\pi(a)$ is locally finite and, by (5), $\{(\operatorname{ad} \pi(a))^n \pi(x) : n \in \mathbb{N}\}$ is finite dimensional.) This shows that $\pi\big((\exp(t \operatorname{ad} a))x\big)$ is locally finite. Taking V to be the adjoint representation, we see that $\big(\exp(t \operatorname{ad} a)\big)\, x \in F_V$.

Let $\mathfrak{s}_V \subset \mathfrak{s}$ be the $\mathbb{C}$-span of F_V. Since

$$\operatorname*{limit}_{t \to 0} \frac{\big(\exp(t \operatorname{ad} a)\big)x - x}{t} = [a, x],$$

we see that $[a, x] \in \mathfrak{s}_V$ (for $a, x \in F_V$). In particular, $\mathfrak{s}_V$ is a Lie subalgebra of $\mathfrak{s}$. This proves (c_1). Now (c_2) follows from (c_1) by the PBW theorem. □

Proposition 1.2.8 *For any $\lambda_c \in \hat{D}$, the $\hat{\mathfrak{g}}$-module $\mathscr{H}(\lambda_c)$ is an integrable highest-weight $\hat{\mathfrak{g}}$-module.*

By Exercise 1.1.E.4, $\mathscr{H}(\lambda_c)$ is nonzero.

Proof We have already seen in Definition 1.2.4(d) that $\mathscr{H}(\lambda_c)$ is a highest-weight $\hat{\mathfrak{g}}$-module. By Lemma 1.2.5, it is locally finite as a $\mathfrak{g}$-module. So, to prove that it is integrable, it suffices to show that it is locally finite as an $\mathfrak{r}$-module.

Apply Lemma 1.2.7(b) in the case $\mathfrak{s} = \tilde{\mathfrak{g}}$, $x = x_\theta[t^{-1}]$ and $V = \mathscr{H}(\lambda_c)$. By Exercise 1.1.E.1, $\mathfrak{s}_x = \mathfrak{s}$. Moreover, clearly $1 \otimes v_+ \in V_x$ and, by Lemma 1.2.7(b), V_x is an $\mathfrak{s}_x = \mathfrak{s}$ submodule of V. Further, the $\mathfrak{s}$-submodule of V generated by $1 \otimes v_+$ is the whole of V. To prove this, observe that the canonical map $j\colon \tilde{M}(\lambda_c) \to \hat{M}(\lambda_c)$ is an isomorphism and, moreover, the canonical map $\pi : \hat{M}(\lambda_c) \to \hat{M}(V(\lambda), c)$ is surjective (cf. Definition 1.2.4(d)). Thus, $V_x = V$, i.e., $x_\theta[t^{-1}]$ acts locally nilpotently on $\mathscr{H}(\lambda_c)$. By the same argument we see that $x_{-\theta}[t]$ acts locally nilpotently on V. Now, any $s\ell_2$-module L such that X and Y act locally nilpotently on L is a locally finite $s\ell_2$-module. This follows, e.g., by Lemma 1.2.7(c_2). Thus, in view of Lemma 1.2.3, the proposition is proved. □

A $(\mathbb{C}d \ltimes \hat{\mathfrak{g}})$-module V is called *integrable* if it is integrable as a $\hat{\mathfrak{g}}$-module. It is called a *highest-weight* $(\mathbb{C}d \ltimes \hat{\mathfrak{g}})$-module if there exists a line $\mathbb{C}v_+ \subset V$ which is stable under $\mathbb{C}d \ltimes \hat{\mathfrak{b}}$ and v_+ generates V as a $(\mathbb{C}d \ltimes \hat{\mathfrak{g}})$-module, where $\hat{\mathfrak{b}}$ is defined by identity (2) of Definition 1.2.2. (The notion of a highest-weight $(\mathbb{C}d \ltimes \tilde{\mathfrak{g}})$-module can, of course, be defined similarly.) With this definition we recall the following important theorem from Kumar (2002, Corollaries 2.2.6, 3.2.10 and Theorem 13.1.3).

Theorem 1.2.9 *Any integrable highest-weight $(\mathbb{C}d \ltimes \tilde{\mathfrak{g}})$-module is irreducible.*

Theorem 1.2.10 *Any integrable highest-weight $\hat{\mathfrak{g}}$-module is isomorphic with a unique $\mathscr{H}(\lambda_c)$, $\lambda_c \in \hat{D}$.*

Thus, $\lambda_c \mapsto \mathscr{H}(\lambda_c)$ sets up a bijective correspondence between $\hat{D}$ and the set of isomorphism classes of integrable highest-weight $\hat{\mathfrak{g}}$-modules.

Moreover, $\mathscr{H}(\lambda_c)$ is an irreducible $\hat{\mathfrak{g}}$-module.

Proof Take an integrable highest-weight $\hat{\mathfrak{g}}$-module V. Let $\mathbb{C}v_+ \subset V$ be a line stable under $\hat{\mathfrak{b}}$ such that the $\hat{\mathfrak{g}}$-submodule generated by v_+ is the whole of V. Let $\lambda_c \in \hat{\mathfrak{h}}^*$ be the character by which $\hat{\mathfrak{b}}$ acts on the line $\mathbb{C}v_+$. Since V is integrable, the $\mathfrak{g}$-submodule V^o generated by v_+ is finite dimensional and so is the $\mathfrak{r}$-submodule V' generated by v_+. Since the Borel subalgebra $\mathfrak{b} \subset \mathfrak{g}$ keeps the line $\mathbb{C}v_+$ stable, from the representation theory of $\mathfrak{g}$ applied to V^o, we get $\lambda \in D$ and $V^o \simeq V(\lambda)$ as $\mathfrak{g}$-modules (cf. (Serre, 1966, Théorème 1 and Proposition 3(d), Chapitre VII)). Moreover, from the $s\ell_2$-representation theory (cf. (Serre, 1966, Corollaire 1, Chapitre IV)) and Lemma 1.2.3, $\lambda_c(C - \theta^\vee) \in \mathbb{Z}_+$, i.e., $\lambda_c \in \hat{D}$.

Since $\hat{\mathfrak{g}}_+$ annihilates v_+ and hence V^o, we get a surjective $\hat{\mathfrak{g}}$-module map

$$\phi \colon \hat{M}\big(V(\lambda), c\big) \to V,$$

taking $I_c(V(\lambda)) \xrightarrow{\sim} V^o$ isomorphically as a $\hat{\mathfrak{r}} = (\mathfrak{g} \oplus \mathbb{C}C)$-module.

Again, using the $s\ell_2$-representation theory (cf. (Serre, 1966, Corollaire 1, Chapitre IV)) and Lemma 1.2.3,

$$(x_\theta[t^{-1}])^{c-\lambda(\theta^\vee)+1} \cdot v_+ = 0 \quad \text{in } V'.$$

Thus, ϕ factors through (as a surjective $\hat{\mathfrak{g}}$-module map)

$$\bar{\phi}: \mathscr{H}(\lambda_c) \to V.$$

For any $\mathfrak{g}$-module L and any $c \in \mathbb{C}$, define the action of d on

$$\hat{M}(L,c) \simeq U(\hat{\mathfrak{g}}_-) \otimes_{\mathbb{C}} I_c(L)$$

via its standard derivation action on $U(\hat{\mathfrak{g}}_-)$ induced from the action on $\hat{\mathfrak{g}}_-$ given in identity (5) of Definition 1.2.1 (d acts trivially on $I_c(L)$). This action of d turns the $\hat{\mathfrak{g}}$-module $\hat{M}(L,c)$ into a $(\mathbb{C}d \ltimes \hat{\mathfrak{g}})$-module. Clearly, this $(\mathbb{C}d \ltimes \hat{\mathfrak{g}})$-module structure on $\hat{M}(V(\lambda),c)$ descends to a $(\mathbb{C}d \ltimes \hat{\mathfrak{g}})$-module structure on the quotient $\mathscr{H}(\lambda_c)$, making it an integrable and a highest-weight $(\mathbb{C}d \ltimes \hat{\mathfrak{g}})$-module (cf. Definition 1.2.4(d) and Proposition 1.2.8); in particular, an integrable and highest-weight $(\mathbb{C}d \ltimes \tilde{\mathfrak{g}})$-module. Thus, by Theorem 1.2.9, it is an irreducible $(\mathbb{C}d \ltimes \tilde{\mathfrak{g}})$-module, and hence an irreducible $(\mathbb{C}d \ltimes \hat{\mathfrak{g}})$-module. We next show that it is irreducible as a $\hat{\mathfrak{g}}$-module.

Let $N \subset \mathscr{H}(\lambda_c)$ be a nonzero $\hat{\mathfrak{g}}$-submodule. Consider the decomposition

$$\mathscr{H}(\lambda_c) = \oplus_{i \in \mathbb{Z}_+} \mathscr{H}(\lambda_c)_i,$$

where

$$\mathscr{H}(\lambda_c)_i := \{v \in \mathscr{H}(\lambda_c) : d \cdot v = -iv\}. \tag{1}$$

Observe that for any $n \in \mathbb{Z}$ and $x \in \mathfrak{g}$,

$$x[t^n] \cdot \mathscr{H}(\lambda_c)_i \subset \mathscr{H}(\lambda_c)_{i-n}. \tag{2}$$

For any nonzero $v \in \mathscr{H}(\lambda_c)$, $v = \sum v_i$ with $v_i \in \mathscr{H}(\lambda_c)_i$, set $|v| = \sum i : v_i \neq 0$. Choose a nonzero $v^o \in N$ such that $|v^o| \leq |v|$ for all nonzero $v \in N$. Then,

$$x[t^n] \cdot v^o = 0 \quad \text{for all } n \geq 1 \text{ and } x \in \mathfrak{g}. \tag{3}$$

For, otherwise, $|x[t^n] \cdot v^o| < |v^o|$, which contradicts the choice of v^o. If $|v^o| > 0$, take a nonzero component $v^o_{i_o}$ with $i_o > 0$. By (1) and (2),

$$x[t^n] \cdot v^o_{i_o} = 0 \quad \text{for all } n \geq 1 \text{ and } x \in \mathfrak{g}.$$

In particular, by the PBW theorem, the $(\mathbb{C}d \ltimes \hat{\mathfrak{g}})$-submodule of $\mathscr{H}(\lambda_c)$ generated by $v^o_{i_o}$ is proper, which contradicts the irreducibility of $\mathscr{H}(\lambda_c)$ as a $(\mathbb{C}d \ltimes \hat{\mathfrak{g}})$-module. Thus, $|v^o| = 0$, i.e., $v^o \in \mathscr{H}(\lambda_c)_0$ and hence, by the PBW theorem, the $(\mathbb{C}d \ltimes \hat{\mathfrak{g}})$-submodule of $\mathscr{H}(\lambda_c)$ generated by v^o is the same as the $\hat{\mathfrak{g}}$-submodule of $\mathscr{H}(\lambda_c)$ generated by v^o. Hence, $N = \mathscr{H}(\lambda_c)$, proving the irreducibility of $\mathscr{H}(\lambda_c)$ as a $\hat{\mathfrak{g}}$-module.

From the irreducibility of $\mathscr{H}(\lambda_c)$ as a $\hat{\mathfrak{g}}$-module, we get that $\bar{\phi}$ is an isomorphism.

So, to complete the proof of the theorem, it suffices to show that for $\lambda_c \neq \mu_{c'} \in \hat{D}$, $\mathscr{H}(\lambda_c)$ and $\mathscr{H}(\mu_{c'})$ are nonisomorphic as $\hat{\mathfrak{g}}$-modules.

Define the $\mathfrak{g}$-submodule

$$\mathscr{H}(\lambda_c)^o = \{v \in \mathscr{H}(\lambda_c) : \hat{\mathfrak{g}}_+ \cdot v = 0\}.$$

Then, clearly, as a $\mathfrak{g}$-submodule of $\mathscr{H}(\lambda_c)$,

$$1 \otimes V(\lambda) = \mathscr{H}(\lambda_c)^o_0 \text{ and } \mathscr{H}(\lambda_c)^o = \bigoplus_{i \geq 0} \mathscr{H}(\lambda_c)^o_i. \tag{4}$$

We claim that, for any $i > 0$,

$$\mathscr{H}(\lambda_c)^o_i = 0. \tag{5}$$

For, if not, the $\hat{\mathfrak{g}}$-submodule of $\mathscr{H}(\lambda_c)$ generated by $\mathscr{H}(\lambda_c)^o_i$ would be proper (again use the PBW theorem), contradicting the irreducibility of the $\hat{\mathfrak{g}}$-module $\mathscr{H}(\lambda_c)$. Thus,

$$\mathscr{H}(\lambda_c)^o = 1 \otimes V(\lambda).$$

So, if $\mathscr{H}(\lambda_c)$ and $\mathscr{H}(\mu_{c'})$ are isomorphic as $\hat{\mathfrak{g}}$-modules, then the $\mathfrak{g}$-modules $V(\lambda)$ and $V(\mu)$ are isomorphic, i.e., $\lambda = \mu$. Moreover, the action of C on $\mathscr{H}(\lambda_c)$ and $\mathscr{H}(\mu_{c'})$ is by the same scalar, i.e., $c = c'$. Thus $\lambda_c = \mu_{c'}$, proving the theorem completely. □

Definition 1.2.11 Recall from the beginning of this section that $\langle \cdot, \cdot \rangle$ is the invariant normalized form on $\mathfrak{g}$. Extend this to an invariant symmetric bilinear form on $\hat{\mathfrak{g}}$, still denoted by $\langle \cdot, \cdot \rangle$, as follows:

$$\langle x[P], y[Q] \rangle = \operatorname*{Res}_{t=0} (t^{-1} PQ) \langle x, y \rangle, \text{ for } x, y \in \mathfrak{g} \text{ and } P, Q \in K, \quad \langle C, \hat{\mathfrak{g}} \rangle = 0.$$

This form clearly descends to a bilinear form on the loop algebra $\mathfrak{g}((t)) = \mathfrak{g} \otimes K$. It is easy to see that this form on $\mathfrak{g}((t))$ is nondegenerate.

Definition 1.2.12 Let W be the Weyl group of $\mathfrak{g}$. Then W can be realized as the subgroup of $\mathrm{Aut}(\mathfrak{h})$ generated by the *simple reflections* $\{s_1, \ldots, s_\ell\}$, where

$$s_i(h) := h - \alpha_i(h)\, \alpha_i^\vee, \text{ for } h \in \mathfrak{h}. \tag{1}$$

Then W is a Coxeter group with Coxeter generators $\{s_1, \ldots, s_\ell\}$.

The dual representation of W in $\mathfrak{h}^*$ is explicitly given by

$$s_i(\lambda) = \lambda - \lambda(\alpha_i^\vee)\, \alpha_i, \text{ for } \lambda \in \mathfrak{h}^*. \tag{2}$$

Let $Q^\vee \in \mathfrak{h}$ be the coroot lattice of $\mathfrak{g}$:

$$Q^\vee := \bigoplus_{i=1}^{\ell} \mathbb{Z}\alpha_i^\vee. \tag{3}$$

Since $\alpha_i(\alpha_j^\vee) \in \mathbb{Z}$ (cf. (Serre, 1966, Chap. V, §11)), $s_i Q^\vee \subset Q^\vee$. Thus, W keeps $Q^\vee$ stable.

Define the *affine Weyl group* to be the semidirect product

$$\hat{W} := W \ltimes Q^\vee. \tag{4}$$

For $q \in Q^\vee$, we denote the corresponding element of $\hat{W}$ by τ_q. By definition, $\hat{W}$ acts on $\mathfrak{h}$ via affine transformations, where W acts linearly on $\mathfrak{h}$ via the standard action (1) and τ_q acts on $\mathfrak{h}$ via translation:

$$\tau_q(h) = q + h. \tag{5}$$

Consider the element $s_0 \in \hat{W}$ defined by

$$s_0 = \tau_{\theta^\vee}\gamma_\theta, \tag{6}$$

where (as in Definition 1.2.2) $\theta^\vee$ is the coroot corresponding to the highest root θ and $\gamma_\theta \in W$ is the reflection through the root plane θ, i.e., $\gamma_\theta h = h - \theta(h)\theta^\vee$.

The following well-known result can be found, e.g., in (Kumar, 2002, Propositions 13.1.7, 1.3.21 and the identity (13.1.1.7)).

Lemma 1.2.13 *The affine Weyl group $\hat{W}$ is a Coxeter group with Coxeter generators $\{s_0, s_1, \ldots, s_\ell\}$. In particular, for any $\hat{w} \in \hat{W}$, we have the notion of its length $\ell(\hat{w})$.*

The Coxeter relations among $\{s_i\}_{1\le i\le \ell}$ together with the following relations provide a complete set of relations for $\hat{W}$:

(a) $s_0^2 = 1$,
(b) $(s_0 s_i)^{m_i} = 1$, for all $1 \le i \le \ell$,

where $m_i = 2, 3, 4, 6$ or ∞ according as $\alpha_i(\theta^\vee)\theta(\alpha_i^\vee) = 0, 1, 2, 3$ or ≥ 4, respectively.

1.2.E Exercises

(1) For any root vector $x_\beta \in \mathfrak{g}_\beta$ and $n \in \mathbb{Z}$, show that $\operatorname{ad}(x_\beta[t^n]) : \tilde{\mathfrak{g}} \to \tilde{\mathfrak{g}}$ is a locally nilpotent transformation.

(2) Show that for any highest-weight $\tilde{\mathfrak{g}}$-module V, its $\tilde{\mathfrak{g}}$-module structure extends to a $\hat{\mathfrak{g}}$-module structure.

(3) For any $\lambda_c \in \hat{D}$, show that the line $\mathbb{C}\big(x_\theta[t^{-1}]^{c-\lambda(\theta^\vee)+1} \otimes v_+\big)$ inside $\hat{M}(V(\lambda), c)$ is stable under the action of $\mathfrak{b}$ and is annihilated by $\hat{\mathfrak{g}}_+$. Thus, the line is stable under $\hat{\mathfrak{b}}$.
(4) Show that, for any $\lambda_c \in \hat{D}$, $\mathscr{H}(\lambda_c)$ is nonzero. *Hint:* Use Exercise 3.
(5) Show that, for any $\lambda_c \in \hat{D}$, the line $\mathbb{C}v_+ \subset \mathscr{H}(\lambda_c)$ is the unique line stable under $\hat{\mathfrak{b}}$. Hence, any $\hat{\mathfrak{g}}$-module endomorphism of $\mathscr{H}(\lambda_c)$ is the identity map up to a scalar multiple. Moreover, $\mathbb{C}v_+ \subset \mathscr{H}(\lambda_c)$ is the unique line annihilated by $\hat{\mathfrak{u}} := (\mathfrak{g} \otimes t\mathbb{C}[[t]]) \oplus \mathfrak{u}$, where $\mathfrak{u}$ is the nil-radical of $\mathfrak{b}$.
(6) Show that, for any $\lambda_c \in \hat{D}$, $\hat{M}(\lambda_c)$ has a unique proper maximal $\hat{\mathfrak{g}}$-submodule. Hence, $\mathscr{H}(\lambda_c)$ is the unique irreducible quotient of $\hat{M}(\lambda_c)$.
(7) For any $f \in K = \mathbb{C}((t))$, any root vector $x_\beta \in \mathfrak{g}_\beta$, and any $\lambda_c \in \hat{D}$, show that $x_\beta[f]$ acts locally nilpotently on $\mathscr{H}(\lambda_c)$.

1.3 Loop Groups and Infinite Grassmannians

We follow the convention from Section 1.1.

As in Definition 1.2.1, let $K = \mathbb{C}((t)) = \mathbb{C}[[t]][t^{-1}]$ be the field of Laurent power series.

For any commutative $\mathbb{C}$-algebra R with identity and affine scheme X over $\mathbb{C}$, let $X(R)$ denote the R-points of X. Then, $X(R)$ can be identified with the set of all the $\mathbb{C}$-algebra homomorphisms $f\colon \mathbb{C}[X] \to R$, where $\mathbb{C}[X]$ is the affine coordinate ring of X (cf. (Mumford, 1988, §II.6, Definition 1 and §II.2, Theorem 1)). We want to realize $G(K)$, $G(\mathbb{C}[t^{-1}])$ as $\mathbb{C}$-points of ind-affine group schemes and $G(\mathbb{C}[[t]])$ as $\mathbb{C}$-points of an affine group scheme.

Recall first that $\operatorname{Spec}(\mathbb{C}[y_1, y_2, \dots])(\mathbb{C}) = \mathbb{C}[[t]]$, where an element $\sum_{n\geq 0} a_n t^n \in \mathbb{C}[[t]]$ corresponds to the unique algebra homomorphism $\mathbb{C}[y_1, y_2, \dots] \to \mathbb{C}$ taking y_n to a_n.

Definition 1.3.1 Let G be any affine algebraic group (of finite type over $\mathbb{C}$). Take a faithful representation $i\colon G \hookrightarrow \mathrm{SL}_N \subset M_N$, where M_N is the vector space of $N \times N$ matrices over $\mathbb{C}$, and let $I_G \subset S(M_N^*)$ be the radical ideal of G inside M_N. For any $1 \leq i, j \leq N$ and an integer n, define the linear function

$$y_n^{i,j}\colon M_N((t)) := M_N \otimes_{\mathbb{C}} \mathbb{C}((t)) \to \mathbb{C},\quad E \otimes \left(\sum_{n\in\mathbb{Z}} a_n t^n\right) \mapsto y^{i,j}(E) a_n,$$

for $E \in M_N$ and $\sum_n a_n t^n \in \mathbb{C}((t))$, where $y^{i,j}\colon M_N \to \mathbb{C}$ is the linear function taking any $E \in M_N$ to its (i, j)th entry.

For any $P \in \mathbb{C}[M_N] = S(M_N^*)$, let $\widehat{P}\colon M_N((t)) \to \mathbb{C}((t))$ be the (polynomial) function obtained from extending the scalars from $\mathbb{C}$ to $\mathbb{C}((t))$. Express $\widehat{P} = \sum_{m\in\mathbb{Z}} \widehat{P}_m t^m$. For any $d \geq 0$, restrict $\widehat{P}$ to $M_N \otimes t^{-d}\mathbb{C}[[t]]$ and denote this restriction by $\widehat{P}^{(d)}$. Then, $\widehat{P}_m^{(d)} = 0$ for $m << 0$ and $\widehat{P}_m^{(d)}$ are polynomial functions on $M_N \otimes t^{-d}\mathbb{C}[[t]]$.

Let $R_N^{(d)}$ be the polynomial ring in the variables $\{y_n^{i,j}\}_{n\geq -d;\, 1\leq i,j\leq N}$ and let $I_G^{(d)}$ be the ideal of $R_N^{(d)}$ generated by $\{\widehat{P}_m^{(d)} : m \in \mathbb{Z}$ and $P \in I_G\}$.

Consider the affine (though non-noetherian) scheme $\bar{G}(t^{-d}\mathbb{C}[[t]])$ associated to the ring $R_N^{(d)}/I_G^{(d)}$, i.e.,

$$\bar{G}(t^{-d}\mathbb{C}[[t]]) := \operatorname{Spec}\left(R_N^{(d)}/I_G^{(d)}\right).$$

In particular, taking $d = 0$, we get the affine scheme

$$\bar{G}[[t]] = \bar{G}(\mathbb{C}[[t]]) := \operatorname{Spec}\left(R_N^{(0)}/I_G^{(0)}\right).$$

Exactly similarly, we can define the scheme

$$\bar{G}\left(\sum_{p=0}^{d}\mathbb{C}t^{-p}\right) := \operatorname{Spec}\left(\left(\mathbb{C}[y_n^{i,j}]_{-d\leq n\leq 0;\,1\leq i,j\leq N}\right)/\left\langle\left(\widehat{P}_m^{(d)}\right)_{|M_N\otimes\sum_{p=0}^{d}\mathbb{C}t^{-p}} : m \in \mathbb{Z} \text{ and } P \in I_G\right\rangle\right).$$

Clearly, the inclusions (for any $d \geq 0$)

$$\bar{G}\left(t^{-d}\mathbb{C}[[t]]\right) \subset \bar{G}\left(t^{-d-1}\mathbb{C}[[t]]\right) \text{ and } \bar{G}\left(\sum_{p=0}^{d}\mathbb{C}t^{-p}\right) \subset \bar{G}\left(\sum_{p=0}^{d+1}\mathbb{C}t^{-p}\right),$$

under the above scheme structures, are closed embeddings. This gives rise to ind-schemes

$$\bar{G}((t)) := \left\{\bar{G}(t^{-d}\mathbb{C}[[t]])\right\}_{d\geq 0} \text{ and}$$
$$\bar{G}[t^{-1}] := \bar{G}\left(\mathbb{C}[t^{-1}]\right) = \left\{\bar{G}\left(\sum_{p=0}^{d}\mathbb{C}t^{-p}\right)\right\}_{d\geq 0}.$$

Observe that $\bar{G}((t))$ is an inductive limit of non-noetherian affine schemes with closed embedding in $\overline{\mathrm{SL}}_N((t))$, whereas $\bar{G}[t^{-1}]$ is an inductive limit of noetherian affine schemes (in fact, affine schemes of finite type over $\mathbb{C}$) with closed embedding in $\overline{\mathrm{SL}}_N[t^{-1}]$.

By virtue of the following Lemma 1.3.2, the (ind)-scheme structures on $\bar{G}[[t]]$, $\bar{G}((t))$ and $\bar{G}[t^{-1}]$ do not depend upon the choice of a faithful representation $G \hookrightarrow \mathrm{SL}_N$.

Lemma 1.3.2 *Let G be any affine algebraic group. Consider the covariant functors $\mathscr{F}_1$, $\mathscr{F}_2$, $\mathscr{F}_3$ from* **Alg** *to* **Set** *by*

$$\mathscr{F}_1(R) = G(R[[t]]),$$

$$\mathscr{F}_2(R) = G(R((t))),$$

$$\mathscr{F}_3(R) = G(R[t^{-1}]).$$

Then all these are representable functors represented respectively by the scheme $\bar{G}[[t]]$ and ind-schemes $\bar{G}((t))$ and $\bar{G}[t^{-1}]$ (with the scheme structure given in Definition 1.3.1).

In particular, the (ind)-scheme structures on these do not depend upon the choice of a faithful representation $i: G \hookrightarrow \mathrm{SL}_N$. Moreover, the $\mathbb{C}$-points of $\bar{G}[[t]]$, $\bar{G}((t))$ and $\bar{G}[t^{-1}]$ coincide with $G[[t]] := G(\mathbb{C}[[t]])$, $G((t)) := G(\mathbb{C}((t)))$ and $G[t^{-1}] := G(\mathbb{C}[t^{-1}])$, respectively.

Further, $\bar{G}[[t]]$ is an affine group scheme, which is a closed subgroup scheme of $\overline{\mathrm{SL}}_N[[t]]$. Similarly, $\bar{G}((t))$ and $\bar{G}[t^{-1}]$ are ind-affine group schemes which are closed ind-subgroup schemes of $\overline{\mathrm{SL}}_N((t))$ and $\overline{\mathrm{SL}}_N[t^{-1}]$, respectively.

Proof We prove the lemma for $\mathscr{F}_1$; the proof for $\mathscr{F}_2$ and $\mathscr{F}_3$ is similar. Let R be a $\mathbb{C}$-algebra. We need to prove that there is a functorial identification

$$\mathrm{Mor}(\mathrm{Spec}\, R, \bar{G}[[t]]) \simeq G(R[[t]]). \tag{1}$$

As at the beginning of the section, since G is an affine variety, there is a canonical bijection

$$G(R[[t]]) \simeq \mathrm{Hom}_{\mathrm{alg}}(\mathbb{C}[G], R[[t]]), \tag{2}$$

where $\mathrm{Hom}_{\mathrm{alg}}(-,-)$ denotes the set of $\mathbb{C}$-algebra homomorphisms. Further, since $\bar{G}[[t]]$ is an affine scheme, there is a canonical bijection

$$\mathrm{Mor}(\mathrm{Spec}\, R, \bar{G}[[t]]) \simeq \mathrm{Hom}_{\mathrm{alg}}(\mathbb{C}[\bar{G}[[t]]], R). \tag{3}$$

Combining (1)–(3), it suffices to prove that there is a canonical bijection

$$\mathrm{Hom}_{\mathrm{alg}}(\mathbb{C}[G], R[[t]]) \simeq \mathrm{Hom}_{\mathrm{alg}}(\mathbb{C}[\bar{G}[[t]]], R). \tag{4}$$

The closed embedding $i: G \hookrightarrow M_N$ gives rise to the closed embedding

$$i_t: \bar{G}[[t]] \hookrightarrow \bar{M}_N[[t]],$$

where $\bar{M}_N[[t]]$ is the scheme Spec $\left(\mathbb{C}[y_n^{i,j}]_{n\geq 0;\, 1\leq i,j\leq N}\right)$.

Clearly, the analogue of (4) for G replaced by M_N is true under the map (following the notation in Definition 1.3.1)

$$\varphi_{M_N}: \mathrm{Hom}_{\mathrm{alg}}(\mathbb{C}[M_N], R[[t]]) \to \mathrm{Hom}_{\mathrm{alg}}(\mathbb{C}[\bar{M}_N[[t]]], R), \quad f \mapsto \bar{f},$$

where $f(y^{i,j}) = \sum_{n\geq 0} \bar{f}(y_n^{i,j})t^n$, for any $1 \leq i, j \leq N$. For any $P \in \mathbb{C}[M_N]$ and any $f \in \mathrm{Hom}_{\mathrm{alg}}(\mathbb{C}[M_N], R[[t]])$, it is easy to see that

$$f(P) = \sum_{m\geq 0} \bar{f}(\widehat{P}_m^{(0)})t^m.$$

From this, it follows that the above bijection φ_{M_N} restricts to a bijection φ_G under the canonical embeddings induced by i:

$$\begin{array}{ccc}
\mathrm{Hom}_{\mathrm{alg}}(\mathbb{C}[G], R[[t]]) & \overset{\varphi_G}{\simeq} & \mathrm{Hom}_{\mathrm{alg}}(\mathbb{C}[\bar{G}[[t]]], R) \\
\big\downarrow \widehat{i} & & \big\downarrow \widehat{i_t} \\
\mathrm{Hom}_{\mathrm{alg}}(\mathbb{C}[M_N], R[[t]]) & \underset{\varphi_{M_N}}{\simeq} & \mathrm{Hom}_{\mathrm{alg}}(\mathbb{C}[\bar{M}_N[[t]]], R).
\end{array}$$

This proves (4) and hence (1).

The 'In particular' part of the lemma follows since the functor $\mathscr{F}_1$ is independent of the choice of an embedding $G \hookrightarrow M_N$ (by (2)) and the representing scheme is unique (cf. Lemma 1.1.1).

To prove that $\bar{G}[[t]]$ is an affine group scheme, since $\mathscr{F}_1$ is representable by $\bar{G}[[t]]$, it suffices to observe (using (Mumford, 1988, Chapter II, §6, Proposition 2)) that the morphism $G \times G \to G$, $(g, h) \mapsto gh^{-1}$, induces a natural map $\mathscr{F}_1(R) \times \mathscr{F}_1(R) \to \mathscr{F}_1(R)$ for any $R \in$ **Alg**. It is a closed subgroup scheme of $\overline{\mathrm{SL}}_N[[t]]$ by construction. The proofs for $\bar{G}((t))$ and $\bar{G}[t^{-1}]$ are identical. □

Corollary 1.3.3 *Let G be any affine algebraic group.*

(a) Consider the morphism $\epsilon(\infty): \bar{G}[t^{-1}] \to G$ induced from the $\mathbb{C}$-algebra homomorphism $R[t^{-1}] \to R$, $t^{-1} \mapsto 0$.

Let $\bar{G}[t^{-1}]^-$ be the (ind)-scheme theoretic fiber of $\epsilon(\infty)$ over 1. Then, it represents the functor $G(R[t^{-1}])^-$: **Alg** $\to$ **Set** *defined as the kernel of the* *homomorphism* $\epsilon_R(\infty)\colon G(R[t^{-1}]) \to G(R)$ *induced from the* $\mathbb{C}$*-algebra homomorphism* $R[t^{-1}] \to R$, $t^{-1} \mapsto 0$.

Since $\bar{G}[t^{-1}] \hookrightarrow \overline{\mathrm{SL}}_N[t^{-1}]$ *is a closed embedding (cf. Definition 1.3.1), it is easy to see that* $\bar{G}[t^{-1}]^- \hookrightarrow \overline{\mathrm{SL}}_N[t^{-1}]^-$ *is a closed embedding.*

(b) Let $H \subset G$ *be a closed subgroup. Consider the morphism* $\epsilon(0)\colon \bar{G}[[t]] \to G$ *induced from the* $\mathbb{C}$*-algebra homomorphism* $R[[t]] \to R$, $t \mapsto 0$.

Let $\bar{G}[[t]]_H$ *be the scheme-theoretic inverse image of* H*. Then it represents the functor* $G(R[[t]])_H$ *defined as the inverse image of* $H(R)$ *under the homomorphism* $\epsilon_R(0)\colon G(R[[t]]) \to G(R)$.

Proof (a) By Lemma 1.3.2, $\bar{G}[t^{-1}]$ represents the functor $G(R[t^{-1}])$ (and, of course, G represents the functor $G(R)$). Now, by Exercise 1.3.E.6, $\bar{G}[t^{-1}]^-$ represents the functor $G(R[t^{-1}])^-$. This proves (a).

The proof of (b) is identical. □

Remark 1.3.4 Even though we do not need to, for any affine scheme X of finite type over $\mathbb{C}$, as in Definition 1.3.1 and Lemma 1.3.2, we can define an affine (non-noetherian) scheme $\bar{X}[[t]]$ which represents the covariant functor $\mathscr{F}_X\colon$ **Alg** $\to$ **Set** defined by

$$\mathscr{F}_X(R) = X(R[[t]]) \cong \mathrm{Hom}_{\mathrm{alg}}(\mathbb{C}[X], R[[t]]).$$

In particular, the $\mathbb{C}$-points of $\bar{X}[[t]] = X(\mathbb{C}[[t]])$.

Exactly the same remark applies to $\mathbb{C}[[t]]$ replaced by $\mathbb{C}[t^{-1}]$ or $\mathbb{C}((t))$.

Definition 1.3.5 (Infinite Grassmannian) For any affine algebraic group G over $\mathbb{C}$, define the *infinite Grassmannian* $\mathscr{X}_G$ as the sheafification of the functor $\mathscr{X}_G^o\colon R \rightsquigarrow G(R((t)))/G(R[[t]])$ (cf. Lemma B.2). Observe that $\mathscr{X}_G^o$ satisfies condition (1) of Lemma B.2 since for any fppf R-algebra R', $R \to R'$ is injective (cf. (Matsumura, 1989, Theorem 7.5)).

In the following, we show that $\mathscr{X}_G$ is representable, represented by an ind-projective scheme $\bar{X}_G$ with $\mathbb{C}$-points $X_G := G((t))/G[[t]]$ for any connected reductive group G. We first consider the case $G = \mathrm{SL}_N$.

Definition 1.3.6 (Representing $\mathscr{X}_{\mathrm{SL}_N}$ by an ind-projective scheme) Denote $V = \mathbb{C}^N$. For any non-negative integer n, define the *nth special lattice functor* $\mathscr{Q}_n = \mathscr{Q}_n^N\colon$ **Alg** $\to$ **Set** by $\mathscr{Q}_n(R) =$ set of projective $R[[t]]$-submodules L^R of $R((t)) \otimes_{\mathbb{C}} V$ satisfying the following two conditions:

(a) $t^n L_o^R \subset L^R \subset t^{-n} L_o^R$, where $L_o^R := R[[t]] \otimes_{\mathbb{C}} V$.

(b) $\wedge^N_{R[[t]]}(L^R) \to \wedge^N_{R((t))}(R((t)) \otimes_{\mathbb{C}} V) \simeq R((t))$ has image, denoted by $\det L^R$, precisely equal to $R[[t]]$.

Now, define the *special lattice functor* $\mathscr{Q} = \mathscr{Q}^N$ by

$$\mathscr{Q}(R) = \cup_{n \geq 0} \mathscr{Q}_n(R).$$

By Exercise 1.3.E.5, the functor $\mathscr{Q}$ is the sheafification $\mathscr{X}_{\mathrm{SL}_N}$ of the functor $R \rightsquigarrow \mathrm{SL}_N(R((t)))/\mathrm{SL}_N(R[[t]])$.

In particular, taking $R = \mathbb{C}$, define

$$Q_n := \mathscr{Q}_n(\mathbb{C}) \text{ and } Q := \mathscr{Q}(\mathbb{C}).$$

(In fact, any $\mathbb{C}[[t]]$-submodule $L^{\mathbb{C}}$ of $V((t)) := \mathbb{C}((t)) \otimes_{\mathbb{C}} V$ satisfying (a) is automatically $\mathbb{C}[[t]]$-free, being a submodule of a free module over a principal ideal domain (PID). Thus, Q_n consists of $\mathbb{C}[[t]]$-submodules L of $V((t))$ such that

$$t^n L_o \subset L \subset t^{-n} L_o, \quad \text{and} \quad \det(L) = \mathbb{C}[[t]],$$

where $L_o := \mathbb{C}[[t]] \otimes_{\mathbb{C}} V$. In fact, in the proof of Theorem 1.3.8, we will see that the condition $\det(L) = \mathbb{C}[[t]]$ can be replaced by the condition $\dim_{\mathbb{C}}(L/t^n L_o) = nN$.)

Recall that for any scheme X and any automorphism f of X, the fixed-point subset X^f acquires a canonical scheme structure as the inverse image subscheme of the diagonal $\Delta(X)$ under the morphism

$$\mathfrak{f}\colon X \to X \times X, \quad x \mapsto (x, f(x)).$$

Consider the complex vector space $V_n := t^{-n} L_o / t^n L_o$ of dimension $2nN$. Then multiplication by t induces a nilpotent endomorphism $\bar{t}_n$ of V_n and hence $1 + \bar{t}_n$ is a unipotent automorphism of V_n. In particular, $1 + \bar{t}_n$ induces an isomorphism (denoted by the same symbol) of the Grassmannian $\mathrm{Gr}(nN, 2nN)$ of nN-dimensional subspaces of the $2nN$-dimensional space V_n. Let $\bar{F}_n = \bar{F}_n^N := \mathrm{Gr}(nN, 2nN)^{1+\bar{t}_n}$ denote its fixed-point projective scheme and let $F_n := \bar{F}_n(\mathbb{C})$ be the $\mathbb{C}$-points of $\bar{F}_n$. Then clearly the map $i_n\colon Q_n \to F_n \subset \mathrm{Gr}(nN, 2nN)$ given by $L \mapsto L/t^n L_o$ is a bijection.

It is easy to see that the inclusion $\theta_n\colon \mathrm{Gr}(nN, 2nN) \hookrightarrow \mathrm{Gr}((n+1)N, 2(n+1)N)$ is a closed embedding, where (denoting $t^k V := t^k \otimes V$) the map θ_n takes $V' \subset t^{-n} L_o / t^n L_o \simeq t^{n-1} V \oplus t^{n-2} V \oplus \cdots \oplus t^{-n} V$ to $t^n V \oplus V'$. Moreover, it is easy to see that θ_n restricts to a closed embedding $\bar{\theta}_n\colon \bar{F}_n \hookrightarrow \bar{F}_{n+1}$.

By virtue of the following lemma, we have a bijection $\beta\colon X_{\mathrm{SL}_N} \to Q^N$.

Extending the scalar, the group $\mathrm{SL}_N((t))$ clearly acts on $V((t))$.

Lemma 1.3.7 *The map*

$$\beta\colon X_{\mathrm{SL}_N} \to Q^N, \quad g\,\mathrm{SL}_N[[t]] \mapsto g L_o, \text{ for } g \in \mathrm{SL}_N((t)),$$

is a bijection.

Proof Let $g \in \mathrm{SL}_N((t))$. It is easy to see that there exists some n (depending upon g) such that

$$t^n L_o \subset g L_o \subset t^{-n} L_o. \tag{1}$$

Of course, gL_o is t-stable. We next calculate the dimension of $gL_o/t^n L_o$.

By the Bruhat decomposition (cf. (Kumar, 2002, Corollary 13.2.10)), we may assume that g is an algebraic group homomorphism $\mathbb{C}^* \to D$, where D is the diagonal subgroup of SL_N. Write

$$g(t) = \begin{pmatrix} t^{n_1} & & O \\ & \ddots & \\ O & & t^{n_N} \end{pmatrix}, \quad \text{for} \quad t \in \mathbb{C}^* \quad \text{and} \quad n_i \in \mathbb{Z}.$$

Then, since $\operatorname{Im} g \subset \mathrm{SL}_N$, we get $\Sigma n_i = 0$. Now

$$\dim \left(gL_o/t^n L_o\right) = \sum_{i=1}^{N} (n - n_i) = Nn - \Sigma n_i = Nn.$$

This proves that $gL_o \in \mathcal{Q}_n$.

Conversely, take $L \in \mathcal{Q}_n$. Since $\mathcal{O} := \mathbb{C}[[t]]$ is a PID and $t^k L_o$ is $\mathcal{O}$-free of rank N (for any $k \in \mathbb{Z}$), we get that L is $\mathcal{O}$-free of rank N. Further, $K \otimes_{\mathcal{O}} L \to V((t))$ is an isomorphism, where $K = \mathbb{C}((t))$. Let $\{e_1, \dots, e_N\}$ be the standard $\mathbb{C}$-basic of V and take a $\mathcal{O}$-basis $\{v_1, \dots, v_N\}$ of L. Now, define the K-linear automorphism g of $V((t))$ by $ge_i = v_i$ $(1 \le i \le N)$. We prove that $\det g$ is a unit of $\mathcal{O}$: write $\det g = t^k u$, where $k \in \mathbb{Z}$ and u is a unit of $\mathcal{O}$. Consider the K-linear automorphism α of $V((t))$ defined by

$$\begin{aligned} \alpha e_i &= e_i, \ \text{ for } \ 1 \le i < N \\ &= t^{-k} u^{-1} e_N, \ \text{ for } \ i = N. \end{aligned}$$

Then $\det(g\alpha) = 1$ and $t^{n+|k|} L_o \subset (g\alpha) L_o \subset t^{-n-|k|} L_o$. Hence, by the earlier part of the proof, we get

$$\dim \left(\frac{g\alpha(L_o)}{t^{n+|k|} L_o} \right) = (n + |k|)N. \tag{2}$$

On the other hand,

$$\begin{aligned} \dim \left(\frac{g\alpha(L_o)}{t^{n+|k|} L_o} \right) &= \dim \left(\frac{gL_o}{t^n L_o} \right) + |k|N + k \\ &= Nn + |k|N + k, \ \text{ since } \ L \in \mathcal{Q}_n. \end{aligned} \tag{3}$$

Combining (2) and (3), we get $k = 0$. Hence, $g\alpha(L_o) = gL_o = L$. This proves the surjectivity of β. The injectivity of β is clear. This proves the lemma. □

Theorem 1.3.8 *For any fixed $N \geq 1$ and $n \geq 0$, the nth special lattice functor $\mathscr{Q}_n = \mathscr{Q}_n^N$ (defined in Definition 1.3.6) is representable, represented by a projective scheme $\bar{H}_n$ (with $\mathbb{C}$-points Q_n), which is a closed subscheme of $\bar{F}_n$ (defined in Definition 1.3.6). Moreover, the inclusion $\bar{H}_n \hookrightarrow \bar{F}_n$ induces an isomorphism of the corresponding reduced schemes $\bar{H}_n^{red} \xrightarrow{\sim} \bar{F}_n^{red}$.*

Further, the canonical morphism $\bar{H}_n \to \bar{H}_{n+1}$ (induced from the inclusion of the functors $\mathscr{Q}_n \subset \mathscr{Q}_{n+1}$) is a closed embedding. Thus, we get an ind-projective scheme $\bar{H} = (\bar{H}_n)_{n\geq 0}$ representing the functor $\mathscr{Q}$, with $\mathbb{C}$-points $Q^N := \bigcup_{n\geq 0} Q_n^N$. Through the bijection β of Lemma 1.3.7, we get the $\mathbb{C}$-points of $\bar{H}$ to be X_{SL_N}.

Thus, by Exercise 1.3.E.5, $\bar{H}$ also represents the functor $\mathscr{X}_{\mathrm{SL}_N}$. In particular, $\mathscr{X}_{\mathrm{SL}_N}(\mathbb{C}) = X_{\mathrm{SL}_N}$.

We denote the ind-scheme $\bar{H}$ by $\bar{X}_{\mathrm{SL}_N}$. Thus, $\bar{X}_{\mathrm{SL}_N}$ represents the functor $\mathscr{X}_{\mathrm{SL}_N}$.

Proof By Eisenbud and Harris (2000, Exercise VI-18), $\mathrm{Gr}(nN, 2nN)$ represents the functor

$$\begin{aligned} R \rightsquigarrow\ & \mathrm{Gr}(nN, 2nN; R) \\ & := \text{set of } R\text{-module direct summands of } R^{2nN} \text{ of rank } nN. \end{aligned}$$

Thus, following the notation of Definition 1.3.6 and Exercise 1.3.E.3, the functor represented by the scheme $\bar{F}_n$ is given by

$$\begin{aligned} \mathscr{F}_n(R) &= \mathrm{Gr}(nN, 2nN; R)^{1+\bar{t}_n} \\ &= \text{set of } R\text{-module direct summands } \tilde{L}^R \text{ of } \frac{t^{-n}L_o^R}{t^n L_o^R} \text{ of rank } nN, \\ &\quad\ \text{which are } (1+\bar{t}_n)\text{-stable.} \end{aligned}$$

Taking the inverse image L^R of $\tilde{L}^R$ under $t^{-n}L_o^R \to t^{-n}L_o^R/t^nL_o^R$, we get that L^R satisfies (a) of Definition 1.3.6 and $\tilde{L}^R$ is $(1+\bar{t}_n)$-stable if and only if L^R is an $R[[t]]$-submodule of $t^{-n}L_o^R$. Further, L^R is a projective $R[[t]]$-module if and only if $\tilde{L}^R$ is an R-module direct summand of $t^{-n}L_o^R/t^nL_o^R$ (cf. Exercise 1.3.E.1).

We next show that when the $\mathbb{C}$-algebra R is a field $k \supset \mathbb{C}$,

$$\mathscr{F}_n(k) = \mathscr{Q}_n(k), \quad \text{for any} \quad n \geq 0. \tag{1}$$

Take any $k[[t]]$-submodule L^k satisfying condition (a) of Definition 1.3.6. By the Elementary Divisor Theorem for free modules over a PID, we get that there exists a $k[[t]]$-basis $\{v_1, \ldots, v_N\}$ of L_o^k such that

$\{t^{-n+d_1}v_1, \ldots, t^{-n+d_N}v_N\}$ is a basis of L^k, for some $d_i \geq 0$. Now, condition (b) of Definition 1.3.6 is equivalent to the condition

$$\sum_{i=1}^{N} -n + d_i = 0. \tag{2}$$

Further,

$$\dim_k \tilde{L}^k = \sum_{i=1}^{N}(2n - d_i) = 2nN - \sum d_i. \tag{3}$$

Comparing (2) and (3), we see that condition (b) is equivalent to the condition that $\dim_k \tilde{L}^k = nN$. This proves (1).

We next show that for any $R \in \mathbf{Alg}$,

$$\mathscr{Q}_n(R) \subset \mathscr{F}_n(R), \quad \text{for any} \quad n \geq 0. \tag{4}$$

Let L^R be a projective $R[[t]]$-submodule satisfying conditions (a) and (b) of Definition 1.3.6. Taking $\mathbb{C}$-algebra homomorphisms $\varphi\colon R \to k$ (where k is a field) and considering $L^k := k[[t]] \otimes_{R[[t]]} L^R$ and using Exercise 1.3.E.2, we get (from the case that R is a field proved earlier) that $\tilde{L}^R$ is of rank nN over R, proving (4).

Conversely, assume that $R \in \mathbf{Alg}$ has no nonzero nilpotents. In this case, we prove that

$$\mathscr{F}_n(R) \subset \mathscr{Q}_n(R). \tag{5}$$

Take $\tilde{L}^R \in \mathscr{F}_n(R)$. Let $R_\mathfrak{p}$ be the localization of R at a prime ideal $\mathfrak{p}$ of R. Since a projective module over a local ring is free (cf. (Matsumura, 1989, Theorem 2.5)), we get that $L^{R_\mathfrak{p}} := R_\mathfrak{p}[[t]] \otimes_{R[[t]]} L^R$ is an $R_\mathfrak{p}[[t]]$-free module of rank N ($R_\mathfrak{p}[[t]]$ is a local ring by Exercise 1.3.E.13). Thus, $\det(L^{R_\mathfrak{p}}) \subset R_\mathfrak{p}((t))$ is given by $t^{-nN}P(t) \cdot R_\mathfrak{p}[[t]]$, where $P(t) \in R_\mathfrak{p}[[t]]$. Now, take any $\mathbb{C}$-algebra homomorphism $\varphi\colon R_\mathfrak{p} \to k$ to a field k. From the case when R is a field proved above as in (1), we get that the image $P^k(t)$ of $P(t)$ in $k[[t]]$ (via φ) is t^{nN} times a unit of $k[[t]]$. Since this is true for any φ and R has no nonzero nilpotents, we get by using Atiyah and Macdonald (1969, Proposition 1.8) that $P(t)$ is t^{nN} times a unit of $R_\mathfrak{p}[[t]]$. (In general, if we allow R to have nilpotents, $P(t)$ would be of the form

$$P(t) = a_0 + a_1 t + \cdots + a_{nN-1}t^{nN-1} + a_{nN}t^{nN} + \cdots, \tag{6}$$

where $a_0, a_1, \ldots, a_{nN-1}$ are nilpotents in $R_\mathfrak{p}$ and a_{nN} is a unit of $R_\mathfrak{p}$.) Reverting to the case when R has no nonzero nilpotents, from the above, we get that

$$\det(L^{R_\mathfrak{p}}) = R_\mathfrak{p}[[t]]. \tag{7}$$

Let $M := \det(L^R) \subset R((t))$. Then, from condition (a) of Definition 1.3.6, we get

$$M = t^{-nN} M_o, \quad \text{where } M_o \text{ is a finitely generated ideal of } R[[t]]. \tag{8}$$

Take $Q(t) = \sum_{i \geq 0} b_i t^i \in M_o$. Then $Q(t)$, considered as an element of $R_\mathfrak{p}[[t]]$, belongs to $t^{nN} \det(L^{R_\mathfrak{p}}) = t^{nN} R_\mathfrak{p}[[t]]$ for any prime ideal $\mathfrak{p}$ of R (by (7)). Thus, $b_1 = \cdots = b_{nN-1} = 0$ as elements of $R_\mathfrak{p}$ (for any $\mathfrak{p}$). Thus, $b_1 = \cdots = b_{nN-1} = 0$ as elements of R (since R has no nonzero nilpotents). Hence, $M_o \subset t^{nN} R[[t]]$. Since $t^n L_o^R \subset L^R$, we have $t^{2nN} R[[t]] \subset M_o$. Consider the quotient R-module

$$A := \frac{t^{nN} R[[t]]}{M_o} \simeq \frac{t^{nN} R[[t]]/t^{2nN} R[[t]]}{M_o/t^{2nN} R[[t]]}.$$

Applying Atiyah and Macdonald (1969, Proposition 3.8) to the R-module A and using (7), we get $A = 0$, i.e., $\det(L^R) = R[[t]]$. Thus, L^R satisfies condition (b) of Definition 1.3.6, proving $L^R \in \mathscr{Q}_n(R)$ by Exercise 1.3.E.1. This proves (5).

Now, we analyze the failure of (5) for general $R \in$ **Alg**. Take any affine open subset $\mathrm{Spec}(S) \subset \bar{F}_n$, for a finitely generated $\mathbb{C}$-algebra S. The inclusion gives rise to the element $\tilde{L}_o^S \in \mathrm{Mor}(\mathrm{Spec}(S), \bar{F}_n) = \mathscr{F}_n(S)$ and hence a projective $S[[t]]$-module L_o^S satisfying (a) of Definition 1.3.6. Take an affine open cover $\{\mathrm{Spec}(S_i)\}_i$ of $\mathrm{Spec}(S)$ so that the $S_i[[t]]$-module

$$L_o^{S_i} := S_i[[t]] \otimes_{S[[t]]} L_o^S \quad \text{is free.} \tag{9}$$

This is possible by Exercise 1.3.E.4. Thus we get, from the proof of (5) given above (see specifically (6)), that

$$\det(L_o^{S_i}) = t^{-nN} P_i(t) S_i[[t]] \subset S_i((t)), \tag{10}$$

where $P_i(t)$ is of the form

$$P_i(t) = a_0^i + a_1^i t + \cdots + a_{nN-1}^i t^{nN-1} + a_{nN}^i t^{nN} + \cdots,$$
$$\text{for some nilpotents } a_0^i, a_1^i, \ldots, a_{nN-1}^i \text{ in } S_i. \tag{11}$$

The nilpotent ideal

$$J_{S_i} = \langle a_0^i, a_1^i, \ldots, a_{nN-1}^i \rangle \subset S_i$$

clearly does not depend upon the choice of the representative $P_i(t)$. In particular, these ideals descend to give a nilpotent ideal $J_S \subset S$. Taking an affine open cover of $\bar{F}_n$ by Spec S, we get a nilpotent ideal sheaf $\mathscr{J} \subset \mathscr{O}_{\bar{F}_n}$. Now, define the closed subscheme $\bar{H}_n$ of $\bar{F}_n$ given by the ideal sheaf $\mathscr{J}$. Thus, their reduced subschemes are isomorphic:

$$\bar{H}_n^{\text{red}} \simeq \bar{F}_n^{\text{red}}. \tag{12}$$

We next prove that the scheme $\bar{H}_n$ represents the functor $\mathscr{Q}_n$, i.e., for any $R \in \mathbf{Alg}$, there is a natural isomorphism

$$\mathscr{Q}_n(R) \xrightarrow{\sim} \text{Mor}(\text{Spec}(R), \bar{H}_n) \hookrightarrow \text{Mor}(\text{Spec}\, R, \bar{F}_n) =: \bar{F}_n(R). \tag{13}$$

By (4), we have an inclusion $\mathscr{Q}_n(R) \subset \mathscr{F}_n(R)$. We claim that the image lands inside $\text{Mor}(\text{Spec}(R), \bar{H}_n)$.

Take $L^R \in \mathscr{Q}_n(R)$. Then, by definition,

$$\det(L^R) = R[[t]]. \tag{14}$$

The element L^R gives rise to a morphism $\tilde{L}^R\colon \text{Spec}(R) \to \bar{F}_n$. Since $\bar{F}_n$ is a scheme of finite type over $\mathbb{C}$, we can assume that R is a finitely generated $\mathbb{C}$-algebra. Take 'small enough' affine open covers $\{\text{Spec}(R_i)\}_i$ of Spec R and $\{\text{Spec}(S_i)\}_i$ of $\bar{F}_n$ such that $\tilde{L}^R$ restricts to $\text{Spec}(R_i) \to \text{Spec}\, S_i$ (i.e., gives a $\mathbb{C}$-algebra homomorphism $f_i\colon S_i \to R_i$) and the $S_i[[t]]$-module $L_o^{S_i}$ is free, where $L_o^{S_i}$ is defined by (9). Thus, by (10), $\det(L_o^{S_i}) = t^{-nN} P_i(t) S_i[[t]]$, where $P_i(t) = \sum_{d \geq 0} a_d^i t^d$ is of the form (11). In particular,

$$\det\left(L^{R_i} := R_i[[t]] \otimes_{S_i[[t]]} L_o^{S_i}\right) = t^{-nN} f_i(P_i(t)) R_i[[t]], \tag{15}$$

where $f_i(P_i(t))$ is obtained from $P_i(t)$ by applying f_i to all the coefficients. But $L^{R_i} = R_i[[t]] \otimes_{R[[t]]} L^R$. Hence, by (14),

$$\det(L^{R_i}) = R_i[[t]]. \tag{16}$$

Comparing (15) and (16), we get

$$f_i(a_d^i) = 0, \quad \text{for all} \quad 0 \leq d < nN, \tag{17}$$

i.e., the homomorphism f_i factors through $S_i/\langle a_0^i, a_1^i, \ldots, a_{nN-1}^i \rangle$. This shows that, from the definition of $\bar{H}_n$, $\tilde{L}^R\colon \text{Spec}(R) \to \bar{H}_n$. Hence, the image of $\mathscr{Q}_n(R)$ (inside $\bar{F}_n(R)$) lands inside $\bar{H}_n(R)$.

Conversely, take $\tilde{L}^R \in \text{Mor}(\text{Spec}\, R, \bar{H}_n)$ and let L^R be the corresponding projective $R[[t]]$-submodule of $R((t)) \otimes_{\mathbb{C}} V$ (which satisfies condition (a) of Definition 1.3.6). Then, by the above calculation (see (15) and (17)), for a 'small enough' open cover $\{\text{Spec}(R_i)\}_i$ of $\text{Spec}(R)$ (since $\tilde{L}^R$ has image inside $\bar{H}_n$), $\det(L^{R_i}) = b_i(t) R_i[[t]]$, for some $b_i(t) = \sum_{d \geq 0} \alpha_d^i t^d \in R_i[[t]]$, where $L^{R_i} := R_i[[t]] \otimes_{R[[t]]} L^R$. Considering $\mathbb{C}$-algebra homomorphisms $\varphi\colon R_i \to k$

(for a field k), from the case when R is a field proved earlier, we get that $\varphi(\alpha_0^i) \neq 0$. Since this is true for any φ, we get that α_0^i is a unit of R_i, i.e.,

$$\det(L^{R_i}) = R_i[[t]]. \tag{18}$$

By (8), $\det(L^R) = t^{-nN} M_o$, for some finitely generated ideal M_o of $R[[t]]$. Since the image of M_o in $R_i[[t]]$ equals $t^{nN} R_i[[t]]$ by (18) (for an affine open cover $\{\mathrm{Spec}(R_i)\}_i$ of Spec R), we first conclude that $M_o \subset t^{nN} R[[t]]$ (and, of course, from the definition of L^R, $M_o \supset t^{2nN} R[[t]]$). Moreover, considering the quotient R-module

$$A := \frac{t^{nN} R[[t]]}{M_o} \simeq \frac{t^{nN} R[[t]]/t^{2nN} R[[t]]}{M_o/t^{2nN} R[[t]]}$$

and using Atiyah and Macdonald (1969, Proposition 3.8) together with the equation (18), we get $\det(L^R) = R[[t]]$. Thus, L^R satisfies condition (b) of Definition 1.3.6 as well, i.e., $L^R \in \mathscr{Q}_n(R)$, proving $\mathrm{Mor}(\mathrm{Spec}\, R, \bar{H}_n) \subset \mathscr{Q}_n(R)$. Thus, $\mathscr{Q}_n(R) \simeq \bar{H}_n(R)$ and hence $\mathscr{Q}_n(R)$ is a representable functor represented by the scheme $\bar{H}_n$ for all $n \geq 0$.

Finally, we have the following commutative diagram of schemes and morphisms between them:

$$\begin{array}{ccc} \bar{H}_n & \overset{i_n}{\hookrightarrow} & \bar{F}_n \\ \Big\downarrow {\scriptstyle j_n} & & \Big\downarrow {\scriptstyle \bar{\theta}_n} \\ \bar{H}_{n+1} & \overset{i_{n+1}}{\hookrightarrow} & \bar{F}_{n+1}, \end{array} \tag{$\mathscr{D}$}$$

where the morphism $j_n : \bar{H}_n \to \bar{H}_{n+1}$ is induced from the canonical inclusion of functors $\mathscr{Q}_n(R) \hookrightarrow \mathscr{Q}_{n+1}(R)$. Since i_n, i_{n+1} are closed embeddings (by definition, $\bar{H}_n \subset \bar{F}_n$ is a closed subscheme) and $\bar{\theta}_n$ is a closed embedding as seen in Definition 1.3.6, we get that j_n is a closed embedding. This completes the proof of the theorem. $\square$

The following example shows that the inclusion $\mathscr{Q}_n(R) \subset \mathscr{F}_n(R)$ (cf. (4) of the proof of Theorem 1.3.8) is proper for some $R \in \mathbf{Alg}$ already for $n = 1$, $N = 2$. In particular, by Theorem 1.3.8, the scheme $\bar{F}_1^2$ is not reduced.

Example 1.3.9 Let $R = \mathbb{C}[\epsilon]/\langle \epsilon^2 \rangle$. Consider the element $g \in \mathrm{GL}_2(R((t)))$ given by

$$g = \begin{pmatrix} \epsilon t^{-1} + 1 & 0 \\ 0 & 1 \end{pmatrix}.$$

Its inverse is $\begin{pmatrix} -\epsilon t^{-1}+1 & 0 \\ 0 & 1 \end{pmatrix}$. Clearly, $tL_o^R \subset gL_o^R \subset t^{-1}L_o^R$. Now $\Lambda^2_{R[[t]]}(gL_o^R) = (\epsilon t^{-1} + 1)R[[t]] \neq R[[t]]$.

But

$$\frac{gL_o^R}{tL_o^R} \simeq \frac{(\epsilon t^{-1} + 1)R[[t]]}{tR[[t]]} \oplus \frac{R[[t]]}{tR[[t]]}$$

is a free R-module of rank 2. To show this, observe that, as an R-module,

$$\frac{(\epsilon t^{-1} + 1)R[[t]]}{tR[[t]]} \simeq \frac{(\epsilon + t)R[[t]]}{(\epsilon + t)(\epsilon - t)R[[t]]} \simeq \frac{R[[t]]}{(\epsilon - t)R[[t]]},$$

where the last isomorphism follows since $\epsilon + t$ is not a zero divisor in $R[[t]]$ as can be seen by multiplying it by $\epsilon - t$.

Define an R-module map

$$\theta\colon R \to R[[t]]/(t-\epsilon)R[[t]] \quad \text{by} \quad r \mapsto r + (t-\epsilon)R[[t]].$$

It is clearly surjective. Moreover, it is injective since if $r + (t-\epsilon)P(t) = 0$, for some $P(t) \in R[[t]]$, then $(t+\epsilon)r + t^2P(t) = 0$. But $t^2P(t)$ has no 't-term,' thus $r = 0$.

Further, it is easy to see that gL_o^R/tL_o^R is an R-module direct summand of $t^{-1}L_o^R/tL_o^R$.

Definition 1.3.10 Recall that $\overline{\mathrm{SL}_N}((t))$ represents the functor $\mathrm{SL}_N(R((t)))$ (cf. Lemma 1.3.2) and $\bar{X}_{\mathrm{SL}_N}$ represents the functor $\mathscr{X}_{\mathrm{SL}_N}$ (cf. Theorem 1.3.8). Also, it is easy to see that the sheafification of the functor $\mathrm{SL}_N(R((t))) \times \mathscr{X}^o_{\mathrm{SL}_N}$ is $\mathrm{SL}_N(R((t))) \times \mathscr{X}_{\mathrm{SL}_N}$ (since $\mathrm{SL}_N(R((t)))$ is representable), where $\mathscr{X}^o_{\mathrm{SL}_N}(R) := \mathrm{SL}_N(R((t)))/\mathrm{SL}_N(R[[t]])$. Thus, the multiplication

$$\mathrm{SL}_N(R((t))) \times \mathscr{X}^o_{\mathrm{SL}_N}(R) \to \mathscr{X}^o_{\mathrm{SL}_N}(R),\ (g, h\bar{o}_R) \mapsto gh\bar{o}_R,$$

gives rise to a $\mathbb{C}$-space functor morphism

$$\mathrm{SL}_N(R((t))) \times \mathscr{X}_{\mathrm{SL}_N} \to \mathscr{X}_{\mathrm{SL}_N},$$

where $\bar{o}_R$ is the base point of $\mathrm{SL}_N(R((t)))/\mathrm{SL}_N(R[[t]])$. This, in turn, gives rise to a morphism of ind-schemes

$$\mu\colon \overline{\mathrm{SL}_N}((t)) \times \bar{X}_{\mathrm{SL}_N} \to \bar{X}_{\mathrm{SL}_N}.$$

We define the 'basic' line bundle $\mathscr{L}$ on $\bar{X}_{\mathrm{SL}_N}$ as follows.

Definition 1.3.11 For any $n \geq 0$, let $\hat{\mathscr{L}_n}$ be the dual of the tautological line bundle over $\mathrm{Gr}(nN, 2nN)$. Recall that the fiber of $\hat{\mathscr{L}_n}$ over any $V' \in \mathrm{Gr}(nN, 2nN)$ is the dual $\Lambda^{nN}(V')^*$. Let $\mathscr{L}_n$ be the pull-back line bundle over $\bar{H}_n$ via the embedding $\bar{i}_n\colon \bar{H}_n \to \mathrm{Gr}(nN, 2nN)$, which is the composite of

$i_n\colon \bar{H}_n \to \bar{F}_n \hookrightarrow \mathrm{Gr}(nN,2nN)$ (cf. Definition 1.3.6 and the proof of Theorem 1.3.8). It is easy to see that $\hat{\mathscr{L}}_{n+1}$ restricts to $\hat{\mathscr{L}}_n$ under the embedding θ_n (cf. Definition 1.3.6). Thus, from the commutative diagram $\mathscr{D}$ in the proof of Theorem 1.3.8, $\mathscr{L}_{n+1}$ restricts to $\mathscr{L}_n$ under the embedding $j_n\colon \bar{H}_n \hookrightarrow \bar{H}_{n+1}$. Hence, we get the 'basic' line bundle $\mathscr{L}$ on $\bar{X}_{\mathrm{SL}_N}$.

It is easy to see that the action of $\overline{\mathrm{SL}}_N[[t]]$ on $\bar{X}_{\mathrm{SL}_N}$ (cf. Definition 1.3.10) canonically lifts to its action on $\mathscr{L}$.

Definition 1.3.12 Let $V_n^- \subset V_n := \frac{t^{-n}L_o}{t^nL_o}$ be the subspace $t^{-1}V \oplus \cdots \oplus t^{-n}V$ under the identification

$$V_n \simeq t^{n-1}V \oplus \cdots \oplus t^0V \oplus t^{-1}V \oplus \cdots \oplus t^{-n}V.$$

Define a section $\hat{\sigma}_n$ of $\hat{\mathscr{L}}_n$ over $\mathrm{Gr}(nN,2nN)$ by defining $\hat{\sigma}_n(L)$ as the linear form

$$\hat{\sigma}_n(L)\colon \Lambda^{nN}(L) \to \Lambda^{nN}\left(\frac{L_o}{t^nL_o}\right), \quad \text{for any } L \in \mathrm{Gr}(nN,2nN),$$

induced from the linear map (obtained from the inclusion $L \subset V_n$):

$$L \to \frac{V_n}{V_n^-} \simeq \frac{L_o}{t^nL_o}.$$

We identify $\Lambda^{nN}(L_o/t^nL_o)$ with $\mathbb{C}$ under the basis

$$\left((t^{n-1}e_1) \wedge \ldots \wedge (t^{n-1}e_N)\right) \wedge \left((t^{n-2}e_1) \wedge \ldots \wedge (t^{n-2}e_N)\right) \wedge \ldots \wedge (e_1 \wedge \ldots \wedge e_N),$$

where $\{e_1, \ldots, e_N\}$ is the standard basis of $V = \mathbb{C}^N$.

It is easy to see that $\hat{\sigma}_{n+1}$ restricts to $\hat{\sigma}_n$ under the embedding $\theta_n\colon \mathrm{Gr}(nN,2nN) \hookrightarrow \mathrm{Gr}((n+1)N,2(n+1)N)$.

Pulling back the sections $\hat{\sigma}_n$ via the embeddings $\bar{i}_n\colon \bar{H}_n \hookrightarrow \bar{F}_n \subset \mathrm{Gr}(nN,2nN)$, we get a section σ of the line bundle $\mathscr{L}$ over $\bar{X}_{\mathrm{SL}_N}$ (cf. diagram ($\mathscr{D}$) in the proof of Theorem 1.3.8).

Let $Z(\sigma) \subset \bar{X}_{\mathrm{SL}_N}$ be the zero set of the section σ.

Lemma 1.3.13 *The open ind-subscheme $\bar{X}_{\mathrm{SL}_N}\backslash Z(\sigma)$ represents the functor*

$$R \rightsquigarrow \mathscr{Q}(R)\backslash Z(\sigma_R),$$

where

$$\mathcal{Q}(R)\backslash Z(\sigma_R) := \Big\{L^R \in \mathcal{Q}(R) : i_{L^R} : L^R \to \big(R((t)) \otimes_{\mathbb{C}} V\big)/\big(t^{-1}R[t^{-1}] \otimes V\big) \text{ is an isomorphism}\Big\},$$

and i_{L^R} *is induced from the inclusion* $L^R \subset R((t)) \otimes_{\mathbb{C}} V$.

Proof We need to prove that for any $R \in \mathbf{Alg}$,

$$\operatorname{Mor}\big(\operatorname{Spec} R, \bar{X}_{\mathrm{SL}_N}\backslash Z(\sigma)\big) \simeq \mathcal{Q}(R)\backslash Z(\sigma_R). \tag{1}$$

Take $f \in \operatorname{Mor}(\operatorname{Spec} R, \bar{X}_{\mathrm{SL}_N}) \simeq \mathcal{Q}(R)$ (by Theorem 1.3.8) and let $L^R = L^R(f) \in \mathcal{Q}_n(R)$ be the corresponding lattice (for some $n \geq 0$). To prove (1), we need to prove that $L^R \in \mathcal{Q}(R)\backslash Z(\sigma_R)$ if and only if

$$\operatorname{Im} f \subset \bar{X}_{\mathrm{SL}_N}\backslash Z(\sigma).$$

Since $\bar{X}_{\mathrm{SL}_N}$ is an ind-scheme filtered by schemes of finite type over $\mathbb{C}$, and for any maximal ideal $\mathfrak{m}$ of a finitely generated algebra S over $\mathbb{C}$, $S/\mathfrak{m} \simeq \mathbb{C}$ (cf. (Atiyah and Macdonald, 1969, Corollary 7.10)), to show (1), it suffices to show that for any $L^R \in \mathcal{Q}_n(R)$, where R is a finitely generated $\mathbb{C}$-algebra,

$$L^R \in \mathcal{Q}_n(R)\backslash Z(\sigma_R) \Longleftrightarrow (R/\mathfrak{m}) \otimes_R L^R \in Q_n\backslash Z(\sigma_{\mathbb{C}}) \quad \text{for all the maximal ideals } \mathfrak{m} \text{ of } R. \tag{2}$$

Now

$$\begin{aligned}
&L^R \in \mathcal{Q}_n(R)\backslash Z(\sigma_R)\\
&\quad \Leftrightarrow i_{L^R} : L^R \to \frac{R((t)) \otimes V}{t^{-1}R[t^{-1}] \otimes V}\\
&\quad\quad \text{is an isomorphism by definition}\\
&\quad \Leftrightarrow \bar{i}_{L^R} : \frac{L^R}{t^n L_o^R} \to \frac{R((t)) \otimes V}{(t^n L_o^R) \oplus (t^{-1}R[t^{-1}] \otimes V)}\\
&\quad\quad \text{is an isomorphism}\\
&\quad \Leftrightarrow \bar{i}_{L^{R_{\mathfrak{m}}}} : \frac{L^{R_{\mathfrak{m}}}}{(R_{\mathfrak{m}} \otimes_R t^n L_o^R)} \to \frac{R_{\mathfrak{m}}((t)) \otimes V}{(t^n L_o^{R_{\mathfrak{m}}}) \oplus (t^{-1}R_{\mathfrak{m}}[t^{-1}] \otimes V)}\\
&\quad\quad \text{is an isomorphism for all the maximal ideals } \mathfrak{m} \subset R\\
&\quad\quad \text{by Atiyah and Macdonald (1969, Proposition 3.9)}
\end{aligned}$$

$$\Leftrightarrow \bar{i}_{L^{R/\mathfrak{m}}}\colon \frac{L^{R/\mathfrak{m}}}{t^n L_o^{R/\mathfrak{m}}} \to \frac{(R/\mathfrak{m})((t)) \otimes V}{(t^n L_o^{R/\mathfrak{m}}) \oplus ((t^{-1}(R/\mathfrak{m})[t^{-1}]) \otimes V)}$$

is an isomorphism by the Nakayama lemma

$$\Leftrightarrow i_{L^{R/\mathfrak{m}}}\colon L^{R/\mathfrak{m}} \to \frac{(R/\mathfrak{m})((t)) \otimes V}{(t^{-1}(R/\mathfrak{m})[t^{-1}]) \otimes V} \text{ is an isomorphism,} \tag{3}$$

where $L^{R_\mathfrak{m}} := R_\mathfrak{m} \otimes_R L^R$ and $L^{R/\mathfrak{m}} := (R_\mathfrak{m}/\mathfrak{m}R_\mathfrak{m}) \otimes_{R_\mathfrak{m}} L^{R_\mathfrak{m}} \simeq (R/\mathfrak{m}) \otimes_R L^R$. (Observe that since $L^R/t^n L_o^R$ is a finitely generated projective R-module by Exercise 1.3.E.1,

$$\left(\frac{R_\mathfrak{m}}{\mathfrak{m}R_\mathfrak{m}}\right) \otimes_{R_\mathfrak{m}} \frac{L^{R_\mathfrak{m}}}{R_\mathfrak{m} \otimes_R t^n L_o^R} \simeq \frac{L^{R/\mathfrak{m}}}{t^n L_o^{R/\mathfrak{m}}}.$$

Moreover, since $\mathfrak{m}$ is a finitely generated ideal, $(R/\mathfrak{m}) \otimes_R t^n L_o^R \simeq t^n L_o^{R/\mathfrak{m}} := t^n (R/\mathfrak{m})[[t]] \otimes_\mathbb{C} V$.)

The equivalence (3) is of course the same as the equivalence (2). This proves the lemma. □

Proposition 1.3.14 *The morphism $\mu_1\colon \overline{\mathrm{SL}}_N[t^{-1}]^- \to \bar{X}_{\mathrm{SL}_N}$, induced from the functor morphism $g \mapsto g \cdot \bar{o}_R$ for $g \in \mathrm{SL}_N(R[t^{-1}])^-$, has its image in $\bar{X}_{\mathrm{SL}_N} \backslash Z(\sigma)$, where $\overline{\mathrm{SL}}_N[t^{-1}]^-$ is defined in Corollary 1.3.3(a) and $\bar{o}_R$ is the base point of $\mathrm{SL}_N(R((t)))/\mathrm{SL}_N(R[[t]])$. Moreover, $\mu_1\colon \overline{\mathrm{SL}}_N[t^{-1}]^- \to \bar{X}_{\mathrm{SL}_N} \backslash Z(\sigma)$ is an isomorphism of ind-schemes.*

Proof By Corollary 1.3.3(a) and Lemma 1.3.13, it suffices to prove that for any $R \in \mathbf{Alg}$, the map

$$\mu_1(R)\colon \mathrm{SL}_N(R[t^{-1}])^- \to \mathcal{Q}(R),\ g \mapsto g L_o^R,$$

gives a bijection onto $\mathcal{Q}(R) \backslash Z(\sigma_R)$. We first show that

$$\operatorname{Im}(\mu_1(R)) \subset \mathcal{Q}(R) \backslash Z(\sigma_R). \tag{1}$$

We show that $g L_o^R \in \mathcal{Q}(R)$ for any $g \in \mathrm{SL}_N(R((t)))$. In fact, we show that for $g \in \mathrm{SL}_N(R((t)))$ and $L^R \in \mathcal{Q}^N(R)$, $gL^R \in \mathcal{Q}^N(R)$, i.e., gL^R satisfies properties (a) and (b) of Definition 1.3.6 for some $n \geq 0$. Let $g \in \mathrm{SL}_N(t^{-d} R[[t]])$ (cf. Definition 1.3.1) and let $L^R \in \mathcal{Q}_m^N(R)$, for some $d, m \geq 0$. Then it is easy to see that

$$gL^R \subset t^{-m-d} L_o^R. \tag{2}$$

Choosing $d' \geq 0$ such that $g^{-1} \in \mathrm{SL}_N(t^{-d'} R[[t]])$, we get from (2) (replacing g by g^{-1}): $g^{-1} L^R \subset t^{-m-d'} L_o^R$, which gives (since $t^m L_o^R \subset L^R$ by assumption)

$$t^{3m+d'} L_o^R \subset gL^R. \tag{3}$$

Combining (3) and (2), we get that gL^R satisfies property (a).

Let $\beta^{L^R}: \Lambda^N_{R[[t]]}(L^R) \to R((t))$ be the map as in (b) of Definition 1.3.6 and let $e_1, \ldots, e_N$ be the standard basis of $V = \mathbb{C}^N$. Taking $v_j = \sum_{i=1}^N p_{ij} e_i \in L^R$, for $p_{ij} \in R((t))$, it is easy to see that

$$\begin{aligned}\beta^{gL^R}(gv_1 \wedge \ldots \wedge gv_N) &= \det g \cdot \beta^{L^R}(v_1 \wedge \ldots \wedge v_N)\\ &= \beta^{L^R}(v_1 \wedge \ldots \wedge v_N), \quad \text{since} \quad g \in \mathrm{SL}_N(R((t))).\end{aligned}$$

Thus, $\det(gL^R) = \det(L^R) = R[[t]]$, proving property (b).

Take $g \in \mathrm{SL}_N(R[t^{-1}])^-$. Then (1) is equivalent to showing that $i_{gL_o^R}: gL_o^R \to \frac{R((t))\otimes V}{t^{-1}R[t^{-1}]\otimes V}$ is an isomorphism. Since $g \in \mathrm{SL}_N(R[t^{-1}])$,

$$g^{-1}(R[t^{-1}] \otimes V) = R[t^{-1}] \otimes V. \tag{4}$$

Thus

$$g^{-1}(t^{-1}R[t^{-1}] \otimes V) = t^{-1}R[t^{-1}] \otimes V. \tag{5}$$

Hence, $i_{gL_o^R}$ is an isomorphism if and only if $i_{L_o^R}: L_o^R \to \frac{R((t))\otimes V}{g^{-1}\cdot(t^{-1}R[t^{-1}]\otimes V)}$ is an isomorphism, which follows from (5).

The injectivity of $\mu_1(R)$ is easy to see.

Finally, take $L^R \in \mathcal{Q}(R)\backslash Z(\sigma_R)$. We first show that the map (induced from the inclusion) $k: L^R \cap (R[t^{-1}] \otimes V) \to L^R/tL^R$ is an isomorphism of R-modules.

From the definition of $\mathcal{Q}(R)\backslash Z(\sigma_R)$,

$$R((t)) \otimes V = L^R \oplus (t^{-1}R[t^{-1}] \otimes V), \tag{6}$$

which gives $R((t)) \otimes V = tL^R \oplus (R[t^{-1}] \otimes V)$. Hence,

$$\begin{aligned}L^R &= L^R \cap (tL^R \oplus (R[t^{-1}] \otimes V))\\ &= tL^R \oplus (L^R \cap (R[t^{-1}] \otimes V)), \ \text{since}\ tL^R \subset L^R.\end{aligned} \tag{7}$$

From (7), we get that k is an isomorphism.

We further claim that the map (induced from the inclusion) $\ell: L^R \cap (R[t^{-1}] \otimes V) \to \frac{R[t^{-1}]\otimes V}{t^{-1}R[t^{-1}]\otimes V}$ is an isomorphism (of R-modules).

From (6), since $t^{-1}R[t^{-1}] \otimes V \subset R[t^{-1}] \otimes V$, we get

$$R[t^{-1}] \otimes V = \left(L^R \cap (R[t^{-1}] \otimes V)\right) \oplus \left(t^{-1}R[t^{-1}] \otimes V\right).$$

This proves that ℓ is an isomorphism.

Since $\frac{R[t^{-1}]\otimes V}{t^{-1}R[t^{-1}]\otimes V}$ is a free R-module of rank N, by virtue of the isomorphism ℓ, we get an R-module basis $\{v_1, \ldots, v_N\}$ of $L^R \cap (R[t^{-1}] \otimes V)$, where

$$v_i = 1 \otimes e_i \pmod{t^{-1}R[t^{-1}] \otimes V}. \tag{8}$$

Since $L^R \in \mathscr{Q}(R)$, by the definition of $\mathscr{Q}$, $\det(L^R) = R[[t]]$. From this and the choice of v_i satisfying (8), we easily get that

$$\beta^{L^R}(v_1 \wedge \ldots \wedge v_N) = 1, \tag{9}$$

where $\beta^{L^R} : \Lambda^N_{R[[t]]}(L^R) \to R((t))$ is the map as in (b) of Definition 1.3.6. Define $g_o \in \mathrm{SL}_N(R[t^{-1}])^-$ as follows:

$$g_o e_i = v_i, \quad \text{for all } 1 \le i \le N. \tag{10}$$

For any $u_1, \ldots, u_N \in R((t)) \otimes V$ and $g \in M_N(R((t)))$, we have

$$gu_1 \wedge \ldots \wedge gu_N = \det g \cdot (u_1 \wedge \ldots \wedge u_N). \tag{11}$$

Thus, from (9)–(11), we get that $\det(g_o) = 1$ and hence, from (8), we get that indeed $g_o \in \mathrm{SL}_N(R[t^{-1}])^-$. From the definition of $\mu_1(R)$ and g_o, we get that

$$\mu_1(R)(g_o) = \sum_{i=1}^{N} R[[t]]v_i \subset L^R.$$

By the isomorphism k,

$$L^R = tL^R + \sum_{i=1}^{N} R[[t]]v_i.$$

Hence, by the Nakayama lemma (Atiyah and Macdonald, 1969, Corollary 2.7), we get that $\sum_{i=1}^{N} R[[t]]v_i = L^R$. (We have used here that any maximal ideal of $R[[t]]$ contains $tR[[t]]$, cf. Exercise 1.3.E.13.) Thus, $\mu_1(R)(g_o) = L^R$. This proves the surjectivity of $\mu_1(R)$ onto $\mathscr{Q}(R)\backslash Z(\sigma_R)$, proving the proposition. □

Following Definition 1.3.10, consider the morphism of ind-schemes:

$$\pi : \overline{\mathrm{SL}}_N((t)) \to \bar{X}_{\mathrm{SL}_N}, \quad \text{induced from } g \mapsto g\bar{o}_R \text{ for } g \in \mathrm{SL}_N(R((t))),$$

where $\bar{o}_R$ is the base point of $\mathrm{SL}_N(R((t)))/\mathrm{SL}_N(R[[t]])$. Let $\tilde{Z}(\sigma)$ be the inverse image of the zero set $Z(\sigma)$ under the above morphism, where σ is the section of the line bundle $\mathscr{L}$ over $\bar{X}_{\mathrm{SL}_N}$ as in Definition 1.3.12.

Corollary 1.3.15 *The morphism*

$$\overline{\mu}: \overline{\mathrm{SL}}_N[t^{-1}]^- \times \overline{\mathrm{SL}}_N[[t]] \to \overline{\mathrm{SL}}_N((t)), \textit{ induced from } (g,h) \mapsto gh$$

for $g \in \mathrm{SL}_N(R[t^{-1}])^-$ *and* $h \in \mathrm{SL}_N(R[[t]])$, *is an isomorphism onto its image* $\overline{\mathrm{SL}}_N((t))\backslash\tilde{Z}(\sigma)$ (*which is an open subset of* $\overline{\mathrm{SL}}_N((t))$).

In particular, π *is a locally trivial principal* $\overline{\mathrm{SL}}_N[[t]]$*-bundle.*

Proof From the representability of the functor $\mathcal{Q}(R)\backslash Z(\sigma_R)$ by $\bar{X}_{\mathrm{SL}_N}\backslash Z(\sigma)$ (cf. Lemma 1.3.13), representability of $\mathrm{SL}_N(R((t)))$ by $\overline{\mathrm{SL}}_N((t))$ (cf. Lemma 1.3.2) and Exercise 1.3.E.6, we get that $\overline{\mathrm{SL}}_N((t))\backslash\tilde{Z}(\sigma)$ represents the functor

$$R \rightsquigarrow \mathrm{SL}_N(R((t)))\backslash\tilde{Z}(\sigma_R)$$

$$:= \left\{g \in \mathrm{SL}_N(R((t))) : i_{gL_o^R} : gL_o^R \to \frac{R((t))\otimes V}{t^{-1}R[t^{-1}]\otimes V} \text{ is an isomorphism}\right\}. \tag{1}$$

So, to prove the corollary, it suffices to show (by Lemma 1.3.2 and Corollary 1.3.3(a)) that the map

$$\overline{\mu}(R)\colon\ \mathrm{SL}_N(R[t^{-1}])^- \times \mathrm{SL}_N(R[[t]]) \to \mathrm{SL}_N(R((t))),\ (g,h)\mapsto gh,$$

gives a bijection onto $\mathrm{SL}_N(R((t)))\backslash\tilde{Z}(\sigma_R)$.

From (1) and (5) of the proof of Proposition 1.3.14,

$$\mathrm{Im}\,(\overline{\mu}(R)) \subset \mathrm{SL}_N(R((t)))\backslash\tilde{Z}(\sigma_R).$$

Conversely, take $g' \in \mathrm{SL}_N(R((t)))\backslash\tilde{Z}(\sigma_R)$. By Proposition 1.3.14, there exists $g \in \mathrm{SL}_N(R[t^{-1}])^-$ such that $gL_o^R = g'L_o^R$. But the isotropy of L_o^R in $\mathrm{SL}_N(R((t)))$ is precisely equal to $\mathrm{SL}_N(R[[t]])$. Thus, $g' = g\cdot h$, for some $h \in \mathrm{SL}_N(R[[t]])$. Hence, $\overline{\mu}(R)$ has image precisely equal to $\mathrm{SL}_N(R((t)))\backslash\tilde{Z}(\sigma_R)$. It is easy to see that $\overline{\mu}(R)$ is injective. This proves the first part of the corollary.

Of course, the assertion that π is a locally trivial principal $\overline{\mathrm{SL}}_N[[t]]$-bundle follows from the first part and Proposition 1.3.14 once we prove that $V = \bar{X}_{\mathrm{SL}_N}$, where $V := \cup_{g\in\mathrm{SL}_N((t))}\, g\left(\bar{X}_{\mathrm{SL}_N}\setminus Z(\sigma)\right)$. But clearly $V(\mathbb{C}) = \bar{X}_{\mathrm{SL}_N}(\mathbb{C}) = \mathrm{SL}_N((t))/\mathrm{SL}_N[[t]]$. Since V and $\bar{X}_{\mathrm{SL}_N}$ have the same $\mathbb{C}$-points and V is open in $\bar{X}_{\mathrm{SL}_N}$, we get that $V = \bar{X}_{\mathrm{SL}_N}$ (since any ind-scheme of ind-finite type has nonempty set of $\mathbb{C}$-points). □

Now, we are ready to show that $\mathscr{X}_G$ is represented by an ind-scheme (cf. Definition 1.3.5). We first prove the following lemma.

Lemma 1.3.16 *For any connected reductive group* G*, the morphism*

$$\overline{\mu}_G\colon \bar{G}[t^{-1}]^- \times \bar{G}[[t]] \to \bar{G}((t)),\ \textit{induced from the morphism}\ (g,h)\mapsto gh,$$

for $g \in G(R[t^{-1}])^-$ *and* $h \in G(R[[t]])$, *is an isomorphism onto an open subset of* $\bar{G}((t))$.

Proof Take a faithful representation $j\colon G \hookrightarrow \mathrm{SL}_N$. This gives rise to the commutative diagram

$$\begin{array}{ccc} \bar{G}[t^{-1}]^- \times \bar{G}[[t]] & \xrightarrow{\overline{\mu}_G} & \bar{G}((t)) \\ \Big\downarrow j_1 & & \Big\downarrow j_2 \\ \overline{\mathrm{SL}}_N[t^{-1}]^- \times \overline{\mathrm{SL}}_N[[t]] & \xrightarrow[\overline{\mu}]{} & \overline{\mathrm{SL}}_N((t)), \end{array} \qquad (\mathscr{D})$$

where $\overline{\mu}$ is as in Corollary 1.3.15 and the vertical maps are induced from the inclusion j. Let $\overline{\mathrm{SL}}_N((t))\backslash\tilde{Z}(\sigma)$ be the open subset of $\overline{\mathrm{SL}}_N((t))$ as in Corollary 1.3.15. Then we assert that

$$\mathrm{Im}\,(\overline{\mu}_G) = j_2^{-1}(\overline{\mathrm{SL}}_N((t))\backslash\tilde{Z}(\sigma)); \qquad (1)$$

in particular, $j_2^{-1}\big(\overline{\mathrm{SL}}_N((t))\backslash\tilde{Z}(\sigma)\big)$ does not depend upon the choice of j. Moreover, we show that $\bar{\mu}_G$ is an isomorphism onto $\mathrm{Im}\,(\overline{\mu}_G)$.

By (1) of Corollary 1.3.15 and Exercise 1.3.E.6 applied to $\bar{G}((t)) \underset{\overline{\mathrm{SL}}_N((t))}{\times} \big(\overline{\mathrm{SL}}_N((t))\backslash\tilde{Z}(\sigma)\big)$, the functor

$$\begin{aligned} R \rightsquigarrow\; & G(R((t)))\backslash\tilde{Z}_G(\sigma_R) \\ := \Big\{ & g \in G(R((t))) : i_{j_2(g)L_o^R} : \\ & j_2(g)L_o^R \to \frac{R((t)) \otimes V}{t^{-1}R[t^{-1}] \otimes V} \text{ is an isomorphism}\Big\} \end{aligned}$$

represents the open subscheme j_2^{-1} $(\overline{\mathrm{SL}}_N((t))\backslash\tilde{Z}(\sigma))$ of the ind-scheme $\bar{G}((t))$.

Moreover, $G(R[t^{-1}])^-$ (resp. $G(R[[t]])$) represents $\bar{G}[t^{-1}]^-$ (resp. $\bar{G}[[t]]$) by Corollary 1.3.3(a) (resp. Lemma 1.3.2). Thus, to prove the lemma, it suffices to show that for any $R \in \mathbf{Alg}$,

$$\overline{\mu}_G(R)\colon G(R[t^{-1}])^- \times G(R[[t]]) \to G(R((t))),\ \ (g_R, h_R) \mapsto g_R \cdot h_R$$

is a bijection onto $G(R((t)))\backslash\tilde{Z}_G(\sigma_R)$.

We have the following commutative diagram $(\mathscr{D}_R)$ (analogue of the diagram $\mathscr{D}$ for any R):

$$\begin{array}{ccc} G(R[t^{-1}])^- \times G(R[[t]]) & \xrightarrow{\overline{\mu}_G(R)} & G(R((t))) \\ \Big\downarrow {\scriptstyle j_1(R)} & & \Big\downarrow {\scriptstyle j_2(R)} \\ \mathrm{SL}_N(R[t^{-1}])^- \times \mathrm{SL}_N(R[[t]]) & \xrightarrow[\overline{\mu}(R)]{} & \mathrm{SL}_N(R((t))), \end{array} \qquad (\mathscr{D}_R)$$

where $\overline{\mu}(R)$ is bijective onto its image $\mathrm{SL}_N(R((t)))\backslash\tilde{Z}(\sigma_R)$ (cf. Proof of Corollary 1.3.15). From this, the injectivity of $\overline{\mu}_G(R)$ follows as well as

$$\mathrm{Im}\,(\overline{\mu}_G(R)) \subset G(R((t)))\backslash\tilde{Z}_G(\sigma_R). \tag{2}$$

To prove the converse, take $x_R \in G(R((t)))\backslash\tilde{Z}_G(\sigma_R)$. Then, by the proof of Corollary 1.3.15, there exists $g_R \in \mathrm{SL}_N(R[t^{-1}])^-$, $h_R \in \mathrm{SL}_N(R[[t]])$ such that

$$g_R \cdot h_R = j_2(R)(x_R). \tag{3}$$

Choose a polynomial representation W over $\mathbb{C}$ of SL_N with a vector $w_o \in W$ such that the scheme-theoretic isotropy subgroup $(\mathrm{SL}_N)_{w_o}$ of w_o in SL_N is precisely equal to G (cf. (Borel, 1991, Chap. II, Theorem 5.1 and §5.5)). Then, for any $S \in \mathbf{Alg}$, by Exercise 1.3.E.6 applied to $\mathrm{SL}_N \underset{W}{\times} w_o$ (for the map $\mathrm{SL}_N \to W$, $g \mapsto gw_o$), we get that

$$G(S) \text{ is precisely the isotropy of } w_o \text{ in } \mathrm{SL}_N(S). \tag{4}$$

Evaluating identity (3) at $w_o \in R((t))\otimes W$ and applying (4) for $S = R((t))$, we get

$$g_R^{-1}(w_o) = h_R(w_o). \tag{5}$$

But $g_R^{-1}(w_o) - w_o \in t^{-1}R[t^{-1}] \otimes W$ and $h_R(w_o) \in R[[t]] \otimes W$. Thus, from (5), we get that

$$h_R(w_o) = w_o = g_R(w_o). \tag{6}$$

Thus, from (4), we get that $(g_R, h_R) \in \mathrm{Im}\,(j_1(R))$. This proves, in view of (2),

$$\mathrm{Im}\,(\overline{\mu}_G(R)) = G(R((t)))\backslash\tilde{Z}_G(\sigma_R),$$

proving the lemma. □

Remark 1.3.17 Lemma 1.3.16 remains valid, more generally, for any closed subgroup $H \subset \mathrm{SL}_N$ (in lieu of G) such that H is the scheme-theoretic stabilizer of a vector w_o in a polynomial representation W of SL_N.

Proposition 1.3.18 *[Realizing $\mathscr{X}_G$ as an ind-scheme] Let G be any group H as in the above remark.*

(a) The functor $\mathscr{X}_G$ as in Definition 1.3.5 is represented by an ind-scheme denoted $\bar{X}_G$ with $\mathbb{C}$-points X_G.

In fact, $\bar{X}_G$ is an ind-projective variety if G is a connected, semisimple algebraic group (cf. Corollary 1.3.19 and Theorem 1.3.23).

(b) The morphism $\bar{G}[t^{-1}]^- \to \bar{X}_G$ induced by the functor morphism

$$G(R[t^{-1}])^- \to \mathscr{X}_G(R),\ g \mapsto g\bar{o}_R,\ \text{for } g \in G(R[t^{-1}])^-,$$

is an open embedding, where $\bar{o}_R$ is the base point of $G(R((t)))/G(R[[t]])$. Moreover, $\{g_o\bar{G}[t^{-1}]^- \cdot \bar{o}\}_{g_o \in G((t))}$ provides an open cover of the ind-scheme $\bar{X}_G$.

(c) We have a morphism

$$\bar{G}((t)) \times \bar{X}_G \to \bar{X}_G$$

induced from the morphism of functors

$$G(R((t))) \times \mathscr{X}_G^o(R) \to \mathscr{X}_G^o(R) \subset \mathscr{X}_G(R),$$
$$(g, h\bar{o}_R) \mapsto gh\bar{o}_R \text{ for } g, h \in G(R((t))).$$

Proof (a) Consider the subfunctor $h_{\bar{G}[t^{-1}]^-} \hookrightarrow \mathscr{X}_G$ which takes (for any $\mathbb{C}$-algebra R) $g \in h_{\bar{G}[t^{-1}]^-}(R) = G(R[t^{-1}])^-$ to $g\bar{o}_R \in \mathscr{X}_G^o(R) = G(R((t)))/G(R[[t]]) \subset \mathscr{X}_G(R)$. Then, $h_{\bar{G}[t^{-1}]^-}$ is an open subfunctor of $\mathscr{X}_G$ (cf. Exercise 1.3.E.7). Thus, for any $g_o \in G((t))$,

$$h_{g_o\bar{G}[t^{-1}]^-} \hookrightarrow \mathscr{X}_G,\ g_o g \mapsto g_o g\bar{o}_R \text{ for } g \in G(R[t^{-1}])^-$$

is an open subfunctor.

We next claim that the collection of subfunctors $\{h_{g_o\bar{G}[t^{-1}]^-}\}_{g_o \in G((t))}$ is an open covering of $\mathscr{X}_G$. To prove this, in view of Eisenbud and Harris (2000, Exercise VI-11), it suffices to show that for any field $k \supset \mathbb{C}$,

$$\cup_{g_o \in G((t))}\, g_o G(k[t^{-1}])^- \bar{o}_k = \mathscr{X}_G(k) = \mathscr{X}_G^o(k) = G(k((t)))/G(k[[t]]), \tag{1}$$

where the second equality in the above equation follows since k is a field (cf. Exercise 1.3.E.7).

To prove (1), equivalently, we need to prove

$$\cup_{g_o \in G((t))}\, g_o G(k[t^{-1}])^- \cdot G(k[[t]]) = G(k((t))). \tag{2}$$

For any $g_o \in G((t))$, by Lemma 1.3.16, the functor $R \rightsquigarrow g_o G(R[t^{-1}])^- \cdot G(R[[t]])$ is an open subfunctor of $G(R((t)))$ represented by an open ind-subscheme of $\bar{G}((t))$ with $\mathbb{C}$-points $g_o G[t^{-1}]^- \cdot G[[t]]$. Consider the sheafification $\mathfrak{F}$ of the functor

$$\mathfrak{F}^o : R \rightsquigarrow \cup_{g_o \in G((t))} g_o G(R[t^{-1}])^- \cdot G(R[[t]]) \subset G(R((t))).$$

Then $\mathfrak{F}$ is an open subfunctor of $G(R((t)))$ represented by an open ind-subscheme denoted V of $\bar{G}((t))$ (cf. Lemma 1.3.2 and Definition B.5(b)) with $\mathbb{C}$-points

$$\cup_{g_o \in G((t))} g_o G[t^{-1}]^- \cdot G[[t]] = G((t)).$$

Also, $\bar{G}((t))$ has $\mathbb{C}$-points $G((t))$. So, both V and $\bar{G}((t))$ have the same set of $\mathbb{C}$-points and hence $V = \bar{G}((t))$ (since any closed ind-subscheme of $\bar{G}((t))$ has nonempty set of $\mathbb{C}$-points, cf. Exercise 1.3.E.8). Thus, for any $\mathbb{C}$-algebra R,

$$V(R) = \bar{G}((t))(R) = G(R((t))). \tag{3}$$

But, k being a field,

$$G(k((t))) = V(k) = \mathfrak{F}(k) = \mathfrak{F}^o(k) = \cup_{g_o \in G((t))} g_o G(k[t^{-1}])^- \cdot G(k[[t]]),$$

where the first equality follows from (3) and the third equality follows since k is a field. This proves (2) and hence (1).

Recall from Lemma 1.3.16 that there is an isomorphism

$$\overline{\mu}_G : \bar{G}[t^{-1}]^- \times \bar{G}[[t]] \to \mathring{V}, \ (g,h) \mapsto gh,$$

where $\mathring{V}$ is an open subset of $\bar{G}((t))$. For any $g_o \in G((t))$ this gives rise to an isomorphism

$$\overline{\mu}_G(g_o) : (g_o \bar{G}[t^{-1}]^-) \times \bar{G}[[t]] \to g_o \mathring{V}, \ (g_o g, h) \mapsto g_o gh.$$

For any $d \geq 0$, recall the closed subscheme $\bar{G}(t^{-d}\mathbb{C}[[t]])$ of $\bar{G}((t))$ from Definition 1.3.1. Then, there exists a closed (affine) subscheme $(g_o \bar{G}[t^{-1}]^-)_d$ of $g_o \bar{G}[t^{-1}]^-$ (of finite type over $\mathbb{C}$) such that $\overline{\mu}_G(g_o)$ restricts to an isomorphism

$$(\overline{\mu}_G(g_o))_d : (g_o \bar{G}[t^{-1}]^-)_d \times \bar{G}[[t]] \to (g_o \mathring{V}) \cap \bar{G}(t^{-d}\mathbb{C}[[t]]).$$

Consider the subfunctor $\mathscr{X}_G^d$ of $\mathscr{X}_G$ (cf. Definition 1.3.5) defined as the sheafification of the functor $R \rightsquigarrow G(t^{-d}R[[t]])/G(R[[t]])$ (cf. Lemma B.2).

Then, parallel to the above proof, we get that the collection of subfunctors $\{h_{(g_o \bar{G}[t^{-1}]^-)_d}\}_{g_o \in G((t))}$ is an open covering of $\mathscr{X}_G^d$. Thus, by Eisenbud and Harris (2000, Theorem VI-14) (since a Zariski cover is an fppf cover by Stacks (2019, Tag 021N)), the functor $\mathscr{X}_G^d$ is represented by a scheme denoted $\bar{X}_G^d$ with $\mathbb{C}$-points $X_G^d = G(t^{-d}\mathbb{C}[[t]])/G[[t]]$. Moreover, the morphism $i_d \colon \mathscr{X}_G^d \to \mathscr{X}_G^{d+1}$ induced from the inclusion gives rise to a morphism $\bar{i}_d \colon \bar{X}_G^d \to \bar{X}_G^{d+1}$. Since $\bar{G}(t^{-d}\mathbb{C}[[t]])$ is a closed subscheme of $\bar{G}(t^{-d-1}\mathbb{C}[[t]])$, we get that $(g_o\bar{G}[t^{-1}]^-)_d$ is a closed subscheme of $(g_o\bar{G}[t^{-1}]^-)_{d+1}$. Moreover, $\{(g_o\bar{G}[t^{-1}]^-)_d\}_{g_o\in G((t))}$ provides an open cover of $\bar{X}_G^d$ (see the proof of (b) below). Thus, the morphism $\bar{i}_d \colon \bar{X}_G^d \to \bar{X}_G^{d+1}$ is a closed embedding. Now, it is easy to see that the ind-scheme

$$\bar{X}_G := (\bar{X}_G^0 \subset \bar{X}_G^1 \subset \bar{X}_G^2 \subset \cdots)$$

represents the functor $\mathscr{X}_G$, once we observe that $\cup_{d\geq 0}\, \mathscr{X}_G^d(R) = \mathscr{X}_G(R)$ for any $\mathbb{C}$-algebra R. This proves (a).

An alternative proof of (a). Fix an embedding $G \subset \mathrm{SL}_N$. Then, by Beilinson and Drinfeld (1994, lemma after Theorem 4.5.1), the functor $\mathscr{X}_G$ is a closed subfunctor of $\mathscr{X}_{\mathrm{SL}_N}$ (for more details of the proof, see Zhu (2017, Proposition 1.2.6)[1], where the notion of a closed subfunctor is parallel to that of an open subfunctor as in Definition B.5(b). Thus, by Theorem 1.3.8, $\mathscr{X}_G$ is a representable functor represented by a closed ind-subscheme of $\bar{X}_{\mathrm{SL}_N}$.

(b) Since $h_{\bar{G}[t^{-1}]^-} \hookrightarrow \mathscr{X}_G$ is an open subfunctor (as observed above), we have that $\bar{G}[t^{-1}]^-\bar{o} \subset \bar{X}_G$ is an open ind-subscheme from the representability of $\mathscr{X}_G$ by $\bar{X}_G$ and Definition B.5(b).

To prove that $\{g_o\bar{G}[t^{-1}]^- \cdot \bar{o}\}_{g_o\in G((t))}$ is an open cover of the ind-scheme $\bar{X}_G$, observe that $U := \cup_{g_o\in G((t))}\, g_o\bar{G}[t^{-1}]^- \cdot \bar{o}$ is an open subset of $\bar{X}_G$, $\bar{X}_G$ is a closed ind-subscheme of $\bar{X}_{\mathrm{SL}_N}$ (cf. Exercise 1.3.E.9) and $\bar{X}_{\mathrm{SL}_N}$ is of ind-finite type (cf. Theorem 1.3.8). Thus, any closed ind-subscheme of $\bar{X}_G$ has nonempty set of $\mathbb{C}$-points. Moreover,

$$U(\mathbb{C}) = \cup_{g_o\in G((t))}\, g_o G[t^{-1}]^- \cdot \bar{o} = \bar{X}_G(\mathbb{C}) = G((t)) \cdot \bar{o}.$$

(c) To prove (c), simply observe that the sheafification of the functor $h_{\bar{G}((t))} \times \mathscr{X}_G^o$ is $h_{\bar{G}((t))} \times \mathscr{X}_G$, since $h_{\bar{G}((t))}$ is representable. □

By the above Proposition 1.3.18 and Lemma 1.3.16, the following corollary follows easily.

Corollary 1.3.19 *Let G be any group H as in Remark 1.3.17. The projection $\pi \colon \bar{G}((t)) \to \bar{X}_G$ is a locally trivial principal $\bar{G}[[t]]$-bundle.*

Further, for any faithful representation $j \colon G \hookrightarrow \mathrm{SL}_N$, $\bar{X}_G \hookrightarrow \bar{X}_{\mathrm{SL}_N}$ is a closed embedding (cf. Exercise 1.3.E.9). Thus, by Theorem 1.3.8, $\bar{X}_G$ is an ind-projective scheme.

[1] I thank X. Zhu for these two references.

Definition 1.3.20 An ind-scheme $X = (X_n)_{n\geq 0}$ is called *reduced* if there exists an equivalent filtration $(Y_n)_{n\geq 0}$ of X (i.e., $\mathrm{Id}\colon X \to X$ is an isomorphism of ind-schemes, where the two copies of X are equipped with the two ind-scheme structures induced from the filtrations X_n and Y_n) such that each Y_n is a reduced scheme.

Lemma 1.3.21 *If $X = (X_n)_{n\geq 0}$ is a reduced ind-scheme, then $(X_n^{\mathrm{red}})_{n\geq 0}$ provides an equivalent filtration of X, where X_n^{red} is the corresponding reduced scheme (cf. (Hartshorne, 1977, Chap. II, Exercise 2.3)). (Of course, as a topological space, $X_n^{\mathrm{red}} = X_n$.)*

Proof Since X is a reduced ind-scheme, for any $n \geq 0$, there exists $k(n) \geq 0$ such that $i_n\colon Y_n \to X_{k(n)}$ is a closed embedding. But, since Y_n is reduced, i_n factors through a closed embedding $\bar{i}_n\colon Y_n \to X_{k(n)}^{\mathrm{red}}$. Thus, the identity map $\mathrm{Id}\colon X \to X^{\mathrm{red}}$ is a morphism of ind-schemes, where X^{red} denotes the ind-scheme obtained from the filtration $(X_n^{\mathrm{red}})_{n\geq 0}$.

Conversely, the closed embedding $X_n^{\mathrm{red}} \to X_n$ clearly shows that $\mathrm{Id}\colon X^{\mathrm{red}} \to X$ is a morphism of ind-schemes. This proves the lemma. □

The following theorem's proof was briefly outlined by G. Faltings (personal communication dated January 26, 2017). B. Conrad provided a detailed proof of the theorem given below (personal communication dated January 27, 2017).

Recall that the Lie algebra $\mathrm{Lie}\,\mathscr{G}$ of an ind-affine group scheme $\mathscr{G}$ is, by definition, the kernel of the group homomorphism $\mathscr{G}(\mathbb{C}(\epsilon)) \to \mathscr{G}(\mathbb{C})$ induced by $\epsilon \mapsto 0$, where $\mathbb{C}(\epsilon) := \mathbb{C}[\epsilon]/\langle \epsilon^2 \rangle$. By Corollary B.21, $\mathrm{Lie}\,\mathscr{G}$ is a Lie algebra.

Let $\mathscr{G} = (\mathscr{G}_n)_{n\geq 0}$ be an ind-affine group scheme. Then $\mathscr{G}^{\mathrm{red}} := (\mathscr{G}_n^{\mathrm{red}})_{n\geq 0}$ is again an ind-affine group scheme. This follows since the multiplication map $\mathscr{G}_n \times \mathscr{G}_m \to \mathscr{G}_l$ clearly restricts to $\mathscr{G}_n^{\mathrm{red}} \times \mathscr{G}_m^{\mathrm{red}} \to \mathscr{G}_l^{\mathrm{red}}$ and so does the morphism induced from the inverse.

Theorem 1.3.22 *Let $\mathscr{G} = (\mathscr{G}_n)_{n\geq 0}$ be an ind-affine group scheme filtered by (affine) finite-type schemes over $\mathbb{C}$ and let $\mathscr{G}^{\mathrm{red}} = (\mathscr{G}_n^{\mathrm{red}})_{n\geq 0}$ be the associated reduced ind-affine group scheme. Assume that the canonical ind-group morphism $i\colon \mathscr{G}^{\mathrm{red}} \to \mathscr{G}$ induces an isomorphism $\dot{i}_e\colon \mathrm{Lie}(\mathscr{G}^{\mathrm{red}}) \xrightarrow{\sim} \mathrm{Lie}\,\mathscr{G}$ of the associated Lie algebras. Then, i is an isomorphism of ind-groups, i.e., $\mathscr{G}$ is a reduced ind-scheme.*

Proof Let $A := \mathbb{C}[\mathscr{G}]$ be the affine coordinate ring of $\mathscr{G}$, i.e., $A := \varprojlim A_n$ with inverse limit topology, where each $A_n := \mathbb{C}[\mathscr{G}_n]$ is given the discrete topology. For any $\mathbb{C}$-algebra B, the set of morphisms $\mathrm{Spec}(B) \to \mathscr{G}$ coincides exactly with the set of continuous $\mathbb{C}$-algebra homomorphisms $A \to B$, where B has discrete topology.

Let $\hat{A}_n$ denote the formal completion of A_n at the identity e, so $\{\hat{A}_n\}_{n\geq 1}$ is an inverse system of complete local noetherian rings with surjective transition maps $\hat{A}_{n+1} \to \hat{A}_n$ with closed kernels for the max-adic topologies by the Artin–Rees Lemma (cf. (Eisenbud, 1995, Lemma 7.15)). Define

$$\hat{A} := \varprojlim \hat{A}_n \tag{1}$$

to be the topological inverse limit of $\hat{A}_n$ equipped with their max-adic topologies. Each $\hat{A}_n$ is itself a topological inverse limit of Artinian local $\mathbb{C}$-algebras, so the same goes for $\hat{A}$. Concretely, viewing the local Artinian algebra quotients of each A_n supported at e as local Artinian algebra quotients of A also, we see that $\hat{A}$ is (as a topological algebra) the inverse limit of all these local Artinian $\mathbb{C}$-algebras. This makes $\hat{A}$ into a (local) pseudo-compact $\mathbb{C}$-algebra. (For an introduction to pseudo-compact rings, we refer to Demazure and Grothendieck (1970, Exp. VII$_B$).) Recall that the class of pseudo-compact $\mathbb{C}$-algebras includes all the complete local noetherian $\mathbb{C}$-algebras with residue field $\mathbb{C}$ and is stable under arbitrary topological inverse limits. The most basic example of a non-noetherian local pseudo-compact $\mathbb{C}$-algebra is the topological ring $\mathbb{C}[[X_i]]_{i\in I}$ of formal power series over $\mathbb{C}$ in an arbitrary infinite set $\{X_i\}_{i\in I}$ of variables, realized as the completion of the polynomial ring $\mathbb{C}[X_i]$ with respect to the system of ideals $(X_j : j \in J)^N + (X_i : i \notin J)$, for finite subsets $J \subset I$ and integers $N \geq 1$.

The tangent space at a $\mathbb{C}$-point x of a pseudo-compact $\mathbb{C}$-algebra is, by definition, the topological $\mathbb{C}$-linear dual of $\mathfrak{m}/\overline{\mathfrak{m}^2}$, where $\mathfrak{m}$ is the (necessarily open) maximal ideal at x and $\overline{\mathfrak{m}^2}$ is the closure of $\mathfrak{m}^2$. Further, any pseudo-compact $\mathbb{C}$-algebra is determined (including its topology) by its functor of points on local Artinian $\mathbb{C}$-algebras (viewed discretely). The continuous maps from $\hat{A}$ (as in (1)) to any discrete $\mathbb{C}$-algebra must factor through one of the $\hat{A}_n$ (even through some local Artinian $\mathbb{C}$-algebra quotient of A_n).

Since $\mathscr{G}$ is an ind-affine group scheme, we get that $\hat{A}$ is a 'Hopf algebra' (in the weaker sense that the coproduct lands in a completed tensor product over $\mathbb{C}$), which makes $\hat{A}$ into a connected formal group. (For a discussion of formal groups, we refer to Fontaine (1977, Chap. I, §9) and also Demazure and Grothendieck (1970, Exp. VII$_B$, §3).) Now, over a field of characteristic 0, any connected formal group is necessarily of the form $\mathbb{C}[[X_i]]$ as an underlying pseudo-compact $\mathbb{C}$-algebra (cf. (Fontaine, 1977, Chap. I, §9.6)).

In exactly the same way

$$\hat{A}^{\text{red}} := \varprojlim \hat{A}_n^{\text{red}}$$

is a formal group, where $A_n^{\text{red}} := \mathbb{C}[\mathscr{G}_n^{\text{red}}]$ and $\hat{A}_n^{\text{red}}$ is the completion of A_n^{red} at e.

The Lie algebra $\mathrm{Lie}(\hat{A})$ of any formal group $\hat{A}$ over $\mathbb{C}$ is defined as the set of points valued in $\mathbb{C}(\epsilon)$ based at e. It is exactly the tangent space of $\hat{A}$ at e.

Assertion I: The canonical map $\mathrm{Lie}(\hat{A}) \to \mathrm{Lie}(\mathscr{G})$ *is an isomorphism and a similar result for* $\mathrm{Lie}(\mathscr{G}^{\mathrm{red}})$. *Thus, the canonical map*

$$\mathrm{Lie}(\hat{A}^{\mathrm{red}}) \to \mathrm{Lie}(\hat{A}) \tag{2}$$

is an isomorphism.

This follows since the definition of $\mathrm{Lie}(\hat{A})$ as $\mathbb{C}(\epsilon)$-points of $\hat{A}$ based at e forces it to factor through a local Artinian algebra quotient of some A_n (based at e). The proof for $\mathrm{Lie}(\mathscr{G}^{\mathrm{red}})$ is identical. So, Assertion I follows from the assumption $\mathrm{Lie}(\mathscr{G}^{\mathrm{red}}) \xrightarrow{\sim} \mathrm{Lie}\,\mathscr{G}$.

Assertion II: The canonical map $\hat{i}: \hat{A} \to \hat{A}^{\mathrm{red}}$ *is surjective.*

We first recall the following general result:

Let $\xi: R' \to R$ *be a continuous homomorphism between pseudo-compact rings such that for every open ideal* J *of* R *(thus* R/J *is an Artinian ring) the map* $R' \to R/J$ *is surjective. Thus, the preimage* J' *of* J *in* R' *is an open ideal with* $R'/J' \simeq R/J$. *The map* $\xi: R' \to R$ *is then identified with the map*

$$\bar{\xi}: R' \to \varprojlim_J R'/J' \simeq \varprojlim_J R/J \simeq R,$$

for J *varying through the full family of open ideals of* R. *Then,* $\bar{\xi}$ *is surjective and hence so is* ξ *(cf. (Demazure and Grothendieck, 1970, Exp.* VII_B*, Cor. 0.2D(ii)(a))).*

We now come to the proof of Assertion II. By the above result, it suffices to show that for any open ideal J of $\hat{A}^{\mathrm{red}}$, the map $\hat{i}_J: \hat{A} \to \hat{A}^{\mathrm{red}}/J$ is surjective, where $\hat{i}_J$ is the map $\hat{i}$ followed by the projection $\hat{A}^{\mathrm{red}} \to \hat{A}^{\mathrm{red}}/J$. But, J being an open ideal, clearly

$$\hat{A}^{\mathrm{red}}/J \simeq \mathbb{C}[\mathscr{G}_n^{\mathrm{red}}]/J_n, \quad \text{for some } n \geq 1 \text{ and some ideal } J_n \text{ of } \mathbb{C}[\mathscr{G}_n^{\mathrm{red}}].$$

Of course, the canonical map $\mathbb{C}[\mathscr{G}_n] \to \mathbb{C}[\mathscr{G}_n^{\mathrm{red}}]/J_n$ is surjective and so is $A \to \mathbb{C}[\mathscr{G}_n]$. Hence, the canonical map $A \to \mathbb{C}[\mathscr{G}_n^{\mathrm{red}}]/J_n$ is surjective. But, since $\hat{i}$ is a continuous map and J is an open ideal, we get that $\hat{i}_j$ is surjective, proving Assertion II.

Assertion III: The canonical map $\hat{i}: \hat{A} \to \hat{A}^{\mathrm{red}}$ *is an isomorphism.*

We first recall the following (simple) general result obtained using the isomorphism of R below with $\mathbb{C}[[X_i]]$ and similarly for R'.

Let R and R' be two connected formal groups over $\mathbb{C}$ and $f: R \to R'$ a continuous surjective homomorphism respecting augmentations to $\mathbb{C}$. Then, f is an isomorphism (necessarily a topological isomorphism) iff the induced map between Lie algebras is a bijection.

Applying the above result to $\hat{i}$ and using Assertions I and II, we get Assertion III.

Assertion IV: For a complete local $\mathbb{C}$-algebra B with residue field $\mathbb{C}$, the canonical map $i: \mathcal{G}^{\text{red}} \to \mathcal{G}$ induces a bijection $i_B: \mathcal{G}^{\text{red}}(B) \simeq \mathcal{G}(B)$.

Let B be any (not necessarily noetherian) $\mathbb{C}$-algebra. By the definition of a morphism to an ind-scheme, any morphism $\varphi\colon$ Spec $B \to \mathcal{G}$ lands inside $\mathcal{G}_n$ (for some $n \geq 1$). From this, we see that $i_B: \mathcal{G}^{\text{red}}(B) \to \mathcal{G}(B)$ is injective (for any B). So, it suffices to prove that for any B as in Assertion IV, i_B is surjective.

Take $g \in \mathcal{G}(B)$. Then, it is represented by an algebra homomorphism $\bar{g}_n: A_n \to B$, for some $n \geq 1$. Since B is a complete local $\mathbb{C}$-algebra with residue field $\mathbb{C}$, $\bar{g}_n$ induces a continuous homomorphism $\hat{g}_n: \hat{A}_n(x) \to B$ and hence a continuous homomorphism $\hat{g}: \hat{A}(x) \to B$, where $\hat{A}_n(x)$ denotes the completion of A_n with respect to some $\mathbb{C}$-point x (not necessarily e) of $\mathcal{G}_n$ and $\hat{A}(x)$ is the inverse limit of $\{\hat{A}_n(x)\}_n$. But, by Assertion III, $\hat{i}: \hat{A} \to \hat{A}^{\text{red}}$ is an isomorphism (and hence so is $\hat{i}(x): \hat{A}(x) \simeq \hat{A}^{\text{red}}(x)$ by translation using $\mathcal{G}(\mathbb{C}) = \mathcal{G}^{\text{red}}(\mathbb{C})$). Thus, we get a continuous homomorphism

$$\hat{g}^{\text{red}} := \hat{g} \circ (\hat{i}(x))^{-1}: \hat{A}^{\text{red}}(x) \to B.$$

Composing $\hat{g}^{\text{red}}$ with the canonical $\mathbb{C}$-algebra homomorphism $A^{\text{red}} \to \hat{A}^{\text{red}}(x)$, we get a $\mathbb{C}$-algebra homomorphism $A^{\text{red}} \to B$. This provides the desired lift of g in $\mathcal{G}^{\text{red}}(B)$. Thus, $\mathcal{G}^{\text{red}}(B) = \mathcal{G}(B)$, proving Assertion IV.

Assertion V: For any local noetherian $\mathbb{C}$-algebra B with residue field $\mathbb{C}$, the canonical map $i: \mathcal{G}^{\text{red}} \to \mathcal{G}$ induces a bijection $i_B: \mathcal{G}^{\text{red}}(B) \simeq \mathcal{G}(B)$.

As observed in the proof of Assertion IV, it suffices to prove that $\mathcal{G}^{\text{red}}(B) \to \mathcal{G}(B)$ is surjective. Since B is noetherian, $B \to \hat{B}$ is injective, where $\hat{B}$ is the completion of B with respect to its unique maximal ideal. Take $g \in \mathcal{G}(B)$ and represent it as $\bar{g}_n: A_n \to B$ (for some n). By Assertion IV, there exists $N \geq n$ such that $\bar{g}_N: A_N \to B \subset \hat{B}$ (obtained from the composition of $\bar{g}_n$ with the canonical map $A_N \to A_n$) descends to $\hat{g}_N: A_N^{\text{red}} \to \hat{B}$. But, since $A_N \to A_N^{\text{red}}$ is surjective, we get that $\hat{g}_N(A_N^{\text{red}}) \subset B$. This proves Assertion V.

With these preparations, we finally come to the proof of the theorem. We need to show that, for any $\mathbb{C}$-algebra B, the canonical map $i: \mathcal{G}^{\text{red}} \to \mathcal{G}$

induces a bijection $i_B\colon \mathscr{G}^{\mathrm{red}}(B) \simeq \mathscr{G}(B)$. As observed earlier, i_B is injective. So, we only need to prove the surjectivity of i_B.

Take $g \in \mathscr{G}(B)$. Then, as shown earlier, $g \in \mathscr{G}_n(B)$ for some $n \geq 1$. Since $\mathscr{G}_n$ is a scheme of finite type over $\mathbb{C}$ (by assumption), we can assume that B is a $\mathbb{C}$-algebra of finite type over $\mathbb{C}$. Let $\mathfrak{m}$ be a maximal ideal of B (so $B/\mathfrak{m} = \mathbb{C}$) and let $B_\mathfrak{m}$ be the localization. Then, by Assertion V, we can find $n(\mathfrak{m}) \geq 1$ and

$$g'_\mathfrak{m} \in \mathscr{G}^{\mathrm{red}}_{n(\mathfrak{m})}(B_\mathfrak{m}) = \mathrm{Mor}\left(\mathbb{C}[\mathscr{G}^{\mathrm{red}}_{n(\mathfrak{m})}], B_\mathfrak{m}\right) \simeq \mathrm{Mor}\left(\mathrm{Spec}(B_\mathfrak{m}), \mathscr{G}^{\mathrm{red}}_{n(\mathfrak{m})}\right)$$

such that

$$i_{B_\mathfrak{m}}(g'_\mathfrak{m}) = g_\mathfrak{m}, \tag{3}$$

where $g_\mathfrak{m}$ is the element of $\mathscr{G}_{n(\mathfrak{m})}(B_\mathfrak{m})$ corresponding to g. Since $i_{B_\mathfrak{m}}$ is injective, $g'_\mathfrak{m}$ is unique, satisfying (3). Further, since $\mathscr{G}^{\mathrm{red}}_{n(\mathfrak{m})}$ is of finite type over $\mathbb{C}$, there exists an affine open set $U_\mathfrak{m} \subset \mathrm{Spec}(B)$ containing the point $\mathfrak{m}$ such that $g'_\mathfrak{m}$ spreads out to

$$g'_{U_\mathfrak{m}} \in \mathscr{G}^{\mathrm{red}}_{n(\mathfrak{m})}(\mathbb{C}[U_\mathfrak{m}]).$$

By the injectivity of $i_{\mathbb{C}[U_\mathfrak{m}]}$, we get the following analogue of (3) (possibly after suitably shrinking $U_\mathfrak{m}$ around $\mathfrak{m}$):

$$i_{\mathbb{C}[U_\mathfrak{m}]}(g'_{U_\mathfrak{m}}) = g_{U_\mathfrak{m}}, \tag{4}$$

where $g_{U_\mathfrak{m}}$ is the element of $\mathscr{G}_{n(\mathfrak{m})}(\mathbb{C}[U_\mathfrak{m}])$ corresponding to g. As $\mathfrak{m}$ runs over the maximal ideals of B, $\{U_\mathfrak{m}\}$ clearly covers $\mathrm{Spec}(B)$. Choose a finite subcover $\{U_{\mathfrak{m}_i}\}_{1\leq i\leq k}$ of $\mathrm{Spec}(B)$ and let $N := \max_i \{n(\mathfrak{m}_i)\}$. From the uniqueness of $g'_{U_{\mathfrak{m}_i}}$ satisfying (4), we get that $g'_{U_{\mathfrak{m}_i}} = g'_{U_{\mathfrak{m}_j}}$ on $U_{\mathfrak{m}_i} \cap U_{\mathfrak{m}_j}$. Thus, we get the element $g' \in \mathscr{G}^{\mathrm{red}}_N(B)$ such that $g'_{|U_{\mathfrak{m}_i}} = g'_{U_{\mathfrak{m}_i}}$ on $U_{\mathfrak{m}_i}$. From this and (3), we get $i_B(g') = g$. This proves the surjectivity of i_B and hence the theorem is proved. □

As a consequence of the above theorem, we deduce the following result.

Theorem 1.3.23 *Let G be a connected semisimple algebraic group. Then the ind-affine group scheme $\bar{G}[t^{-1}]$ is reduced and hence so is $\bar{G}[t^{-1}]^-$.*

Thus, the infinite Grassmannian $\bar{X}_G$ is a reduced ind-scheme.

Proof We first show that $\mathscr{G} := \bar{G}[t^{-1}]$ is reduced. By Theorem 1.3.22, it suffices to show that

$$\mathrm{Lie}(\mathscr{G}^{\mathrm{red}}) = \mathrm{Lie}(\mathscr{G}). \tag{1}$$

Take an embedding $G \hookrightarrow \mathrm{SL}_N \subset M_N$. This gives rise to a closed embedding of groups: $\mathscr{G} \subset \overline{\mathrm{SL}}_N[t^{-1}]$ (cf. Lemma 1.3.2).

In particular,

$$\mathrm{Lie}(\mathscr{G}) \subset \mathrm{Lie}(\overline{\mathrm{SL}}_N[t^{-1}]) = sl_N \otimes \mathbb{C}[t^{-1}],$$

where for the last equality, see Exercise 1.3.E.12. Considering the evaluation homomorphisms for any $\alpha \in \mathbb{P}^1(\mathbb{C})\backslash\{0\}$ (induced from the $\mathbb{C}$-algebra homomorphisms $R[t^{-1}] \to R$, $t \mapsto \alpha$), $\epsilon(\alpha)\colon \mathscr{G} \to G$ and $\overline{\epsilon}(\alpha)\colon \overline{\mathrm{SL}}_N[t^{-1}] \to \mathrm{SL}_N$, it is easy to see (using Exercise 1.3.E.12 again) that

$$\mathrm{Lie}(\mathscr{G}) \subset \mathfrak{g} \otimes \mathbb{C}[t^{-1}], \quad \text{where} \quad \mathfrak{g} := \mathrm{Lie}\, G. \tag{2}$$

For any root $\beta \in \Delta$ of $\mathfrak{g}$, consider the root subgroup $U_\beta \subset G$ with Lie algebra the root space $\mathfrak{g}_\beta$ (cf. (Jantzen, 2003, Part II, §1.2)). This gives rise to a closed embedding of ind-affine group schemes $j\colon \bar{U}_\beta[t^{-1}] \hookrightarrow \mathscr{G}$. Since $U_\beta \simeq \mathbb{A}^1$, clearly, $\bar{U}_\beta[t^{-1}]$ is reduced. In particular, the embedding $j\colon \bar{U}_\beta[t^{-1}] \hookrightarrow \mathscr{G}$ factors through $\bar{U}_\beta[t^{-1}] \hookrightarrow \mathscr{G}^{\mathrm{red}}$. Further, under the differential $\dot{j}$, similar to (2), we get

$$\mathrm{Lie}(\bar{U}_\beta[t^{-1}]) \subset \mathfrak{g}_\beta \otimes \mathbb{C}[t^{-1}]. \tag{3}$$

In fact, identifying the group U_β with the additive group $G_a \simeq \mathbb{C}$, it is easy to see that

$$\mathrm{Lie}(\bar{U}_\beta[t^{-1}]) = \mathfrak{g}_\beta \otimes \mathbb{C}[t^{-1}]. \tag{4}$$

Thus,

$$\mathrm{Lie}(\mathscr{G}^{\mathrm{red}}) \supset \sum_{\beta\in\Delta} (\mathfrak{g}_\beta \otimes \mathbb{C}[t^{-1}]).$$

But, since $\mathrm{Lie}(\mathscr{G}^{\mathrm{red}})$ is a Lie algebra which is a Lie subalgebra of $sl_N \otimes \mathbb{C}[t^{-1}]$ under the standard bracket as in Exercise 1.3.E.12 and $\sum_{\beta\in\Delta} \mathfrak{g}_\beta$ generates the Lie algebra $\mathfrak{g}$ (this is where we have used the assumption that G is semisimple), we get that

$$\mathrm{Lie}(\mathscr{G}^{\mathrm{red}}) \supset \mathfrak{g} \otimes \mathbb{C}[t^{-1}]. \tag{5}$$

Combining (2) and (5), we get

$$\mathrm{Lie}(\mathscr{G}^{\mathrm{red}}) = \mathrm{Lie}(\mathscr{G}) = \mathfrak{g} \otimes \mathbb{C}[t^{-1}].$$

This proves (1) and hence $\mathscr{G}$ is reduced by Theorem 1.3.22.

The evaluation $\epsilon(\infty)\colon \mathscr{G} \to G$ admits a group splitting obtained from the inclusion $G \to \mathscr{G}$ (which is induced from the $\mathbb{C}$-algebra homomorphism $R \hookrightarrow R[t^{-1}]$). This gives rise to an isomorphism of ind-schemes (cf. Corollary 1.3.3(a)):

$$\mathscr{G} \simeq \mathscr{G}^- \times G, \quad \text{where} \quad \mathscr{G}^- := \bar{G}[t^{-1}]^-. \tag{6}$$

Now, since $\mathscr{G}$ is reduced, so is $\mathscr{G}^-$.

Finally, by Proposition 1.3.18, the infinite Grassmannian $\bar{X}_G$ has an open cover isomorphic with the ind-scheme $\mathscr{G}^-$. Thus, $\bar{X}_G$ is a reduced ind-scheme. □

Recall the ind-projective variety X_G^r (in particular, reduced) with closed points X_G from Kumar (2002, §§13.2.12, 13.2.13 and 13.2.15). Also, recall that the structure X_G^r coincides with the ind-variety structure obtained via the representation theory (cf. (Kumar, 2002, Proposition 13.2.18)).

Proposition 1.3.24 *Let G be a connected, simply-connected simple algebraic group. Then the ind-scheme $\bar{X}_G$ as in Proposition 1.3.18 coincides with the ind-projective variety X_G^r.*

Proof We first prove the proposition for $G = \mathrm{SL}_N$. In this case, following the notation as in Theorem 1.3.8, by definition $\bar{X}_{\mathrm{SL}_N}$ is the ind-scheme given by the filtration $(\bar{H}_n)_{n\geq 0}$. By Theorem 1.3.23 and Lemma 1.3.21, since $\bar{X}_{\mathrm{SL}_N}$ is reduced, $(\bar{H}_n^{\mathrm{red}})_{n\geq 0}$ gives an equivalent filtration with $\bar{H}_n(\mathbb{C}) = \bar{H}_n^{\mathrm{red}}(\mathbb{C}) = Q_n^N$. Hence, $\bar{X}_{\mathrm{SL}_N}$ is an ind-variety. By Theorem 1.3.8 and Kumar (2002, §13.2.13), we get that $\bar{X}_{\mathrm{SL}_N}$ coincides with $X_{\mathrm{SL}_N}^r$.

We now come to the general G. Fix an embedding $G \hookrightarrow \mathrm{SL}_N$. This induces closed embeddings (cf. Corollary 1.3.19 for $\bar{X}_G$ and Kumar (2002, §13.2.15) for X_G^r):

$$\bar{X}_G \hookrightarrow \bar{X}_{\mathrm{SL}_N} \quad \text{and} \quad X_G^r \hookrightarrow X_{\mathrm{SL}_N}^r.$$

Now, since both $\bar{X}_G$ and X_G^r are reduced, by Kumar (2002, Lemma 4.1.2), we see that the identity map $\mathrm{Id}\colon \bar{X}_G \to X_G^r$ is an isomorphism of ind-varieties. □

Unlike noetherian group schemes over $\mathbb{C}$ (which are always reduced by a result due to Cartier), ind-affine group schemes over $\mathbb{C}$ are, in general, not reduced.

Example 1.3.25 The ind-affine group scheme $\mathscr{H} := \bar{H}[t]$ is *not* reduced for $H = \mathbb{C}^*$.

Consider the embedding

$$\mathbb{C}^* \hookrightarrow \mathrm{SL}_2,\ z \mapsto \begin{pmatrix} z & 0 \\ 0 & z^{-1} \end{pmatrix}.$$

Then the set of $\mathbb{C}$-points of $\mathscr{H}$ is given by

$$\left\{ \begin{pmatrix} P(t) & 0 \\ 0 & Q(t) \end{pmatrix} : P(t), Q(t) \in \mathbb{C}[t] \quad \text{and} \quad PQ = 1 \right\} \simeq \mathbb{C}^*.$$

Since the set of $\mathbb{C}$-points of $\mathscr{H}$ coincides with that of $\mathscr{H}^{\text{red}}$, we see that the ind-variety

$$\mathscr{H}^{\text{red}} \simeq \mathbb{C}^*.$$

In particular, for any $R \in \mathbf{Alg}$,

$$\begin{aligned}\operatorname{Mor}(\operatorname{Spec} R, \mathscr{H}^{\text{red}}) &= \mathbb{C}^*(R) \\ &\simeq \text{set of invertible elements in } R. \qquad (1)\end{aligned}$$

On the other hand, by Lemma 1.3.2,

$$\begin{aligned}\operatorname{Mor}(\operatorname{Spec} R, \mathscr{H}) &\simeq \operatorname{Hom}_{\text{alg}}(\mathbb{C}[x, x^{-1}], R[t]) \\ &\simeq \text{set of invertible elements in } R[t]. \qquad (2)\end{aligned}$$

If R has a nonzero nilpotent element a, then $1 - at \in R[t]$ is invertible. Thus, by comparing (1) and (2), we get that

$$\operatorname{Mor}(\operatorname{Spec} R, \mathscr{H}^{\text{red}}) \subsetneq \operatorname{Mor}(\operatorname{Spec} R, \mathscr{H}).$$

This shows that $\mathscr{H}$ is not reduced. Thus, the infinite Grassmannian $\bar{X}_{\mathbb{C}^*}$ is not reduced (cf. Proposition 1.3.18).

Remark 1.3.26 (a) Similar to the above example, one can see that for any algebraic group H with a surjective algebraic group homomorphism $H \to \mathbb{C}^*$, $\bar{H}[t]$ is *not* reduced.

(b) Any (not necessarily noetherian) affine group scheme $\mathscr{G}$ over a field of characteristic 0 is reduced (extension of Cartier's result to non-noetherian group schemes). We refer to (Oort, 1966) in combination with (Waterhouse, 1979, §3.3)[2] for a short proof.

In particular, for any affine algebraic group H, the affine group scheme $\bar{H}[[t]]$ (cf. Lemma 1.3.2) is reduced.

Thus, combining Theorem 1.3.23 with Lemma 1.3.16, we get that for any connected semisimple group G, the ind-affine group scheme $\bar{G}((t))$ is reduced.

1.3.E Exercises

(1) Let $R \in \mathbf{Alg}$. For any positive integer N, let $L_o^R = L_o^R(N) := R[[t]] \otimes_{\mathbb{C}} V$ be as in Definition 1.3.6, where $V = \mathbb{C}^N$. Let L^R be an $R[[t]]$-submodule of $R((t)) \otimes_{\mathbb{C}} V$ satisfying

$$t^n L_o^R \subset L^R \subset t^{-n} L_o^R, \quad \text{for some } n \geq 0.$$

[2] We thank B. Conrad for providing this reference.

Then show that L^R is an $R[[t]]$-projective module if and only if $\tilde{L}^R := L^R/t^n L_o^R$ is an R-module direct summand of the R-module $t^{-n}L_o^R/t^n L_o^R$.

(2) Let $R \in \mathbf{Alg}$ and let L^R be a projective $R[[t]]$-submodule of $R((t)) \otimes_{\mathbb{C}} V$ satisfying

$$t^n L_o^R \subset L^R \subset t^{-n} L_o^R, \quad \text{for some } n \geq 0.$$

Then show that for any $\mathbb{C}$-algebra homomorphism $R \to R'$, the above inclusions induce the inclusions

$$t^n L_o^{R'} \subset L^{R'} \subset t^{-n} L_o^{R'}, \text{ where } L^{R'} := R'[[t]] \otimes_{R[[t]]} L^R.$$

Show further that if $L^R/t^n L_o^R$ is a direct summand of $t^{-n}L_o^R/t^n L_o^R$ as R-modules, then

$$R' \otimes_R \frac{L^R}{t^n L_o^R} \xrightarrow{\sim} \frac{L^{R'}}{t^n L_o^{R'}}.$$

(3) Let X be a scheme and let f be an automorphism of X. Then show that the fixed subscheme X^f (which is defined as the scheme-theoretic inverse image of the diagonal under the morphism $\theta_f : X \to X \times X$, $x \mapsto (x, f(x))$) represents the functor $\mathscr{X}^f : \mathbf{Alg} \to \mathbf{Set}$ defined by

$$\mathscr{X}^f(R) = X(R)^{f_R},$$

where f_R is the induced automorphism of $X(R) := \text{Mor}(\text{Spec } R, X)$.

(4) Let S be a noetherian algebra over $\mathbb{C}$ and let P be a finitely generated projective $S[[t]]$-module. Show that there exists an affine open cover $\{\text{Spec}(S_i)\}_i$ of the scheme $\text{Spec}(S)$ such that $S_i[[t]] \otimes_{S[[t]]} P$ is a free $S_i[[t]]$-module, for each i.

Hint (due to N. Mohan Kumar): Consider the projective S-module $P_o := P/tP$. Then show that $P \simeq \hat{P}_o$, where $\hat{P}_o := P_o \otimes_S S[[t]]$. To show this, using the projectivity of P and $\hat{P}_o$ as $S[[t]]$-modules, get an $S[[t]]$-module lift $\theta : \hat{P}_o \to P$ of the S-module isomorphism $\hat{P}_o/t\hat{P}_o \simeq P/tP$. Prove that θ is an isomorphism by observing that a finitely generated module over $S[[t]]$ is complete with respect to t and is zero if it is zero mod t.

(5) Show that the special lattice functor $\mathscr{Q} = \mathscr{Q}^N$ as in Definition 1.3.6 is the sheafification (cf. Lemma B.2) of the functor $R \rightsquigarrow \text{SL}_N(R((t)))/\text{SL}_N(R[[t]])$.

Hint: Use the fact proved in Proposition 1.3.14 that for $g \in \text{SL}_N(R((t)))$ and $L^R \in \mathscr{Q}^N(R)$, $gL^R \in \mathscr{Q}^N(R)$, i.e., gL^R satisfies properties (a) and (b) of Definition 1.3.6 for some $n \geq 0$.

(6) Let $f\colon X \to Z$ and $g\colon Y \to Z$ be two morphisms of ind-schemes. Then their fiber product $X \underset{Z}{\times} Y$ represents the functor

$$R \rightsquigarrow X(R) \underset{Z(R)}{\times} Y(R),$$

where

$$X(R) \underset{Z(R)}{\times} Y(R) := \{(x, y) \in X(R) \times Y(R) : \ f_R(x) = g_R(y)\}$$

and $f_R\colon X(R) \to Z(R)$ is the map induced from f.

(7) Show that the functor $h_{\bar{G}[t^{-1}]^-}$ as in the proof of Proposition 1.3.18 is an open subfunctor of $\mathscr{X}_G$.

Moreover, show that $\mathscr{X}_G(k) = \mathscr{X}_G^o(k)$ for any field $k \supset \mathbb{C}$, where the functors $\mathscr{X}_G$ and $\mathscr{X}_G^o$ are defined in Definition 1.3.5.

(8) Let R be a $\mathbb{C}$-algebra generated (as a $\mathbb{C}$-algebra) by countably many elements. Then show that for any maximal ideal $\mathfrak{m}$ of R, $R/\mathfrak{m} \simeq \mathbb{C}$ (as $\mathbb{C}$-algebras).

In particular, any closed ind-subscheme of $\bar{G}((t))$ (for any affine algebraic group G) has nonempty set of $\mathbb{C}$-points.

Hint: We can assume that $R = \mathbb{C}[x_1, x_2, x_3, \dots]$. Now, $R/\mathfrak{m}$ is a field extension k of $\mathbb{C}$. In particular, $\mathbb{C}$ being algebraically closed, $k \supset \mathbb{C}(x)$, where $\mathbb{C}(x)$ is the quotient field of the polynomial ring $\mathbb{C}[x]$. Show that $\mathbb{C}(x)$ as a vector space over $\mathbb{C}$ is of uncountable dimension, whereas clearly R and hence $R/\mathfrak{m}$ is of countable dimension over $\mathbb{C}$.

(9) Show that for any G as H in Remark 1.3.17, the canonical map $\bar{X}_G \hookrightarrow \bar{X}_{\mathrm{SL}_N}$ (induced by an embedding $G \hookrightarrow \mathrm{SL}_N$) is a closed embedding.

(10) For any integer $n \geq 1$, consider the covariant group functor $\mathscr{F}_n$ from **Alg** to **Set** defined by

$$\mathscr{F}_n(R) = G\left(\frac{R[[t]]}{\langle t^n \rangle}\right),$$

where G is an affine algebraic group. Then show that $\mathscr{F}_n$ is a representable functor represented by an affine group scheme of finite type over $\mathbb{C}$ (i.e., an affine algebraic group) $\bar{G}\left(\frac{\mathbb{C}[[t]]}{\langle t^n \rangle}\right)$ with $\mathbb{C}$-points $G\left(\frac{\mathbb{C}[[t]]}{\langle t^n \rangle}\right)$. Since it is a variety, we denote it by $G\left(\frac{\mathbb{C}[[t]]}{\langle t^n \rangle}\right)$ itself.

Hint: Follow the proof of Lemma 1.3.2.

(11) Let G be a connected reductive group and let $P \subset G$ be a parabolic subgroup. Define the (parahoric) closed subgroup scheme $\mathcal{P} \subset \bar{G}((t))$ by $\mathcal{P} := ev_0^{-1}(P)$, under the evaluation map $ev_0\colon \bar{G}[[t]] \to G$ at $t = 0$

equipped with the scheme-theoretic inverse image structure. Then $\mathcal{P}$ is reduced (cf. Remark 1.3.26(b)). Moreover, by Exercise 6 and Lemma 1.3.2, $\mathcal{P}$ represents the functor $R \rightsquigarrow \tilde{\mathcal{P}}(R)$, where $\tilde{\mathcal{P}}(R) := (ev_0^R)^{-1}(P(R))$, under the evaluation map $ev_0^R : G(R[[t]]) \to G(R)$ at $t = 0$.

Consider the functor

$$\mathbf{Alg} \to \mathbf{Set}, \quad R \rightsquigarrow G(R((t)))/\tilde{\mathcal{P}}(R).$$

Show that its sheafification (cf. Lemma B.2) is a representable functor represented by an ind-projective scheme denoted $\bar{X}_G(P)$ (with $\mathbb{C}$-points $G((t))/\tilde{\mathcal{P}}(\mathbb{C})$). Moreover, show that this ind-scheme is, in fact, an ind-variety if G is semisimple.

Show further that for any connected reductive group G,

$$\bar{G}((t)) \to \bar{X}_G(P)$$

is a locally trivial principal $\mathcal{P}$-bundle and

$$\bar{X}_G(P) \to \bar{X}_G$$

is a locally trivial G/P-fibration, where the ind-scheme $\bar{X}_G$ is as in Proposition 1.3.18.

(12) Show that the Lie algebra Lie $(\overline{\mathrm{SL}}_N[t^{-1}])$ of the ind-affine group scheme $\overline{\mathrm{SL}}_N[t^{-1}]$ is isomorphic with the Lie algebra $sl_N \otimes \mathbb{C}[t^{-1}]$ under the bracket $[x \otimes P, y \otimes Q] = [x, y] \otimes PQ$, for $x, y \in sl_N$ and P, $Q \in \mathbb{C}[t^{-1}]$.

Moreover, the evaluation homomorphism $\epsilon(\alpha)\colon \overline{\mathrm{SL}}_N[t^{-1}] \to \mathrm{SL}_N$, induced from $t \mapsto \alpha$ for any $\alpha \in \mathbb{P}^1(\mathbb{C})\backslash\{0\}$, induces the Lie algebra homomorphism (cf. Lemma B.19):

$$\dot{\epsilon}(\alpha)_1 : sl_N \otimes \mathbb{C}[t^{-1}] \to sl_N,\ x \otimes P \mapsto P(\alpha)x,$$
$$\text{for } x \in sl_N \text{ and } P \in \mathbb{C}[t^{-1}].$$

(13) For any algebra $R \in \mathbf{Alg}$, any maximal ideal of $R[[t]]$ contains $t\ R[[t]]$.

1.4 Central Extension of Loop Groups

Let $\mathfrak{g}$ be a finite-dimensional simple Lie algebra over $\mathbb{C}$ and let G be the connected, simply-connected complex algebraic group with Lie algebra $\mathfrak{g}$. For $\lambda_c \in \hat{D}$, let $\mathscr{H}(\lambda_c)$ be the integrable highest-weight $\hat{\mathfrak{g}}$-module with highest weight λ_c (cf. Theorem 1.2.10). Recall the definition of $\mathbb{C}$-space and $\mathbb{C}$-group functors from Definition B.1.

1.4.1

Consider the $\mathbb{C}$-group functor $\mathscr{L}_G(R) := G(R((t)))$, which is represented by the ind-affine group scheme $\bar{G}((t))$ (cf. Lemma 1.3.2). In particular, it satisfies the property (E) (cf. Exercise B.E.4). Recall that $\mathbf{PGL}_{\mathscr{H}(\lambda_c)}$ is the projective linear group functor (cf. Example B.4(2)) with the tangent space at 1 given by $\mathrm{End}_R(\mathscr{H}(\lambda_c)_R)/R \cdot \mathrm{Id}_{\mathscr{H}(\lambda_c)_R}$, where $\mathscr{H}(\lambda_c)_R := \mathscr{H}(\lambda_c) \otimes R$ (cf. Lemma B.13). By Lemma B.13, **PGL** satisfies the condition (E). Of course, since G satisfies the condition (L) (cf. Exercise B.E.5), thinking of $\mathscr{L}_G(R)$ as $R((t))$-points of G,

$$T_1(\mathscr{L}_G)_R = \mathfrak{g} \otimes R((t)), \quad \text{for any } R \in \mathbf{Alg}.$$

Moreover, the Lie algebra bracket in $T_1(\mathscr{L}_G)_R$ coincides with the standard Lie algebra bracket in $\mathfrak{g} \otimes_{\mathbb{C}} R((t))$, as can easily be seen from Definition B.17(c). The functor $\mathscr{L}_G$ satisfies the condition (L) finitely (cf. Definition B.15(b)). Also, by Exercise B.E.6, $\mathbf{PGL}_{\mathscr{H}(\lambda_c)}$ satisfies the condition (L) finitely.

Definition 1.4.2 (Adjoint action of $\bar{G}((t))$) Define the R-linear *adjoint action* of the group functor $\mathscr{L}_G(R)$ on the Lie-algebra functor $\hat{\mathfrak{g}}(R) := \mathfrak{g} \otimes R((t)) \oplus R.C$ (where the R-linear bracket in $\hat{\mathfrak{g}}(R)$ is defined by the same formula as (4) of Definition 1.2.1) by

$$(\mathbf{Ad}_C\gamma)(x[P] \oplus sC) = \gamma x[P]\gamma^{-1} + \left(s + \operatorname*{Res}_{t=0}\langle \gamma^{-1}d\gamma, x[P]\rangle\right) C,$$

for $\gamma \in \mathscr{L}_G(R)$, $x \in \mathfrak{g}$, $P \in R((t))$ and $s \in R$, where $\langle , \rangle$ is the $R((t))$-bilinear extension of the normalized invariant form on $\mathfrak{g}$ and taking an embedding $i \colon G \hookrightarrow GL_N$ we view $G(R((t)))$ as a subgroup of $N \times N$ invertible matrices over the ring $R((t))$. Observe that for the group functor $GL_N(R((t)))$, the adjoint action (defined in Definition B.17) is given by

$$(\mathbf{Ad}\gamma) \cdot M = \gamma M \gamma^{-1}, \quad \text{for } \gamma \in GL_N(R((t))) \text{ and } M \in M_N(R((t))).$$

From the functoriality of **Ad** (cf. (1) of Definition B.17), $\gamma x[P]\gamma^{-1} \in \mathfrak{g} \otimes R((t))$ (for $\gamma \in \mathscr{L}_G(R)$ and $x[P] \in \mathfrak{g} \otimes R((t))$) and it does not depend upon the choice of the embedding i. A similar remark applies to $\gamma^{-1}d\gamma$. Here $d\gamma$ for $\gamma = (\gamma_{i,j}) \in M_N(R((t)))$ denotes $d\gamma := \big(d\gamma_{i,j}/dt\big)$.

It is easy to check that for any $\gamma \in \mathscr{L}_G(R)$, $\mathbf{Ad}_C\gamma \colon \hat{\mathfrak{g}}(R) \to \hat{\mathfrak{g}}(R)$ is an R-linear Lie algebra homomorphism. Moreover,

$$\mathbf{Ad}_C(\gamma_1\gamma_2) = \mathbf{Ad}_C(\gamma_1) \circ \mathbf{Ad}_C(\gamma_2). \tag{1}$$

Using Lemma B.18, one easily sees that for any finite-dimensional $\mathbb{C}$-algebra R and $x \in \mathfrak{g} \otimes R((t))$,

$$\dot{\mathbf{Ad}}_C(x)(y) = [x, y], \quad \text{for any } y \in \hat{\mathfrak{g}}(R). \tag{2}$$

It is easy to see that the representation $\mathscr{H}(\lambda_c)$ of $\hat{\mathfrak{g}}$ extends R-linearly to a representation $\bar{\rho}_R$ in $\mathscr{H}(\lambda_c)_R := \mathscr{H}(\lambda_c) \underset{\mathbb{C}}{\otimes} R$ of $\hat{\mathfrak{g}}(R)$.

A proof of the following result due to Faltings can be found in Beauville and Laszlo (1994, Lemma A.3) for $G = \mathrm{SL}_n$. The proof for general G is identical.

Proposition 1.4.3 *With the notation as above, for any $R \in$ **Alg** and $\gamma \in \mathscr{L}_G(R)$, locally over Spec R, there exists an R-linear automorphism $\hat{\rho}(\gamma)$ of $\mathscr{H}(\lambda_c)_R$ uniquely determined up to an invertible element of R satisfying*

$$\hat{\rho}(\gamma)\bar{\rho}_R(x)\hat{\rho}(\gamma)^{-1} = \bar{\rho}_R(\mathbf{Ad}_C(\gamma) \cdot x), \quad \textit{for any } x \in \hat{\mathfrak{g}}(R), \tag{1}$$

where the adjoint representation of $\mathscr{L}_G(R)$ on $\hat{\mathfrak{g}}(R)$ is defined in the previous Definition 1.4.2.

As a corollary of the above Proposition 1.4.3, we get the following.

Theorem 1.4.4 *With the notation and assumptions as at the beginning of this section, there exists a homomorphism $\rho\colon \mathscr{L}_G \to \mathbf{PGL}_{\mathscr{H}(\lambda_c)}$ of group functors such that*

$$\dot{\rho} = \dot{\rho}(\mathbb{C})\colon T_1(\mathscr{L}_G)_{\mathbb{C}} = \mathfrak{g} \otimes \mathbb{C}((t)) \to \mathit{End}_{\mathbb{C}}(\mathscr{H}(\lambda_c))/\mathbb{C} \cdot \mathit{Id}_{\mathscr{H}(\lambda_c)} \tag{1}$$

coincides with the projective representation $\mathscr{H}(\lambda_c)$ of $\mathfrak{g} \otimes \mathbb{C}((t))$ (cf. Lemmas B.13 and B.14).

By Exercise 1.4.E.1, in fact, $\dot{\rho}_R\colon \mathfrak{g} \otimes R((t)) \to \mathit{End}_R(\mathscr{H}(\lambda_c)_R)/R \cdot \mathit{Id}_{\mathscr{H}(\lambda_c)_R}$ coincides with the projective representation $\mathscr{H}(\lambda_c)_R$ of $\mathfrak{g} \otimes R((t))$.

Proof Fix $\gamma \in \mathscr{L}_G(R)$. As guaranteed by the existence of an R-linear automorphism $\hat{\rho}(\gamma)$ of $\mathscr{H}(\lambda_c)_R$ locally in Spec R and its uniqueness up to an invertible element of R, we get an fppf R-algebra S_γ (depending upon γ) (cf. (Stacks, 2019, Tag 021N)) and a unique element (obtained by glueing locally obtained $\hat{\rho}(\gamma)$) $\bar{\rho}_{S_\gamma}(\gamma) \in \mathrm{PGL}_{S_\gamma}(\mathscr{H}(\lambda_c)_{S_\gamma}) := \mathrm{Aut}_{S_\gamma}(\mathscr{H}(\lambda_c)_{S_\gamma})/S_\gamma^* I$, where S_γ^* denotes the set of invertible elements in S_γ. Consider the exact sequence (cf. Definition B.1):

$$\mathscr{L}_G(R) \xrightarrow{i_{S_\gamma}} \mathscr{L}_G(S_\gamma) \rightrightarrows \mathscr{L}_G\left(S_\gamma \underset{R}{\otimes} S_\gamma\right).$$

Since $i_{S_\gamma}(\gamma)$ goes to the same element in $\mathscr{L}_G(S_\gamma \underset{R}{\otimes} S_\gamma)$ under the above two homomorphisms $\mathscr{L}_G(S_\gamma) \rightrightarrows \mathscr{L}_G(S_\gamma \underset{R}{\otimes} S_\gamma)$, we get that $\bar{\rho}_{S_\gamma}(\gamma) \in \mathrm{PGL}_{S_\gamma}(\mathscr{H}(\lambda_c)_{S_\gamma})$ goes to the same element under the two maps

$$\mathrm{PGL}_{S_\gamma}(\mathscr{H}(\lambda_c)_{S_\gamma}) \rightrightarrows \mathrm{PGL}_{S_\gamma \underset{R}{\otimes} S_\gamma}\left(\mathscr{H}(\lambda_c)_{S_\gamma \otimes_R S_\gamma}\right).$$

Hence, $\bar{\rho}_{S_\gamma}(\gamma) \in K_R(S_\gamma)$ for the functor $\mathbf{PGL}_{\mathscr{H}(\lambda_c)}$ (cf. proof of Lemma B.2 for the notation $K_R(S_\gamma)$). Finally, define $\rho(\gamma) \in \mathbf{PGL}_{\mathscr{H}(\lambda_c)}(R)$ as the image of $\bar{\rho}_{S_\gamma}(\gamma)$ under the canonical map $K_R(S_\gamma) \to \mathbf{PGL}_{\mathscr{H}(\lambda_c)}(R)$. From the uniqueness of R-linear automorphisms $\hat{\rho}(\gamma)$ locally in Spec R up to an invertible element in R, we get that $\rho(\gamma)$ is well defined (i.e., it does not depend upon the choice of S_γ).

Again using the uniqueness of $\hat{\rho}(\gamma)$ (up to invertible elements in R locally) satisfying (1) of Proposition 1.4.3 and using (1) of Definition 1.4.2 and Exercise 1.4.E.4, we get that ρ is a group homomorphism and, in fact, it is a morphism from the group functor $\mathscr{L}_G$ to the group functor $\mathbf{PGL}_{\mathscr{H}(\lambda_c)}$.

We now prove (1). Take $x \in \mathfrak{g}((t))$ and $y \in \hat{\mathfrak{g}}$. Then, by Proposition 1.4.3 applied to $R = \mathbb{C}(\epsilon)$, we get

$$\hat{\rho}(e^{\epsilon x})\bar{\rho}_R(y)\hat{\rho}(e^{-\epsilon x}) = \bar{\rho}_R(\mathbf{Ad}_C(e^{\epsilon x}) \cdot y), \tag{2}$$

where for the notation $e^{\epsilon x} \in \mathscr{L}_G(R)$ see Definition B.15(a). By Lemma B.18 applied to the representation $\mathbf{Ad}_C$ of $\mathscr{L}_G$, and identity (2) of Definition 1.4.2,

$$\bar{\rho}_R(\mathbf{Ad}_C(e^{\epsilon x})y) = \bar{\rho}_R(y + \epsilon[x, y]). \tag{3}$$

Similarly, fixing a lift of $\dot{\rho}(x)$ in $\mathrm{End}_{\mathbb{C}}(\mathscr{H}(\lambda_c))$, for $v \in \mathscr{H}(\lambda_c)$, by Lemma B.18,

$$\begin{aligned} &\hat{\rho}(e^{\epsilon x})\bar{\rho}_R(y)\hat{\rho}(e^{-\epsilon x})v \\ &\quad = \hat{\rho}(e^{\epsilon x})\bar{\rho}_R(y)\left(v - \epsilon\dot{\rho}(x)v - \epsilon\lambda_x v\right), \quad \text{for some } \lambda_x \in \mathbb{C} \\ &\quad = \hat{\rho}(e^{\epsilon x})\left(\bar{\rho}(y)v - \epsilon\bar{\rho}(y)\dot{\rho}(x)v - \epsilon\lambda_x\bar{\rho}(y)v\right) \\ &\quad = \bar{\rho}(y)v - \epsilon\bar{\rho}(y)\dot{\rho}(x)v - \epsilon\lambda_x\bar{\rho}(y)v + \epsilon\dot{\rho}(x)\bar{\rho}(y)v + \epsilon\lambda_x\bar{\rho}(y)v. \end{aligned} \tag{4}$$

Combining the equations (2)–(4), we get

$$\bar{\rho}[x, y] = \left[\dot{\rho}(x), \bar{\rho}(y)\right],$$

i.e.,

$$\left[\bar{\rho}(x) - \dot{\rho}(x), \bar{\rho}(y)\right] = 0, \text{ for all } x \in \mathfrak{g}((t)) \text{ and } y \in \hat{\mathfrak{g}}.$$

Thus, by Exercise 1.2.E.5,

$$\bar{\rho}(x) - \dot{\rho}(x) = \mu_x \operatorname{Id}_{\mathscr{H}(\lambda_c)}, \quad \text{for some } \mu_x \in \mathbb{C}.$$

This proves the theorem. □

Definition 1.4.5 (Central extensions of loop groups) Following the notation and assumptions at the beginning of this section, take any $\lambda_c \in \hat{D}$. By Theorem 1.4.4, we have a homomorphism of group functors:

$$\rho\colon \mathscr{L}_G \to \mathbf{PGL}_{\mathscr{H}(\lambda_c)}.$$

Also, there is a canonical homomorphism of group functors (cf. Example B.4(2)):

$$\pi : \mathbf{GL}_{\mathscr{H}(\lambda_c)} \to \mathbf{PGL}_{\mathscr{H}(\lambda_c)}.$$

All these $\mathbb{C}$-group functors $\mathscr{L}_G$, $\mathbf{GL}_{\mathscr{H}(\lambda_c)}$ and $\mathbf{PGL}_{\mathscr{H}(\lambda_c)}$ satisfy the condition (L) finitely (cf. §1.4.1 for $\mathscr{L}_G$ and Exercise B.E.6 for **GL** and **PGL**). Thus, by Exercise B.E.7, we get the fiber product group functor $\hat{\mathscr{G}}_{\lambda_c}$ satisfying the condition (L) finitely:

$$\hat{\mathscr{G}}_{\lambda_c} := \mathscr{L}_G \underset{\mathbf{PGL}_{\mathscr{H}(\lambda_c)}}{\times} \mathbf{GL}_{\mathscr{H}(\lambda_c)}.$$

By the definition, we get homomorphisms of group functors

$$p\colon \hat{\mathscr{G}}_{\lambda_c} \to \mathscr{L}_G \quad \text{and} \quad \hat{\rho}\colon \hat{\mathscr{G}}_{\lambda_c} \to \mathbf{GL}_{\mathscr{H}(\lambda_c)}$$

making the following diagram commutative:

$$\begin{array}{ccc} \hat{\mathscr{G}}_{\lambda_c} & \xrightarrow{\hat{\rho}} & \mathbf{GL}_{\mathscr{H}(\lambda_c)} \\ {\scriptstyle p}\downarrow & & \downarrow{\scriptstyle \pi} \\ \mathscr{L}_G & \xrightarrow[\rho]{} & \mathbf{PGL}_{\mathscr{H}(\lambda_c)} \end{array}.$$

By Exercise B.E.7, the Lie algebra Lie $\hat{\mathscr{G}}_{\lambda_c}(R) := T_1(\hat{\mathscr{G}}_{\lambda_c})_R$ is identified with the fiber product Lie algebra (cf. §1.4.1, Example B.12 and Lemma B.13)

$$\hat{\mathfrak{G}}_{\lambda_c}(R) = \mathfrak{g} \otimes R((t)) \underset{\operatorname{End}_R(\mathscr{H}(\lambda_c)_R)/R.\operatorname{Id}}{\times} \operatorname{End}_R(\mathscr{H}(\lambda_c)_R),$$

for any finite-dimensional $\mathbb{C}$-algebra R.

Lemma 1.4.6 *The Lie algebra $\hat{\mathfrak{g}}_{\lambda_c} :=$ Lie $\hat{\mathscr{G}}_{\lambda_c}(\mathbb{C})$ can canonically be identified with the affine Lie algebra $\hat{\mathfrak{g}}$.*

Proof Let $\bar{\rho}: \hat{\mathfrak{g}} \to \mathrm{End}_{\mathbb{C}}(\mathscr{H}(\lambda_c))$ denote the representation. Define

$$\psi: \hat{\mathfrak{g}} \to \hat{\mathfrak{g}}_{\lambda_c}, \quad x[P] + zC \mapsto (x[P], \bar{\rho}(x[P]) + zc\,\mathrm{Id}),$$
$$\text{for } x \in \mathfrak{g}, P \in K \text{ and } z \in \mathbb{C}.$$

From the definition of the bracket in $\hat{\mathfrak{g}}$ and Theorem 1.4.4, ψ is an isomorphism of Lie algebras. □

Combining Theorem 1.4.4, Definition 1.4.5 and Lemma 1.4.6, we get the following.

Corollary 1.4.7 *For any $\lambda_c \in \hat{D}$, we have a homomorphism of group functors*

$$\hat{\rho}: \hat{\mathscr{G}}_{\lambda_c} \to \mathbf{GL}_{\mathscr{H}(\lambda_c)}$$

such that its derivative for $R = \mathbb{C}$

$$\dot{\hat{\rho}}: \hat{\mathfrak{g}}_{\lambda_c} \to \mathrm{End}_{\mathbb{C}}(\mathscr{H}(\lambda_c))$$

under the identification of Lemma 1.4.6 coincides with the Lie algebra representation

$$\bar{\rho}: \hat{\mathfrak{g}} \to \mathrm{End}_{\mathbb{C}}(\mathscr{H}(\lambda_c)).$$

Moreover, for any $\mathbb{C}$-algebra R, $\hat{\gamma} \in \hat{\mathscr{G}}_{\lambda_c}(R)$ and $x \in \hat{\mathfrak{g}}(R)$,

$$\hat{\rho}(\hat{\gamma})\bar{\rho}_R(x)\hat{\rho}(\hat{\gamma})^{-1} = \bar{\rho}_R(\mathbf{Ad}_C(p(\hat{\gamma}))x), \text{ as operators on } \mathscr{H}(\lambda_c)_R. \quad (1)$$

The following lemma is trivial to verify.

Lemma 1.4.8 *Let V be a vector space over $\mathbb{C}$ and let $v_+ \in V$ be a nonzero vector and $V' \subset V$ a subspace such that $V = \mathbb{C}v_+ \oplus V'$. Then the following subgroup functors of $\mathbf{GL}_V$:*

$$\mathbf{GL}^\circ_V(R) = \{T \in \mathrm{GL}_R(V_R) : Tv_+ = v_+\} \quad \text{and}$$
$$\mathbf{GL}'_V(R) = \left\{T \in \mathrm{GL}_R(V_R) : Tv_+ - v_+ \in V'_R \text{ and } T(V'_R) \subset V'_R\right\}$$

are $\mathbb{C}$-group functors, i.e., they satisfy the sheaf condition for the fppf topology.

Moreover, the projection homomorphism $\pi: \mathbf{GL}_V \to \mathbf{PGL}_V$ is an isomorphism of group functors restricted to either of $\mathbf{GL}^\circ_V$ or $\mathbf{GL}'_V$ onto their images.

Lemma 1.4.9 *Let Y be a connected variety (over $\mathbb{C}$). Then any morphism $f: Y \to \mathbb{C}^*$, which is null-homotopic in the topological category with the analytic topology Y^{an} on Y, is constant.*

Observe that if the singular cohomology $H^1(Y^{an}, \mathbb{Z}) = 0$, then any continuous map $Y^{an} \to \mathbb{C}^*$ is null-homotopic since $\mathbb{C}^*$ is a $K(\mathbb{Z}, 1)$-space (cf. (Spanier, 1966, Chap. 8, §1, Theorem 8)).

Proof Assume, if possible, that there exists a null-homotopic nonconstant morphism $f: Y \to \mathbb{C}^*$. Since f is a morphism, there exists $N_o > 0$ such that the number of irreducible components of $f^{-1}(z) \leq N_o$, for any $z \in \mathbb{C}^*$. Now take the N-sheeted covering $\pi_N: \mathbb{C}^* \to \mathbb{C}^*$, $z \mapsto z^N$, for any $N > N_o$. Since f is null-homotopic, there exists a lift as a morphism $\tilde{f}: Y \to \mathbb{C}^*$ (cf. (Serre, 1958, Proposition 20)) making the following diagram commutative:

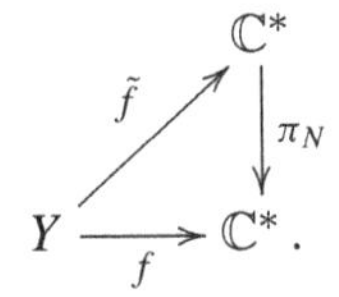

Since $\tilde{f}$ is a morphism and nonconstant, by Chevalley's Theorem (cf. (Hartshorne, 1977, Chap. II, Exercise 3.19)) Im $\tilde{f}$ (being a constructible and connected set) misses only finitely many points of $\mathbb{C}^*$. In particular, there exists $z_o \in \mathbb{C}^*$ (in fact, an open dense set of points) such that $\pi_N^{-1}(z_o) \subset \text{Im}\, \tilde{f}$. But then the number of irreducible components of $f^{-1}(z_o) = \tilde{f}^{-1}\pi_N^{-1}(z_o) \geq N > N_o$, which is a contradiction to the choice of N. This proves the lemma. □

1.4.10

Recall the homomorphism of group functors $p: \hat{\mathscr{G}}_{\lambda_c} \to \mathscr{L}_G$ (for any $\lambda_c \in \hat{D}$) from Definition 1.4.5. Since

$$\mathbf{GL}_{\mathscr{H}(\lambda_c)}(R) \to \mathbf{PGL}_{\mathscr{H}(\lambda_c)}(R)$$

has kernel R^* for any $R \in \mathbf{Alg}$, we get an exact sequence of group functors (i.e., for any R, the following sequence specialized at R is exact):

$$1 \to \mathbb{G}_m \xrightarrow{i} \hat{\mathscr{G}}_{\lambda_c} \xrightarrow{p} \mathscr{L}_G, \tag{1}$$

where i is given by $r \mapsto (1, r\,\text{Id})$, for any $r \in R^*$. Of course, $\mathbb{G}_m$ is central in $\hat{\mathscr{G}}_{\lambda_c}$ (for any $R \in \mathbf{Alg}$). By Exercise 1.4.E.2, $\hat{\mathscr{G}}_{\lambda_c}(R) \to \mathscr{L}_G(R)$ is surjective for any field $R \supset \mathbb{C}$. Combining Lemma 1.4.8 and Proposition 1.4.3, we get the following.

Theorem 1.4.11 *Take any $\lambda_c \in \hat{D}$. Then the homomorphism of group functors $p\colon \hat{\mathscr{G}}_{\lambda_c} \to \mathscr{L}_G$ splits over the subgroup functors*

$$\mathscr{L}_G^+(R) := G(R[[t]])_e,\ \mathscr{L}_G^-(R) := G(R[t^{-1}])^- \ \text{and}\ \mathscr{G}(R) := G(R);$$

where $G(R[[t]])_e$ and $G(R[t^{-1}])^-$ are defined in Corollary 1.3.3 with $H = (e)$.

Thus, p also splits over the group functor $\mathscr{F}_1\colon R \rightsquigarrow G(R[[t]])$ (cf. Lemma 1.3.2).

Similarly, p also splits over the group functor $\mathscr{F}_3\colon R \rightsquigarrow G(R[t^{-1}])$.

Proof We first prove the theorem for $\mathscr{L}_G^+$. Fix $v_+ \in \mathscr{H}(\lambda_c)$ a highest-weight vector (which is unique up to a scalar multiple). We claim that for any $\gamma \in \mathscr{L}_G^+(R)$,

$$\hat{\rho}(\gamma_S)v_+ \in S^* v_+, \tag{1}$$

for any fppf R-algebra $S = S_\gamma$ such that $\rho(\gamma_S)$ lifts to an element $\hat{\rho}(\gamma_S)$ of $\mathrm{Aut}_S(\mathscr{H}(\lambda_c)_S)$, where γ_S is the image of γ in $\mathscr{L}_G^+(S)$. By (1) of Proposition 1.4.3,

$$\bar{\rho}_S(x)\hat{\rho}(\gamma_S)v_+ = \hat{\rho}(\gamma_S)\bar{\rho}_S(\mathbf{Ad}_C(\gamma_S^{-1})x)v_+,\quad \text{for any } x \in \hat{\mathfrak{u}}_S := \mathfrak{g} \otimes tS[[t]] \oplus (S \otimes \mathfrak{u}), \tag{2}$$

where $\mathfrak{u}$ is the nil-radical of $\mathfrak{b}$. But, by definition,

$$\mathbf{Ad}_C(\gamma_S^{-1})(\hat{\mathfrak{u}}_S) \subset \hat{\mathfrak{u}}_S. \tag{3}$$

Moreover,

$$\bar{\rho}_S(\hat{\mathfrak{u}}_S) \cdot v_+ = 0, \ \text{since } v_+ \text{ is a highest-weight vector.} \tag{4}$$

Thus, combining (2)–(4), we get

$$\bar{\rho}_S(x)(\hat{\rho}(\gamma_S)v_+) = 0, \quad \text{for all } x \in \hat{\mathfrak{u}}_S. \tag{5}$$

By Exercise 1.4.E.4, we get that

$$\hat{\rho}(\gamma_S)v_+ \in Sv_+, \quad \text{and hence} \quad \hat{\rho}(\gamma_S)v_+ \in S^* \cdot v_+$$

($\hat{\rho}(\gamma_S)$ being represented by an invertible S-linear map). This proves (1).

Thus, by Exercise B.E.9,

$$\rho\left(\mathscr{L}_G^+(R)\right) \subset \pi\left(\mathbf{GL}^0_{\mathscr{H}(\lambda_c)}(R)\right),$$

where π is the canonical morphism $\mathbf{GL}_{\mathscr{H}(\lambda_c)} \to \mathbf{PGL}_{\mathscr{H}(\lambda_c)}$. Therefore, by Lemma 1.4.8, we get the splitting of p over the subgroup functor $\mathscr{L}_G^+$.

We next come to the case of $\mathscr{L}_G^-$. We claim that for any $\gamma \in \mathscr{L}_G^-(R)$,

$$[\hat{\rho}(\gamma_S)v_+]_+ \subset S^*, \tag{6}$$

for any fppf R-algebra $S = S_\gamma$ such that $\rho(\gamma_S)$ lifts to an element $\hat{\rho}(\gamma_S)$ of $\mathrm{Aut}_S(\mathscr{H}(\lambda_c)_S)$, where $[\hat{\rho}(\gamma_S)v_+]_+$ is the coefficient of v_+ in the component of $\hat{\rho}(\gamma_S)v_+$ under the decomposition

$$\mathscr{H}(\lambda_c) \otimes S = Sv_+ \oplus (\mathscr{H}'(\lambda_c) \otimes S),$$

where $\mathscr{H}'(\lambda_c)$ is the sum of weight spaces of $\mathscr{H}(\lambda_c)$ of weights $< \lambda_c$. Applying (1) of Proposition 1.4.3 to $\hat{\rho}(\gamma_S)v_+$, we get

$$\hat{\rho}(\gamma_S)\,(\bar{\rho}_S(x_1)\dots\bar{\rho}_S(x_n)v_+) = \bar{\rho}_S(\mathbf{Ad}_C(\gamma_S)x_1)\dots\bar{\rho}_S(\mathbf{Ad}_C(\gamma_S)x_n)\hat{\rho}(\gamma_S)v_+, \tag{7}$$

for any $x_i \in \hat{\mathfrak{u}}_S^- := (\mathfrak{g} \otimes t^{-1}S[t^{-1}]) \oplus (S \otimes \mathfrak{u}^-)$, where $\mathfrak{u}^-$ is the nil-radical of the opposite Borel $\mathfrak{b}^-$. From the definition of $\mathbf{Ad}_C$,

$$\mathbf{Ad}_C(\gamma_S)\cdot\left(\hat{\mathfrak{u}}_S^-\right) \subset \hat{\mathfrak{u}}_S^-.$$

Thus, from (7) we get (since $\hat{\mathfrak{u}}_{\mathbb{C}}^- \cdot \mathscr{H}(\lambda_c) \subset \mathscr{H}'(\lambda_c)$)

$$\hat{\rho}(\gamma_S)\,(\bar{\rho}_S(x_1)\dots\bar{\rho}_S(x_n)v_+) \in \mathscr{H}'(\lambda_c) \otimes S, \quad \text{for any } n \geq 1. \tag{8}$$

But, since $\mathscr{H}(\lambda_c)$ is an irreducible $\hat{\mathfrak{g}}$-module and $\hat{\mathfrak{u}}_S$ annihilates v_+, the span of $\bar{\rho}_S(x_1)\dots\bar{\rho}_S(x_n)v_+$, as x_i run over $\hat{\mathfrak{u}}_S^-$, is equal to $\mathscr{H}'(\lambda_c) \otimes S$. Thus, from (8), we get

$$\hat{\rho}(\gamma_S)(\mathscr{H}'(\lambda_c) \otimes S) \subset \mathscr{H}'(\lambda_c) \otimes S. \tag{9}$$

From this we immediately obtain (6) by applying $\hat{\rho}(\gamma_S^{-1})$ to (9). Thus, by (6) and (9), and Exercise B.E.9,

$$\rho\left(\mathscr{L}_G^-(R)\right) \subset \pi\left(\mathbf{GL}'_{\mathscr{H}(\lambda_c)}(R)\right)$$

under the decomposition $\mathscr{H}(\lambda_c) = \mathbb{C}v_+ \oplus \mathscr{H}'(\lambda_c)$. Again applying Lemma 1.4.8, we get the splitting of p over the subgroup functor $\mathscr{L}_G^-$.

Since the action of $\mathfrak{g}$ on $\mathscr{H}(\lambda_c)$ decomposes as a direct sum of finite-dimensional $\mathfrak{g}$-submodules V_i:

$$\mathscr{H}(\lambda_c) = \bigoplus_i V_i$$

and G is simply connected, the action of $\mathfrak{g}$ on any V_i integrates to give an action of G on V_i. This gives a representation of G in $\mathrm{GL}_{\mathbb{C}}(\mathscr{H}(\lambda_c))$. From this we get a homomorphism of group functors

$$\mathscr{G} \to \mathbf{GL}_{\mathscr{H}(\lambda_c)},$$

which provides a splitting of p over $\mathscr{G}$.

We finally prove that p splits over the subgroup functor $\mathscr{F}_1(R) := G(R[[t]])$ (cf. Lemma 1.3.2) of $\mathscr{L}_G$.

First of all, from the definition of $\mathscr{L}_G^+(R)$ as the kernel of

$$\epsilon_R(0) \colon G(R[[t]]) \to G(R),\ t \mapsto 0 \quad \text{(cf. Corollary 1.3.3)},$$

and the splitting of $\epsilon_R(0)$ obtained from the embedding $G(R) \hookrightarrow G(R[[t]])$, induced from the embedding $R \hookrightarrow R[[t]]$, we see that there is a semidirect product decomposition of the group functor

$$\mathscr{F}_1 = \mathscr{L}_G^+ \ltimes \mathscr{G}. \tag{10}$$

Take splittings σ_+ and σ_0 of p over $\mathscr{L}_G^+$ and $\mathscr{G}$, respectively. Now define (for any $R \in \mathbf{Alg}$ and $g \in \mathscr{F}_1(R)$ uniquely written as $g = g_+ g_0$, with $g_+ \in \mathscr{L}_G^+(R)$ and $g_0 \in \mathscr{G}(R)$)

$$\sigma(g) = \sigma_+(g_+) \cdot \sigma_0(g_0). \tag{11}$$

It is clear that σ is a $\mathbb{C}$-space functor section of p. We now prove that σ, in fact, is a homomorphism of group functors. To prove this, since $\mathscr{G}$ normalizes $\mathscr{L}_G^+$, it suffices to show

$$\sigma_+(g_0 g_+ g_0^{-1}) = \sigma_0(g_0)\sigma_+(g_+)\sigma_0(g_0)^{-1}, \quad \text{for } g_0 \in \mathscr{G}(R) \text{ and } g_+ \in \mathscr{L}_G^+(R). \tag{12}$$

Consider the morphism of $\mathbb{C}$-space functors

$$\psi : \mathscr{G} \times \mathscr{L}_G^+ \to \hat{\mathscr{G}}_{\lambda_c},\ (g_0, g_+) \mapsto \sigma_+(g_0 g_+ g_0^{-1})\sigma_0(g_0)\sigma_+(g_+)^{-1}\sigma_0(g_0)^{-1}.$$

Its image clearly lands in $\mathbb{G}_m$ under the sequence (1) of §1.4.10. Thus, the morphism ψ gives rise to a morphism of $\mathbb{C}$-space functors:

$$\bar{\psi}: \mathscr{G} \times \bar{\mathscr{L}}_G^+ \to \mathbb{G}_m \text{ such that } \bar{\psi}(1, g_+) = 1, \text{ for any } g_+ \in \bar{\mathscr{L}}_G^+.$$

Since $\mathscr{L}_G^+$ is a representable functor (cf. Corollary 1.3.3) represented by a scheme denoted $\bar{L}_G^+$ with $\mathbb{C}$-points L_G^+ (the kernel of $G[[t]] \to G, t \mapsto 0$), $\bar{\psi}$ is induced from a morphism of schemes:

$$\bar{\psi}_o: G \times \bar{L}_G^+ \to \mathbb{G}_m, \text{ such that } \bar{\psi}_o(1, g_+) = 1, \text{ for any } g_+ \in \bar{L}_G^+.$$

But any morphism $f: G \to \mathbb{G}_m$ is a constant (cf. Lemma 1.4.9). Thus, $\bar{\psi}_o \equiv 1$ and hence so is ψ. This proves (12) and hence we obtain a splitting of p over $\mathscr{F}_1$.

The proof for $\mathscr{F}_3$ is identical to that of $\mathscr{F}_1$. This proves the theorem. □

Proposition 1.4.12 *For $\lambda_c \in \hat{D}$, the group functor $\hat{\mathscr{G}}_{\lambda_c}$ is represented by a reduced ind-affine group scheme denoted $\bar{\hat{G}}_{\lambda_c}$ (with $\mathbb{C}$-points $\hat{G}_{\lambda_c} = \hat{\mathscr{G}}_{\lambda_c}(\mathbb{C})$). This gives rise to an exact sequence of ind-group schemes:*

$$1 \to \mathbb{G}_m \to \bar{\hat{G}}_{\lambda_c} \xrightarrow{\bar{p}} \bar{G}((t)) \to 1. \tag{1}$$

Moreover, $\bar{p}$ admits a regular section over $N := \bar{G}[t^{-1}]^- \times \bar{G}[[t]]$ (cf. Lemma 1.3.16).

Thus, $\bar{\hat{G}}_{\lambda_c} \to \bar{G}((t))$ is a Zariski locally trivial principal $\mathbb{G}_m$-bundle.

Proof We first show that the group functor $\hat{\mathscr{G}}_{\lambda_c}$ is represented by an ind-scheme. Consider the open cover of $\bar{G}((t))$:

$$\bar{G}((t)) = \bigcup_{g \in G((t))} gN. \tag{2}$$

Then, the subfunctors $\{h_{gN}\}_{g \in G((t))}$ are an open covering of $h_{\bar{G}((t))}$ (cf. Definition B.5). To prove the above equality, observe that

$$\bar{G}((t))(\mathbb{C}) = G((t)) = \cup_{g \in G((t))}\, gN(\mathbb{C}).$$

(Now, use the fact that any closed ind-subscheme of $\bar{G}((t))$ has nonempty set of $\mathbb{C}$-points as observed in Exercise 1.3.E.8.) Thus, by Exercise B.E.8, $\{p^{-1}h_{gN}\}_{g \in G((t))}$ is an open cover consisting of subfunctors of $\hat{\mathscr{G}}_{\lambda_c}$. (Observe that $p^{-1}h_{gN}$ indeed satisfies the sheaf condition by using condition (1) of Exercise B.E.8 and using the fact that Spec $R' \to$ Spec R is surjective for any faithfully flat homomorphism $R \to R'$, cf. (Matsumura, 1989, Theorem 7.3).)

Recall from Theorem 1.4.11 that the homomorphism of group functors $p\colon \hat{\mathscr{G}}_{\lambda_c} \to \mathscr{L}_G$ admits splittings over the subgroup functors $G(R[[t]])$ and $G(R[t^{-1}])^-$. Combining them we get a morphism s from the $\mathbb{C}$-space functor $N(R) = G(R[[t]]) \times G(R[t^{-1}])^-$ to $\hat{\mathscr{G}}_{\lambda_c}$ such that $p \circ s = \mathrm{Id}$, i.e., a functorial section s of p over $N(R)$. For any $g \in G((t))$, choosing $\hat{g} \in \hat{G}_{\lambda_c}$ over g (which is possible by the surjectivity of $\hat{G}_{\lambda_c} \to G((t))$, cf. §1.4.10), we get a functorial section $\hat{g}s$ of p over h_{gN}. This gives rise to an isomorphism of $\mathbb{C}$-space functors

$$f\colon h_{gN} \times \mathbb{G}_m \simeq p^{-1}(h_{gN})$$

given by (for any $R \in \mathbf{Alg}$)

$$h_{gN}(R) \times R^* \xrightarrow{\sim} \left(p^{-1}h_{gN}\right)(R), \quad (\theta, r) \mapsto \hat{g}s(\theta) \cdot r,$$
$$\text{for } \theta \in h_{gN}(R) \text{ and } r \in R^*, \tag{3}$$

making the following diagram commutative:

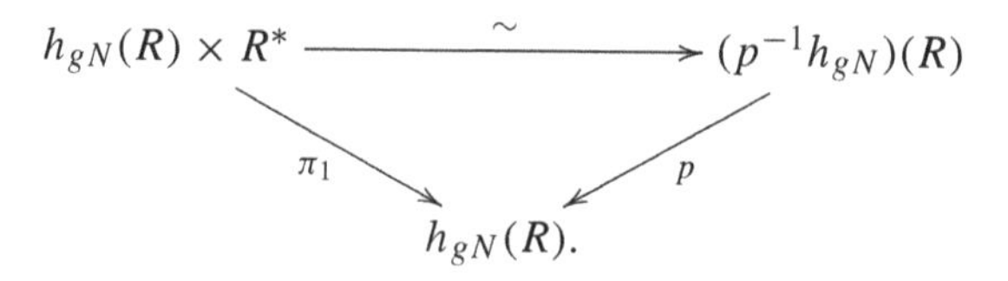

In particular, this implies that the open subfunctor $p^{-1}h_{gN}$ of $\hat{\mathscr{G}}_{\lambda_c}$ (for any $g \in G((t))$) over h_{gN} is represented by the ind-scheme $gN \times \mathbb{G}_m \to gN$. We now show that $\hat{\mathscr{G}}_{\lambda_c}$ is a representable functor with $\mathbb{C}$-points $\hat{G}_{\lambda_c}$.

Since $\{p^{-1}h_{gN}\}_{g \in G((t))}$ is an open cover consisting of subfunctors of $\hat{\mathscr{G}}_{\lambda_c}$ and $p^{-1}h_{gN}$ is represented by the ind-scheme $gN \times \mathbb{G}_m$ (by the isomorphism (3)), we get that the functor $\hat{\mathscr{G}}_{\lambda_c}$ is represented by an ind-scheme denoted $\bar{\hat{G}}_{\lambda_c}$ (cf. the proof of Proposition 1.3.18(a) using (Eisenbud and Harris, 2000, Theorem VI-14 and Exercise VI-11); since a Zariski cover is an fppf cover by Stacks (2019, Tag 021N)). Since $\hat{\mathscr{G}}_{\lambda_c}$ is a group functor, we get that $\bar{\hat{G}}_{\lambda_c}$ is an ind-group scheme giving rise to the exact sequence (1) of ind-group schemes. Moreover, the morphism $\bar{p}\colon \bar{\hat{G}}_{\lambda_c} \to \bar{G}((t))$ admits regular sections over gN (by (3)) for any $g \in G((t))$.

We now show that $\bar{\hat{G}}_{\lambda_c}$ is a reduced ind-affine group scheme. Let $\{(\bar{G}((t)))^n\}_{n \geq 0}$ be an increasing filtration of $\bar{G}((t))$ by reduced closed affine subschemes (cf. Remark 1.3.26(b) and Lemma 1.3.21), giving an ind-affine group scheme structure. Then, under the inverse image ind-scheme structure,

$$(\bar{\hat{G}}_{\lambda_c})^n := \bar{p}^{-1}\left((\bar{G}((t)))^n\right)$$

is a closed subset of $\bar{\hat{G}}_{\lambda_c}$ acquiring a reduced scheme structure by virtue of the isomorphism (3). Moreover, $(\bar{\hat{G}}_{\lambda_c})^n \hookrightarrow (\bar{\hat{G}}_{\lambda_c})^{n+1}$ is a closed embedding since closed embedding is preserved under base change (cf. (Hartshorne, 1977, Chap. II, Exercise 3.11(a))). Thus, $\bar{\hat{G}}_{\lambda_c}$ is a reduced ind group scheme.

Since a (Zariski locally trivial) principal $\mathbb{G}_m$-bundle over an affine scheme is affine (being an affine morphism), from the affineness of $(\bar{G}((t)))^n$ and (2), we get that $\bar{\hat{G}}_{\lambda_c}$ is ind-affine. This completes the proof of the proposition. □

Remark 1.4.13 As proved later (cf. Corollary 8.2.3), the (group) splittings of $\bar{p}\colon \bar{\hat{G}}_{\lambda_c} \to \bar{G}((t))$ over either of $\bar{G}[[t]]$ or $\bar{G}[t^{-1}]^-$ are unique (cf. Theorem 1.4.11).

Moreover, for any two regular sections s_1, s_2 of $\bar{p}$ over $N = \bar{G}[t^{-1}]^- \times \bar{G}[[t]]$ (cf. Proposition 1.4.12 and Corollary 8.2.3),

$$s_1 = s_2 z, \quad \text{for a fixed } z \in \mathbb{G}_m.$$

1.4.E Exercises

(1) With the notation as in Theorem 1.4.4, show that

$$\dot{\rho}_R : \mathfrak{g} \otimes R((t)) \to \mathrm{End}_R(\mathscr{H}(\lambda_c)_R)/R.\,\mathrm{Id}_{\mathscr{H}(\lambda_c)_R}$$

coincides with the projective representation of $\mathfrak{g} \otimes R((t))$ in $\mathscr{H}(\lambda_c)_R$, for any $R \in \mathbf{Alg}$.

(2) For any $\mathbb{C}$-algebra R which is a field, show that $\mathbf{GL}_V(R) \to \mathbf{PGL}_V(R)$ is surjective for any (not necessarily finite dimensional) $\mathbb{C}$-vector space V.

Hint: Let R be any noetherian $\mathbb{C}$-algebra. Let S be an fppf R-algebra. In particular, S is a noetherian $\mathbb{C}$-algebra. By Stacks (2019, Tag 0311), there is an embedding of rings

$$S \hookrightarrow \Pi_{i=1}^N S_{\mathfrak{p}_i},$$

where $\mathfrak{p}_1, \ldots, \mathfrak{p}_N$ are the associated prime ideals of S.

(a) Show that for any injective $\mathbb{C}$-algebra homomorphism $T \hookrightarrow T'$,

$$\mathrm{PGL}_V(T) \subset \mathrm{PGL}_V(T'), \quad \text{where } \mathrm{PGL}_V(T) := \mathrm{GL}_T(V_T)/T^* \cdot \mathrm{Id}.$$

Moreover, if $R \subset T \hookrightarrow T'$ and $K_R(T') = \mathrm{PGL}_V(R)$ for the functor $\mathbf{PGL}_V$ (cf. proof of Lemma B.2 for the notation $K_R(T)$), show that $K_R(T) = \mathrm{PGL}_V(R)$.

(b) Let $R \subset T$, where T is a local ring such that R is an R-module direct summand of T and T is R-flat. Show that $K_R(T) = \mathrm{PGL}_V(R)$.

(c) Let $R \subset T_1$ and $R \subset T_2$ be two $\mathbb{C}$-algebras such that T_2 is flat over R. Assume further that

$$K_R(T_1) = K_R(T_2) = \mathrm{PGL}_V(R).$$

Show that $K_R(T_1 \times T_2) = \mathrm{PGL}_V(R)$. Combining (a)–(c), the exercise follows.

(3) Show that, for any $\lambda_c \in \hat{D}$, there is an isomorphism of ind-group schemes

$$\hat{\bar{G}}_{\lambda_c} \xrightarrow{\psi} \hat{\bar{G}}_{0_c}$$

making the following diagram commutative:

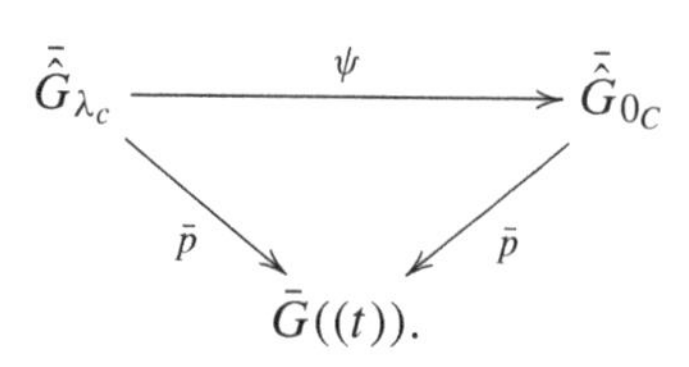

Hint: To prove this, show first that the $\mathbb{G}_m$-bundles $\hat{\bar{G}}_{\lambda_c}$ and $\hat{\bar{G}}_{0_C}$ are isomorphic. Then show that a $\mathbb{G}_m$-bundle isomorphism which takes 1 to 1 over $1 \in \bar{G}((t))$ is automatically a group homomorphism.

(4) For any $\lambda_c \in \hat{D}$ and any $\mathbb{C}$-algebra R, show that $Rv_+ \subset \mathscr{H}(\lambda_c)_R$ is the unique line annihilated by $\hat{\mathfrak{u}}(R) := (\mathfrak{g} \otimes tR[[t]]) \oplus (R \otimes \mathfrak{u})$, where $\mathfrak{u}$ is the nil-radical of $\mathfrak{b}$.

Hence, show that any $\hat{\mathfrak{g}}(R)$-module endomorphism of $\mathscr{H}(\lambda_c)_R$ is the identity map up to a scalar multiple.

Hint: Use Exercise 1.2.E.5.

1.C Comments

The content of Section 1.2 is fairly standard (cf. (Kac, 1990, Chaps. 7 and 12 and Lemma 9.10) and (Kumar, 2002, Chaps. 1 and 13)).

The content of Section 1.3 is also fairly standard by now. Lemma 1.3.2, Theorem 1.3.8, Corollary 1.3.15 for $G = \mathrm{SL}_n, \mathrm{GL}_N$; Corollary 1.3.19 for $G = \mathrm{SL}_N$; and Lemma 1.3.21 are proved in Beauville and Laszlo (1994). The proof of Theorem 1.3.8 is an elaboration of Beauville and Laszlo (1994, Proof of Proposition 2.4) (with some help from P. Belkale). The approach we have taken in this section is largely derived from Faltings (2003) (though we have supplied here significantly more details). For example, Proposition 1.3.14, Corollary 1.3.15 and Lemma 1.3.16 are taken from Faltings (2003, §2).

As mentioned before, the detailed proof of Theorem 1.3.22 given here was provided by B. Conrad (and a brief outline was given earlier by G. Faltings in a private communication). Theorem 1.3.23 and Proposition 1.3.24 are given in Laszlo and Sorger (1997, Propositions 4.6 and 4.7). However, the proof of Theorem 1.3.23 outlined in Laszlo and Sorger (1997, Proposition 4.6) is incorrect (since it wrongly uses an incorrect theorm of Shafarevich). For a different representation-theoretic approach to many of the results in this Section 1.3 and the next, see Kumar (2002, Chap. 13.2).

Lemma 1.4.9 is taken from Kumar, Narasimhan and Ramanathan (1994, Lemma 2.5). Theorem 1.4.11 is taken from Laszlo and Sorger (1997). There is an alternative construction of the central extension of $\mathrm{SL}_N((t))$ via the Fredholm group in Beauville and Laszlo (1994, §4) (also see Kumar (2002, Theorem 13.2.8)).

2

Space of Vacua and its Propagation

Let $\mathfrak{g}$ be a simple Lie algebra over the complex numbers and let $(\Sigma, \vec{p})$ be an s-pointed curve (for any $s \geq 1$) defined in Definition 2.1.1, where $\vec{p} = (p_1, \ldots, p_s)$. We fix a positive integer c called the *level* or *central charge*. Let D_c be the set of dominant integral weights of $\mathfrak{g}$ of level $\leq c$ and let $\vec{\lambda} = (\lambda_1, \ldots, \lambda_s)$ be an s-tuple of weights with $\lambda_i \in D_c$ attached to the point p_i. To this data, there is associated the *space of vacua* (also called the *space of conformal blocks*) $\mathscr{V}_\Sigma^\dagger(\vec{p}, \vec{\lambda})$ and its dual space $\mathscr{V}_\Sigma(\vec{p}, \vec{\lambda})$ called the *space of covacua* (or the space of *dual conformal blocks*), which are fundamental objects of this book (cf. Definition 2.1.1). The definition of these spaces requires choosing a local parameter of the curve Σ at the points p_i. However, we show that, up to a natural isomorphism, these spaces do not depend upon the choice of the local parameters (cf. Corollary 2.1.3). It is further shown in Lemma 2.1.4 that these spaces are finite-dimensional (the dimensions of which are given by the Verlinde dimension formula given and proved in Chapter 4).

In Section 2.2, we prove the 'propagation of vacua' (cf. Theorem 2.2.2), which roughly gives a natural isomorphism between the space of covacua for $n + m$ points on the curve Σ and certain space of covariants of the tensor product of n copies of integrable highest-weight modules of $\hat{\mathfrak{g}}$ with m copies of irreducible $\mathfrak{g}$-modules. As a corollary of this result, one obtains that the space of covacua (and the space of vacua) does not change by adding additional smooth points on the curve if we attach the 0 weight to these points (cf. Corollary 2.2.3).

In Section 2.3, we study the spaces of vacua and covacua for the curve $\Sigma = \mathbb{P}^1$. Denote the tensor product $\mathfrak{g}$-module

$$V(\vec{\lambda}) := V(\lambda_1) \otimes \cdots \otimes V(\lambda_s).$$

Then, in this case (i.e., when $\Sigma = \mathbb{P}^1$) the space of covacua (resp. vacua) is identified as a quotient (resp. subspace) of $V(\vec{\lambda})$ (resp. $V(\vec{\lambda}^*)$) in terms of a certain operator φ on $V(\vec{\lambda})$ (cf. Definition 2.3.1 and Theorem 2.3.2). The space of vacua for $\Sigma = \mathbb{P}^1$ with three marked points (i.e., $s = 3$) and $\mathfrak{g} = sl_2$ for any level c is determined in Proposition 2.3.3. This allows us to prove that the space of vacua for $\Sigma = \mathbb{P}^1$ with three marked points and any $\mathfrak{g}$ for any level c is given explicitly in terms of the decomposition of any irreducible $\mathfrak{g}$-module with respect to the $sl_2 \subset \mathfrak{g}$ passing through the highest root space of $\mathfrak{g}$ (cf. Corollary 2.3.5).

2.1 Space of Vacua

We follow the notation from Section 1.2. In particular, $\mathfrak{g}$ is a simple Lie algebra. We also fix a positive integer c called the *level* or *central charge*.

Definition 2.1.1 Let Σ be a reduced (but not necessarily connected) projective curve with at worst only nodal singularity. (Recall that a point $P \in \Sigma$ is called a *node* if analytically a neighborhood of P in Σ is isomorphic with an analytic neighborhood of $(0,0)$ in the curve $xy = 0$ in $\mathbb{A}^2$.) For any $s \geq 1$, by an *s-pointed curve* we mean the pair $(\Sigma, \vec{p} = (p_1, \ldots, p_s))$ consisting of distinct and smooth points $\{p_1, \ldots, p_s\}$ of Σ, such that the following condition is satisfied:

(a) Each irreducible component of Σ contains at least one point p_i.

(This condition ensures that $\Sigma \setminus \vec{p}$ is affine.)

An s-pointed curve $(\Sigma, \vec{p})$ is called a (Deligne–Mumford) *stable s-pointed curve* if Σ is connected and it satisfies the following condition:

(b) The automorphism group $\mathrm{Aut}_{\vec{p}}(\Sigma)$ of Σ fixing all the points p_i is finite.

(Actually, Deligne–Mumford do not require the condition (a) for their stability.)

Recall that the above condition (b) (in the presence of (a)) is equivalent to the requirement that if an irreducible component Σ_k of Σ is $\mathbb{P}^1$, then the number of points p_i's on Σ_k together with the number of points in $\Sigma_k \cap \Sigma_k'$ is at least three, where Σ_k' is the union of other irreducible components of Σ.

As in Definition 1.2.6, let $D \subset \mathfrak{h}^*$ be the set of dominant integral weights for $\mathfrak{g}$. Define the subset

$$D_c := \left\{\lambda \in D : \lambda(\theta^\vee) \leq c\right\}, \tag{1}$$

where θ is the highest root of $\mathfrak{g}$ and $\theta^\vee \in \mathfrak{h}$ is the corresponding coroot.

Let $(\Sigma, \vec{p} = (p_1, \ldots, p_s))$ be an s-pointed curve and let $\vec{\lambda} = (\lambda_1, \ldots, \lambda_s)$ be an s-tuple of weights with each $\lambda_i \in D_c$, where we think of λ_i 'attached' to the point p_i. To this data, there is associated the *space of vacua* (also called the *space of conformal blocks*) $\mathscr{V}_\Sigma^\dagger(\vec{p}, \vec{\lambda})$ and its dual space $\mathscr{V}_\Sigma(\vec{p}, \vec{\lambda})$ called the *space of covacua* (or the space of *dual conformal blocks*) defined as follows.

Fix *formal parameters* $(t_1, \ldots, t_s)$ of Σ at the points $p_1, \ldots, p_s$, respectively (i.e., $\varprojlim \mathscr{O}_\Sigma/\mathfrak{m}_{p_i}^n \simeq \mathbb{C}[[t_i]]$, where $\mathfrak{m}_{p_i}$ is the maximal ideal of Σ at p_i). Let $\mathfrak{g}[\Sigma \backslash \vec{p}]$ denote the space of morphisms $f\colon \Sigma \backslash \vec{p} \to \mathfrak{g}$, i.e., $\mathfrak{g}[\Sigma \backslash \vec{p}] := \mathfrak{g} \otimes \mathbb{C}[\Sigma \backslash \vec{p}]$, where $\vec{p}$ also denotes the set $\{p_1, \ldots, p_s\}$. Then, $\mathfrak{g}[\Sigma \backslash \vec{p}]$ is a Lie algebra under the pointwise bracket:

$$[x[f], y[g]] = [x, y][fg], \quad \text{for } f, g \in \mathbb{C}[\Sigma \backslash \vec{p}] \text{ and } x, y \in \mathfrak{g}, \tag{2}$$

where, as earlier, we denote $x \otimes f$ by $x[f]$. For any $\lambda \in D_c$, let

$$\lambda_c := (\lambda, c) \in \hat{D}, \tag{3}$$

where $\hat{D}$ is as in Definition 1.2.6 and $(\lambda, c) \in \hat{\mathfrak{h}}^*$ denotes the element such that $(\lambda, c)_{|\mathfrak{h}} = \lambda$ and $(\lambda, c)(C) = c$ (cf. identity (1) of Definition 1.2.2). Since we have fixed the level, we will abbreviate $\mathscr{H}(\lambda_c)$ by $\mathscr{H}(\lambda)$ for any $\lambda \in D_c$. Set

$$\mathscr{H}(\vec{\lambda}) := \mathscr{H}(\lambda_1) \otimes \cdots \otimes \mathscr{H}(\lambda_s). \tag{4}$$

Define an action of the Lie algebra $\mathfrak{g}[\Sigma \backslash \vec{p}]$ on $\mathscr{H}(\vec{\lambda})$ as follows:

$$\begin{aligned} x[f] \cdot (v_1 \otimes \cdots \otimes v_s) &= \Sigma_{i=1}^s v_1 \otimes \cdots \otimes x[f_{p_i}] \cdot v_i \otimes \cdots \otimes v_s, \\ &\text{for } f \in \mathbb{C}[\Sigma \backslash \vec{p}], x \in \mathfrak{g} \text{ and } v_i \in \mathscr{H}(\lambda_i), \end{aligned} \tag{5}$$

where f_{p_i} denotes the Laurent series expansion of f at p_i with respect to the formal parameter t_i.

By the Residue Theorem (Hartshorne, 1977, Chap. III, Theorem 7.14.2), taking the normalization of Σ,

$$\sum_{i=1}^s \operatorname{Res}_{p_i}(f dg) = 0, \quad \text{for any} \quad f, g \in \mathbb{C}[\Sigma \backslash \vec{p}]. \tag{6}$$

Thus, the action (5) indeed is an action of the Lie algebra $\mathfrak{g}[\Sigma \backslash \vec{p}]$ on $\mathscr{H}(\vec{\lambda})$. Moreover, by the next lemma, the action of $\mathfrak{g}[\Sigma \backslash \vec{p}]$ on $\mathscr{H}(\vec{\lambda})$ does not depend upon the choice of the formal parameters t_i at p_i up to a natural isomorphism of $\mathscr{H}(\vec{\lambda})$.

Now, we are ready to define the *space of vacua*

$$\mathscr{V}_\Sigma^\dagger(\vec{p}, \vec{\lambda}) := \operatorname{Hom}_{\mathfrak{g}[\Sigma \backslash \vec{p}]}(\mathscr{H}(\vec{\lambda}), \mathbb{C}), \tag{7}$$

and the *space of covacua*

$$\mathscr{V}_\Sigma(\vec{p},\vec{\lambda}) := [\mathscr{H}(\vec{\lambda})]_{\mathfrak{g}[\Sigma\backslash\vec{p}]}, \tag{8}$$

where $\mathbb{C}$ is considered as the trivial module under the action of $\mathfrak{g}[\Sigma\backslash\vec{p}]$, and $[\mathscr{H}(\vec{\lambda})]_{\mathfrak{g}[\Sigma\backslash\vec{p}]}$ denotes the space of covariants $\mathscr{H}(\vec{\lambda})/\big(\mathfrak{g}[\Sigma\backslash\vec{p}]\cdot\mathscr{H}(\vec{\lambda})\big)$. Clearly,

$$\mathscr{V}^\dagger_\Sigma(\vec{p},\vec{\lambda}) \simeq \mathscr{V}_\Sigma(\vec{p},\vec{\lambda})^*. \tag{9}$$

For any power series $\omega(t) \in \mathbb{C}[[t]]$ with $\omega(0) = 0$ and $\omega'(0) \neq 0$, consider the automorphism of $K = \mathbb{C}((t))$, $f \mapsto f \circ \omega$. This automorphism gives rise to an automorphism φ_ω of $\hat{\mathfrak{g}}$, taking $x[f] \mapsto x[f \circ \omega]$ and $C \to C$. Observe that for any two formal parameters t and s of Σ at any smooth point p, s is of the form $\omega(t)$, for some ω as above.

Lemma 2.1.2 *For any $\lambda \in D_c$, there is a natural linear isomorphism $\varphi_\omega(\lambda)\colon \mathscr{H}(\lambda) \to \mathscr{H}(\lambda)$ satisfying $\varphi_\omega(\lambda)(X \cdot h) = \varphi_\omega(X)\cdot(\varphi_\omega(\lambda)h)$, for $X \in \hat{\mathfrak{g}}$ and $h \in \mathscr{H}(\lambda)$.*

Proof Twist the $\hat{\mathfrak{g}}$-module structure of $\mathscr{H}(\lambda)$ via the automorphism φ_ω. To distinguish, we denote $\mathscr{H}(\lambda)$ with this twisted $\hat{\mathfrak{g}}$-module structure by $\mathscr{H}(\lambda)'$. Since φ_ω keeps the Borel subalgebra $\hat{\mathfrak{b}}$ stable, $\mathscr{H}(\lambda)'$ continues to be a highest-weight module with the same highest-weight subspace $\mathbb{C}h_+$. Moreover, since $\varphi_{\omega|\hat{\mathfrak{h}}}$ is the identity map, $\hat{\mathfrak{h}}$ acts on the highest-weight subspace by the same weight λ_c. In particular, $\mathscr{H}(\lambda)'$ is a quotient of the Verma module $\hat{M}(\lambda_c)$. Further, φ_ω being an automorphism of $\hat{\mathfrak{g}}$, $\mathscr{H}(\lambda)'$ is an irreducible $\hat{\mathfrak{g}}$-module. In particular, $\mathscr{H}(\lambda)'$ is isomorphic with $\mathscr{H}(\lambda)$ as $\hat{\mathfrak{g}}$-modules (cf. Exercise 1.2.E.6). Moreover, by Exercise 1.2.E.5, there is a unique (up to a nonzero scalar multiple) $\hat{\mathfrak{g}}$-module isomorphism $\varphi_\omega(\lambda)\colon \mathscr{H}(\lambda) \to \mathscr{H}(\lambda)'$. In fact, since $\mathbb{C}h_+$ is the unique $\hat{\mathfrak{b}}$-stable line in both of $\mathscr{H}(\lambda)$ and $\mathscr{H}(\lambda)'$, we can canonically choose $\varphi_\omega(\lambda)$ so that $\varphi_\omega(\lambda)h_+ = h_+$. This proves the lemma. □

As an immediate consequence of the above lemma, we get the following:

Corollary 2.1.3 *Up to a natural isomorphism, spaces $\mathscr{V}^\dagger_\Sigma(\vec{p},\vec{\lambda})$ and $\mathscr{V}_\Sigma(\vec{p},\vec{\lambda})$ do not depend upon the choice of formal parameters $(t_1,\ldots,t_s)$ of Σ at the points $(p_1,\ldots,p_s)$.*

Lemma 2.1.4 *With the notation and assumptions as in Definition 2.1.1, the space of covacua $\mathscr{V}_\Sigma(\vec{p},\vec{\lambda})$ is finite dimensional and hence by identity* (9) *of Definition 2.1.1, so is the space of vacua $\mathscr{V}^\dagger_\Sigma(\vec{p},\vec{\lambda})$.*

Proof Define a Lie algebra bracket on

$$\hat{\mathfrak{g}}^{(s)} := \left(\bigoplus_{i=1}^{s} (\mathfrak{g} \otimes \mathbb{C}((t_i))) \right) \oplus \mathbb{C}C, \tag{1}$$

by declaring C to be a central element and

$$\left[\sum_{i=1}^{s} x_i[f_i], \sum_{i=1}^{s} y_i[g_i] \right] = \sum_{i=1}^{s} [x_i, y_i][f_i g_i] + \left(\sum_{i=1}^{s} \langle x_i, y_i \rangle \operatorname{Res}_{p_i}(g_i df_i) \right) C,$$

for $f_i, g_i \in \mathbb{C}((t_i)), x_i, y_i \in \mathfrak{g}$.

Now, define an embedding of Lie algebras:

$$\beta \colon \mathfrak{g}[\Sigma \backslash \vec{p}] \to \hat{\mathfrak{g}}^{(s)}, \ x[f] \mapsto \sum_{i=1}^{s} x[f_{p_i}], \ \text{ for } f \in \mathbb{C}[\Sigma \backslash \vec{p}] \text{ and } x \in \mathfrak{g}.$$

By the Residue Theorem (Hartshorne, 1977, Chap. III, Theorem 7.14.2), β is indeed a Lie algebra homomorphism. Moreover, by using the Riemann–Roch theorem for curves (Hartshorne, 1977, Chap. IV, Theorem 1.3), $\operatorname{Im} \beta + \left(\bigoplus_{i=1}^{s} \mathfrak{g} \otimes \mathbb{C}[[t_i]] \right)$ has finite codimension in $\hat{\mathfrak{g}}^{(s)}$. (Here we have used assumption (a) of Definition 2.1.1.) Further, define the following surjective Lie algebra homomorphism from the direct sum Lie algebra:

$$\pi : \bigoplus_{i=1}^{s} \hat{\mathfrak{g}}_{t_i} \to \hat{\mathfrak{g}}^{(s)}, \ \sum_{i=1}^{s} x_i[f_i]$$
$$\mapsto \sum_{i=1}^{s} x_i[f_i], \ C_i \to C, \ \text{ for } f_i \in \mathbb{C}((t_i)) \text{ and } x_i \in \mathfrak{g},$$

here C_i is the center C of $\hat{\mathfrak{g}}_{t_i}$.

Now, the lemma follows from Kumar (2002, Lemma 10.2.2) by applying Lemma 10.2.2 from the same (following its notation) to the Kac–Moody Lie algebra $\mathfrak{g} := \bigoplus_{i=1}^{s} \hat{\mathfrak{g}}_{t_i}$, $\mathfrak{n}$ the nil-radical of the standard Borel subalgebra and $\mathfrak{s} := \pi^{-1}(\operatorname{Im} \beta)$. □

2.2 Propagation of Vacua

We follow the notation and assumptions as in Section 2.1.

Definition 2.2.1 Let $\vec{p} = (p_1, \ldots, p_s)$ and $\vec{q} = (q_1, \ldots, q_a)$ be two disjoint nonempty sets of smooth and distinct points in Σ such that $(\Sigma, \vec{p})$ is an s-pointed curve and let $\vec{\lambda} = (\lambda_1, \ldots, \lambda_s)$, $\vec{\mu} = (\mu_1, \ldots, \mu_a)$ be a collection of weights in D_c.

Denote the tensor product $\mathfrak{g}$-module

$$V(\vec{\mu}) := V(\mu_1) \otimes \cdots \otimes V(\mu_a). \tag{1}$$

Define a $\mathfrak{g}[\Sigma\backslash\vec{p}]$-module structure on $V(\vec{\mu})$ as follows:

$$x[f] \cdot (v_1 \otimes \cdots \otimes v_a) = \sum_{j=1}^{a} v_1 \otimes \cdots \otimes f(q_j)x \cdot v_j \otimes \cdots \otimes v_a, \tag{2}$$

for $v_j \in V(\mu_j)$, $x \in \mathfrak{g}$ and $f \in \mathbb{C}[\Sigma\backslash\vec{p}]$. Thus, we get the tensor product $\mathfrak{g}[\Sigma\backslash\vec{p}]$-module structure on $\mathscr{H}(\vec{\lambda}) \otimes V(\vec{\mu})$.

Theorem 2.2.2 *For $\vec{p}$, $\vec{q}$, $\vec{\lambda}$, $\vec{\mu}$ as above, the natural map*

$$\theta : \left[\mathscr{H}(\vec{\lambda}) \otimes V(\vec{\mu})\right]_{\mathfrak{g}[\Sigma\backslash\vec{p}]} \to \mathscr{V}_\Sigma\left((\vec{p},\vec{q}),(\vec{\lambda},\vec{\mu})\right)$$

is an isomorphism, where $\mathscr{V}_\Sigma$ is the space of covacua defined by identity (8) *of Definition 2.1.1 and the map θ is induced from the $\mathfrak{g}[\Sigma\backslash\vec{p}]$-module embedding*

$$\mathscr{H}(\vec{\lambda}) \otimes V(\vec{\mu}) \hookrightarrow \mathscr{H}(\vec{\lambda},\vec{\mu}),$$

with $V(\mu_j)$ identified as a $\mathfrak{g}$-submodule of $\mathscr{H}(\mu_j)$ generated by its highest-weight line. (Observe that the subspace $V(\mu_j) \subset \mathscr{H}(\mu_j)$ is annihilated by $\hat{\mathfrak{g}}_+$, and hence the embedding $V(\mu_j) \subset \mathscr{H}(\mu_j)$ is indeed a $\mathfrak{g}[\Sigma\backslash\vec{p}]$-module embedding. Here $\hat{\mathfrak{g}}_+$ is defined by (4) *of Definition 1.2.2.)*

Proof Let $\mathscr{H} = \mathscr{H}(\vec{\lambda}) \otimes V(\mu_1) \otimes \cdots \otimes V(\mu_{a-1})$. By induction on a, it suffices to show that the inclusion $V(\mu_a) \hookrightarrow \mathscr{H}(\mu_a)$ induces an isomorphism (abbreviating μ_a by μ and q_a by q)

$$[\mathscr{H} \otimes V(\mu)]_{\mathfrak{g}[\Sigma^o]} \xrightarrow{\sim} [\mathscr{H} \otimes \mathscr{H}(\mu)]_{\mathfrak{g}[\Sigma^o\backslash q]}, \tag{1}$$

where $\Sigma^o := \Sigma\backslash\vec{p}$.

We first prove (1) replacing $\mathscr{H}(\mu)$ by the generalized Verma module $\hat{M}(V(\mu),c)$ (cf. identity (4) of Definition 1.2.4), i.e.,

$$[\mathscr{H} \otimes V(\mu)]_{\mathfrak{g}[\Sigma^o]} \xrightarrow{\sim} [\mathscr{H} \otimes \hat{M}(V(\mu),c)]_{\mathfrak{g}[\Sigma^o\backslash q]}. \tag{2}$$

Choose a parameter z around q so that $z^{-1} \in \mathbb{C}[\Sigma^o\backslash q]$ (this is possible since each irreducible component of Σ contains a point of $\vec{p}$). With this choice of the local parameter z,

$$\mathfrak{g}[\Sigma^o\backslash q] = \mathfrak{g}[\Sigma^o] \oplus \hat{\mathfrak{g}}_-, \text{ where, as in Definition 1.2.2, } \hat{\mathfrak{g}}_- := \oplus_{n\geq 1}\, \mathfrak{g} \otimes z^{-n}. \tag{3}$$

Consider the Lie algebra

$$\mathfrak{s} := \mathfrak{g}[\Sigma^o\backslash q] \oplus \mathbb{C}C,$$

where C is central in $\mathfrak{s}$ and

$$[x[f], y[g]] = [x, y][fg] + \langle x, y\rangle \operatorname*{Res}_q (gdf)\, C, \quad \text{for } f, g \in \mathbb{C}[\Sigma^o \backslash q], x, y \in \mathfrak{g}.$$

Let $\mathfrak{g}[\Sigma^o\backslash q]$ act on $\mathscr{H}$ by the same formula as (5) of Definition 2.1.1 and identity (2) of Definition 2.2.1 and let C act on $\mathscr{H}$ by the scalar $-c$. By the Residue Theorem (Hartshorne, 1977, Chap. III, Theorem 7.14.2), these actions combine to make $\mathscr{H}$ into an $\mathfrak{s}$-module.

Consider the embedding of the Lie algebra

$$\mathfrak{s} \hookrightarrow \hat{\mathfrak{g}}$$

by taking $C \mapsto C$ and any $x[f] \mapsto x[f_q]$, where the Laurent series expansion f_q of f at q is taken with respect to the parameter z. Thus, the action of $C \in \mathfrak{s}$ on the tensor product $\mathscr{H} \otimes \hat{M}(V(\mu), c)$ is trivial. Now, by the definition of $\hat{M}(V(\mu), c)$ and (3) (in the following, $\mathfrak{g}[\Sigma^o]$ acts on $V(\mu)$ via its evaluation at q and C acts via the scalar c)

$$\begin{aligned}
\left[\mathscr{H} \otimes \hat{M}(V(\mu), c)\right]_{\mathfrak{g}[\Sigma^o\backslash q]} &= \left[\mathscr{H} \otimes \hat{M}(V(\mu), c)\right]_{\mathfrak{s}}, \quad \text{since } C \text{ acts trivially}\\
&\simeq \mathscr{H} \otimes_{U(\mathfrak{s})} \hat{M}(V(\mu), c)\\
&\simeq \mathscr{H} \otimes_{U(\mathfrak{s})} \left(U(\mathfrak{s}) \otimes_{U(\mathfrak{g}[\Sigma^o]\oplus \mathbb{C}C)} V(\mu)\right)\\
&\simeq \mathscr{H} \otimes_{U(\mathfrak{g}[\Sigma^o]\oplus \mathbb{C}C)} V(\mu)\\
&= \mathscr{H} \otimes_{U(\mathfrak{g}[\Sigma^o])} V(\mu)\\
&= [\mathscr{H} \otimes V(\mu)]_{\mathfrak{g}[\Sigma^o]}.
\end{aligned}$$

This proves (2).

Now we come to the proof of (1).

Let $K(\mu)$ be the kernel of the canonical projection $\hat{M}(V(\mu), c) \twoheadrightarrow \mathscr{H}(\mu)$. In view of (2), to prove (1), it suffices to show that the image of

$$\iota\colon\ [\mathscr{H} \otimes K(\mu)]_{\mathfrak{g}[\Sigma^o\backslash q]} \to \left[\mathscr{H} \otimes \hat{M}(V(\mu), c)\right]_{\mathfrak{g}[\Sigma^o\backslash q]}$$

is zero. By identity (3), the map $x[f] \mapsto x[f_q]$ and $C \to C$ (for $f \in \mathbb{C}[\Sigma^o\backslash q]$ and $x \in \mathfrak{g}$) induces a surjection:

$$\mathfrak{s} \to \hat{\mathfrak{g}}/\hat{\mathfrak{g}}_+.$$

In particular,

$$\hat{\mathfrak{g}} = \mathfrak{s} + \hat{\mathfrak{g}}_+$$

and hence, by the PBW theorem, $U(\hat{\mathfrak{g}})$ is the span of elements of the form

$$Y_1 \ldots Y_m \cdot X_1 \ldots X_n, \quad \text{for } \ Y_i \in \mathfrak{s}, X_j \in \hat{\mathfrak{g}}_+ \ \text{ and } \ m, n \geq 0.$$

Thus, to prove the vanishing of the map ι, from the definition of $K(\mu)$ (cf. Definition 1.2.6), it suffices to show that (putting $Y = x_\theta[z^{-1}]$)

$$\iota\left(h \otimes (X_1 \dots X_n \cdot Y^{n_\mu} \cdot v_+)\right) = 0, \tag{4}$$

for $h \in \mathscr{H}$, any $n \geq 0$ and $X_j \in \hat{\mathfrak{g}}_+$, where $n_\mu := c + 1 - \mu(\theta^\vee)$ and v_+ is a highest-weight vector of $\hat{M}(V(\mu), c)$. But, by Exercise 1.2.E.3,

$$\hat{\mathfrak{g}}_+ \cdot (Y^{n_\mu} \cdot v_+) = 0.$$

Thus, to prove (4), it suffices to show that

$$\iota(h \otimes (Y^{n_\mu} \cdot v_+)) = 0, \quad \text{for any} \quad h \in \mathscr{H}. \tag{5}$$

Take $f \in \mathbb{C}[\Sigma^o]$ such that

$$f_q \equiv z \,(\text{mod } z^2).$$

By Exercise 1.2.E.7,

$$(x_{-\theta}[f])^N \cdot h = 0, \quad \text{for large enough } N. \tag{6}$$

We next show that, as elements of $\hat{M}(V(\mu), c)$, for any $N \geq 0$,

$$Y^{n_\mu} \cdot v_+ = \alpha X^N Y^{n_\mu + N} \cdot v_+, \quad \text{for some } \alpha \in \mathbb{C}, \tag{7}$$

where $X := x_{-\theta}[f_q] \in \hat{\mathfrak{g}}$. We can (and do) choose $x_{-\theta}$ so that $\langle x_{-\theta}, x_\theta \rangle = 1$. Put $H = [X, Y] \in \hat{\mathfrak{g}}$. Then $H = C - \theta^\vee[z^{-1} f_q]$, $[Y, H] = 2x_\theta[z^{-2} f_q]$. In particular, $[Y, H]$ commutes with Y.

For any associative algebra A and element $y \in A$, define the operators $L_y(x) = yx$, $R_y(x) = xy$ and $ad(y) = L_y - R_y$.

Considering the Binomial Theorem for the operator $L_y^n = (ad(y) + R_y)^n$, we get

$$y^n x = \sum_{j=0}^{n} \binom{n}{j} \left((ad(y))^j x\right) y^{n-j}. \tag{8}$$

(Observe that R_y and $ad(y)$ commute.) Applying this identity for $y = Y$ and $x = H$ in the enveloping algebra $U(\hat{\mathfrak{g}})$, we get

$$Y^n H = H Y^n + n Y^{n-1}[Y, H], \quad \text{since } Y \text{ commutes with } [Y, H]. \tag{9}$$

Now,

$$\begin{aligned} [Y, H] \cdot v_+ &= 2x_\theta[z^{-2} f_q] \cdot v_+ \\ &= 2x_\theta[z^{-1}] \cdot v_+, \quad \text{since } \hat{\mathfrak{g}}_+ \oplus (\mathfrak{u} \otimes t^0) \text{ annihilates } v_+ \\ &= 2Y \cdot v_+, \end{aligned} \tag{10}$$

where $\mathfrak{u}$ is the nil-radical of $\mathfrak{b}$. Further,

$$\begin{aligned} H \cdot v_+ &= \left(C - \theta^\vee[z^{-1} f_q]\right) \cdot v_+ \\ &= cv_+ - \theta^\vee \cdot v_+, \ \text{ since } \hat{\mathfrak{g}}_+ \oplus (\mathfrak{u} \otimes t^0) \text{ annihilates } v_+ \\ &= (n_\mu - 1)v_+, \ \text{ since } n_\mu := c + 1 - \mu(\theta^\vee). \end{aligned} \tag{11}$$

Hence, by (9)–(11), we get

$$\begin{aligned} HY^n \cdot v_+ &= Y^n H \cdot v_+ - nY^{n-1}[Y, H] \cdot v_+ \\ &= (n_\mu - 1)Y^n \cdot v_+ - 2nY^n \cdot v_+ \\ &= (n_\mu - 1 - 2n)Y^n \cdot v_+. \end{aligned} \tag{12}$$

Applying identity (8) for $y = Y$ and $x = X$, we get

$$Y^n X = XY^n - nHY^{n-1} - \binom{n}{2} Y^{n-2}[Y, H], \ \text{ since } Y \text{ commutes with } [Y, H].$$

Thus,

$$\begin{aligned} XY^n \cdot v_+ &= nHY^{n-1} \cdot v_+ + \binom{n}{2} Y^{n-2}[Y, H] \cdot v_+ \\ &= n(n_\mu + 1 - 2n)Y^{n-1} \cdot v_+ + n(n-1)Y^{n-1} \cdot v_+, \ \text{ by (10) and (12)} \\ &= n(n_\mu - n)Y^{n-1} \cdot v_+. \end{aligned}$$

Thus, by induction on m, for any $0 \leq m \leq n$,

$$\begin{aligned} X^m Y^n \cdot v_+ = &(n(n-1)\dots(n-m+1)) \\ &\times \left((n_\mu - n)(n_\mu - n + 1)\dots(n_\mu + m - 1 - n)\right) Y^{n-m} \cdot v_+. \end{aligned}$$

This proves identity (7). By (7),

$$\begin{aligned} \iota\left(h \otimes (Y^{n_\mu} \cdot v_+)\right) &= \alpha\, \iota\left(h \otimes X^N Y^{n_\mu + N} \cdot v_+\right) \\ &= (-1)^N \alpha\, \iota\left((x_{-\theta}[f])^N \cdot h \otimes Y^{n_\mu + N} \cdot v_+\right) \\ &= 0, \ \text{ by (6).} \end{aligned}$$

This proves (5) and hence completes the proof of the theorem. □

The following result is usually referred to as 'propagation of vacua.'

Corollary 2.2.3 *Let $(\Sigma, \vec{p})$ be an s-pointed curve. Then, for any smooth point $q \in \Sigma \backslash \vec{p}$, there are natural isomorphisms*

(a) $\mathscr{V}_\Sigma(\vec{p},\vec{\lambda}) \simeq \mathscr{V}_\Sigma((\vec{p},q),(\vec{\lambda},0))$.
(b) *If* Σ *is irreducible,* $\mathscr{V}_\Sigma(\vec{p},\vec{\lambda}) \simeq [\mathscr{H}(0) \otimes V(\vec{\lambda})]_{\mathfrak{g}[\Sigma\setminus q]}$, *where the point* q *is assigned weight* 0.

Proof (a) Apply Theorem 2.2.2 for the case $\vec{q} = (q)$ and $\vec{\mu} = (0)$.

(b) Follows from Theorem 2.2.2 and the (a)-part. (In Theorem 2.2.2 replace $\vec{p}$ by the singleton (q), $\vec{\lambda}$ by (0), $\vec{q}$ by $\vec{p}$ and $\vec{\mu}$ by $\vec{\lambda}$.) □

2.2.E Exercises

(1) Let $(\Sigma, \vec{p} = (p_1, \ldots, p_s))$ be an s-pointed irreducible curve and let $\vec{\lambda} = (\lambda_1, \ldots, \lambda_s)$ be a collection of weights in D_c. Let $\mathcal{V}_1, \ldots, \mathcal{V}_s$ be nonzero quotients of the generalized Verma modules $\hat{M}(V(\lambda_1), c), \ldots, \hat{M}(V(\lambda_s), c)$, respectively, such that at least one of $\mathcal{V}_i$ is equal to $\mathscr{H}(\lambda_i)$. Then, show that the quotient map

$$\mathcal{V}_1 \otimes \cdots \otimes \mathcal{V}_s \to \mathscr{H}(\vec{\lambda})$$

induces an isomorphism

$$\left[\mathcal{V}_1 \otimes \cdots \otimes \mathcal{V}_s\right]_{\mathfrak{g}[\Sigma\setminus\vec{p}]} \to \mathscr{V}_\Sigma(\vec{p},\vec{\lambda}).$$

Hint: Use ideas similar to the proof of Theorem 2.2.2.

2.3 A Description of the Space of Vacua on $\mathbb{P}^1$ via Representations of $\mathfrak{g}$

We follow the notation from Section 2.1. In particular, let $\mathfrak{g}$ be a simple (finite-dimensional) Lie algebra over $\mathbb{C}$ and let $c \geq 1$ be the level. In this section, we take $\Sigma = \mathbb{P}^1$, where $\mathbb{P}^1$ is thought of as $\mathbb{A}^1 \cup \{\infty\}$. Let $\vec{p} = (p_1, \ldots, p_s)$ be an s-tuple of distinct points in $\mathbb{A}^1$ and let $\vec{\lambda} = (\lambda_1, \ldots, \lambda_s)$ be an s-tuple of weights with each $\lambda_i \in D_c$ (we take $s \geq 1$).

Definition 2.3.1 Let θ be the highest root of $\mathfrak{g}$ and let $x_\theta \in \mathfrak{g}_\theta$ be a nonzero (highest) root vector. Define the operator $\varphi = \varphi_{(\vec{p},\vec{\lambda})} : V(\vec{\lambda}) \to V(\vec{\lambda})$ by

$$\varphi(v_1 \otimes \cdots \otimes v_s) = \sum_{i=1}^{s} p_i v_1 \otimes \cdots \otimes x_\theta \cdot v_i \otimes \cdots \otimes v_s, \quad \text{for} \quad v_i \in V(\lambda_i),$$

where $V(\vec{\lambda})$ is defined by identity (1) of Definition 2.2.1.

Let $sl_2(\theta) \subset \mathfrak{g}$ be the subalgebra spanned by $\{\mathfrak{g}_\theta, \mathfrak{g}_{-\theta}, \theta^\vee\}$. Then $sl_2(\theta)$ is isomorphic with sl_2 under the map $X \mapsto x_\theta$, $Y \mapsto x_{-\theta}$, $H \mapsto \theta^\vee$, where $x_{-\theta} \in \mathfrak{g}_{-\theta}$ is chosen so that $[x_\theta, x_{-\theta}] = \theta^\vee$.

Decompose $V(\lambda_i)$ under the action of $sl_2(\theta)$:

$$V(\lambda_i) = \bigoplus_{n \geq 0} V(\lambda_i)_{(n)},$$

where $V(\lambda_i)_{(n)}$ denotes the isotypic component of $V(\lambda_i)$ (under the action of $sl_2(\theta)$) with highest weight the non-negative integer n. (The irreducible representation of sl_2 with highest weight n is of dimension $n+1$.)

Theorem 2.3.2 *With the notation and assumptions as above:*

(a) $\mathscr{V}_{\mathbb{P}^1}(\vec{p}, \vec{\lambda}) \simeq V(\vec{\lambda})/(\mathfrak{g} \cdot V(\vec{\lambda}) + Im\, \varphi^{c+1})$.
In particular, for $c >> 0$ (for fixed $\vec{\lambda}$), $\mathscr{V}_{\mathbb{P}^1}(\vec{p}, \vec{\lambda})$ coincides with the space of covariants $V(\vec{\lambda})/(\mathfrak{g} \cdot V(\vec{\lambda}))$.

Similarly:

(b) $\mathscr{V}^\dagger_{\mathbb{P}^1}(\vec{p}, \vec{\lambda}) \simeq \{\mathfrak{g}$*-module maps* $f : V(\vec{\lambda}) \to \mathbb{C} : f \circ \varphi^{c+1} = 0\}$.

Proof Let $q = \infty \in \mathbb{P}^1$. Applying Corollary 2.2.3(b) for $\Sigma = \mathbb{P}^1$, we get

$$\mathscr{V}_{\mathbb{P}^1}(\vec{p}, \vec{\lambda}) \simeq \left[\mathscr{H}(0) \otimes V(\vec{\lambda})\right]_{\mathfrak{g}[\mathbb{A}^1]}. \tag{1}$$

Let $K(0)$ be the kernel of the canonical projection $\hat{M}(V(0), c) \to \mathscr{H}(0)$. Then we have the exact sequence

$$[K(0) \otimes V(\vec{\lambda})]_{\mathfrak{g}(\mathbb{A}^1)} \xrightarrow{i} \left[\hat{M}(V(0), c) \otimes V(\vec{\lambda})\right]_{\mathfrak{g}(\mathbb{A}^1)} \\ \to \left[\mathscr{H}(0) \otimes V(\vec{\lambda})\right]_{\mathfrak{g}(\mathbb{A}^1)} \to 0, \tag{*}$$

where the generalized Verma module $\hat{M}(V(0), c)$ is considered as a $\mathfrak{g}(\mathbb{A}^1)$-module with respect to the parameter $t = z^{-1}$ on $\mathbb{P}^1$ at ∞. Since $\hat{M}(V(0), c)$ is a free $U(\hat{\mathfrak{g}}_-)$-module generated by v_+ and $\hat{\mathfrak{g}}_-$ is an ideal in $\mathfrak{g} \otimes \mathbb{C}[t^{-1}]$, it is easy to see that (setting $\mathfrak{g}_0(\mathbb{A}^1)$ to be the kernel of the evaluation map $\mathfrak{g}(\mathbb{A}^1) \to \mathfrak{g}$ at 0, thus $\mathfrak{g}_0(\mathbb{A}^1) = \hat{\mathfrak{g}}_-$)

$$\left[\hat{M}(V(0), c) \otimes V(\vec{\lambda})\right]_{\mathfrak{g}(\mathbb{A}^1)} \simeq \left[\left[\hat{M}(V(0), c) \otimes V(\vec{\lambda})\right]_{\mathfrak{g}_0(\mathbb{A}^1)}\right]_{\mathfrak{g}}$$

$$\simeq [v_+ \otimes V(\vec{\lambda})]_{\mathfrak{g}},$$

by (Kumar, 2002, Proposition 3.1.10)

$$\simeq v_+ \otimes \frac{V(\vec{\lambda})}{\mathfrak{g} \cdot V(\vec{\lambda})}. \tag{2}$$

Moreover, by the definition of $\mathscr{H}(0)$ (cf. Definition 1.2.6) and Exercise 1.2.E.3, we get

$$K(0) = U\left(\mathfrak{g} \otimes \mathbb{C}[t^{-1}]\right) \cdot (x_\theta[t^{-1}])^{c+1} \cdot v_+ \subset \hat{M}(V(0), c).$$

Thus, the image of $[K(0) \otimes V(\vec{\lambda})]_{\mathfrak{g}(\mathbb{A}^1)}$ in $[\hat{M}(V(0), c) \otimes V(\vec{\lambda})]_{\mathfrak{g}(\mathbb{A}^1)}$ under i is the same as the image of $(x_\theta[t^{-1}])^{c+1} \cdot v_+ \otimes V(\vec{\lambda})$. The latter, under identification (2), is clearly equal to

$$\begin{aligned}
&(-1)^{c+1} v_+ \otimes \frac{\left((x_\theta[t^{-1}])^{c+1} \cdot V(\vec{\lambda}) + \mathfrak{g} \cdot V(\vec{\lambda})\right)}{\mathfrak{g} \cdot V(\vec{\lambda})} \\
&= v_+ \otimes \frac{\left(\varphi^{c+1}(V(\vec{\lambda})) + \mathfrak{g} \cdot V(\vec{\lambda})\right)}{\mathfrak{g} \cdot V(\vec{\lambda})}, \quad \text{by the definition of } \varphi.
\end{aligned}$$

Thus, from the exact sequence $(*)$ and (2), we get that

$$[\mathscr{H}(0) \otimes V(\vec{\lambda})]_{\mathfrak{g}(\mathbb{A}^1)} \simeq \frac{V(\vec{\lambda})}{\left(\mathfrak{g} \cdot V(\vec{\lambda}) + \varphi^{c+1}(V(\vec{\lambda}))\right)}.$$

This proves the (a)-part of the theorem, by using (1). Since $\mathscr{V}^\dagger_{\mathbb{P}^1}(\vec{p}, \vec{\lambda})$ is dual to $\mathscr{V}_{\mathbb{P}^1}(\vec{p}, \vec{\lambda})$ (cf. isomorphism (9) of Definition 2.1.1), the (b)-part follows immediately from the (a)-part. □

We now determine the space of vacua for $\Sigma = \mathbb{P}^1$ with three marked points (i.e., $s = 3$) and $\mathfrak{g} = sl_2$ for any level c. Denote the irreducible sl_2-representation of dimension $n + 1$ by $W(n)$, i.e., $W(n) = S^n(\mathbb{C}^2)$, under the standard action of sl_2 on $\mathbb{C}^2$.

Proposition 2.3.3 *(a) For any $\vec{\lambda} = (n_1, n_2, n_3) \in \mathbb{Z}^3_+$, the space $W(\vec{\lambda})/ sl_2 \cdot W(\vec{\lambda})$ is at most 1-dimensional, where $W(\vec{\lambda}) := W(n_1) \otimes W(n_2) \otimes W(n_3)$, and $\mathbb{Z}_+ := \mathbb{Z}_{\geq 0}$.*

Moreover, it is 1-dimensional if and only if

$$n_1 + n_2 + n_3 \in 2\mathbb{Z}_+ \tag{1}$$

and the sum of any two of $\{n_1, n_2, n_3\}$ is at least as much as the third.

(b) Let $\mathfrak{g} = sl_2$, $\vec{\lambda} = (n_1, n_2, n_3) \in [c]^3$ and $\vec{p} = (p_1, p_2, p_3)$ with distinct $p_i \in \mathbb{A}^1$, where $[c] := \{0, 1, \ldots, c\}$. Then the space of covacua $\mathscr{V}_{\mathbb{P}^1}(\vec{p}, \vec{\lambda})$ (equivalently, the space of vacua) is at most 1-dimensional. Moreover, it is 1-dimensional if and only if the above condition (1) is satisfied together with the following condition:

$$n_1 + n_2 + n_3 \leq 2c. \tag{2}$$

Proof (a) Of course, it follows from the Clebsch–Gordan formula for the tensor product decomposition of sl_2-representations (Bröcker and tom Dieck, 1985, Proposition 5.5).

(b) We prove the (b)-part for the space of vacua $V^{\dagger}_{\mathbb{P}^1}(\vec{p},\vec{\lambda})$. Since $V^{\dagger}_{\mathbb{P}^1}(\vec{p},\vec{\lambda})$ is a subspace of $\mathrm{Hom}_{sl_2}(W(\vec{\lambda}),\mathbb{C})$ by Theorem 2.3.2(b), it is at most 1-dimensional by the (a)-part.

Further, by the (a)-part again, if $\vec{\lambda}$ does not satisfy condition (1), then $W(\vec{\lambda}) = sl_2 \cdot W(\vec{\lambda})$ and hence $\mathscr{V}^{\dagger}_{\mathbb{P}^1}(\vec{p},\vec{\lambda}) = 0$. So, assume now that $\vec{\lambda} \in [c]^3$ satisfies condition (1). Thus, $\mathrm{Hom}_{sl_2}(W(\vec{\lambda}),\mathbb{C})$ is 1-dimensional.

Let $\{e_1,e_2\}$ be the standard basis of $\mathbb{C}^2$. Then $\mathrm{Hom}_{\mathbb{C}}(W(\vec{\lambda}),\mathbb{C})$ can be identified with the space of polynomials P in the variables $\{x_i\}_{1\le i\le 3}$ which are (not necessarily homogeneous) of degree $\le n_i$ in the variable x_i for each $i = 1,2,3$. Under this correspondence $f \in \mathrm{Hom}_{\mathbb{C}}(W(\vec{\lambda}),\mathbb{C}) \mapsto P$, where

$$P(x_1,x_2,x_3) = f\left((e_1 + x_1e_2)^{\otimes n_1} \otimes (e_1 + x_2e_2)^{\otimes n_2} \otimes (e_1 + x_3e_3)^{\otimes n_3}\right).$$

Now, letting $n = (n_1 + n_2 + n_3)/2$, the 1-dimensional space $\mathrm{Hom}_{sl_2}(W(\vec{\lambda}),\mathbb{C})$ corresponds to the polynomial

$$P_o(x_1,x_2,x_3) = (x_2 - x_3)^{n-n_1}(x_3 - x_1)^{n-n_2}(x_1 - x_2)^{n-n_3}$$

(up to a nonzero scalar multiple). Moreover, as in Exercise 2.3.E.1, for any $m \in \mathbb{Z}_+$, $(P_o \circ \varphi^m)(x_1,x_2,x_3)$ corresponds to the coefficient of $\epsilon^m/m!$ in the expansion of

$$\begin{aligned} P_o(x_1 + p_1\epsilon, x_2 + p_2\epsilon, x_3 + p_3\epsilon) &= (x_2 - x_3 + \epsilon(p_2 - p_3))^{n-n_1} \\ &\times (x_3 - x_1 + \epsilon(p_3 - p_1))^{n-n_2}(x_1 - x_2 + \epsilon(p_1 - p_2))^{n-n_3}. \end{aligned} \tag{3}$$

But P_o is a polynomial of degree equal to $3n - (n_1 + n_2 + n_3) = n$. Thus, $P_o \circ \varphi^{c+1} = 0$ if $n < c+1$, i.e., $n_1 + n_2 + n_3 \le 2c$. Conversely, if $n \ge c+1$, it can be seen from (3) that $P_o \circ \varphi^{c+1} \ne 0$, since the points p_i's are distinct.

Thus, the (b)-part follows from Theorem 2.3.2(b). □

Remark 2.3.4 In Proposition 2.3.3(b), we assume that $p_i \in \mathbb{A}^1$. In fact, the same result is true with any $\vec{p}$ in $\mathbb{P}^1$ consisting of distinct points, as we can use an automorphism of $\mathbb{P}^1$ to move ∞ to another point.

As a corollary of Theorem 2.3.2 and Proposition 2.3.3, we get the following.

Corollary 2.3.5 *Let the assumptions be as at the beginning of this section and the notation as in Definition 2.3.1, and take $s = 3$.*

(a) The space $\mathscr{V}_{\mathbb{P}^1}(\vec{p},\vec{\lambda})$ *is naturally isomorphic to the quotient of* $[V(\vec{\lambda})]_{\mathfrak{g}}$ *by the image of the subspace*

$$V(\vec{\lambda})_F := \bigoplus_{n_1+n_2+n_3>2c} V(\vec{\lambda})_{(n_1,n_2,n_3)}$$

under the canonical projection $V(\vec{\lambda}) \to [V(\vec{\lambda})]_{\mathfrak{g}}$, *where* $V(\vec{\lambda})_{(n_1,n_2,n_3)} := V(\lambda_1)_{(n_1)} \otimes V(\lambda_2)_{(n_2)} \otimes V(\lambda_3)_{(n_3)}$.

Similarly:
(b) $\mathscr{V}^{\dagger}_{\mathbb{P}^1}(\vec{p},\vec{\lambda}) \simeq \{\mathfrak{g}\text{-module maps } f\colon V(\vec{\lambda}) \to \mathbb{C} : f \text{ vanishes on } V(\vec{\lambda})_F\}$.

Proof (a) By virtue of Theorem 2.3.2(a), it suffices to show that

$$\varphi^{c+1}(V(\vec{\lambda})) + \mathfrak{g}\cdot V(\vec{\lambda}) = V(\vec{\lambda})_F + \mathfrak{g}\cdot V(\vec{\lambda}). \tag{1}$$

Decompose $V(\vec{\lambda})$ under $sl_2(\theta)$:

$$V(\vec{\lambda}) = \left(\bigoplus_{n_1+n_2+n_3>2c} V(\vec{\lambda})_{(n_1,n_2,n_3)}\right) \oplus \left(\bigoplus_{n_1+n_2+n_3\le 2c} V(\vec{\lambda})_{(n_1,n_2,n_3)}\right). \tag{2}$$

The operator φ clearly keeps any $V(\vec{\lambda})_{(n_1,n_2,n_3)}$ stable.

Further, if $n_1+n_2+n_3 > 2c$, by Theorem 2.3.2(a) and Proposition 2.3.3(b),

$$\varphi^{c+1}\left(V(\vec{\lambda})_{(n_1,n_2,n_3)}\right) + sl_2(\theta)\cdot V(\vec{\lambda})_{(n_1,n_2,n_3)} = V(\vec{\lambda})_{(n_1,n_2,n_3)}. \tag{3}$$

If $n_1+n_2+n_3 \le 2c$, by Proposition 2.3.3 and Theorem 2.3.2(a),

$$\varphi^{c+1}\left(V(\vec{\lambda})_{(n_1,n_2,n_3)}\right) \subset sl_2(\theta)\cdot V(\vec{\lambda})_{(n_1,n_2,n_3)}. \tag{4}$$

Combining (2)–(4), we get

$$V(\vec{\lambda})_F + \mathfrak{g}\cdot V(\vec{\lambda}) = \operatorname{Im}(\varphi^{c+1}) + \mathfrak{g}\cdot V(\vec{\lambda}).$$

This proves (1) and hence the (a)-part of the corollary is proved.

The (b)-part clearly follows from the (a)-part and identity (9) of Definition 2.1.1. □

2.3.E Exercises

(1) Following the notation as in the proof of Proposition 2.3.3, show that $(P \circ \varphi^m)(x_1,x_2,x_3)$ corresponds to the coefficient of $\epsilon^m/m!$ in the expansion of $P(x_1+p_1\epsilon, x_2+p_2\epsilon, x_3+p_3\epsilon)$, for any $m \ge 0$ and any

polynomial P in the variables $\{x_i\}_{1\leq i\leq 3}$ which are of degree $\leq n_i$ in the variables x_i for each $i = 1,2,3$.

Hint: Show that the action of $x_\theta \otimes 1 \otimes 1$ on P is the derivation with respect to x_1 and similarly for $1 \otimes x_\theta \otimes 1$ and $1 \otimes 1 \otimes x_\theta$.

(2) Let $\mathfrak{g}$ be any simple Lie algebra and let $c > 0$ be any level.

(a) Show that for any point $p \in \mathbb{A}^1 \subset \mathbb{P}^1$ and any $\lambda \in D_c$,

$$\begin{aligned}\mathscr{V}_{\mathbb{P}^1}(p,\lambda) &= 0, \text{ if } \lambda \neq 0 \\ &\simeq \mathbb{C}, \text{ if } \lambda = 0.\end{aligned}$$

(b) For any distinct points $\vec{p} = (p_1, p_2) \subset \mathbb{A}^1$ and any $\lambda, \mu \in D_c$,

$$\begin{aligned}\mathscr{V}_{\mathbb{P}^1}(\vec{p},(\lambda,\mu)) &= 0, \text{ if } \lambda \neq \mu^* \\ &\simeq \mathbb{C}, \text{ if } \lambda = \mu^*,\end{aligned}$$

where μ^* is the highest weight of the dual representation $V(\mu)^*$. In fact, the result is true for any $p \in \mathbb{P}^1$ in (a) and any $\vec{p} \subset \mathbb{P}^1$ in (b).

Hint: Use, e.g., Theorem 2.3.2.

2.C Comments

Most of the results in this chapter are due to Tsuchiya, Ueno and Yamada (1989) (see also Ueno (2008, Chap. 3 and §6.1) for a more detailed account of their work). However, our treatment in this chapter follows more closely Beauville (1996). Theorem 2.3.2 is originally due to Tsuchiya and Kanie (1988) in the language of vertex operators.

There is another proof of the finite-dimensionality of the space of vacua (Lemma 2.1.4) due to Suzuki (1990). Exercise 2.3.E.2 is taken from Ueno (2008, Corollary 6.2).

Belkale, Gibney and Mukhopadhyay (2016) have determined a bound on c such that the space $\mathscr{V}_{\mathbb{P}^1}(\vec{p},\vec{\lambda})$ coincides with the space of covariants (cf. Theorem 2.3.2).

3

Factorization Theorem for Space of Vacua

Let $\mathfrak{g}$ be a simple Lie algebra and let $(\Sigma, \vec{p})$ be an s-pointed curve (for any $s \geq 1$), where $\vec{p} = (p_1, \ldots, p_s)$. We fix a level $c \geq 1$ and let $\vec{\lambda} = (\lambda_1, \ldots, \lambda_s)$ be an s-tuple of weights with $\lambda_i \in D_c$ attached to the point p_i, where D_c is the set of dominant integral weights of $\mathfrak{g}$ of level $\leq c$. As in Section 2.1, there is associated the space of covacua $\mathscr{V}_\Sigma(\vec{p}, \vec{\lambda})$. Let $q_o \in \Sigma$ be a node and let $\tilde{\Sigma}$ be the curve obtained from Σ by the normalization $\pi : \tilde{\Sigma} \to \Sigma$ at only the point q_o. Thus, $\pi^{-1} q_o$ consists of two points say q_o', q_o'', which are smooth points in $\tilde{\Sigma}$ and, moreover, $\tilde{\Sigma} \backslash \{q_o', q_o''\} \to \Sigma \backslash \{q_o\}$ is an isomorphism. We fix formal local parameters z' and z'' at q_o' and q_o'', respectively.

In Section 3.1, we prove the *Factorization Theorem*, which asserts that there is an explicit 'canonical' isomorphism between the spaces of covacua for Σ and $\tilde{\Sigma}$:

$$F: \mathscr{V}_\Sigma(\vec{p}, \vec{\lambda}) \to \bigoplus_{\mu \in D_c} \mathscr{V}_{\tilde{\Sigma}}\left((\vec{p}, q_o', q_o''), \left(\vec{\lambda}, \mu^*, \mu \right) \right).$$

This section is devoted to its proof.

In Section 3.2, we recall the definition of *Sugawara elements* $\{L_n\}_{n \in \mathbb{Z}}$, which are elements in a certain completion of the enveloping algebra of the affine Lie algebra $\tilde{\mathfrak{g}}$ (cf. Definition 3.2.1). The commutation of these elements 'essentially' gives the commutation of the standard basis elements of the Virasoro algebra Vir (for a more precise statement, see Proposition 3.2.2(b)). We further show (cf. Lemma 3.2.4) that the action of the Sugawara elements L_n on a highest-weight $\tilde{\mathfrak{g}}$-module V (more generally, any smooth $\tilde{\mathfrak{g}}$-module) gives rise to the structure of a Vir-module on V, if the central element $C \in \tilde{\mathfrak{g}}$ acts via a scalar $c \neq -h^\vee$ on V, where $h^\vee$ is the dual Coxeter number of $\mathfrak{g}$. Moreover, if V is a unitary $\tilde{\mathfrak{g}}$-module, then it is also a unitary Vir-module under the same positive-definite Hermitian form on V. Let $\mathfrak{s} \hookrightarrow \mathfrak{g}$ be an embedding of simple Lie algebras and let V be a smooth $\tilde{\mathfrak{g}}$-module. Then we extend the definition of the action of the Virasoro algebra on V in this setting of the pair $(\mathfrak{g}, \mathfrak{s})$ (cf. Exercises 3.2.E.2 and 3.2.E.3).

In Section 3.3, we sheafify the definition of conformal blocks. Let $\mathfrak{F}_T = \left(\Sigma_T, \vec{p}, \vec{t}\right)$ be a *family of s-pointed curves with formal parameters* parameterized by an irreducible smooth variety T (cf. Definition 3.3.1). Associated to this family, one defines the sheaf of covacua $\mathscr{V}_{\mathfrak{F}_T}(\vec{p}, \vec{\lambda})$ and the sheaf of vacua $\mathscr{V}^{\dagger}_{\mathfrak{F}_T}(\vec{p}, \vec{\lambda})$ over T (cf. Definitions 3.3.2 and 3.3.5). It is shown that these sheaves are coherent $\mathscr{O}_T$-modules (cf. Proposition 3.3.4 and Exercise 3.3.E.1). Further, the fiber of $\mathscr{V}_{\mathfrak{F}_T}(\vec{p}, \vec{\lambda})$ at any $b \in T$ coincides with the space of covacua for the curve over b (cf. Lemma 3.3.3). We also sheafify the Virasoro algebra and its action on the integrable highest-weight modules (cf. Definition 3.3.6).

In Section 3.4, the above Virasoro action is crucially used to define a functorial flat projective connection on the sheaf of covacua $\mathscr{V}_{\mathfrak{F}_T}(\vec{p}, \vec{\lambda})$ associated to the family $\mathfrak{F}_T$ if T parameterizes a family of smooth curves (cf. Theorem 3.4.2). In the process we prove that the sheaf $\mathscr{V}_{\mathfrak{F}_T}(\vec{p}, \vec{\lambda})$ of $\mathscr{O}_T$-modules is locally free. We explicitly determine this connection for a specific (and important) family of $\mathbb{P}^1$s, which turns out to be the Knizhnik–Zamolodchikov connection (cf. Example 3.4.3 and Remark 3.4.4).

In Section 3.5, the local freeness of the sheaf $\mathscr{V}_{\mathfrak{F}_T}(\vec{p}, \vec{\lambda})$ of $\mathscr{O}_T$-modules (proved in Theorem 3.4.2), when T parameterizes a family of smooth curves, is extended to a more general family. Let $(\Sigma_o, \vec{p}_o)$ be an s-pointed curve with a node at q (and possibly other nodes). Let $\tilde{\Sigma}_o$ be the normalization of Σ_o at the point q. There exists a 'canonical' smoothing deformation of $(\Sigma_o, \vec{p}_o)$ over the formal disc $\mathbb{D}_\tau := Spec\, \mathbb{C}[[\tau]]$ with special fiber $\tilde{\Sigma}_o$ (cf. Lemma 3.5.1). We prove the Factorization Theorem over the smoothing deformation in Theorem 3.5.6 by making crucial use of a 'gluing' tensor element (cf. Definition 3.5.3). For any genus $g \geq 0$ and $s \geq 1$, let $\bar{\mathcal{M}}_{g,s}$ be the stack whose objects over any scheme $T \in \mathfrak{S}$ are families of *stable* s-pointed curves of genus g over T. Then, by a classical result of Deligne–Mumford extended by Knudsen, $\bar{\mathcal{M}}_{g,s}$ is a proper and smooth Deligne–Mumford stack of finite type such that the substack $\bar{\mathcal{S}}_{g,s}$ consisting of singular curves is a divisor with normal crossings (cf. Theorem 3.5.7). Using the stack $\bar{\mathcal{M}}_{g,s}$, we prove the following (cf. Proposition 3.5.8).

Proposition Let $(\Sigma, \vec{p})$ be a smooth connected s-pointed projective curve $(s \geq 1)$ of genus $g \geq 0$ such that $2g - 2 + s > 0$. Then, for any $\vec{\lambda} \in D_c^s$, $\dim \mathscr{V}_\Sigma(\vec{p}, \vec{\lambda})$ only depends on g and $\vec{\lambda}$. Denote this dimension by $F_g(\vec{\lambda})$.

Being a Deligne–Mumford smooth stack of finite type, $\bar{\mathcal{M}}_{g,s}$ has an atlas $\phi \colon X \to \bar{\mathcal{M}}_{g,s}$, where X is a smooth scheme of finite type over $\mathbb{C}$.

Let $(\Sigma_X, \vec{p}_X)$ be the family of s-pointed curves over X associated to $\phi: X \to \bar{\mathcal{M}}_{g,s}$. By taking an open cover of X (if needed), we can assume that $(\Sigma_X, \vec{p}_X)$ is a family $\mathcal{F}_X = (\Sigma_X, \vec{p}_X, \vec{t}_X)$ of s-pointed curves with formal parameters. Then, we have the following extension of part of Theorem 3.4.2 (cf. Theorem 3.5.9).

Theorem For any genus $g \geq 0$ and $s \geq 1$ such that $2g - 2 + s > 0$ and any set of dominant weights $\vec{\lambda} = (\lambda_1, \ldots, \lambda_s)$ with $\lambda_i \in D_c$, the sheaf of conformal blocks $\mathcal{V}_{\mathcal{F}_X}(\vec{p}_X, \vec{\lambda})$ is locally free over X of the same rank over each of the components of X.

As an application of the local freeness of $\mathcal{V}_{\mathcal{F}_X}(\vec{p}_X, \vec{\lambda})$ and the Factorization Theorem, we deduce the following corollary (cf. Corollary 3.5.10).

Corollary (a) For any genus $g \geq 1$, $s \geq 1$ and any $\vec{\lambda} \in D_c^s$,

$$F_g(\vec{\lambda}) = \sum_{\mu \in D_c} F_{g-1}(\vec{\lambda}, \mu^*, \mu).$$

(b) For any genus $g = g' + g''$ (where $g', g'' \geq 0$) and any $\vec{\lambda}' \in D_c^{s'}$, $\vec{\lambda}'' \in D_c^{s''}$ (with $s', s'' \geq 1$),

$$F_g(\vec{\lambda}', \vec{\lambda}'') = \sum_{\mu \in D_c} F_{g'}(\vec{\lambda}', \mu^*) F_{g''}(\vec{\lambda}'', \mu).$$

As a consequence of the above corollary, we deduce that $F_g(0) \geq (\sharp D_c)^g$ (cf. Corollary 3.5.11).

3.1 Factorization Theorem

We continue to follow the notation from Chapter 2. In particular, $(\Sigma, \vec{p})$ is an s-pointed curve, $\mathfrak{g}$ is a simple Lie algebra and we fix a level $c \geq 1$. We do *not* assume that Σ is irreducible.

Definition 3.1.1 For any highest-weight $\hat{\mathfrak{g}}$-module $\mathcal{V}$ (cf. Definition 1.2.4), define the $\hat{\mathfrak{g}}$-module $D(\mathcal{V})$ as the restricted dual $\mathcal{V}^\vee$ of $\mathcal{V}$ (where by the *restricted dual* we mean the direct *sum* of the homogeneous components of $\mathcal{V}$ as in (1) of the proof of Theorem 1.2.10) with the following twisted action of $\hat{\mathfrak{g}}$:

$$x[n] \odot f = x[-n] \cdot f, \quad \text{for } x \in \mathfrak{g}, n \in \mathbb{Z} \text{ and } f \in \mathcal{V}^\vee,$$

and

$$C \odot f = -C \cdot f,$$

where $x[n] := x[t^n]$. It is easy to see that if $\mathcal{V}$ is an integrable highest-weight $\hat{\mathfrak{g}}$-module of the form $\mathscr{H}(\lambda)$ with central charge c, then $D(\mathcal{V})$ is of the form $\mathscr{H}(\lambda^*)$ with the same central charge c, where λ^* is the highest weight of the dual $\mathfrak{g}$-module $V(\lambda)^*$.

Let $q_o \in \Sigma$ be a node, and let $\tilde{\Sigma}$ be the curve obtained from Σ by the normalization $\pi : \tilde{\Sigma} \to \Sigma$ at only the point q_o. Thus, $\pi^{-1}q_o$ consists of two points say q'_o, q''_o, which are smooth points in $\tilde{\Sigma}$ and

$$\pi_{|\tilde{\Sigma}\backslash\{q'_o,q''_o\}} : \tilde{\Sigma}\backslash\{q'_o,q''_o\} \to \Sigma\backslash\{q_o\}$$

is an isomorphism. We fix formal local parameters z' and z'' at q'_o and q''_o, respectively.

The map π on restriction gives rise to a map

$$\pi_{|\tilde{\Sigma}\backslash\{\vec{p},q'_o,q''_o\}} : \tilde{\Sigma}\backslash\{\vec{p},q'_o,q''_o\} \to \Sigma\backslash\{\vec{p}\},$$

which, in turn, gives rise to a Lie algebra homomorphism

$$\pi^* : \mathfrak{g}[\Sigma\backslash\vec{p}] \to \mathfrak{g}[\tilde{\Sigma}\backslash\{\vec{p},q'_o,q''_o\}],$$

where the Lie algebra bracket in $\mathfrak{g}[\Sigma\backslash\vec{p}]$ is given by (2) of Definition 2.1.1 and similarly for $\mathfrak{g}[\tilde{\Sigma}\backslash\{\vec{p},q'_o,q''_o\}]$.

Define the linear map

$$\hat{F} : \mathscr{H}(\vec{\lambda}) \to \mathscr{H}(\vec{\lambda}) \otimes \left(\bigoplus_{\mu\in D_c} D(\mathscr{H}(\mu)) \otimes \mathscr{H}(\mu) \right),$$

$$h \mapsto h \otimes \sum_{\mu\in D_c} I_\mu, \quad \text{for } h \in \mathscr{H}(\vec{\lambda}),$$

where I_μ is the identity map thought of as an element of $V(\mu)^* \otimes V(\mu) \simeq \mathrm{End}_{\mathbb{C}}(V(\mu))$, and $V(\mu)$ (resp. $V(\mu)^*$) sits inside $\mathscr{H}(\mu)$ (resp. $D(\mathscr{H}(\mu))$) as the zeroth homogeneous component.

Realize $\mathscr{H}(\vec{\lambda}) \otimes D(\mathscr{H}(\mu)) \otimes \mathscr{H}(\mu)$ as a $\mathfrak{g}[\tilde{\Sigma}\backslash\{\vec{p},q'_o,q''_o\}]$-module via the Laurent series expansion at the points $\vec{p}, q'_o, q''_o$, respectively. Then, I_μ being a $\mathfrak{g}$-invariant and since $(\pi^* f)(q'_o) = (\pi^* f)(q''_o)$ for any $f \in \mathbb{C}[\Sigma\backslash\vec{p}]$, $\hat{F}$ is a $\mathfrak{g}[\Sigma\backslash\vec{p}]$-module map, where we realize the range as a $\mathfrak{g}[\Sigma\backslash\vec{p}]$-module via the Lie algebra homomorphism π^*. Hence, $\hat{F}$ induces a linear map

$$F : \mathscr{V}_\Sigma(\vec{p},\vec{\lambda}) \to \bigoplus_{\mu\in D_c} \mathscr{V}_{\tilde{\Sigma}}\left((\vec{p},q'_o,q''_o), \left(\vec{\lambda},\mu^*,\mu\right) \right).$$

Theorem 3.1.2 *With the notation and assumptions as above, the map*

$$F : \mathscr{V}_\Sigma(\vec{p},\vec{\lambda}) \to \bigoplus_{\mu \in D_c} \mathscr{V}_{\tilde{\Sigma}}\left((\vec{p},q_o',q_o''),\left(\vec{\lambda},\mu^*,\mu\right)\right)$$

is an isomorphism.

Dualizing the map F, we get an isomorphism

$$F^* : \bigoplus_{\mu \in D_c} \mathscr{V}^\dagger_{\tilde{\Sigma}}\left((\vec{p},q_o',q_o''),\left(\vec{\lambda},\mu^*,\mu\right)\right) \xrightarrow{\sim} \mathscr{V}^\dagger_\Sigma(\vec{p},\vec{\lambda}).$$

Proof We first prove the surjectivity of F. For $n \geq 1$, pick $g_n \in \mathbb{C}_{q_o''}[\tilde{\Sigma}\backslash\{\vec{p},q_o'\}]$ such that

$$(g_n)_{q_o'} - (z')^{-n} \equiv 0 \mod z',$$

where

$$\mathbb{C}_{q_o''}[\tilde{\Sigma}\backslash\{\vec{p},q_o'\}] := \{f \in \mathbb{C}[\tilde{\Sigma}\backslash\{\vec{p},q_o'\}] : f(q_o'') = 0\}.$$

Then, for any $x \in \mathfrak{g}, h \in \mathscr{H}(\vec{\lambda}), v_1 \in V(\mu)^*, v_2 \in V(\mu)$,

$$x[g_n]\cdot(h \otimes v_1 \otimes v_2) = (x[g_n]\cdot h) \otimes v_1 \otimes v_2 + h \otimes (x[-n]\cdot v_1) \otimes v_2.$$

Thus, for any $x \in \mathfrak{g}, n \in \mathbb{N} := \{1,2,3,\dots\}$ and $h \in \mathscr{H}(\vec{\lambda})$, as elements of $Q_\mu := [\mathscr{H}(\vec{\lambda},\mu^*,\mu)]_{\mathfrak{g}[\tilde{\Sigma}\backslash\{\vec{p},q_o',q_o''\}]}$,

$$(x[g_n]\cdot h) \otimes v_1 \otimes v_2 = -h \otimes (x[-n]\cdot v_1) \otimes v_2. \tag{1}$$

Similarly, taking a function $f_n \in \mathbb{C}[\tilde{\Sigma}\backslash\{\vec{p},q_o',q_o''\}]$ such that

$$(f_n)_{q_o''} - (z'')^{-n} \equiv 0 \mod z'',$$

we see that, for any $x \in \mathfrak{g}, n \in \mathbb{N}, h \in \mathscr{H}(\vec{\lambda}), h_1 \in D(\mathscr{H}(\mu))$, and $v_2 \in V(\mu)$, as elements of Q_μ,

$$h \otimes h_1 \otimes x[-n]\cdot v_2 = -(x[f_n]\cdot h) \otimes h_1 \otimes v_2 - h \otimes (x[(f_n)_{q_o'}]\cdot h_1) \otimes v_2. \tag{2}$$

Finally, take a function $f \in \mathbb{C}[\tilde{\Sigma}\backslash\vec{p}]$ such that

$$f(q_o') = 1, \quad f(q_o'') = 0. \tag{3}$$

Then, for $x \in \mathfrak{g}, h \in \mathscr{H}(\vec{\lambda}), v \in \oplus_{\mu \in D_c}\left(V(\mu)^* \otimes V(\mu)\right)$, as elements of $Q := \oplus_{\mu \in D_c} Q_\mu$,

$$-h \otimes (x \odot v) = (x[f]\cdot h) \otimes v, \tag{4}$$

where the action $\odot$ of $\mathfrak{g}$ on $V(\mu^*) \otimes V(\mu)$ is via its action on the first factor only. In particular, as elements of Q,

$$-h \otimes \beta(x) = (x[f] \cdot h) \otimes \sum_{\mu \in D_c} I_\mu, \tag{5}$$

where β is the map defined by

$$\beta \colon U(\mathfrak{g}) \to \bigoplus_{\mu \in D_c} V(\mu^*) \otimes V(\mu), \quad \beta(a) = a \odot \sum_{\mu} I_\mu. \tag{6}$$

Observe that $\operatorname{Im}(\beta)$ is $\mathfrak{g} \oplus \mathfrak{g}$-stable under the component-wise action of $\mathfrak{g} \oplus \mathfrak{g}$ on $V(\mu^*) \otimes V(\mu)$, since I_μ is $\mathfrak{g}$-invariant under the diagonal action of $\mathfrak{g}$; $V(\mu^*) \otimes V(\mu)$ is an irreducible $\mathfrak{g} \oplus \mathfrak{g}$-module with highest weight (μ^*, μ); and $\operatorname{Im}(\beta)$ has a nonzero component in each $V(\mu^*) \otimes V(\mu)$. Thus, β is surjective.

From the surjectivity of β, we get that the map F is surjective by combining (1), (2) and (5).

We next show that F is injective. Equivalently, we show that the dual map

$$F^* \colon \bigoplus_{\mu \in D_c} \mathscr{V}_{\tilde{\Sigma}}^\dagger \left((\vec{p}, q_o', q_o''), (\vec{\lambda}, \mu^*, \mu) \right) \to \mathscr{V}_{\Sigma}^\dagger (\vec{p}, \vec{\lambda})$$

is surjective.

From the definition of $\mathscr{V}_\Sigma^\dagger$ (as in identity (7) of Definition 2.1.1) and identifying the domain of F^* via Theorem 2.2.2, we think of F^* as the map

$$F^* \colon \operatorname{Hom}_{\mathfrak{g}[\tilde{\Sigma} \backslash \vec{p}]} \left(\mathscr{H}(\vec{\lambda}) \otimes \left(\oplus_{\mu \in D_c} V(\mu^*) \otimes V(\mu) \right), \mathbb{C} \right)$$
$$\to \operatorname{Hom}_{\mathfrak{g}[\Sigma \backslash \vec{p}]} (\mathscr{H}(\vec{\lambda}), \mathbb{C})$$

induced from the inclusion

$$i \colon \mathscr{H}(\vec{\lambda}) \to \mathscr{H}(\vec{\lambda}) \otimes \left(\oplus_{\mu \in D_c} V(\mu^*) \otimes V(\mu) \right),$$
$$h \mapsto h \otimes \sum_{\mu \in D_c} I_\mu, \quad \text{for} \quad h \in \mathscr{H}(\vec{\lambda}).$$

(Since I_μ is $\mathfrak{g}$-invariant under the diagonal action of $\mathfrak{g}$, it is easy to see that i is a $\mathfrak{g}[\Sigma \backslash \vec{p}]$-module map, where we identify $\mathfrak{g}[\Sigma \backslash \vec{p}]$ as a Lie subalgebra of $\mathfrak{g}[\tilde{\Sigma} \backslash \vec{p}]$ induced from the morphism $\tilde{\Sigma} \backslash \vec{p} \to \Sigma \backslash \vec{p}$ and $V(\mu^*)$, $V(\mu)$ are thought of as $\mathfrak{g}[\tilde{\Sigma} \backslash \vec{p}]$-modules via the evaluations at q_o', q_o'', respectively.)

Let $\mathbb{C}_{q_o}[\Sigma \backslash \vec{p}] \subset \mathbb{C}_{q_o''}[\tilde{\Sigma} \backslash \vec{p}] \subset \mathbb{C}[\tilde{\Sigma} \backslash \vec{p}]$ be the ideals of $\mathbb{C}[\tilde{\Sigma} \backslash \vec{p}]$:

$$\mathbb{C}_{q_o}[\Sigma \backslash \vec{p}] := \{ f \in \mathbb{C}[\Sigma \backslash \vec{p}] : f(q_o) = 0 \},$$

and

$$\mathbb{C}_{q_o''}[\tilde{\Sigma} \backslash \vec{p}] = \{ f \in \mathbb{C}[\tilde{\Sigma} \backslash \vec{p}] : f(q_o'') = 0 \}.$$

(Observe that, under the canonical inclusion $\mathbb{C}[\Sigma\backslash\vec{p}] \subset \mathbb{C}[\tilde{\Sigma}\backslash\vec{p}]$, $\mathbb{C}_{q_o}[\Sigma\backslash\vec{p}]$ is an ideal of $\mathbb{C}[\tilde{\Sigma}\backslash\vec{p}]$ consisting of those functions vanishing at both of q'_o and q''_o.) Now, define the Lie ideals of $\mathfrak{g}[\tilde{\Sigma}\backslash\vec{p}]$:

$$\mathfrak{g}_{q_o}[\Sigma\backslash\vec{p}] := \mathfrak{g} \otimes \mathbb{C}_{q_o}[\Sigma\backslash\vec{p}], \text{ and } \mathfrak{g}_{q''_o}[\tilde{\Sigma}\backslash\vec{p}] := \mathfrak{g} \otimes \mathbb{C}_{q''_o}[\tilde{\Sigma}\backslash\vec{p}]. \tag{7}$$

Define the Lie algebra homomorphism

$$\mathfrak{g} \to \mathfrak{g}_{q''_o}\left[\tilde{\Sigma}\backslash\vec{p}\right]/\mathfrak{g}_{q_o}[\Sigma\backslash\vec{p}], \quad x \mapsto x[f] + \mathfrak{g}_{q_o}[\Sigma\backslash\vec{p}],$$

where $f \in \mathbb{C}_{q''_o}[\tilde{\Sigma}\backslash\vec{p}]$ is any function satisfying (3). (Clearly, the above map is independent of the choice of f satisfying (3).) Let

$$\varphi\colon U(\mathfrak{g}) \to U\left(\mathfrak{g}_{q''_o}[\tilde{\Sigma}\backslash\vec{p}]/\mathfrak{g}_{q_o}[\Sigma\backslash\vec{p}]\right)$$

be the induced homomorphism of the enveloping algebras.

To prove the surjectivity of F^*, take $\Phi \in \mathrm{Hom}_{\mathfrak{g}[\Sigma\backslash\vec{p}]}(\mathscr{H}(\vec{\lambda}), \mathbb{C})$ and define the linear map

$$\tilde{\Phi}\colon \mathscr{H}(\vec{\lambda}) \otimes \left(\bigoplus_{\mu \in D_c} V(\mu^*) \otimes V(\mu)\right) \to \mathbb{C}$$

via

$$\tilde{\Phi}(h \otimes \beta(a)) = \Phi(\varphi(a^t) \cdot h), \text{ for } h \in \mathscr{H}(\vec{\lambda}) \text{ and } a \in U(\mathfrak{g}),$$

where $t\colon U(\mathfrak{g}) \to U(\mathfrak{g})$ is the anti-automorphism taking $x \mapsto -x$ for $x \in \mathfrak{g}$, β is the map defined by (6) and φ is defined above. (Observe that even though $\varphi(a) \cdot h$ is not well defined, $\Phi(\varphi(a) \cdot h)$ is well defined, i.e., it does not depend upon the choice of the coset representatives in $\mathfrak{g}_{q''_o}[\tilde{\Sigma}\backslash\vec{p}]/\mathfrak{g}_{q_o}[\Sigma\backslash\vec{p}]$.)

To show that $\tilde{\Phi}$ is well defined, we need to show that for any $a \in \operatorname{Ker}\beta$ and $h \in \mathscr{H}(\vec{\lambda})$,

$$\Phi(\varphi(a^t) \cdot h) = 0. \tag{8}$$

This will be proved in Lemma 3.1.4.

We next show that $\tilde{\Phi}$ is a $\mathfrak{g}[\tilde{\Sigma}\backslash\vec{p}]$-module map. Take $x \in \mathfrak{g}$ and $f \in \mathbb{C}[\tilde{\Sigma}\backslash\vec{p}]$ satisfying (3). Then, for any $h \in \mathscr{H}(\vec{\lambda})$ and $a \in U(\mathfrak{g})$,

$$\begin{aligned}\tilde{\Phi}(x[f] \cdot (h \otimes \beta(a))) &= \tilde{\Phi}\left((x[f] \cdot h) \otimes \beta(a) + h \otimes \beta(xa)\right)\\ &= \Phi\left(\varphi(a^t)x[f] \cdot h - \varphi(a^t)x[f] \cdot h\right)\\ &= 0.\end{aligned} \tag{9}$$

Next, take $x \in \mathfrak{g}$ and $g \in \mathbb{C}[\tilde{\Sigma}\backslash\vec{p}]$ such that $g(q_o') = 0$ and $g(q_o'') = 1$. Then, for any $h \in \mathscr{H}(\vec{\lambda})$ and $a \in U(\mathfrak{g})$,

$$\begin{aligned}
\tilde{\Phi}(x[g] \cdot (h \otimes \beta(a))) &= \tilde{\Phi}\left((x[g] \cdot h) \otimes \beta(a) - h \otimes \beta(ax)\right) \\
&= \Phi\left(\varphi(a^t)x[g] \cdot h + \varphi(x)\varphi(a^t) \cdot h\right) \\
&= \Phi\left(\varphi(a^t)x[g] \cdot h - x[g]\varphi(a^t) \cdot h + x[g+f]\varphi(a^t) \cdot h\right), \\
&\qquad \text{for any } \; f \in \mathbb{C}[\tilde{\Sigma}\backslash\vec{p}] \; \text{ satisfying (3)} \\
&= -\Phi\left((\operatorname{ad} x[g](\varphi(a^t)) \cdot h\right), \text{ since } x[g+f] \in \mathfrak{g}[\Sigma\backslash\vec{p}] \\
&= 0, \text{ since } \left[x[g], \mathfrak{g}_{q_o''}[\tilde{\Sigma}\backslash\vec{p}]\right] \subset \mathfrak{g}_{q_o}[\Sigma\backslash\vec{p}]. \qquad (10)
\end{aligned}$$

Combining (9) and (10) we see that $\tilde{\Phi}$ is a $\mathfrak{g}[\tilde{\Sigma}\backslash\vec{p}]$-module map.

From the definition of $\tilde{\Phi}$, it is clear that $F^*(\tilde{\Phi}) = \Phi$. This proves the surjectivity of F^* (and hence the injectivity of F) modulo the next lemma. Thus, the theorem is proved (modulo the next lemma). □

Definition 3.1.3 For any $\mu \in D$, consider the algebra homomorphism

$$\overline{\beta}_\mu : U(\mathfrak{g}) \to \operatorname{End}_{\mathbb{C}}(V(\mu)),$$

defined by

$$\overline{\beta}_\mu(a)(v) = a \cdot v, \text{ for any } a \in U(\mathfrak{g}) \text{ and } v \in V(\mu).$$

Let K_μ be the kernel of $\overline{\beta}_\mu$, which is a two-sided ideal of $U(\mathfrak{g})$ (called a primitive ideal). From the definition of β (cf. (6) of the proof of Theorem 3.1.2), it is easy to see that, under the identification of $V(\mu)^* \otimes V(\mu)$ with $\operatorname{End}_{\mathbb{C}}(V(\mu))$,

$$\beta(a)(v) = a^t \cdot v, \text{ for any } a \in U(\mathfrak{g}) \text{ and } v \in V(\mu). \qquad (1)$$

Thus,

$$\operatorname{Ker} \beta = \bigcap_{\mu \in D_c} K_\mu^t. \qquad (2)$$

It is easy to see that $\operatorname{Ker}\beta$ is stable under t. From the definition of $\overline{\beta}_\mu$, it follows immediately that for any left ideal $K \subset U(\mathfrak{g})$ such that $U(\mathfrak{g})/K$ is an integrable $\mathfrak{g}$-module and the $\mathfrak{g}$-module $U(\mathfrak{g})/K$ has isotypic components of highest weights $\{\mu_i\}_{i\in\Lambda} \subset D$, then

$$K \supset \bigcap_{i \in \Lambda} K_{\mu_i}. \qquad (3)$$

We are now ready to prove the following lemma.

Lemma 3.1.4 *With the notation as in the proof of Theorem 3.1.2 (cf. identity (8)), for any* $a \in Ker\,\beta$, $\Phi \in \mathrm{Hom}_{\mathfrak{g}[\Sigma\backslash\vec{p}]}(\mathscr{H}(\vec{\lambda}), \mathbb{C})$, *and* $h \in \mathscr{H}(\vec{\lambda})$,

$$\Phi(\varphi(a^t) \cdot h) = 0. \tag{1}$$

Proof Let $\hat{\mathfrak{g}}_o = \hat{\mathfrak{g}}_o[\tilde{\Sigma}\backslash\{\vec{p} \cup q'_o\}]$ be the Lie algebra

$$\hat{\mathfrak{g}}_o := \mathfrak{g} \otimes \mathbb{C}_{q''_o}[\tilde{\Sigma}\backslash\{\vec{p} \cup q'_o\}] \oplus \mathbb{C}\,C,$$

where $\mathbb{C}_{q''_o}[\tilde{\Sigma}\backslash\{\vec{p} \cup q'_o\}] \subset \mathbb{C}[\tilde{\Sigma}\backslash\{\vec{p} \cup q'_o\}]$ is the ideal consisting of functions vanishing at q''_o, with the Lie bracket in $\hat{\mathfrak{g}}_o$ defined by

$$[x[f], y[g]] = [x, y][fg] + \operatorname*{Res}_{q'_o}((df)g)\,\langle x, y\rangle C, \text{ for } f, g \in \mathbb{C}_{q''_o}[\tilde{\Sigma}\backslash\{\vec{p} \cup q'_o\}],$$
$$\text{and } x, y \in \mathfrak{g}; \tag{2}$$

and C is central in $\hat{\mathfrak{g}}_o$.

Let $t = z'$ be a local parameter on $\tilde{\Sigma}$ at the point q'_o. Then, there is a Lie algebra embedding

$$\hat{\mathfrak{g}}_o \hookrightarrow \hat{\mathfrak{g}},\ x[f] \mapsto x[f_{q'_o}] \quad \text{and} \quad C \mapsto C. \tag{3}$$

Let $\mathscr{H}(\vec{\lambda})^*$ be the full vector space dual of $\mathscr{H}(\vec{\lambda})$. The Lie algebra $\hat{\mathfrak{g}}_o$ acts on $\mathscr{H}(\vec{\lambda})$ via

$$x[f] \cdot (v_1 \otimes \cdots \otimes v_s) = \sum_{i=1}^{s} v_1 \otimes \cdots \otimes x[f_{p_i}] \cdot v_i \otimes \cdots \otimes v_s, \quad \text{for } v_i \in \mathscr{H}(\lambda_i),$$

and $C(v_1 \otimes \cdots \otimes v_s) = -c(v_1 \otimes \cdots \otimes v_s)$. By the Residue Theorem, it is indeed a Lie algebra action.

This gives rise to the (dual) action of $\hat{\mathfrak{g}}_o$ on $\mathscr{H}(\vec{\lambda})^*$. Let $M \subset \mathscr{H}(\vec{\lambda})^*$ be the $\hat{\mathfrak{g}}_o$-submodule generated by $\Phi \in \mathscr{H}(\vec{\lambda})^*$. We claim that the action of $\hat{\mathfrak{g}}_o$ on M extends to a $\hat{\mathfrak{g}}$-module structure on M via the embedding (3). Let $\hat{\mathfrak{g}}_o^+ \subset \hat{\mathfrak{g}}_o$ be the subalgebra $\mathfrak{g}_{q_o}(\Sigma\backslash\vec{p})$ defined by identity (7) of the proof of Theorem 3.1.2. Then, by the definition of Φ,

$$\hat{\mathfrak{g}}_o^+ \cdot \Phi = 0. \tag{4}$$

Define an increasing filtration $\{\mathscr{F}_d(M)\}_{d \geq 0}$ of M by

$$\mathscr{F}_d(M) = \text{span of } \{(x_1[f_1] \ldots x_k[f_k]) \cdot \Phi : f_i \in \mathbb{C}_{q''_o}[\tilde{\Sigma}\backslash\{\vec{p}, q'_o\}], x_i \in \mathfrak{g}$$
$$\text{and } \sum_{i=1}^{k} o(f_i) \leq d\},$$

where $o(f_i)$ is the order of pole of f_i at q'_o. (If f_i is regular at q'_o, we set $o(f_i) = 0$.) From (4), it is easy to see that for any $\Psi \in \mathscr{F}_d(M)$, any $x \in \mathfrak{g}$ and any $g \in \mathbb{C}_{q_o}[\Sigma \backslash \vec{p}] \hookrightarrow \mathbb{C}_{q_o''}[\tilde{\Sigma} \backslash \{\vec{p}, q_o'\}]$ such that the order of zero of g at q_o' is at least $d+1$,

$$x[g] \cdot \Psi = 0. \tag{5}$$

Now, for any $\Psi \in \mathscr{F}_d(M)$ and $x[P] \in \hat{\mathfrak{g}}$ (i.e., $x \in \mathfrak{g}$ and $P \in K := \mathbb{C}((z')))$, pick $g \in \mathbb{C}_{q_o''}[\tilde{\Sigma} \backslash \{\vec{p}, q_o'\}]$ such that

$$g_{q_o'} - P \equiv 0 \,(\text{mod } (z')^{d+1}) \tag{6}$$

and define

$$x[P] \cdot \Psi = x[g] \cdot \Psi, \tag{7}$$

$$C \cdot \Psi = c\Psi. \tag{8}$$

From (5), it follows that (7) gives a well-defined action $x[P] \cdot \Psi$ (i.e., it does not depend upon the choice of g satisfying (6)). Observe that, taking $g = 0$,

$$x \left[\sum_{i \geq d+1} a_i (z')^i \right] \cdot \Psi = 0. \tag{9}$$

Of course, the action of $\hat{\mathfrak{g}}$ on M defined by (7) and (8) extends the action of $\hat{\mathfrak{g}}_o$ on M.

We next show that this action indeed makes M into a module for the Lie algebra $\hat{\mathfrak{g}}$. To show this, it suffices to show that, for $x, y \in \mathfrak{g}$, $P, Q \in K$, and $\Psi \in \mathscr{F}_d(M)$,

$$\begin{aligned} &x[P] \cdot (y[Q] \cdot \Psi) - y[Q] \cdot (x[P] \cdot \Psi) \\ &\quad = [x, y][PQ] \cdot \Psi + \langle x, y \rangle \operatorname*{Res}_{q_o'}((dP)Q)c\Psi. \end{aligned} \tag{10}$$

Take $f, g \in \mathbb{C}_{q_o''}[\tilde{\Sigma} \backslash \{\vec{p}, q_o'\}]$ such that

$$f_{q_o'} - P \quad \text{and} \quad g_{q_o'} - Q \equiv 0 \,(\text{mod } (z')^{d+1+o(P)+o(Q)}),$$

where, as earlier, $o(P)$ is the order of the pole of P (defined to be 0 if P is regular at 0). Using the definition (7), it is easy to see that (10) is equivalent to the same identity with P replaced by f and Q by g. The latter of course follows since M is a representation of $\hat{\mathfrak{g}}_o$. As a special case of (9), we get

$$\hat{\mathfrak{g}}_+ \cdot \Phi = 0, \tag{11}$$

where $\hat{\mathfrak{g}}_+$ is defined by identity (4) of Definition 1.2.2. We next show that M is an integrable $\hat{\mathfrak{g}}$-module. To prove this, it suffices to show that for any root vector $x_\beta \in \mathfrak{g}$ (corresponding to any nonzero root β) and $P \in K$, $x_\beta[P]$ acts locally nilpotently on M. Since M is generated by Φ as a $\hat{\mathfrak{g}}$-module, by

Lemma 1.2.7 and Kumar (2002, Corollary 1.3.4), it suffices to show that $x_\beta[P]$ acts nilpotently on Φ.

For any positive integer N, let $\mathbb{C}_N[\Sigma\backslash\vec{p}] \subset \mathbb{C}[\Sigma\backslash\vec{p}]$ be the ideal consisting of those $g \in \mathbb{C}[\Sigma\backslash\vec{p}]$ such that its pull-back π^*g to $\hat{\Sigma}\backslash\vec{p}$ has a zero of order $\geq N$ at q_o'. Let $\mathfrak{g}_N[\Sigma\backslash\vec{p}] \subset \mathfrak{g}[\Sigma\backslash\vec{p}]$ be the Lie subalgebra defined as $\mathfrak{g} \otimes \mathbb{C}_N[\Sigma\backslash\vec{p}]$. By the same proof as that of Lemma 2.1.4, $\mathfrak{g}_N[\Sigma\backslash\vec{p}] \cdot \mathscr{H}(\vec{\lambda})$ is of finite codimension in $\mathscr{H}(\vec{\lambda})$. Pick $f \in \mathbb{C}_{q_o''}[\tilde{\Sigma}\backslash\{\vec{p}\cup q_o'\}]$ such that $f_{q_o'} - P \equiv 0$ (mod z'). Then, for any $k \geq 1$,

$$x_\beta[P]^k \cdot \Phi = x_\beta[f]^k \cdot \Phi, \tag{12}$$

since $x_\beta[P]$ and $x_\beta[f]$ commute. Pick $N_o > 0$ such that $(\operatorname{ad} x_\beta)^{N_o}(\mathfrak{g}) = 0$. Let V be a finite-dimensional complement of $\mathfrak{g}_{o(P)N_o+1}[\Sigma\backslash\vec{p}] \cdot \mathscr{H}(\vec{\lambda})$ in $\mathscr{H}(\vec{\lambda})$. Since $x_\beta(f)$ acts locally nilpotently on $\mathscr{H}(\vec{\lambda})$ (cf. Exercise 1.2.E.7) and V is finite dimensional, there exists N (which we take $\geq N_o$) such that

$$x_\beta[f]^N \cdot V = 0. \tag{13}$$

Take any $x[g] \in \mathfrak{g}_{o(P)N_o+1}[\Sigma\backslash\vec{p}]$. By identity (8) of Theorem 2.2.2 in the enveloping algebra $U(\hat{\mathfrak{g}}_o)$,

$$x_\beta[f]^N \cdot x[g] = \sum_{j=0}^{N_o-1} \binom{N}{j} \Big((\operatorname{ad} x_\beta)^j(x)\Big)[f^j g] x_\beta[f]^{N-j}.$$

Thus,

$$x_\beta[f]^N \cdot \Big(\mathfrak{g}_{o(P)N_o+1}[\Sigma\backslash\vec{p}] \cdot \mathscr{H}(\vec{\lambda})\Big) \subset \mathfrak{g}(\Sigma\backslash\vec{p}) \cdot \mathscr{H}(\vec{\lambda}). \tag{14}$$

Combining (13) and (14), we get that

$$x_\beta[f]^N \cdot \Phi = 0.$$

But then, by (12),

$$x_\beta[P]^N \cdot \Phi = 0.$$

This proves that $M \subset \mathscr{H}(\vec{\lambda})^*$ is an integrable $\hat{\mathfrak{g}}$-module (generated by Φ). Let $M_o \subset M$ be the $\mathfrak{g}$-submodule generated by Φ. Decompose M_o into irreducible components:

$$M_o = \bigoplus_{\mu \in D} V(\mu)^{\oplus n_\mu}.$$

Take any highest-weight vector v_o in any irreducible $\mathfrak{g}$-submodule $V(\mu)$ of M_o. Since $\hat{\mathfrak{g}}_+$ annihilates M_o (use (11)), v_o generates an integrable highest-weight $\hat{\mathfrak{g}}$-submodule of M of highest weight μ_c. In particular, any $V(\mu)$

appearing in M_o satisfies $\mu \in D_c$ (cf. Theorem 1.2.10). Thus, from (2) and (3) of Definition 3.1.3, applied to the map

$$U(\mathfrak{g}) \to M_o, \quad a \mapsto a \cdot \Phi,$$

we get that for any $a \in \operatorname{Ker}\beta$, $a^t \cdot \Phi = 0$. But, as in Definition 3.1.3, $\operatorname{Ker}\beta$ is stable under t. Thus, for any $a \in \operatorname{Ker}\beta$, $a \cdot \Phi = 0$, i.e., $\Phi(\varphi(a^t) \cdot h) = 0$, for any $h \in \mathscr{H}(\vec{\lambda})$. This proves the lemma and hence Theorem 3.1.2 is fully established. □

3.2 Sugawara Operators

Let $\mathfrak{g}$ be a simple Lie algebra over $\mathbb{C}$. We begin with the definition of *Sugawara elements* $\{L_n\}_{n\in\mathbb{Z}}$. We follow the notation from Section 1.2. In particular, $\tilde{\mathfrak{g}}$ is the uncompleted affine Kac–Moody Lie algebra as in (1) of Definition 1.2.1.

Definition 3.2.1 Let $\{e_i\}_{i\in I}$ be a basis of $\mathfrak{g}$ and let $\{e^i\}_{i\in I}$ be the dual basis of $\mathfrak{g}$ with respect to the normalized form $\langle , \rangle$ on $\mathfrak{g}$. Let $\mathfrak{n} \subset \mathfrak{b}$ be the nil-radical of the Borel subalgebra $\mathfrak{b}$ of $\mathfrak{g}$ and let $\mathfrak{b}_-$ be the opposite Borel subalgebra (i.e., $\mathfrak{b}_-$ is a Borel subalgebra of $\mathfrak{g}$ such that $\mathfrak{b} \cap \mathfrak{b}_- = \mathfrak{h}$). Define

$$\tilde{\mathfrak{n}} := \left(\mathfrak{g} \otimes t\mathbb{C}[t]\right) \oplus \left(\mathfrak{n} \otimes t^0\right), \quad \tilde{\mathfrak{b}}_- := \left(\mathfrak{g} \otimes t^{-l}\mathbb{C}[t^{-1}]\right) \oplus \left(\mathfrak{b}_- \otimes t^0\right) \oplus \mathbb{C}C. \tag{1}$$

As in Kumar (2002, Section 1.5.8), define a certain completion $\hat{U}(\tilde{\mathfrak{g}})$ of the enveloping algebra $U(\tilde{\mathfrak{g}})$ by

$$\hat{U}(\tilde{\mathfrak{g}}) := \Pi_{d\geq 0}\left(U(\tilde{\mathfrak{b}}_-) \otimes_{\mathbb{C}} U_d(\tilde{\mathfrak{n}})\right), \tag{2}$$

where $U_d(\tilde{\mathfrak{n}}) \subset U(\tilde{\mathfrak{n}})$ is the set of homogeneous elements of principal degree d. (Recall that, for an affine root $\alpha = \sum_{i=0}^{\ell} n_i\alpha_i$, its principal degree $|\alpha|$ is defined as $|\sum_{i=0}^{\ell} n_i|$.) The algebra structure on $U(\tilde{\mathfrak{g}})$ canonically extends to an algebra structure on $\hat{U}(\tilde{\mathfrak{g}})$.

For any $n \in \mathbb{Z}$, define the *Sugawara element* $L_n \in \hat{U}(\tilde{\mathfrak{g}})$ by

$$L_n := \frac{1}{2}\sum_{j\in\mathbb{Z}}\sum_{i\in I} : e_i[-j]e^i[j+n] : \in \hat{U}(\tilde{\mathfrak{g}}), \tag{3}$$

where the normal ordering $: x[a]y[b] :$ means

$$\begin{cases} x[a]y[b], & \text{if } a \leq b \\ y[b]x[a], & \text{if } a > b. \end{cases}$$

It is easy to see that L_n does not depend upon the choice of the basis $\{e_i\}$ of $\mathfrak{g}$.

A proof of the following proposition can be found in Kac, Raina and Rozhkovskaya (2013, §10.1).

Proposition 3.2.2 *(a) For any $x \in \mathfrak{g}$ and $m, n \in \mathbb{Z}$, as elements of $\hat{U}(\tilde{\mathfrak{g}})$,*

$$[x[m], L_n] = (C + h^\vee) m \, x[m+n],$$

where $h^\vee$ is the dual Coxeter number of $\mathfrak{g}$.

(b) As elements of $\hat{U}(\tilde{\mathfrak{g}})$,

$$[L_m, L_n] = (C + h^\vee)(m-n)L_{m+n} + \delta_{m,-n} \frac{m^3 - m}{12} \dim(\mathfrak{g}) C (C + h^\vee).$$

Definition 3.2.3 (a) A representation V of $\tilde{\mathfrak{g}}$ is called *smooth* if for any $v \in V$ and $x \in \mathfrak{g}$, there exists an integer d (depending upon v) such that

$$x[m] \cdot v = 0, \quad \text{for all} \quad m \geq d. \tag{1}$$

Any highest-weight $\tilde{\mathfrak{g}}$-module (cf. Definition 1.2.4(c)) is clearly smooth. Clearly, L_n acts on any smooth $\tilde{\mathfrak{g}}$-module.

(b) Let us recall the definition of the *Virasora algebra* Vir. It is the Lie algebra over $\mathbb{C}$ with basis $\{d_n; \bar{C}\}_{n \in \mathbb{Z}}$ and the commutation relation is given by

$$[d_m, d_n] = (m-n)d_{m+n} + \delta_{m,-n} \frac{m^3 - m}{12} \bar{C}; \quad [d_m, \bar{C}] = 0. \tag{1}$$

A representation V of Vir endowed with a positive-definite Hermitian form $\{,\}$ is called *unitary* if

$$\{\bar{C}v, w\} = \{v, \bar{C}w\}; \; \{d_m v, w\} = \{v, d_{-m} w\}, \quad \text{for all} \;\; v, w \in V \;\; \text{and} \;\; m \in \mathbb{Z}.$$

(c) Let $\mathfrak{k}$ be a *maximal compact subalgebra* of $\mathfrak{g}$ (i.e., the Lie subalgebra of a maximal compact subgroup K of G, G being the connected, simply-connected complex algebraic group with Lie algebra $\mathfrak{g}$). Then, as vector space over $\mathbb{R}$,

$$\mathfrak{g} = \mathfrak{k} \oplus i\mathfrak{k}, \tag{2}$$

(cf. (Helgason, 1978, Chap. III, Corollary 7.5)). This decomposition gives rise to a conjugate-linear (under (2)) Lie algebra anti-automorphism $\sigma_o = \sigma_o(\mathfrak{k})$ of $\mathfrak{g}$ such that $(\sigma_o)_{|\mathfrak{k}} = -\operatorname{Id}_{\mathfrak{k}}$ and $(\sigma_o)_{|i\mathfrak{k}} = \operatorname{Id}_{i\mathfrak{k}}$.

A representation V of $\tilde{\mathfrak{g}}$ endowed with a positive-definite Hermitian form $\{,\}$ is called *unitary* (with respect to $\mathfrak{k}$) if

$$\{C\, v, w\} = \{v, Cw\}; \; \{x[m]v, w\} = \{v, \sigma_o(x)[-m]w\},$$
$$\text{for all} \;\; v, w \in V, x \in \mathfrak{g} \;\; \text{and} \;\; m \in \mathbb{Z}.$$

Lemma 3.2.4 *Let V be a smooth representation of $\tilde{\mathfrak{g}}$. Assume that the central element $C \in \tilde{\mathfrak{g}}$ acts via a scalar $c \neq -h^\vee$ on V. Then V is a module for the Lie algebra Vir under the following action:*

$$\bar{C} \mapsto \left(\frac{c \dim \mathfrak{g}}{c+h^\vee}\right) I_V; \; d_n \mapsto \frac{L_n}{c+h^\vee}, \quad \textit{for any} \quad n \in \mathbb{Z}. \tag{1}$$

Moreover, if V is a unitary $\tilde{\mathfrak{g}}$-module, then it is also a unitary Vir-module under the same positive-definite Hermitian form.

Proof The assertion that V is a Vir-module under (1) follows immediately from Proposition 3.2.2(b).

We now prove the unitarity of V as a Vir-module. Since V is a unitary $\tilde{\mathfrak{g}}$-module, $c \in \mathbb{R}$. By Exercise 3.2.E.1, for $n \neq 0$,

$$L_n = \frac{1}{2} \sum_{j \in \mathbb{Z}} \sum_{i \in I} e_i[-j] e^i[j+n].$$

Choose an orthonormal $\mathbb{R}$-basis $\{e_i\}_{i \in I}$ of $\mathfrak{k}$ so that $e^i = -e_i$ for all $i \in I$. Then, for any $n \neq 0$, $v, w \in V$,

$$\begin{aligned}
\{d_n v, w\} &= \frac{1}{c+h^\vee} \{L_n v, w\} \\
&= -\frac{1}{2(c+h^\vee)} \sum_{j \in \mathbb{Z}} \sum_{i \in I} \{e_i[-j] e_i[j+n] v, w\} \\
&= -\frac{1}{2(c+h^\vee)} \sum_{j \in \mathbb{Z}} \sum_{i \in I} \{v, e_i[-j-n] e_i[j] w\}, \\
&\quad \text{by the unitarity of } V \text{ as a } \tilde{\mathfrak{g}}\text{-module} \\
&= \{v, d_{-n} w\}.
\end{aligned} \tag{2}$$

Similarly,

$$\begin{aligned}
\{d_0 v, w\} &= -\frac{1}{2(c+h^\vee)} \left\{ \left(\sum_{i \in I} e_i^2 + 2 \sum_{j \geq 1} \sum_{i \in I} e_i[-j] e_i[j] \right) v, w \right\} \\
&= -\frac{1}{2(c+h^\vee)} \left\{ v, \left(\sum_{i \in I} e_i^2 + 2 \sum_{j \geq 1} \sum_{i \in I} e_i[-j] e_i[j] \right) w \right\} \\
&= \{v, d_0 w\}.
\end{aligned} \tag{3}$$

Also,

$$\{\bar{C} v, w\} = \frac{c \dim \mathfrak{g}}{c+h^\vee} \{v, w\} = \{v, \bar{C} w\}, \quad \text{since } c \in \mathbb{R}. \tag{4}$$

Combining (2), (3) and (4), we get that V is unitary as a Vir-module. This proves the lemma. □

3.2.E Exercises

(1) Recall the definition of the Sugawara elements $L_n \in \hat{U}(\tilde{\mathfrak{g}})$ from identity (3) of Definition 3.2.1. Show that, for $n \neq 0$,

$$L_n = \frac{1}{2} \sum_{j \in \mathbb{Z}} \sum_{i \in I} e_i[-j] e^i[j+n],$$

i.e., we can dispense with the normal order in the expression of L_n, for $n \neq 0$.

Hint: Show that $\sum_{i \in I} [e_i, e^i] = 0$ and thus $\sum_{i \in I} [e_i[j], e^i[k]] = j \delta_{j,-k} \dim \mathfrak{g}\, C$.

(2) Let $\mathfrak{s}$ and $\mathfrak{g}$ be simple Lie algebras with an embedding $\varphi\colon \mathfrak{s} \hookrightarrow \mathfrak{g}$. This gives rise to an embedding $\tilde{\varphi}\colon \tilde{\mathfrak{s}} \to \tilde{\mathfrak{g}}$ taking $x[n] \mapsto \varphi(x)[n]$ (for $x \in \mathfrak{s}$, $n \in \mathbb{Z}$), and $C_{\tilde{\mathfrak{s}}} \mapsto d_\varphi C_{\tilde{\mathfrak{g}}}$, where $C_{\tilde{\mathfrak{s}}}$ (resp. $C_{\tilde{\mathfrak{g}}}$) is the canonical central element of $\tilde{\mathfrak{s}}$ (resp. $\tilde{\mathfrak{g}}$) and d_φ is the Dynkin index of φ (cf. Definition A.1 and Lemma A.12).

Let V be a smooth representation of $\tilde{\mathfrak{g}}$ such that the central element $C_{\tilde{\mathfrak{g}}}$ acts via the scalar c. Then show that V considered as a representation of $\tilde{\mathfrak{s}}$ via $\tilde{\varphi}$ is a smooth $\tilde{\mathfrak{s}}$-module such that $C_{\tilde{\mathfrak{s}}}$ acts via the scalar cd_φ.

(3) Let the notation and assumptions be as in Exercise 2. Assume further that $c + h^\vee_{\mathfrak{g}} \neq 0$ and $cd_\varphi + h^\vee_{\mathfrak{s}} \neq 0$, where $h^\vee_{\mathfrak{g}}$ is the dual Coxeter number of $\mathfrak{g}$. Recall the definition of the Sugawara elements $L_n = L^{\mathfrak{g}}_n \in \hat{U}(\tilde{\mathfrak{g}})$ from identity (3) of Definition 3.2.1. Similarly, define the Sugawara elements $L^{\mathfrak{s}}_n \in \hat{U}(\tilde{\mathfrak{s}})$. Then, show that

$$d_n \mapsto \frac{L^{\mathfrak{g}}_n}{c + h^\vee_{\mathfrak{g}}} - \frac{L^{\mathfrak{s}}_n}{cd_\varphi + h^\vee_{\mathfrak{s}}}; \quad \bar{C} \mapsto \frac{c \dim \mathfrak{g}}{c + h^\vee_{\mathfrak{g}}} - \frac{cd_\varphi \dim \mathfrak{s}}{cd_\varphi + h^\vee_{\mathfrak{s}}}$$

acting on V is a module for Vir.

Let $\mathfrak{k} \subset \mathfrak{g}$ be a maximal compact subalgebra such that $\varphi^{-1}(\mathfrak{k})$ is a maximal compact subalgebra of $\mathfrak{s}$. Prove that if V is unitary $\tilde{\mathfrak{g}}$-module (with respect to $\mathfrak{k}$), then this Vir-module is also unitary.

In particular, if V is unitary and $\frac{c \dim \mathfrak{g}}{c + h^\vee_{\mathfrak{g}}} = \frac{cd_\varphi \dim \mathfrak{s}}{cd_\varphi + h^\vee_{\mathfrak{s}}}$, then show that V is a trivial Vir-module.

Hint: Show that for any $x \in \mathfrak{s}$ and $k \in \mathbb{Z}$,

$$\left[\frac{L_n^{\mathfrak{g}}}{c+h_{\mathfrak{g}}^{\vee}} - \frac{L_n^{\mathfrak{s}}}{cd_{\varphi}+h_{\mathfrak{s}}^{\vee}}, x[k]\right] = 0, \quad \text{as operator on } V.$$

Thus, conclude that

$$\left[\frac{L_n^{\mathfrak{g}}}{c+h_{\mathfrak{g}}^{\vee}} - \frac{L_n^{\mathfrak{s}}}{cd_{\varphi}+h_{\mathfrak{s}}^{\vee}}, \frac{L_n^{\mathfrak{s}}}{cd_{\varphi}+h_{\mathfrak{s}}^{\vee}}\right] = 0, \quad \text{as operator on } V.$$

(4) Show that any smooth $\tilde{\mathfrak{g}}$-module V, such that C acts by a scalar c with $c+h^{\vee} \neq 0$, admits an extension of its $\tilde{\mathfrak{g}}$-module structure to a $(\mathbb{C}d \ltimes \tilde{\mathfrak{g}})$-module structure (cf. identity (5) of Definition 1.2.1) by requiring d to act via $(-L_0/c+h^{\vee})$.

Hint: Use Proposition 3.2.2(a).

3.3 Sheaf of Conformal Blocks

As in earlier sections, let $s \geq 1$ and let $c \geq 1$ be the level.

Definition 3.3.1 (Family of s-pointed curves with formal parameters) Let T be an irreducible smooth variety and let $\pi : \Sigma_T \to T$ be a proper flat morphism (in particular, π is of finite type) such that each scheme-theoretic fiber Σ_b is a reduced (but not necessarily irreducible) curve. In addition, we are given the following:

(a) Mutually non-intersecting sections $\{p_i\}_{1 \leq i \leq s}$ of π.

(b) For any $b \in T$, $(\Sigma_b, p_1(b), \ldots, p_s(b))$ is an s-pointed curve; i.e., it is reduced, $p_i(b)$ are smooth points in Σ_b, it has at worst nodal singularity and each irreducible component of Σ_b contains at least one point from $\{p_i(b)\}_{1 \leq i \leq s}$. Thus, by Mumford and Oda (2015, Criterion 5.4.8 and Proposition 5.3.2), $\pi : \Sigma_T \to T$ is a smooth morphism in a neighborhood of $\bigcup_i p_i(T)$.

Moreover, we require that the restriction map $\pi_{|\Sigma_T^o} : \Sigma_T^o \to T$ is an affine morphism, where $\Sigma_T^o = \Sigma_T \setminus \bigcup_{i=1}^s p_i(T)$.[1] (Observe that $p_i(T)$ is closed in Σ_T since p_i is a section of π.)

[1] By a result appearing in 'mathoverflow' due to R. van Dobben de Bruyn dated April 25, 2018, the restriction map $\pi_{|\Sigma_T^o} : \Sigma_T^o \to T$ is an affine morphism if we assume that the geometric fibers of π are connected.

(c) Formal local parameters $\{t_i\}_{1\le i\le s}$ along $p_i(T)$, i.e., an element t_i of the formal completion $\hat{\mathscr{O}}_{\Sigma_T, p_i(T)}$ of Σ_T along $p_i(T)$ satisfying a ring isomorphism:

$$\hat{\mathscr{O}}_{\Sigma_T, p_i(T)} \simeq \mathscr{O}_T[[t_i]], \tag{1}$$

where the isomorphism is induced from $\mathscr{O}_T \xrightarrow{\pi^*} \mathscr{O}_{\Sigma_T} \to \hat{\mathscr{O}}_{\Sigma_T, p_i(T)}$.

The triple $\mathfrak{F}_T = \left(\Sigma_T, \vec{p}, \vec{t}\right)$ is called a *family of s-pointed curves with formal parameters over T*, where $\vec{p} = (p_1, \dots, p_s)$ and $\vec{t} = (t_1, \dots, t_s)$. The pair $(\Sigma_T, \vec{p})$ is called a *family of s-pointed curves* if it satisfies the above conditions (a)–(b).

Let T' be a smooth irreducible variety and let $f: T' \to T$ be a morphism. Then we can pull back the triple $(\Sigma_T, \vec{p}, \vec{t})$ to T' to get a family of s-pointed curves with formal parameters

$$\mathfrak{F}_{T'} = \left(f^*(\Sigma_T), f^*\vec{p}, f^*\vec{t}\right)$$

over T'.

Definition 3.3.2 (Sheaf of conformal blocks) Let $\mathfrak{F}_T = (\Sigma_T, \vec{p}, \vec{t})$ be a family of s-pointed curves with formal parameters over T as in the above Definition 3.3.1. Let $\vec{\lambda} = (\lambda_1, \dots, \lambda_s) \in D_c^s$. Recall the definition of $\mathscr{H}(\vec{\lambda})$ from identity (4) of Definition 2.1.1 and the Lie algebra $\hat{\mathfrak{g}}^{(s)}$ from identity (1) of Lemma 2.1.4. Then, as in the proof of Lemma 2.1.4, there is a Lie algebra homomorphism from the direct sum Lie algebra

$$\oplus_{i=1}^s \hat{\mathfrak{g}}_{t_i} \to \hat{\mathfrak{g}}^{(s)}.$$

Clearly, the tensor product $\mathscr{H}(\vec{\lambda})$ is a module for $\bigoplus_{i=1}^s \hat{\mathfrak{g}}_{t_i}$ under componentwise action. Moreover, since each $\mathscr{H}(\lambda_i)$ has the same central charge c, the $\bigoplus_{i=1}^s \hat{\mathfrak{g}}_{t_i}$-module structure on $\mathscr{H}(\vec{\lambda})$ descends to give a $\hat{\mathfrak{g}}^{(s)}$-module structure on $\mathscr{H}(\vec{\lambda})$. Now, let us consider the sheaf of $\mathscr{O}_T$-modules:

$$\mathscr{H}(\vec{\lambda})_T := \mathscr{O}_T \otimes_{\mathbb{C}} \mathscr{H}(\vec{\lambda}) \quad \text{and} \tag{1}$$

$$\hat{\mathfrak{g}}_T^{(s)} := \left(\bigoplus_{i=1}^s \mathfrak{g} \otimes_{\mathbb{C}} \mathscr{O}_T((t_i))\right) \bigoplus \mathscr{O}_T C, \tag{2}$$

where $\mathscr{O}_T((t_i)) := \mathscr{O}_T\left([[t_i]][t_i^{-1}]\right)$. Define a $\mathscr{O}_T$-linear bracket in $\hat{\mathfrak{g}}_T^{(s)}$ by

$$\left[\sum_{i=1}^s x_i \otimes f_i, \sum_{i=1}^s y_i \otimes g_i\right] = \sum_{i=1}^s [x_i, y_i] \otimes f_i g_i + \left(\sum_{i=1}^s \langle x_i, y_i\rangle \underset{t_i=0}{Res}\,(g_i d_{t_i} f_i)\right) C, \tag{3}$$

for $x_i, y_i \in \mathfrak{g}$, $f_i, g_i \in \mathscr{O}_T((t_i))$; and C is central in $\hat{\mathfrak{g}}_T^{(s)}$. Then, $\hat{\mathfrak{g}}_T^{(s)}$ is a sheaf of $\mathscr{O}_T$-Lie algebras. Similar to the action of $\hat{\mathfrak{g}}^{(s)}$ on $\mathscr{H}(\vec{\lambda})$, we have a $\mathscr{O}_T$-linear action of the sheaf $\hat{\mathfrak{g}}_T^{(s)}$ of $\mathscr{O}_T$-Lie algebras on the sheaf $\mathscr{H}(\vec{\lambda})_T$ of $\mathscr{O}_T$-modules. Also, define the $\mathscr{O}_T$-module

$$\hat{\mathfrak{g}}(\mathfrak{F}_T) := \mathfrak{g} \otimes_{\mathbb{C}} \pi_*^o \left(\mathscr{O}_{\Sigma_T^o}\right), \quad \text{where } \pi^o := \pi_{|\Sigma_T^o}. \tag{4}$$

Then $\hat{\mathfrak{g}}(\mathfrak{F}_T)$ is a sheaf of $\mathscr{O}_T$-Lie algebras under the bracket

$$[x \otimes f, y \otimes g] = [x, y] \otimes fg, \quad \text{for} \quad x, y \in \mathfrak{g} \quad \text{and} \quad f, g \in \pi_*^o \left(\mathscr{O}_{\Sigma_T^o}\right). \tag{5}$$

There is an embedding of sheaves of $\mathscr{O}_T$-Lie algebras

$$\varphi\colon \hat{\mathfrak{g}}(\mathfrak{F}_T) \hookrightarrow \hat{\mathfrak{g}}_T^{(s)}, \quad x \otimes f \mapsto \sum_{i=1}^{s} x \otimes f_{p_i}, \quad \text{for } x \in \mathfrak{g} \text{ and } f \in \pi_*^o \left(\mathscr{O}_{\Sigma_T^o}\right), \tag{6}$$

where f_{p_i} denotes the image of f in $\hat{\mathscr{O}}_{\Sigma_T, p_i(T)}[t_i^{-1}] \simeq \mathscr{O}_T((t_i))$. By the Residue Theorem, φ is indeed a Lie algebra embedding. (By assumptions (a) and (b) of Definition 3.3.1, we conclude that no irreducible component of Σ_T is disjoint from $\cup_i p_i(T)$. Now, using assumption (c), the injectivity of φ follows.)

Finally, define the *sheaf of covacua* $\mathscr{V}_{\mathfrak{F}_T}(\vec{p}, \vec{\lambda})$ over T as the quotient sheaf of $\mathscr{O}_T$-modules

$$\mathscr{V}_{\mathfrak{F}_T}(\vec{p}, \vec{\lambda}) = \mathscr{H}(\vec{\lambda})_T \Big/ \hat{\mathfrak{g}}(\mathfrak{F}_T) \cdot \mathscr{H}(\vec{\lambda})_T, \tag{7}$$

where $\hat{\mathfrak{g}}(\mathfrak{F}_T)$ acts on $\mathscr{H}(\vec{\lambda})_T$ via the embedding φ (given by (6)) and $\hat{\mathfrak{g}}(\mathfrak{F}_T) \cdot \mathscr{H}(\vec{\lambda})_T \subset \mathscr{H}(\vec{\lambda})_T$ denotes the image sheaf under the sheaf homomorphism

$$\theta\colon \hat{\mathfrak{g}}(\mathfrak{F}_T) \otimes_{\mathscr{O}_T} \mathscr{H}(\vec{\lambda})_T \to \mathscr{H}(\vec{\lambda})_T \tag{8}$$

induced from the action of $\hat{\mathfrak{g}}(\mathfrak{F}_T)$ on $\mathscr{H}(\vec{\lambda})_T$.

Lemma 3.3.3 *Let $\mathfrak{F}_T = (\Sigma_T, \vec{p}, \vec{t})$ be a family of s-pointed curves with formal parameters over T. Then, for any $b \in T$, and any $\vec{\lambda} = (\lambda_1, \ldots, \lambda_s) \in D_c^s$, there are canonical isomorphisms:*

(a) $\mathbb{C}_b \bigotimes_{\mathscr{O}_T} \mathscr{H}(\vec{\lambda})_T \simeq \mathscr{H}(\vec{\lambda})$, where $\mathbb{C}_b = \mathscr{O}_{T,b}/\mathfrak{m}_{T,b}$. (Here $\mathscr{O}_{T,b}$ is the local ring of T at b and $\mathfrak{m}_{T,b}$ is the unique maximal ideal of $\mathscr{O}_{T,b}$.)

(b) $\mathbb{C}_b \bigotimes_{\mathscr{O}_T} \hat{\mathfrak{g}}_T^{(s)} \simeq \hat{\mathfrak{g}}^{(s)}$.

(c) $\mathbb{C}_b \bigotimes_{\mathscr{O}_T} \mathscr{V}_{\mathfrak{F}_T}(\vec{p}, \vec{\lambda}) \simeq \mathscr{V}_{\Sigma_b}(\vec{p}(b), \vec{\lambda})$.

Proof (a) From the definition of $\mathscr{H}(\vec{\lambda})_T$, (a) is clear.

(b) To prove (b), it suffices to observe that

$$\mathbb{C}((t_i)) \xrightarrow{\sim} \mathbb{C}_b \otimes_{\mathscr{O}_T} \mathscr{O}_T((t_i)), \quad \text{under the map } f \mapsto 1 \otimes f. \tag{1}$$

(To prove (1), use the fact that $\mathfrak{m}_{T,b}$ is a finitely generated $\mathscr{O}_{T,b}$-module.)

(c) By the definition of $\mathscr{V}_{\mathfrak{F}_T}(\vec{p},\vec{\lambda})$, we have an exact sequence of sheaves:

$$\hat{\mathfrak{g}}(\mathfrak{F}_T) \otimes_{\mathscr{O}_T} \mathscr{H}(\vec{\lambda})_T \xrightarrow{\theta} \mathscr{H}(\vec{\lambda})_T \to \mathscr{V}_{\mathfrak{F}_T}(\vec{p},\vec{\lambda}) \to 0,$$

where θ is as in (8) of Definition 3.3.2. Since $\otimes$ is a right exact functor, we get the following diagram with exact top row:

$$\begin{array}{ccccccc}
\mathbb{C}_b \underset{\mathscr{O}_T}{\bigotimes} \left(\hat{\mathfrak{g}}(\mathfrak{F}_T) \underset{\mathscr{O}_T}{\bigotimes} \mathscr{H}(\vec{\lambda})_T\right) & \xrightarrow{\mathrm{Id}\otimes\theta} & \mathbb{C}_b \underset{\mathscr{O}_T}{\bigotimes} \mathscr{H}(\vec{\lambda})_T & \longrightarrow & \mathbb{C}_b \underset{\mathscr{O}_T}{\bigotimes} \mathscr{V}_{\mathfrak{F}_T}(\vec{p},\vec{\lambda}) & \longrightarrow & 0 \\
\wr\!\!\mid & & \downarrow\wr & & \Vert & & \\
\left(\mathbb{C}_b \underset{\mathscr{O}_T}{\bigotimes} \hat{\mathfrak{g}}(\mathfrak{F}_T)\right) \underset{\mathbb{C}}{\bigotimes} \mathscr{H}(\vec{\lambda}) & \longrightarrow & \mathscr{H}(\vec{\lambda}) & \longrightarrow & \mathbb{C}_b \underset{\mathscr{O}_T}{\bigotimes} \mathscr{V}_{\mathfrak{F}_T}(\vec{p},\vec{\lambda}) & \longrightarrow & 0.
\end{array} \tag{$\mathfrak{D}$}$$

Observe next that, by definition (cf. (4) of Definition 3.3.2),

$$\begin{aligned}
\mathbb{C}_b \otimes_{\mathscr{O}_T} \hat{\mathfrak{g}}(\mathfrak{F}_T) &= \mathbb{C}_b \otimes_{\mathscr{O}_T} \left(\mathfrak{g} \otimes_{\mathbb{C}} \pi^o_*(\mathscr{O}_{\Sigma^o_T})\right) \\
&\simeq \mathfrak{g} \otimes_{\mathbb{C}} \left(\mathbb{C}_b \otimes_{\mathscr{O}_T} \pi^o_*(\mathscr{O}_{\Sigma^o_T})\right) \\
&\simeq \mathfrak{g} \otimes_{\mathbb{C}} \mathbb{C}[\Sigma_b \backslash \vec{p}(b)] \\
&= \mathfrak{g}[\Sigma_b \backslash \vec{p}(b)].
\end{aligned} \tag{2}$$

Moreover, under the identifications of the above diagram ($\mathfrak{D}$) and (2), the map

$$\mathrm{Id} \otimes \theta \colon \mathfrak{g}[\Sigma_b \backslash \vec{p}(b)] \otimes \mathscr{H}(\vec{\lambda}) \to \mathscr{H}(\vec{\lambda})$$

corresponds to the $\mathfrak{g}[\Sigma_b \backslash \vec{p}(b)]$-module structure of $\mathscr{H}(\vec{\lambda})$ as in identity (5) of Definition 2.1.1. This proves the (c)-part of the lemma. □

Proposition 3.3.4 *Let $\mathfrak{F}_T = (\Sigma_T, \vec{p}, \vec{t})$ be as in Definition 3.3.1. Then $\mathscr{V}_{\mathfrak{F}_T}(\vec{p},\vec{\lambda})$ is a coherent sheaf of $\mathscr{O}_T$-modules.*

Proof Recall the embedding $\varphi\colon \hat{\mathfrak{g}}(\mathfrak{F}_T) \hookrightarrow \hat{\mathfrak{g}}^{(s)}_T$ of $\mathscr{O}_T$-Lie algebras from (6) of Definition 3.3.2. Also, consider the subsheaf

$$\hat{\mathfrak{p}}^{(s)}_T := \left(\oplus_{i=1}^s \mathfrak{g} \otimes_{\mathbb{C}} \mathscr{O}_T([[t_i]])\right) \oplus \mathscr{O}_T C$$

of $\hat{\mathfrak{g}}_T^{(s)}$ and let $\hat{\mathfrak{g}}(\mathfrak{F}_T)+\hat{\mathfrak{p}}_T^{(s)}$ be the subsheaf of $\hat{\mathfrak{g}}_T^{(s)}$ of $\mathcal{O}_T$-modules generated by $\operatorname{Im}\varphi$ and $\hat{\mathfrak{p}}_T^{(s)}$. Then, as can be seen, the quotient sheaf $\hat{\mathfrak{g}}_T^{(s)}/\big(\hat{\mathfrak{g}}(\mathfrak{F}_T)+\hat{\mathfrak{p}}_T^{(s)}\big)$ is a coherent (in fact, a locally free) $\mathcal{O}_T$-module (cf. (Looijenga, 2005, Lemma 3.6)). Thus, we can find a finite set of elements $\{x_j\}$ of $\hat{\mathfrak{g}}^{(s)}$ such that each x_j acts locally finitely on $\mathscr{H}(\vec{\lambda})$ and

$$\hat{\mathfrak{g}}_T^{(s)}=\hat{\mathfrak{g}}(\mathfrak{F}_T)+\hat{\mathfrak{p}}_T^{(s)}+\sum_j \mathcal{O}_T x_j$$

(cf. (Kumar, 2002, proof of Lemma 10.2.2)).

Now, following the proof of Lemma 2.1.4 and recalling that the PBW theorem holds for any Lie algebra $\mathfrak{s}$ over a commutative ring R such that $\mathfrak{s}$ is free as an R-module (cf. (Cartan and Eilenberg, 1956, Chap. XIII, Theorem 3.1)), we get the proposition. □

Definition 3.3.5 Let $\mathfrak{F}_T=(\Sigma_T,\vec{p},\vec{t})$ be a family of s-pointed curves with formal parameters over T as in Definition 3.3.1. Let $\vec{\lambda}=(\lambda_1,\ldots,\lambda_s)\in D_c^s$. Then, similar to the definition of the sheaf of covacua (cf. identity (7) of Definition 3.3.2), define the *sheaf of vacua* (also called the *sheaf of conformal blocks*):

$$\mathscr{V}^{\dagger}_{\mathfrak{F}_T}(\vec{p},\vec{\lambda})=\mathbf{Hom}_{\hat{\mathfrak{g}}(\mathfrak{F}_T)}(\mathscr{H}(\vec{\lambda})_T,\mathcal{O}_T),$$

where **Hom** denotes the sheaf Hom. Then, as in Exercise 3.3.E.1, $\mathscr{V}^{\dagger}_{\mathfrak{F}_T}(\vec{p},\vec{\lambda})$ is a coherent sheaf of $\mathcal{O}_T$-modules.

We also need to consider a sheafified and 'completed' version of the Virasoro algebra. Let T be any scheme and let $\vec{\lambda}=(\lambda_1,\ldots,\lambda_s)\in D_c^s$.

Definition 3.3.6 Define the $\mathcal{O}_T$-module

$$\mathrm{Vir}_T^{(s)}:=\big(\oplus_{i=1}^s\mathcal{O}_T((t_i))\partial_{t_i}\big)\oplus\mathcal{O}_T\bar{C},$$

where we have abbreviated ∂/∂_{t_i} by ∂_{t_i}. Define a $\mathcal{O}_T$-linear bracket in $\mathrm{Vir}_T^{(s)}$ by declaring $\bar{C}$ to be central and defining the bracket

$$\left[\sum_{i=1}^s f_i\partial_{t_i},\sum_{i=1}^s g_i\partial_{t_i}\right]=\sum_{i=1}^s(f_i\partial_{t_i}(g_i)-g_i\partial_{t_i}(f_i))\partial_{t_i}+\left(\sum_{i=1}^s\operatorname*{Res}_{t_i=0}\left((\partial_{t_i}^3 f_i)g_i\right)\right)\frac{\bar{C}}{12},\tag{1}$$

for $f_i,\ g_i\in\mathcal{O}_T((t_i))$. (Of course, ∂_{t_i} annihilates $\mathcal{O}_T$.) Observe that this bracket corresponds to the bracket of the Virasoro algebra defined in Definition 3.2.3(b) if we take $s=1, d_m=-t_1^{m+1}\partial_{t_1}$ and $T=\operatorname{Spec}\mathbb{C}$.

The action of Vir on $\mathscr{H}(\lambda_i)$ given by Lemma 3.2.4 gives rise to a $\mathscr{O}_T$-linear Lie algebra action of $\mathrm{Vir}_T^{(s)}$ on $\mathscr{H}(\vec{\lambda})_T$ as follows.

Let $\bar{C}$ act on $\mathscr{H}(\vec{\lambda})_T$ by the scalar $\frac{c \dim \mathfrak{g}}{c+h^\vee}$. Further, for any $f \in \mathscr{O}_T$, $v = v_1 \otimes \cdots \otimes v_s \in \mathscr{H}(\vec{\lambda})$, with $v_i \in \mathscr{H}(\lambda_i)$, the action is given by

$$\left(f t_i^{n+1} \partial_{t_i}\right) \cdot v = \frac{-f}{c+h^\vee} \left(v_1 \otimes \cdots \otimes L_n v_i \otimes \cdots \otimes v_s\right), \tag{2}$$

where $L_n \in \hat{U}(\tilde{\mathfrak{g}})$ is the element defined by identity (3) of Definition 3.2.1. Observe that for any $v \in \mathscr{H}(\vec{\lambda})$, there exists a large enough N (depending upon v) such that $(f t_i^{n+1} \partial_{t_i}) \cdot v = 0$, for $n \geq N$. Thus, the action (2) indeed extends to an action of $\mathscr{O}_T((t_i))\partial_{t_i}$. We denote the action of $\mathrm{Vir}_T^{(s)}$ on $\mathscr{H}(\vec{\lambda})_T$ by $\gamma_{\vec{\lambda}}$.

Definition 3.3.7 Let $\mathfrak{F}_T = (\Sigma_T, \vec{p}, \vec{t})$ be a family of s-pointed curves with formal parameters over T as in Definition 3.3.1 such that $\pi : \Sigma_T \to T$ is a smooth morphism. Let $\Theta_{\Sigma_T^o}$ be the sheaf of vector fields on Σ_T^o. For any $\tilde{\theta} \in \Theta_{\Sigma_T^o}$ and $1 \leq i \leq s$, in a formal punctured neighborhood of $p_i(T)$ (i.e., $\mathrm{Spec}\,(\mathscr{O}_T((t_i)))$ under identification (1) of Definition 3.3.1), we can uniquely write

$$\tilde{\theta} = \tilde{\theta}_i^o + \tilde{\theta}_i^h, \tag{1}$$

where

$$\tilde{\theta}_i^h(t_i) = 0 \quad \text{and} \quad \tilde{\theta}_i^o \in \mathscr{O}_T((t_i))\partial_{t_i}.$$

(Of course, the decomposition (1) depends upon the choice of the formal local parameter t_i.) This gives rise to a $\mathscr{O}_T$-linear map

$$\beta : \pi_*^o(\Theta_{\Sigma_T^o}) \to \mathrm{Vir}_T^{(s)}, \ \tilde{\theta} \mapsto \sum_{i=1}^s \tilde{\theta}_i^o.$$

We caution that β is *not* a Lie algebra homomorphism. The map β together with the Lie algebra action $\gamma_{\vec{\lambda}}$ of $\mathrm{Vir}_T^{(s)}$ on $\mathscr{H}(\vec{\lambda})_T$ gives rise to a $\mathscr{O}_T$-linear action $\tilde{\gamma}_{\vec{\lambda}}$ of $\pi_*^o(\Theta_{\Sigma_T^o})$ on $\mathscr{H}(\vec{\lambda})_T$:

$$\tilde{\gamma}_{\vec{\lambda}}(\tilde{\theta}) \cdot v = \gamma_{\vec{\lambda}}(\beta(\tilde{\theta})) \cdot v. \tag{2}$$

The standard action of $\pi_*^o(\Theta_{\Sigma_T^o})$ on $\pi_*^o(\mathscr{O}_{\Sigma_T^o})$ clearly extends to an action of $\pi_*^o(\Theta_{\Sigma_T^o})$ on $\hat{\mathfrak{g}}(\mathfrak{F}_T)$ by demanding the action to be trivial on the $\mathfrak{g}$-component.

We have the following commutation relation between the action $\tilde{\gamma}_{\vec{\lambda}}$ and the action of $\hat{\mathfrak{g}}(\mathfrak{F}_T)$ on $\mathscr{H}(\vec{\lambda})_T$.

Lemma 3.3.8 *Let the assumption be as in the above Definition 3.3.7. Then, for any* $\tilde{\theta} \in \pi_*^o(\Theta_{\Sigma_T^o})$, $x \in \mathfrak{g}$ *and* $\tilde{g} \in \pi_*^o(\mathscr{O}_{\Sigma_T^o})$, *we have*

$$\left[\tilde{\gamma}_{\vec{\lambda}}(\tilde{\theta}), x \otimes \tilde{g}\right] = \sum_{i=1}^{s} x \otimes \left(\tilde{\theta}_i^o(\tilde{g}_{p_i})\right) \quad \text{as operators on} \quad \mathscr{H}(\vec{\lambda})_T, \tag{1}$$

where the action of $x \otimes \tilde{g} \in \hat{\mathfrak{g}}(\mathfrak{F}_T)$ *on* $\mathscr{H}(\vec{\lambda})_T$ *is defined via its embedding* φ *(cf. (6) of Definition 3.3.2).*

Proof Since all the operators are $\mathscr{O}_T$-linear, it suffices to check (1) acting on $\mathscr{H}(\vec{\lambda})$. Take $v = v_1 \otimes \cdots \otimes v_s \in \mathscr{H}(\vec{\lambda})$ with $v_i \in \mathscr{H}(\lambda_i)$. Then, writing $\tilde{\theta}_i^o = \sum_{n_i \geq -N} f_{n_i} t_i^{n_i+1} \partial_{t_i} \in \mathscr{O}_T((t_i))\partial_{t_i}$ and $\tilde{g}_{p_i} = \sum_{m_i \geq -N} g_{m_i} t_i^{m_i} \in \mathscr{O}_T((t_i))$, we get

$$\begin{aligned}
&\left[\tilde{\gamma}_{\vec{\lambda}}(\tilde{\theta}), x \otimes \tilde{g}\right] v \\
&= \sum_{i=1}^{s} v_1 \otimes \cdots \otimes v_{i-1} \otimes \left[\gamma_{\vec{\lambda}}(\tilde{\theta}_i^o), x \otimes \tilde{g}_{p_i}\right] v_i \otimes v_{i+1} \otimes \cdots \otimes v_s \\
&= -\frac{1}{c+h^\vee} \sum_{i=1}^{s} \sum_{n_i \geq -N} f_{n_i} v_1 \otimes \cdots \otimes v_{i-1} \otimes [L_{n_i}, x \otimes \tilde{g}_{p_i}] v_i \otimes v_{i+1} \\
&\quad \otimes \cdots \otimes v_s, \quad \text{by (6) of Definition 3.3.2 and (2) of Definition 3.3.6} \\
&= \sum_{i=1}^{s} \sum_{n_i, m_i \geq -N} m_i f_{n_i} g_{m_i} v_1 \otimes \cdots \otimes v_{i-1} \otimes x[m_i + n_i] v_i \\
&\quad \otimes v_{i+1} \otimes \cdots \otimes v_s, \quad \text{by Proposition 3.2.2(a).}
\end{aligned} \tag{2}$$

Now

$$\begin{aligned}
(x \otimes \tilde{\theta}_i^o(\tilde{g}_{p_i})) \cdot v = \sum_{n_i, m_i \geq -N} & v_1 \otimes \cdots \otimes v_{i-1} \\
& \otimes f_{n_i} g_{m_i} m_i x[n_i + m_i] v_i \otimes v_{i+1} \otimes \cdots \otimes v_s.
\end{aligned} \tag{3}$$

Combing (2) and (3), we get the lemma. □

3.3.E Exercises

(1) Let $\mathscr{V}_{\mathfrak{F}_T}^{\dagger}(\vec{p}, \vec{\lambda})$ be the sheaf of vacua as in Definition 3.3.5. Show that it is a coherent sheaf of $\mathscr{O}_T$-modules.

(2) Let T' be a smooth irreducible variety and let $f : T' \to T$ be a morphism. Let $\mathfrak{F}_T$ be a family of s-pointed curves with formal

parameters over T. Then, as in Definition 3.3.1, we have an induced family $\mathfrak{F}_{T'}$ over T'. Let $\vec{\lambda} = (\lambda_1, \ldots, \lambda_s) \in D_c^s$.

Prove the analogue of Lemma 3.3.3 for the morphism f, i.e., prove the following:

(a) $\mathcal{O}_{T'} \otimes_{\mathcal{O}_T} \mathcal{H}(\vec{\lambda})_T \simeq \mathcal{H}(\vec{\lambda})_{T'}$,
(b) $\mathcal{O}_{T'} \otimes_{\mathcal{O}_T} \hat{\mathfrak{g}}_T^{(s)} \simeq \hat{\mathfrak{g}}_{T'}^{(s)}$,
(c) $\mathcal{O}_{T'} \otimes_{\mathcal{O}_T} \mathcal{V}_{\mathfrak{F}_T}(\vec{p}, \vec{\lambda}) \simeq \mathcal{V}_{\mathfrak{F}_{T'}}(f^*\vec{p}, \vec{\lambda})$.

3.4 Flat Projective Connection on the Sheaf of Conformal Blocks

Let $\mathfrak{F}_T = (\Sigma_T, \vec{p}, \vec{t})$ be a family of s-pointed curves with formal parameters over a smooth irreducible variety T as in Section 3.3. In addition, assume that $\pi : \Sigma_T \to T$ is a smooth morphism (in particular, Σ_T is a smooth variety). Let $\vec{\lambda} = (\lambda_1, \ldots, \lambda_s) \in D_c^s$. Recall the definition of the sheaf of covacua $\mathcal{V}_{\mathfrak{F}_T}(\vec{p}, \vec{\lambda})$ over T from identity (7) of Definition 3.3.2.

Definition 3.4.1 Let $\mathcal{V} \to S$ be an (algebraic) vector bundle of rank n over a smooth variety S. By a *projective connection* over $\mathcal{V}$ we mean a PGL_n-connection over the corresponding projective bundle $\mathbb{P}(\mathcal{V}) \to S$. We only consider algebraic connections.

More explicitly, a projective connection ∇ over $\mathcal{V}$ consists of connections ∇^i over an open cover $\{S_i\}$ of S (i.e., ∇^i is a connection over $\mathcal{V}_{|S_i}$) such that for any pair i, j, any point $p \in S_i \cap S_j$ and any tangent vector $v \in T_p(S_i \cap S_j)$, $(\nabla^i_v - \nabla^j_v)$ as a linear operator on the fiber $\mathcal{V}_p$ is a scalar operator $z(v)\,\mathrm{Id}_{\mathcal{V}_p}$ (for some $z(v) \in \mathbb{C}$). (Observe that by the Leibnitz rule, $(\nabla^i_X - \nabla^j_X)$ is an $\mathcal{O}_{S_i \cap S_j}$-linear operator acting on the sheaf of sections of $\mathcal{V}_{|S_i \cap S_j}$ as well as, of course, in the X-variable.)

A projective connection ∇ is called a *flat projective connection* if each ∇^i is a projectively flat connection over $\mathcal{V}_{|S_i}$, i.e., the curvature of ∇^i is a scalar operator.

The following is the main result of this section.

Theorem 3.4.2 *With the notation and assumptions as above, at the beginning of this section, the sheaf $\mathcal{V}_{\mathfrak{F}_T}(\vec{p}, \vec{\lambda})$ over T is a locally free $\mathcal{O}_T$-module. Moreover, it carries a flat projective connection $\nabla^{(\mathfrak{F}_T, \vec{\lambda})}$.*

Further, the projective connection ∇ is functorial in $\mathfrak{F}_T$, i.e., for a smooth variety T' and a morphism $f: T' \to T$, the pull-back

$$f^*\nabla^{(\mathfrak{F}_T,\vec{\lambda})} = \nabla^{(\mathfrak{F}_{T'},\vec{\lambda})}, \tag{1}$$

where the family $\mathfrak{F}_{T'}$ is defined in Definition 3.3.1.

Proof We first claim that any vector field on T can locally (in T) be lifted to a vector field on $\Sigma_T^o := \Sigma_T \setminus \bigcup_{i=1}^s p_i(T)$. To prove this, consider the short exact sequence of sheaves over Σ_T^o:

$$0 \to \Theta_{\Sigma_T^o/T} \to \Theta_{\Sigma_T^o} \xrightarrow{d\pi^o} (\pi^o)^*\Theta_T \to 0, \tag{2}$$

where Θ_Y is the tangent sheaf of Y, π^o is the restriction of $\pi: \Sigma_T \to T$ to Σ_T^o and $\Theta_{\Sigma_T^o/T}$ denotes the subsheaf of $\Theta_{\Sigma_T^o}$ consisting of the vertical vector fields, i.e., the vector fields which project to 0 via $d\pi^o$. (Observe that the sheaf map $d\pi^o$ is surjective since we have assumed π to be a smooth morphism.)

Since π^o is an affine morphism (cf. Definition 3.3.1), the sheaf sequence (2) gives rise to the sheaf exact sequence of $\mathcal{O}_T$-modules:

$$0 \to \pi_*^o\left(\Theta_{\Sigma_T^o/T}\right) \to \pi_*^o\left(\Theta_{\Sigma_T^o}\right) \to \pi_*^o(\pi^o)^*\left(\Theta_T\right) \to 0. \tag{3}$$

Since $(\pi^o)^{-1}(\Theta_T) \subset (\pi^o)^*(\Theta_T)$, the claim, that any vector field θ on T can locally be lifted to a vector field $\tilde{\theta}$ on Σ_T^o, is established. (In fact, if T is affine, then it can be globally lifted.) For any $\tilde{\theta} \in \Theta_{\Sigma_T^o}$ and $1 \leq i \leq s$, in a formal punctured neighborhood of $p_i(T)$, we can uniquely write (cf. Definition 3.3.7)

$$\tilde{\theta} = \tilde{\theta}_i^o + \tilde{\theta}_i^h, \tag{4}$$

where

$$\tilde{\theta}_i^h(t_i) = 0 \quad \text{and} \quad \tilde{\theta}_i^o \in \mathcal{O}_T((t_i))\partial_{t_i}.$$

For any $\tilde{\theta} \in (d\pi^o)^{-1}(\pi^o)^{-1}(\Theta_T)$, i.e., $\tilde{\theta}$ is a lift of a vector field $\theta \in \Theta_T$ and $g \otimes v \in \mathscr{H}(\vec{\lambda})_T$ (for $g \in \mathcal{O}_T$ and $v \in \mathscr{H}(\vec{\lambda})$), define the covariant derivative

$$\nabla_{\tilde{\theta}}(g \otimes v) = \theta(g) \otimes v + g\tilde{\gamma}_{\vec{\lambda}}(\tilde{\theta}) \cdot v, \tag{5}$$

where the $\mathcal{O}_T$-linear operator $\tilde{\gamma}_{\vec{\lambda}}(\tilde{\theta})$ is defined by identity (2) of Definition 3.3.7. (Observe that if $\theta \in \Theta_T$ is defined on an affine open subset $U \subset T$, then θ admits a lift $\tilde{\theta} \in \Theta_{\Sigma_T^o}$ defined over $(\pi^o)^{-1}(U)$. If $U \subset T$ is not affine, then $\tilde{\theta}$ may not be defined on $(\pi^o)^{-1}(U)$ though. Also, observe that if we choose formal local coordinates $\{x_k\}$ at $b \in T$, then $\{x_k, t_i\}$ gives a formal local coordinate for Σ_T at $p_i(b)$. Write $\theta \in \Theta_T$ locally (in a formal neighborhood of b) as $\theta = \sum_k f_k(x)\partial_{x_k}$. Then, any lift $\tilde{\theta}$ in a formal neighborhood of $p_i(b)$ can be written as $\tilde{\theta} = g(x,t_i)\partial_{t_i} + \sum_k f_k(x)\partial_{x_k}$.)

The operator $\nabla_{\tilde{\theta}}$ satisfies the following properties. For any $f,\ g \in \mathcal{O}_T$, $v \in \mathscr{H}(\vec{\lambda})$, $\hat{y} \in \hat{\mathfrak{g}}(\mathfrak{F}_T)$ (where $\hat{\mathfrak{g}}(\mathfrak{F}_T)$ is defined by identity (4) of Definition 3.3.2),

$$\nabla_{\tilde{\theta}}(f \cdot (g \otimes v)) = \theta(f) g \otimes v + f \nabla_{\tilde{\theta}}(g \otimes v), \tag{6}$$

$$\nabla_{f\tilde{\theta}}(g \otimes v) = f \nabla_{\tilde{\theta}}(g \otimes v), \tag{7}$$

$$[\nabla_{\tilde{\theta}}, \hat{y}] = \tilde{\theta}(\hat{y}), \quad \text{as operators on } \mathscr{H}(\vec{\lambda})_T, \tag{8}$$

where $\hat{y}$ acts on $\mathscr{H}(\vec{\lambda})_T$ via the embedding φ given in (6) of Definition 3.3.2 and $\tilde{\theta}(\hat{y})$ denotes the standard action of the vector field on the second factor of $\hat{\mathfrak{g}}(\mathfrak{F}_T) := \mathfrak{g} \otimes \pi_*^o(\mathcal{O}_{\Sigma_T^o})$. Of course, (6) and (7) are clear from definition (5). We now prove (8).

Consider the operator $\mathfrak{X}_\theta$ on $\mathscr{H}(\vec{\lambda})_T$ defined by

$$\mathfrak{X}_\theta(g \otimes v) = \theta(g) \otimes v, \quad \text{for} \quad g \in \mathcal{O}_T \quad \text{and} \quad v \in \mathscr{H}(\vec{\lambda}).$$

Then

$$\nabla_{\tilde{\theta}} = \mathfrak{X}_\theta + \tilde{\gamma}_{\vec{\lambda}}(\tilde{\theta}). \tag{9}$$

Thus, for $\hat{y} = x \otimes \tilde{g}$ with $x \in \mathfrak{g}$ and $\tilde{g} \in \pi_*^o(\mathcal{O}_{\Sigma_T^o})$, writing

$$\tilde{\theta}_i^o = \sum_{n_i \geq -N} f_{n_i} t_i^{n_i+1} \partial_{t_i} \in \mathcal{O}_T((t_i))\partial_{t_i} \ \text{ and } \tilde{g}_{p_i} = \sum_{m_i \geq -N} g_{m_i} t_i^{m_i} \in \mathcal{O}_T((t_i)),$$

we get (for $g \in \mathcal{O}_T$ and $v = v_1 \otimes \cdots \otimes v_s \in \mathscr{H}(\vec{\lambda})$ with $v_i \in \mathscr{H}(\lambda_i)$)

$$\begin{aligned}
[\nabla_{\tilde{\theta}}, \hat{y}](g \otimes v) &= [\mathfrak{X}_\theta, \hat{y}](g \otimes v) + [\tilde{\gamma}_{\vec{\lambda}}(\tilde{\theta}), \hat{y}](g \otimes v) \\
&= \mathfrak{X}_\theta(\hat{y}(g \otimes v)) - \hat{y}(\mathfrak{X}_\theta(g \otimes v)) \\
&\quad + \sum_{i=1}^s \left(x \otimes \tilde{\theta}_i^o(\tilde{g}_{p_i})\right)(g \otimes v), \quad \text{by Lemma 3.3.8} \\
&= \sum_{i=1}^s \sum_{m_i \geq -N} \left(\theta(g g_{m_i}) - \theta(g) g_{m_i}\right) \\
&\quad \times (v_1 \otimes \cdots \otimes v_{i-1} \otimes x[m_i] \cdot v_i \otimes v_{i+1} \otimes \cdots \otimes v_s)
\end{aligned}$$

$$+\sum_{i=1}^{s}\sum_{n_i,m_i\geq -N} gf_{n_i}g_{m_i}m_i(v_1\otimes\cdots\otimes v_{i-1}\otimes x[n_i+m_i]\cdot v_i$$

$$\otimes v_{i+1}\otimes\cdots\otimes v_s),$$

by identity (3) of the proof of Lemma 3.3.8

$$=\sum_{i=1}^{s}\sum_{m_i\geq -N} g\theta(g_{m_i})$$

$$\times(v_1\otimes\cdots\otimes v_{i-1}\otimes x[m_i]\cdot v_i\otimes v_{i+1}\otimes\cdots\otimes v_s)$$

$$+\sum_{i=1}^{s}\sum_{n_i,m_i\geq -N} gf_{n_i}g_{m_i}m_i(v_1\otimes\cdots\otimes v_{i-1}\otimes x[n_i+m_i]\cdot v_i$$

$$\otimes v_{i+1}\otimes\cdots\otimes v_s). \tag{10}$$

Further (with the same notation as above),

$$\tilde{\theta}(\hat{y})(g\otimes v)=\sum_{i=1}^{s} gv_1\otimes\cdots\otimes v_{i-1}\otimes(x\otimes\tilde{\theta}(\tilde{g})_{p_i})\cdot v_i\otimes v_{i+1}\otimes\cdots\otimes v_s$$

$$=\sum_{i=1}^{s}\sum_{n_i,m_i\geq -N} gf_{n_i}g_{m_i}m_i(v_1\otimes\cdots\otimes v_{i-1}\otimes x[n_i+m_i]\cdot v_i$$

$$\otimes v_{i+1}\otimes\cdots\otimes v_s)$$

$$+\sum_{i=1}^{s}\sum_{m_i\geq -N} gv_1\otimes\cdots\otimes v_{i-1}\otimes\tilde{\theta}_i^h(g_{m_i})x[m_i]\cdot v_i$$

$$\otimes v_{i+1}\otimes\cdots\otimes v_s,\quad \text{by (4)}. \tag{11}$$

But, since $\tilde{\theta}$ is a lift of a vector field $\theta\in\Theta_T$,

$$\tilde{\theta}_i^h(g_{m_i})=\theta(g_{m_i}). \tag{12}$$

Combining (10)–(12), we get (8).

From (8), we see that

$$\nabla_{\tilde{\theta}}\left(\hat{\mathfrak{g}}(\mathfrak{F}_T)\cdot\mathscr{H}(\vec{\lambda})_T\right)\subset\hat{\mathfrak{g}}(\mathfrak{F}_T)\cdot\mathscr{H}(\vec{\lambda})_T.$$

In particular, $\nabla_{\tilde{\theta}}$ descends to an operator on the sheaf of covacua $\mathscr{V}_{\mathfrak{F}_T}(\vec{p},\vec{\lambda})$ defined by (7) of Definition 3.3.2.

We next calculate the curvature $\mathcal{K}$ of ∇. Let $\theta, \delta \in \Theta_T$ and let $\tilde{\theta}, \tilde{\delta} \in \Theta_{\Sigma_T^o}$ be their lifts. Write, for any $1 \le i \le s$,

$$\tilde{\theta}_i^o = \sum_{n_i \ge -N} f_{n_i} t_i^{n_i+1} \partial_{t_i}, \quad \tilde{\delta}_i^o = \sum_{n_i \ge -N} h_{n_i} t_i^{n_i+1} \partial_{t_i}.$$

Then, for any $g \otimes v \in \mathscr{H}(\lambda)_T$ with $g \in \mathscr{O}_T$ and $v = v_1 \otimes \cdots \otimes v_s \in \mathscr{H}(\vec{\lambda})$,

$$\begin{aligned}
\mathcal{K}(\tilde{\delta},\tilde{\theta})(g \otimes v) &:= \nabla_{\tilde{\delta}} \nabla_{\tilde{\theta}}(g \otimes v) - \nabla_{\tilde{\theta}} \nabla_{\tilde{\delta}}(g \otimes v) - \nabla_{[\tilde{\delta},\tilde{\theta}]}(g \otimes v) \\
&= \nabla_{\tilde{\delta}} \left(\theta(g) \otimes v + g \sum_{i=1}^{s} \tilde{\theta}_i^o \cdot v \right) \\
&\quad - \nabla_{\tilde{\theta}} \left(\delta(g) \otimes v + g \sum_{i=1}^{s} \tilde{\delta}_i^o \cdot v \right) - [\delta,\theta](g) \otimes v \\
&\quad - g \sum_{i=1}^{s} [\tilde{\delta}_i^o, \tilde{\theta}_i^o] \cdot v \\
&\quad - g \sum_{i=1}^{s} \sum_{n_i \ge -N} \big(\delta(f_{n_i}) - \theta(h_{n_i})\big)(t_i^{n_i+1} \partial_{t_i}) \cdot v, \quad \text{by (4)} \\
&= \delta\big(\theta(g)\big) \otimes v + \theta(g) \sum_{i=1}^{s} \tilde{\delta}_i^o \cdot v \\
&\quad - \frac{1}{c+h^\vee} \sum_{i=1}^{s} \sum_{n_i \ge -N} \delta(g f_{n_i}) v_1 \otimes \cdots \otimes L_{n_i} v_i \otimes \cdots \otimes v_s \\
&\quad + g \sum_{i=1}^{s} \tilde{\delta}_i^o (\tilde{\theta}_i^o \cdot v) - \theta(\delta(g)) \otimes v - \delta(g) \sum_{i=1}^{s} \tilde{\theta}_i^o \cdot v \\
&\quad + \frac{1}{c+h^\vee} \sum_{i=1}^{s} \sum_{n_i \ge -N} \theta(g h_{n_i}) v_1 \otimes \cdots \otimes L_{n_i} v_i \otimes \cdots \otimes v_s \\
&\quad - g \sum_{i=1}^{s} \tilde{\theta}_i^o (\tilde{\delta}_i^o \cdot v) - [\delta,\theta](g) \otimes v - g \sum_{i=1}^{s} [\tilde{\delta}_i^o, \tilde{\theta}_i^o] \cdot v \\
&\quad + \frac{g}{c+h^\vee} \sum_{i=1}^{s} \sum_{n_i \ge -N} \big(\delta(f_{n_i}) - \theta(h_{n_i})\big) \\
&\quad \times (v_1 \otimes \cdots \otimes v_{i-1} \otimes L_{n_i} v_i \otimes v_{i+1} \otimes \cdots \otimes v_s), \\
&\qquad \text{by identity (2) of Definition 3.3.6}
\end{aligned}$$

$$= g\beta(\tilde{\delta},\tilde{\theta})v - \frac{1}{c+h^\vee}\sum_{i=1}^{s}$$

$$\sum_{n_i\geq -N} \delta(g) f_{n_i} v_1 \otimes \cdots \otimes L_{n_i} v_i \otimes \cdots \otimes v_s$$

$$+ \frac{1}{c+h^\vee}\sum_{i=1}^{s}\sum_{n_i\geq -N} \theta(g) h_{n_i} v_1 \otimes \cdots \otimes L_{n_i} v_i \otimes \cdots \otimes v_s$$

$$+ \theta(g)\sum_{i=1}^{s} \tilde{\delta}_i^o \cdot v - \delta(g)\sum_{i=1}^{s}\tilde{\theta}_i^o \cdot v,$$

for some $\beta(\tilde{\delta},\tilde{\theta}) \in \mathcal{O}_T$ depending only upon $\tilde{\delta},\tilde{\theta}$

using Proposition 3.2.2(b)

$$= g\beta(\tilde{\delta},\tilde{\theta})v, \quad \text{by identity (2) of Definition 3.3.6.} \tag{13}$$

This shows that the curvature $\mathcal{K}(\tilde{\delta},\tilde{\theta})$ is a scalar operator.

We next show that the sheaf $\mathcal{V}_{\mathfrak{F}_T}(\vec{p},\vec{\lambda})$ is a locally free $\mathcal{O}_T$-module. It suffices to show that for any $b \in T$, the stalk $\mathcal{V}_{\mathfrak{F}_T}(\vec{p},\vec{\lambda})_b$ of $\mathcal{V}_{\mathfrak{F}_T}(\vec{p},\vec{\lambda})$ at b is a free $\mathcal{O}_{T,b}$-module, where $\mathcal{O}_{T,b}$ is the local ring of T at b (cf. (Hartshorne, 1977, Chap. II, Exercise 5.7)). Take $v_1,\ldots,v_n \in \mathcal{V}_{\mathfrak{F}_T}(\vec{p},\vec{\lambda})_b$ such that their evaluation $v_1(b),\ldots,v_n(b) \in \mathbb{C}_b \bigotimes_{\mathcal{O}_T} \mathcal{V}_{\mathfrak{F}_T}(\vec{p},\vec{\lambda})$ is a $\mathbb{C}$-basis. Then, by Nakayama's lemma, $v_1,\ldots,v_n$ generates $\mathcal{V}_{\mathfrak{F}_T}(\vec{p},\vec{\lambda})_b$ over the local ring $\mathcal{O}_{T,b}$. We show that $v_1,\ldots,v_n$ are linearly independent elements of $\mathcal{V}_{\mathfrak{F}_T}(\vec{p},\vec{\lambda})_b$ over $\mathcal{O}_{T,b}$. For otherwise, choose a nontrivial relation

$$f_1v_1 + \cdots + f_nv_n = 0, \quad \text{with} \quad f_i \in \mathcal{O}_{T,b}, \tag{14}$$

satisfying the following:

(a) For some $1 \leq i_o \leq n$, $f_{i_o} \in \mathfrak{m}_{T,b}^k \backslash \mathfrak{m}_{T,b}^{k+1}$, and $f_i \in \mathfrak{m}_{T,b}^k$ for all $1 \leq i \leq n$. (Observe that since $v_1(b),\ldots,v_n(b)$ are linearly independent, $k \geq 1$.)

(b) For any nontrivial relation

$$g_1v_1 + \cdots + g_nv_n = 0, \quad \text{with} \quad g_i \in \mathcal{O}_{T,b},$$

we have $g_i \in \mathfrak{m}_{T,b}^k$, for all $1 \leq i \leq n$.

Now, choose a vector field $\theta_o \in \Theta_{T,b}$ such that $\theta_o(f_{i_o}) \in \mathfrak{m}_{T,b}^{k-1}\backslash\mathfrak{m}_{T,b}^k$, where $\Theta_{T,b}$ is the stalk of the sheaf Θ_T at b, and lift it to $\tilde{\theta}_o \in \Theta_{\Sigma_T^o}$. Then, by (6),

$$0 = \nabla_{\tilde{\theta}_o}(f_1 v_1 + \cdots + f_n v_n) = \theta_o(f_{i_o})v_{i_o} + \sum_{i \neq i_o} \theta_o(f_i)v_i + \sum_{i,j=1}^{n} f_i \beta_{i,j} v_j,$$

for some $\beta_{i,j} \in \mathcal{O}_{T,b}$.

Since each $f_i \in \mathfrak{m}_{T,b}^k$ by (a), the above relation contradicts (b). Hence $\{v_1, \ldots, v_n\}$ is a basis of $\mathscr{V}_{\mathfrak{F}_T}(\vec{p}, \vec{\lambda})_b$ over $\mathcal{O}_{T,b}$ and thus $\mathscr{V}_{\mathfrak{F}_T}(\vec{p}, \vec{\lambda})$ is a locally free $\mathcal{O}_T$-module.

We finally show that the operator $\nabla_{\tilde{\theta}}$ acting on $\mathscr{V}_{\mathfrak{F}_T}(\vec{p}, \vec{\lambda})$ projectively does not depend upon the choice of the lift $\tilde{\theta} \in \Theta_{\Sigma_T^o}$ of $\theta \in \Theta_T$ (though $\nabla_{\tilde{\theta}}$ does depend upon $\tilde{\theta}$, even projectively, as an operator on $\mathscr{H}(\vec{\lambda})_T$).

Consider the $\mathcal{O}_{\Sigma_T^o}$-locally free sheaf $\Theta_{\Sigma_T^o/T}$ of vertical vector fields for the smooth morphism $\pi^o \colon \Sigma_T^o \to T$. Then, any two lifts of θ differ (locally in T) by an element of $\pi_*^o(\Theta_{\Sigma_T^o/T})$ (and conversely). Thus, it suffices to show that for any $\tilde{\theta} \in \pi_*^o(\Theta_{\Sigma_T^o/T})$, $\nabla_{\tilde{\theta}}$ acts by a scalar operator on $\mathscr{V}_{\mathfrak{F}_T}(\vec{p}, \vec{\lambda})$ in the precise sense given below. Observe first that $\pi_*^o(\Theta_{\Sigma_T^o/T})$ is a $\mathcal{O}_T$-Lie algebra under the standard bracket of vector fields. By the definition, for any $\tilde{\theta} \in \pi_*^o(\Theta_{\Sigma_T^o/T})$, $\nabla_{\tilde{\theta}}$ is an $\mathcal{O}_T$-linear operator. Thus, for any $b \in T$, $\nabla_{\tilde{\theta}}$ induces a $\mathbb{C}$-linear operator

$$\nabla_{\tilde{\theta}}(b) \colon \mathbb{C}_b \otimes_{\mathcal{O}_T} \mathscr{V}_{\mathfrak{F}_T}(\vec{p}, \vec{\lambda}) \to \mathbb{C}_b \otimes_{\mathcal{O}_T} \mathscr{V}_{\mathfrak{F}_T}(\vec{p}, \vec{\lambda}).$$

Hence, we get a linear map (by using (7))

$$\nabla(b) \colon \mathbb{C}_b \otimes_{\mathcal{O}_T} \pi_*^o(\Theta_{\Sigma_T^o/T}) \to \operatorname{End}_{\mathbb{C}}\left(\mathbb{C}_b \otimes_{\mathcal{O}_T} \mathscr{V}_{\mathfrak{F}_T}(\vec{p}, \vec{\lambda})\right).$$

Moreover, $\nabla(b)$ is a projective representation of the Lie algebra $\mathbb{C}_b \otimes_{\mathcal{O}_T} \pi_*^o(\Theta_{\Sigma_T^o/T})$ by (13) (i.e.,

$$\nabla(b) \colon \mathbb{C}_b \otimes_{\mathcal{O}_T} \pi_*^o(\Theta_{\Sigma_T^o/T}) \to \operatorname{End}_{\mathbb{C}}\left(\mathbb{C}_b \otimes_{\mathcal{O}_T} \mathscr{V}_{\mathfrak{F}_T}(\vec{p}, \vec{\lambda})\right) / \mathbb{C}\,\mathrm{Id}$$

is a Lie algebra homomorphism). Further, $\mathbb{C}_b \otimes_{\mathcal{O}_T} \pi_*^o(\Theta_{\Sigma_T^o/T})$ can easily be seen to be the Lie algebra of vector fields on the fiber $\Sigma_b \backslash \vec{p}(b)$ of the affine and smooth morphism π^o over b. Hence, $\mathbb{C}_b \otimes_{\mathcal{O}_T} \pi_*^o(\Theta_{\Sigma_T^o/T})$ is a finite direct sum of infinite-dimensional simple Lie algebras, summands coming from components of $\Sigma_b \backslash \vec{p}(b)$ (cf. (Beilinson, Feigin and Mazur, 1990)). Also, since $\mathscr{V}_{\mathfrak{F}_T}(\vec{p}, \vec{\lambda})$ is a coherent sheaf, $\mathbb{C}_b \otimes_{\mathcal{O}_T} \mathscr{V}_{\mathfrak{F}_T}(\vec{p}, \vec{\lambda})$ is a finite-dimensional complex vector space. Thus, $\nabla(b)$ being a projective representation for any $b \in T$, $\nabla_{\tilde{\theta}}(b)$ is a scalar operator

$$z(b, \tilde{\theta})\,\mathrm{Id}_{(\mathbb{C}_b \otimes_{\mathcal{O}_T} \mathscr{V}_{\mathfrak{F}_T}(\vec{p}, \vec{\lambda}))}$$

for any $\tilde{\theta} \in \pi^o_*(\Theta_{\Sigma^o_T/T})$ and $b \in T$ (for some $z(b,\tilde{\theta}) \in \mathbb{C}$). Since $\mathscr{V}_{\mathfrak{F}_T}(\vec{p},\vec{\lambda})$ is a locally free $\mathscr{O}_T$-module, we get that $\nabla_{\tilde{\theta}}$ acting on $\mathscr{V}_{\mathfrak{F}_T}(\vec{p},\vec{\lambda})$ is a scalar operator for any $\tilde{\theta} \in \pi^o_*(\Theta_{\Sigma^o_T/T})$ (see Exercise 3.4.E.1 for a counter-example when local freeness is not available).

Thus, using (6), (7) and (13), we get that the locally defined connections $\nabla_{\tilde{\theta}}$ as $\tilde{\theta}$ ranges over $(d\pi^o)^{-1}(\pi^o)^{-1}(\Theta_T)$ (i.e., $\tilde{\theta}$ is a lift of a vector field $\theta \in \Theta_T$) patch-up to give a flat projective connection $\nabla^{(\mathfrak{F}_T,\vec{\lambda})}$ over $\mathscr{V}_{\mathfrak{F}_T}(\vec{p},\vec{\lambda})$.

The functoriality of the projective connection ∇ is clear from its definition. This completes the proof of the theorem. □

Example 3.4.3 We consider a particular (important) example of Theorem 3.4.2. Let $\Sigma = \mathbb{P}^1 = \mathbb{C} \cup \{\infty\}$ and let $T = \mathbb{C}^s \backslash D$, where D is the multi-diagonal

$$D := \left\{\vec{z} = (z_1, \dots, z_s) \in \mathbb{C}^s : z_i = z_j \quad \text{for some} \quad i \neq j\right\}.$$

Define the family $\mathfrak{F}_T = (\Sigma_T, \vec{p}, \vec{t})$ by the projection $\pi : \Sigma_T := \mathbb{P}^1 \times T \to T$, with the sections (for $1 \leq i \leq s$) $p_i(\vec{z}) = (z_i, \vec{z})$, for $\vec{z} = (z_1, \dots, z_s) \in T$ and the local parameters t_i along $p_i(T)$ defined by

$$t_i(x,\vec{z}) = x - z_i, \quad \text{for} \quad x \in \mathbb{P}^1 \quad \text{and} \quad z \in T.$$

In this case, the connection ∇ of Theorem 3.4.2 is given by $\Big($for $\theta = \sum_{j=1}^s f_j \partial_{z_j} \in \Theta_T$, $g \in \mathscr{O}_T$ and $v = v_1 \otimes \cdots \otimes v_s \in \mathscr{H}(\vec{\lambda})\Big)$

$$\nabla_\theta(g \otimes v) = \theta(g) \otimes v + \frac{g}{c + h^\vee} \sum_{i=1}^s f_i v_1 \otimes \cdots \otimes v_{i-1} \otimes L_{-1} v_i \otimes v_{i+1} \otimes \cdots \otimes v_s, \tag{1}$$

where $\vec{\lambda} = (\lambda_1, \dots, \lambda_s) \in D^s_c$ and the Sugawara element $L_n \in \hat{U}(\tilde{\mathfrak{g}})$ is defined by identity (3) of Definition 3.2.1.

To prove (1), lift the vector field θ to the vector field $\tilde{\theta}$ on Σ_T by

$$\tilde{\theta} = \sum_{j=1}^s f_j \partial_{z_j} + 0 \cdot \partial_x \,,$$

where $\{\partial_x, \partial_{z_j}\}_j$ is taken with respect to the standard coordinates $(x, z_1, \dots, z_s)$ on $\mathbb{C} \times T$. Then, with respect to the coordinates $(t_i, z_1, \dots, z_s)$ in a neighborhood of $p_i(T)$, $\tilde{\theta}$ is given by

$$\tilde{\theta} = \sum_{j=1}^s f_j \tilde{\partial}_{z_j} - f_i \tilde{\partial}_{t_i},$$

where the system of vector fields $\{\tilde{\partial}_{t_i}, \tilde{\partial}_{z_j}\}_{1\leq j\leq s}$ in a neighborhood of $p_i(T)$ is taken with respect to the coordinates $(t_i, z_1, \ldots, z_s)$. Clearly, under decomposition (4) of the proof of Theorem 3.4.2,

$$\tilde{\theta}_i^h = \sum_{j=1}^{s} f_j \tilde{\partial}_{z_j} \quad \text{and} \quad \tilde{\theta}_i^o = -f_i \tilde{\partial}_{t_i}. \tag{2}$$

Now, (1) follows from the definition of ∇ given by (5) of the proof of Theorem 3.4.2 (in view of identity (2) of Definition 3.3.7 and identity (2) of Definition 3.3.6). Observe that ∇ is a connection on $\mathscr{H}(\vec{\lambda})_T$ and descends to a connection on $\mathscr{V}_{\mathfrak{F}_T}(\vec{p}, \vec{\lambda})$ (as opposed to a projective connection in general). From expression (1), it is easy to see that the connection ∇ in this case is actually flat.

Remark 3.4.4 Let $T = \mathbb{C}^s \backslash D$ and $\mathfrak{F}_T$ be as in Example 3.4.3, and let $\vec{\lambda} = (\lambda_1, \ldots, \lambda_s) \in D_c^s$. Then Knizhnik–Zamolodchikov gave the following connection $\overline{\nabla}$ (known as the *KZ-connection*) on the $\mathscr{O}_T$-module $\mathscr{O}_T \otimes_{\mathbb{C}} V(\vec{\lambda})$, where $V(\vec{\lambda}) := V(\lambda_1) \otimes \cdots \otimes V(\lambda_s)$:

$$\overline{\nabla}_{\partial_{z_i}}(g \otimes v) = (\partial_{z_i} g) \otimes v + \frac{g}{c + h^\vee} \otimes \sum_{j \neq i} \frac{\Omega_{i,j}}{z_j - z_i}(v), \tag{1}$$

for $g \in \mathscr{O}_T$ and $v \in V(\vec{\lambda})$ where $\Omega_{i,j}$ denotes the Casimir operator acting (only) on the (i, j)th factor $V(\lambda_i) \otimes V(\lambda_j)$. Under the identification of the conformal block bundle $\mathscr{V}_{\mathfrak{F}_T}(\vec{p}, \vec{\lambda})$ with a $\mathscr{O}_T$-module quotient of $\mathscr{O}_T \otimes_{\mathbb{C}} V(\vec{\lambda})$ (cf. Theorem 2.3.2), the KZ-connection $\overline{\nabla}$ descends to give a connection on $\mathscr{V}_{\mathfrak{F}_T}(\vec{p}, \vec{\lambda})$. Moreover, this connection on $\mathscr{V}_{\mathfrak{F}_T}(\vec{p}, \vec{\lambda})$ coincides with the connection of Example 3.4.3 (cf. (Feigin, Schechtman and Varchenko, 1990, Theorem 1) and (Looijenga, 2005, §3)).

3.4.E Exercises

(1) Let $X = \operatorname{Spec}(\mathbb{C}[t])$ and consider the $\mathbb{C}[t]$-module $M = \mathbb{C}[t]/(t^2)$. The module M canonically gives rise to a coherent sheaf $\mathcal{M}$ over X. Consider the $\mathbb{C}[t]$-module map

$$M \to M, \quad f \mapsto ft,$$

and let $\varphi \colon \mathcal{M} \to \mathcal{M}$ be the corresponding $\mathscr{O}_X$-module map. Show that the induced fiber map

$$\varphi(x) \colon \mathbb{C}_x \otimes_{\mathscr{O}_X} \mathcal{M} \to \mathbb{C}_x \otimes_{\mathscr{O}_X} \mathcal{M}$$

is zero for all $x \in X$. But, of course, φ is a nonzero map.

(2) Let Σ be a smooth projective curve and let $\xi : T \to \Sigma^o$ be a holomorphic isomorphism from an analytic open subset $T \subset \mathbb{C}$ onto an analytic open subset Σ^o of Σ. Consider the holomorphic family $\mathfrak{F}_T = (\Sigma_T, p, t)$ for $s = 1$, where $\pi : \Sigma_T := \Sigma \times T \to T$ is the projection, $p(z) = (\xi(z), z)$ and $t(x,z) = \xi^{-1}(x) - z$, for $(x,z) \in \Sigma^o \times T$.

Calculate the holomorphic connection ∇ (as in Theorem 3.4.2) in this case and show that it is flat.

Generalize this to any $s \geq 1$, replacing T by $T^s \backslash D$, where D is the multi-diagonal.

Hint: Follow Example 3.4.3.

3.5 Local Freeness of the Sheaf of Conformal Blocks

As in earlier sections, let $s \geq 1$ and let $c \geq 1$ be the level. Let $\vec{\lambda} = (\lambda_1, \ldots, \lambda_s) \in D_c^s$. For preliminaries on stacks, we refer to Appendix C. In this section, we consider families of s-pointed curves with formal parameters over $\mathbb{D}_\tau := Spec\, \mathbb{C}[[\tau]]$ instead of over smooth irreducible varieties as in Section 3.3.

Let $(\Sigma_o, \vec{p}_o)$ be an s-pointed curve with a node at q (and possibly other nodes). Let $\tilde{\Sigma}_o$ be the normalization of Σ_o at the point q. The nodal point q splits into two smooth points q', q'' in $\tilde{\Sigma}_o$. The following lemma shows that there exists a 'canonical' smoothing deformation of $(\Sigma_o, \vec{p}_o)$ over the formal disc $\mathbb{D}_\tau$. We denote by $\mathbb{D}_\tau^\times$ the associated punctured formal disc $Spec\, \mathbb{C}((\tau))$.

A proof of the following lemma is sketched in Looijenga (2013, §6). For more details, we refer to Damiolini (2020, §6.1).

Lemma 3.5.1 *There exist families of s-pointed curves with formal parameters $\mathfrak{F}_\Sigma = (\Sigma, \vec{p}, \vec{t}), \mathfrak{F}_{\tilde{\Sigma}} = (\tilde{\Sigma}, \vec{\tilde{p}}, \vec{\tilde{t}})$ over $\mathbb{D}_\tau$ and a morphism $\zeta : \tilde{\Sigma} \to \Sigma$ of families of s-pointed curves with formal parameters over $\mathbb{D}_\tau$ (in particular, $\zeta \circ \tilde{p}_i = p_i$ for all $1 \leq i \leq s$), such that the following properties hold:*

(1) Over the closed point $o \in \mathbb{D}_\tau$, $\zeta|_o : \tilde{\Sigma}_o \to \Sigma_o$ is the normalization of Σ_o at (only) the point q, $\vec{p}(o) = \vec{p}_o$.

(2) The completed local ring $\hat{\mathcal{O}}_{\Sigma,q}$ of $\mathcal{O}_\Sigma$ at q is isomorphic to $\mathbb{C}[[z', z'', \tau]]/\langle \tau - z'z'' \rangle \simeq \mathbb{C}[[z', z'']]$. Moreover, $(z', \tau/z')$ (resp. $(z'', \tau/z'')$) gives a formal parameter around q' (resp. q'') in $\tilde{\Sigma}$, where we still denote by z' (resp. z'') the function around q' (resp. q'') by pulling back z' (resp. z'') via ζ.

(3) There exists an isomorphism of $\mathbb{C}[[\tau]]$-algebras

$$\kappa : \hat{\mathcal{O}}_{\tilde{\Sigma}\backslash\{q',q''\}, \tilde{\Sigma}_o\backslash\{q',q''\}} \simeq \mathcal{O}_{\tilde{\Sigma}_o\backslash\{q',q''\}}[[\tau]], \tag{1}$$

where $\hat{\mathcal{O}}_{\tilde{\Sigma}\backslash\{q',q''\},\tilde{\Sigma}_o\backslash\{q',q''\}}$ *is the completion of* $\mathcal{O}_{\tilde{\Sigma}\backslash\{q',q''\}}$ *along* $\tilde{\Sigma}_o\backslash\{q',q''\}$ *and the* $\mathbb{C}[[\tau]]$*-algebra structure on* $\hat{\mathcal{O}}_{\tilde{\Sigma}\backslash\{q',q''\},\tilde{\Sigma}_o\backslash\{q',q''\}}$ *is obtained from the projection* $\tilde{\Sigma}\backslash\{q',q''\}\to\mathbb{D}_\tau$*. Moreover,* $\vec{\tilde{p}}$ *in* $\tilde{\Sigma}$ *and* $\vec{\tilde{p}}_o$ *in* $\tilde{\Sigma}_o$ *are compatible under this isomorphism, i.e., the points* $\tilde{p}_{o,i}\in\tilde{\Sigma}_o\backslash\{q',q''\}$ *(for* $i=1,\dots,s$*) identified with the algebra homomorphisms* $\beta_i:\mathcal{O}_{\tilde{\Sigma}_o\backslash\{q',q''\}}\to\mathbb{C}$ *extended to*

$$\beta_i^\tau:\hat{\mathcal{O}}_{\tilde{\Sigma}\backslash\{q',q''\},\tilde{\Sigma}_o\backslash\{q',q''\}}\to\mathbb{C}[[\tau]]$$

under the identification κ *correspond to the sections* $\tilde{p}_i$ *of* $\tilde{\Sigma}\to\mathbb{D}_\tau$*.*

We will use $\vec{p}$ *to denote the sections* $\vec{\tilde{p}}$ *in* $\tilde{\Sigma}$ *if there is no confusion.*

Let $\hat{\mathfrak{g}}'$ (resp. $\hat{\mathfrak{g}}''$) be the affine Kac–Moody Lie algebra attached to the point q' (resp. q'') in $\tilde{\Sigma}_o$ with respect to the parameters z' and z'', respectively. For any $\mu\in D_c$, let $\mathscr{H}'(\mu^*)$ (resp. $\mathscr{H}''(\mu)$) be the highest-weight integrable representation of $\hat{\mathfrak{g}}'$ (resp. $\hat{\mathfrak{g}}''$) with highest weights μ^* (resp. μ). We will abbreviate $\mathscr{H}'(\mu^*)$ and $\mathscr{H}''(\mu)$, respectively by $\mathscr{H}(\mu^*)$ and $\mathscr{H}(\mu)$ and understand that $\mathscr{H}(\mu^*)$ (resp. $\mathscr{H}(\mu)$) is a representation of $\hat{\mathfrak{g}}'$ (resp. $\hat{\mathfrak{g}}''$).

Lemma 3.5.2 *There exists a nondegenerate pairing* $\langle\,,\rangle:\mathscr{H}(\mu^*)\times\mathscr{H}(\mu)\to\mathbb{C}$ *such that for any* $h_1\in\mathscr{H}(\mu^*)$, $h_2\in\mathscr{H}(\mu)$, *and* $x[z'^n]\in\hat{\mathfrak{g}}'$,

$$\langle x[z'^n]\cdot h_1,h_2\rangle+\langle h_1,x[z''^{-n}]\cdot h_2\rangle=0. \tag{1}$$

Proof This follows immediately from the explicit construction of $\mathscr{H}(\mu^*)$ as in Definition 3.1.1 by taking the standard pairing $\mathscr{H}(\mu)^\vee\times\mathscr{H}(\mu)\to\mathbb{C}$. □

Definition 3.5.3 There exist direct sum decompositions by the negative of the d-degree (cf. identity (1) of the proof of Theorem 1.2.10) (putting the d-degree of the highest-weight vectors at 0):

$$\mathscr{H}(\mu^*)=\bigoplus_{k=0}^\infty\mathscr{H}(\mu^*)_k,\ \mathscr{H}(\mu)=\bigoplus_{k=0}^\infty\mathscr{H}(\mu)_k.$$

Choose any basis $\{f_j^k\}_{j\in S_\mu^k}$ of $\mathscr{H}(\mu^*)_k$ and let $\{v_j^k\}_{j\in S_\mu^k}$ be the dual basis of $\mathscr{H}(\mu)_k$ under the above pairing $\langle\,,\rangle$. (Observe that $\langle\mathscr{H}(\mu^*)_k,\mathscr{H}(\mu)_{k'}\rangle=0$ unless $k=k'$.) Set, for any $k\geq 0$,

$$\Delta_{\mu,k}:=\sum_{j\in S_\mu^k}f_j^k\otimes v_j^k\in\mathscr{H}(\mu^*)_k\otimes\mathscr{H}(\mu)_k.$$

For $k<0$, we set $\Delta_{\mu,k}=0$.

Then, $\Delta_{\mu,0} = \mathrm{I}_\mu$, where I_μ is the element introduced in Definition 3.1.1. In view of Lemma 3.5.2, $\Delta_{\mu,k}$ satisfies the following property (for any $k, n \in \mathbb{Z}$):

$$(x[z'^n] \otimes 1) \cdot \Delta_{\mu,k+n} + (1 \otimes x[z''^{-n}]) \cdot \Delta_{\mu,k} = 0, \text{ for any } x[z'^n] \in \hat{\mathfrak{g}}'. \quad (1)$$

To prove (1), take any $f^{k+n} \in \mathscr{H}(\mu^*)_{k+n}$ and $v^k \in \mathscr{H}(\mu)_k$. Then, under the standard tensor product bilinear form,

$$\begin{aligned}
&\langle (x[z'^n] \otimes 1) \cdot \Delta_{\mu,k+n} + (1 \otimes x[z''^{-n}]) \cdot \Delta_{\mu,k}, v^k \otimes f^{k+n} \rangle \\
&= \sum_{j \in S_\mu^{k+n}} \langle x[z'^n] \cdot f_j^{k+n}, v^k \rangle \langle f^{k+n}, v_j^{k+n} \rangle \\
&\quad + \sum_{j_1 \in S_\mu^k} \langle f_{j_1}^k, v^k \rangle \langle f^{k+n}, x[z''^{-n}] \cdot v_{j_1}^k \rangle \\
&= - \sum_{j \in S_\mu^{k+n}} \langle f_j^{k+n}, x[z''^{-n}] \cdot v^k \rangle \langle f^{k+n}, v_j^{k+n} \rangle \\
&\quad - \sum_{j_1 \in S_\mu^k} \langle f_{j_1}^k, v^k \rangle \langle x[z'^n] \cdot f^{k+n}, v_{j_1}^k \rangle, \quad \text{by Lemma 3.5.2} \\
&= -\langle f^{k+n}, x[z''^{-n}] \cdot v^k \rangle - \langle x[z'^n] \cdot f^{k+n}, v^k \rangle \\
&= 0, \text{ by Lemma 3.5.2.}
\end{aligned}$$

This proves (1).

We now construct the following 'gluing' tensor element:

$$\Delta_\mu := \sum_{k \geq 0} \Delta_{\mu,k} \tau^k \in \big(\mathscr{H}(\mu^*) \otimes \mathscr{H}(\mu)\big)[[\tau]].$$

Let θ', θ'' be the maps of pulling-back local functions via the map $\zeta : \tilde{\Sigma} \to \Sigma$:

$$\theta' : \hat{\mathscr{O}}_{\Sigma,q} \to \hat{\mathscr{O}}_{\tilde{\Sigma},q'} \subset \mathbb{C}((z'))[[\tau]] \text{ and } \theta'' : \hat{\mathscr{O}}_{\Sigma,q} \to \hat{\mathscr{O}}_{\tilde{\Sigma},q''} \subset \mathbb{C}((z''))[[\tau]],$$

where, as earlier, $\hat{\mathscr{O}}_{\Sigma,q}$ is the completion of $\mathscr{O}_\Sigma$ along q, and $\hat{\mathscr{O}}_{\tilde{\Sigma},q'}$ and $\hat{\mathscr{O}}_{\tilde{\Sigma},q''}$ are defined similarly. Then, for any $f(z', z'') = \sum_{i \geq 0, j \geq 0} a_{i,j} z'^i z''^j \in \hat{\mathscr{O}}_{\Sigma,q}$ (cf. Lemma 3.5.1(2)), we have

$$\theta'(f) = f(z', \tau/z') = \sum_{j \geq 0} \left(\sum_{i \geq 0} a_{i,j} z'^{i-j} \right) \tau^j$$

and

$$\theta''(f) = f(\tau/z'', z'') = \sum_{i\geq 0}\left(\sum_{j\geq 0} a_{i,j} z''^{j-i}\right)\tau^i.$$

The morphisms θ', θ'' induce a $\mathbb{C}[[\tau]]$-linear Lie algebra morphism

$$\theta : \mathfrak{g}\otimes \hat{\mathscr{O}}_{\Sigma,q} \to \mathfrak{g}((z'))[[\tau]] \oplus \mathfrak{g}((z''))[[\tau]],\ x\otimes f \mapsto x\otimes \theta'(f) + x\otimes \theta''(f),$$

where τ acts on $\hat{\mathscr{O}}_{\Sigma,q}$ via

$$\tau \cdot f(z', z'') = z'z'' f(z', z'').$$

Thus, we get an injective map of $\mathfrak{g}\otimes \hat{\mathscr{O}}_{\Sigma,q}$ into $\hat{\mathfrak{g}}'[[\tau]] \oplus \hat{\mathfrak{g}}''[[\tau]]$ (but *not* a Lie algebra homomorphism). The latter acts canonically on $(\mathscr{H}(\mu^*)\otimes \mathscr{H}(\mu))\,[[\tau]]$. Thus, we get a $\mathbb{C}[[\tau]]$-linear projective representation of $\mathfrak{g}\otimes \hat{\mathscr{O}}_{\Sigma,q}$ in $(\mathscr{H}(\mu^*)\otimes \mathscr{H}(\mu))\,[[\tau]]$.

Lemma 3.5.4 *The element $\Delta_\mu \in (\mathscr{H}(\mu^*)\otimes \mathscr{H}(\mu))\,[[\tau]]$ defined above is annihilated by $\mathfrak{g}\otimes \hat{\mathscr{O}}_{\Sigma,q}$ via the morphism θ defined as above.*

Proof For any $x[z'^i z''^j] \in \mathfrak{g}\otimes \hat{\mathscr{O}}_{\Sigma,q}$,

$$\begin{aligned}
x[z'^i z''^j]\cdot \Delta_\mu &= \sum_{k\in\mathbb{Z}} (x[z'^{i-j}]\otimes 1)\cdot \Delta_{\mu,k}\tau^{k+j} \\
&\quad + \sum_{k\in\mathbb{Z}} (1\otimes x[z''^{j-i}])\cdot \Delta_{\mu,k}\tau^{k+i} \\
&= -\sum_{k\in\mathbb{Z}} (1\otimes x[z''^{j-i}])\cdot \Delta_{\mu,k+j-i}\tau^{k+j} \\
&\quad + \sum_{k\in\mathbb{Z}} (1\otimes x[z''^{j-i}])\cdot \Delta_{\mu,k}\tau^{k+i}, \text{ by (1) of Definition 3.5.3} \\
&= 0.
\end{aligned}$$

From this it is easy to see that $x[f]\cdot \Delta_\mu = 0$ for any $x[f] \in \mathfrak{g}\otimes \hat{\mathscr{O}}_{\Sigma,q}$. This proves the lemma. □

Definition 3.5.5 We follow the notation from Lemma 3.5.1. Let $\mathscr{H}(\lambda_i)$ denote the integrable $\hat{\mathfrak{g}}$-module with highest-weight λ_i and let $\mathscr{H}(\vec{\lambda})$ denote their tensor product. Define

$$\mathscr{H}(\vec{\lambda})_{\mathbb{D}_\tau} := \mathscr{H}(\vec{\lambda})[[\tau]].$$

Then, as in Definition 3.3.2, we get an action of $\mathfrak{g}[\Sigma\backslash\vec{p}]$ on $\mathscr{H}(\vec{\lambda})_{\mathbb{D}_\tau}$ (induced from the embedding φ as in (6) of Definition 3.3.2).

We now construct a morphism of $\mathbb{C}[[\tau]]$-modules:

$$F_{\vec{\lambda}}: \mathscr{H}(\vec{\lambda})_{\mathbb{D}_\tau} \longrightarrow \bigoplus_{\mu \in D_c} \Big(\mathscr{H}(\vec{\lambda}) \otimes \mathscr{H}(\mu^*) \otimes \mathscr{H}(\mu)\Big)[[\tau]]$$

given by

$$\sum_{j=0}^{\infty} h_j \tau^j \mapsto \sum_{j,k=0}^{\infty} (h_j \otimes \Delta_{\mu,k})\tau^{j+k}, \ \text{ for } h_j \in \mathscr{H}(\vec{\lambda}).$$

Consider the following canonical homomorphisms (obtained by pull-back and restrictions):

$$\mathscr{O}_{\Sigma\backslash\vec{p}} \to \mathscr{O}_{\tilde{\Sigma}\backslash\vec{p}} \to \mathscr{O}_{\tilde{\Sigma}\backslash\{\vec{p},q',q''\}} \to \hat{\mathscr{O}}_{\tilde{\Sigma}\backslash\{\vec{p},q',q''\}, \tilde{\Sigma}_o\backslash\{\vec{p}_o,q',q''\}}$$
$$\simeq \mathscr{O}_{\tilde{\Sigma}_o\backslash\{\vec{p}_o,q',q''\}}[[\tau]],$$

where the last isomorphism is obtained from the isomorphism κ of Lemma 3.5.1 (see isomorphism (1) there). This gives rise to a Lie algebra homomorphism (depending upon the isomorphism κ):

$$\kappa_{\vec{p}}: \mathfrak{g}[\Sigma\backslash\vec{p}] \to \Big(\mathfrak{g} \otimes \mathscr{O}_{\tilde{\Sigma}_o\backslash\{\vec{p}_o,q',q''\}}\Big)[[\tau]].$$

Hence, the Lie algebra $\mathfrak{g}[\Sigma\backslash\vec{p}]$ acts on $\big(\mathscr{H}(\vec{\lambda}) \otimes \mathscr{H}(\mu^*) \otimes \mathscr{H}(\mu)\big)[[\tau]]$ via the action of $\mathfrak{g} \otimes \mathscr{O}_{\tilde{\Sigma}_o\backslash\{\vec{p}_o,q',q''\}}$ on $\mathscr{H}(\vec{\lambda}) \otimes \mathscr{H}(\mu^*) \otimes \mathscr{H}(\mu)$ at the points $\{\vec{p}_o, q', q''\}$ (cf. Definition 3.1.1) and extending it $\mathbb{C}[[\tau]]$-linearly.

Theorem 3.5.6 *We have the following:*

(i) *The morphism $F_{\vec{\lambda}}$ is a $\mathbb{C}[[\tau]]$-linear $\mathfrak{g}[\Sigma\backslash\vec{p}]$-module map under the action of $\mathfrak{g}[\Sigma\backslash\vec{p}]$ defined above.*
(ii) *The morphism $F_{\vec{\lambda}}$ induces an isomorphism of sheaf of covacua over $\mathbb{D}_\tau$:*

$$\bar{F}_{\vec{\lambda}}: \mathscr{V}_{\mathfrak{F}_\Sigma}(\vec{p},\vec{\lambda}) \simeq \bigoplus_{\mu \in D_c} \mathscr{V}_{\tilde{\Sigma}_o}\Big((\vec{p}_o, q', q''), (\vec{\lambda}, \mu^*, \mu)\Big)[[\tau]], \quad (1)$$

where $\mathfrak{F}_\Sigma$ is the family as in Lemma 3.5.1.

Proof By Lemma 3.5.4, the morphism

$$F_{\vec{\lambda}}: \mathscr{H}(\vec{\lambda})_{\mathbb{D}_\tau} \to \bigoplus_{\mu \in D_c} \Big(\mathscr{H}(\vec{\lambda}) \otimes \mathscr{H}(\mu^*) \otimes \mathscr{H}(\mu)\Big)[[\tau]]$$

given by $h \mapsto \sum_{\mu \in D_c} h \otimes \Delta_\mu$, is a morphism of $\mathfrak{g}[\Sigma\backslash\vec{p}]$-modules. This proves part (i) of the theorem.

We now proceed to prove part (ii) of the theorem. Using part (i) of the theorem and the morphism $\kappa_{\vec{p}}$, we get the $\mathbb{C}[[\tau]]$-module morphism (1). Taking quotient by τ, by the Factorization Theorem (Theorem 3.1.2) together with Lemma 3.3.3, the morphism $\bar{F}_{\vec{\lambda}}$ gives rise to an isomorphism

$$\mathscr{V}_{\Sigma_o}(\vec{p}_o, \vec{\lambda}) \to \bigoplus_{\mu \in D_c} \mathscr{V}_{\tilde{\Sigma}_o}\left((\vec{p}_o, q', q''), (\vec{\lambda}, \mu^*, \mu)\right).$$

As a consequence of the Nakayama Lemma (cf. (Atiyah and Macdonald, 1969, Chap. 2, Exercise 10)), $\bar{F}_{\vec{\lambda}}$ is surjective. (Observe that by Proposition 3.3.4, both the domain and the range of $\bar{F}_{\vec{\lambda}}$ are finitely generated $\mathbb{C}[[\tau]]$-modules.) Now, since the range of $\bar{F}_{\vec{\lambda}}$ is a free $\mathbb{C}[[\tau]]$-module, we get that $\bar{F}_{\vec{\lambda}}$ splits over $\mathbb{C}[[\tau]]$. Thus, applying the Nakayama lemma (cf. (Atiyah and Macdonald, 1969, Proposition 2.6)) again to the kernel K of $\bar{F}_{\vec{\lambda}}$, we get that $K = 0$. Thus, $\bar{F}_{\vec{\lambda}}$ is an isomorphism, proving part (ii) of the theorem. □

For any genus $g \geq 0$ and $s \geq 1$, let $\bar{\mathcal{M}}_{g,s}$ be the category over $\mathfrak{S}$ (where $\mathfrak{S}$ is the category of schemes as in Section 1.1) whose objects over any scheme $T \in \mathfrak{S}$ are families of *stable* s-pointed connected curves of genus g over T, though we do not require that each irreducible component of the geometric fibers contains at least one marked point (cf. Definition 3.3.1). Even though in Definition 3.3.1 we assumed T to be smooth, its extension to any scheme T is straightforward (cf. (Knudsen, 1983a, Definition 1.1)).

We recall the following result due to Knudsen (1983a, Theorem 2.7). This is an extension of the classical result of Deligne–Mumford for $\bar{\mathcal{M}}_{g,0}$ (cf. (Deligne and Mumford, 1969)).

Theorem 3.5.7 *Assume that $2g - 2 + s > 0$. Then the category $\bar{\mathcal{M}}_{g,s}$ is a proper and smooth Deligne–Mumford stack such that the substack $\bar{\mathcal{S}}_{g,s}$ consisting of singular curves is a divisor with normal crossings.*

Proposition 3.5.8 *Let $(\Sigma, \vec{p})$ be a smooth connected s-pointed projective curve ($s \geq 1$) of genus $g \geq 0$ such that $2g - 2 + s > 0$. Then, for any $\vec{\lambda} \in D_c^s$, $\dim \mathscr{V}_\Sigma(\vec{p}, \vec{\lambda})$ only depends on g and $\vec{\lambda}$.*

We denote this dimension by $F_g(\vec{\lambda})$.

Proof By Knudsen (1983b, proof of Theorem 6.1), there exists a proper normal (irreducible) variety $X_{g,s}$ and a finite surjective morphism $q: X_{g,s} \to \bar{\mathcal{M}}_{g,s}$. Thus, there exists a family $\mathcal{F}_{X_{g,s}} = (\Sigma_{X_{g,s}}, \vec{p}_{X_{g,s}})$ of s-pointed stable curves of genus g over $X_{g,s}$. Let

$$X^0_{g,s} = \{x \in X_{g,s} : \Sigma_x \text{ is a smooth curve}\},$$

where $\Sigma_x := \Sigma_{X_{g,s}|x}$. Then $X^0_{g,s}$ is an open (nonempty) subset of irreducible variety $X_{g,s}$. In particular, $X^0_{g,s}$ is also irreducible. Taking a desingularization

$\tilde{X}^0_{g,s}$ of $X^0_{g,s}$ and pulling back the family $\mathcal{F}_{X_{g,s}}$ to $\tilde{X}^0_{g,s}$, we get a family $\mathcal{F}_{\tilde{X}^0_{g,s}} = (\Sigma_{\tilde{X}^0_{g,s}}, \vec{p}_{\tilde{X}^0_{g,s}})$ of s-pointed smooth irreducible curves of genus g over $\tilde{X}^0_{g,s}$. Thus, by Hartshorne (1977, Chap. III, Theorem 10.2), $\pi : \Sigma_{\tilde{X}^0_{g,s}} \to \tilde{X}^0_{g,s}$ is a smooth morphism. Moreover, by taking a small enough open cover $\{U_i\}$ of $\tilde{X}^0_{g,s}$, we can enlarge the family $\mathcal{F}_{\tilde{X}^0_{g,s}|U_i}$ to a family $(\Sigma_{U_i}, \vec{p}_{U_i}, \vec{t}_{U_i})$ of s-pointed genus g-curves with formal parameters over U_i, where Σ_{U_i} is the restriction of $\Sigma_{\tilde{X}^0_{g,s}}$ to U_i and similarly for $\vec{p}_{U_i}$ (cf. Exercise 3.5.E.2). Now, by Theorem 3.4.2 and Lemma 3.3.3, we conclude that $\dim \mathcal{V}_{\Sigma_x}(\vec{p}_x, \vec{\lambda})$ is independent of $x \in U_i$, where Σ_x is the restriction of Σ_{U_i} to x and similarly for $\vec{p}_x$. Since $\tilde{X}^0_{g,s}$ is irreducible (in particular, connected), we conclude that $\dim \mathcal{V}_{\Sigma_x}(\vec{p}_x, \vec{\lambda})$ is independent of $x \in \tilde{X}^0_{g,s}$. From the surjectivity of $X_{g,s} \to \bar{\mathcal{M}}_{g,s}$, we get the proposition. □

Being a Deligne–Mumford stack, $\bar{\mathcal{M}}_{g,s}$ has an étale presentation $\phi : X \to \bar{\mathcal{M}}_{g,s}$. Moreover, $\bar{\mathcal{M}}_{g,s}$ being smooth, X is a smooth (but not necessarily connected) scheme of finite type over $\mathbb{C}$.

Let $(\Sigma_X, \vec{p}_X)$ be the family of s-pointed curves over X associated to $\phi : X \to \bar{\mathcal{M}}_{g,s}$. By taking an open cover $\{V_j\}_j$ of X (if needed) with connected V_j, we can assume that (abbreviating V_j by V) $(\Sigma_V, \vec{p}_V)$ can be enlarged to a family $\mathcal{F}_V = (\Sigma_V, \vec{p}_V, \vec{t}_V)$ of s-pointed curves with formal parameters, where $\Sigma_V := \Sigma_{X|V}$. As in Definition 3.3.2, we get the corresponding sheaf of covacua $\mathcal{V}_{\mathcal{F}_V}(\vec{p}_V, \vec{\lambda})$ over V, which is coherent by Proposition 3.3.4.

Theorem 3.5.9 *For any genus $g \geq 0$ and $s \geq 1$ such that $2g - 2 + s > 0$ and any set of dominant weights $\vec{\lambda} = (\lambda_1, \ldots, \lambda_s)$ with $\lambda_i \in D_c$, the sheaf of conformal blocks $\mathcal{V}_{\mathcal{F}_V}(\vec{p}_V, \vec{\lambda})$ is locally free over V.*

Moreover, the rank of $\mathcal{V}_{\mathcal{F}_V}(\vec{p}_V, \vec{\lambda})$ is independent of $V = V_j$ and it is equal to $F_g(\vec{\lambda})$.

Proof We introduce a filtration on $\bar{\mathcal{M}}_{g,s}$:

$$\bar{\mathcal{M}}^0_{g,s} \subset \bar{\mathcal{M}}^1_{g,s} \subset \cdots \subset \bar{\mathcal{M}}^k_{g,s} = \bar{\mathcal{M}}_{g,s},$$

where $\bar{\mathcal{M}}^i_{g,s}$ is the open substack of $\bar{\mathcal{M}}_{g,s}$ with each geometric fiber consisting of at most i many nodal points. (By Knudsen (1983b, Theorem 2.7), $\bar{\mathcal{M}}^0_{g,s}$ is open in $\bar{\mathcal{M}}_{g,s}$. Moreover, following Harris and Morrison (1998, p. 50), $\bar{\mathcal{M}}_{g,s} \backslash \bar{\mathcal{M}}^i_{g,s}$ is closed in $\bar{\mathcal{M}}_{g,s}$, as stability forbids different nodes to collide in a limit. So, the only possibility in the limit is for a component to break up and create a new node.[2]) Note that $\bar{\mathcal{M}}^0_{g,s}$ consists of smooth s-pointed connected curves. For any fixed genus g, there exists $k = k_g \geq 0$ such that the number of

[2] I thank Prakash Belkale and Sándor Kovács for their input and references.

nodal points on a (connected) stable curve of genus g is bounded by k. (In fact, $k_g = 3g - 3$ by Harris and Morrison (1998, Exercise 2.20).) This filtration of $\bar{\mathcal{M}}_{g,s}$ induces an open filtration on V via ϕ:

$$V^0 \subset V^1 \subset \cdots \subset V^k = V.$$

We now prove inductively that the coherent sheaf $\mathcal{V}_{\mathcal{F}_{V^i}}(\vec{p}_{V^i}, \vec{\lambda})$ is locally free, where $\mathcal{F}_{V^i}$ (resp. $\vec{p}_{V^i}$) is the restriction of $\mathcal{F}_V$ (resp. $\vec{p}_V$) to V^i. When $i = 0$, in view of Theorem 3.4.2, $\mathcal{V}_{\mathcal{F}_{V^0}}(\vec{p}_{V^0}, \vec{\lambda})$ is locally free. (Observe that by Hartshorne (1977, Chap. III, Theorem 10.2), $\Sigma_{V^0} \to V^0$ is a smooth morphism.) Assume that $\mathcal{V}_{\Sigma_{V^{i-1}}}(\vec{p}_{V^{i-1}}, \vec{\lambda})$ is locally free where $i \geq 1$. By the smoothing construction in Lemma 3.5.1, for any $\mathbb{C}$-point $x \in V^i \backslash V^{i-1}$, there exists a morphism $\beta_x \colon \mathbb{D}_\tau \to \bar{\mathcal{M}}^i_{g,s}$ such that $\beta_x(o) = \phi(x)$ and $\beta_x(g_\tau) \in \bar{\mathcal{M}}^{i-1}_{g,s} \backslash \bar{\mathcal{M}}^{i-2}_{g,s}$, where g_τ is the generic point of $\mathbb{D}_\tau$. Since $\phi \colon X \to \bar{\mathcal{M}}_{g,s}$ is étale and surjective, β_x can be lifted to $\beta'_x \colon \mathbb{D}_\tau \to V \subset X$ such that $\phi \circ \beta'_x = \beta_x$ and $\beta'_x(o) = x$. It follows that $\beta'_x(g_\tau) \in V^{i-1} \backslash V^{i-2}$. By Lemma 3.3.3 and Theorem 3.5.6, the rank of $\mathcal{V}_{\mathcal{F}_{V^{i-1}}}(\vec{p}_{V^{i-1}}, \vec{\lambda})$ (which is locally free by induction) is equal to the dimension of $\mathcal{V}_{\Sigma_x}(\vec{p}_x, \vec{\lambda})$. It follows that $\mathcal{V}_{\mathcal{F}_{V^i}}(\vec{p}_{V^i}, \vec{\lambda})$ being coherent, it is locally free (cf. (Hartshorne, 1977, Chap. II, Exercise 5.8(c))). It concludes the proof that $\mathcal{V}_{\mathcal{F}_V}(\vec{p}_V, \vec{\lambda})$ is locally free over V.

We next prove that it is of the same rank $F_g(\vec{\lambda})$ over each of the V_j. We first claim that V^0 is nonempty for any $V = V_j$. For, otherwise, $\phi(V) \subset \bar{\mathcal{S}}_{g,s}$. But $\bar{\mathcal{S}}_{g,s}$ is a divisor in $\bar{\mathcal{M}}_{g,s}$ and $\phi(V)$ is open in $\bar{\mathcal{M}}_{g,s}$. Since $\bar{\mathcal{M}}_{g,s}$ is irreducible (cf. proof of Proposition 3.5.8), this is a contradiction. Hence, V^0 is nonempty. Thus, by Proposition 3.5.8, we get that $\mathcal{V}_{\mathcal{F}_V}(\vec{p}_V, \vec{\lambda})$ (which is locally free over V as already proved) is of rank $F_g(\vec{\lambda})$. This proves the theorem. □

As a consequence of Theorem 3.5.9 and the Factorization Theorem 3.1.2, we get the following.

Corollary 3.5.10 *(a) For any genus $g \geq 1$, $s \geq 1$ and any $\vec{\lambda} \in D^s_c$,*

$$F_g(\vec{\lambda}) = \sum_{\mu \in D_c} F_{g-1}(\vec{\lambda}, \mu^*, \mu),$$

where $F_g(\vec{\lambda})$ is defined in Proposition 3.5.8.

(b) For any genus $g = g' + g''$ (where $g', g'' \geq 0$) and any $\vec{\lambda}' \in D^{s'}_c$, $\vec{\lambda}'' \in D^{s''}_c$ (with $s', s'' \geq 1$),

$$F_g(\vec{\lambda}', \vec{\lambda}'') = \sum_{\mu \in D_c} F_{g'}(\vec{\lambda}', \mu^*) F_{g''}(\vec{\lambda}'', \mu).$$

Proof (a) This follows from Theorems 3.1.2, 3.5.9 and Lemma 3.3.3 by observing that the normalization of an irreducible nodal curve of genus

$g \geq 1$ (with a single node) is a (smooth irreducible) curve of genus $g - 1$ (cf. (Hartshorne, 1977, Chap. IV, Exercise 1.8)).

(b) Let $(\Sigma', \vec{p}\,')$ (resp. $(\Sigma'', \vec{p}\,'')$) be a smooth irreducible s'-pointed (resp. s''-pointed) curve of genus g' (resp. g'') and let Σ be the curve obtained from joining Σ' and Σ'' at one point p disjoint from the marked points on either. Then, considering the long exact cohomology sequences obtained from the sheaf exact sequences for the subvarieties $\Sigma' \subset \Sigma$ and $\{p\} \subset \Sigma''$, we get that the genus g of Σ equals $g' + g''$. We mark Σ as an $s = s' + s''$-pointed curve by taking the markings as $\vec{p} = (\vec{p}\,', \vec{p}\,'')$. Now, assuming $s' \geq 2$ if $g' = 0$ and $s'' \geq 2$ if $g'' = 0$, we get that $(\Sigma, \vec{p})$ is an s-pointed connected stable curve of genus g. Thus, (b) follows from Theorems 3.1.2, 3.5.9 and Lemma 3.3.3 by observing that the normalization of Σ is the smooth curve $\Sigma' \sqcup \Sigma''$.

If $g' = 0, g'' > 0$ and $s' = 1$ (similarly, if $g' > 0, g'' = 0$ and $s'' = 1$), introduce an extra marked point $p'_o \in \Sigma' = \mathbb{P}^1$ and attach weight 0 to p'_o. Then, $(\Sigma, (\vec{p}\,', p'_o, \vec{p}\,''))$ is an $s+1$-pointed stable connected curve of genus g. Thus, we get

$$F_g(\vec{\lambda}\,', 0, \vec{\lambda}\,'') = \sum_{\mu \in D_c} F_{g'}(\vec{\lambda}\,', 0, \mu^*) F_{g''}(\vec{\lambda}\,'', \mu).$$

Using Corollary 2.2.3(a), we get (b) in this case as well. The proof in the case $g' = g'' = 0$ and s' or $s'' < 2$ is similar. □

As a corollary of the above corollary and Corollary 2.2.3(a), we get the following.

Corollary 3.5.11 *Let $c \geq 1$ be the central charge and let $\vec{\lambda} = (0) \in D_c$ be the zero weight. Then, for any genus $g \geq 0$,*

$$F_g(0) \geq (\sharp D_c)^g.$$

Proof For $g = 0$, $F_g(0) = 1$ (cf. Exercise 2.3.E.2). So, assume that $g \geq 1$. Then

$$\begin{aligned} F_g(0) &= \sum_{\mu_1, \ldots, \mu_g \in D_c} F_0\left(\mu_1^*, \mu_1, \ldots, \mu_g^*, \mu_g\right), \\ &\qquad \text{by Corollaries 3.5.10(a) and 2.2.3(a)} \\ &\geq \sum_{\mu_1, \ldots, \mu_g \in D_c} F_0(\mu_1^*, \mu_1) F_0(\mu_2^*, \mu_2) \ldots F_0(\mu_g^*, \mu_g), \\ &\qquad \text{by Corollaries 3.5.10(b) and 2.2.3(a)} \\ &= (\sharp D_c)^g, \quad \text{by Exercise 2.3.E.2(b).} \end{aligned}$$

This proves the corollary. □

Remark 3.5.12 (a) There is an intrinsic way to define the sheaf of covacua attached to any family of s-pointed curves (without the choice of the local parameters) and $\vec{\lambda} = (\lambda_1, \ldots, \lambda_s) \in D_c^s$. This allows us to define a coherent sheaf of covacua $\mathscr{V}_{\mathcal{F}}(\vec{p}_{\bar{\mathcal{M}}_{g,s}}, \vec{\lambda})$ over $\bar{\mathcal{M}}_{g,s}$, where $\mathcal{F} = \mathcal{F}_{\bar{\mathcal{M}}_{g,s}}$. Similar to the above proof, one obtains that this sheaf $\mathscr{V}_{\mathcal{F}}(\vec{p}_{\bar{\mathcal{M}}_{g,s}}, \vec{\lambda})$ over $\bar{\mathcal{M}}_{g,s}$ is locally free. Moreover, its pull-back to V under ϕ is isomorphic to the vector bundle $\mathscr{V}_{\mathcal{F}_V}(\vec{p}_V, \vec{\lambda})$ over $V = V_j$ as in Theorem 3.5.9. The details can be found in Tsuchimoto (1993) and in a more general setting can be found in Hong and Kumar (2019, Theorem 8.9).

(b) Let $\mathcal{F}_T = (\Sigma_T, \vec{p}, \vec{t})$ be a family of s-pointed stable curves with formal parameters such that the family $\mathcal{F}_T^o = (\Sigma_T, \vec{p})$ is a versal family of s-pointed stable curves. Let $D = \{x \in T : \Sigma_x \text{ is singular}\}$. Then, there exists a flat projective connection on the sheaf of covacua $\mathscr{V}_{\mathcal{F}_{T\setminus D}}(\vec{p}_{|T\setminus D}, \vec{\lambda})$ (for any $\vec{\lambda} \in D_c^s$) for the family $\mathcal{F}_T$ restricted to $T\setminus D$ such that the connection extends to the whole of $\mathscr{V}_{\mathcal{F}_T}(\vec{p}, \vec{\lambda})$ with regular singularities along D (cf. (Ueno, 2008, Proposition 5.4 and Theorem 5.5)).

3.5.E Exercises

(1) Let Σ be an irreducible curve with smooth base point p and let $\mathcal{S}$ be a coherent sheaf over Σ such that $\mathcal{S}_{|\Sigma\setminus p}$ and $\mathcal{S}_{|\mathbb{D}_p}$ are locally free, where $\mathbb{D}_p$ is the formal disc in Σ centered at p. Then, show that $\mathcal{S}$ is locally free over Σ.

(2) Let $\pi : X \to T$ be a morphism between smooth varieties such that π admits a section $\sigma : T \to X$. Then, show that there exists an open cover $\{U_i\}_i$ of X such that the completion $\hat{\mathcal{O}}_{U_i,V_i}$ of U_i along $V_i := U_i \cap \sigma(T)$ admits the following ring isomorphism:

$$\hat{\mathcal{O}}_{U_i,V_i} \simeq \mathcal{O}_{V_i}[[\tau]]$$

induced from the pull-back morphism $\sigma^* : \mathcal{O}_{V_i} \to \mathcal{O}_{U_i}$, where we identify $\mathcal{O}_{V_i} \simeq \mathcal{O}_{\pi(V_i)}$.

Hint: On a small enough open set U_i of X, there exists a morphism $f : U_i \to \mathbb{C}$ such that the zero scheme $Z(f) = V_i$. Now, define the map $\mathcal{O}_{V_i}[[\tau]] \to \hat{\mathcal{O}}_{U_i,V_i}$ by $\tau \mapsto f$.

3.C Comments

Most of the main results in Sections 3.1, 3.3–3.5 (including the Factorization Theorem 3.1.2; Lemma 3.3.3; Proposition 3.3.4; Theorems 3.4.2, 3.5.6

and 3.5.9 (for versal family); and Corollary 3.5.10) are originally due to Tsuchiya, Ueno and Yamada (1989) (see also (Ueno, 2008, Chaps. 4 and 5) for a more detailed account of their work). The proofs presented here are adaptations of the proofs given in Tsuchiya, Ueno and Yamada (1989), Bakalov and Kirillov (2001, §7.4) and Ueno (2008). It may be mentioned that some of the points in the proof of the Factorization Theorem given in Bakalov and Kirillov (2001, §7.7) are unclear to me. A geometric proof of the Factorization Theorem can be found in Faltings (1994). We have taken Example 3.4.3 from Bakalov and Kirillov (2001, §7.4).

For the material in Section 3.2 (including Lemma 3.2.4), we refer to Kac, Raina and Rozhkovskaya (2013, Lecture 10). Exercise 3.2.E.3 is known as the Goddard–Kent–Olive construction (cf. (Goddard, Kent and Olive, 1985)).

A coordinate-free approach to the description of conformal blocks and flat projective connection on the bundle of conformal blocks is developed in Tsuchimoto (1993) (where the projective connection was explicitly given in terms of the Atiyah algebra) and also in Bakalov and Kirillov (2001, §7.5) and Looijenga (2005). Hitchin (1990) has constructed a projectively flat connection on a vector bundle over the Teichmüller space $\mathbb{T}$. The fiber of the vector bundle over $\Sigma \in \mathbb{T}$ consists of the global sections of a power of the determinant line bundle (called the generalized theta functions) on the moduli space of stable vector bundles over Σ, which via Theorem 8.4.15 coincides with the space of vacua $\mathscr{V}_{\Sigma}^{\dagger}(p, 0_d)$ (see also (Faltings, 1993)). Laszlo (1998a) has shown that the Hitchin connection corresponds to the flat projective connection as in Theorem 3.4.2 under the above identification of the space of vacua with that of generalized theta functions. We also refer to the works of Beilinson and Schechtman (1988), Ginzburg (1995, §10), van Geemen and de Jong (1998), Sun and Tsai (2004), Belkale (2012b), Mukhopadhyay and Wentworth (2019) and Baier et al. (2020) on variants and generalizations of Hitchin's connection.

It was observed in Sorger (1999a, §2.3) that $F_g(0) \geq 1$ (cf. Corollary 3.5.11). Boysal (2008) has proved that for $\mathfrak{g} = sl_n$, any $g \geq 1$ and $\vec{\lambda} = (\lambda_1, \ldots, \lambda_s) \in D_c^s$ such that $\lambda_1 + \cdots + \lambda_s$ is in the root lattice, $F_g(\vec{\lambda}) \geq (\sharp D_c)^{g-1}$.

4
Fusion Ring and Explicit Verlinde Formula

The aim of this chapter is to prove the celebrated Verlinde formula giving an explicit expression for the dimension of the space of conformal blocks.

To facilitate this, we develop an algebraic formalism for general fusion rules and the corresponding associated ring. By a *fusion rule* on a finite set A together with an involution $*$, one means a map $F\colon \mathbb{Z}_+[A] \to \mathbb{Z}$ satisfying certain properties (cf. Definition 4.1.1). It is shown that given a nondegenerate fusion rule on A, the free $\mathbb{Z}$-module $\mathbb{Z}[A]$ with basis A acquires a natural associative and commutative ring structure together with a trace form $t\colon \mathbb{Z}[A] \to \mathbb{Z}$. This ring $\mathbb{Z}[A]$ together with the trace form is called the *fusion ring* associated to the fusion rule F on A. Further, the involution $*$ extends to a ring isomorphism of $\mathbb{Z}[A]$ and $\mathbb{Z}[A]$ is a Gorenstein ring (cf. Proposition 4.1.2). The trace form t and the involution $*$ together give rise to a natural positive-definite symmetric bilinear form $\langle\, ,\, \rangle$ on $\mathbb{Z}[A]$ (cf. Definition 4.1.4). We further show (cf. Lemma 4.1.5) that the complexified algebra $\mathbb{C}[A] := \mathbb{C} \otimes_{\mathbb{Z}} \mathbb{Z}[A]$ is a (finite-dimensional) reduced algebra (i.e., it has no nonzero nilpotent elements). Hence, we get the decomposition (as $\mathbb{C}$-algebras):

$$\varphi_A\colon \mathbb{C}[A] \simeq \mathbb{C}^{S_A}, \quad \varphi_A(x) = (\chi(x))_{\chi \in S_A}, \quad \text{for } x \in \mathbb{C}[A],$$

where S_A is the set of all the $\mathbb{C}$-algebra homomorphisms from $\mathbb{C}[A]$ to $\mathbb{C}$ (cf. Lemma 4.1.5).

The fusion rule F should be thought of as the 'genus 0 fusion rule.' Motivated from the Factorization Theorem, the fusion rule F is extended to a fusion map $F_g\colon \mathbb{Z}_+[A] \to \mathbb{Z}$ for any genus $g \geq 0$ (cf. Definition 4.1.6). The map F_g is explicitly determined in Proposition 4.1.7 and Corollary 4.1.8 using the trace t, a 'Casimir element' $\Omega \in \mathbb{Z}[A]$ defined as

$$\Omega = \sum_{\lambda \in A} \lambda \cdot \lambda^* \in \mathbb{Z}[A],$$

and the set S_A. The matrix encoding the fusion product is diagonalized in Exercise 4.1.E.3.

Having developed the algebraic machinery of fusion rules and fusion ring in Section 4.1, the aim of Section 4.2 is to state and give a proof of the Verlinde dimension formula. For any simple Lie algebra $\mathfrak{g}$ and central charge $c > 0$, recall the set D_c consisting of dominant integral weights of $\mathfrak{g}$ with level $\leq c$. Then, the finite set D_c has a natural involution $\lambda \mapsto \lambda^* := -w_o\lambda$, where w_o is the longest element of the Weyl group of $\mathfrak{g}$. Define the function $F_c : \mathbb{Z}_+[D_c] \to \mathbb{Z}_+$ by $F_c(0) = 1$ and, for any $s \geq 1$ and $\lambda_i \in D_c$,

$$F_c(\lambda_1 + \cdots + \lambda_s) = \dim \mathscr{V}_{\mathbb{P}^1}(\vec{p}, \vec{\lambda}),$$

where $\vec{p} = (p_1, \ldots, p_s)$ are any distinct points in $\mathbb{P}^1$ and $\vec{\lambda} = (\lambda_1, \ldots, \lambda_s)$. (By Proposition 3.5.8 and Exercise 2.3.E.2, $F_c(\lambda_1 + \cdots + \lambda_s)$ does not depend upon the choice of the points $\vec{p}$.) These F_c will be our most important examples of fusion rule (cf. Example 4.2.1). The corresponding fusion ring $\mathbb{Z}[D_c]$ is called the *fusion ring of* $\mathfrak{g}$ *at level* c and will be denoted by $R_c(\mathfrak{g})$. We denote the basis of the fusion ring $R_c(\mathfrak{g})$ by $\{[V(\lambda)]\}_{\lambda \in D_c}$ and denote the fusion product (i.e., the product in $\mathbb{Z}[D_c]$) by $\otimes^c$. Proposition 4.2.3 determines the fusion product $[V(\lambda)] \otimes^c [V(\mu)]$ in terms of the action of $sl_2(\theta)$ (sl_2 passing through the highest root space of $\mathfrak{g}$) on the components. This proposition is an easy consequence of Corollary 2.3.5. As an application of this proposition, we show that $[V(\lambda)] \otimes^c [V(\mu)]$ coincides with the usual tensor product $[V(\lambda)] \otimes [V(\mu)]$ in the case $\lambda + \mu \in D_c$. We also determine $[V(\lambda)] \otimes^c [V(\mu)]$ when $(\lambda + \mu)(\theta^\vee) = c + 1$ or $c + 2$ (cf. Corollary 4.2.4 and Exercise 4.2.E.1).

We give another definition of the fusion product $\otimes^c_F$ in terms of the $\mathfrak{g}$-equivariant Euler–Poincaré characteristic of certain vector bundles on the infinite Grassmannian $\bar{X}_G$ (cf. Definition 4.2.11). In Definition 4.2.7, we define a $\mathbb{Z}$-module map

$$\xi_c : R(\mathfrak{g}) \to R_c(\mathfrak{g}),$$

where $R(\mathfrak{g})$ is the representation ring of $\mathfrak{g}$. It is shown to be a (surjective) ring homomorphism if we endow $R_c(\mathfrak{g})$ with the fusion product $\otimes^c_F$ (cf. identity (8) of Subsection 4.2.18). Its kernel is determined in Lemma 4.2.8.

Then, using a result of Teleman on the Lie algebra homology of $\hat{\mathfrak{g}}_- := \mathfrak{g} \otimes t^{-1}\mathbb{C}[t^{-1}]$ with coefficient in the tensor product of an integrable highest-weight $\hat{\mathfrak{g}}$-module with a finite-dimensional $\mathfrak{g}$-module (cf. Theorem 4.2.16), we show that the above two fusion products $\otimes^c$ and $\otimes^c_F$ coincide (cf. Corollary 4.2.17). This identification, together with the combinatorics of the affine Weyl group W_c (cf. Definition 4.2.5), gives rise to another (closed) expression for the fusion product (cf. Lemma 4.2.12):

For $\lambda, \mu \in D_c$,

$$[V(\lambda)] \otimes^c [V(\mu)] = \sum_{\nu \in D_c} \sum_{w \in W'_c} (-1)^{\ell(w)} n^{\lambda,\mu}_{w^{-1}*\nu} [V(\nu)],$$

where W'_c is defined in Definition 4.2.7 and $n^{\lambda,\mu}_{w^{-1}*\nu} := \dim(\mathrm{Hom}_{\mathfrak{g}}(V(w^{-1}*\nu), V(\lambda) \otimes V(\mu)))$. Further, by using the equality of the two fusion products $\otimes^c$ and $\otimes^c_F$, we of course get that ξ_c is a ring homomorphism with respect to the usual fusion product $\otimes^c$ (cf. Theorem 4.2.9). The equality of $\otimes^c$ and $\otimes^c_F$ can be more easily deduced for any simple $\mathfrak{g}$ not of type $E_\bullet$ and F_4 (cf. Exercise 4.2.E.4).

The surjective ring homomorphism ξ_c allows us to determine the set of algebra homomorphisms $R_c(\mathfrak{g}) \to \mathbb{C}$ and identify them with T_c^{reg}/W (cf. Corollary 4.2.18), where W is the (finite) Weyl group of $\mathfrak{g}$ and T_c^{reg} is a certain finite subgroup of the maximal torus T of the simply-connected algebraic group G (with Lie algebra $\mathfrak{g}$) (cf. Definition 4.2.5). Now, the stage is set to state and prove the following Verlinde formula giving an explicit expression for the dimension of the space of conformal blocks (cf. Theorem 4.2.19). This is one of the most important results of the book.

Theorem Let $\mathfrak{g}$ be any simple Lie algebra and let $c > 0$ be any central charge. Let $(\Sigma, \vec{p} = (p_1, \ldots, p_s))$ be an irreducible smooth s-pointed curve of any genus $g \geq 0$ (where $s \geq 1$) and let $\vec{\lambda} = (\lambda_1, \ldots, \lambda_s)$ be a collection of weights in D_c. Then, for the space $\mathscr{V}_\Sigma(\vec{p}, \vec{\lambda})$ of covacua,

$$\dim \mathscr{V}_\Sigma(\vec{p}, \vec{\lambda}) = |T_c|^{g-1} \sum_{\mu \in D_c} \Bigg(\left(\Pi_{i=1}^s (\mathrm{ch}_{t_\mu}([V(\lambda_i)]))\right) \cdot \Pi_{\alpha \in \Delta_+} \left(2 \sin\left(\frac{\pi}{c + h^\vee} \langle \mu + \rho, \alpha \rangle\right)\right)^{2-2g} \Bigg),$$

where Δ_+ is the set of positive roots of $\mathfrak{g}$, $\kappa \colon \mathfrak{h}^* \to \mathfrak{h}$ is the isomorphism induced from the normalized invariant form, T_c is defined in Definition 4.2.5, $h^\vee$ is the dual Coxeter number of $\mathfrak{g}$ and $t_\mu := \mathrm{Exp}\left(\frac{2\pi i \kappa(\mu+\rho)}{c+h^\vee}\right) \in T_c$.

In particular, if $g = 1$, then $\dim \mathscr{V}_\Sigma(p, 0) = |D_c|$.

In Exercise 4.2.E.8, $\dim \mathscr{V}_\Sigma(p, 0)$ is determined for any g if $\mathfrak{g}$ is simply-laced and $c = 1$.

4.1 General Fusion Rules and the Associated Ring

Definition 4.1.1 Let A be a finite set with an involution $*$ (i.e., a bijection of order 2). Let $\mathbb{Z}_+[A] := \bigoplus_{a\in A} \mathbb{Z}_+ a$ be the free monoid generated by A, where $\mathbb{Z}_+ := \mathbb{Z}_{\geq 0}$. The involution $*$ of A clearly extends to an involution of $\mathbb{Z}_+[A]$.

By a *fusion rule* on A, we mean a map $F\colon \mathbb{Z}_+[A] \to \mathbb{Z}$ satisfying the following conditions:

(f_1) $F(0) = 1$
(f_2) $F(a) > 0$, for some $a \in A$
(f_3) $F(x) = F(x^*)$, for $x \in \mathbb{Z}_+[A]$
(f_4) $F(x+y) = \sum_{\lambda\in A} F(x+\lambda)F(y+\lambda^*)$, for $x, y \in \mathbb{Z}_+[A]$.
The fusion rule F is said to be *nondegenerate* if it satisfies
(f_5) For any $a \in A$, there exists $\lambda_a \in A$ such that $F(a+\lambda_a) \neq 0$.

Proposition 4.1.2 *Let $F\colon \mathbb{Z}_+[A] \to \mathbb{Z}$ be a nondegenerate fusion rule. Then the abelian group $\mathbb{Z}[A] := \bigoplus_{a\in A} \mathbb{Z}$ acquires a product (defined by* (1) *below), making $\mathbb{Z}[A]$ into a commutative ring with identity.*

Moreover, it admits a unique linear form $t\colon \mathbb{Z}[A] \to \mathbb{Z}$ (called the trace) satisfying the following properties:

(a) $t(a\cdot b^*) = \delta_{a,b}$ for $a, b \in A$.
(b) $t\left(\Pi_{a\in A} a^{n_a}\right) = F\left(\sum_{a\in A} n_a a\right)$, for any $n_a \in \mathbb{Z}_+$.

Further, $*$ is a ring homomorphism of $\mathbb{Z}[A]$ and $\mathbb{Z}[A]$ is a Gorenstein ring.

The ring $\mathbb{Z}[A]$ together with the trace form t is called the *fusion ring associated to the fusion rule F*. Observe that t is $*$-invariant.

Proof Define the multiplication (called the *fusion product*)

$$a\cdot b = \sum_{\lambda\in A} F(a+b+\lambda^*)\lambda, \quad \text{for } a,b\in A, \tag{1}$$

and extend bilinearly on $\mathbb{Z}[A]$. By definition, the product is commutative. We next show that it is associative.

Take $a, b, c \in A$. Then

$$\begin{aligned}(a\cdot b)\cdot c &= \sum_{\lambda,\mu\in A} F(a+b+\lambda^*)F(\lambda+c+\mu^*)\mu \\ &= \sum_{\mu\in A} F(a+b+c+\mu^*)\mu, \quad \text{by } (f_4) \text{ of Definition 4.1.1} \qquad (2) \\ &= a\cdot(b\cdot c).\end{aligned}$$

This proves the associativity.

We next show that $*$ is a ring homomorphism; i.e.,

$$*(a \cdot b) = (*a) \cdot (*b), \quad \text{for} \quad a, b \in A.$$

Now

$$\begin{aligned} *(a \cdot b) &= \sum_{\lambda \in A} F(a + b + \lambda^*)\lambda^* \\ &= \sum_{\lambda \in A} F(a^* + b^* + \lambda)\lambda^*, \quad \text{by } (f_3) \text{ of Definition 4.1.1} \\ &= (*a) \cdot (*b). \end{aligned}$$

We next show that there exists a unique $\mathbb{1} \in A$ such that

$$F(\mathbb{1}) = 1, \quad F(a) = 0, \quad \text{for all} \quad a \in A, a \neq \mathbb{1}. \tag{3}$$

Applying (f_1) and (f_4) of Definition 4.1.1 to $x = y = 0$, we get

$$\begin{aligned} 1 &= \sum_{\lambda \in A} F(\lambda)F(\lambda^*) \\ &= \sum_{\lambda \in A} F(\lambda)^2, \quad \text{by} \quad (f_3) \quad \text{of Definition 4.1.1.} \end{aligned}$$

From this, we get (3). (Observe that we have used (f_2) of Definition 4.1.1 here.) In particular, by (f_3) of Definition 4.1.1,

$$\mathbb{1}^* = \mathbb{1}. \tag{4}$$

Applying (f_4) of Definition 4.1.1 to $x = a \in A$, $y = a^*$, we get

$$\begin{aligned} F(a + a^*) &= \sum_{\lambda \in A} F(a + \lambda)F(a^* + \lambda^*) \\ &= \sum_{\lambda \in A} F(a + \lambda)^2, \quad \text{by } (f_3) \text{ of Definition 4.1.1.} \end{aligned} \tag{5}$$

In particular, for $a \in A$,

$$F(a + a^*) \geq F(a + a^*)^2.$$

This forces

$$F(a + a^*) = 0 \quad \text{or} \quad 1. \tag{6}$$

Further, by (5),

$$F(a + \lambda) = 0, \quad \text{for all} \quad \lambda \neq a^*, \ \lambda \in A. \tag{7}$$

Since F is nondegenerate, (6) and (7) force

$$F(a+a^*) = 1. \tag{8}$$

Combining (7) and (8), we get

$$F(a+b^*) = \delta_{a,b}, \quad \text{for} \quad a,b \in A. \tag{9}$$

Now, define the linear form $t\colon \mathbb{Z}[A] \to \mathbb{Z}$ by

$$t(a) = F(a), \quad \text{for } a \in A. \tag{10}$$

Of course, this definition is forced by property (b). Now, by the definition of the product, for $a, b \in A$,

$$\begin{aligned} t(a \cdot b^*) &= \sum_{\lambda \in A} F(a+b^*+\lambda^*)F(\lambda) \\ &= F(a+b^*), \quad \text{by } (f_4) \text{ of Definition 4.1.1} \\ &= \delta_{a,b}, \quad \text{by (9).} \end{aligned}$$

This proves property (a).

We next show that $\mathbb{1}$ is the multiplicative identify of $\mathbb{Z}[A]$. For any $x \in \mathbb{Z}_+[A]$,

$$F(x) = F(x+0) = \sum_{\lambda \in A} F(x+\lambda)F(\lambda^*) = F(x+\mathbb{1}), \quad \text{by (3) and (4).} \tag{11}$$

Now, for any $a \in A$,

$$\begin{aligned} \mathbb{1} \cdot a &= \sum_{\lambda \in A} F(\mathbb{1}+a+\lambda^*)\lambda \\ &= \sum_{\lambda \in A} F(a+\lambda^*)\lambda, \quad \text{by (11)} \\ &= a, \quad \text{by (8) and (9).} \end{aligned}$$

This proves that $\mathbb{1}$ is the multiplicative identity of $\mathbb{Z}[A]$.

Similar to the derivation of (2), by induction on n, it is easy to see that for any $a_1, \ldots, a_n \in A$,

$$a_1 \cdot a_2 \cdots a_n = \sum_{\mu \in A} F(a_1 + \cdots + a_n + \mu)\mu^*.$$

Thus,

$$
\begin{aligned}
t(a_1 \ldots a_n) &= \sum_{\mu \in A} F(a_1 + \cdots + a_n + \mu) t(\mu^*) \\
&= \sum_{\mu \in A} F(a_1 + \cdots + a_n + \mu) F(\mu^*), \quad \text{by (10)} \\
&= F(a_1 + \cdots + a_n), \quad \text{by } (f_4) \text{ of Definition 4.1.1.}
\end{aligned}
$$

This proves property (b).

Finally, we show that $\mathbb{Z}[A]$ is a Gorenstein ring over $\mathbb{Z}$.

Consider the $\mathbb{Z}$-linear map

$$\varphi\colon \mathbb{Z}[A] \to \mathrm{Hom}_{\mathbb{Z}}(\mathbb{Z}[A], \mathbb{Z}),$$

defined by

$$\varphi(x)(y) = t(x \cdot y), \quad \text{for} \quad x, y \in \mathbb{Z}[A],$$

where $\mathrm{Hom}_{\mathbb{Z}}(\mathbb{Z}[A], \mathbb{Z})$ denotes the $\mathbb{Z}$-module of $\mathbb{Z}$-linear maps from $\mathbb{Z}[A]$ to $\mathbb{Z}$. By the property (a) of t, φ is a $\mathbb{Z}$-linear isomorphism. Put a $\mathbb{Z}[A]$-module structure on $\mathrm{Hom}_{\mathbb{Z}}(\mathbb{Z}[A], \mathbb{Z})$ by

$$(x \cdot \alpha)(y) = \alpha(xy), \quad \text{for } x, y \in \mathbb{Z}[A] \text{ and } \alpha \in \mathrm{Hom}_{\mathbb{Z}}(\mathbb{Z}[A], \mathbb{Z}).$$

Then φ is a $\mathbb{Z}[A]$-module isomorphism (under the multiplication action of $\mathbb{Z}[A]$ on itself). The exact sequence

$$0 \to \mathbb{Z} \to \mathbb{Q} \to \mathbb{Q}/\mathbb{Z} \to 0$$

gives rise to the exact sequence of $\mathbb{Z}[A]$-modules ($\mathbb{Z}[A]$ being a free $\mathbb{Z}$-module):

$$0 \to \mathrm{Hom}_{\mathbb{Z}}(\mathbb{Z}[A], \mathbb{Z}) \to \mathrm{Hom}_{\mathbb{Z}}(\mathbb{Z}[A], \mathbb{Q}) \to \mathrm{Hom}_{\mathbb{Z}}(\mathbb{Z}[A], \mathbb{Q}/\mathbb{Z}) \to 0.$$

Now, $\mathrm{Hom}_{\mathbb{Z}}(\mathbb{Z}[A], \mathbb{Q})$ and $\mathrm{Hom}_{\mathbb{Z}}(\mathbb{Z}[A], \mathbb{Q}/\mathbb{Z})$ are injective $\mathbb{Z}[A]$-modules (cf. (Lang, 1965, Chap. III, Exercise 9(d))). Thus, $\mathbb{Z}[A] \simeq \mathrm{Hom}_{\mathbb{Z}}(\mathbb{Z}[A], \mathbb{Z})$ has finite injective dimension. Hence, $\mathbb{Z}[A]$ is a Gorenstein ring (by one of the equivalent definitions of Gorenstein rings). This completes the proof of the proposition. □

The converse of Proposition 4.1.2 is also true. Specifically, we have the following.

Lemma 4.1.3 *Let R be a commutative ring with identity $\mathbb{1}$ endowed with a ring involution $*$ and a $*$-invariant linear form $t\colon R \to \mathbb{Z}$. Assume that R admits a finite orthonormal $\mathbb{Z}$-basis A with respect to the (symmetric) bilinear* *form* $\langle , \rangle : R \times R \to \mathbb{Z}$, $\langle x, y \rangle := t(x \cdot y^*)$. *Assume further that* $\mathbb{1} \in A$ *and* A *is stable under* $*$. *Then the map*

$$F : \mathbb{Z}_+[A] \to \mathbb{Z}, \quad F\left(\sum_{a \in A} n_a a\right) = t\left(\Pi_{a \in A} a^{n_a}\right), \quad \textit{for} \quad n_a \in \mathbb{Z}_+, \tag{1}$$

is a nondegenerate fusion rule such that the associated fusion ring with trace coincides with (R, t).

Proof Since $\mathbb{1} \in A$, $t(\mathbb{1}) = 1$. Thus, $F(0) = t(\mathbb{1}) = F(\mathbb{1}) = 1$. Since t is invariant under $*$, $F(x) = F(x^*)$, for any $x \in \mathbb{Z}_+[A]$. For any $a \in A$,

$$F(a + a^*) = t(a \cdot a^*) = \langle a, a \rangle = 1.$$

Finally, take $x = \sum_{a \in A} n_a a$, $y = \sum_{a \in A} m_a a \in \mathbb{Z}_+[A]$. Then

$$\begin{aligned}
F(x + y) &= t\left(\left(\Pi_{a \in A} a^{n_a}\right) \cdot \left(\Pi_{a \in A} a^{m_a}\right)\right) \\
&= \left\langle \Pi_{a \in A} a^{n_a}, \ \Pi_{a \in A} (a^*)^{m_a} \right\rangle \\
&= \sum_{b \in A} \left\langle \Pi_{a \in A} a^{n_a}, b \right\rangle \left\langle b, \Pi_{a \in A} (a^*)^{m_a} \right\rangle, \\
&\quad \text{since } A \text{ is an orthonormal basis of } R \\
&= \sum_{b \in A} \left\langle \Pi_{a \in A} a^{n_a}, b \right\rangle \left\langle b^*, \Pi_{a \in A} a^{m_a} \right\rangle, \quad \text{since } t \text{ is } *\text{-invariant} \\
&= \sum_{b \in A} F(x + b^*) F(y + b).
\end{aligned}$$

Thus, F satisfies all the defining properties of a nondegenerate fusion rule.

From the definition of the induced product in $\mathbb{Z}[A]$ as in identity (1) of the proof of Proposition 4.1.2 (denoted by $\cdot$), we get that, for $a, b \in A$,

$$\begin{aligned}
a \cdot b &= \sum_{\lambda \in A} F(a + b + \lambda^*) \lambda \\
&= \sum_{\lambda \in A} t(ab\lambda^*) \lambda \\
&= \sum_{\lambda \in A} \langle ab, \lambda \rangle \lambda \\
&= ab, \quad \text{since } A \text{ is an orthonormal basis.}
\end{aligned}$$

This shows that the fusion product in $\mathbb{Z}[A] = R$ induced from F coincides with the original product in R.

The induced trace form clearly coincides with t by identity (10) of the proof of Proposition 4.1.2 and (1). This proves the lemma. □

Definition 4.1.4 Let F be a nondegenerate fusion rule on a finite set A (cf. Definition 4.1.1) and let $\mathbb{Z}[A]$ be the corresponding fusion ring with trace $t\colon \mathbb{Z}[A] \to \mathbb{Z}$ (cf. Proposition 4.1.2). Define the positive-definite symmetric bilinear form $\langle , \rangle$ on $\mathbb{Z}[A]$ by

$$\langle x, y\rangle = t(xy^*). \tag{1}$$

(It is positive definite by the defining property (a) of Proposition 4.1.2.)

Let $\mathbb{R}[A]$ be the $\mathbb{R}$-algebra $\mathbb{Z}[A] \otimes_{\mathbb{Z}} \mathbb{R}$ obtained by extending the scalars. Clearly $\langle , \rangle$ extends to a positive-definite symmetric $\mathbb{R}$-bilinear form on $\mathbb{R}[A]$. Also, the involution $*$ on $\mathbb{Z}[A]$ extends to an $\mathbb{R}$-algebra involution on $\mathbb{R}[A]$.

Let $S_A := \operatorname{Spec}(\mathbb{R}[A])$ be the (finite) set of $\mathbb{R}$-algebra homomorphisms $f\colon \mathbb{R}[A] \to \mathbb{C}$, which is the same as the set of $\mathbb{C}$-algebra homomorphisms $\mathbb{C}[A] \to \mathbb{C}$, where $\mathbb{C}[A] := \mathbb{R}[A] \otimes_{\mathbb{R}} \mathbb{C}$.

Lemma 4.1.5 *With the assumptions and notation as in the above Definition 4.1.4, the $\mathbb{R}$-algebra $\mathbb{R}[A]$ is reduced. Further, the $\mathbb{C}$-algebra homomorphism $\varphi_A\colon \mathbb{C}[A] \to \mathbb{C}^{S_A}$ into the product algebra given by*

$$\varphi_A(x) = (\chi(x))_{\chi\in S_A}, \quad \text{for} \quad x \in \mathbb{C}[A], \tag{1}$$

is an isomorphism.

Also, for any $x \in \mathbb{R}[A]$,

$$\varphi_A(x^*) = \overline{\varphi_A(x)}. \tag{2}$$

Proof We first show that $\mathbb{R}[A]$ is reduced, i.e., it has no nonzero nilpotent elements. It clearly suffices to show that for $x \in \mathbb{R}[A]$ such that $x^2 = 0$, we have $x = 0$. Now,

$$\langle xx^*, xx^*\rangle = t(x^2 x^{*^2}) = 0.$$

Thus, $xx^* = 0$, which gives $\langle x, x\rangle = 0$ and hence $x = 0$. Thus, $\mathbb{C}[A]$ is a reduced algebra as well, which implies that φ_A is an isomorphism.

We now prove (2).

Since $\mathbb{R}[A]$ is reduced, we have a canonical decomposition (as $\mathbb{R}$-algebras):

$$\mathbb{R}[A] \simeq \mathbb{R}_1 \times \cdots \times \mathbb{R}_m \times \mathbb{C}_1 \times \cdots \times \mathbb{C}_n$$

obtained from the indecomposable idempotents of $\mathbb{R}[A]$, where each $\mathbb{R}_i$ is the $\mathbb{R}$-algebra $\mathbb{R}$ and each $\mathbb{C}_j$ is the $\mathbb{R}$-algebra $\mathbb{C}$. The decomposition being canonical, for any $1 \le i \le m$ and $1 \le j \le n$,

$$* \mathbb{R}_i = \mathbb{R}_{i'} \quad \text{and} \quad * \ \mathbb{C}_j = \mathbb{C}_{j'},$$

for some $1 \leq i' \leq m$ and $1 \leq j' \leq n$. We next claim that $i' = i$ and $j' = j$. For, otherwise, if $i' \neq i$ for some i, then, for any $x \in \mathbb{R}_i$, $x \cdot x^* = 0$, which gives $\langle x, x \rangle = 0$. This is a contradiction since $\langle , \rangle$ on $\mathbb{R}[A]$ is positive definite.

By the same argument, we see that $j' = j$, for all j, i.e., $*$ keeps each of the factors $\mathbb{R}_i$ and $\mathbb{C}_j$ stable. Of course, $*$ being an $\mathbb{R}$-algebra homomorphism on each factor, $*$ is the identify map on each $\mathbb{R}_i$ factor.

We next claim that $*$ is the complex conjugation on each $\mathbb{C}_j$ factor. For, if not, $*$ would be the identity map on some $\mathbb{C}_j$. This would give

$$t(x^2) = t(xx^*) = \langle x, x \rangle \geq 0, \quad \text{for all} \quad x \in \mathbb{C}_j.$$

This is a contradiction since t is a $\mathbb{R}$-linear map and $\{x^2 : x \in \mathbb{C}_j\} = \mathbb{C}_j$.

Finally, the canonical $\mathbb{C}$-algebra isomorphism $\mathbb{C}_j \otimes_{\mathbb{R}} \mathbb{C} \simeq \mathbb{C} \times \mathbb{C}$ is given by $z \otimes w \mapsto (wz, w\bar{z})$. From this (2) follows easily. □

Definition 4.1.6 Let F be a nondegenerate fusion rule on a finite set A. Then, for any 'genus' $g \geq 0$, define the map (by induction on g)

$$F_g : \mathbb{Z}_+[A] \to \mathbb{Z}$$

by

$$F_0 = F, \text{ and, for any } g \geq 1, \ F_g(x) = \sum_{\lambda \in A} F_{g-1}(x + \lambda + \lambda^*), \ x \in \mathbb{Z}_+[A].$$

Also, define the 'Casimir' element

$$\Omega = \sum_{\lambda \in A} \lambda \cdot \lambda^* \in \mathbb{Z}[A]. \tag{1}$$

For any $x \in \mathbb{Z}[A]$, let $\mu_x : \mathbb{Z}[A] \to \mathbb{Z}[A]$ be the multiplication by $x : y \mapsto xy$. Let $\mathrm{Tr}(x)$ denote the trace of μ_x. Then, by Proposition 4.1.2(a) and Definition 4.1.4,

$$\mathrm{Tr}(x) = \sum_{\lambda \in A} \langle x \cdot \lambda, \lambda \rangle = \sum_{\lambda \in A} t(x \cdot \lambda \cdot \lambda^*) = t(x \cdot \Omega). \tag{2}$$

Proposition 4.1.7 *For $g \geq 1$ and any nondegenerate fusion rule on a finite set A and any $a_1, \ldots, a_n \in A$,*

$$F_g(a_1 + \cdots + a_n) = t(a_1 \ldots a_n \cdot \Omega^g) = Tr(a_1 \ldots a_n \cdot \Omega^{g-1}), \tag{1}$$

where t is the trace as in Proposition 4.1.2.

In fact, (1) *remains true for $g = 0$ as well by Lemma 4.1.9.*

Proof By the definition of F_g,

$$F_g(a_1 + \cdots + a_n) = \sum_{\lambda_1,\ldots,\lambda_g \in A} F_0(a_1 + \cdots + a_n + \lambda_1 + \lambda_1^* + \cdots + \lambda_g + \lambda_g^*)$$

$$= \sum_{\lambda_1,\ldots,\lambda_g \in A} t(a_1 \ldots a_n \cdot \lambda_1 \cdot \lambda_1^* \ldots \lambda_g \cdot \lambda_g^*),$$

by the defining property 4.1.2(b)

$$= t(a_1 \ldots a_n \cdot \Omega^g).$$

This proves the first equality in (1). The second equality in (1) of course follows from the identity (2) of Definition 4.1.6. □

As a consequence of Lemma 4.1.5 and Proposition 4.1.7, we have the following.

Corollary 4.1.8 *With the notation and assumption as in Proposition 4.1.7, for any* $a_1, \ldots, a_n \in A$ *and* $g \geq 0$,

$$F_g(a_1 + \cdots + a_n) = \sum_{\chi \in S_A} \chi(a_1) \ldots \chi(a_n) \chi(\Omega)^{g-1}. \tag{1}$$

Moreover, for any $\chi \in S_A$,

$$\chi(\Omega) = \sum_{\lambda \in A} |\chi(\lambda)|^2. \tag{2}$$

Proof For any $x \in \mathbb{C}[A]$, the multiplication map $\mu_x \colon \mathbb{C}[A] \to \mathbb{C}[A]$, under the identification φ_A of Lemma 4.1.5, is given by the diagonal matrix $(\chi(x))_{\chi \in S_A}$ (in the standard coordinate basis of $\mathbb{C}^{S_A}$). Thus,

$$\text{Tr}(x) = \sum_{\chi \in S_A} \chi(x).$$

Combining identity (1) of Proposition 4.1.7 with the above, we get

$$F_g(a_1 + \cdots + a_n) = \text{Tr}(a_1 \ldots a_n \cdot \Omega^{g-1})$$

$$= \sum_{\chi \in S_A} \chi(a_1) \ldots \chi(a_n) \chi(\Omega)^{g-1}.$$

This proves (1). To prove (2), use identity (2) of Lemma 4.1.5. □

The following lemma is due to Jiuzu Hong.

Lemma 4.1.9 *The element* $\Omega \in \mathbb{Z}[A]$ *(cf. identity* (1) *of Definition 4.1.6) is an invertible element in* $\mathbb{C}[A]$.

In particular, identity (1) *of Proposition 4.1.7 remains true for* $g = 0$ *as well.*

Proof By Lemma 4.1.5 it suffices to prove that, for any $\chi \in S_A$, $\chi(\Omega)$ is nonzero. Now,

$$\begin{aligned}\chi(\Omega) &= \sum_{\lambda \in A} |\chi(\lambda)|^2, \text{ by identity (2) of Corollary 4.1.8} \\ &> 0.\end{aligned}$$

This proves the lemma. □

4.1.E Exercises

In the following Exercises 1–3, F is a nondegenerate fusion rule on a finite set A.

(1) Let $\mathbb{Q}[A] := \mathbb{Z}[A] \otimes_{\mathbb{Z}} \mathbb{Q}$ be the associated fusion algebra over $\mathbb{Q}$. Show that there is a canonical decomposition (as a $\mathbb{Q}$-algebra)

$$\mathbb{Q}[A] \simeq \Pi_{i=1}^{m} E_i \times \Pi_{j=1}^{n} F_j,$$

where each E_i is a (finite) totally real extension of $\mathbb{Q}$ and each F_j is an imaginary quadratic extension of a (finite) totally real extension F'_j of $\mathbb{Q}$.

Moreover, each E_i and F_j are stable under $*$ with $*$ acting on each E_i and F'_j via the identity map and $*$ on each F_j acts via the nontrivial automorphism of F_j over F'_j.

Hint: Follow the proof of Lemma 4.1.5.

(2) Show that, for any $g, h \in \mathbb{Z}_+$ and $x, y \in \mathbb{Z}_+[A]$,

$$F_{g+h}(x + y) = \sum_{\lambda \in A} F_g(x + \lambda) F_h(y + \lambda^*).$$

Observe that this is a higher-genus analogue of the condition (f_4) of Definition 4.1.1.

(3) By definition, A is a basis of $\mathbb{C}[A]$. For $a \in A$, let $F^a = (F^a_{b,c})_{b,c \in A}$ be the matrix of the multiplication μ_a in the A-basis. Then, by the definition of multiplication in $\mathbb{C}[A]$ (cf. identity (1) of the proof of Proposition 4.1.2),

$$F^a_{b,c} = F(a + c + b^*).$$

Also, under the identification of $\mathbb{C}[A]$ with $\mathbb{C}^{S_A}$ (as in Lemma 4.1.5), the standard coordinate basis of $\mathbb{C}^{S_A}$ gives rise to a basis of $\mathbb{C}[A]$

parameterized by S_A. For $a \in A$, let D^a be the matrix of μ_a in the S_A-basis. Then, clearly $D^a = (D^a_\chi)_{\chi \in S_A}$ is a $S_A \times S_A$ diagonal matrix with

$$D^a_\chi = \chi(a).$$

Show that there exists a unitary matrix $\Sigma = (\Sigma_{\chi,a})_{\chi \in S_A, a \in A}$ such that

$$F^a = \Sigma^{-1} \cdot D^a \cdot \Sigma, \quad \text{for all } a \in A.$$

In fact, Σ can be taken so that its entries

$$\Sigma_{\chi,a} = \frac{\chi(a)}{\left(\sum_{\lambda \in A} |\chi(\lambda)|^2\right)^{\frac{1}{2}}}.$$

Prove further that for any $a_1, \dots, a_n \in A$ and $g \geq 0$,

$$F_g(a_1 + \cdots + a_n) = \sum_{\chi \in S_A} \frac{\Sigma_{\chi,a_1} \cdots \Sigma_{\chi,a_n}}{(\Sigma_{\chi,\mathbb{1}})^{n+2g-2}},$$

where $\mathbb{1}$ is the multiplicative identity of $\mathbb{Z}[A]$ (cf. the proof of Proposition 4.1.2).

4.2 Fusion Ring of a Simple Lie Algebra and an Explicit Verlinde Dimension Formula

In this section $\mathfrak{g}$ is a simple Lie algebra over $\mathbb{C}$. We fix a level $c > 0$. Associated to the pair $(\mathfrak{g}, c)$, there is a fusion rule F_c giving rise to the fusion ring $R_c(\mathfrak{g})$. We will draw upon the general results proved in Section 4.1 to study $R_c(\mathfrak{g})$ in this section. This is our most important example of fusion rings, which leads to an explicit Verlinde dimension formula. We continue to follow the notation from Section 1.2, often without explanation.

Example 4.2.1 Let D_c be as defined in (1) of Definition 2.1.1. Define the function $F_c \colon \mathbb{Z}_+[D_c] \to \mathbb{Z}_+$ by $F_c(0) = 1$ (for the zero element of $\mathbb{Z}_+[D_c]$) and, for any $s \geq 1$ and $\lambda_i \in D_c$,

$$F_c(\lambda_1 + \cdots + \lambda_s) = \dim \mathscr{V}_{\mathbb{P}^1}(\vec{p}, \vec{\lambda}),$$

where $\vec{p} = (p_1, \dots, p_s)$ are any distinct points in $\mathbb{P}^1$, $\vec{\lambda} = (\lambda_1, \dots, \lambda_s)$ and $\mathscr{V}_{\mathbb{P}^1}(\vec{p}, \vec{\lambda})$ denotes the space of covacua on $\mathbb{P}^1$ with respect to the weights $\vec{\lambda}$ attached to the points $\vec{p}$. By Proposition 3.5.8 and Exercise 2.3.E.2, $F_c(\lambda_1 + \cdots + \lambda_s)$ does not depend upon the choice of the points $\vec{p}$.

Define the involution $*: D_c \to D_c$ by $\lambda^* := -w_o\lambda$, where w_o is the longest element of the Weyl group of $\mathfrak{g}$. Thus, λ^* is the highest weight of the dual module $V(\lambda)^*$.

By Exercise 2.3.E.2, F_c satisfies the properties (f_2) and (f_5) of Definition 4.1.1 for $A = D_c$. The property (f_4) follows from Corollary 3.5.10(b) for $g = 0$. (If $x = 0$ as an element of $\mathbb{Z}_+[D_c]$, it follows from Exercise 2.3.E.2.) The property (f_3) follows from the following lemma. Thus, F_c is indeed a nondegenerate fusion rule.

The corresponding fusion ring $\mathbb{Z}[D_c]$ (as given by Proposition 4.1.2) is called the *fusion ring of* $\mathfrak{g}$ *at level* c. Henceforth, it will be denoted by $R_c(\mathfrak{g})$. By definition, as a $\mathbb{Z}$-module, it is freely generated by the isomorphism classes $\{[V(\lambda)]\}_{\lambda \in D_c}$ and the fusion product $\otimes^c$ (at level c) is given by (cf. identity (1) of the proof of Proposition 4.1.2):

$$[V(\lambda)] \otimes^c [V(\mu)] := \sum_{\nu \in D_c} \dim \mathscr{V}_{\mathbb{P}^1}((\lambda, \mu, \nu^*))[V(\nu)], \tag{1}$$

where $\mathscr{V}_{\mathbb{P}^1}((\lambda, \mu, \nu^*))$ denotes the space of covacua on $\mathbb{P}^1$ with respect to the weights (λ, μ, ν^*) attached to any three distinct points in $\mathbb{P}^1$.

Lemma 4.2.2 *Let $s \geq 1$. For any s-pointed curve $(\Sigma, \vec{p})$ and any $\vec{\lambda} = (\lambda_1, \dots, \lambda_s)$ with each $\lambda_i \in D_c$, there is an isomorphism*

$$\mathscr{V}_\Sigma(\vec{p}, \vec{\lambda}) \simeq \mathscr{V}_\Sigma(\vec{p}, \vec{\lambda}^*),$$

where $\vec{\lambda}^* := (\lambda_1^*, \dots, \lambda_s^*)$.

Proof Recall first that there exists an automorphism $\beta: \mathfrak{g} \to \mathfrak{g}$ such that, for any $\lambda \in D$, the $\mathfrak{g}$-module $V(\lambda)^\beta$ is isomorphic with $V(\lambda^*)$, where $V(\lambda)^\beta$ is the same vector space as $V(\lambda)$ but the $\mathfrak{g}$-module structure on $V(\lambda)^\beta$ is twisted via

$$x \odot v = \beta^{-1}(x) \cdot v, \quad \text{for } x \in \mathfrak{g} \text{ and } v \in V(\lambda) \tag{1}$$

(cf. (Bourbaki, 2005, Chap. VIII, §7, no. 6, Remark 1)).

The automorphism β clearly gives rise to an automorphism $\hat{\beta}$ of the affine Lie algebra $\hat{\mathfrak{g}}$ defined by

$$\hat{\beta}(x[f]) = \beta(x)[f], \quad \text{for} \quad x \in \mathfrak{g},\ f \in K = \mathbb{C}((t)) \text{ and } \hat{\beta}(C) = C.$$

Now, for any $\lambda \in D_c$, the twisted $\hat{\mathfrak{g}}$-module $\mathscr{H}(\lambda)^{\hat{\beta}}$ (with the same space as $\mathscr{H}(\lambda)$ and the $\hat{\mathfrak{g}}$-module structure twisted by the same formula as (1)) is isomorphic with $\mathscr{H}(\lambda^*)$. To see this, let $v_+ \in \mathscr{H}(\lambda)$ be a highest-weight vector. Then, $(\mathfrak{g} \otimes t\mathbb{C}[[t]]) \odot v_+ = 0$ in $\mathscr{H}(\lambda)^{\hat{\beta}}$. Moreover, the $\mathfrak{g}$-submodule of $\mathscr{H}(\lambda)^{\hat{\beta}}$ generated by v_+ is the same as $V(\lambda)^\beta \simeq V(\lambda^*)$. Thus, $\mathscr{H}(\lambda)^{\hat{\beta}}$ is an irreducible quotient of the generalized Verma module $\hat{M}(V(\lambda^*), c)$ (cf. (4) of Definition 1.2.4) and hence an irreducible quotient of the Verma module $\hat{M}(\lambda_c^*)$. Now, apply Exercise 1.2.E.6 to get that $\mathscr{H}(\lambda)^{\hat{\beta}} \simeq \mathscr{H}(\lambda^*)$.

Choose a $\hat{\mathfrak{g}}$-module isomorphism $\mathscr{H}(\lambda)^{\hat{\beta}} \simeq \mathscr{H}(\lambda^*)$ (which is unique up to a scalar multiple by Exercise 1.2.E.5) and let

$$\hat{\beta}_\lambda : \mathscr{H}(\lambda) \to \mathscr{H}(\lambda^*)$$

be the set-theoretic identity map under the above identification. Then, clearly

$$\hat{\beta}_\lambda(\hat{x} \cdot v) = \hat{\beta}(\hat{x}) \cdot \hat{\beta}_\lambda(v), \quad \text{for} \quad \hat{x} \in \hat{\mathfrak{g}}, v \in \mathscr{H}(\lambda). \tag{2}$$

Now, we are ready to prove the lemma. Let

$$\hat{\beta}_{\vec{\lambda}} : \mathscr{H}(\vec{\lambda}) := \mathscr{H}(\lambda_1) \otimes \cdots \otimes \mathscr{H}(\lambda_s) \to \mathscr{H}(\vec{\lambda}^*)$$

be the linear isomorphism $\hat{\beta}_{\vec{\lambda}} := \hat{\beta}_{\lambda_1} \otimes \cdots \otimes \hat{\beta}_{\lambda_s}$.

From property (2), it is easy to see $\hat{\beta}_{\vec{\lambda}}$ induces an isomorphism $\mathscr{V}_\Sigma(\vec{p}, \vec{\lambda}) \simeq \mathscr{V}_\Sigma(\vec{p}, \vec{\lambda}^*)$. This proves the lemma. □

As a consequence of Corollary 2.3.5 and the definition of the fusion product $\otimes^c$ (as in (1) of Example 4.2.1), we get the following result.

Proposition 4.2.3 *For any $\lambda, \mu \in D_c$, $[V(\lambda)] \otimes^c [V(\mu)]$ is the isomorphism class of the quotient $Q_{\lambda,\mu}$ of $V(\lambda) \otimes V(\mu)$ by the $\mathfrak{g}$-submodule $K_{\lambda,\mu}$ generated by*

$$\bigoplus_{p+q+r>2c} \big(V(\lambda)_{(p)} \otimes V(\mu)_{(q)}\big)_{(r)},$$

where $\big(V(\lambda)_{(p)} \otimes V(\mu)_{(q)}\big)_{(r)}$ denotes the isotypic component of $V(\lambda)_{(p)} \otimes V(\mu)_{(q)}$ corresponding to the irreducible representation indexed by r of $sl_2(\theta)$ (cf. Definition 2.3.1).

In particular, $Q_{\lambda,\mu}$ has no components $V(\nu)$ with $\nu \notin D_c$.

Proof Observe first that for any $\nu \notin D_c$, $V(\nu)$ does not occur in $Q_{\lambda,\mu}$:

If $V(\nu)$ does not occur in $V(\lambda) \otimes V(\mu)$, there is nothing to prove. So, assume that $\nu(\theta^\vee) \geq c+1$ and there is a copy $V(\nu) \subset V(\lambda) \otimes V(\mu)$. Consider the $sl_2(\theta)$-submodule V_1 of $V(\nu)$ passing through the highest-weight vector of $V(\nu)$. Then,

$$V_1 \subset \bigoplus_{p,q \geq 0} \big(V(\lambda)_{(p)} \otimes V(\mu)_{(q)}\big)_{(\nu(\theta^\vee))}.$$

But, for $\big(V(\lambda)_{(p)} \otimes V(\mu)_{(q)}\big)_{(\nu(\theta^\vee))}$ to be nonzero, we must have $p + q \geq \nu(\theta^\vee)$. Thus, from the definition of $K_{\lambda,\mu}$, $V_1 \subset K_{\lambda,\mu}$ hence so is $V(\nu) \subset K_{\lambda,\mu}$. This proves that $V(\nu)$ does not occur in $Q_{\lambda,\mu}$.

We next show that, for any $\nu \in D_c$, the multiplicity $m^{\nu}_{\lambda,\mu} := \dim \mathscr{V}_{\mathbb{P}^1}(\lambda, \mu, \nu^*)$ of $[V(\nu)]$ in $[V(\lambda)] \otimes^c [V(\mu)]$ coincides with the multiplicity $n^{\nu}_{\lambda,\mu}$ of $V(\nu)$ in $Q_{\lambda,\mu}$.

By Corollary 2.3.5,

$$\begin{aligned} m^{\nu}_{\lambda,\mu} &= \dim \Big\{ \mathfrak{g}\text{-module maps } f \colon V(\lambda) \otimes V(\mu) \otimes V(\nu^*) \to \mathbb{C} \\ &\qquad \text{such that } f \text{ vanishes on } \bigoplus_{p+q+r>2c} V(\lambda)_{(p)} \otimes V(\mu)_{(q)} \otimes V(\nu^*)_{(r)} \Big\} \\ &= \dim \{\mathfrak{g}\text{-module maps } \bar{f} \colon V(\lambda) \otimes V(\mu) \to V(\nu) \text{ such that } \bar{f} \\ &\qquad \text{vanishes on } [V(\lambda)_{(p)} \otimes V(\mu)_{(q)}]_{(r)} \text{ with } p+q+r > 2c\} \\ &= n^{\nu}_{\lambda,\mu}. \end{aligned}$$

This proves the proposition. □

Corollary 4.2.4 *With the notation as in Proposition 4.2.3,*

(a) $[V(\lambda)] \otimes^c [V(\mu)] = [V(\lambda) \otimes V(\mu)]$, *if* $\lambda + \mu \in D_c$.

(b) If $(\lambda+\mu)(\theta^\vee) = c+1$, *then* $[V(\lambda)] \otimes^c [V(\mu)]$ *is obtained from* $V(\lambda) \otimes V(\mu)$ *by removing all the components* $V(\nu)$ *with* $\nu(\theta^\vee) \geq c+1$. *(In fact, in this case* $V(\lambda) \otimes V(\mu)$ *cannot have any component* $V(\nu)$ *with* $\nu(\theta^\vee) > c+1$, *since* $(\lambda+\mu)(\theta^\vee) = c+1$.*)*

Proof (a) In this case, clearly $K_{\lambda,\mu} = 0$, where $K_{\lambda,\mu}$ is as defined in Proposition 4.2.3. This proves (a) by Proposition 4.2.3.

(b) Let $V(\nu) \subset V(\lambda) \otimes V(\mu)$ be a component with $\nu(\theta^\vee) = c+1$. Then

$$V(\nu)_{(\nu(\theta^\vee))} \subset V(\lambda)_{(\lambda(\theta^\vee))} \otimes V(\mu)_{(\mu(\theta^\vee))}.$$

Thus

$$V(\nu)_{(\nu(\theta^\vee))} \subset K_{\lambda,\mu} \quad \text{and hence} \quad V(\nu) \subset K_{\lambda,\mu}.$$

Conversely, take p, q, r such that $p+q+r \geq 2c+1$ and $(V(\lambda)_{(p)} \otimes V(\mu)_{(q)})_{(r)} \neq 0$. Then $p = \lambda(\theta^\vee)$, $q = \mu(\theta^\vee)$ and $r = (\lambda+\mu)(\theta^\vee)$. But any $\mathfrak{g}$-component $V(\nu)$ of the $\mathfrak{g}$-submodule of $V(\lambda) \otimes V(\mu)$ generated by $\big(V(\lambda)_{(\lambda(\theta^\vee))} \otimes V(\mu)_{(\mu(\theta^\vee))}\big)_{((\lambda+\mu)(\theta^\vee))}$ clearly satisfies $\nu(\theta^\vee) \geq (\lambda+\mu)(\theta^\vee) = c+1$. Thus, $K_{\lambda,\mu}$ does not contain any component $V(\nu)$ with $\nu(\theta^\vee) \leq c$. This proves (b) and hence the corollary is proved. □

Definition 4.2.5 Let $\mathfrak{g}$ be a simple Lie algebra over $\mathbb{C}$ and let W be its Weyl group. Then W acts naturally on the weight lattice $P \subset \mathfrak{h}^*$, where

$$P := \{\lambda \in \mathfrak{h}^* : \lambda(\alpha_i^\vee) \in \mathbb{Z} \text{ for all the simple coroots } \alpha_i^\vee\},$$

and hence also on $P_{\mathbb{R}} := P \otimes_{\mathbb{Z}} \mathbb{R}$.

Let $Q \subset P$ be the root lattice and let $Q_{lg} \subset Q$ be the sublattice generated by the long roots (if all the root lengths are equal, we call them long roots). Let $h^\vee := 1 + \rho(\theta^\vee)$ be the *dual Coxeter number*, where ρ is the half sum of positive roots of $\mathfrak{g}$ and θ is the highest root.

Let G be the connected simply-connected complex algebraic group with Lie algebra $\mathfrak{g}$ and let $T \subset G$ be the maximal torus with Lie algebra $\mathfrak{h}$. (Here we have deviated from our usual convention to denote the Lie algebra of a group by the corresponding Gothic character.)

Let W_c be the *affine Weyl group* of $\mathfrak{g}$ at level c, which is, by definition, the group of affine transformations of $P_{\mathbb{R}}$ generated by W and the translation $\lambda \mapsto \lambda + (c + h^\vee)\theta$. Since each long root is W-conjugate to θ, W_c is the semi-direct product of W by the lattice $(c + h^\vee)Q_{lg}$. For any $\alpha \in \Delta$, where Δ is the set of all the roots of $\mathfrak{g}$, and $n \in \mathbb{Z}$, define the *affine wall*

$$H_{\alpha,n} = \{\lambda \in P_{\mathbb{R}} : \langle \lambda, \alpha \rangle = n(c + h^\vee)\}.$$

Let

$$H := \bigcup_{\alpha \in \Delta,\, n \in \mathbb{Z}} H_{\alpha,n}.$$

The connected components of $P_{\mathbb{R}} \backslash H$ are called *alcoves*. Then, the closure of any alcove is a fundamental domain for the action of W_c on $P_{\mathbb{R}}$ and W_c acts simply transitively on the set of alcoves (cf. (Bourbaki, 2002, Chap. VI, §2, no. 1)). In particular, W_c acts freely on $P_{\mathbb{R}} \backslash H$. Moreover, W_c is a Coxeter group (cf. (Bourbaki, 2002, Chap. V, §3, no. 2)). The *fundamental alcove* is defined by

$$A^\circ := \{\lambda \in P_{\mathbb{R}} : \lambda(\alpha_i^\vee) > 0 \text{ for all the simple coroots } \alpha_i^\vee \text{ and } \lambda(\theta^\vee) < c + h^\vee\}.$$

Its closure in $P_{\mathbb{R}}$ is clearly given by

$$A = \{\lambda \in P_{\mathbb{R}} : \lambda(\alpha_i^\vee) \geq 0 \text{ for all the simple coroots } \alpha_i^\vee \text{ and } \lambda(\theta^\vee) \leq c + h^\vee\}.$$

Then, it is easy to see that

$$A^\circ = A \backslash H. \tag{1}$$

Define the shifted action of W_c on $P_{\mathbb{R}}$ by

$$w * \lambda = w(\lambda + \rho) - \rho, \quad \text{for } w \in W_c \text{ and } \lambda \in P_{\mathbb{R}}. \tag{2}$$

It is easy to see that the map

$$D_c \to A^{\circ} \cap P, \quad \mu \mapsto \mu + \rho, \tag{3}$$

is a bijection. Since W_c keeps P stable and, moreover, it acts simply transitively on the set of alcoves, for any $\lambda \in P \backslash H$, there exists a unique $w \in W_c$ and $\mu \in D_c$ such that

$$\lambda - \rho = w * \mu. \tag{4}$$

Conversely, for any $w \in W_c$ and $\mu \in D_c$,

$$(w * \mu) + \rho \in P \backslash H. \tag{5}$$

Further, for any $\lambda \in H_{\alpha,n}$ with $\alpha \in \Delta$ and $n \in \mathbb{Z}$,

$$\tau_{\frac{2n(c+h^\vee)\alpha}{\langle\alpha,\alpha\rangle}} \cdot s_\alpha(\lambda) = s_\alpha \cdot \tau_{\frac{-2n(c+h^\vee)\alpha}{\langle\alpha,\alpha\rangle}}(\lambda) = \lambda, \tag{6}$$

where τ_β denotes the translation by β and $s_\alpha(\lambda) := \lambda - \lambda(\alpha^\vee)\alpha \in W$ is the reflection corresponding to the root α. (Observe that for any $\alpha \in \Delta$,

$$\frac{2\alpha}{\langle\alpha,\alpha\rangle} \in Q_{lg}, \tag{7}$$

which can easily be seen from the tables in (Bourbaki, 2002, Plates II–IX).)

Denote by $T_c \subset T$ the subgroup

$$T_c = \{t \in T : e^{\beta}(t) = 1, \text{ for all } \beta \in (c + h^\vee)Q_{lg}\},$$

and let T_c^{reg} be the subset of T_c consisting of regular elements, i.e.,

$$\begin{aligned} T_c^{\mathrm{reg}} &:= \{t \in T_c : w \cdot t \neq t \text{ for any } w \neq 1 \in W\} \\ &= \{t \in T_c : e^{\alpha}(t) \neq 1 \text{ for any root } \alpha\}. \end{aligned}$$

Let $\kappa \colon \mathfrak{h}^* \to \mathfrak{h}$ be the isomorphism induced from the invariant form normalized by $\langle\theta,\theta\rangle = 2$.

Lemma 4.2.6 *(a) The map $\varphi \colon \lambda \mapsto \operatorname{Exp}\left(\frac{2\pi i\kappa(\lambda)}{c+h^\vee}\right)$ induces an isomorphism of groups:*

$$\overline{\varphi} \colon P/(c + h^\vee)Q_{lg} \xrightarrow{\sim} T_c.$$

(b) The map $\lambda \mapsto \operatorname{Exp}\left(\frac{2\pi i\kappa(\lambda+\rho)}{c+h^\vee}\right)$ induces a bijection:

$$D_c \xrightarrow{\sim} T_c^{reg}/W.$$

Proof (a) For any long root α and $\lambda \in P$,

$$\begin{aligned} e^{(c+h^\vee)\alpha}\left(\operatorname{Exp}\left(\frac{2\pi i\kappa(\lambda)}{c+h^\vee}\right)\right) &= e^{2\pi i\langle\lambda,\alpha\rangle} \\ &= e^{2\pi i\lambda(\alpha^\vee)}, \quad \text{since } \alpha \text{ is a long root} \\ &= 1. \end{aligned}$$

Thus, $\operatorname{Im}\varphi \subset T_c$.

Let $\lambda = (c+h^\vee)\alpha$, for a long root α. Then

$$\operatorname{Exp}\left(\frac{2\pi i\kappa(\lambda)}{c+h^\vee}\right) = \operatorname{Exp}(2\pi i\kappa(\alpha)) = 1,$$

since

$$\kappa(\Delta_{lg}) \subset Q^\vee \tag{1}$$

as the following calculation shows, where Δ_{lg} denotes the set of long roots. For $\alpha \in \Delta_{lg}$,

$$\omega_i(\kappa(\alpha)) = \langle\omega_i,\alpha\rangle = \langle\omega_i,\alpha^\vee\rangle \in \mathbb{Z},$$

where $\{\omega_1,\ldots,\omega_\ell\} \subset \mathfrak{h}^*$ are the fundamental weights. Thus, φ factors through $(c+h^\vee)Q_{lg}$. We next show that $\overline{\varphi}$ is injective.

Take $\lambda \in P$ such that $\operatorname{Exp}\left(\frac{2\pi i\kappa(\lambda)}{c+h^\vee}\right) = 1$. Then, $\frac{\kappa(\lambda)}{c+h^\vee} \in Q^\vee$, which gives

$$\lambda \in (c+h^\vee)\kappa^{-1}(Q^\vee). \tag{2}$$

For any simple root α_i,

$$\kappa^{-1}(\alpha_i^\vee) = \frac{2\alpha_i}{\langle\alpha_i,\alpha_i\rangle}.$$

Taking a simple root α_k in the W-orbit of non-long α_i such that $\langle\alpha_k,\alpha_j\rangle \neq 0$ for a long root α_j and considering $s_k(\alpha_j)$, we see that

$$\kappa^{-1}(Q^\vee) \subset Q_{lg}.$$

Combining this with (1), we get

$$\kappa^{-1}(Q^\vee) = Q_{lg}. \tag{3}$$

In particular, W_c is canonically isomorphic with the affine Weyl group defined in Definition 1.2.12. Combining (2) and (3), we get that $\overline{\varphi}$ is injective.

Take $t \in T_c$ and choose $\lambda \in \mathfrak{h}^*$ such that $\operatorname{Exp}\left(\frac{2\pi i\kappa(\lambda)}{c+h^\vee}\right) = t$. Then

$$e^\beta\left(\operatorname{Exp}\left(\frac{2\pi i\kappa(\lambda)}{c+h^\vee}\right)\right) = 1 \quad \text{for all} \quad \beta \in (c+h^\vee)Q_{lg}.$$

This gives

$$\lambda(\kappa(\alpha)) \in \mathbb{Z}, \quad \text{for all} \quad \alpha \in Q_{lg}.$$

Hence, by (3), we get

$$\lambda(Q^\vee) \subset \mathbb{Z} \Rightarrow \lambda \in P.$$

This proves the surjectivity of $\overline{\varphi}$ and hence (a) is proved.

(b) As in Definition 4.2.5 (see (3) of Definition 4.2.5), the map

$$D_c \xrightarrow{\sim} A^\circ \cap P, \ \lambda \mapsto \lambda + \rho, \tag{4}$$

is a bijection. Moreover, by the (a) part, the W-equivariant map

$$\overline{\varphi}\colon P/(c+h^\vee)Q_{lg} \xrightarrow{\sim} T_c$$

is an isomorphism. Thus, $\overline{\varphi}$ induces a bijection

$$P/W_c \xrightarrow{\sim} T_c/W. \tag{5}$$

Now, for any $\lambda \in H_{\alpha,n}$ (with $\alpha \in \Delta$ and $n \in \mathbb{Z}$), by identity (6) of Definition 4.2.5,

$$s_\alpha \cdot \tau_{\frac{-2n(c+h^\vee)\alpha}{\langle\alpha,\alpha\rangle}}(\lambda) = \lambda.$$

Of course, by (7) of Definition 4.2.5, for any $\lambda \in P$,

$$\overline{\varphi}(\lambda) = \overline{\varphi}(\tau_{\frac{-2n(c+h^\vee)\alpha}{\langle\alpha,\alpha\rangle}}\lambda).$$

Thus, for $\lambda \in H_{\alpha,n}$,

$$s_\alpha(\overline{\varphi}(\lambda)) = \overline{\varphi}(\lambda); \quad \text{in particular,} \quad \overline{\varphi}(\lambda) \in T_c \backslash T_c^{\text{reg}}. \tag{6}$$

Further, since W_c acts freely on $P\backslash H$, by (5) and (6),

$$(P\backslash H)/W_c \xrightarrow{\sim} T_c^{\text{reg}}/W. \tag{7}$$

Finally, since A° is an alcove and W_c acts simply transitively on the set of alcoves (cf. Definition 4.2.5), the canonical inclusion induces a bijection:

$$A^\circ \cap P \xrightarrow{\sim} (P\backslash H)/W_c. \tag{8}$$

Combining the bijections (4), (8) and (7), we get the (b)-part of the lemma. □

Definition 4.2.7 Let $R(\mathfrak{g})$ be the representation ring of $\mathfrak{g}$. Define the $\mathbb{Z}$-linear map

$$\xi_c \colon R(\mathfrak{g}) \to R_c(\mathfrak{g})$$

as follows. For $\lambda \in D$ (where D is as in Definition 1.2.6) such that $\lambda + \rho$ lies on an affine wall $H_{\alpha,n}$ (for some $\alpha \in \Delta$ and $n \in \mathbb{Z}$), define

$$\xi_c([V(\lambda)]) = 0.$$

Otherwise, there is a unique $\mu \in D_c$ and $w \in W'_c$ such that $\lambda = w^{-1} * \mu$, where W'_c is the set of minimal-length coset representatives in W_c/W (cf. (4) of Definition 4.2.5). (Observe that, for $w \in W_c$ and $\mu \in D_c$, $w^{-1} * \mu \in D$ if and only if $w \in W'_c$, see Kostant (2004, Remark 1.3).) In this case, we define

$$\xi_c([V(\lambda)]) = \epsilon(w)[V(\mu)],$$

where $\epsilon(w)$ is the sign of the Coxeter group element w.

The following lemma follows easily from the definition of the map ξ_c since $R(\mathfrak{g}) = \bigoplus_{\lambda \in D} \mathbb{Z}[V(\lambda)]$.

Lemma 4.2.8 *The kernel of $\xi_c \colon R(\mathfrak{g}) \to R_c(\mathfrak{g})$ is the $\mathbb{Z}$-submodule of $R(\mathfrak{g})$ spanned by*

(a) $[V(\lambda)]$, $\lambda \in D$ such that $\lambda + \rho \in H$, and
(b) $[V(w^{-1} * \mu)] - \epsilon(w)[V(\mu)]$, for $\mu \in D_c$ and $w \in W'_c$.

Proof Take $\lambda \in D$ such that $\lambda + \rho \notin H$. Then, W_c acts freely on $\lambda + \rho$ (cf. Definition 4.2.5). From this the lemma follows easily. □

The following is a crucial result used in the proof of an explicit Verlinde dimension formula (cf. Theorem 4.2.19).

Theorem 4.2.9 *For any simple Lie algebra $\mathfrak{g}$ and any level $c > 0$, the map*

$$\xi_c \colon R(\mathfrak{g}) \to R_c(\mathfrak{g})$$

is a surjective ring homomorphism.

In particular, the fundamental representations $\{[V(\omega_i)]\}_{1 \le i \le \ell}$ generate the fusion ring $R_c(\mathfrak{g})$.

Before we come to the proof of this theorem, we give a different geometric definition of a product in $R_c(\mathfrak{g})$ which (as shown below in Corollary 4.2.17) coincides with the product $\otimes^c$. The proof of Theorem 4.2.9 is given in Subsection 4.2.18.

As earlier, we fix a level $c > 0$. Let G be the (simple) simply-connected complex algebraic group with Lie algebra $\mathfrak{g}$ and let $\bar{\hat{G}} = \bar{\hat{G}}_{0_1}$ be the corresponding affine Kac–Moody group (corresponding to the basic weight $0_1 \in \hat{D}$), which is a $\mathbb{G}_m$ central extension of the formal loop group $\bar{G}((t))$ (cf. Proposition 1.4.12). Let $\hat{\mathcal{P}} \subset \bar{\hat{G}}$ be the parahoric subgroup, which is the inverse image of $\bar{G}[[t]]$ under the central extension $\bar{p}\colon \bar{\hat{G}} \to \bar{G}((t))$. The central extension uniquely splits over $\bar{G}[[t]]$ (cf. Remark 1.4.13) and hence

$$\hat{\mathcal{P}} \simeq \mathbb{G}_m \times \bar{G}[[t]].$$

Recall the infinite Grassmannian $\bar{X}_G$, which is an ind-projective variety (cf. Propositions 1.3.18 and 1.3.24) with $\mathbb{C}$-points

$$\bar{X}_G(\mathbb{C}) = G((t))/G[[t]].$$

Then, $\bar{\hat{G}}$ acts on $\bar{X}_G$ through $\bar{p}$ via the action of $\bar{G}((t))$ on $\bar{X}_G$ (cf. Proposition 1.3.18(c)).

Lemma 4.2.10 *(a) Let V be a finite-dimensional representation of G. Then there exists a $\bar{\hat{G}}$-equivariant vector bundle $\mathscr{L}_c(V)$ over $\bar{X}_G$ such that the fiber of $\mathscr{L}_c(V)$, over the base point $\bar{o} \in \bar{X}_G$, which is a module for $(\bar{p})^{-1}(\bar{G}[[t]]) \simeq \mathbb{G}_m \times \bar{G}[[t]]$ is acted by the $\mathbb{G}_m$-factor via the character $z \mapsto z^{-c}$ and $\bar{G}[[t]]$ acts through the representation V^* of G under the evaluation map $\bar{G}[[t]] \to G, t \mapsto 0$.*

Moreover,

(b) For any $\lambda \in D$ such that $\lambda + \rho \in H$, $H^i(\bar{X}_G, \mathscr{L}_c(V(\lambda))) = 0$ for all $i \geq 0$, where H is defined in Definition 4.2.5.

And,

(c) For $\mu \in D_c$ and $w \in W'_c$, $H^i(\bar{X}_G, \mathscr{L}_c(V(w^{-1} * \mu))) = 0$ unless $i = \ell(w)$, and

$$H^{\ell(w)}(\bar{X}_G, \mathscr{L}_c(V(w^{-1} * \mu))) \simeq \mathscr{H}(\mu_c)^*$$

as a module of Lie $\bar{\hat{G}} = \hat{\mathfrak{g}}$ (cf. Definition B.22 and Lemma 1.4.6).

Thus,

$$H^{\ell(w)}(\bar{X}_G, \mathscr{L}_c(V(w^{-1} * \mu)))^{\hat{\mathfrak{g}}_-} \simeq V(\mu)^*,$$

where, as in (4) of Definition 1.2.2, $\hat{\mathfrak{g}}_- := \mathfrak{g} \otimes (t^{-1}\mathbb{C}[t^{-1}])$.

Proof To prove the existence of $\mathscr{L}_c(V)$, it clearly suffices to assume that $V = V(\lambda)$ for $\lambda \in D$. Consider the affine full-flag ind-variety $\bar{X}_G(B)$ defined in Exercise 1.3.E.11, where $B \subset G$ is the Borel subgroup with Lie algebra $\mathfrak{b}$ (as at the beginning of Section 1.2). Then, parallel to $\bar{X}_G$, $\hat{\bar{G}}$ acts on $\bar{X}_G(B)$ through $\bar{p}: \hat{\bar{G}} \to \bar{G}((t))$. Let $\hat{\mathcal{B}}$ be the (affine) Borel subgroup of $\hat{\bar{G}}$, which is the inverse image of $\mathcal{B} \subset \bar{G}((t))$ defined in Exercise 1.3.E.11. Recall that $\mathcal{B} := ev_0^{-1}(B)$ is the closed subgroup scheme under the evaluation map $ev_0: \bar{G}[[t]]) \to G, t \mapsto 0$.

Now, for any $\lambda \in D$, there exists a $\hat{\bar{G}}$-equivariant line bundle $\mathscr{L}_c^B(\lambda)$ over $\bar{X}_G(B)$ such that the fiber of $\mathscr{L}_c^B(\lambda)$ over the base point $\bar{o}_B \in \bar{X}_G(B)$ (which is a module for $\hat{\mathcal{B}} \simeq \mathbb{G}_m \times \mathcal{B}$) is acted by the $\mathbb{G}_m$-factor via the character $z \mapsto z^{-c}$ and $\mathcal{B}$ acts through the character $e^{-\lambda}$ of B under the evaluation map $\mathcal{B} \to B, t \mapsto 0$. More generally, as we will need in Subsection 4.2.18, for any finite-dimensional B-module M, there exists a $\hat{\bar{G}}$-equivariant vector bundle $\mathscr{L}_c^B(M)$ over $\bar{X}_G(B)$ such that the fiber of $\mathscr{L}_c^B(M)$ over the base point $\bar{o}_B \in \bar{X}_G(B)$ is acted by the $\mathbb{G}_m$-factor via the character $z \mapsto z^{-c}$ and $\mathcal{B}$ acts through the representation M^* of B. We denote this representation of $\hat{\mathcal{B}}$ by M_c^*. We now show the existence of such $\mathscr{L}_c^B(M)$.

Since $\bar{G}((t)) \to \bar{X}_G(B)$ is a locally trivial principal $\mathcal{B}$-bundle (cf. Exercise 1.3.E.11) and $\hat{\bar{G}} \xrightarrow{\bar{p}} \bar{G}((t))$ is a locally trivial principal $\mathbb{G}_m$-bundle trivial over $\bar{G}[t^{-1}]^- \times \bar{G}[[t]]$ (cf. Proposition 1.4.12), we get that the composite $\beta: \hat{\bar{G}} \to \bar{X}_G(B)$ is a locally trivial principal $\hat{\mathcal{B}}$-bundle. Consider the $\hat{\mathcal{B}}$-equivariant vector bundle $\theta: \hat{\bar{G}} \times M_c^* \to \hat{\bar{G}}$ under the projection θ, where $\hat{\mathcal{B}}$ acts on $\hat{\bar{G}} \times M_c^*$ via

$$b \cdot (g,v) = (gb^{-1}, b \cdot v), \quad \text{for } b \in \hat{\mathcal{B}}, g \in \hat{\bar{G}} \text{ and } v \in M_c^*.$$

Thus, by Theorem C.17, θ descends to give the vector bundle $\mathscr{L}_c^B(M)$ over $\bar{X}_G(B)$ as above. Recall from the proof of Theorem C.17 that, thought of as a $\mathbb{C}$-space functor, $\mathscr{L}_c^B(M)$ is the sheafification of the functor

$$S \rightsquigarrow \left(\hat{\bar{G}}(S) \times (M_c^*)(S)\right) / \hat{\mathcal{B}}(S)$$

(also see Kumar (2002, Corollary 8.2.5) for another construction of $\mathscr{L}_c^B(M)$).

Now, define

$$\mathscr{L}_c(V(\lambda)) := \pi_* \left(\mathscr{L}_c^B(\lambda)\right)$$

for the locally trivial G/B-fibration $\pi : \bar{X}_G(B) \to \bar{X}_G$ (cf. Exercise 1.3.E.11). By the classical BWB theorem, $\mathscr{L}_c(V(\lambda))$ satisfies the property (a). Further, from the degenerate Leray spectral sequence applied to the fibration π with respect to the line bundle $\mathscr{L}_c^B(\lambda)$, for any $i \geq 0$ and $\lambda \in D$,

$$H^i\left(\bar{X}_G(B), \mathscr{L}_c^B(\lambda)\right) \simeq H^i\left(\bar{X}_G, \mathscr{L}_c(V(\lambda))\right).$$

Thus, (b) and (c)-parts of the lemma follow from the affine analogue of the BWB theorem (cf. (Kumar, 2002, Corollary 8.3.12)). (Observe that we have used here Proposition 1.3.24.) □

Definition 4.2.11 For any $\lambda, \mu \in D_c$, define the following (*a priori* different from $\otimes^c$) product $\otimes^c_F$:

$$[V(\lambda)] \otimes^c_F [V(\mu)] = \chi_{\mathfrak{g}}\left(\bar{X}_G, \mathscr{L}_c(V(\lambda) \otimes V(\mu))\right)^*, \tag{1}$$

where for any finite-dimensional G-module V, we define the virtual G-module

$$\chi_{\mathfrak{g}}(\bar{X}_G, \mathscr{L}_c(V)) := \sum_{i \geq 0} (-1)^i \left[H^i(\bar{X}_G, \mathscr{L}_c(V))^{\hat{\mathfrak{g}}_-}\right] \in R_c(\mathfrak{g}). \tag{2}$$

As shown below in the following Lemma 4.2.12, the above sum is a finite sum and it is determined there. In particular, it lies in $R_c(\mathfrak{g})$.

We can rewrite (1) as follows. Let

$$H^i\left(\bar{X}_G, \mathscr{L}_c(V(\lambda) \otimes V(\mu))\right) \simeq \bigoplus_{\nu \in D_c} d_i^{\lambda,\mu}(\nu)\, \mathscr{H}(\nu)^*,$$

as $\hat{\mathfrak{g}}$-modules, for some (unique) $d_i^{\lambda,\mu}(\nu) \in \mathbb{Z}_+$ (cf. Lemma 4.2.10). Then

$$[V(\lambda)] \otimes^c_F [V(\mu)] = \sum_{\nu \in D_c} \left(\sum_{i \geq 0} (-1)^i d_i^{\lambda,\mu}(\nu)\right) [V(\nu)]. \tag{3}$$

Lemma 4.2.12 *For any* $\lambda, \mu \in D_c$ *,*

$$[V(\lambda)] \otimes^c_F [V(\mu)] = \sum_{\nu \in D_c} \sum_{w \in W'_c} (-1)^{\ell(w)} n^{\lambda,\mu}_{w^{-1} * \nu} [V(\nu)], \tag{1}$$

where

$$n^{\lambda,\mu}_{w^{-1} * \nu} := \dim\left(\operatorname{Hom}_{\mathfrak{g}}\left(V(w^{-1} * \nu), V(\lambda) \otimes V(\mu)\right)\right).$$

Proof Decompose as $\mathfrak{g}$-modules (cf. Definition 4.2.7):

$$V(\lambda) \otimes V(\mu) \simeq \left(\bigoplus_{\nu \in D_c} \bigoplus_{w \in W'_c} n^{\lambda,\mu}_{w^{-1} * \nu} V(w^{-1} * \nu)\right) \bigoplus \bigoplus_{\gamma \in D \cap (H - \rho)} d_\gamma^{\lambda,\mu} V(\gamma), \tag{2}$$

for some $d_\gamma^{\lambda,\mu} \in \mathbb{Z}_+$.

Then, from decomposition (2) and Lemma 4.2.10, we get

$$
\begin{aligned}
H^i(\bar{X}_G, \mathscr{L}_c(V(\lambda) \otimes V(\mu))) \\
&\simeq \left(\bigoplus_{\nu \in D_c} \bigoplus_{w \in W'_c} n^{\lambda,\mu}_{w^{-1}*\nu} H^i \left(\bar{X}_G, \mathscr{L}_c(V(w^{-1} * \nu)) \right) \right) \\
&\quad \bigoplus \bigoplus_{\gamma \in D \cap (H-\rho)} d^{\lambda,\mu}_{\gamma} H^i \left(\bar{X}_G, \mathscr{L}_c(V(\gamma)) \right) \\
&\simeq \bigoplus_{\nu \in D_c} \left(\bigoplus_{\substack{w \in W'_c \\ \ell(w)=i}} n^{\lambda,\mu}_{w^{-1}*\nu} \right) \mathscr{H}(\nu)^*, \quad \text{by Lemma 4.2.10.} \qquad (3)
\end{aligned}
$$

Thus,

$$
\chi_{\mathfrak{g}} \left(\bar{X}_G, \mathscr{L}_c(V(\lambda) \otimes V(\mu)) \right) = \sum_{\nu \in D_c} \sum_{w \in W'_c} (-1)^{\ell(w)} n^{\lambda,\mu}_{w^{-1}*\nu} [V(\nu)^*].
$$

From the definition of $[V(\lambda)] \otimes^c_F [V(\mu)]$ as in (1) of Definition 4.2.11, we get (1). This proves the lemma. □

Definition 4.2.13 For any $\mu \in D$ and nonzero $z \in \mathbb{C}$, realize $V(\mu)$ as a module for the affine Lie algebra $\tilde{\mathfrak{g}} := (\mathfrak{g} \otimes \mathcal{A}) \oplus \mathbb{C}C$ (cf. (1) of Definition 1.2.1) via the evaluation at z:

$$
ev_z : (\mathfrak{g} \otimes \mathcal{A}) \oplus \mathbb{C}C \to \mathfrak{g}, \quad C \mapsto 0 \text{ and } x[f] \mapsto f(z)x, \quad \text{for } x \in \mathfrak{g} \text{ and } f \in \mathcal{A}.
$$

We denote this $\tilde{\mathfrak{g}}$-module by $V_z(\mu)$.

For any $\nu \in D_c$ and $\mu \in D$, we give a resolution of $\mathscr{H}(\nu) \otimes V_z(\mu)$.

First, recall the BGG resolution consisting of $\hat{\mathfrak{g}}$-modules and $\hat{\mathfrak{g}}$-module maps (cf. (Kumar, 2002, Theorem 9.1.3)):

$$
\cdots \xrightarrow{\delta_2} F_1 \xrightarrow{\delta_1} F_0 \xrightarrow{\epsilon} \mathscr{H}(\nu) \to 0,
$$

where

$$
F_p := \bigoplus_{\substack{w \in W'_c \\ \ell(w)=p}} \hat{M}(V(w^{-1} * \nu), c).
$$

Tensoring with $V_z(\mu)$, we get the resolution

$$
\cdots \to F_1 \otimes V_z(\mu) \to F_0 \otimes V_z(\mu) \to \mathscr{H}(\nu) \otimes V_z(\mu) \to 0.
$$

Recall from Definition 1.2.2 that $\hat{\mathfrak{g}}_- := \mathfrak{g} \otimes (t^{-1}\mathbb{C}[t^{-1}])$.

Lemma 4.2.14 *For any $\nu \in D_c$, $\mu \in D$ and $z \in \mathbb{C}$, the Lie algebra homology $H_*(\hat{\mathfrak{g}}_-, \mathscr{H}(\nu) \otimes V_z(\mu))$ as a $\mathfrak{g}$-module is given by the homology of the following complex consisting of $\mathfrak{g}$-modules and $\mathfrak{g}$-module maps:*

$$\cdots \to \hat{F}_p \to \cdots \xrightarrow{\hat{\delta}_2} \hat{F}_1 \xrightarrow{\hat{\delta}_1} \hat{F}_0 \to 0,$$

where

$$\hat{F}_p := \bigoplus_{\substack{w \in W'_c \\ \ell(w)=p}} \left(V(w^{-1} * \nu) \otimes V(\mu)\right).$$

Proof By the proof of Lemma 1.2.5, for any $\mathfrak{g}$-module V, as $U(\hat{\mathfrak{g}}_-)$-modules,

$$\hat{M}(V,c) \simeq U(\hat{\mathfrak{g}}_-) \otimes_{\mathbb{C}} V,$$

where $U(\hat{\mathfrak{g}}_-)$ acts on the right-hand side via the left multiplication on the first factor. Thus, by the Hopf principle (cf. (Kumar, 2002, Proposition 3.1.10)), as $U(\hat{\mathfrak{g}})$-modules,

$$\hat{M}(V,c) \otimes V_z(\mu) \simeq U(\hat{\mathfrak{g}}_-) \otimes_{\mathbb{C}} (V \otimes V_z(\mu)). \tag{1}$$

In particular, $\hat{M}(V,c) \otimes V_z(\mu)$ is free as a $U(\hat{\mathfrak{g}}_-)$-module. Thus, the homology $H_*(\hat{\mathfrak{g}}_-, \mathscr{H}(\nu) \otimes V_z(\mu))$ is given by the homology of the complex:

$$\cdots \to \mathbb{C} \otimes_{U(\hat{\mathfrak{g}}_-)} \left(F_1 \otimes V_z(\mu)\right) \to \mathbb{C} \otimes_{U(\hat{\mathfrak{g}}_-)} \left(F_0 \otimes V_z(\mu)\right) \to 0. \tag{2}$$

By (1), there exists a $\mathfrak{g}$-module isomorphism (for any $\mathfrak{g}$-module V):

$$\mathbb{C} \otimes_{U(\hat{\mathfrak{g}}_-)} \left(\hat{M}(V,c) \otimes V_z(\mu)\right) \simeq V \otimes V_z(\mu) \simeq V \otimes V(\mu). \tag{3}$$

Combining (2) and (3), we get the lemma. □

Proposition 4.2.15 *The products $\otimes^c$ and $\otimes^c_F$ in $R_c(\mathfrak{g})$ coincide if and only if for all λ, μ, $\nu \in D_c$,*

$$\overline{\chi}_{\mathfrak{g}}(\lambda, \mu, \nu) = 0,$$

where

$$\overline{\chi}_{\mathfrak{g}}(\lambda, \mu, \nu) := \sum_{i \geq 1} (-1)^i \dim \left(\operatorname{Hom}_{\mathfrak{g}}(V(\lambda), H_i(\hat{\mathfrak{g}}_-, \mathscr{H}(\nu) \otimes V_1(\mu)))\right).$$

Proof By Lemma 4.2.12, for $\lambda, \mu \in D_c$,

$$
\begin{aligned}
[V(\lambda)] \otimes_F^c [V(\mu)] &= \sum_{\nu \in D_c} \sum_{w \in W'_c} (-1)^{\ell(w)} n^{\lambda,\mu}_{w^{-1}*\nu} [V(\nu)] \\
&= \sum_{\nu \in D_c} \sum_{w \in W'_c} (-1)^{\ell(w)} \dim \Big(\mathrm{Hom}_{\mathfrak{g}}(V(\lambda), V(w^{-1} * \nu) \\
&\quad \otimes V(\mu^*)) \Big) [V(\nu)] \\
&= \sum_{\nu \in D_c} \sum_{i \geq 0} (-1)^i \dim \Big(\mathrm{Hom}_{\mathfrak{g}}(V(\lambda), H_i(\hat{\mathfrak{g}}_-, \mathscr{H}(\nu) \\
&\quad \otimes V_1(\mu^*)) \Big) [V(\nu)], \qquad (1)
\end{aligned}
$$

by Lemma 4.2.14 using the Euler–Poincaré principle. Further, by the definition of $\otimes^c$ (cf. (1) of Example 4.2.1),

$$
[V(\lambda)] \otimes^c [V(\mu)] = \sum_{\nu \in D_c} \dim \mathscr{V}_{\mathbb{P}^1}(\lambda, \mu, \nu^*)[V(\nu)], \qquad (2)
$$

where (λ, μ, ν^*) are attached to the points $(\infty, 1, 0)$ on $\mathbb{P}^1$, respectively.

By Lemma 4.2.2,

$$
\begin{aligned}
\mathscr{V}_{\mathbb{P}^1}(\lambda, \mu, \nu^*) &\simeq \mathscr{V}_{\mathbb{P}^1}(\lambda^*, \mu^*, \nu) \\
&\simeq [\mathscr{H}(\nu) \otimes V_\infty(\lambda^*) \otimes V_1(\mu^*)]_{\mathfrak{g} \otimes \mathbb{C}[t^{-1}]}, \text{ by Theorem 2.2.2.} \\
&\simeq \mathrm{Hom}_{\mathfrak{g}}(V(\lambda), H_0(\hat{\mathfrak{g}}_-, \mathscr{H}(\nu) \otimes V_1(\mu^*))). \qquad (3)
\end{aligned}
$$

Combining (1)–(3) and replacing μ by μ^*, we get the proposition. □

We now recall the following result from Teleman (1995, Theorem 0), the proof of which is omitted due to its length. Actually, he proves a more general result, but the following version is sufficient for our purposes.

Theorem 4.2.16 *For any λ, μ, $\nu \in D_c$ and any $i \geq 1$, $V(\lambda)$ does not occur in $H_i(\hat{\mathfrak{g}}_-, \mathscr{H}(\nu) \otimes V_1(\mu))$ as a $\mathfrak{g}$-module.* □

From the above theorem, one can completely determine $H_i(\hat{\mathfrak{g}}_-, \mathscr{H}(\nu) \otimes V_1(\mu))$ as a $\mathfrak{g}$-module, provided one knows the $\mathfrak{g}$-module $H_0(\hat{\mathfrak{g}}_-, \mathscr{H}(\nu) \otimes V_1(\mu))$ (cf. Exercise 4.2.E.6).

As an immediate consequence of Lemma 4.2.12, Proposition 4.2.15 and Theorem 4.2.16, we get the following.

Corollary 4.2.17 *The two products $\otimes^c$ and $\otimes_F^c$ in $R_c(\mathfrak{g})$ coincide for any simple Lie algebra $\mathfrak{g}$ and any central charge $c > 0$.*

Hence, for any λ, $\mu \in D_c$,

$$[V(\lambda)] \otimes^c [V(\mu)] = \sum_{\nu \in D_c} \sum_{w \in W'_c} (-1)^{\ell(w)} \times \dim \left(\mathrm{Hom}_{\mathfrak{g}}(V(w^{-1} * \nu), V(\lambda) \otimes V(\mu)) \right) [V(\nu)].$$

Now, we are ready to prove Theorem 4.2.9.

4.2.18 Proof of Theorem 4.2.9

In view of Corollary 4.2.17, it suffices to show that the map $\xi_c \colon R(\mathfrak{g}) \to R_c(\mathfrak{g})$ is a ring homomorphism with respect to the product $\otimes_F^c$ in $R_c(\mathfrak{g})$. As an immediate consequence of ξ_c being a ring homomorphism with respect to the product $\otimes_F^c$ in $R_c(\mathfrak{g})$, we get that $\otimes_F^c$ is associative (since ξ_c is surjective).

For any finite-dimensional G-module V, define the virtual $\hat{\mathfrak{g}}$-module

$$\chi\left(\bar{X}_G, \mathscr{L}_c(V)\right) := \sum_{i \geq 0} (-1)^i H^i\left(\bar{X}_G, \mathscr{L}_c(V)\right).$$

By the definition of ξ_c and Lemma 4.2.10, for any $\lambda \in D$,

$$\xi_c([V(\lambda)]) = \chi_{\mathfrak{g}}\left(\bar{X}_G, \mathscr{L}_c(V(\lambda))\right)^*, \tag{1}$$

where $\chi_{\mathfrak{g}}$ is defined by (2) of Definition 4.2.11.

We next show that for any $\lambda \in D$, $w \in W_c$ with $w^{-1} * \lambda \in D$ and finite-dimensional G-module V,

$$\chi\left(\bar{X}_G, \mathscr{L}_c(V(\lambda) \otimes V)\right) = (-1)^{\ell(w)} \chi\left(\bar{X}_G, \mathscr{L}_c\left(V(w^{-1} * \lambda) \otimes V\right)\right). \tag{2}$$

In particular,

$$\chi_{\mathfrak{g}}\left(\bar{X}_G, \mathscr{L}_c(V(\lambda) \otimes V)\right) = (-1)^{\ell(w)} \chi_{\mathfrak{g}}\left(\bar{X}_G, \mathscr{L}_c\left(V(w^{-1} * \lambda) \otimes V\right)\right). \tag{3}$$

Since V is a G-module, from the Leray spectral sequence applied to the locally trivial G/B-fibration $\bar{X}_G(B) \to \bar{X}_G$ (cf. Exercise 1.3.E.11) with respect to the vector bundle $\mathscr{L}_c^B(\mathbb{C}_\lambda \otimes V)$ and the classical BWB theorem, we get

$$\begin{aligned}\chi\left(\bar{X}_G, \mathscr{L}_c(V(\lambda) \otimes V)\right) &= \chi\left(\bar{X}_G(B), \mathscr{L}_c^B(\mathbb{C}_\lambda \otimes V)\right) \\ &= \sum_{\beta \in P_V} n_\beta\, \chi\left(\bar{X}_G(B), \mathscr{L}_c^B(\lambda + \beta)\right),\end{aligned} \tag{4}$$

where the vector bundle $\mathscr{L}_c^B(M)$ over $\bar{X}_G(B)$ for any finite-dimensional B-module M is as in the proof of Lemma 4.2.10, $\mathbb{C}_\lambda$ denotes the 1-dimensional representation of B corresponding to the character e^λ, P_V is the set of weights of V and the character

$$\mathrm{ch}(V) = \sum_{\beta \in P_V} n_\beta e^\beta.$$

Similarly,

$$\begin{aligned}
&\chi(\bar{X}_G, \mathscr{L}_c(V(w^{-1} * \lambda) \otimes V)) \\
&\quad = \sum_{\beta \in P_V} n_\beta \, \chi\left(\bar{X}_G(B), \mathscr{L}_c^B(w^{-1} * \lambda + \beta)\right) \\
&\quad = \sum_{\beta \in P_V} n_\beta \, \chi\left(\bar{X}_G(B), \mathscr{L}_c^B(w^{-1} * (\lambda + \beta)\right), \text{ since } n_\beta = n_{v\beta}, \text{ for any } v \in W \\
&\quad = (-1)^{\ell(w)} \sum_{\beta \in P_V} n_\beta \, \chi\left(\bar{X}_G(B), \mathscr{L}_c^B(\lambda + \beta)\right),
\end{aligned}$$

by Kumar (2002, Corollary 8.3.12). (5)

Combining (4) and (5), we get (2).

Specializing (2) to the case when $\lambda \in D$ is such that $\lambda + \rho \in H_{\alpha,n}$ (with $\alpha \in \Delta$ and $n \in \mathbb{Z}$), we get from (6) of Definition 4.2.5 (since $\tau_{\frac{2n(c+h^\vee)\alpha}{\langle \alpha,\alpha \rangle}}$ has even length by Kumar (2002, Exercise 13.1.E.3) and $(s_\alpha \cdot \tau_{\frac{-2n(c+h^\vee)\alpha}{\langle \alpha,\alpha \rangle}})(\lambda+\rho) = \lambda+\rho$ by (6) of Definition 4.2.5),

$$\chi(\bar{X}_G, \mathscr{L}_c(V(\lambda) \otimes V)) = -\chi(\bar{X}_G, \mathscr{L}_c(V(\lambda) \otimes V))$$

and hence

$$\chi(\bar{X}_G, \mathscr{L}_c(V(\lambda) \otimes V)) = 0. \tag{6}$$

In particular,

$$\chi_{\mathfrak{g}}(\bar{X}_G, \mathscr{L}_c(V(\lambda) \otimes V)) = 0. \tag{7}$$

We are now ready to prove that ξ_c is a ring homomorphism with respect to the product $\otimes_F^c$ in $R_c(\mathfrak{g})$, i.e., for $\lambda, \mu \in D$,

$$\xi_c([V(\lambda) \otimes V(\mu)]) = \xi_c\left([V(\lambda)]\right) \otimes_F^c \xi_c\left([V(\mu)]\right). \tag{8}$$

If at least one of λ or μ (say λ) is such that $\lambda + \rho \in H$, then by (1) and (7), both sides of (8) are zero. So, assume that both of $\lambda + \rho$ and $\mu + \rho$ lie in $P \backslash H$, i.e., there exists $v, w \in W_c'$ and $\lambda_o, \mu_o \in D_c$ such that $\lambda = v^{-1} * \lambda_o$ and $\mu = w^{-1} * \mu_o$.

Then, by (1) and (3),

$$\begin{aligned}\xi_c([V(\lambda) \otimes V(\mu)]) &= \chi_{\mathfrak{g}}\Big(\bar{X}_G, \mathscr{L}_c\,(V(\lambda) \otimes V(\mu))\Big)^* \\ &= (-1)^{\ell(v)+\ell(w)} \chi_{\mathfrak{g}}\left(\bar{X}_G, \mathscr{L}_c(V(\lambda_o) \otimes V(\mu_o))\right)^* \\ &= (-1)^{\ell(v)+\ell(w)} \xi_c\left([V(\lambda_o)]\right) \otimes_F^c \xi_c\left([V(\mu_o)]\right), \\ &\quad \text{by (1) of Definition 4.2.11} \\ &= \xi_c\left([V(\lambda)]\right) \otimes_F^c \xi_c\left([V(\mu)]\right).\end{aligned}$$

This completes the proof of (8) and hence Theorem 4.2.9 is proved. □

For any $t \in T$, we get an algebra homomorphism

$$\mathrm{ch}_t : R(\mathfrak{g}) \to \mathbb{C},$$

where for any representation δ of G in a finite-dimensional vector space V,

$$\mathrm{ch}_t([V]) := \mathrm{trace}_V \delta(t).$$

As a consequence of the Weyl character formula, Lemmas 4.1.5, 4.2.6, 4.2.8 and Theorem 4.2.9, we get the following.

Corollary 4.2.18 *For any $t \in T_c^{reg}$ (cf. Definition 4.2.5), the character $ch_t : R(\mathfrak{g}) \to \mathbb{C}$ factors through $R_c(\mathfrak{g})$ via ξ_c (cf. Definition 4.2.7) to give an algebra homomorphism*

$$ch_{t,c} : R_c(\mathfrak{g}) \to \mathbb{C}.$$

Moreover, $\{ch_{t,c}\}_{t \in T_c^{reg}/W}$ bijectively parameterizes the set S_{D_c} of algebra homomorphisms of $R_c(\mathfrak{g})$ to $\mathbb{C}$.

Proof To prove that, for any $t \in T_c^{\mathrm{reg}}$, ch_t factors through $R_c(\mathfrak{g})$, by Lemma 4.2.6, we can assume that

$$t = \mathrm{Exp}\left(\frac{2\pi i \kappa(\mu + \rho)}{c + h^\vee}\right), \quad \text{for some } \mu \in D_c.$$

By the Weyl character formula, for any $\lambda \in D$,

$$\mathrm{ch}_t([V(\lambda)]) = \frac{\sum\limits_{w \in W} \epsilon(w) e^{w(\lambda+\rho)}(t)}{\sum\limits_{w \in W} \epsilon(w) e^{w\rho}(t)}.$$

Since $t \in T$ is regular, the denominator is nonzero.

So, by Lemma 4.2.8, it suffices to prove that

$$\sum_{w \in W} \epsilon(w) e^{\frac{2\pi i}{c+h^\vee}\langle w(\lambda+\rho), \mu+\rho\rangle} = \epsilon(v\tau_\beta) \sum_{w \in W} \epsilon(w) e^{\frac{2\pi i}{c+h^\vee}\langle wv(\lambda+\rho+\beta), \mu+\rho\rangle}, \quad (1)$$

for any $v \in W$, $\lambda, \mu \in D_c$ and $\beta \in (c + h^\vee)Q_{lg}$, and

$$\sum_{w \in W} \epsilon(w) e^{\frac{2\pi i}{c+h^\vee} \langle w(\lambda+\rho), \mu+\rho \rangle} = 0, \text{ for } \lambda \in D \text{ with } \lambda + \rho \in H \text{ and any } \mu \in D_c. \tag{2}$$

For any $\beta \in (c + h^\vee)Q_{lg}$, $\epsilon(\tau_\beta) = 1$, by (3) of the proof of Lemma 4.2.6 and Kumar (2002, Exercise 13.1.E.3). Moreover, for any $\beta \in (c + h^\vee)Q_{lg}$ (and hence $w \cdot \beta \in (c + h^\vee)Q_{lg}$),

$$\langle \beta, \mu + \rho \rangle \in (c + h^\vee)\mathbb{Z}, \text{ by (3) of Lemma 4.2.6.} \tag{3}$$

This proves (1).

To prove (2), let $\lambda + \rho \in H_{\alpha,n}$ for some $\alpha \in \Delta$ and $n \in \mathbb{Z}$. Then, by identity (6) of Definition 4.2.5, $s_\alpha \cdot \tau_{\frac{-2n(c+h^\vee)\alpha}{\langle \alpha,\alpha \rangle}}(\lambda + \rho) = \lambda + \rho$. Thus,

$$\begin{aligned}
\sum_{w \in W} \epsilon(w) e^{\frac{2\pi i}{c+h^\vee} \langle w(\lambda+\rho), \mu+\rho \rangle} &= \sum_{w \in W} \epsilon(w) e^{\frac{2\pi i}{c+h^\vee} \langle w s_\alpha (\lambda+\rho - \frac{2n(c+h^\vee)\alpha}{\langle \alpha,\alpha \rangle}),\, \mu+\rho \rangle} \\
&= -\sum_{w \in W} \epsilon(w) e^{\frac{2\pi i}{c+h^\vee} \langle w (\lambda+\rho - \frac{2n(c+h^\vee)\alpha}{\langle \alpha,\alpha \rangle}),\, \mu+\rho \rangle} \\
&= -\sum_{w \in W} \epsilon(w) e^{\frac{2\pi i}{c+h^\vee} \langle w(\lambda+\rho), \mu+\rho \rangle}
\end{aligned}$$

and hence

$$\sum_{w \in W} \epsilon(w) e^{\frac{2\pi i}{c+h^\vee} \langle w(\lambda+\rho), \mu+\rho \rangle} = 0,$$

proving (2).

Since $\xi_c \colon R(\mathfrak{g}) \to R_c(\mathfrak{g})$ is a surjective algebra homomorphism and $\mathrm{ch}_t \colon R(\mathfrak{g}) \to \mathbb{C}$ is an algebra homomorphism, we get that $\mathrm{ch}_{t,c} \colon R_c(\mathfrak{g}) \to \mathbb{C}$ is an algebra homomorphism.

By Lemma 4.1.5,

$$|S_{D_c}| = |D_c|. \tag{4}$$

Further, by Lemma 4.2.6,

$$\left|T_c^{\mathrm{reg}}/W\right| = |D_c|. \tag{5}$$

Since

$$\gamma \colon R(\mathfrak{g}) \otimes_{\mathbb{Z}} \mathbb{C} \xrightarrow{\sim} \mathbb{C}[T/W]$$

is an isomorphism (cf. (Bröcker and tom Dieck, 1985, Chap. VI, Proposition 2.1)), where

$$\gamma([V])(t) = \mathrm{ch}_t([V]), \quad \text{for} \quad t \in T/W,$$

we get that $\{\mathrm{ch}_{t,c}\}_{t\in T_c^{\mathrm{reg}}/W}$ are all distinct. Thus, by (4) and (5), $\{\mathrm{ch}_{t,c}\}_{t\in T_c^{\mathrm{reg}}/W}$ bijectively parameterizes S_{D_c}. This proves the corollary. □

We now come to the following *Verlinde formula*, which is one of the most important results of the book.

Theorem 4.2.19 *Let $\mathfrak{g}$ be any simple Lie algebra and let $c > 0$ be any central charge. Let $(\Sigma, \vec{p} = (p_1, \ldots, p_s))$ be an irreducible smooth s-pointed curve of any genus $g \geq 0$ (where $s \geq 1$) and let $\vec{\lambda} = (\lambda_1, \ldots, \lambda_s)$ be a collection of weights in D_c. Then*

$$\dim \mathscr{V}_\Sigma(\vec{p}, \vec{\lambda}) = |T_c|^{g-1} \sum_{\mu \in D_c} \left(\left(\Pi_{i=1}^s (\mathrm{ch}_{t_\mu}([V(\lambda_i)]))\right) \cdot \Pi_{\alpha\in\Delta_+} \left(2 \sin\left(\frac{\pi}{c+h^\vee}\langle \mu + \rho, \alpha\rangle\right)\right)^{2-2g} \right), \tag{1}$$

where Δ_+ is the set of positive roots of $\mathfrak{g}$, $h^\vee$ is the dual Coxeter number, T_c is as in Definition 4.2.5, $\kappa : \mathfrak{h}^* \to \mathfrak{h}$ is the isomorphism induced from the normalized invariant form and $t_\mu := \mathrm{Exp}\left(\frac{2\pi i \kappa(\mu+\rho)}{c+h^\vee}\right) \in T_c$.

Moreover,

$$|T_c| = (c + h^\vee)^\ell |P/Q|\ |Q/Q_{lg}|, \tag{2}$$

where ℓ is the rank of $\mathfrak{g}$, P (resp. Q) is the weight (resp. root) lattice and Q_{lg} is the sublattice of Q generated by the long roots.

In particular, for $g = 1, s = 1$ and $\vec{\lambda} = (0)$,

$$\dim \mathscr{V}_\Sigma(\vec{p}, \vec{\lambda}) = |D_c|.$$

Proof Let F_c be the fusion rule as in Example 4.2.1. By Corollaries 3.5.10(a) and 4.1.8,

$$\dim \mathscr{V}_\Sigma(\vec{p}, \vec{\lambda}) = \sum_{\chi\in S_{D_c}} \chi([V(\lambda_1)]) \ldots \chi([V(\lambda_s)])\chi(\Omega)^{g-1}, \tag{3}$$

where S_{D_c} is the set of algebra homomorphisms $R_c(\mathfrak{g}) \to \mathbb{C}$ and

$$\chi(\Omega) = \sum_{\nu\in D_c} |\chi([V(\nu)])|^2. \tag{4}$$

By Corollary 4.2.18 and Lemma 4.2.6,

$$S_{D_c} = \{\mathrm{ch}_{t_\mu,c}\}_{\mu \in D_c}. \tag{5}$$

We now determine $\chi(\Omega)$ for any $\chi = \mathrm{ch}_{t_\mu,c}$.
Let $L^2(T_c)$ be the space of $\mathbb{C}$-valued functions on T_c with inner product

$$\langle f, g \rangle = \frac{1}{|T_c|} \sum_{t \in T_c} f(t)\overline{g(t)}.$$

For any $\mu \in D_c$, consider the function on T_c:

$$J_\mu(t) = \sum_{w \in W} \epsilon(w) e^{w(\mu+\rho)}(t).$$

Since J_μ is W-anti-invariant, i.e.,

$$J_\mu(vt) = \epsilon(v) J_\mu(t), \quad \text{for any } v \in W \text{ and } t \in T_c,$$

J_μ vanishes on $T_c \backslash T_c^{\mathrm{reg}}$. Of course, $|J_\mu|^2$ is W-invariant. Thus, by Lemma 4.2.6(b),

$$\sum_{\nu \in D_c} |J_\mu(t_\nu)|^2 = \frac{|T_c|}{|W|} ||J_\mu||^2. \tag{6}$$

Now, we claim that for any $\mu \in D_c$, $\{e^{w(\mu+\rho)}\}_{w \in W}$ are distinct characters of T_c, i.e., for $w \neq 1$,

$$e^{w(\mu+\rho)}{}_{|T_c} \neq e^{\mu+\rho}{}_{|T_c}.$$

By Lemma 4.2.6(a), it is equivalent to the assertion that

$$\langle w(\mu+\rho) - (\mu+\rho), \lambda \rangle \notin (c + h^\vee)\mathbb{Z}, \quad \text{for some} \quad \lambda \in P.$$

If not, assuming $\langle w(\mu+\rho) - (\mu+\rho), \lambda \rangle \in (c + h^\vee)\mathbb{Z}$, for all $\lambda \in P$, we get from (3) of the proof of Lemma 4.2.6 that

$$w(\mu+\rho) - (\mu+\rho) = \beta, \quad \text{for some} \quad \beta \in (c + h^\vee) Q_{lg}.$$

Thus,

$$\tau_{-\beta} \cdot w(\mu+\rho) = \mu + \rho,$$

which is a contradiction, since $\mu + \rho \in A^\circ$ by (3) of Definition 4.2.5. This proves that $\{e^{w(\mu+\rho)}{}_{|T_c}\}_{w \in W}$ are distinct characters.

Thus, by the orthogonality relation for the finite group T_c, we get

$$||J_\mu||^2 = |W|. \tag{7}$$

Taking $\chi = \mathrm{ch}_{t_\mu, c}$ in (4), we get

$$\begin{aligned}
\chi(\Omega) &= \sum_{\nu \in D_c} |\,\mathrm{ch}_{t_\mu, c}([V(\nu)])|^2 \\
&= \frac{\sum_{\nu \in D_c} |J_\nu(t_\mu)|^2}{\prod_{\alpha \in \Delta}(e^\alpha(t_\mu) - 1)}, \quad \text{by the Weyl character formula} \\
&= \frac{\sum_{\nu \in D_c} |J_\mu(t_\nu)|^2}{\prod_{\alpha \in \Delta}(e^\alpha(t_\mu) - 1)}, \quad \text{by the definition of } J_\mu \\
&= \frac{|T_c|}{\prod_{\alpha \in \Delta_+}\left(2 \sin \frac{\pi}{c+h^\vee}\langle \mu + \rho, \alpha\rangle\right)^2}, \quad \text{by (6) and (7).} \qquad (8)
\end{aligned}$$

Combining (3), (5) and (8), we get (1).

By Lemma 4.2.6(a),

$$|T_c| = (c + h^\vee)^\ell |P/Q|\, |Q/Q_{lg}|.$$

This proves (2) and hence the theorem is proved. □

Remarks 4.2.20 (a) The expression for $\dim \mathscr{V}_\Sigma(\vec{p}, \vec{\lambda})$ as in (1) of Theorem 4.2.19 remains valid for any s-pointed stable curve Σ by Theorem 3.5.9 and Lemma 3.3.3.

(b) The number $|P/Q|$ is called the *index of connection*. It is the order of the fundamental group π_1 of the corresponding adjoint group. Its values are given by (cf. (Bourbaki, 2002, Plates I–IX)):

- $A_\ell (\ell \geq 1) : \ell + 1$
- $B_\ell (\ell \geq 2)$, $C_\ell (\ell \geq 2)$, $E_7 : 2$
- $D_\ell (\ell \geq 4) : 4$
- $E_6 : 3$
- G_2, F_4, $E_8 : 1$

(c) The order $|Q/Q_{lg}|$ of course is 1 for simply-laced $\mathfrak{g}$. This order for non-simply-laced $\mathfrak{g}$ is given as follows:

- $B_\ell (\ell \geq 2) : 2$
- $C_\ell (\ell \geq 2) : 2^{\ell - 1}$
- $F_4 : 4$
- $G_2 : 6$

The above values can be read off from Bourbaki (2002, Plates I–IX).

4.2.E Exercises

(1) Show that for any $\lambda, \mu \in D_c$ such that $(\lambda + \mu)(\theta^\vee) = c + 2$, $[V(\lambda)] \otimes^c [V(\mu)]$ is obtained from $V(\lambda) \otimes V(\mu)$ by removing all the components $V(\nu)$ with $\nu(\theta^\vee) = c + 2$ or $c + 1$ along with those components $V(\nu)$ with $\nu(\theta^\vee) = c$ that intersect $V(\lambda)_{(\lambda(\theta^\vee))} \otimes V(\mu)_{(\mu(\theta^\vee))}$ nontrivially.

Hint: Use Proposition 4.2.3.

(2) Following the notation of Lemma 4.2.14, it is easy to see that

$$\hat{F}_1 = V(\nu + m\theta) \otimes V(\mu), \quad \text{where} \quad m := c + 1 - \nu(\theta^\vee).$$

Show that the differential $\hat{\delta}_1 : \hat{F}_1 \to \hat{F}_0 = V(\nu) \otimes V(\mu)$ is the composite map $\eta \circ (j \otimes I)$ given as follows:

$$V(\nu + m\theta) \otimes V(\mu) \overset{j \otimes I}{\hookrightarrow} (V(\nu) \otimes V(\theta)^{\otimes m}) \otimes V(\mu)$$

and

$$\eta : (V(\nu) \otimes V(\theta)^{\otimes m}) \otimes V(\mu) \to V(\nu) \otimes V(\mu)$$

is given by

$$\eta(v \otimes (x_1 \otimes \cdots \otimes x_m) \otimes w)$$
$$= v \otimes (x_m \ldots x_1 \cdot w), \quad \text{for } v \in V(\nu), x_i \in V(\theta) = \mathfrak{g}, w \in V(\mu).$$

Hint: Use the definition of $\mathscr{H}(\nu)$ as in Definition 1.2.6 to describe the $\hat{\mathfrak{g}}$-module map $\delta_1 : F_1 \to F_0$ as in Definition 4.2.13. Now, use the explicit identification (1) of the proof of Lemma 4.2.14.

(3) Simply using Corollary 4.2.4 (and not using Theorem 4.2.9), show that for any $c > 0$, $R_c(\mathfrak{g})$ under the product $\otimes^c$ is generated by $\{[V(\omega_i)] : \omega_i \in D_c\}$, where $\{\omega_i\}_{1 \le i \le \ell}$ are the fundamental weights.

Hint: Choose an element $H \in Q^\vee$ such that $\alpha_i(H) > 0$ for each simple root α_i. Now, use the induction on $\lambda(H)$ to show that $[V(\lambda)]$ lies in the ring generated by $\{[V(\omega_i)] : \omega_i \in D_c\}$.

Thus, to show that $\otimes^c = \otimes^c_F$ in $R_c(\mathfrak{g})$, it suffices to prove that

$$[V(\lambda)] \otimes^c [V(\omega_i)] = [V(\lambda)] \otimes^c_F [V(\omega_i)], \quad \text{for any } \lambda, \omega_i \in D_c.$$

(4) Show that for any $\mathfrak{g}$ of type A_ℓ, B_ℓ, C_ℓ, D_ℓ or G_2 and any central charge $c > 0$, $\overline{\chi}_{\mathfrak{g}}(\lambda, \omega_i, \nu) = 0$, for any $\lambda, \nu, \omega_i \in D_c$, where $\overline{\chi}_g$ is defined in Proposition 4.2.15.

Thus, in view of the above Exercise 3 and Proposition 4.2.15, this gives an alternative (and much simpler) proof of the equality $\otimes^c = \otimes^c_F$ in $R_c(\mathfrak{g})$ for $\mathfrak{g}$ of any type other that $E_\bullet$ and F_4 (cf. Corollary 4.2.17 for any $\mathfrak{g}$).

Hint: Use some 'partial' determination of $H_*(\hat{\mathfrak{g}}_-, \mathscr{H}(\nu) \otimes V_1(\omega_i))$ for those ω_i such that $\omega_i(\theta^\vee) \leq 2$ by observing that any irreducible $\mathfrak{g}$-submodule of the tensor product $V(\lambda) \otimes V(\mu)$ has highest weight of the form $\lambda + \beta$ for some weight β of $V(\mu)$ and using the following Exercise (7). Further, any fundamental weight ω_i of level 1 cannot belong to Q_{lg}

(5) Show that for any $\mu, \nu \in D_c$, if a $\mathfrak{g}$-module $V(\lambda)$ occurs in $H_0(\hat{\mathfrak{g}}_-, \mathscr{H}(\nu) \otimes V_1(\mu))$, then $\lambda \in D_c$.

(6) For any $\mu, \nu \in D_c$, consider the decomposition as $\mathfrak{g}$-modules (cf. the above Exercise 5):

$$H_0(\hat{\mathfrak{g}}_-, \mathscr{H}(\nu) \otimes V_1(\mu)) \simeq \bigoplus_{\lambda \in D_c} m_\lambda^{\mu,\nu} V(\lambda).$$

Then, assuming the validity of Theorem 4.2.16, show that, for any $p \geq 0$, as $\mathfrak{g}$-modules,

$$H_p\left(\hat{\mathfrak{g}}_-, \mathscr{H}(\nu) \otimes V_1(\mu)\right) \simeq \bigoplus_{\lambda \in D_c} m_\lambda^{\mu,\nu} \left(\bigoplus_{\substack{w \in W'_c \\ \ell(w) = p}} V(w^{-1} * \lambda) \right).$$

Hint: Use the Hochschild–Serre spectral sequence for Lie algebra homology.

For $\mu = 0$, this is a result due to Garland and Lepowsky (1976).

(7) Show that

$$W'_c = \{v\tau_\alpha : v \text{ is the shortest coset representative in } W/W_\alpha \text{ and } \alpha \in (c + h^\vee)Q_{lg} \text{ is anti-dominant weight}\},$$

where $W_\alpha \subset W$ is the stabilizer of α.

(8) Show that for any simply-laced $\mathfrak{g}$ and $c = 1$,

$$\dim \mathscr{V}_\Sigma(p, 0) = |Z(G)|^g,$$

where g is the genus of the smooth irreducible projective curve Σ, p is any point of Σ and $Z(G)$ is the center of simply-connected G with Lie algebra $\mathfrak{g}$.

4.C Comments

We refer to the Bourbaki talk (Sorger, 1994a) for a brief survey of the Verlinde formula and its proof.

The content of Section 4.1 (including Exercises 4.1.E) is largely taken from Beauville (1996) (barring Lemma 4.1.9 which is a personal communication due to Jiuzu Hong). Also, Exercise 4.1.E.3 is well known (see, e.g., (Verlinde, 1988), (Moore and Seiberg, 1989), (Kac, 1990, Exercise 13.34) and (Beauville, 1996, §6.4)). We also refer to Szenes (1995) and Ueno (2008, §5.5) for the contents of this section.

Section 4.2 is largely taken from Faltings (1994, §§5, 6) and Beauville (1996). Proposition 4.2.3 is contained in Tsuchiya, Ueno and Yamada (1989, Example 2.2.8). Corollary 4.2.18 (equivalently Theorem 4.2.9) was conjectured in Faltings (1994, Conjecture 5.1), wherein it was proved for all the classical $\mathfrak{g}$ and $\mathfrak{g}$ of type G_2. The uniform proof (of Theorem 4.2.9) given in Subsection 4.2.18 uses another definition of the fusion product $\otimes_F^c$ given in terms of the $\mathfrak{g}$-equivariant Euler–Poincaré characteristic of certain vector bundles on the infinite Grassmannian $\bar{X}_G$ (cf. Definition 4.2.11) and its coincidence with the usual fusion product $\otimes^c$ defined by (1) of Example 4.2.1. The definition of the fusion product $\otimes_F^c$ (as in Definition 4.2.11) and the result that $\xi_c\colon R(\mathfrak{g}) \to R_c(\mathfrak{g})$ is a ring homomorphism with respect to the fusion product $\otimes_F^c$ (cf. Section 4.2.18), as well as (then) conjectural equality of $\otimes_F^c$ with $\otimes^c$ is due to Kumar (1997b) (apparently the equality of $\otimes_F^c$ with $\otimes^c$ was also conjectured by Bott, as mentioned in Teleman (1995)). Lemma 4.2.14 and Proposition 4.2.15 are also taken from Kumar (1997b). Now, as in Corollary 4.2.17, the equality of $\otimes_F^c$ with $\otimes^c$ follows from a result due to Teleman (1995) (cf. Theorem 4.2.16). In fact, Teleman (1995) determines the Lie algebra cohomology of the pair $(\mathfrak{g}[t^{-1}], \mathfrak{g})$ with coefficients in the tensor product $\mathscr{H}(\lambda_c) \otimes \vec{V}(\vec{\lambda})$ of an integrable highest-weight $\hat{\mathfrak{g}}$-module with finite-dimensional evaluation modules. More generally, Teleman has determined the Lie algebra cohomology of the pair $(\mathfrak{g} \otimes \mathbb{C}[\Sigma], \mathfrak{g})$ with coefficients in $\mathscr{H}(\lambda_c) \otimes \vec{V}(\vec{\lambda})$ and proved its rigidity under nodal deformations of Σ (cf. (Teleman, 1996)).

The Verlinde formula (Theorem 4.2.19) was, in some form, conjectured by Verlinde (1988). **A very significant part of the proof of the formula (in the precise form of Theorem 4.2.19) was done by** Tsuchiya, Ueno and Yamada (1989).

Exercises 4.2.E.1 and 4.2.E.4 are essentially due to Faltings (1994). Exercises 4.2.E.2 and 4.2.E.6 are due to Kumar (1997b) and Exercise 4.2.E.8 is due to Faltings (2009) (also see (Zhu, 2017, Corollary 4.2.5)).

There are several (geometric) proofs of the Verlinde formula for the dimension of the space of generalized theta functions for $G = \mathrm{SL}_2$ (or more generally for GL_2 with fixed determinant) as in Kirwan (1992), Szenes (1993), Bertram (1993), Bertram and Szenes (1993), Narasimhan and Ramadas (1993), Daskalopoulos and Wentworth (1993, 1996), Thaddeus (1994) and Zagier (1995) (and possibly more). Jeffrey and Kirwan (1998) contains a proof of the Verlinde formula for GL_N using Witten's formula (Witten, 1991) for the symplectic volume of the moduli space. For the volume computation of moduli spaces, we refer, in addition, to the papers Pantev (1994), Beauville (1997), Boysal and Vergne (2010), Oprea (2011), Krepski and Meinrenken (2013) and Baldoni, Boysal and Vergne (2015). Alekseev, Meinrenken and Woodward (2000) gave a generalization of the Verlinde formula for some non-simply-connected groups. Bismut and Labourie (1999) gave a symplectic geometry proof of the Verlinde formula for $c \gg 0$ and any G. As proved later in Chapter 8, the space of generalized theta functions is isomorphic with the space of conformal blocks.

Fuchs and Schweigert (1997) gave a proof of the Verlinde formula for genus $g = 0$ using Theorem 4.2.9 but without invoking the Factorization Theorem in $g = 0$ case.

Following the works of Moore and Seiberg (1988, 1989), Huang formulated and proved a generalization of the Verlinde conjecture in the framework of the theory of vertex operator algebras using the results on the duality and modular invariance of genus 0 and 1 correlation functions (cf. (Huang, 2008)). For more general conformal blocks arising from vertex operator algebras, under some natural assumptions, factorization, local freeness and computation of Chern classes has been done in Damiolini, Gibney and Tarasca (2019, 2020).

An explicit residue formula for $\dim \mathscr{V}_\Sigma(p, 0_c)$ for $G = \mathrm{SL}_N$ is given in Szenes (1995). Zagier (1996) contains several number-theoretical and combinatorial aspects of the Verlinde formula for GL_n (especially for $n = 2, 3$). Further, an explicit formula for $\dim \mathscr{V}_\Sigma(p, 0_c)$ for the classical groups can be found in Oxbury and Wilson (1996).

Fakhruddin (2012) gave a formula for the Chern classes of the Verlinde bundle (i.e., bundle of the conformal blocks) over the moduli stack $\bar{\mathcal{M}}_{0,s}$ of stable s-pointed curves of genus 0 as well as the first Chern class of the Verlinde bundle over the moduli stack $\bar{\mathcal{M}}_{g,s}$ of stable s-pointed curves of any genus g (see also (Mukhopadhyay, 2016c)). Marian et al. (2017) extended this work by giving an explicit formula in terms of the tautological classes for the total Chern character of the Verlinde bundle over the moduli stack $\bar{\mathcal{M}}_{g,s}$.

5
Moduli Stack of Quasi-parabolic G-Bundles and its Uniformization

Let X be a scheme and let G be a (not necessarily reductive) affine algebraic group. Let $\mathbf{Bun}_G(X)$ be the groupoid fibration over the category of schemes $\mathfrak{S}$ whose objects over $S \in \mathfrak{S}$ are (principal) G-bundles E_S over $X \times S$ (cf. Definition 5.1.1). Then, $\mathbf{Bun}_G(X)$ is an algebraic stack if X is a projective variety and G is any affine algebraic group. Moreover, if X is a projective curve, then $\mathbf{Bun}_G(X)$ is a smooth stack. Further, if X is a smooth irreducible projective curve of genus g and G is connected, reductive, then

$$\dim \mathbf{Bun}_G(X) = (g-1)\dim G$$

(cf. Theorem 5.1.3).

Fix an integer $s \geq 1$ and let $(\Sigma, \vec{p})$ be an s-pointed projective curve. Let G be a connected reductive algebraic group. Label the points $\vec{p} = \{p_1, \ldots, p_s\}$ by standard parabolic subgroups $\vec{P} = \{P_1, \ldots, P_s\}$, respectively. We define the *quasi-parabolic moduli stack* $\mathbf{Parbun}_G(\Sigma, \vec{P})$ whose objects over $S \in \mathfrak{S}$ are the pairs $(E, \vec{\sigma})$, where E is a G-bundle over $\Sigma \times S$ (for any scheme $S \in \mathfrak{S}$) and $\vec{\sigma} = (\sigma_1, \ldots, \sigma_s)$ consists of sections σ_i of $E_{|p_i \times S}/P_i$ (cf. Definition 5.1.4). Then, $\mathbf{Parbun}_G(\Sigma, \vec{P})$ is a smooth (algebraic) stack. Moreover, if Σ is a smooth irreducible curve of genus g,

$$\dim \mathbf{Parbun}_G(\Sigma, \vec{P}) = (g-1)\dim G + \sum_{i=1}^{s} \dim G/P_i$$

(cf. Theorem 5.1.5).

Section 5.2 is devoted to a 'uniformization theorem' for $\mathbf{Bun}_G(\Sigma)$ as well as $\mathbf{Parbun}_G(\Sigma, \vec{P})$, for a semisimple, connected algebraic group G and a smooth irreducible projective curve Σ. Let Σ^* be the curve Σ with a single puncture at any fixed base point p_o and let $\mathbb{D} := \operatorname{Spec} \mathbb{C}[[t]]$ be the formal disc at p_o of Σ for a local parameter t. We consider the functor

$$\mathscr{F}_{G,\Sigma^*,\mathbb{D}}\colon \mathbf{Alg} \to \mathbf{Set},\ \mathscr{F}_{G,\Sigma^*,\mathbb{D}}(R) = \{\mathscr{E}_R = (E_R, \sigma_R, \mu_R)\} \text{ modulo isomorphisms},$$

where **Alg** is the category of $\mathbb{C}$-algebras, E_R is a principal G-bundle over $\Sigma_R := \Sigma \times \operatorname{Spec} R$, σ_R is a section of E_R over $\Sigma_R^* := \Sigma^* \times \operatorname{Spec} R$ and μ_R is a section over the formal disc $\mathbb{D}_R := \operatorname{Spec} R[[t]]$. It is shown in Lemma 5.2.2 that $\mathscr{F}_{G,\Sigma^*,\mathbb{D}}$ is a representable functor represented by the formal loop group $\bar{G}((t))$, the proof of which relies on the 'descent' lemma (cf. Lemma 5.2.3). Similarly, the functor $\mathscr{F}_{G,\Sigma^*}$, where we drop the section μ_R from the definition, is represented by the infinite Grassmannian $\bar{X}_G$ with $\mathbb{C}$-points $G((t))/G[[t]]$ (cf. Proposition 5.2.7). We show the existence of a 'tautological bundle,' which is a G-bundle $\mathbf{U} \to \Sigma \times \bar{X}_G$ (cf. Proposition 5.2.4). Let $\Gamma := \operatorname{Mor}(\Sigma^*, G)$ be the set of morphisms from Σ^* to G. It is shown that $\Gamma = \bar{\Gamma}(\mathbb{C})$ for a reduced ind-affine group scheme $\bar{\Gamma}$ which is of ind-finite type over $\mathbb{C}$ (cf. Definition 5.2.9 and Lemma 5.2.10). It is shown that the tautological bundle $\mathbf{U}$ is, in fact, $\bar{\Gamma}$-equivariant (cf. Lemma 5.2.12).

Finally, we prove the uniformization theorem (cf. Theorem 5.2.14) asserting that the stack $\mathbf{Bun}_G(\Sigma)$ is isomorphic with the quotient stack $[\bar{\Gamma}\backslash\bar{X}_G]$ under the left multiplication of $\bar{\Gamma}$ on $\bar{X}_G$. Similarly, there is a corresponding parabolic analogue (cf. Theorem 5.2.16) asserting that

$$\mathbf{Parbun}_G\left(\Sigma, \vec{P}\right) \simeq \left[\bar{\Gamma}\backslash\left(\bar{X}_G \times \Pi_{i=1}^s (G/P_i)\right)\right],$$

where $\mathbf{Parbun}_G(\Sigma, \vec{P})$ is the parabolic moduli stack as above, $\vec{p} = (p_1, \ldots, p_s)$ for $s \geq 1$ is a set of distinct marked points on Σ distinct from the chosen base point p_o and $\bar{\Gamma}$ acts on $\bar{X}_G \times \Pi_{i=1}^s (G/P_i)$ from the left via

$$\gamma \cdot (\bar{g}, \bar{g}_1, \ldots, \bar{g}_s) = \left(\overline{(\gamma)_{p_o} g}, \overline{\gamma(p_1) g_1}, \ldots, \overline{\gamma(p_s) g_s}\right),$$

for $\gamma \in \bar{\Gamma}(R) = \operatorname{Mor}(\Sigma_R^*, G)$, $g \in G(R((t)))$ and $g_i \in G(R)$, where $(\gamma)_{p_o}$ is the restriction of γ to the punctured disc $\mathbb{D}_R^* = \operatorname{Spec} R((t)), \bar{g} := gG(R[[t]])$ and $\bar{g}_i := g_i P_i(R)$. In particular, the set $\operatorname{Bun}_G(\Sigma)$ of isomorphism classes of G-bundles over Σ is naturally bijective with the quotient set $\Gamma\backslash X_G$ (cf. Corollary 5.2.15). For a similar parabolic analogue of this corollary, see Corollary 5.2.17.

As outlined in Exercise 5.2.E.3, the stack $\mathbf{Parbun}_G(\Sigma, \vec{P})$ is also isomorphic with the quotient stack $\left[\bar{\Gamma}_{\vec{p}}\backslash(\Pi_{i=1}^s \bar{X}_G(P_i))\right]$ (see the same for the unexplained notation). As asserted in Exercise 5.2.E.4,

$$\bar{X}_G \times \Pi_{i=1}^s (G/P_i) \to \left[\bar{\Gamma}\backslash\left(\bar{X}_G \times \Pi_{i=1}^s (G/P_i)\right)\right]$$

is a locally trivial $\bar{\Gamma}$-torsor in the étale topology.

5.1 Moduli Stack of Quasi-parabolic G-Bundles

Let $X \in \mathfrak{S}$ be a scheme and let G be a (not necessarily reductive) affine algebraic group, where $\mathfrak{S}$ (as in Section 1.1) is the category of quasi-compact separated schemes over $\mathbb{C}$ and morphisms between them. We freely follow the notation and definitions from Appendix C. We first discuss the moduli stack of G-bundles over X.

Definition 5.1.1 Let $\mathbf{Bun}_G(X)$ be the groupoid fibration over $\mathfrak{S}$ (cf. Definition C.5) whose objects are G-bundles E_S over $X \times S$ (cf. Example C.4(d) for the definition of G-bundles). By a morphism between two G-bundles E_S (over $X \times S$) and $E'_{S'}$ (over $X \times S'$), we mean a G-equivariant morphism $f: E_S \to E'_{S'}$ and a morphism $\bar{f}: S \to S'$ making the following diagram commutative:

$$\begin{array}{ccc} E_S & \xrightarrow{f} & E'_{S'} \\ \downarrow & & \downarrow \\ X \times S & \xrightarrow[\mathrm{Id}_X \times \bar{f}]{} & X \times S'. \end{array}$$

Then, $\mathbf{Bun}_G(X)$ is a category over $\mathfrak{S}$ taking a G-bundle E_S (over $X \times S$) to S and morphism $f: E_S \to E'_{S'}$ (as above) to the morphism $\bar{f}: S \to S'$.

Let G be any affine algebraic group and $H \subset G$ an algebraic subgroup. Let $E \to X$ be a G-bundle over any scheme X. By a H-subbundle E_H of E, we mean a H-stable subset E_H of E such that $E_H \to X$ is a H-bundle.

We recall the following well-known result for its subsequent use.

Lemma 5.1.2 *The set of H-subbundles E_H of E is in bijective correspondence with the sections of the G/H-fibration $\pi_H: E \times^G G/H = E/H \to X$. Under this correspondence, a section σ of π_H gives rise to the subbundle*

$$E_H(\sigma) := \coprod_{x \in X} \sigma(x) H \subset E.$$

Proof The lemma follows easily if we show that $E_H(\sigma)$ is a (locally isotrivial) H-bundle. To prove this, use the fact that $G \to G/H$ is an H-bundle (cf. (Serre, 1958, Proposition 3)). □

The following is one of the most important results about $\mathbf{Bun}_G(X)$. For a complete proof of (a), (b)-parts, see Wang (2011, Theorem 1.0.1 and Proposition 6.0.18). For a sketch of the proof of the (c)-part, see Sorger (1999b, Proposition 3.6.8).

Theorem 5.1.3 (a) *For a projective variety X and any affine algebraic group G, $\mathbf{Bun}_G(X)$ is an algebraic stack.*

(b) *Moreover, if X is a projective curve, then $\mathbf{Bun}_G(X)$ is a smooth stack.*

(c) *If X is a smooth irreducible projective curve of genus g and G is reductive and connected, then*

$$\dim \mathbf{Bun}_G(X) = (g-1)\dim G.$$

We now consider the moduli stack of *quasi-parabolic G-bundles* over curves. Fix an integer $s \geq 1$ and let $(\Sigma, \vec{p})$ be an s-pointed projective curve as in Definition 2.1.1. Let G be a connected reductive algebraic group over $\mathbb{C}$ with a fixed Borel subgroup B and a maximal torus $T \subset B$. Label the points $\vec{p} = \{p_1, \ldots, p_s\}$ by standard parabolic subgroups $\vec{P} = \{P_1, \ldots, P_s\}$, respectively.

Definition 5.1.4 (Quasi-parabolic moduli stack) A *quasi-parabolic G-bundle of type $\vec{P}$ over $(\Sigma, \vec{p})$* is, by definition, a G-bundle E over Σ together with sections σ_i of E/P_i at p_i, i.e., elements of the fiber E_{p_i}/P_i.

The *quasi-parabolic moduli stack* $\mathbf{Parbun}_G(\Sigma, \vec{P})$ is the category over $\mathfrak{S}$ whose objects are the pairs $(E, \vec{\sigma})$, where E is a G-bundle over $\Sigma \times S$ (for any scheme $S \in \mathfrak{S}$) and $\vec{\sigma} = (\sigma_1, \ldots, \sigma_s)$ consists of sections σ_i of $E_{|p_i \times S}/P_i$.

Let $(E', \vec{\sigma}')$ be another such pair, where E' is a G-bundle over $\Sigma \times S'$. A morphism from $(E, \vec{\sigma})$ to $(E', \vec{\sigma}')$ consists of a pair $(f, \bar{f})$ making the following diagram commutative, where $f \colon E \to E'$ is a G-equivariant morphism and $\bar{f} \colon S \to S'$ is a morphism,

$$\begin{array}{ccc} E & \xrightarrow{f} & E' \\ \pi \downarrow & & \downarrow \pi' \\ \Sigma \times S & \xrightarrow[\mathrm{Id}_\Sigma \times \bar{f}]{} & \Sigma \times S'. \end{array}$$

Moreover, we require that, for all $1 \leq i \leq s$,

$$\sigma'_i \circ (\mathrm{Id}_\Sigma \times \bar{f})_{|p_i \times S} = f_{p_i} \circ \sigma_i,$$

where $f_{p_i} \colon E/P_i \to E'/P_i$ is the induced map from f.

It is easy to see that $\mathbf{Parbun}_G(\Sigma, \vec{P})$ is a groupoid fibration over $\mathfrak{S}$ where $(E, \vec{\sigma})$ is mapped to S and $(f, \bar{f})$ is mapped to $\bar{f}$.

Analogous to Theorem 5.1.3, we have the following result, which, in fact, follows from the same.

Theorem 5.1.5 *Let $(\Sigma, \vec{p})$ be an s-pointed (for any $s \geq 1$) projective curve and let $\vec{P} = (P_1, \ldots, P_s)$ be standard parabolic subgroups of a connected, reductive algebraic group G. Then* $\mathbf{Parbun}_G(\Sigma, \vec{P})$ *is a smooth (algebraic) stack.*

Moreover, if Σ is smooth and irreducible,

$$\dim \mathbf{Parbun}_G(\Sigma, \vec{P}) = (g-1)\dim G + \sum_{i=1}^{s} \dim G/P_i,$$

where g is the genus of Σ.

5.2 Uniformization of Moduli Stack of Quasi-parabolic G-Bundles

Let G be a connected semisimple algebraic group over $\mathbb{C}$ and let Σ be a smooth irreducible projective curve. *This will be our tacit assumption in this section unless otherwise stated explicitly.*

We fix a base point $p_o \in \Sigma$ and let $\Sigma^* := \Sigma \backslash \{p_o\}$. Let $\hat{\mathscr{O}}_{p_o}$ be the completion of the local ring $\mathscr{O}_{p_o}$ of Σ at p_o and let

$$\mathbb{D} := \operatorname{Spec}\left(\hat{\mathscr{O}}_{p_o}\right)$$

be the *formal disc* in Σ centered at p_o. Let $\mathbb{D}^*$ be the *formal punctured disc* in Σ centered at p_o, defined as the fiber product

$$\begin{array}{ccc} \mathbb{D}^* & \xrightarrow{i^*} & \mathbb{D} \\ {\scriptstyle j^*}\downarrow & & \downarrow{\scriptstyle j} \\ \Sigma^* & \xrightarrow[i]{} & \Sigma. \end{array} \qquad (\mathscr{D})$$

We fix a local parameter t for Σ at p_o. Then, $\hat{\mathscr{O}}_{p_o} \simeq \mathbb{C}[[t]]$ and $\mathbb{D}^* = \operatorname{Spec}(\mathbb{C}((t)))$. For any $R \in \mathbf{Alg}$ (cf. Section 1.1), define

$$\Sigma_R := \Sigma \times \operatorname{Spec}(R), \quad \Sigma_R^* = \Sigma^* \times \operatorname{Spec} R,$$
$$\mathbb{D}_R := \operatorname{Spec}(R[[t]]), \quad \mathbb{D}_R^* := \operatorname{Spec}(R((t))).$$

Then, we have the fiber diagram

$$\begin{array}{ccc} \mathbb{D}_R^* & \xrightarrow{i_R^*} & \mathbb{D}_R \\ {\scriptstyle j_R^*}\downarrow & & \downarrow{\scriptstyle j_R} \\ \Sigma_R^* & \xrightarrow[i_R]{} & \Sigma_R . \end{array} \qquad (\mathscr{D}_R)$$

Let **Set** be the category of sets (cf. Section 1.1).

Definition 5.2.1 Define the functor

$$\begin{aligned} &\mathscr{F}_{G,\Sigma^*,\mathbb{D}} : \mathbf{Alg} \to \mathbf{Set}, \ \ \mathscr{F}_{G,\Sigma^*,\mathbb{D}}(R) \\ &\quad = \{\mathscr{E}_R = (E_R, \sigma_R, \mu_R) : E_R \text{ is a principal } G\text{-bundle over } \Sigma_R, \sigma_R \\ &\qquad\qquad \text{is a section of } E_R \text{ over } \Sigma_R^* \text{ and } \mu_R \text{ is a section over } \mathbb{D}_R\}/\sim, \end{aligned}$$

where $\mathscr{E}_R \sim \mathscr{E}'_R$ if there exists an isomorphism θ_R of G-bundles

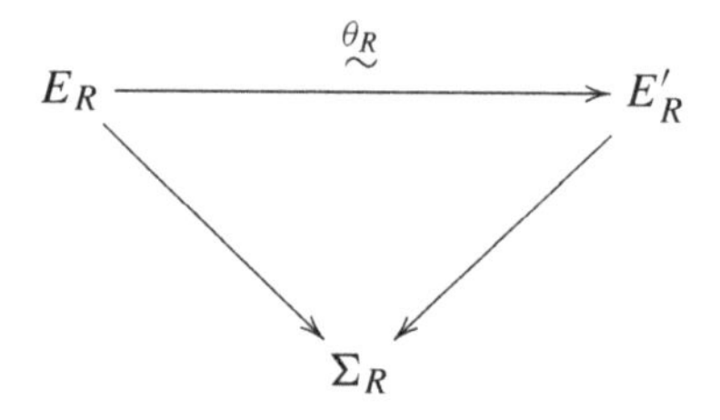

such that $\theta_R \circ \sigma_R = \sigma'_R$ and $\theta_R \circ \mu_R = \mu'_R$.

Lemma 5.2.2 *The functor* $\mathscr{F} = \mathscr{F}_{G,\Sigma^*,\mathbb{D}}$ *is a representable functor represented by the formal loop group* $\bar{G}((t))$ *(cf. Definition 1.3.1).*

Proof In view of Lemma 1.3.2, we need to prove that for any $R \in \mathbf{Alg}$, $\mathscr{F}(R)$ is canonically isomorphic with $G(R((t)))$.

Define a map $\alpha : \mathscr{F}(R) \to G(R((t)))$ as follows. For $\mathscr{E}_R = (E_R, \sigma_R, \mu_R) \in \mathscr{F}(R)$, let $\sigma_R^*, \mu_R^* : \mathbb{D}_R^* \to E_R$ be the sections

$$\sigma_R^* := \sigma_R \circ j_R^*, \ \mu_R^* = \mu_R \circ i_R^*, \qquad (1)$$

where j_R^* and i_R^* are as in the diagram ($\mathscr{D}_R$) at the beginning of this section. Now, define $\alpha_{\mathscr{E}_R} : \mathbb{D}_R^* \to G$ by

$$\mu_R^* = \sigma_R^* \cdot \alpha_{\mathscr{E}_R}.$$

Thus, $\alpha_{\mathscr{E}_R} \in G(R((t)))$. It is easy to see that if $\mathscr{E}_R \sim \mathscr{E}'_R$, then $\alpha_{\mathscr{E}_R} = \alpha_{\mathscr{E}'_R}$. Thus, we can view α as a map

$$\mathscr{F}(R) \to G(R((t))), \quad [\mathscr{E}_R] \mapsto \alpha_{\mathscr{E}_R},$$

where $[\mathscr{E}_R]$ denotes the equivalence class of $\mathscr{E}_R$.

Conversely, by the following 'descent' lemma, given any element $\gamma \in G(R((t)))$, there exists a triple $\mathscr{E}_R^\gamma = (E_R^\gamma, \sigma_R^\gamma, \mu_R^\gamma) \in \mathscr{F}(R)$ (which is unique up to a unique isomorphism) such that $\alpha_{\mathscr{E}_R^\gamma} = \gamma$. Thus, the map $\gamma \mapsto [\mathscr{E}_R^\gamma]$ provides the inverse of α. □

We recall the following 'descent' lemma. Even though it is proved in Beauville and Laszlo (1995) for vector bundles, the same proof works for G-bundles. (It may be mentioned that it does *not* follow from Grothendieck's descent lemma since for non-noetherian $\mathbb{C}$-algebras R, $\mathbb{D}_R \to \Sigma_R$ is *not* flat.)

Lemma 5.2.3 *For any $R \in$ **Alg** and any $\gamma \in G(R((t)))$, there exists a principal G-bundle E_R over Σ_R together with sections σ_R over Σ_R^* and μ_R over $\mathbb{D}_R$ such that*

$$\mu_R^* = \sigma_R^* \cdot \gamma, \tag{1}$$

where σ_R^* and μ_R^* are defined by (1) *of Lemma 5.2.2.*

Moreover, the triple (E_R, σ_R, μ_R) satisfying (1) *is unique up to a unique isomorphism.* □

By a (principal) G-bundle over an ind-scheme $X = (X_n)_{n \geq 0}$ we mean a collection of G-bundles $(\pi_n \colon E_n \to X_n)_{n \geq 0}$ (with right G-action) together with G-bundle closed embeddings $E_n \hookrightarrow E_{n+1}$ over $X_n \hookrightarrow X_{n+1}$. We think of the ind-scheme $E = (E_n)_{n \geq 0}$ with G-action as the G-bundle $E \to X$ built from the G-bundles $E_n \to X_n$.

The following proposition guarantees the existence of the 'tautological' bundle over Σ parameterized by the infinite Grassmannian $\bar{X}_G$ with $\mathbb{C}$-points $X_G := G((t))/G[[t]]$ (cf. Proposition 1.3.18).

Proposition 5.2.4 *There exists a G-bundle $\mathbf{U} \to \Sigma \times \bar{X}_G$ together with a section $\sigma_{\bar{X}_G}$ over $\Sigma^* \times \bar{X}_G$ such that, for any $g \in G((t))$, the section over $\mathbb{D}^* \times \bar{g}$:*

$$\sigma_{\bar{X}_G}^*(-, \bar{g}) \cdot g \quad \textit{extends to a section over } \mathbb{D} \times \bar{g}, \tag{1}$$

where $\bar{g} := gG[[t]]$ and $\sigma_{\bar{X}_G}^*$ is defined by (1) *of Lemma 5.2.2.*

(As we will see later in Corollary 5.2.8, the pair $(\mathbf{U}, \sigma_{\bar{X}_G})$ is unique up to a unique isomorphism.)

Proof For any $g \in G((t))$, let $\bar{V}_g \subset \bar{X}_G$ be the open subset $g \cdot \bar{G}[t^{-1}]^- \cdot \bar{o}$, where $\bar{o} \in \bar{X}_G$ is the base point (cf. Proposition 1.3.18(b)). The identification

$\bar{V}_g \simeq g \cdot \bar{G}[t^{-1}]^-$ gives rise to a section i_g of the locally trivial principal $\bar{G}[[t]]$-bundle $\bar{G}((t)) \to \bar{X}_G$ over $\bar{V}_g$ (cf. Exercise 1.3.E.11).

Let $\{F^n\}_{n\geq 0}$ be a filtration of $\bar{G}[t^{-1}]^-$ by affine schemes of finite type over $\mathbb{C}$ (cf. Definition 1.3.1 and Corollary 1.3.3). By Grothendieck's standard faithfully flat descent lemma applied to the flat cover $\{\Sigma^* \times F^n_g, \mathbb{D}_{F^n_g}\}$ of $\Sigma \times F^n_g$ (cf. (Grothendieck, 1971, VIII 5.1, 1.1 et 1.2)), we have a G-bundle $\mathbf{U}^n_g$ over $\Sigma \times F^n_g$ with sections σ^n_g over $\Sigma^* \times F^n_g$ and μ^n_g over $\mathbb{D}_{F^n_g}$ such that

$$\sigma^{n*}_g \cdot i^n_g = \mu^{n*}_g, \tag{2}$$

where σ^{n*}_g and μ^{n*}_g are defined by (1) of Lemma 5.2.2, $F^n_g := g \cdot F^n \cdot \bar{o}$ and $i^n_g := i_g|_{F^n_g}$ is an element of $\mathrm{Mor}(F^n_g, \bar{G}((t)))$. Moreover, the triple $(\mathbf{U}^n_g, \sigma^n_g, \mu^n_g)$ satisfying (2) is unique up to a unique isomorphism. From this it is easy to see that $(\mathbf{U}^n_g, \sigma^n_g, \mu^n_g)_{n\geq 0}$ give rise to a G-bundle $\mathbf{U}_g$ over $\Sigma \times \bar{V}_g$ together with sections σ_g over $\Sigma^* \times \bar{V}_g$ and μ_g over $\mathbb{D}_{\bar{V}_g} := \bigcup_{n\geq 0} \mathbb{D}_{F^n_g}$ such that

$$\sigma^*_g \cdot i_g = \mu^*_g. \tag{3}$$

For $g, h \in G((t))$, there clearly exists a morphism $\theta_{g,h}: \bar{V}_{g,h} \to \bar{G}[[t]]$ such that

$$i_g = i_h \cdot \theta_{g,h} \quad \text{over} \quad \bar{V}_{g,h} := \bar{V}_g \cap \bar{V}_h. \tag{4}$$

Now, by Lemma 5.2.2 (rather its immediate extension to $\operatorname{Spec} R$ replaced by ind-affine schemes) and the identities (3) and (4), we have a unique isomorphism of the triples over $\bar{V}_{g,h}$:

$$\left(\mathbf{U}_g|_{\Sigma\times\bar{V}_{g,h}}, \sigma_g|_{\Sigma^*\times\bar{V}_{g,h}}, \mu_g|_{\mathbb{D}_{\bar{V}_{g,h}}}\right) \sim \left(\mathbf{U}_h|_{\Sigma\times\bar{V}_{g,h}}, \sigma_h|_{\Sigma^*\times\bar{V}_{g,h}}, \mu_h|_{\mathbb{D}_{\bar{V}_{g,h}}} \cdot \theta_{g,h}\right). \tag{5}$$

In particular, there is an isomorphism of pairs

$$\beta_{g,h}: \left(\mathbf{U}_g|_{\Sigma\times\bar{V}_{g,h}}, \sigma_g|_{\Sigma^*\times\bar{V}_{g,h}}\right) \sim \left(\mathbf{U}_h|_{\Sigma\times\bar{V}_{g,h}}, \sigma_h|_{\Sigma^*\times\bar{V}_{g,h}}\right). \tag{6}$$

From (5) we see that $\beta_{g,h}$ satisfy the cocycle condition, i.e., for g, h, $k \in G((t))$,

$$\beta_{h,k} \circ \beta_{g,h} = \beta_{g,k} \quad \text{over} \quad \Sigma \times (\bar{V}_g \cap \bar{V}_h \cap \bar{V}_k). \tag{7}$$

Thus, the bundles $\mathbf{U}_g$ over $\Sigma \times \bar{V}_g$ and the sections σ_g over $\Sigma^* \times \bar{V}_g$ patch up to give a G-bundle $\mathbf{U}$ over $\Sigma \times \bar{X}_G$ and a section $\sigma_{\bar{X}_G}$ over $\Sigma^* \times \bar{X}_G$. Moreover, by (3), (1) is satisfied. This proves the proposition. □

We recall the following result (due to (Drinfeld and Simpson, 1995, Theorem 3 and Remark 2(e))) without proof.

Theorem 5.2.5 *Let G and Σ be as at the beginning of this section and let $E \to \Sigma \times S$ be a G-bundle, where S is a (not necessarily noetherian) scheme over $\mathbb{C}$. Then, there is an étale cover $\pi : \tilde{S} \to S$ such that the pull-back of E to $\Sigma^* \times \tilde{S}$ is trivial.*

In fact, the above result holds more generally for any connected affine algebraic group G such that $Hom(G, \mathbb{G}_m) = \{1\}$.

Similar to Definition 5.2.1, we have the following.

Definition 5.2.6 Define the functor

$$\mathscr{F}_{G,\Sigma^*} : \mathbf{Alg} \to \mathbf{Set} \quad \text{by}$$

$\mathscr{F}_{G,\Sigma^*}(R) = \{(E_R, \sigma_R) : E_R$ is a principal G-bundle over Σ_R and σ_R is a section of E_R over $\Sigma_R^*\}/\sim$, where $(E_R, \sigma_R) \sim (E'_R, \sigma'_R)$ if there exists an isomorphism θ_R of G-bundles:

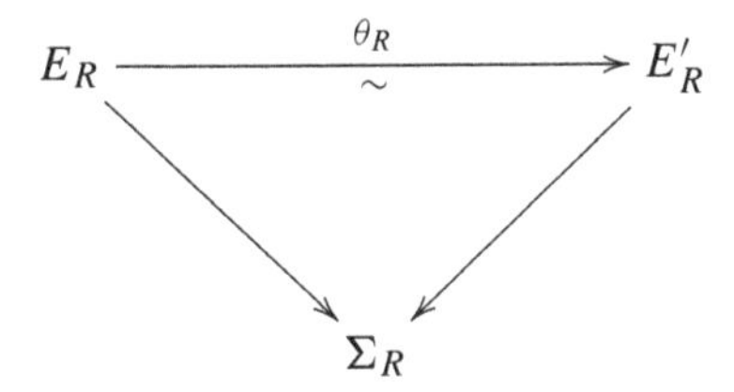

such that $\theta_R \circ \sigma_R = \sigma'_R$. We denote the equivalence class of (E_R, σ_R) by $[E_R, \sigma_R]$.

Proposition 5.2.7 *The functor $\mathscr{F}_{G,\Sigma^*}$ is a representable functor represented by the infinite Grassmannian $\bar{X}_G$.*

Moreover, for any $[E_R, \sigma_R] \in \mathscr{F}_{G,\Sigma^*}(R)$,

$$[\bar{f}^* \mathbf{U}, \bar{f}^* \sigma_{\bar{X}_G}] = [E_R, \sigma_R], \tag{1}$$

where f: Spec$R \to \bar{X}_G$ is the morphism representing $[E_R, \sigma_R]$, $\bar{f} := Id_\Sigma \times f$ and $(\mathbf{U}, \sigma_{\bar{X}_G})$ is as in Proposition 5.2.4.

Proof We need to prove that for any $R \in \mathbf{Alg}$, $\mathscr{F}_{G,\Sigma^*}(R)$ is canonically isomorphic with $\mathscr{X}_G(R) = \mathrm{Mor}(\mathrm{Spec}\, R, \bar{X}_G)$.

Define the map $\alpha \colon \mathscr{F}_{G,\Sigma^*}(R) \to \mathrm{Mor}(\mathrm{Spec}\, R, \bar{X}_G)$ as follows. Let $[E_R, \sigma_R] \in \mathscr{F}_{G,\Sigma^*}(R)$. Recall first that there exists an algebra $R' \in \mathbf{Alg}$ and a surjective étale morphism $\varphi \colon \mathrm{Spec}\, R' \to \mathrm{Spec}\, R$ such that the G-bundle $\bar{\varphi}^*(E_{R|\mathbb{D}_R})$ is trivial (where $\bar{\varphi} := \mathrm{Id}_\Sigma \times \varphi \colon \Sigma_{R'} \to \Sigma_R$), i.e., we have a section

$\mu_{R'}$ of $E_{R'|\mathbb{D}_{R'}}$, where $E_{R'} := \bar{\varphi}^*(E_R)$. (This follows, e.g., from Theorem 5.2.5.) Let $\sigma_{R'}$ be the pull-back of the section σ_R. Write

$$\mu^*_{R'} = \sigma^*_{R'} \cdot \gamma, \quad \text{for} \quad \gamma \in G(R'((t))),$$

where $\sigma^*_{R'}$, $\mu^*_{R'}$ are defined by (1) of Lemma 5.2.2. Choose a different section $\mu'_{R'}$ of $E_{R'|\mathbb{D}_{R'}}$ and write

$$\mu'^*_{R'} = \sigma^*_{R'} \cdot \gamma', \quad \text{for} \quad \gamma' \in G(R'((t))).$$

Then, clearly, $\gamma'^{-1}\gamma \in G(R'[[t]])$. Thus, the element $\gamma G(R'[[t]]) \in G(R'((t))) / G(R'[[t]])$ is independent of the choice of trivialization of $E_{R'|\mathbb{D}_{R'}}$. By virtue of Proposition 1.3.18, we have a canonical map

$$G(R'((t)))/G(R'[[t]]) \to \mathrm{Mor}(\mathrm{Spec}\, R', \bar{X}_G).$$

Let $\hat{\gamma}$ be the image of $\gamma G(R'[[t)]])$ in $\mathrm{Mor}(\mathrm{Spec}\, R', \bar{X}_G)$. Then, $\hat{\gamma}$ is independent of the trivialization of $E_{R'|\mathbb{D}_{R'}}$. Consider now the fiber product

$$\mathrm{Spec}\, R' \underset{\mathrm{Spec}\, R}{\times} \mathrm{Spec}\, R' = \mathrm{Spec}\, R'', \quad \text{where} \quad R'' := R' \otimes_R R'.$$

Let $(E^{(1)}_{R''}, \sigma^{(1)}_{R''})$ and $(E^{(2)}_{R''}, \sigma^{(2)}_{R''})$ be the pull-backs of $(E_{R'}, \sigma_{R'})$ induced from the two projections $q_1, q_2\colon \mathrm{Spec}\, R'' \to \mathrm{Spec}\, R'$. Then, both being pull-backs of (E_R, σ_R), they are isomorphic. In particular, $\hat{\gamma} \circ q_1 = \hat{\gamma} \circ q_2$. Thus, $\hat{\gamma}$ descends to give a morphism $\alpha([E_R, \sigma_R]) \in \mathrm{Mor}(\mathrm{Spec}\, R, \bar{X}_G)$. This is our map α.

We now define (the inverse) map $\beta\colon \mathrm{Mor}(\mathrm{Spec}\, R, \bar{X}_G) \to \mathscr{F}_{G,\Sigma^*}(R)$ by taking

$$\beta(f) = [\bar{f}^*\mathbf{U}, \bar{f}^*\sigma_{\bar{X}_G}], \quad \text{for} \quad f\colon \mathrm{Spec}\, R \to \bar{X}_G,$$

where $(\mathbf{U}, \sigma_{\bar{X}_G})$ is the pair as in Proposition 5.2.4 and $\bar{f} := \mathrm{Id}_\Sigma \times f\colon \Sigma_R \to \Sigma \times \bar{X}_G$.

It is easy to see that $\alpha \circ \beta = \mathrm{Id}$ by using the defining property of $(\mathbf{U}, \sigma_{\bar{X}_G})$ as in (1) of Proposition 5.2.4. To prove that $\beta \circ \alpha = \mathrm{Id}$, we equivalently need to prove that α is injective. Take $[E_R, \sigma_R]$, $[E'_R, \sigma'_R] \in \mathscr{F}_{G,\Sigma^*}(R)$ such that $\alpha([E_R, \sigma_R]) = \alpha([E'_R, \sigma'_R])$. We need to prove that they are isomorphic (and the isomorphism is unique).

Observe first that if there exists an isomorphism between (E_R, σ_R) and (E'_R, σ'_R), then it is unique (since, by definition, the isomorphism is unique over the open subset $\Sigma^* \times \mathrm{Spec}\, R$ of the base). We now come to the existence. Take a surjective étale morphism $\varphi\colon \mathrm{Spec}\, R' \to \mathrm{Spec}\, R$ such that both of $\bar{\varphi}^*(E_{R|\mathbb{D}_R})$

and $\bar{\varphi}^*(E'_{R|_{\mathbb{D}_R}})$ have sections $\mu_{R'}$ and $\mu'_{R'}$, respectively, where $\bar{\varphi} := \mathrm{Id}_\Sigma \times \varphi$. Write

$$\mu^*_{R'} = \sigma^*_{R'} \cdot \gamma \quad \text{and} \quad \mu'^*_{R'} = \sigma'^*_{R'} \cdot \gamma', \quad \text{for} \quad \gamma, \gamma' \in G(R'((t))), \tag{2}$$

where $\sigma_{R'}$ (resp. $\sigma'_{R'}$) is the pull-back of σ_R (resp. σ'_R) via $\bar{\varphi}$. Since $\alpha([E_R, \sigma_R]) = \alpha([E'_R, \sigma'_R])$, there exists $\theta \in G(R'[[t]])$ such that

$$\gamma = \gamma' \cdot \theta. \tag{3}$$

(To guarantee the existence of $\theta \in G(R'[[t]])$, we may have to go to an fppf R'-algebra, which we still denote by R', cf. proof of Lemma B.2.)

Now, consider the triples over $\Sigma \times \operatorname{Spec} R'$:

$$(E_{R'}, \sigma_{R'}, \mu_{R'}) \quad \text{and} \quad (E'_{R'}, \sigma'_{R'}, \mu'_{R'} \cdot \theta).$$

By Lemma 5.2.2 and (2) and (3), these two triples are isomorphic under a unique isomorphism; in particular, the pairs $(E_{R'}, \sigma_{R'})$ and $(E'_{R'}, \sigma'_{R'})$ are isomorphic under a unique isomorphism. Considering the pull-backs of this isomorphism induced from the two projections

$$q_1, q_2 \colon \operatorname{Spec} R' \underset{\operatorname{Spec} R}{\times} \operatorname{Spec} R' \to \operatorname{Spec} R',$$

it is easy to see that isomorphism $(E_{R'}, \sigma_{R'}) \simeq (E'_{R'}, \sigma'_{R'})$ being unique descends to give an isomorphism $(E_R, \sigma_R) \simeq (E'_R, \sigma'_R)$. This proves that α is injective, proving the first part of the proposition.

To prove the isomorphism (1), observe that $\alpha[E_R, \sigma_R] = \alpha[\bar{f}^*\mathbf{U}, \bar{f}^*\sigma_{\bar{X}_G}]$. Thus, from the injectivity of α, we get the isomorphism (1). This proves the proposition. □

The following result follows easily from the above proposition (cf. Exercise 5.2.E.2).

Corollary 5.2.8 *For the bundle $\mathbf{U} \to \Sigma \times \bar{X}_G$ and its section $\sigma_{\bar{X}_G}$ over $\Sigma^* \times \bar{X}_G$ given in Proposition 5.2.4, the pair $(\mathbf{U}, \sigma_{\bar{X}_G})$ satisfying (1) of Proposition 5.2.4 is unique up to a unique isomorphism.* □

Definition 5.2.9 (Ind-scheme structure on $\bar{\Gamma}$) Let G be any affine algebraic group (of finite type over $\mathbb{C}$). Let $\Sigma^* = \Sigma \backslash \bar{q}$ for any set of points $\bar{q} := \{q_1, \dots, q_s\}$ ($s \geq 1$) and let $\Gamma = \Gamma_{\bar{q}} := \mathrm{Mor}(\Sigma^*, G)$. We define an ind-scheme $\bar{\Gamma} = \bar{\Gamma}_{\bar{q}}$ with $\mathbb{C}$-points $\bar{\Gamma}(\mathbb{C}) = \Gamma$ as follows. We follow the construction as in Definition 1.3.1.

Fix an embedding $G \hookrightarrow \mathrm{SL}_N \subset M_N$ and let $I_G \subset \mathbb{C}[M_N]$ be the (radical) ideal of G inside M_N. For any $n \geq 0$, let $\mathrm{Mor}_{\leq n}(\Sigma^*, M_N)$ be the set of all the morphisms $f\colon \Sigma^* \to M_N$, $f = (f_{ij})_{1\le i,\, j\le N}$, such that the order of poles at each of the points of $\bar{q}$ of each $f_{ij} \le n$. Then, being equal to $H^0(\Sigma, \mathscr{L}) \otimes M_N$ for a line bundle $\mathscr{L}$ on Σ, $\mathrm{Mor}_{\le n}(\Sigma^*, M_N)$ is a finite-dimensional vector space over $\mathbb{C}$ of dimension (say) d_n. Choose a basis $\{f_k\}_{1\le k\le d_n}$ of $\mathrm{Mor}_{\le n}(\Sigma^*, M_N)$ inductively so that $\{f_k\}_{1\le k\le d_{n-1}}$ is a basis of $\mathrm{Mor}_{\le n-1}(\Sigma^*, M_N)$. Now, set

$$\mathrm{Mor}_{\le n}\left(\Sigma^*, G\right) := \mathrm{Mor}\left(\Sigma^*, G\right) \cap \mathrm{Mor}_{\le n}\left(\Sigma^*, M_N\right).$$

Using the coordinates $\{z_k\}_{1\le k\le d_n}$ of $\mathrm{Mor}_{\le n}(\Sigma^*, M_N)$ determined by the basis $\{f_k\}$ and the ideal I_G, we define the ideal $I_G^{\Sigma^*}(n) \subset \mathbb{C}[\mathbf{z}]$ generated by $\{Q_{P,w}(\mathbf{z})\}_{P\in I_G,\, w\in\Sigma^*}$, where $\mathbf{z} = (z_1, \ldots, z_{d_n})$ and

$$Q_{P,w}(\mathbf{z}) := P\left(\sum_{k=1}^{d_n} z_k f_k(w)\right). \tag{1}$$

Now, define the affine scheme of finite type over $\mathbb{C}$:

$$\bar{\Gamma}^{(n)} = \bar{\Gamma}_{\bar{q}}^{(n)} := \mathrm{Spec}\left(\mathbb{C}[\mathbf{z}]/I_G^{\Sigma^*}(n)\right).$$

Clearly, $\bar{\Gamma}^{(n)} \hookrightarrow \bar{\Gamma}^{(n+1)}$ is a closed embedding. Thus, we get an ind-affine scheme $\bar{\Gamma} = \bar{\Gamma}_{\bar{q}}$ of ind-finite type given by the filtration

$$\bar{\Gamma}^{(0)} \subset \bar{\Gamma}^{(1)} \subset \cdots.$$

By definition, $\bar{\Gamma}$ is a closed ind-subscheme of $\mathrm{Mor}(\Sigma^*, M_N)$, where we think of $\mathrm{Mor}(\Sigma^*, M_N)$ as an ind-affine scheme filtered by the affine spaces $\mathrm{Mor}_{\le n}(\Sigma^*, M_N)$. Clearly,

$$\bar{\Gamma}^{(n)}(\mathbb{C}) = \mathrm{Mor}_{\le n}(\Sigma^*, G) \text{ and thus } \bar{\Gamma}(\mathbb{C}) = \Gamma = \mathrm{Mor}(\Sigma^*, G).$$

Lemma 5.2.10 *Let G be any affine algebraic group. Consider the covariant functor $\mathscr{F}_\Gamma\colon$ **Alg** $\to$ **Set** by*

$$\mathscr{F}_\Gamma(R) = \mathrm{Mor}\left(\Sigma_R^*, G\right), \quad \textit{where, as earlier,} \quad \Sigma_R^* := \Sigma^* \times \mathrm{Spec}\, R.$$

Then, the ind-scheme $\bar{\Gamma}$ (defined above) represents the functor $\mathscr{F}_\Gamma$.

In particular, the ind-scheme structure on $\bar{\Gamma}$ is independent of the choice of the embedding $G \hookrightarrow \mathrm{SL}_N$ and the choice of the basis $\{f_k\}$ of $\mathrm{Mor}(\Sigma^*, M_N)$.

Further, $\bar{\Gamma}$ is an ind-affine group scheme of ind-finite type.

Moreover, if G is a connected semisimple group, $\bar{\Gamma}$ is reduced and thus $\bar{\Gamma}$ is an ind-affine group variety in this case.

Proof We need to prove that there is a natural bijection

$$\mathrm{Mor}\left(\Sigma_R^*, G\right) \simeq \mathrm{Mor}\left(\mathrm{Spec}\, R, \bar{\Gamma}\right). \tag{1}$$

Take an embedding $G \hookrightarrow \mathrm{SL}_N \subset M_N$. We first prove an analogue of (1) for G replaced by M_N, i.e.,

$$\mathrm{Mor}\left(\Sigma_R^*, M_N\right) \overset{\varphi_{M_N}}{\simeq} \mathrm{Mor}\left(\mathrm{Spec}\, R, \mathrm{Mor}\left(\Sigma^*, M_N\right)\right). \tag{2}$$

To prove (2) for any M_N, it suffices to prove it for $N = 1$, where it is easy to see by observing that any morphism $f\colon \mathrm{Spec}\, R \to \mathrm{Mor}(\Sigma^*, M_N)$ lands inside $\mathrm{Mor}_{\leq n}(\Sigma^*, M_N)$ for some n.

From the bijection (2), and the description of the ideal $I_G^{\Sigma^*}(n)$ as in (1) of Definition 5.2.9, it follows that the restriction of the isomorphism φ_{M_N} to $\mathrm{Mor}(\Sigma_R^*, G)$ gives a bijection φ_G making the following diagram commutative:

$$\begin{array}{ccc} \mathrm{Mor}\left(\Sigma_R^*, G\right) & \overset{\varphi_G}{\simeq} & \mathrm{Mor}\left(\mathrm{Spec}\, R, \bar{\Gamma}\right) \\ \big\downarrow & & \big\downarrow \\ \mathrm{Mor}\left(\Sigma_R^*, M_N\right) & \underset{\varphi_{M_N}}{\simeq} & \mathrm{Mor}\left(\mathrm{Spec}\, R, \mathrm{Mor}\left(\Sigma^*, M_N\right)\right), \end{array}$$

where the vertical maps are induced by the embedding $G \hookrightarrow M_N$ and $\bar{\Gamma} \hookrightarrow \mathrm{Mor}(\Sigma^*, M_N)$, respectively. This proves (1).

To prove that $\bar{\Gamma}$ is an ind-affine group scheme, since $\mathscr{F}_\Gamma$ is representable by $\bar{\Gamma}$, it suffices to observe (using Lemma 1.1.1) that the morphism $G \times G \to G$, $(g,h) \mapsto gh^{-1}$, induces a natural map $\mathscr{F}_\Gamma(R) \times \mathscr{F}_\Gamma(R) \to \mathscr{F}_\Gamma(R)$ for any $R \in \mathbf{Alg}$.

Exactly the same proof, that of the result that $\bar{G}[t^{-1}]$ is reduced (cf. Theorem 1.3.23), gives that $\bar{\Gamma}$ is reduced in the case G is connected and semisimple. This proves the lemma. □

Corollary 5.2.11 *Let G be any affine algebraic group. Choose a local parameter t for Σ at p_o, where $p_o \in \bar{q}$. Then the morphism $\mathbb{D}^* = \mathbb{D}^*_{p_o} \to \Sigma^*$ gives rise to a morphism $\bar{\Gamma} \to \bar{G}((t))$, where $\bar{\Gamma} := \bar{\Gamma}_{\bar{q}}$.*

Proof For any $R \in \mathbf{Alg}$, the morphism $j_R^*\colon \mathbb{D}_R^* \to \Sigma_R^*$ (see the beginning of this section diagram $\mathscr{D}_R$) induces the map θ

$$\begin{array}{ccc} \mathrm{Mor}\left(\Sigma_R^*, G\right) & \xrightarrow{\ \theta\ } & \mathrm{Mor}(\mathbb{D}_R^*, G) \\ \wr\big\downarrow & & \wr\big\downarrow \\ \mathrm{Mor}(\mathrm{Spec}\, R, \bar{\Gamma}) & \xrightarrow{\ \hat{\theta}\ } & \mathrm{Mor}(\mathrm{Spec}\, R, \bar{G}((t))), \end{array}$$

where the left (resp. right) vertical bijection follows from Lemma 5.2.10 (resp. Lemma 1.3.2) and $\hat{\theta}$ is defined to be the map making the above diagram commutative. The map $\hat{\theta}$ gives rise to the morphism $\bar{\Gamma} \to \bar{G}((t))$ via Lemma 1.1.1. □

We take now $\bar{q}$ to consist of a single point p_o and abbreviate $\bar{\Gamma}_{\bar{q}} = \bar{\Gamma}_{p_o}$ by $\bar{\Gamma}$ till the end of this Section 5.2.

Lemma 5.2.12 *The principal G-bundle $\mathbf{U} \to \Sigma \times \bar{X}_G$, as in Proposition 5.2.4, acquires the structure of a $\bar{\Gamma}$-equivariant bundle with respect to the (left) action of $\bar{\Gamma}$ on $\bar{X}_G$ via the morphism $\bar{\Gamma} \to \bar{G}((t))$ as in the above Corollary 5.2.11.*

Further, for any $\gamma \in \Gamma = \bar{\Gamma}(\mathbb{C})$,

$$(\gamma \cdot \sigma_{\bar{X}_G})(z, \bar{g}) = \sigma_{\bar{X}_G}(z, \bar{g})\gamma(z), \text{ for } z \in \Sigma^* \text{ and } \bar{g} \in \bar{X}_G, \tag{1}$$

where $\sigma_{\bar{X}_G}$ is the section over $\Sigma^* \times \bar{X}_G$ as in Proposition 5.2.4.

Proof The pair $(\mathbf{U}, \sigma_{\bar{X}_G})$ guaranteed by Proposition 5.2.4 is unique up to a unique isomorphism satisfying (1) of the same (cf. Corollary 5.2.8). For any $\gamma \in \Gamma$, define the pair $(\mathbf{U}, \sigma^{\gamma}_{\bar{X}_G})$, where $\sigma^{\gamma}_{\bar{X}_G}$ is the section over $\Sigma^* \times \bar{X}_G$ defined by

$$\sigma^{\gamma}_{\bar{X}_G}(z, \bar{g}) = \sigma_{\bar{X}_G}(z, \bar{g})\gamma(z)^{-1}, \quad \text{for} \quad z \in \Sigma^* \quad \text{and} \quad \bar{g} \in \bar{X}_G. \tag{2}$$

Then, the pair $\left(\mathbf{U}, \sigma^{\gamma}_{\bar{X}_G}\right)$, under the correspondence of Proposition 5.2.7 (rather its extension to Spec R replaced by ind-schemes) corresponds to the morphism

$$L_\gamma : \bar{X}_G \to \bar{X}_G, \ \bar{g} \mapsto \gamma \cdot \bar{g}, \quad \text{for} \quad \bar{g} \in \bar{X}_G$$

(use (1) of Proposition 5.2.4).

Further, the pair $[\bar{L}^*_\gamma \mathbf{U}, \bar{L}^*_\gamma \sigma_{\bar{X}_G}]$ also corresponds to the morphism L_γ, where $\bar{L}_\gamma := \mathrm{Id}_\Sigma \times L_\gamma$. Hence, by Proposition 5.2.7 (and Corollary 5.2.8) there exists a unique isomorphism between the pairs $(\mathbf{U}, \sigma^{\gamma}_{\bar{X}_G})$ and $(\bar{L}^*_\gamma \mathbf{U}, \bar{L}^*_\gamma \sigma_{\bar{X}_G})$. This unique isomorphism; in particular, the isomorphism $\mathbf{U} \simeq \bar{L}^*_\gamma \mathbf{U}$

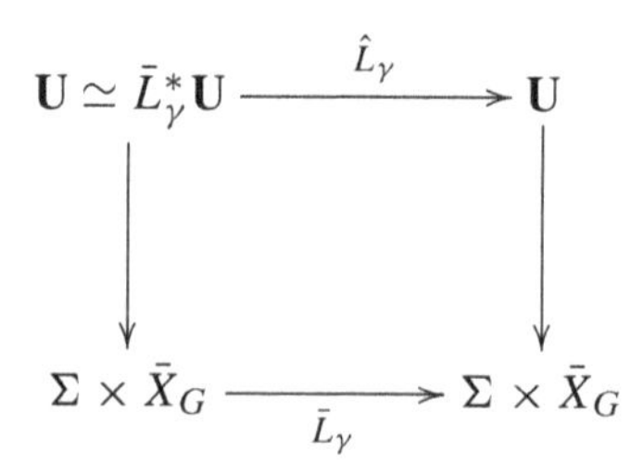

gives rise to an action $\hat{L}_\gamma$ of γ on $\mathbf{U}$ for any $\gamma \in \Gamma$. By the uniqueness of the isomorphism $(\mathbf{U}, \sigma^{\gamma}_{\bar{X}_G})$ and $(\bar{L}^*_\gamma \mathbf{U}, \bar{L}^*_\gamma \sigma_{\bar{X}_G})$, we get $\hat{L}_\gamma \cdot \hat{L}_{\gamma'} = \hat{L}_{\gamma\gamma'}$. Now,

$\bar{X}_G, \bar{\Gamma}$ and hence **U** being ind-varieties (cf. Proposition 1.3.18 and Lemma 5.2.10), the action of Γ extends to an action of $\bar{\Gamma}$.

Now, the identity (1) follows from the identity (2). This proves the lemma. □

Remark 5.2.13 I do not know if there is a unique $\bar{\Gamma}$-equivariant bundle structure on $\mathbf{U} \to \bar{X}_G$.

We are ready to prove the following *Uniformization Theorem.*

Theorem 5.2.14 *Let G and Σ be as at the beginning of this section. Then there is an isomorphism of stacks*

$$\mathbf{Bun}_G(\Sigma) \simeq [\bar{\Gamma} \backslash \bar{X}_G],$$

where $\mathbf{Bun}_G(\Sigma)$ *is the moduli stack (cf. Definition 5.1.1 and Theorem 5.1.3),* $\bar{\Gamma}$ *is the ind-group scheme as in Definition 5.2.9 with* $\mathbb{C}$*-points* $\bar{\Gamma}(\mathbb{C}) = \Gamma := Mor(\Sigma^*, G), \Sigma^* := \Sigma \backslash \{p_o\}$ *and* $[\bar{\Gamma} \backslash \bar{X}_G]$ *is the quotient stack (cf. Example C.18(b)) under the left action of* $\bar{\Gamma}$ *on* $\bar{X}_G$.

Proof We first define a functor $\mathbf{f}\colon \mathbf{Bun}_G(\Sigma) \to [\bar{\Gamma} \backslash \bar{X}_G]$. Let $E = E_S \to \Sigma \times S$ be a G-bundle, for any scheme S. We define a $\mathbb{C}$-space functor $\mathscr{E}_S$ as follows. For any $\mathbb{C}$-algebra R and an element in $S(R)$, i.e., a morphism $\varphi\colon \operatorname{Spec} R \to S$, define

$$\mathscr{E}_S^o(\varphi) := \text{set of trivializations of } E_{\varphi|\Sigma_R^*},$$

where $\Sigma_R^* := \Sigma^* \times \operatorname{Spec} R$ and E_φ denotes the pull-back bundle $(\mathrm{Id}_\Sigma \times \varphi)^*(E)$. Now, for any $\mathbb{C}$-algebra R, set

$$\mathscr{E}_S^o(R) := \sqcup_{\varphi \in S(R)} \mathscr{E}_S^o(\varphi).$$

Then, for an fppf R-algebra R', $\mathscr{E}_S^o(R) \to \mathscr{E}_S^o(R')$ is injective since $\operatorname{Spec}(R') \to \operatorname{Spec}(R)$ is surjective by Matsumura (1989, Theorem 7.3). Let $\mathscr{E}_S$ be the sheafification of $\mathscr{E}_S^o$ (cf. Lemma B.2). Then we get a morphism

$$p\colon \mathscr{E}_S \to S,$$

where we have abbreviated the $\mathbb{C}$-space functor $h_S(R) = S(R)$ associated to S by S itself. Further, for any $\varphi \in S(R)$, using Lemma 5.2.10, there is a canonical action $\mathscr{E}_S^o(\varphi) \times \bar{\Gamma}(R) \to \mathscr{E}_S^o(\varphi)$ giving rise to an action

$$\mathscr{E}_S \times \bar{\Gamma} \to \mathscr{E}_S. \tag{1}$$

Take an affine étale cover S' of S such that $E'_{|\Sigma_{S'}^*} \to \Sigma_{S'}^*$ is trivial (cf. Theorem 5.2.5), where E' is the pull-back of E to $\Sigma_{S'}$. Then $\mathscr{E}_{S'} = S' \times_S \mathscr{E}_S$ is isomorphic with $S' \times \bar{\Gamma}$ such that the induced action of $\bar{\Gamma}$ on $\mathscr{E}_{S'}$ (as in (1)) corresponds to the right multiplication on the $\bar{\Gamma}$-factor. Thus, $p\colon \mathscr{E}_S \to S$ is a $\bar{\Gamma}$-torsor with the right action of $\bar{\Gamma}$ (cf. Definition C.16).

We further define a $\bar{\Gamma}$-equivariant morphism $\mathscr{E}_S \to \bar{X}_G$ as follows. For any $\varphi \in S(R)$, map any element of $\mathscr{E}^o_S(\varphi)$ to the pair $(E_\varphi, \sigma_\varphi)$, where σ_φ is the corresponding section of $E_{\varphi|\Sigma^*_R}$. The sections σ_φ (for $\varphi \in S(R)$) give rise to the section σ_o of $\bar{p}^*(E_S)$ over $\Sigma^* \times \mathscr{E}_S$, where $\bar{p}\colon \Sigma \times \mathscr{E}_S \to \Sigma \times S$ is the morphism $\mathrm{Id}_\Sigma \times p$. We call σ_o the *tautological section*. Now, the equivalence class $[E_\varphi, \sigma_\varphi] \in \mathscr{F}_{G,\Sigma^*}(R)$ (cf. Definition 5.2.6) corresponds to an element in $\bar{X}_G(R)$ (cf. Proposition 5.2.7). This gives us the desired morphism

$$\beta\colon \mathscr{E}_S \to \bar{X}_G.$$

It is clearly $\bar{\Gamma}$-equivariant, where $\bar{\Gamma}$ acts on $\bar{X}_G$ via the ind-affine group morphism $\bar{\Gamma} \to \bar{G}((t))$ as in Corollary 5.2.11 and we switch the right action of $\bar{\Gamma}$ on $\mathscr{E}_S$ to the left action by the standard procedure:

$$\gamma \cdot x = x \cdot \gamma^{-1}, \quad \text{for } \gamma \in \bar{\Gamma} \text{ and } x \in \mathscr{E}_S.$$

By definition, the functor $\mathbf{f}$ associates to $E_S \to \Sigma_S$, the $\bar{\Gamma}$-torsor $p\colon \mathscr{E}_S \to S$ and the $\bar{\Gamma}$-equivariant morphism $\varphi\colon \mathscr{E}_S \to \bar{X}_G$.

We now define a functor $\mathbf{g}\colon [\bar{\Gamma}\backslash\bar{X}_G] \to \mathbf{Bun}_G(\Sigma)$ as follows. Take a $\bar{\Gamma}$-torsor (with left $\bar{\Gamma}$-action) $p\colon \mathscr{E}_S \to S$ (over a scheme S) and a $\bar{\Gamma}$-equivariant morphism $\varphi\colon \mathscr{E}_S \to \bar{X}_G$. Let $\mathbf{U} \to \Sigma \times \bar{X}_G$ be the tautological bundle as in Proposition 5.2.4. Then, by Lemma 5.2.12, $\mathbf{U}$ is a $\bar{\Gamma}$-equivariant G-bundle with respect to the left action of $\bar{\Gamma}$ on the $\bar{X}_G$-factor. Thus, $\bar{\varphi}^*\mathbf{U}$ is a $\bar{\Gamma}$-equivariant G-torsor over $\Sigma \times \mathscr{E}_S$ with the right action of G, where $\bar{\varphi} := \mathrm{Id}_\Sigma \times \varphi$ (cf. Definition C.16). Since $p\colon \mathscr{E}_S \to S$ is a $\bar{\Gamma}$-torsor, the $\bar{\Gamma}$-equivariant G-torsor $\bar{\varphi}^*\mathbf{U}$ over $\Sigma \times \mathscr{E}_S$ descends to a G-bundle E_S over $\Sigma \times S$ (cf. Theorem C.17). (As observed in the proof of Lemma C.19, E_S is indeed a scheme.) This is our functor $\mathbf{g}$. We need to prove that

(a) $\mathbf{g} \circ \mathbf{f} \simeq \mathrm{Id}_{\mathbf{Bun}_G(\Sigma)}$, and
(b) $\mathbf{f} \circ \mathbf{g} \simeq \mathrm{Id}_{[\bar{\Gamma}|\bar{X}_G]}$.

To prove (a), from the definitions of $\mathbf{f}$ and $\mathbf{g}$, it suffices to prove that for any G-bundle $\pi\colon E_S \to \Sigma_S$ with the trivial $\bar{\Gamma}$-equivariant structure (for a scheme S),

$$\bar{\beta}^*(\mathbf{U}) \simeq \bar{p}^*(E_S), \quad \text{as } \bar{\Gamma}\text{-equivariant } G\text{-torsors}, \tag{2}$$

where $p\colon \mathscr{E}_S \to S$ is the $\bar{\Gamma}$-torsor obtained as earlier and $\beta\colon \mathscr{E}_S \to \bar{X}_G$ is the $\bar{\Gamma}$-equivariant morphism obtained from E_S via the functor $\mathbf{f}$ as above.

Now, the G-torsors over $\Sigma \times \mathscr{E}_S$ and sections over $\Sigma^* \times \mathscr{E}_S$: $\left(\bar{\beta}^*(\mathbf{U}), \bar{\beta}^*\sigma_{\bar{X}_G}\right)$ and $(\bar{p}^*(E_S), \sigma_o)$ correspond to the same morphism $\beta\colon \mathscr{E}_S \to \bar{X}_G$ under the correspondence of Proposition 5.2.7, where $\sigma_{\bar{X}_G}$ is the section of $\mathbf{U}$ as in Proposition 5.2.4 and σ_o is the tautological section as above. Thus, there is a unique isomorphism

$$\theta\colon \left(\bar{\beta}^*(\mathbf{U}), \bar{\beta}^*\sigma_{\bar{X}_G}\right) \xrightarrow{\sim} \left(\bar{p}^*(E_S), \sigma_o\right).$$

We next claim that the isomorphism $\theta\colon \bar{\beta}^*(\mathbf{U}) \to \bar{p}^*(E_S)$ is $\bar{\Gamma}$-equivariant.

For $\gamma \in \bar{\Gamma}$, let L_γ be the left action of γ on $\mathscr{E}_S$ (which is the same as the right action of γ^{-1} on $\mathscr{E}_S$) extended to $\Sigma \times \mathscr{E}_S$ with the trivial action on the Σ-factor. Then, it is easy to see that, for $\gamma \in \Gamma = \bar{\Gamma}(\mathbb{C})$,

$$(L_\gamma^*\sigma_o)(z,\sigma_\varphi) = \sigma_o(z,\sigma_\varphi)ev_z(\gamma)^{-1}, \quad \text{for } z \in \Sigma^* \text{ and } \sigma_\varphi \in \mathscr{E}_S, \tag{3}$$

where $ev_z\colon \bar{\Gamma} \to G$ is the morphism induced from $\bar{\Gamma}(R) = \mathrm{Mor}(\Sigma_R^*, G) \to G(R)$ by $f \mapsto f_{|z\times \mathrm{Spec}\, R}$ via Lemma 5.2.10. Further, by the proof of Lemma 5.2.12,

$$(\bar{L}_\gamma^*\sigma_{\bar{X}_G})(z,\bar{g}) = \sigma_{\bar{X}_G}(z,\bar{g})ev_z(\gamma)^{-1}, \quad \text{for } z \in \Sigma^* \text{ and } \bar{g} \in \bar{X}_G. \tag{4}$$

But, since $\bar{\Gamma}$ is an ind-variety, any morphism $\bar{\Gamma} \to G$ is determined by its restriction to Γ (cf. (Hartshorne, 1977, Chap. II, Proposition 2.6)). Thus (3) and (4) hold for any $\gamma \in \bar{\Gamma}$. Combining (3) and (4), we easily get that the unique isomorphism θ is $\bar{\Gamma}$-equivariant. This proves (2) and hence (a).

To prove (b), it suffices to show that for any $\bar{\Gamma}$-torsor $\pi : \mathscr{F}_S \to S$ over a scheme S with the left action of $\bar{\Gamma}$ on $\mathscr{F}_S$ and a $\bar{\Gamma}$-equivariant morphism $\beta\colon \mathscr{F}_S \to \bar{X}_G$, there is a natural isomorphism of $\bar{\Gamma}$-torsors:

$$\eta\colon \mathscr{F}_S \xrightarrow{\sim} \mathscr{E}_S, \tag{5}$$

such that the sections $\bar{\beta}^*(\sigma_{\bar{X}_G})$ of $\bar{\beta}^*(\mathbf{U})$ over $\Sigma^* \times \mathscr{F}_S$ and the tautological section σ_o of $\bar{p}^*(E_S)$ over $\Sigma^* \times \mathscr{E}_S$ correspond, where E_S is, by definition, the G-bundle over $\Sigma \times S$ such that (denoting $\bar{\pi} := Id_\Sigma \times \pi$)

$$\bar{\pi}^*(E_S) \simeq \bar{\beta}^*(\mathbf{U}), \quad \text{as } \bar{\Gamma}\text{-equivariant } G\text{-torsors over } \Sigma \times \mathscr{F}_S \tag{6}$$

and $p\colon \mathscr{E}_S \to S$ is the bundle of sections of $E_{S|_{\Sigma^*\times x}}$ as above. For $e_x \in \mathscr{F}_S(R)$ (for a $\mathbb{C}$-algebra R) with $\pi(e_x) = x \in S(R)$, let $\bar{\beta}^*(\sigma_{\bar{X}_G})_{e_x}$ be the section $\bar{\beta}^*(\sigma_{\bar{X}_G})_{|\Sigma^*(R)\times e_x}$. This section corresponds to an element $\eta(e_x)$ of $\mathscr{E}_S(R)$ over x due to the isomorphism (6). Using (4), it is easy to see that η is an isomorphism of $\bar{\Gamma}$-torsors. By the definition of η, the sections $\bar{\beta}^*(\sigma_{\bar{X}_G})$ and σ_o correspond under η. This proves (5) and hence (b) is proved. This completes the proof of the theorem. □

As an immediate corollary of Theorem 5.2.14 and Exercise C.E.11, we get the following.

Corollary 5.2.15 *Let G and Σ be as at the beginning of this section and let $\mathrm{Bun}_G(\Sigma)$ be the set of isomorphism classes of G-bundles over Σ. Then there is a natural set-theoretic bijection*

$$\mathrm{Bun}_G(\Sigma) \simeq \Gamma \backslash X_G.$$

Let G and Σ be as at the beginning of this section. Let $\vec{p} = (p_1, \ldots, p_s)$, for $s \geq 1$, be a set of distinct marked points in Σ distinct from the chosen base point p_o. We fix a Borel subgroup B of G. We label the points $\vec{p}$ with standard parabolic subgroups $\vec{P} = (P_1, \ldots, P_s)$, respectively. Let $\bar{X}_G$ be the infinite Grassmannian with respect to a local parameter t of Σ at p_o. Then, the ind-group variety $\bar{\Gamma} = \bar{\Gamma}_{p_o}$ (where $\Sigma^* = \Sigma \backslash p_o$) acts on $\bar{X}_G \times \Pi_{i=1}^s (G/P_i)$ from the left via

$$\gamma \cdot (x, \bar{g}_1, \ldots, \bar{g}_s) = \left(\gamma \cdot x, \overline{ev_{p_1}(\gamma) g_1}, \ldots, \overline{ev_{p_s}(\gamma) g_s}\right),$$
$$\text{for } \gamma \in \bar{\Gamma}, x \in \bar{X}_G \text{ and } g_i \in G,$$

where the action of $\bar{\Gamma}$ on $\bar{X}_G$ is through the morphism $\bar{\Gamma} \to \bar{G}((t))$ as in Corollary 5.2.11.

Recall the $\bar{\Gamma}$-equivariant tautological G-bundle $\mathbf{U} \to \Sigma \times \bar{X}_G$ and a section $\sigma_{\bar{X}_G}$ over $\Sigma^* \times \bar{X}_G$ from Proposition 5.2.4 and Lemma 5.2.12. Let

$$\delta \colon \bar{X}_G \times \Pi_{i=1}^s (G/P_i) \to \bar{X}_G$$

be the $\bar{\Gamma}$-equivariant projection on the first factor and let $\bar{\delta} := \mathrm{Id}_\Sigma \times \delta$. Let $\bar{\delta}^*(\mathbf{U})$ be the $\bar{\Gamma}$-equivariant pull-back G-bundle over $\Sigma \times \bar{X}_G \times \Pi_{i=1}^s (G/P_i)$. For any $1 \leq i \leq s$, define a section $\hat{\sigma}_i$ of $\bar{\delta}^*(\mathbf{U})/P_i$ over $p_i \times \bar{X}_G \times \Pi_{i=1}^s (G/P_i)$ by

$$\hat{\sigma}_i(p_i, x, \bar{g}_1, \ldots, \bar{g}_s) = \bar{\delta}^* \left(\sigma_{\bar{X}_G}(p_i, x)\right) \cdot g_i P_i, \quad \text{for } x \in \bar{X}_G \text{ and } g_i \in G,$$

where $\bar{g}_i := g_i P_i$. Then, $\hat{\sigma}_i$ is $\Gamma = \bar{\Gamma}(\mathbb{C})$-invariant from (1) of Lemma 5.2.12.

Define a morphism of stacks $\left[\bar{\Gamma} \backslash \left(\bar{X}_G \times \Pi_{i=1}^s (G/P_i)\right)\right] \to \mathbf{Parbun}_G(\Sigma, \vec{P})$ as follows.

Take any pair (π, φ) where $\pi : \mathbf{E} \to S$ is a $\bar{\Gamma}$-torsor with the left action of $\bar{\Gamma}$ on a $\mathbb{C}$-space functor $\mathbf{E}$ over base scheme $S \in \mathfrak{S}$ and $\varphi \colon \mathbf{E} \to \bar{X}_G \times \Pi_{i=1}^s (G/P_i)$ is a $\bar{\Gamma}$-equivariant morphism. Now, associated to (π, φ), define the quasi-parabolic G-bundle $\mathcal{E}$ over $\Sigma \times S$ as follows. Take the $\bar{\Gamma}$-equivariant pull-back G-torsor $\mathcal{E} := \bar{\varphi}^*(\bar{\delta}^*(\mathbf{U}))$ over $\Sigma \times \mathbf{E}$, where $\bar{\varphi} := \mathrm{Id}_\Sigma \times \varphi$. Since π is a $\bar{\Gamma}$-torsor, $\mathcal{E}$ descends to a G-bundle $\mathcal{E}^o$ over $\Sigma \times S$ (cf. Theorem C.17). Moreover, the sections $\hat{\sigma}_i$ of $\bar{\delta}^*(\mathbf{U})/P_i$ over $p_i \times \bar{X}_G \times \Pi_{i=1}^s (G/P_i)$ being Γ-invariant (and hence $\bar{\Gamma}$-invariant from the proof of Lemma B.23), give rise to sections (via pull-back and descent) $\hat{\sigma}_i^o$ of $\mathcal{E}^o/P_i$ over $p_i \times S$ (cf. Theorem C.17). The functor $(\pi, \varphi) \rightsquigarrow (\mathcal{E}^o, \hat{\sigma}_i^o)_{1\le i\le s}$ gives the following parabolic analogue of Theorem 5.2.14. Its proof is similar to that of the proof of Theorem 5.2.14 (and hence omitted).

Theorem 5.2.16 *With the above notation and assumptions, there is an isomorphism of stacks*

$$\mathbf{Parbun}_G\left(\Sigma, \vec{P}\right) \simeq \left[\bar{\Gamma}\backslash\left(\bar{X}_G \times \Pi_{i=1}^s (G/P_i)\right)\right],$$

where $\mathbf{Parbun}_G(\Sigma, \vec{P})$ *is the quasi-parabolic moduli stack as defined in Definition 5.1.4.*

Similar to Corollary 5.2.15, we get the following from the above theorem.

Corollary 5.2.17 *With the notation and assumptions as in the above theorem, there is a natural set-theoretic bijection*

$$\mathrm{Parbun}_G(\Sigma, \vec{P}) \simeq \Gamma\backslash\left(X_G \times \Pi_{i=1}^s (G/P_i)\right),$$

where $\mathrm{Parbun}_G(\Sigma, \vec{P})$ *denotes the set of isomorphism classes of quasi-parabolic G-bundles of type $\vec{P}$ over $(\Sigma, \vec{p})$ (cf. Definition 5.1.4).*

5.2.E Exercises

Let Σ and G be as at the beginning of this section.

(1) Let $\Sigma^* = \Sigma\backslash\bar{q}$, for a nonempty finite subset $\bar{q}$ of Σ. Then, show that for any affine variety X, the functor **Alg** $\to$ **Set**:

$$R \rightsquigarrow \mathrm{Mor}\left(\Sigma_R^*, X\right)$$

is represented by an ind-affine scheme $\bar{\Gamma}(X)$ with $\mathbb{C}$-points $\mathrm{Mor}(\Sigma^*, X)$. *Hint:* Follow the construction as in Definition 5.2.9 and the proof of Lemma 5.2.10.

(2) Give details of the proof of Corollary 5.2.8.

(3) Prove the following analogue of Proposition 5.2.4, Lemma 5.2.12 and Theorem 5.2.16 (under the same assumptions and notation as of Theorem 5.2.16).

First some more notation. Denote $\Sigma^* := \Sigma \setminus \{p_1, \ldots, p_s\}$, the ind-group variety $\bar{\Gamma}_{\vec{p}}$ with $\mathbb{C}$-points $\Gamma_{\vec{p}} = \mathrm{Mor}\,(\Sigma^*, G)$ is as in Lemma 5.2.10, t_i is a local parameter of Σ at p_i, $ev_0\colon \bar{G}[[t_i]] \to G$ is the evaluation map at $t_i = 0$, closed subgroup scheme $\mathcal{P}_i := ev_0^{-1}(P_i)$ and $\gamma \in \bar{\Gamma}_{\vec{p}}$ acts on $\bar{X}_G$ and $\bar{X}_G(P_i)$ based at $p_i \in \Sigma$ with respect to the parameter t_i (cf. Exercise 1.3.E.11) via the left multiplication by the Laurent series expansion of γ at p_i (see Corollary 5.2.11).

Prove that there exists a $\bar{\Gamma}_{\vec{p}}$-equivariant G-bundle $\mathbf{U}_s \to \Sigma \times \Pi_{i=1}^s \bar{X}_G$ together with a section σ_s over $\Sigma^* \times \Pi_{i=1}^s \bar{X}_G$ such that, for any $g_i \in G((t_i))$, the section over $\mathbb{D}^*_{p_j} \times (\bar{g}_1, \ldots, \bar{g}_s)$ for $1 \leq j \leq s$:

$$\sigma_s\left(-, (\bar{g}_1, \ldots, \bar{g}_s)\right) \cdot g_j \tag{1}$$

extends to a section $\mu_j^{(g_1,\ldots,g_s)}$ of $(\mathbf{U}_s)_{|\mathbb{D}_{p_j} \times (\bar{g}_1,\ldots,\bar{g}_s)}$, where $\bar{g}_i := g_i G[[t_i]] \in \bar{X}_G(\mathbb{C})$ and $\mathbb{D}_{p_j}$ (resp. $\mathbb{D}^*_{p_j}$) is the formal disc (resp. punctured disc) in Σ around p_j.

The section σ_s satisfies

$$(\gamma \cdot \sigma_s)(z, (\bar{g}_1, \ldots, \bar{g}_s)) = \sigma_s(z, (\bar{g}_1, \ldots, \bar{g}_s))\, \gamma(z),$$
$$\text{for any } \gamma \in \bar{\Gamma}_{\vec{p}}(\mathbb{C}), z \in \Sigma^* \text{ and } \bar{g}_i \in \bar{X}_G.$$

Moreover, the pair $(\mathbf{U}_s, \sigma_s)$ is unique up to a unique isomorphism with the above extension property (1) of σ_s.

Consider the $\bar{\Gamma}_{\vec{p}}$-equivariant projection

$$q \colon \Pi_{i=1}^s \bar{X}_G(P_i) \to \Pi_{i=1}^s \bar{X}_G$$

and let $\bar{q} := \mathrm{Id}_\Sigma \times q$. Then $\bar{q}^*(\mathbf{U}_s)$ is a $\bar{\Gamma}_{\vec{p}}$-equivariant G-bundle over $\Sigma \times \Pi_{i=1}^s \bar{X}_G(P_i)$. Prove that, for any $1 \leq j \leq s$, the bundle $\left(\bar{q}^*(\mathbf{U}_s)_{|p_j \times \Pi_{i=1}^s \bar{X}_G(P_i)}\right)/P_j$ admits a $\bar{\Gamma}_{\vec{p}}$-invariant section $\bar{\mu}_j$. The section $\bar{\mu}_j$ satisfies the following for any $g_i \in G((t_i))$:

$$\bar{\mu}_j(p_j, (g_1\mathcal{P}_1(\mathbb{C}), \ldots, g_s\mathcal{P}_s(\mathbb{C}))) = \mu_j^{(g_1,\ldots,g_s)}(p_j, (\bar{g}_1, \ldots, \bar{g}_s))P_j. \tag{$*$}$$

Following the construction given above Theorem 5.2.16, we can now define a functor from the stack $\left[\bar{\Gamma}_{\vec{p}} \backslash \Pi_{i=1}^s \bar{X}_G(P_i)\right] \to \mathbf{Parbun}_G(\Sigma, \vec{P})$. Show that this functor gives an isomorphism of stacks

$$\mathbf{Parbun}_G\left(\Sigma, \vec{P}\right) \simeq \left[\bar{\Gamma}_{\vec{p}} \backslash \Pi_{i=1}^s \bar{X}_G(P_i)\right].$$

(4) Show that under the assumptions and notation of Theorem 5.2.16, the map of stacks

$$\bar{X}_G \times \Pi_{i=1}^s (G/P_i) \to \left[\bar{\Gamma} \backslash \left(\bar{X}_G \times \Pi_{i=1}^s (G/P_i)\right)\right]$$

is a locally trivial $\bar{\Gamma}$-torsor even in the étale topology (i.e., for any morphism $S \to [\bar{\Gamma} \backslash Y]$ of stacks for $S \in \mathfrak{S}$, the fiber product

$S \times_{[\bar{\Gamma}\backslash Y]} Y \to S$ is an étale locally trivial $\bar{\Gamma}$-torsor, cf. Definitions C.13 and C.16), where we identify the ind-variety $Y = \bar{X}_G \times \Pi_{i=1}^{s}(G/P_i)$ with the corresponding stack $\mathfrak{S}_Y$.

(5) Follow the notation of Lemma 5.2.12. For $x \in \bar{X}_G(\mathbb{C}) = X_G$, let Γ_x be the isotropy subgroup under the action of $\Gamma = \bar{\Gamma}(\mathbb{C})$ on X_G. Then, show that (as groups) $\Gamma_x \simeq \mathrm{Aut}(\mathbf{U}_{|\Sigma \times x})$.

5.C Comments

Theorem 5.1.3 seems to be well known (see (Laszlo and Sorger, 1997, Proposition 3.4) and also (Behrend, 1991), (Olsson, 2006)). Its (a)-part for $G = \mathrm{GL}_N$ is originally due to Laumon and Moret-Bailly (1999). As mentioned earlier, a complete proof of Theorem 5.1.3(a,b) appeared in Wang (2011) and a sketch of the proof of the (c)-part appeared in Sorger (1999b).

Lemma 5.2.2, Proposition 5.2.7 and the Uniformization Theorem (Theorem 5.2.14) are due to Beauville and Laszlo (1994) for $G = \mathrm{SL}_N$ (and also for GL_N). Theorem 5.2.16 for $G = \mathrm{SL}_N$ is due to Pauly (1996). Lemma 5.2.2 for any G is obtained in Laszlo and Sorger (1997, Proposition 3.8). Proposition 5.2.7 for any G is obtained in Kumar, Narasimhan and Ramanathan (1994, Proposition 2.8(b)) and also in Laszlo and Sorger (1997, Proposition 3.10). Theorem 5.2.14 for any G follows essentially from results in Drinfeld and Simpson (1995) (see also (Faltings, 2003, Corollary 16)). The construction of the tautological bundle $\mathbf{U}$ as in Proposition 5.2.4 is taken from Kumar, Narasimhan and Ramanathan (1994, Proposition 2.8). Exercise 5.2.E.4 is taken from Laszlo and Sorger (1997, Theorem 8.5) (for $G = \mathrm{SL}_N$ in the non-parabolic case it is due to Beauville and Laszlo (1994, Proposition 3.4)). There is an alternative proof of the reducedness of $\bar{\Gamma}$ in Laszlo and Sorger (1997, §5) (cf. Lemma 5.2.10).

A generalization of Theorem 5.2.5 for singular curves Σ is proved in Belkale and Fakhruddin (2019).

6

Parabolic G-Bundles and Equivariant G-Bundles

Let G be a simple, simply-connected algebraic group with maximal torus H and let $(\Sigma, \vec{p} = (p_1, \ldots, p_s))$ be an s-pointed smooth projective irreducible curve (of any genus g). Fix a maximal compact subgroup K of G. Then the set of K-orbits $K/\operatorname{Ad} K$ in K under the adjoint action is parameterized by the fundamental alcove Φ_o (cf. Lemma 6.1.1). Recall that a parabolic G-bundle $(E, \vec{\tau}, \vec{\sigma})$ consists of a principal G-bundle $E \to \Sigma$ together with markings $\vec{\tau} = (\tau_1, \ldots, \tau_s)$, for $\tau_j \in \Phi_o$, and a section σ_j of E_{p_j}/P_j over p_j, for each $1 \leq j \leq s$, where $P_j := P(\tau_j)$ is the standard parabolic subgroup such that its Levi subgroup $L(\tau_j)$ containing H has for its simple roots $S_{\tau_j} := \{\alpha_i : \alpha_i(\tau_j) = 0\}$. We define the parabolic semistability (and parabolic stability) of $(E, \vec{\tau}, \vec{\sigma})$ in Definition 6.1.4(d). This definition generalizes the standard definition of parabolic semistability (and stability) for parabolic vector bundles (cf. Exercise 6.1.E.7). In particular, when $s = 0$, we recover the definition of semistability and stability of the G-bundle $E \to \Sigma$ (cf. Definition 6.1.4(b)) generalizing the corresponding notion for vector bundles (cf. Definition 6.1.4(a)). We show that a G-bundle $E \to \Sigma$ is semistable if and only if its adjoint bundle $\operatorname{ad} E$ is semistable (cf. Lemma 6.1.5).

For an algebra R over $\mathbb{C}$, let $\mathbb{D}_R = \operatorname{Spec} R[[t]]$ denote the formal disc. Let a finite group A act on $\mathbb{D} := \operatorname{Spec} \mathbb{C}[[t]]$ and let $\mathscr{E} \to \mathbb{D}_R$ be an A-equivariant principal G-bundle, which is trivial as a G-bundle, where A acts on $\mathbb{D}_R$ with the trivial action on R. Then, as proved in Theorem 6.1.9, there exists a G-bundle trivialization of $\mathscr{E}$ in which the A-action is the product action, i.e., there exists an A-equivariant G-bundle isomorphism inducing the identity on the base: $\mathscr{E} \underset{\sim}{\overset{\varphi}{\to}} \mathbb{D}_R \times G$ such that the action of A on $\mathbb{D}_R \times G$ is given by

$$\gamma \odot (x, g) = (\gamma x, \theta_\gamma(x(0))g),$$

where $x(0)$ is the image of x in Spec R and θ_γ: Spec $R \to G$ is a morphism. Moreover, for any $x^o \in$ Spec R, the group homomorphism $\theta(x^o)\colon A \to G$, $\gamma \mapsto \theta_\gamma(x^o)$, is unique up to a conjugation, which is called the *type of* $\mathscr{E}$ *over* x^o. The proof of Theorem 6.1.9 uses non-abelian group cohomology. In fact, Theorem 6.1.9 is true for any connected affine algebraic group.

Let $\vec{\tau} = (\tau_1, \dots, \tau_s)$ be a set of rational markings, i.e., $\tau_j = \bar{\tau}_j/d_j$, for some positive integers d_j and $\mathrm{Exp}(2\pi i \bar{\tau}_j) = 1$. As in Theorem 6.1.8, we fix a Galois cover $\pi\colon \hat{\Sigma} \to \Sigma$ with signature the pair $\vec{p}$ and the sequence $\vec{d} = (d_1, \dots, d_s)$ with finite Galois group A. We also fix inverse images $\{\hat{p}_j \in \pi^{-1}(p_j)\}_{1\le j\le s}$ and generators $\vec{\gamma} = (\gamma_1, \dots, \gamma_s)$ of the cyclic isotropy groups $(A_{\hat{p}_1}, \dots, A_{\hat{p}_s})$. Thus, $A_{\hat{p}_j}$ is of order d_j. As earlier in Section 1.1, let **Alg** be the category of algebras over $\mathbb{C}$ and **Set** the category of sets. Define the functor $\mathscr{F}^{A,\vec{\tau}}_{G,\hat{\Sigma}^*}$: **Alg** $\to$ **Set** by

$$\begin{aligned}\mathscr{F}^{A,\vec{\tau}}_{G,\hat{\Sigma}^*}(R) = \{(\hat{E}_R, \hat{\sigma}_R) : \;& \hat{E}_R \text{ is an } A\text{-equivariant } G\text{-bundle over } \hat{\Sigma}_R \\ & \text{such that } \hat{E}_{R_{|\hat{\Sigma}\times x}} \text{ has local type } \vec{\tau} \text{ for any } x \in \mathrm{Spec}\, R \text{ and } \hat{\sigma}_R \text{ is an} \\ & A\text{-equivariant section of } \hat{E}_R \text{ over } (\hat{\Sigma}^*)_R\}/\text{ isomorphisms,}\end{aligned}$$

where $\Sigma^* := \Sigma\backslash\vec{p}$, $\hat{\Sigma}^* := \pi^{-1}(\Sigma^*)$, A acts trivially on R and $\hat{\Sigma}_R := \hat{\Sigma} \times$ Spec R.

For any parabolic subgroup P of G, consider the parahoric subgroup scheme $\mathcal{P} \subset \bar{G}((t))$ defined by $\mathcal{P} := ev_0^{-1}(P)$, under the evaluation map $ev_0\colon \bar{G}[[t]] \to G$ at $t = 0$ (cf. Exercise 1.3.E.11). Let t_j be the formal parameter at $p_j \in \Sigma$ defined by identity (1) of Definition 6.1.11. Then, we prove (cf. Theorem 6.1.12) that, if $\theta(\tau_j) < 1$ for all j (for the highest root θ), the functor $\mathscr{F}^{A,\vec{\tau}}_{G,\hat{\Sigma}^*}$ is representable, represented by the ind-scheme $\bar{X}_{\vec{P}} = \Pi^s_{j=1}\bar{X}_G(P_j)$, where $P_j := P(\tau_j)$ is defined in the first paragraph.

Similar to the definition of the stack $\mathbf{Bun}_G(\Sigma)$ as in Chapter 5, define the groupoid fibration over $\mathfrak{S}$ of A-equivariant G-bundles $\mathbf{Bun}^{A,\vec{\tau}}_G(\hat{\Sigma})$ of local type $\vec{\tau}$, whose objects are A-equivariant G-bundles E_S over $\hat{\Sigma} \times S$ (with the trivial action of A on S) such that $E_{S_{|\hat{\Sigma}\times t}}$ (for any $t \in S$) is of local type $\vec{\tau}$ (cf. Definition 6.1.14). Let $\bar{X}_{\vec{P}} := \Pi^s_{j=1}\bar{X}_G(P_j)$ and let $\bar{\Gamma}$ be the ind-affine group variety with $\mathbb{C}$-points $\Gamma := \mathrm{Mor}(\Sigma^*, G)$, where Σ^* and P_j are as in the above paragraph. Then $\bar{\Gamma}$ acts on $\bar{X}_{\vec{P}}$ by the left multiplication on each factor via its Laurent series expansion in the formal coordinates t_j. With this notation, there exists an equivalence of categories between $\mathbf{Bun}^{A,\vec{\tau}}_G(\hat{\Sigma})$ and the quotient stack $\left[\bar{\Gamma}\backslash\bar{X}_{\vec{P}}\right]$ (cf. Theorem 6.1.15). In particular, $\mathbf{Bun}^{A,\vec{\tau}}_G(\hat{\Sigma})$ is isomorphic to the stack $\mathbf{Parbun}_G(\Sigma, \vec{P})$ of quasi-parabolic G-bundles over $(\Sigma, \vec{p})$ of type

$\vec{P} := (P_1, \dots, P_s)$ (defined in Chapter 5) and hence it is a smooth (algebraic) stack. Specializing this result to the fiber over a point, we get (cf. Theorem 6.1.17) that there is a natural set-theoretic bijection between the set $\text{Bun}_G^{A,\vec{\tau}}(\hat{\Sigma})$ of isomorphism classes of A-equivariant G-bundles over $\hat{\Sigma}$ of local type $\vec{\tau}$ and the set $\text{Parbun}_G(\Sigma, \vec{P})$ of isomorphism classes of quasi-parabolic G-bundles of type $\vec{P}$ over $(\Sigma, \vec{p})$. Under this bijection, A-semistable (resp. A-stable) G-bundles over $\hat{\Sigma}$ correspond to the parabolic semistable (resp. parabolic stable) bundles over Σ with respect to the markings $\vec{\tau}$. This reduces the problem of studying the quasi-parabolic moduli stack (resp. parabolic semistable moduli *space*, resp. parabolic stable moduli *space*) of parabolic G-bundles over $(\Sigma, \vec{p})$ to that of the moduli stack (resp. semistable moduli *space*, resp. stable moduli *space*) of (non-parabolic) A-equivariant G-bundles over a cover $\hat{\Sigma}$ of Σ with Galois group A.

Let us assume now that G, more generally, is a connected reductive group and Σ continues to be a smooth irreducible projective curve. In Section 6.2, we prove the existence and uniqueness of Harder–Narasimhan (for short HN) reduction of a G-bundle over Σ. Let $\pi : E \to \Sigma$ be a G-bundle. Then, a P-subbundle $E_P \subset E$ for a standard parabolic subgroup P of G is called a *Harder–Narasimhan reduction* if the associated L-bundle $E_P(L)$, obtained from the P-bundle E_P via the extension of the structure group $P \to P/U \simeq L$, is semistable, where L is the Levi subgroup of P containing H and U is the unipotent radical of P. Moreover, we require that for any nontrivial character λ of P such that $\lambda \in \bigoplus_{i=1}^{\ell} \mathbb{Z}_+\alpha_i$ (in particular, λ is trivial restricted to the identity component of the center of G),

$$\deg\left(E_P \times^P \mathbb{C}_\lambda\right) > 0.$$

By virtue of Theorem 6.2.3, such a reduction exists and is unique. Moreover, for a G-bundle E over Σ, and an embedding of connected reductive groups $G \hookrightarrow G'$, the HN reduction of E coincides with the HN reduction of $E(G')$ intersected with E (cf. Theorem 6.2.6 for a more precise statement). As a consequence, it is shown (cf. Corollary 6.2.7) that if $E(G')$ is semistable, then so is E. Further, if E is semistable and G is not contained in any proper (not necessarily standard) parabolic subgroup of G', then $E(G')$ is semistable. As another consequence of HN reduction, an A-equivariant G-bundle over $\hat{\Sigma}$ is A-semistable if and only if it is semistable (cf. Exercise 6.2.E.4). By virtue of Exercise 6.1.E.15, a vector bundle over Σ is polystable (where polystability is defined in Definition 6.1.4(c)) if and only if it is a direct sum of stable vector bundles of the same slope. In Exercise 6.2.E.2 the HN reduction of vector bundles is discussed.

Section 6.3 is devoted to the classical result of Narasimhan–Seshadri on topological construction of stable and polystable vector bundles over Σ and its generalization to any connected reductive G. For any homomorphism from the fundamental group $\rho\colon \pi_1(\Sigma) \to G$, we get a holomorphic G-bundle

$$E_\rho := \tilde{\Sigma} \times^{\pi_1(\Sigma)} G,$$

where $\tilde{\Sigma}$ is the simply-connected cover of Σ. By the Serre's GAGA principle, E_ρ is an algebraic G-bundle over Σ. If $\operatorname{Im}\rho$ lies in a compact subgroup of G, then ρ is called a *unitary* homomorphism and E_ρ is called a *unitary G-bundle*. The homomorphism ρ is called *irreducible* if $\operatorname{Im}\rho$ is not contained in any proper (not necessarily standard) parabolic subgroup of G. Then, by Proposition 6.3.4, E_ρ is a semistable G-bundle if ρ is unitary. Further, for a unitary ρ, E_ρ is a stable G-bundle if and only if ρ is irreducible. In fact, we prove a generalization of these results for equivariant bundles. It is shown that for a unitary representation V of $\pi_1(\Sigma)$, the subspace $V^{\pi_1(\Sigma)}$ of $\pi_1(\Sigma)$-invariants in V is canonically isomorphic with the space of global sections of the corresponding vector bundle over Σ (cf. Lemma 6.3.6 for its equivariant generalization). This leads to the result that for two unitary homomorphisms ρ, ρ', the corresponding bundles E_ρ and $E_{\rho'}$ are isomorphic if and only if ρ is conjugate to ρ' (cf. Corollary 6.3.7 for its equivariant generalization). A classification of topological G-bundles over Σ is obtained in Lemma 6.3.10. For a unitary representation ρ of $\pi_1(\Sigma)$, the dimension of the group cohomology $H^1(\pi_1(\Sigma), \operatorname{ad}\rho)$ is calculated in Corollary 6.3.14.

Let K be a compact connected Lie group (which we take to be a maximal compact subgroup of G). For any integer $g \geq 1$, let F_g be the free group on the symbols $\{a_1, b_1, a_2, b_2, \ldots, a_g, b_g\}$. Define the map

$$\beta : K^{2g} \to [K,K], \ \big((h_1,k_1),(h_2,k_2),\ldots,(h_g,k_g)\big) \mapsto \Pi_{i=1}^{g}[h_i,k_i].$$

Any $\bar{\rho} = \big((h_1,k_1),\ldots,(h_g,k_g)\big) \in K^{2g}$ determines a group homomorphism $\tilde{\rho}\colon F_g \to K$ taking $a_i \mapsto h_i$ and $b_i \mapsto k_i$. If $\bar{\rho} \in \beta^{-1}(e)$, then the homomorphism $\tilde{\rho}$ descends to a group homomorphism $\rho\colon \pi_1(\Sigma) \to K$, where g is the genus of Σ. For any $\bar{\rho} \in \beta^{-1}(e)$, $\operatorname{Ker}((d\beta)_{\bar{\rho}})$ is determined in Proposition 6.3.15 and identified with the space of 1-cocycles of $\pi_1(\Sigma)$ with coefficients in $\operatorname{ad}\rho$. As a corollary, we get that $M_g(K) := \{\bar{\rho} \in \beta^{-1}(e) : \rho \text{ is irreducible}\}$ is an $\mathbb{R}$-analytic (smooth) manifold of dimension $(2g-1)\dim K + \dim \mathfrak{z}$, where $\mathfrak{z}$ is the center of $\mathfrak{g}$ (cf. Corollary 6.3.16). Moreover, $M_g(K)$ parameterizes an $\mathbb{R}$-analytic family of holomorphic G-bundles over Σ. It is shown in Proposition 6.3.18 that the infinitesimal deformation map for this family is surjective. In particular, this family is complete at each of its points (cf. Theorem 6.3.20).

As proved in Proposition 6.3.30, let $\mathscr{F} \to \Sigma \times T$ be a $\mathbb{C}$-analytic family of stable G-bundles over Σ parameterized by a $\mathbb{C}$-analytic space T. Then, the subset

$$T_u := \{t \in T : \mathscr{F}_t \simeq E_\rho \text{ for some unitary representation } \rho \text{ of } \pi_1(\Sigma) \text{ in } G\}$$

is a closed subset of T. Moreover, for any $\mathbb{C}$-analytic family $\mathscr{F}' \to \Sigma \times T$ of G-bundles, by Lemma 6.3.31 (resp. Exercise 6.3.E.9), the subset $T_s := \{t \in T : \mathscr{F}'_t \text{ is a stable } G\text{-bundle}\}$ (resp. T_{ss} defined as T_s by replacing 'stable' by 'semistable') is an open subset which is complement of a (closed) $\mathbb{C}$-analytic subset of T. Further, for any $\mathbb{R}$-analytic family $\mathscr{F}$ of G-bundles over Σ parameterized by an $\mathbb{R}$-analytic space T,

$$T_o := \{t \in T : \mathscr{F}_t \simeq E_\rho \text{ for some irreducible representation } \rho \text{ of } \pi_1(\Sigma) \text{ in } K\}$$

is an open subset of T (cf. Corollary 6.3.21). The above results lead finally to the following fundamental Theorem 6.3.35.

Theorem Let G be a connected reductive group and let E be a holomorphic G-bundle over a smooth irreducible projective curve Σ of genus $g \geq 2$. Then E is polystable of degree 0 (i.e., $E \times^G \mathbb{C}_\chi$ has degree 0 for any character χ of G) if and only if $E \simeq E_\rho$ (as holomorphic G-bundles) for a unitary representation $\rho : \pi_1(\Sigma) \to G$.

We further have the following equivariant generalization of the Narasimhan–Seshadri Theorem 6.3.35 (cf. Theorem 6.3.41).

Theorem Let $\hat{\Sigma}$ be an irreducible smooth projective curve with faithful action of a finite group A such that $\Sigma := \hat{\Sigma}/A$ has genus $g \geq 2$. Then an A-equivariant G-bundle $\hat{E}$ over $\hat{\Sigma}$ is A-unitary if and only it is A-polystable of degree 0.

In particular, an A-equivariant G-bundle over $\hat{\Sigma}$ is A-polystable if and only if it is polystable.

We also prove the following result (cf. Proposition 6.3.42).

Proposition Let $\hat{E}$ be an A-equivariant G-bundle over $\hat{\Sigma}$ such that $\hat{\Sigma}/A$ has genus ≥ 2 and let $\theta : G \to \mathrm{GL}_V$ be a representation with finite kernel, where G is a connected semisimple group. Then the vector bundle $\hat{E}(V)$ is A-unitary if and only if $\hat{E}$ is A-unitary.

6.1 Identification of Parabolic G-Bundles with Equivariant G-Bundles

Let G be a simple, connected, simply-connected algebraic group over $\mathbb{C}$ and let $(\Sigma, \vec{p})$ be an s-pointed (for any $s \geq 1$) smooth projective irreducible curve (of any genus g), where $\vec{p} = (p_1, \ldots, p_s)$. *Unless otherwise stated to the contrary, this will be our tacit assumption during this Section 6.1.* Fix a maximal compact subgroup K of G. Following the notation from Section 1.2, define the *fundamental alcove*:

$$\Phi_o = \{h \in \mathfrak{h} : \alpha_i(h) \geq 0 \quad \text{and} \quad \theta(h) \leq 1, \quad \text{for all the simple roots} \quad \alpha_i\},$$

where θ is the highest root.

For any semisimple element $x \in \mathfrak{g}$, define the corresponding *Kempf's parabolic subalgebra*

$$\mathfrak{p}(x) := \{v \in \mathfrak{g} : \lim_{t \to -\infty} \mathrm{Ad}(\mathrm{Exp}(tx)) \cdot v \text{ exists in } \mathfrak{g}\},$$

and let $P(x)$ be the corresponding parabolic subgroup of G.

Then, for $h \in \Phi_o$, $P(h)$ is the standard parabolic subgroup such that its Levi subgroup $L(h)$ containing H has for its simple roots $S_h := \{\alpha_i : \alpha_i(h) = 0\}$.

We recall the following well-known result (cf. (Helgason, 1978, Chap. VII, Theorem 7.9)).

Lemma 6.1.1 *The map*

$$\Phi_o \to K/\mathrm{Ad}K, \quad h \mapsto [\mathrm{Exp}(2\pi i h)],$$

is a bijection, where $K/\mathrm{Ad}K$ denotes the set of K-orbits in K under the adjoint action and $[\mathrm{Exp}(2\pi i h)]$ denotes the K-orbit of $\mathrm{Exp}(2\pi i h)$.

Definition 6.1.2 Let $E \to \Sigma$ be a principal G-bundle (cf. Example C.4(d)). A *parabolic structure* on E (with respect to the pointed curve $(\Sigma, \vec{p})$) consists of:

(a) Markings (called *parabolic weights*) $\vec{\tau} = (\tau_1, \ldots, \tau_s)$, for $\tau_j \in \Phi_o$, where τ_j is 'attached' to the point p_j, and
(b) A section σ_j of E_{p_j}/P_j over p_j, for each $1 \leq j \leq s$, where $P_j := P(\tau_j)$ and E_{p_j} is the fiber of E over p_j.
Denote $\vec{\sigma} := (\sigma_1, \ldots, \sigma_s)$.

A G-bundle $E \to \Sigma$ with the above additional structures (a) and (b) is called a *parabolic G-bundle over $(\Sigma, \vec{p})$ with markings $\vec{\tau}$* and denoted by $(E,\vec{\tau},\vec{\sigma})$. Thus, a parabolic G-bundle over $(\Sigma,\vec{p})$ is nothing but a quasi-parabolic G-bundle over $(\Sigma,\vec{p})$ of type $\vec{P} := (P_1,\dots,P_s)$ (cf. Definition 5.1.4) together with the markings $\vec{\tau}$.

Similarly, a family of parabolic G-bundles parameterized by a scheme S is a G-bundle $\mathscr{E}$ over $\Sigma \times S$ consisting of:

(a′) markings $\vec{\tau} = (\tau_1,\dots,\tau_s)$ as in (a), and
(b′) a section σ_j^S of $(\mathscr{E}_{|p_j\times S})/P_j$, for each $1 \le j \le s$.

Let $\mathscr{E}_1$ and $\mathscr{E}_2$ be two families of parabolic G-bundles with the same markings $\vec{\tau}$ (parameterized by schemes S_1 and S_2, respectively). By a morphism $\varphi\colon \mathscr{E}_1 \to \mathscr{E}_2$ of families of parabolic G-bundles, we simply mean a morphism of the underlying quasi-parabolic G-bundles (cf. Definition 5.1.4).

Definition 6.1.3 (a) Let P be a standard parabolic subgroup with the Levi subgroup $L = L_P$ containing the maximal torus H (with Lie algebra $\mathfrak{h}$). Let $S_P \subset \{\alpha_1,\dots,\alpha_\ell\}$ be the set of simple roots for L. Then the set $X(P)$ of characters of P (i.e., algebraic group homomorphisms $P \to \mathbb{G}_m$) can be identified with

$$\mathfrak{h}^*_{\mathbb{Z},P} := \{\lambda \in \mathfrak{h}^* : \lambda(\alpha_i^\vee) \in \mathbb{Z}\ \forall \text{ simple roots } \alpha_i \text{ and } \lambda(\alpha_i^\vee) = 0\ \forall \alpha_i \in S_P\} \tag{1}$$

under $\chi \mapsto \dot{\chi}(1)|_{\mathfrak{h}}$. We often identify χ with $\dot{\chi}(1)|_{\mathfrak{h}}$ and write it additively.

Let $\{\omega_1,\dots,\omega_\ell\}$ denote the set of fundamental weights, i.e.,

$$\omega_i(\alpha_j^\vee) = \delta_{i,j}, \quad 1 \le i,j \le \ell. \tag{2}$$

Then

$$\mathfrak{h}^*_{\mathbb{Z},P} = \bigoplus_{\alpha_i \notin S_P} \mathbb{Z}\omega_i.$$

Recall that the standard maximal parabolic subgroups Q_k are parameterized by $1 \le k \le \ell$, where Q_k is the unique standard parabolic subgroup with

$$S_{Q_k} := \{\alpha_1,\dots,\hat{\alpha}_k,\dots,\alpha_\ell\}.$$

For a standard parabolic subgroup P, let $W_P \subset W$ be the Weyl group of its Levi subgroup L_P.

(b) Let $E \to \Sigma$ be a principal G-bundle and let $f\colon G \to G'$ be a homomorphism of algebraic groups. Then, by $E(G')$ we mean the principal G'-bundle $E \times^G G' \to \Sigma$, where G acts on G' via the left multiplication through the morphism f and G' acts on $E \times^G G'$ via the right multiplication on the G'-factor.

(c) Let $E \to \Sigma$ be a principal G-bundle. For any parabolic subgroup P of G and $\chi \in X(P)$, define the line bundle over E/P:

$$\mathscr{L}_P(\chi) = \mathscr{L}_P(\chi, E) = E \times^P \mathbb{C}_{\chi^{-1}} \to E/P,$$

where $\mathbb{C}_{\chi-1}$ is the 1-dimensional representation of P associated to the character χ^{-1}.

Let $T_v(E/P)$ be the relative tangent bundle of E/P over Σ (consisting of tangent vectors of E/P along the fibers of the bundle $E/P \to \Sigma$). Then, $T_v(E/P)$ can canonically be identified with the vector bundle $E \times^P (\mathfrak{g}/\mathfrak{p})$ over E/P, where $\mathfrak{p} := \text{Lie } P$ and P acts on $\mathfrak{g}/\mathfrak{p}$ via the adjoint action.

Definition 6.1.4 (Semistable bundles) (a) A vector bundle $\mathscr{V}$ over Σ is defined to be *semistable* (resp. *stable*) if for any subbundle $(0) \subsetneq \mathscr{W} \subsetneq \mathscr{V}$,

$$\mu(\mathscr{W}) \le \mu(\mathscr{V}) \text{ (resp. } \mu(\mathscr{W}) < \mu(\mathscr{V})), \tag{1}$$

where the slope $\mu(\mathscr{V}) := \deg(\mathscr{V})/\text{rank}(\mathscr{V})$ and deg denotes the first Chern class.

Thus, a vector bundle $\mathscr{V}$ is semistable (resp. stable) if and only if $\mathscr{V} \otimes \mathscr{L}$ is semistable (resp. stable) for any line bundle $\mathscr{L}$ over Σ.

(b) A G-bundle $E \to \Sigma$ is called *semistable* (resp. *stable*) if for any standard maximal parabolic subgroup Q_k of G $(1 \le k \le \ell)$ and any section μ of $E/Q_k \to \Sigma$,

$$\deg \mu^* \left(\mathscr{L}_{Q_k}(-\omega_k)\right) \le 0 \left(\text{resp. } \deg \mu^* \left(\mathscr{L}_{Q_k}(-\omega_k)\right) < 0\right). \tag{2}$$

Observe that the trivial bundle $\Sigma \times G \to \Sigma$ is semistable.

Alternatively, a G-bundle $E \to \Sigma$ is called *semistable* (resp. *stable*) if for any standard proper parabolic subgroup P of G and any section μ of $E/P \to \Sigma$,

$$\deg \mu^* \left(T_v(E/P)\right) \ge 0 \text{ (resp. } > 0).$$

By Exercise 6.1.E.4, these two definitions are equivalent.

These alternative definitions remain valid for any connected reductive group G provided we take the fundamental weights ω_k to vanish on the center $Z(\mathfrak{g})$ $(\subset \mathfrak{h})$ of $\mathfrak{g}$ and we replace ω_k by some positive multiple $d\omega_k$ so that $d\omega_k$ is a character of T.

By Exercise 6.1.E.5, a vector bundle $\mathscr{V}$ over Σ is semistable (resp. stable) if and only if the associated frame bundle $F(\mathscr{V})$ (which is a principal GL_n-bundle for $n = \text{rank } \mathscr{V}$) is semistable (resp. stable).

(c) As in (b), let G be a connected reductive group. Then, a G-bundle E over Σ is called *polystable* if it has a reduction E_L to a Levi subgroup L

(of a parabolic subgroup P of G) such that the L-bundle E_L is stable and for any character χ of L which is trivial restricted to the center of G, we have

$$\deg \left(E_L \times^L \mathbb{C}_\chi\right) = 0,$$

where $\mathbb{C}_\chi$ is the 1-dimensional representation of L given by the character χ.

A vector bundle $\mathscr{V}$ over Σ of rank r is called *polystable* if the associated frame bundle $F(\mathscr{V})$ is polystable as a GL_r-bundle. By Exercise 6.1.E.15, $\mathscr{V}$ is polystable if and only if it is a direct sum of stable vector bundles all of which have the same slope.

By Theorem 6.1.7, $\operatorname{ad} E$ is polystable if E is so, where $\operatorname{ad} E := E \times^G \mathfrak{g}$. Thus, by Exercise 6.1.E.15, $\operatorname{ad} E$ is semistable and hence E is semistable by Lemma 6.1.5.

(d) Let $(E, \vec{\tau}, \vec{\sigma})$ be a parabolic G-bundle over $(\Sigma, \vec{p})$. Then, it is called *parabolic semistable* (resp. *parabolic stable*) if for any standard maximal parabolic subgroup Q_k $(1 \leq k \leq \ell)$ and any section μ of $E/Q_k \to \Sigma$, we have

$$\deg \mu^* \left(\mathscr{L}_{Q_k}(-\omega_k)\right) + \sum_{j=1}^{s} \omega_k(w_j^{-1}\tau_j) \leq 0 \quad (\text{resp. } < 0), \tag{3}$$

where $\bar{w}_j := W_{P_j} w_j W_{Q_k} \in W_{P_j} \backslash W / W_{Q_k}$ is the unique element such that taking any $e_j \in E_{p_j}$ and writing $\sigma_j = e_j g_j P_j$ and $\mu(p_j) = e_j h_j Q_k$, for some $g_j, h_j \in G$, we have

$$h_j \in g_j P_j w_j Q_k. \tag{4}$$

(It is easy to see that $\bar{w}_j$ does not depend upon the choices of e_j, g_j and h_j. This $\bar{w}_j$ is called the *relative position* of μ with respect to the quasi-parabolic structure at p_j.)

The number on the left side of (3) is called the *parabolic degree* (denoted $\operatorname{pardeg} \mu^* \mathscr{L}_{Q_k}(-\omega_k)$) of the parabolic bundle E with respect to the section μ and the line bundle $\mathscr{L}_{Q_k}(-\omega_k)$ for the parabolic markings $\vec{\tau} = (\tau_1, \ldots, \tau_s)$.

An equivalent characterization of parabolic semistability (resp. parabolic stability) for vector bundles is given in Exercise 6.1.E.7.

Lemma 6.1.5 *Let G be a connected reductive group and let $E \to \Sigma$ be a G-bundle. If the adjoint vector bundle*

$$\operatorname{ad} E := E \times^G \mathfrak{g}$$

is semistable (resp. stable), then so is E.

In fact, by Theorem 6.1.7, we see that if E is semistable, then so is $\operatorname{ad} E$. *Thus, semistability of E is equivalent to that of* $\operatorname{ad} E$.

In general E being stable does not necessarily imply that $\operatorname{ad} E$ *is stable even when G is a simple group (cf. Exercise 6.3.E.10).*

Proof Let P be a standard proper parabolic subgroup of G and let $E_P \subset E$ be a P-subbundle obtained from a section μ of $E/P \to \Sigma$ (cf. Lemma 5.1.2). Now, by definition, $\mu^* (T_v(E/P))$ is a quotient of the adjoint bundle $\operatorname{ad} E$. But, $\deg(\operatorname{ad} E) = 0$ (since G acts trivially on $\wedge^{\mathrm{top}}(\mathfrak{g})$ under the adjoint action) and since $\operatorname{ad} E$ is semistable (resp. stable), by assumption, we get $\deg \mu^* (T_v(E/P)) \geq 0$ (resp. > 0). This proves the lemma. □

Remark 6.1.6 A G-bundle can be thought of as a parabolic G-bundle for $s = 0$. Further, in this case, parabolic semistable (resp. stable) bundle is nothing but a semistable (resp. stable) bundle.

We recall the following result without proof from Ramanan and Ramanathan (1984, Theorem 3.18). The proof in the same has a gap, but a modified proof is given in Balaji and Parameswaran (2003, Proposition 6 and Remarks 17, 18).

Theorem 6.1.7 *Let $f : G \to G'$ be a homomorphism between connected reductive groups such that $f(Z^o(G)) \subset Z^o(G')$, where $Z^o(G)$ denotes the identity component of the center of G. Then, if $E \to \Sigma$ is a semistable (resp. polystable) G-bundle, then so is $E(G')$ obtained from E by extension of the structure group to G' (cf. Definition 6.1.3(b)).*

In particular, for any semistable (resp. polystable) G-bundle E, $\operatorname{ad} E$ *is a semistable (resp. polystable) vector bundle (cf. Exercise 6.1.E.5).*

We recall the following result. To prove the result, by Selberg (1960, Lemma 8), any finitely generated linear group Γ has a normal torsion-free subgroup Γ_o of finite index in Γ. Moreover, observe that if Γ acts faithfully on the upper half plane $\mathbb{H}$ (resp. $\mathbb{A}^1(\mathbb{C})$) with all its Γ-orbits closed and the action of Γ is properly discontinuous on a nonempty Γ-stable open subset, then Γ_o acts fixed point freely on $\mathbb{H}$ (resp. $\mathbb{A}^1(\mathbb{C})$). To prove this, realize $\mathbb{H} = \mathrm{SL}_2(\mathbb{R})/\mathrm{SO}_2$ and thus $\Gamma \subset \mathrm{PSL}_2(\mathbb{R})$ in this case. In the case of $\mathbb{A}^1(\mathbb{C})$, observe that the group of variety automorphisms of the affine line:

$$\operatorname{Aut}(\mathbb{A}^1(\mathbb{C})) = \left\{ \begin{bmatrix} a & b \\ 0 & 1 \end{bmatrix} : a \in \mathbb{C}^*, b \in \mathbb{C} \right\}$$

acting on $\mathbb{A}^1(\mathbb{C}) = \{[z : 1] : z \in \mathbb{C}\}$ as a subset of $\mathbb{P}^1(\mathbb{C})$. An element $\gamma = \begin{bmatrix} a & b \\ 0 & 1 \end{bmatrix} \in \operatorname{Aut}(\mathbb{A}^1(\mathbb{C}))$ is of infinite order if and only if either $a = 1$ and $b \neq 0$ or $a \in \mathbb{C}^*$ is of infinite order (in the multiplicative group). Now, using the results from Serre (1992, §6.4) the following result is obtained.

Theorem 6.1.8 *Let $(\Sigma, \vec{p})$ be a smooth irreducible projective s-pointed curve for $s \geq 1$ and let $\vec{d} = (d_1, \dots, d_s)$ be a set of integers $d_i \geq 2$ attached to $\vec{p}$.*

$$\text{We assume that if } \Sigma = \mathbb{P}^1, \text{ then } s \geq 3. \tag{1}$$

Then there exists a smooth irreducible projective curve $\hat{\Sigma}$ and a Galois cover $\pi : \hat{\Sigma} \to \Sigma$ with finite Galois group A such that A acts freely on $\hat{\Sigma} \backslash \pi^{-1}(\bar{p})$ and the isotropy subgroup $A_{\hat{p}_i}$ (for any $\hat{p}_i \in \pi^{-1}(p_i)$) is cyclic of order d_i, where $\bar{p} \subset \Sigma$ denotes the subset $\{p_1, \dots, p_s\}$. The set $\vec{p}$ together with $\vec{d}$ is called signature on Σ.

Conversely, any smooth irreducible projective curve $\hat{\Sigma}$ with faithful action of a finite group A gives rise to such an example by taking $\Sigma = \hat{\Sigma}/A$ and $\vec{p} = (p_1, \dots, p_s)$ in Σ consists of ramification points. Here, $\vec{d} = (d_1, \dots, d_s)$ is the set of integers ≥ 2 such that d_i is the order of the isotropy group for any point $\hat{p}_i \in \hat{\Sigma}$ over p_i.

Even though, given $(\Sigma, \vec{p})$ and $\vec{d}$, $\hat{\Sigma}$ is *not* unique, we will fix one such $\hat{\Sigma}$ in the sequel.

Let A be a finite group acting on the formal disc $\mathbb{D} := \mathrm{Spec}(\mathbb{C}[[t]])$ and let $R \in \mathbf{Alg}$ (cf. Section 1.1). Then, A acts on $\mathbb{D}_R := \mathrm{Spec}(R[[t]])$ with the trivial action of A on R, by observing that $R[[t]] = \varprojlim_n \big(R \otimes_{\mathbb{C}} (\mathbb{C}[[t]]/\langle t^n \rangle)\big)$.

Theorem 6.1.9 *Let G be any connected affine algebraic group (not necessarily semisimple) and let $\mathscr{E} \to \mathbb{D}_R$ be an A-equivariant principal G-bundle, which is trivial as a G-bundle. Then, there exists a G-bundle trivialization of $\mathscr{E}$ in which the A-action is the product action, in the sense that there exists an A-equivariant G-bundle isomorphism inducing the identity on the base:*

$$\mathscr{E} \underset{\sim}{\xrightarrow{\varphi}} \mathbb{D}_R \times G$$

such that the action of A on $\mathbb{D}_R \times G$ is given by

$$\gamma \odot (x, g) = (\gamma x, \theta_\gamma(x(0))g), \text{ for } \gamma \in A, x \in \mathbb{D}_R \text{ and } g \in G, \tag{1}$$

where $x(0)$ is the image of x in $\mathrm{Spec}\, R$ induced from the embedding $R \to R[[t]]$ and $\theta_\gamma : \mathrm{Spec}\, R \to G$ is a morphism.

Moreover, for any $x^o \in \mathrm{Spec}\, R$, the group homomorphism $\theta(x^o) : A \to G$, $\gamma \mapsto \theta_\gamma(x^o)$, is unique up to a conjugation, which is called the type of $\mathscr{E}$ over x^o.

If $R = \mathbb{C}$, so that $\mathrm{Spec}\, R$ is a point, we simply call θ the type of $\mathscr{E}$.

Proof Pick any G-bundle trivialization $\mathscr{E} \xrightarrow[\sim]{\beta} \mathbb{D}_R \times G$. Then, the action of A transports via β to an action given by

$$\gamma \cdot (x,g) = (\gamma x, \alpha_\gamma(\gamma x) g), \quad \text{for } \gamma \in A, x \in \mathbb{D}_R, g \in G, \tag{2}$$

where $\alpha_\gamma : \mathbb{D}_R \to G$ is a morphism. Since $\mathscr{E}$ is an A-equivariant G-bundle, we get

$$\alpha_{\gamma_1\gamma_2}(x) = \alpha_{\gamma_1}(x)\alpha_{\gamma_2}(\gamma_1^{-1}x), \quad \text{for } \gamma_1, \gamma_2 \in A, x \in \mathbb{D}_R. \tag{3}$$

Thus, thinking of α_γ as an element of $G(R[[t]])$, we get a 1-cochain $\alpha : A \to G(R[[t]])$, $\gamma \mapsto \alpha_\gamma$, for the group A with coefficients in $G(R[[t]])$, with the trivial action of A on G. Moreover, α is a 1-cocycle by (3) (cf. (Serre, 1997, Chap. I, §5.1)).

Evaluation at $t = 0$ gives rise to an A-equivariant algebra homomorphism $R[[t]] \to R$ and hence an A-equivariant group homomorphism

$$e^o : G(R[[t]]) \to G(R).$$

Composing $e^o \circ \alpha$, we get a 1-cocycle

$$\alpha^o : A \to G(R) \hookrightarrow G(R[[t]]).$$

Let $G(R[[t]])^+$ be the kernel of e^o. Clearly, e^o is surjective (due to the inclusion $G(R) \hookrightarrow G(R[[t]])$). The exact sequence

$$1 \to G(R[[t]])^+ \to G(R[[t]]) \xrightarrow{e^o} G(R) \to 1$$

gives rise to an exact sequence of pointed sets in non-abelian group cohomology (cf. (Serre, 1997, Proposition 38, §5.5)):

$$H^1(A, G(R[[t]])^+) \to H^1(A, G(R[[t]])) \xrightarrow{\hat{e^o}} H^1(A, G(R)). \tag{4}$$

We next show that $\hat{e}^o$ is a one-to-one map. To prove this, by Serre (1997, Chap. I, Corollary 2, §5.5), it suffices to show that for any 1-cocycle $\beta : A \to G(R[[t]])$,

$$H^1(A, G(R[[t]])^+_\beta) \text{ is trivial,} \tag{5}$$

where $G(R[[t]])^+_\beta$ denotes the same group $G(R[[t]])^+$ but with a twisted action of A via β:

$$\gamma \odot_\beta f = \beta(\gamma)(\gamma \cdot f)\beta(\gamma)^{-1}, \text{ for } \gamma \in A \text{ and } f \in G(R[[t]])^+.$$

We first prove by induction on $n \geq 1$ that

$$H^1\left(A, G\left(R[[t]]/\langle t^n\rangle\right)^+_\beta\right) \quad \text{is trivial,} \tag{6}$$

where $G\left(R[[t]]/\langle t^n\rangle\right)^+$ is the kernel of the surjective homomorphism $G\left(R[[t]]/\langle t^n\rangle\right) \to G(R)$ and $G\left(R[[t]]/\langle t^n\rangle\right)^+_\beta$ denotes the same group with twisted action of A via the image of β in $G\left(R[[t]]/\langle t^n\rangle\right)$. Clearly, (6) for $n = 1$ is trivial. Now, consider the exact sequence of A-groups:

$$1 \to (K_n(R))_\beta \to G\left(R[[t]]/\langle t^{n+1}\rangle\right)^+_\beta \xrightarrow{\pi_n^R} G\left(R[[t]]/\langle t^n\rangle\right)^+_\beta \to 1, \tag{7}$$

where $K_n(R)$ is the kernel of π_n^R. By Exercise 6.1.E.1, π_n^R is surjective with kernel isomorphic (as a group) to the $\mathbb{C}$-vector space $R \otimes_{\mathbb{C}} \left(\mathfrak{g} \otimes_{\mathbb{C}} \frac{t^n\mathbb{C}[[t]]}{t^{n+1}\mathbb{C}[[t]]}\right)$. Next, observe that any element $\gamma \in A$ acts on $(K_n(R))_\beta$ via a $\mathbb{C}$-linear isomorphism. Thus, by Hochschild and Serre (1953, Proposition 6),

$$H^1(A, (K_n(R))_\beta) = 0. \tag{8}$$

From the cohomology sequence (analogue of (4)) associated to the coefficient sequence (7) of A-groups, and using (6) (valid by the induction hypothesis) and (8), we get that

$$H^1\left(A, G\left(R[[t]]/\langle t^{n+1}\rangle\right)^+_\beta\right) = 0,$$

completing the induction and hence (6) is proved for all $n \geq 1$ and any 1-cocycle $\beta\colon A \to G(R[[t]])$.

Since $M_N(R[[t]]) \simeq \varprojlim_n M_N\left(R[[t]]/\langle t^n\rangle\right)$, by considering an embedding $G \hookrightarrow M_N$ and the equations defining G, it is easy to see that

$$G(R[[t]])^+ \simeq \varprojlim G\left(R[[t]]/\langle t^n\rangle\right)^+. \tag{9}$$

Consider the isomorphism of varieties induced from the exponential map (cf. Exercise 6.1.E.1):

$$\operatorname{Exp}\colon \mathfrak{g} \otimes \left(t\mathbb{C}[[t]]/\langle t^n\rangle\right) \to G\left(\mathbb{C}[[t]]/\langle t^n\rangle\right)^+.$$

It induces a bijection

$$\begin{aligned}\mathfrak{g} \otimes \left(tR[[t]]/\langle t^n\rangle\right) &\simeq \operatorname{Mor}\left(\operatorname{Spec} R, \mathfrak{g} \otimes \left(t\mathbb{C}[[t]]/\langle t^n\rangle\right)\right)\\ &\simeq \operatorname{Mor}\left(\operatorname{Spec} R, G\left(\mathbb{C}[[t]]/\langle t^n\rangle\right)^+\right)_\beta, \quad f \mapsto \operatorname{Exp} \circ f\\ &\overset{\theta}{\underset{\sim}{\to}} G\left(R[[t]]/\langle t^n\rangle\right)^+_\beta,\end{aligned}$$

where the bijection θ is obtained by using Exercises 1.3.E.10 and 1.3.E.6.

The bijection θ allows us to transport the action of A on $G\,(R[[t]]/\langle t^n\rangle)^+_\beta$ to that on $\mathfrak{g}\otimes(tR[[t]]/\langle t^n\rangle)$. Moreover, it is easy to see that any $\gamma\in A$ acts on $\mathfrak{g}\otimes(tR[[t]]/\langle t^n\rangle)$ via a $\mathbb{C}$-linear isomorphism. Thus, $\left[\mathfrak{g}\otimes(tR[[t]]/\langle t^n\rangle)\right]^A$ is a linear subspace. From this we immediately see that the canonical map

$$\left[G\left(R[[t]]/\langle t^{n+1}\rangle\right)^+_\beta\right]^A\to\left[G\left(R[[t]]/\langle t^n\rangle\right)^+_\beta\right]^A\quad\text{is surjective.}$$

Thus, by Exercise 6.1.E.2, (6) and (9), we get

$$H^1\left(A,G(R[[t]])^+_\beta\right)=0,$$

for any 1-cocycle $\beta\colon A\to G(R[[t]])$. This proves (5) and hence the map (cf. (4))

$$\hat{e}^o\colon H^1(A,G(R[[t]]))\to H^1(A,G(R))\quad\text{is one-to-one.}$$

We return to the 1-cocycle α as at the beginning of the proof. Clearly, $\hat{e}^o([\alpha])=\hat{e}^o([\alpha^o])$, where $[\alpha]$, $[\alpha^o]\in H^1(A,G(R[[t]])$ denote the cohomology classes of α and α^o, respectively. Since $\hat{e}^o$ is one-to-one, we get

$$[\alpha]=[\alpha^o],\tag{10}$$

i.e., there exists a $\tau\in G(R[[t]])=\mathrm{Mor}(\mathbb{D}_R,G)$ such that

$$\begin{aligned}\tau(\gamma x)^{-1}\alpha_\gamma(\gamma x)\tau(x)=\alpha_\gamma(x(0)),\quad&\text{for all }x\in\mathbb{D}_R,\gamma\in A,\\ \text{since }(\gamma x)(0)=x(0)&\end{aligned}\tag{11}$$

(cf. (Serre, 1997, Chap. I, §5.1)).

Define a G-bundle isomorphism

$$\mathbb{D}_R\times G\xrightarrow{\hat{\tau}}\mathbb{D}_R\times G,\quad(x,g)\mapsto(x,\tau(x)^{-1}g).$$

Then, the action of A on the range transported via $\hat{\tau}$ (to be denoted $\odot$) becomes (cf. (2))

$$\begin{aligned}\gamma\odot(x,g)&=\hat{\tau}(\gamma\cdot\hat{\tau}^{-1}(x,g))\\&=\hat{\tau}(\gamma\cdot(x,\tau(x)g))\\&=\hat{\tau}(\gamma x,\alpha_\gamma(\gamma x)\tau(x)g)\\&=(\gamma x,\tau(\gamma x)^{-1}\alpha_\gamma(\gamma x)\tau(x)g)\\&=(\gamma x,\alpha_\gamma(x(0))g),\quad\text{by (11).}\end{aligned}$$

Taking $\theta_\gamma=e^o(\alpha_\gamma)$, we get the first part of the theorem.

To prove the uniqueness of $\theta(x^o)$ up to a conjugation for any $x^o \in \operatorname{Spec} R$, let

$$\mathbb{D}_R \times G \xrightarrow{\delta} \mathbb{D}_R \times G, \ \ \delta(x,g) = (x, \bar{\delta}(x)g), \quad \text{for} \quad x \in \mathbb{D}_R, g \in G,$$

be an A-equivariant G-bundle isomorphism such that A acts on the domain by (1) and on the range by

$$\gamma \odot' (x,g) = (\gamma x, \theta'_\gamma(x(0))g), \quad \text{for} \quad \gamma \in A, x \in \mathbb{D}_R \quad \text{and} \quad g \in G,$$

where $\bar{\delta}\colon \mathbb{D}_R \to G$ is a morphism. In particular, for $x^o \in \operatorname{Spec} R \subset \operatorname{Spec} R[[t]]$,

$$\begin{aligned} \delta(\gamma \odot (x^o, g)) &= \delta(\gamma x^o, \theta_\gamma(x^o)g) \\ &= (\gamma x^o, \bar{\delta}(\gamma x^o)\theta_\gamma(x^o)g) \\ &= (x^o, \bar{\delta}(x^o)\theta_\gamma(x^o)g), \end{aligned} \tag{12}$$

since A acts trivially on $\operatorname{Spec} R$. On the other hand, from the A-equivariance of δ, we get

$$\begin{aligned} \delta(\gamma \odot (x^o, g)) &= \gamma \odot' \delta((x^o, g)) \\ &= \gamma \odot' (x^o, \bar{\delta}(x^o)g) \\ &= (x^o, \theta'_\gamma(x^o)\bar{\delta}(x^o)g). \end{aligned} \tag{13}$$

Comparing (12) and (13), we get

$$\bar{\delta}(x^o)\theta_\gamma(x^o)\bar{\delta}(x^o)^{-1} = \theta'_\gamma(x^o).$$

Thus, $\theta'(x^o)\colon A \to G$ is a conjugate of $\theta(x^o)$, proving the theorem. □

The above theorem justifies the following.

Definition 6.1.10 Let G be as at the beginning of this section.

(a) Let $(\Sigma, \vec{p})$ be an s-pointed curve as in Theorem 6.1.8 (in particular, it satisfies (1) of Theorem 6.1.8) and let $\vec{d} = (d_1, \ldots, d_s)$ be a set of positive integers attached to $\vec{p}$. Fix a Galois cover $\pi : \hat{\Sigma} \to \Sigma$ with Galois group A as guaranteed by Theorem 6.1.8. We also fix preimages $\vec{\hat{p}} = (\hat{p}_1, \ldots, \hat{p}_s)$ in $\hat{\Sigma}$ of $\vec{p}$ and generators $\vec{\gamma} = (\gamma_1, \ldots, \gamma_s)$ of the isotropy groups $(A_{\hat{p}_1}, \ldots, A_{\hat{p}_s})$. Observe that $A_{\hat{p}_i}$ are cyclic groups, being subgroups of $\operatorname{Aut}(T_{\hat{p}_i}(\hat{\Sigma}))$.

For any A-equivariant principal G-bundle $\hat{E}$ over $\hat{\Sigma}$, $\hat{E}_{|\mathbb{D}_{\hat{p}_j}}$ is trivial as a G-bundle (e.g., by Theorem 5.2.5), where $\mathbb{D}_{\hat{p}_j} \subset \hat{\Sigma}$ is the formal disc around $\hat{p}_j$. Since $\hat{p}_j$ is fixed by $A_{\hat{p}_j}$ (in particular, it acts on $\mathbb{D}_{\hat{p}_j}$), $\hat{E}_{|\mathbb{D}_{\hat{p}_j}}$ is an $A_{\hat{p}_j}$-equivariant trivial G-bundle. Thus, by Theorem 6.1.9, we get a homomorphism (the type of $\hat{E}_{|\mathbb{D}_{\hat{p}_j}}$) $\theta_j\colon A_{\hat{p}_j} \to G$ (unique up to a conjugation). Moreover, any conjugate of θ_j can be realized as θ_j with respect to some G-bundle trivialization of $\hat{E}_{|\mathbb{D}_{\hat{p}_j}}$. Let $\tau_j \in \Phi_o$ be the unique element such that $\operatorname{Exp}(2\pi i \tau_j)$ is conjugate to $\theta_j(\gamma_j)$ (cf. Lemma 6.1.1). Define the *local type* of $\hat{E}$ to be the sequence

$$\vec{\tau} = (\tau_1, \ldots, \tau_s).$$

Observe that τ_j *does* depend upon the choice of the generator γ_j of $A_{\hat{p}_j}$.

(b) Let $(\Sigma, \vec{p})$ be an s-pointed curve as in Theorem 6.1.8 (in particular, it satisfies (1) of Theorem 6.1.8). Let $\vec{\tau} = (\tau_1, \ldots, \tau_s)$ be a set of markings (cf. Definition 6.1.2) with τ_j rational points of Φ_o, i.e., we can write $\tau_j = \bar{\tau}_j / d_j$, for some positive integers d_j and $\operatorname{Exp}(2\pi i \bar{\tau}_j) = 1$.

As in Theorem 6.1.8, we fix a Galois cover $\pi : \hat{\Sigma} \to \Sigma$ with finite Galois group A associated to $(\Sigma, \vec{p})$ and the sequence $\vec{d} = (d_1, \ldots, d_s)$, ignoring those p_i with $d_i = 1$. We also fix inverse images $\{\hat{p}_j \in \pi^{-1}(p_j)\}_{1 \le j \le s}$ and generators $\vec{\gamma} = (\gamma_1, \ldots, \gamma_s)$ of the cyclic isotropy groups $(A_{\hat{p}_1}, \ldots, A_{\hat{p}_s})$.

We make the following definition similar to Definition 5.2.6.

Definition 6.1.11 With the above notation; in particular, $s \ge 1$ and $s \ge 3$ if $\Sigma = \mathbb{P}^1$, define the functor $\mathscr{F}^{A,\vec{\tau}}_{G,\hat{\Sigma}^*} : \mathbf{Alg} \to \mathbf{Set}$ by

$$\begin{aligned}\mathscr{F}^{A,\vec{\tau}}_{G,\hat{\Sigma}^*}(R) = \{(\hat{E}_R, \hat{\sigma}_R) :\ & \hat{E}_R \text{ is an } A\text{-equivariant principal } G\text{-bundle over } \hat{\Sigma}_R \\ & \text{such that } \hat{E}_{R_{|\hat{\Sigma} \times x}} \text{ has local type } \vec{\tau} \text{ for any } x \in \operatorname{Spec} R \text{ and } \hat{\sigma}_R \text{ is an} \\ & A\text{-equivariant section of } \hat{E}_R \text{ over } (\hat{\Sigma}^*)_R\}/\sim,\end{aligned}$$

where $\Sigma^* := \Sigma \backslash \{p_1, \ldots, p_s\}$, $\hat{\Sigma}^* := \pi^{-1}(\Sigma^*)$, A acts trivially on R and $(\hat{E}_R, \hat{\sigma}_R) \sim (\hat{E}'_R, \hat{\sigma}'_R)$ if there exists an isomorphism $\hat{\theta}_R$ of A-equivariant G-bundles:

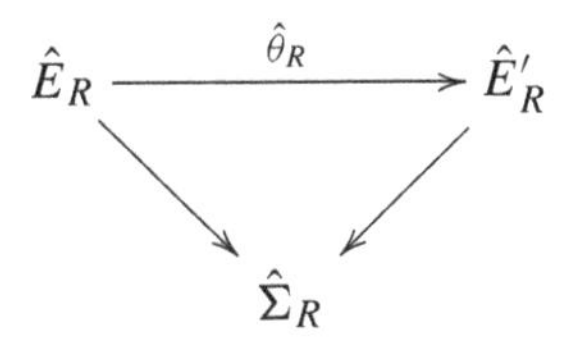

such that $\hat{\theta}_R \circ \hat{\sigma}_R = \hat{\sigma}'_R$. We denote the equivalence class of $(\hat{E}_R, \hat{\sigma}_R)$ by $[\hat{E}_R, \hat{\sigma}_R]$.

Choose a local parameter $\hat{t}_j$ of $\hat{\Sigma}$ around $\hat{p}_j$ such that the generator γ_j of the isotropy group $A_{\hat{p}_j}$ acts on the function $\hat{t}_j$ via

$$\gamma_j \cdot \hat{t}_j = e^{-\frac{2\pi i}{d_j}} \hat{t}_j \quad \text{and} \quad t_j := (\hat{t}_j)^{d_j} \tag{1}$$

is a local parameter for Σ at p_j. Such a local parameter $\hat{t}_j$ exists.

For any one-parameter subgroup $\tau\colon \mathbb{G}_m \to G$, recall the definition of *Kempf's parabolic subgroup*

$$Q(\tau) := \left\{ g \in G : \lim_{z\to 0} \tau(z) g \tau(z)^{-1} \text{ exists in } G \right\}. \tag{2}$$

If semisimple $x \in \mathfrak{g}$ is such that $\mathrm{Exp}(2\pi i d x) = 1$ for some positive integer d, then

$$\bar{\sigma}_x : \mathbb{C} \to G, \ t \mapsto \mathrm{Exp}(dtx)$$

descends to a one-parameter subgroup $\sigma_x\colon \mathbb{C}^* \to G$, where $\mathbb{C}^* := \mathbb{C}/2\pi i \mathbb{Z}$. In this case, it is easy to see that $\mathrm{Lie}(Q(\sigma_x)) = \mathfrak{p}(x)$, where $\mathfrak{p}(x)$ is defined at the beginning of this section.

For any parabolic subgroup P of G, define the parahoric subgroup scheme $\mathcal{P} \subset \bar{G}((t))$ as in Exercise 1.3.E.11 by

$$\mathcal{P} := ev_0^{-1}(P), \ \text{ under the evaluation map } \ ev_0\colon \bar{G}[[t]] \to G \ \text{ at } t = 0. \tag{3}$$

Analogous to Proposition 5.2.7, we have the following.

Theorem 6.1.12 *Let the notation and assumptions be as in the above definition. Assume further that $\theta(\tau_j) < 1$ for each $1 \le j \le s$, where θ is the highest root of G. Then, the functor $\mathscr{F}^{A,\vec{\tau}}_{G,\hat{\Sigma}^*}$ is representable, represented by the ind-scheme (cf. Exercise 1.3.E.11)*

$$\bar{X}_{\vec{P}} = \Pi_{j=1}^{s} \bar{X}_G(P_j),$$

where P_j is the standard parabolic subgroup $P(\tau_j)$ of G as at the beginning of this section and $\bar{X}_G(P_j)$ is the partial infinite flag variety, which is an ind-projective variety as in Exercise 1.3.E.11.

Proof We need to prove that for any $R \in \mathbf{Alg}$, $\mathscr{F}^{A,\vec{\tau}}_{G,\hat{\Sigma}^*}(R)$ is canonically isomorphic with $\bar{X}_{\vec{P}}(R) = \mathrm{Mor}(\mathrm{Spec}\, R, \bar{X}_{\vec{P}})$.

Define the map $\mathfrak{H}\colon \mathscr{F}^{A,\vec{\tau}}_{G,\hat{\Sigma}^*}(R) \to \bar{X}_{\vec{P}}(R)$ as follows. Let $\hat{\mathbb{D}}_j := \mathrm{Spec}\, \mathbb{C}[[\hat{t}_j]]$ be the formal disc around $\hat{p}_j$. Let $[\hat{E}_R, \hat{\sigma}_R] \in \mathscr{F}^{A,\vec{\tau}}_{G,\hat{\Sigma}^*}(R)$. Recall that there exists an algebra $R' \in \mathbf{Alg}$ and a surjective étale morphism $\varphi\colon \mathrm{Spec}\, R' \to \mathrm{Spec}\, R$ such that the G-bundle $\hat{E}_{R'}{}_{|(\hat{\mathbb{D}}_j)_{R'}}$ is trivial for each $1 \le j \le s$, where $\bar{\varphi} := \mathrm{Id}_{\hat{\Sigma}} \times \varphi\colon \hat{\Sigma}_{R'} \to \hat{\Sigma}_R$ and $\hat{E}_{R'} := \bar{\varphi}^*(\hat{E}_R)$ (cf. Theorem 5.2.5). Moreover, by Theorem 6.1.9, we can assume that the action of $A_{\hat{p}_j}$ on $\hat{E}_{R'}{}_{|(\hat{\mathbb{D}}_j)_{R'}}$ is the 'product action' in the sense that there exists a section $\hat{\mu}_j = \hat{\mu}_{j,R'}$ of $\hat{E}_{R'}{}_{|(\hat{\mathbb{D}}_j)_{R'}}$ such that the generator γ_j of the stabilizer $A_{\hat{p}_j} \subset A$ acts on $\hat{\mu}_j$ via

$$\gamma_j \cdot \hat{\mu}_j = \hat{\mu}_j \cdot \mathrm{Exp}(2\pi i \tau_j). \tag{1}$$

Write as sections

$$\hat{\mu}_j^* = \hat{\sigma}_{R'}^* \cdot \hat{\beta}_j, \quad \text{for} \quad \hat{\beta}_j \in G(R'((\hat{t}_j))), \tag{2}$$

where $\hat{\sigma}_{R'}$ is the section of $\hat{E}_{R'_{|(\hat{\Sigma}^*)_{R'}}}$ obtained from the pull-back of $\hat{\sigma}_R$, and

$$\hat{\sigma}_{R'}^* := \hat{\sigma}_{R'_{|\operatorname{Spec} R'((\hat{t}_j))}}, \quad \hat{\mu}_j^* := \hat{\mu}_{j|\operatorname{Spec} R'((\hat{t}_j))}.$$

By (1), it is easy to see that

$$\gamma_j \cdot \hat{\beta}_j = \hat{\beta}_j \cdot \mathrm{Exp}(2\pi i \tau_j). \tag{3}$$

Define the transition function

$$\beta_j := \hat{\beta}_j \cdot (\hat{t}_j)^{\bar{\tau}_j} \in G(R'((\hat{t}_j))). \tag{4}$$

From identity (1) of Definition 6.1.11 and identity (3), it is easy to see that

$$\gamma_j \cdot \beta_j = \beta_j. \tag{5}$$

Thus, β_j descends to an element of $G(R'((t_j)))$. If we take a different section $\hat{\mu}'_j$ of $\hat{E}_{R'_{|(\hat{\mathbb{D}}_j)_{R'}}}$, then we can write

$$\hat{\mu}'_j = \hat{\mu}_j \cdot \hat{f}_j, \quad \text{for some} \quad \hat{f}_j \in G(R'[[\hat{t}_j]]).$$

Hence,

$$\hat{\beta}_j \hat{f}_j = \hat{\beta}'_J.$$

Moreover, if $\hat{\mu}'_j$ also satisfies (1), then we see that

$$\gamma_j \cdot \hat{f}_j = \mathrm{Exp}(2\pi i \tau_j)^{-1} \cdot \hat{f}_j \cdot \mathrm{Exp}(2\pi i \tau_j). \tag{6}$$

Conversely, for any $\hat{f}_j \in G(R'[[\hat{t}_j]])$ satisfying (6), the section $\hat{\mu}_j \cdot \hat{f}_j$ of $\hat{E}_{R'_{|(\hat{\mathbb{D}}_j)_{R'}}}$ satisfies condition (1). Let

$$f_j := (\hat{t}_j)^{-\bar{\tau}_j} \cdot \hat{f}_j \cdot (\hat{t}_j)^{\bar{\tau}_j}.$$

Then, by (6) and identity (1) of Definition 6.1.11,

$$\gamma_j \cdot f_j = f_j. \tag{7}$$

Thus, $f_j \in G(R'((t_j)))$. We next claim that

$$f_j \in \mathcal{P}_j(R') := ev_0^{-1}(P_j(R')), \tag{8}$$

where $ev_0\colon G(R'[[t_j]]) \to G(R')$ is the map induced from the evaluation at $t_j = 0$ (cf. Exercise 1.3.E.11). Let

$$\hat{f}_j^o := \widehat{ev}_0(\hat{f}_j), \quad \text{where} \quad \widehat{ev}_0 : G(R'[[\hat{t}_j]]) \to G(R').$$

Then, by (6), $\hat{f}_j^o \in Z_{\mathrm{Exp}(2\pi i\tau_j)}(R')$, where $Z_{\mathrm{Exp}(2\pi i\tau_j)}$ is the centralizer scheme of $\mathrm{Exp}(2\pi i\tau_j)$ in G and $Z_{\mathrm{Exp}(2\pi i\tau_j)}(R')$ is its R'-rational points. Now, $Z_{\mathrm{Exp}(2\pi i\tau_j)}$ is the Levi subgroup L_j of G containing H with roots $\beta \in \Delta$ such that $\beta(\tau_j) = 0$, where $\Delta \subset \mathfrak{h}^*$ is the set of roots of G. (We have used the assumption here that $|\beta(\tau_j)| < 1$ for all $\beta \in \Delta$) Thus, we get

$$(\hat{t}_j)^{-\bar{\tau}_j} \cdot \hat{f}_j^o \cdot (\hat{t}_j)^{\bar{\tau}_j} = \hat{f}_j^o \quad \text{and} \quad \hat{f}_j^o \in L_j(R') \subset P_j(R'). \tag{9}$$

Think of $\hat{f}_j^o \in G(R') \subset G(R'[[\hat{t}_j]])$. Then $\hat{\zeta}_j := (\hat{f}_j^o)^{-1} \cdot \hat{f}_j\colon \mathrm{Spec}(R'[[\hat{t}_j]]) \to G$ has image inside the big cell $H \times \Pi_{\alpha\in\Delta} U_\alpha$ (fixing an ordering of Δ so that all the positive roots appear first and then all the negative roots or vice versa), where U_α is the one-parameter unipotent subgroup corresponding to the root α. (To prove this observe that $((\hat{f}_j^o)^{-1} \cdot \hat{f}_j)_{|\mathrm{Spec}\,R'}$ is the constant map going to $e \in G$.) Decompose the morphism $\hat{\zeta}_j = (\hat{\zeta}_j(0), \hat{\zeta}_j(\alpha))_{\alpha\in\Delta}$, where $\hat{\zeta}_j(0)$ (resp. $\hat{\zeta}_j(\alpha)$) is the component of $\hat{\zeta}_j$ in H (resp. U_α). Then, for any $\alpha \in \Delta$,

$$\zeta_j(\alpha) := (\hat{t}_j)^{-\bar{\tau}_j} \cdot \hat{\zeta}_j(\alpha) \cdot (\hat{t}_j)^{\bar{\tau}_j} \in (\hat{t}_j)^{-\alpha(\bar{\tau}_j)+1} R'[[\hat{t}_j]], \tag{10}$$

where we have identified $\epsilon_\alpha\colon \mathbb{G}_a \xrightarrow{\sim} U_\alpha$ satisfying $h\epsilon_\alpha(z)h^{-1} = \epsilon_\alpha(\alpha(h)z)$, for any $z \in \mathbb{G}_a$ and $h \in H$ (cf. (Jantzen, 2003, Part II, §1.2)). (Observe that the '+1' in the exponent of $\hat{t}_j$ in (10) appears due to the fact that $\hat{\zeta}_{j|\mathrm{Spec}\,R'}$ is the constant map with image e and hence $\hat{\zeta}_j(\alpha) \in \hat{t}_j R'[[\hat{t}_j]]$.)

By (7) and (9) (since $\gamma_j \cdot \hat{f}_j^o = \hat{f}_j^o$) we get

$$\gamma_j \cdot \zeta_j = \zeta_j, \quad \text{where} \quad \zeta_j := (\hat{t}_j)^{-\bar{\tau}_j} \cdot \hat{\zeta}_j \cdot (\hat{t}_j)^{\bar{\tau}_j}.$$

In particular,

$$\gamma_j \cdot \zeta_j(\alpha) = \zeta_j(\alpha) \ \ (\text{i.e., } \zeta_j(\alpha) \in R'((t_j))) \ \text{and} \quad \gamma_j \cdot \zeta_j(0) = \zeta_j(0) = \hat{\zeta}_j(0). \tag{11}$$

By the assumption $\theta(\tau_j) < 1$, we get (since $\alpha(\bar{\tau}_j) \in \mathbb{Z}$)

$$-(d_j - 2) \le -\alpha(\bar{\tau}_j) + 1 \le d_j, \quad \text{for any root} \ \ \alpha \in \Delta. \tag{12}$$

Moreover, for any (negative) root α which is not a root of P_j,

$$2 \le -\alpha(\bar{\tau}_j) + 1. \tag{13}$$

By (1) of Definition 6.1.11, since $t_j = (\hat{t}_j)^{d_j}$, the exponents of $\hat{t}_j$ in $\zeta_j(\alpha)$ (for any $\alpha \in \Delta$) are multiples of d_j by (11). Hence, by (10)–(13), $\zeta_j \in \mathcal{P}_j(R')$ and hence so is $f_j \in \mathcal{P}_j(R')$ by (9). This proves (8). Thus, by (8), associated to $[\hat{E}_{R'}, \hat{\sigma}_{R'}]$, we get a well-defined element

$$\bar{\beta} = (\bar{\beta}_1, \ldots, \bar{\beta}_s) \in \Pi_{j=1}^s \left(G(R'((t_j)))/\mathcal{P}_j(R')\right), \tag{14}$$

i.e., it does not depend upon the choice of the trivializations $(\hat{\mu}_j)_{1\leq j\leq s}$ satisfying (1), where $\bar{\beta}_j := \beta_j \cdot \mathcal{P}_j(R')$ and β_j is defined by (4).

Consider the canonical injective map (cf. Exercise 1.3.E.11):

$$i_j(R') \colon G(R'((t_j)))/\mathcal{P}_j(R') \to \mathrm{Mor}\left(\mathrm{Spec}\, R', \bar{X}_G(P_j)\right).$$

Let $\bar{\beta}'$ be the image of $\bar{\beta}$ in $\mathrm{Mor}(\mathrm{Spec}\, R', \bar{X}_{\vec{P}})$.

Considering $\mathrm{Spec}\, R' \underset{\mathrm{Spec}\, R}{\times} \mathrm{Spec}\, R'$ as in the proof of Proposition 5.2.7, from the uniqueness of $\bar{\beta}'$ we get a well-defined element in $\bar{X}_{\vec{P}}(R) := \mathrm{Mor}(\mathrm{Spec}\, R, \bar{X}_{\vec{P}})$. This gives our sought-after map $\mathfrak{H}\colon \mathscr{F}^{A,\vec{\tau}}_{G,\hat{\Sigma}^*}(R) \to \bar{X}_{\vec{P}}(R)$.

We now prove that $\mathfrak{H}$ is a bijection. We first prove that $\mathfrak{H}$ is injective. Take $(\hat{E}_R, \hat{\sigma}_R)$, $(\hat{E}'_R, \hat{\sigma}'_R) \in \mathscr{F}^{A,\vec{\tau}}_{G,\hat{\Sigma}^*}(R)$ such that their images under $\mathfrak{H}$ coincide. Choose a surjective étale morphism $\varphi\colon \mathrm{Spec}\, R' \to \mathrm{Spec}\, R$ such that both the G-bundles $\hat{E}_{R'_{|(\hat{\mathbb{D}}_j)_{R'}}}$ and $\hat{E}'_{R'|(\hat{\mathbb{D}}_j)_{R'}}$ are trivial for each $1 \leq j \leq s$, where $\hat{E}_{R'}$, $(\hat{\mathbb{D}}_j)_{R'}$ are as at the beginning of this proof. Taking a section $\hat{\mu}_j$ (resp. $\hat{\mu}'_j$) of $\hat{E}_{R'_{|(\hat{\mathbb{D}}_j)_{R'}}}$ (resp. $\hat{E}'_{R'_{|(\hat{\mathbb{D}}_j)_{R'}}}$) satisfying (1), we get $\hat{\beta}_j$ (resp. $\hat{\beta}'_j$) defined by (2). From the injectivity of $i_j(R')$, we get that

$$\hat{\beta}'_j \cdot (\hat{t}_j)^{\bar{\tau}_j} \in \hat{\beta}_j \cdot (\hat{t}_j)^{\bar{\tau}_j}\mathcal{P}_j(R'), \quad \text{for all} \quad 1 \leq j \leq s,$$

i.e., there exists $f_j \in \mathcal{P}_j(R')$ such that

$$\hat{\beta}'_j = \hat{\beta}_j \cdot \hat{f}_j, \quad \text{where} \quad \hat{f}_j := (\hat{t}_j)^{\bar{\tau}_j} \cdot f_j \cdot (\hat{t}_j)^{-\bar{\tau}_j}. \tag{15}$$

It is easy to see from (3) that

$$\gamma_j \cdot \hat{f}_j = \mathrm{Exp}(2\pi i \tau_j)^{-1} \cdot \hat{f}_j \cdot \mathrm{Exp}(2\pi i \tau_j). \tag{16}$$

We next claim that $\hat{f}_j \in G(R'[[\hat{t}_j]])$. Similar to $\hat{f}^o_j$, consider $f^o_j := ev_0(f_j) \in P_j(R')$ under the evaluation map $G(R'[[t_j]]) \to G(R')$. Considering $\zeta_j := (f^o_j)^{-1} \cdot f_j$, it is easy to see (similar to the case of $\hat{\zeta}_j$ considered earlier) that

$$\hat{\zeta}_j := (\hat{t}_j)^{\bar{\tau}_j} \cdot \zeta_j \cdot (\hat{t}_j)^{-\bar{\tau}_j} \in G(R'[[\hat{t}_j]]). \tag{17}$$

Further,

$$\hat{f}_j = (\hat{t}_j)^{\bar{\tau}_j} \cdot f^o_j \cdot (\hat{t}_j)^{-\bar{\tau}_j} \cdot \hat{\zeta}_j. \tag{18}$$

Since $(\hat{t}_j)^{\bar{\tau}_j}$ commutes with $L_j(R')$ (where L_j is the Levi subgroup of P_j containing H), it is easy to see that

$$(\hat{t}_j)^{\bar{\tau}_j} \cdot f_j^o \cdot (\hat{t}_j)^{-\bar{\tau}_j} \in G(R'[[\hat{t}_j]]). \tag{19}$$

Combining (17)–(19), we get that

$$\hat{f}_j \in G(R'[[\hat{t}_j]]), \quad \text{for all} \quad 1 \leq j \leq s. \tag{20}$$

For any $1 \leq j \leq s$, choose a set of coset representatives:

$$\left\{a_j^1 A_{\hat{p}_j}, \dots, a_j^{q_j} A_{\hat{p}_j}\right\} \quad \text{of} \quad A/A_{\hat{p}_j}.$$

For any $1 \leq k \leq q_j$, consider the formal disc $a_j^k \cdot \hat{\mathbb{D}}_j$ in $\hat{\Sigma}$ centered at $a_j^k \cdot \hat{p}_j$. Identify the disc $a_j^k \cdot \hat{\mathbb{D}}_j$ with $\hat{\mathbb{D}}_j$ under the action of a_j^k and transport the local parameter $\hat{t}_j$ of $\hat{\mathbb{D}}_j$ to $a_j^k \cdot \hat{\mathbb{D}}_j$ (still denoted by $\hat{t}_j$) under this identification.

Take the section $\hat{\mu}_j(k)$ (resp. $\hat{\mu}'_j(k)$) of $\hat{E}_{R'_{|(a_j^k \cdot \hat{\mathbb{D}}_j)_{R'}}}$ $\left(\text{resp. } \hat{E}'_{R'_{|(a_j^k \cdot \hat{\mathbb{D}}_j)_{R'}}}\right)$ defined by

$$\hat{\mu}_j(k)(a_j^k \cdot x) := a_j^k \cdot (\hat{\mu}_j(x)), \quad \text{for any} \quad x \in (\hat{\mathbb{D}}_j)_{R'},$$

and similarly for $\hat{\mu}'_j(k)$, where $\hat{\mu}_j$ and $\hat{\mu}'_j$ are any sections of $\hat{E}_{R'_{|(\hat{\mathbb{D}}_j)_{R'}}}$ and $\hat{E}'_{R'_{|(\hat{\mathbb{D}}_j)_{R'}}}$ respectively satisfying (1). Then, it is easy to see (since $\hat{\sigma}_{R'}$ is A-equivariant) that $\hat{\mu}_j(k)^* = \hat{\sigma}^*_{R'} \cdot \hat{\beta}_j$ as sections over $(a_j^k \hat{\mathbb{D}}_j^*)_{R'}$ for any $1 \leq k \leq q_j$ and similarly for $\hat{\mu}'_j(k)^*$. Thus, by the analogue of Proposition 5.2.7 with several punctures (for Σ^* replaced by $\hat{\Sigma}^*$) and using (15) and (20), we get that there exists a G-bundle isomorphism

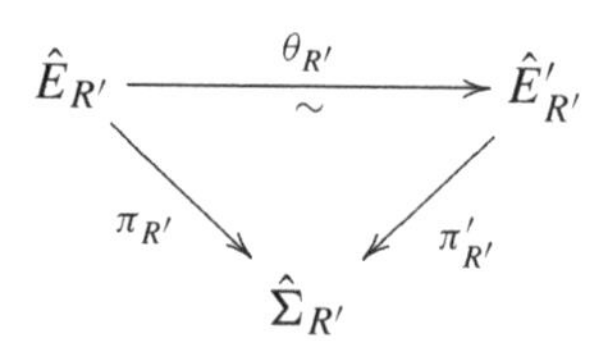

$$\begin{array}{ccc} \hat{E}_{R'} & \xrightarrow[\sim]{\theta_{R'}} & \hat{E}'_{R'} \\ & {\scriptstyle \pi_{R'}} \searrow \quad \swarrow {\scriptstyle \pi'_{R'}} & \\ & \hat{\Sigma}_{R'} & \end{array}$$

taking $\hat{\sigma}_{R'}$ to $\hat{\sigma}'_{R'}$. Since $\hat{\sigma}_{R'}$ and $\hat{\sigma}'_{R'}$ are A-equivariant over $(\hat{\Sigma}^*)_{R'}$ (by assumption) and $\pi_{R'}^{-1}((\hat{\Sigma}^*)_{R'})$ is dense in $\hat{E}_{R'}$, we conclude that $\theta_{R'}$ is A-equivariant. From the uniqueness of $\theta_{R'}$ (since it is uniquely determined on $\pi_{R'}^{-1}((\hat{\Sigma}^*)_{R'})$), following the same argument as in the last part of the proof of Proposition 5.2.7, by considering the fiber product

$$\operatorname{Spec} R' \underset{\operatorname{Spec} R}{\times} \operatorname{Spec} R',$$

we conclude that $(\hat{E}_R, \hat{\sigma}_R)$ is isomorphic with $(\hat{E}'_R, \hat{\sigma}'_R)$ as A-equivariant G-bundles. This proves that $\mathfrak{H}$ is one-to-one.

We next prove that $\mathfrak{H}$ is surjective. Take a morphism $\delta\colon \operatorname{Spec} R \to \bar{X}_{\vec{p}}$. Then by Exercise 1.3.E.11 and the proof of Lemma B.2, there exists an fppf cover $\varphi\colon \operatorname{Spec} R' \to \operatorname{Spec} R$ such that the morphism $\delta_{R'} := \delta \circ \varphi\colon \operatorname{Spec} R' \to \bar{X}_{\vec{p}}$ lifts to a morphism $\hat{\delta}_{R'}\colon \operatorname{Spec} R' \to \Pi_{j=1}^s \bar{G}((t_j))$ giving rise to the elements $\beta_j \in G(R'((t_j)))$ by taking the projection of $\hat{\delta}_{R'}$ to the jth factor and using Lemma 1.3.2. Define

$$\hat{\beta}_j := \beta_j \cdot (\hat{t}_j)^{-\bar{\tau}_j} \in G(R'((\hat{t}_j))). \tag{21}$$

Consider the trivial G-bundle $E'_{R'}$ over $(\hat{\Sigma}^*)_{R'}$ with the trivial A-action, i.e.,

$$E'_{R'} = (\hat{\Sigma}^*)_{R'} \times G \to (\hat{\Sigma}^*)_{R'}$$

with

$$a \cdot (x,g) = (a \cdot x, g), \quad \text{for} \quad a \in A, x \in (\hat{\Sigma}^*)_{R'},\ g \in G.$$

Further, consider the $A_{\hat{p}_j}$-equivariant trivial G-bundle $E^j_{R'} = (\hat{\mathbb{D}}_j)_{R'} \times G \to (\hat{\mathbb{D}}_j)_{R'}$ with the action of the generator γ_j of $A_{\hat{p}_j}$ given by

$$\gamma_j \cdot (x,g) = (\gamma_j \cdot x,\ \operatorname{Exp}(2\pi i \tau_j) g), \quad \text{for } x \in (\hat{\mathbb{D}}_j)_{R'} \text{ and } g \in G.$$

There is an A-equivariant isomorphism of schemes

$$A \times^{A_{\hat{p}_j}} (\hat{\mathbb{D}}_j)_{R'} \to (\hat{F}_j)_{R'}, \quad [a,x] \mapsto a \cdot x,$$

where $\hat{F}_j := A \cdot \hat{\mathbb{D}}_j = \coprod_{k=1}^{q_j} (a_j^k \cdot \hat{\mathbb{D}}_j)$, $\{a_j^1, \ldots, a_j^{q_j}\}$ is a set of coset representatives of $A/A_{\hat{p}_j}$ (as earlier) and $A_{\hat{p}_j}$ acts on $A \times (\hat{\mathbb{D}}_j)_{R'}$ diagonally:

$$\gamma \cdot (a,x) = (a \cdot \gamma^{-1}, \gamma \cdot x), \text{ for } a \in A, \gamma \in A_{\hat{p}_j} \text{ and } x \in (\hat{\mathbb{D}}_j)_{R'}.$$

Hence, an $A_{\hat{p}_j}$-equivariant G-bundle on $(\hat{\mathbb{D}}_j)_{R'}$ extends uniquely (unique up to a unique isomorphism) to an A-equivariant G-bundle on $(\hat{F}_j)_{R'}$ (cf. (Chriss and Ginzburg, 1997, §5.2.16)). In particular, the $A_{\hat{p}_j}$-equivariant G-bundle $E^j_{R'}$ extends uniquely to an A-equivariant G-bundle $\hat{E}^j_{R'}$ over $(\hat{F}_j)_{R'}$.

Identify the $A_{\hat{p}_j}$-equivariant bundles $E'_{R'}$ and $E^j_{R'}$ over the intersection $(\hat{\mathbb{D}}^*_j)_{R'} = (\hat{\mathbb{D}}_j)_{R'} \cap (\hat{\Sigma}^*)_{R'}$ via

$$\begin{aligned}\theta_j\colon E^j_{R'}{}_{|(\hat{\mathbb{D}}^*_j)_{R'}} &= (\hat{\mathbb{D}}^*_j)_{R'} \times G \to E'_{R'}{}_{|(\hat{\mathbb{D}}^*_j)_{R'}} \\ &= (\hat{\mathbb{D}}^*_j)_{R'} \times G,\ (x,g) \mapsto (x, \hat{\beta}_j(x) g), \text{for } x \in (\hat{\mathbb{D}}^*_j)_{R'} \text{ and } g \in G,\end{aligned}$$

where $\hat{\beta}_j \in G(R'((\hat{t}_j)))$ is defined by (21).

Clearly, θ_j is an $A_{\hat{p}_j}$-equivariant isomorphism of G-bundles and hence gives rise to a unique A-equivariant isomorphism of G-bundles

$$\hat{\theta}_j\colon \hat{E}^j_{R'_{|(\hat{F}^*_j)_{R'}}} \to E'_{R'_{|(\hat{F}^*_j)_{R'}}}, \quad \text{where } (\hat{F}^*_j)_{R'} := (A\cdot \hat{\mathbb{D}}^*_j)_{R'}.$$

The A-equivariant G-bundles $E'_{R'}$ and $\{\hat{E}^j_{R'}\}_{1\le j\le s}$ and the above isomorphisms allow us to get an A-equivariant G-bundle $\hat{E}_{R'}$ over $\hat{\Sigma}_{R'}$ via the 'descent' lemma (cf. the analogue of Lemma 5.2.3 for several punctures in $\hat{\Sigma}$). By the definition, $\hat{E}_{R'}$ is of local type $\{\tau_j\}_{1\le j\le s}$ which comes equipped with an A-equivariant section $\hat{\sigma}_{R'}$ over $(\hat{\Sigma}^*)_{R'}$ given by

$$\hat{\sigma}_{R'}(x) = (x,1) \quad \text{in} \quad E'_{R'}, \quad \text{for } x\in(\hat{\Sigma}^*)_{R'}.$$

Further, from the definition of $\mathfrak{H}$,

$$\mathfrak{H}\left(\left[\hat{E}_{R'},\hat{\sigma}_{R'}\right]\right) = \delta_{R'}. \tag{22}$$

From the injectivity of $\mathfrak{H}$, $(\hat{E}_{R'},\hat{\sigma}_{R'})$ satisfying (22) is unique (up to a unique isomorphism) and hence considering (as earlier) the fiber product

$$\operatorname{Spec} R'' := \operatorname{Spec} R' \underset{\operatorname{Spec} R}{\times} \operatorname{Spec} R'$$

with the two projection to $\operatorname{Spec} R'$, we get (e.g., applying the analogue of Proposition 5.2.7 for $\hat{\Sigma}$ with several punctures) that $(\hat{E}_{R'},\hat{\sigma}_{R'})$ descends to a G-bundle $(\hat{E}_R,\hat{\sigma}_R)$ over $\hat{\Sigma}_R$ with section over $(\hat{\Sigma}^*)_R$. Moreover, it is easy to see by considering $(\hat{E}_{R''},\hat{\sigma}_{R''})$ that the A-equivariant structure on $\hat{E}_{R'}$ also descends to give an A-equivariant structure on $\hat{E}_R$ such that $\hat{\sigma}_R$ is A-equivariant . Thus, $(\hat{E}_R,\hat{\sigma}_R)\in \mathscr{F}^{A,\hat{\tau}}_{G,\hat{\Sigma}^*}(R)$, which maps to δ under $\mathfrak{H}$. This proves the surjectivity of $\mathfrak{H}$ and hence the theorem is fully established. □

Remark 6.1.13 In Balaji and Seshadri (2015), the restriction $\theta(\tau_j) < 1$ in Theorem 6.1.12 plays no role since by considering general parahoric subgroups of $G((t_j))$, their work is independent of the location of the weights τ_j in the fundamental alcove. However, the proof, in the case when $\theta(\tau_j)$ is allowed to be 1, is very similar to the proof given above.

It might be remarked that for any semisimple group G, the 'parahoric viewpoint' is a natural one since the 'unit group' of A-invariant local sections is a parahoric subgroup of a general kind.

Definition 6.1.14 Similar to the definition of the stack $\mathbf{Bun}_G(\Sigma)$ as in Definition 5.1.1, define the groupoid fibration of A-equivariant G-bundles $\mathbf{Bun}^{A,\vec{\tau}}_G(\hat{\Sigma})$ of local type $\vec{\tau}$ over the category $\mathfrak{S}$, whose objects are A-equivariant G-bundles E_S over $\hat{\Sigma}\times S$ (with the trivial action of A on S) such that $E_{S|_{\hat{\Sigma}\times t}}$ (for any $t\in S$) is of local type $\vec{\tau}$. By a morphism between two such bundles E_S (over $\hat{\Sigma} \times S$) and $E'_{S'}$ (over $\hat{\Sigma} \times S'$), we mean an $A \times G$-equivariant morphism $f: E_S \to E'_{S'}$ and a morphism $\bar{f}: S \to S'$ making the following diagram commutative:

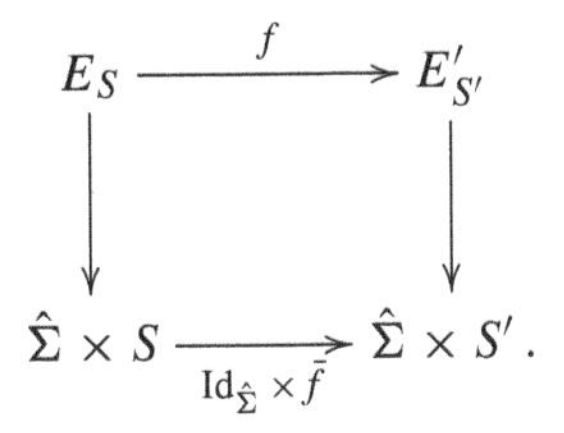

$$\begin{array}{ccc} E_S & \xrightarrow{f} & E'_{S'} \\ \downarrow & & \downarrow \\ \hat{\Sigma} \times S & \xrightarrow[\mathrm{Id}_{\hat{\Sigma}} \times \bar{f}]{} & \hat{\Sigma} \times S'. \end{array}$$

The functor $\mathbf{Bun}_G^{A,\vec{\tau}}(\hat{\Sigma}) \to \mathfrak{S}$ takes $E_S \rightsquigarrow S$ and $f \rightsquigarrow \bar{f}$.

Let $\bar{X}_{\vec{P}} := \Pi_{j=1}^{s} \bar{X}_G(P_j)$ be as in Theorem 6.1.12 and let $\bar{\Gamma}$ be the ind-affine group variety as in Definition 5.2.9 and Lemma 5.2.10 with $\mathbb{C}$-points $\Gamma := \mathrm{Mor}(\Sigma^*, G)$. Then $\bar{\Gamma}$ acts on $\bar{X}_{\vec{P}}$ by the left multiplication on each factor via its Laurent series expansion in the coordinates t_j (cf. Corollary 5.2.11).

With the notation as above, we have the following.

Theorem 6.1.15 *Let $s \geq 1$ and $\vec{\tau}$ be as in Theorem 6.1.12. Then there exists an equivalence of categories over $\mathfrak{S}$ between the groupoid fibration $\mathbf{Bun}_G^{A,\vec{\tau}}(\hat{\Sigma})$ over $\mathfrak{S}$ of A-equivariant G-bundles of local type $\vec{\tau}$ and the quotient stack $\left[\bar{\Gamma}\backslash\bar{X}_{\vec{P}}\right]$ (cf. Example C.18(b)).*

In particular, $\mathbf{Bun}_G^{A,\vec{\tau}}(\hat{\Sigma})$ is isomorphic to the stack $\mathbf{Parbun}_G(\Sigma, \vec{P})$ of quasi-parabolic G-bundles over $(\Sigma, \vec{p})$ of type $\vec{P} := (P_1, \ldots, P_s)$ (cf. Definition 5.1.4) and hence it is a smooth (algebraic) stack.

(Even though $\mathbf{Parbun}_G(\Sigma, \vec{P})$ only depends upon $\vec{P}$, its isomorphism with $\mathbf{Bun}_G^{A,\vec{\tau}}(\hat{\Sigma})$ does depend upon the choice of $\vec{\tau}$.)

Proof The proof is parallel to the proof of Theorem 5.2.14. We first define a functor $\zeta: \mathbf{Bun}_G^{A,\vec{\tau}}(\hat{\Sigma}) \to [\bar{\Gamma}\backslash\bar{X}_{\vec{P}}]$. Let $\hat{E} = \hat{E}_S \to \hat{\Sigma} \times S \in \mathbf{Bun}_G^{A,\vec{\tau}}(\hat{\Sigma})$ (for any scheme $S \in \mathfrak{S}$). Define a $\mathbb{C}$-space functor $\hat{\mathscr{E}}_S$ as follows. For any $\mathbb{C}$-algebra R and an element in $S(R)$, i.e., a morphism φ: Spec $R \to S$, define

$$\hat{\mathscr{E}}_S^o(\varphi) := \text{ set of } A\text{-equivariant sections of } \hat{E}_{\varphi_{|\hat{\Sigma}_R^*}},$$

where $\hat{E}_\varphi$ denotes the pull-back bundle $(\mathrm{Id}_{\hat{\Sigma}} \times \varphi)^*(\hat{E})$. Now, for any $\mathbb{C}$-algebra R, set

$$\hat{\mathscr{E}}_S^o(R) := \sqcup_{\varphi \in S(R)} \hat{\mathscr{E}}_S^o(\varphi).$$

Then, for an fppf R-algebra R', $\hat{\mathscr{E}}_S^o(R) \to \hat{\mathscr{E}}_S^o(R')$ is injective. Let $\hat{\mathscr{E}}_S$ be the sheafification of $\hat{\mathscr{E}}_S^o$ (cf. Lemma B.2). Then, we get a morphism

$$\hat{p}: \hat{\mathscr{E}}_S \to S.$$

Further, for any $\varphi \in S(R)$, using Lemma 5.2.10, there is a canonical action $\hat{\mathscr{E}}^o_S(\varphi) \times \bar{\Gamma}(R) \to \hat{\mathscr{E}}^o_S(\varphi)$ giving rise to an action

$$\hat{\mathscr{E}}_S \times \bar{\Gamma} \to \hat{\mathscr{E}}_S. \tag{1}$$

(Here, we have used the identification $\mathrm{Mor}(\Sigma^*_R, G) \simeq \mathrm{Mor}_A(\hat{\Sigma}^*_R, G)$, where $\mathrm{Mor}_A(\hat{\Sigma}^*_R, G)$ denotes the set of A-invariant morphisms $\hat{\Sigma}^*_R \to G$.) Since A acts freely on $\hat{\Sigma}^*_S$ with quotient Σ^*_S, the A-equivariant G-bundle $\hat{E}_{S_{|\hat{\Sigma}^*_S}}$ is the pull-back of a unique G-bundle E_S over Σ^*_S with trivial A-action. Take an affine étale cover $S' \to S$ such that $E_{S'}$ over $\Sigma^*_{S'}$ is trivial (cf. Theorem 5.2.5), where $E_{S'}$ is the pull-back of E_S to $\Sigma^*_{S'}$. Then it is easy to see that $\hat{E}_{S'}$ (which is the pull-back of $\hat{E}_S$ to $\hat{\Sigma} \times S'$) satisfies $\hat{E}_{S'_{|\hat{\Sigma}^*_{S'}}} \simeq q^*(E_{S'})$ as A-equivariant G-bundles, where $q\colon \hat{\Sigma}^*_{S'} \to \Sigma^*_{S'}$ is the standard quotient map induced by $\pi : \hat{\Sigma} \to \Sigma$. In particular, $\hat{E}_{S'_{|\hat{\Sigma}^*_{S'}}}$ admits an A-equivariant section. Then $\hat{\mathscr{E}}_{S'} = S' \times_S \hat{\mathscr{E}}_S$ is isomorphic with $S' \times \bar{\Gamma}$ such that the induced action of $\bar{\Gamma}$ on $\hat{\mathscr{E}}_{S'}$ (as in (1)) corresponds to the right multiplication on the $\bar{\Gamma}$-factor. Thus, $\hat{p}\colon \hat{\mathscr{E}}_S \to S$ is a $\bar{\Gamma}$-torsor with the right action of $\bar{\Gamma}$ (cf. Definition C.16).

We further define a $\bar{\Gamma}$-equivariant morphism $\hat{\beta}\colon \hat{\mathscr{E}}_S \to \bar{X}_{\vec{P}}$ as follows. For any $\varphi \in S(R)$, map any element of $\hat{\mathscr{E}}^o_S(\varphi)$ to the pair $(\hat{E}_\varphi, \hat{\sigma}_\varphi)$, where $\hat{\sigma}_\varphi$ is the corresponding section of $\hat{E}_{\varphi_{|\hat{\Sigma}^*_R}}$. The sections $\hat{\sigma}_\varphi$ (for $\varphi \in S(R)$) give rise to the section $\hat{\sigma}_o$ of $\bar{\hat{p}}^*(\hat{E}_S)$ over $\hat{\Sigma}^* \times \hat{\mathscr{E}}_S$, where $\bar{\hat{p}}\colon \hat{\Sigma} \times \hat{\mathscr{E}}_S \to \hat{\Sigma} \times S$ is the morphism $\mathrm{Id}_{\hat{\Sigma}} \times \hat{p}$. We call $\hat{\sigma}_o$ the *A-equivariant tautological section*. Now, the equivalence class $[\hat{E}_\varphi, \hat{\sigma}_\varphi] \in \mathscr{F}^{A,\vec{\tau}}_{G,\hat{\Sigma}^*}(R)$ (cf. Definition 6.1.11) corresponds to an element in $\bar{X}_{\vec{P}}(R)$ (cf. Theorem 6.1.12). This gives us the desired morphism

$$\hat{\beta}\colon \hat{\mathscr{E}}_S \to \bar{X}_{\vec{P}}.$$

It is easy to see that $\hat{\beta}$ is $\bar{\Gamma}$-equivariant, where we switch the right action of $\bar{\Gamma}$ on $\hat{\mathscr{E}}_S$ to the left action by the standard procedure:

$$\gamma \cdot x = x \cdot \gamma^{-1}, \quad \text{for } \gamma \in \bar{\Gamma} \text{ and } x \in \hat{\mathscr{E}}_S.$$

The functor ζ takes $\hat{E}_S \in \mathbf{Bun}^{A,\vec{\tau}}_G(\hat{\Sigma})$ to the pair $(\hat{p}, \hat{\beta})$.

Conversely, we define a functor $\eta\colon [\bar{\Gamma}\backslash\bar{X}_{\vec{P}}] \to \mathbf{Bun}^{A,\vec{\tau}}_G(\hat{\Sigma})$ as follows.

Take a $\bar{\Gamma}$-torsor (with the left action of $\bar{\Gamma}$) $\hat{p}\colon \hat{\mathscr{E}}_S \to S$ (over a scheme $S \in \mathfrak{S}$) and a $\bar{\Gamma}$-equivariant morphism $\hat{\beta}\colon \hat{\mathscr{E}}_S \to \bar{X}_{\vec{P}}$. The identity morphism $\mathrm{Id}\colon \bar{X}_{\vec{P}} \to \bar{X}_{\vec{P}}$ gives rise (via Theorem 6.1.12) to an A-equivariant G-bundle $\mathfrak{U}(\vec{\tau})$ over $\hat{\Sigma} \times \bar{X}_{\vec{P}}$ (with A acting trivially on $\bar{X}_{\vec{P}}$) of local type $\vec{\tau}$ restricted to any $\hat{\Sigma} \times x$ (for $x \in \bar{X}_{\vec{P}}$) together with an A-equivariant section $\hat{\sigma}_{\bar{X}_{\vec{P}}}$ of

$\mathfrak{U}(\vec{\tau})$ over $\hat{\Sigma}^* \times \bar{X}_{\vec{p}}$. Moreover, the pair $(\mathfrak{U}(\vec{\tau}), \hat{\sigma}_{\bar{X}_{\vec{p}}})$ is unique up to a unique isomorphism. (Even though Theorem 6.1.12 only guarantees a G-bundle over $\hat{\Sigma} \times \operatorname{Spec} R$, for $R \in \mathbf{Alg}$, but the uniqueness insures its extension to $\hat{\Sigma} \times \bar{X}_{\vec{p}}$.) We fix one such pair in its isomorphism class. By the same proof as that of Lemma 5.2.12, the G-bundle $\mathfrak{U}(\vec{\tau})$ acquires the structure of a $\bar{\Gamma}$-equivariant G-bundle over $\hat{\Sigma} \times \bar{X}_{\vec{p}}$ commuting with the A-equivariant structure. The $\bar{\Gamma}$-equivariant morphism $\hat{\beta}\colon \hat{\mathscr{E}}_S \to \bar{X}_{\vec{p}}$ gives rise to a $\bar{\Gamma}$-equivariant G-bundle $\bar{\hat{\beta}}^*(\mathfrak{U}(\vec{\tau}))$ over $\hat{\Sigma} \times \hat{\mathscr{E}}_S$ via pull-back through $\bar{\hat{\beta}} := \operatorname{Id}_{\hat{\Sigma}} \times \hat{\beta}\colon \hat{\Sigma} \times \hat{\mathscr{E}}_S \to \hat{\Sigma} \times \bar{X}_{\vec{p}}$. Since $\hat{p}\colon \hat{\mathscr{E}}_S \to S$ is a $\bar{\Gamma}$-torsor, the $\bar{\Gamma}$-equivariant bundle $\bar{\hat{\beta}}^*(\mathfrak{U}(\vec{\tau}))$ descends to give a G-bundle (denoted) $\hat{E}(\hat{p}, \hat{\beta}) \to \hat{\Sigma} \times S$ (cf. Lemma C.17). Since $\mathfrak{U}(\vec{\tau})$ is of local type $\vec{\tau}$, so is $\hat{E}(\hat{p}, \hat{\beta})$. Further, since $\mathfrak{U}(\vec{\tau})$ has an A-equivariant structure commuting with the $\bar{\Gamma}$-equivariant structure and $\bar{\hat{\beta}}$ is an $A \times \bar{\Gamma}$-equivariant morphism with A acting only on $\hat{\Sigma}$ (acting trivially on $\hat{\mathscr{E}}_S$ and $\bar{X}_{\vec{p}}$ and $\bar{\Gamma}$ acting trivially on $\hat{\Sigma}$), $\hat{E}(\hat{p}, \hat{\beta})$ is an A-equivariant G-bundle over $\hat{\Sigma} \times S$. This is our map $\eta\colon [\bar{\Gamma} \backslash \bar{X}_{\vec{p}}] \to \mathbf{Bun}_G^{A,\vec{\tau}}(\hat{\Sigma})$, taking $(\hat{p}, \hat{\beta}) \mapsto \hat{E}(\hat{p}, \hat{\beta})$.

The proof that $\eta \circ \zeta \simeq \operatorname{Id}_{\mathbf{Bun}_G^{A,\vec{\tau}}(\hat{\Sigma})}$ and $\zeta \circ \eta \simeq \operatorname{Id}_{[\bar{\Gamma} \backslash \bar{X}_{\vec{p}}]}$ is similar to the one given in the proof of Theorem 5.2.14 and hence is left to the reader. This proves the first part of the theorem.

The 'In particular' part of the theorem follows from the first part and Exercise 5.2.E.3 together with Theorem 5.1.5. □

Following Definition 6.1.14, let $\operatorname{Bun}_G^{A,\vec{\tau}}(\hat{\Sigma})$ be the set of isomorphism classes of A-equivariant G-bundles over $\hat{\Sigma}$ of local type $\vec{\tau}$. Similarly, $\operatorname{Parbun}_G(\Sigma, \vec{P})$ is as defined in Corollary 5.2.17.

Definition 6.1.16 Let G be a connected reductive group. An A-equivariant G-bundle $\hat{E}$ over $\hat{\Sigma}$ is called *A-semistable* (resp. *A-stable*) if condition (2) of Definition 6.1.4(b) is satisfied for any standard maximal parabolic subgroup Q_k of G and any A-equivariant section μ of $\hat{E}/Q_k \to \hat{\Sigma}$.

Similarly, following Definition 6.1.4(c), $\hat{E}$ is called *A-polystable* if it has an A-equivariant reduction $\hat{E}_L$ to a Levi subgroup L such that the L-bundle $\hat{E}_L$ is A-stable and for any character χ of L which is trivial restricted to the center of G, we have

$$\deg\left(\hat{E}_L \times^L \mathbb{C}_\chi\right) = 0.$$

Similar to the definition of semistable and stable vector bundles as in Definition 6.1.4(a), an A-equivariant vector bundle $\mathscr{V}$ over $\hat{\Sigma}$ is called *A-semistable* (resp. *A-stable*) if the inequality (1) of Definition 6.1.4(a) is satisfied for any A-stable subbundle $(0) \subsetneq \mathscr{W} \subsetneq \mathscr{V}$.

Similar to Exercise 6.1.E.5, an A-equivariant vector bundle $\mathscr{V}$ is A-semistable (resp. A-stable) if and only if the corresponding frame bundle $F(\mathscr{V})$ is so.

An A-equivariant vector bundle $\mathscr{V}$ over $\hat{\Sigma}$ of rank r is called *A-polystable* if the associated frame bundle $F(\mathscr{V})$ is A-polystable as a GL_r-bundle.

By Exercise 6.1.E.16, an A-equivariant vector bundle $\mathscr{V}$ over $\hat{\Sigma}$ is A-polystable if and only if we can write

$$\mathscr{V} = \bigoplus_i \mathscr{V}_i,$$

where each $\mathscr{V}_i$ is an A-stable vector bundle all of which have the same slope.

Similar to Corollaries 5.2.15 and 5.2.17, we get the following result from Theorem 6.1.15.

Theorem 6.1.17 *With the notation and assumptions as in Theorem 6.1.15, we have a natural set-theoretic bijection*

$$\mathrm{Bun}_G^{A,\vec{\tau}}(\hat{\Sigma}) \simeq \mathrm{Parbun}_G(\Sigma, \vec{P}). \tag{1}$$

In fact, there is a similar set-theoretic natural bijection as (1) *with* $\hat{\Sigma}$ *replaced by* $\hat{\Sigma} \times S$ *(for any ind-scheme* S *as parameter space).*

Under the bijection (1), *A-semistable (resp. A-stable) G-bundles over* $\hat{\Sigma}$ *correspond to the parabolic semistable (resp. stable) bundles over* Σ *with respect to the markings* $\vec{\tau}$ *(cf. Definition 6.1.4(d)). In fact, a more precise result is true (cf. identity* (17) *in the proof).*

Proof The bijection (1) (resp. its extension to $\hat{\Sigma} \times S$) follows immediately from Theorem 6.1.15 by specializing the equivalence of the groupoid fibrations $\mathbf{Bun}_G^{A,\vec{\tau}}(\hat{\Sigma})$ and $\mathbf{Parbun}_G(\Sigma, \vec{P})$ over a point (resp. over S).

We now prove the assertion about the correspondence of semistable and stable bundles. Take any standard maximal parabolic subgroup Q_k ($1 \leq k \leq \ell$) of G. Let $\hat{E} \to \hat{\Sigma}$ be an A-equivariant G-bundle over $\hat{\Sigma}$ of local type $\vec{\tau}$ and let $E \to \Sigma$ be the corresponding quasi-parabolic G-bundle over Σ of type $\vec{P}$ given by the correspondence (1). Then, by definition (given in the proofs of Theorems 6.1.12 and 6.1.15 following the notation therein which we follow freely) as A-equivariant G-bundles (with the trivial A-action on E):

$$\hat{E}_{|\hat{\Sigma}^*} := \pi^*(E_{|\Sigma^*}). \tag{2}$$

From this we see that the pull-back of sections provides a bijective correspondence between the sections of $(E_{|\Sigma^*})/Q_k$ and A-equivariant sections of $(\hat{E}_{|\hat{\Sigma}^*})/Q_k$. Moreover, since $Y_k := G/Q_k$ is a projective variety and Σ is a curve, this correspondence extends to give a

$$\text{bijective correspondence } \psi \text{ between the sections of } E/Q_k \text{ and } A\text{-equivariant sections of } \hat{E}/Q_k. \tag{3}$$

(In fact, this bijective correspondence holds for any parabolic subgroup Q of G.) Take a section θ of E/Q_k and let $\hat{\theta} = \psi(\theta)$ be the corresponding section of $\hat{E}/Q_k$. For any $1 \leq j \leq s$, we assert that there exists a section $\hat{\mu}_j$ of $\hat{E}_{|\hat{\mathbb{D}}_j}$ satisfying the following two conditions (writing $\hat{\tau}_j := \text{Exp}(2\pi i \tau_j)$):

$$\gamma_j \cdot \hat{\mu}_j = \hat{\mu}_j \cdot \hat{\tau}_j, \text{ cf. (1) of the proof of Theorem 6.1.12,} \tag{4}$$

and

$$\hat{\theta}_{|\hat{\mathbb{D}}_j} = \hat{\mu}_j \cdot \bar{w}_j Q_k, \text{ for some } \bar{w}_j \in N(H), \tag{5}$$

where $N(H)$ is the normalizer of H in G. To prove the existence of such a $\hat{\mu}_j$, take any $\hat{\mu}'_j$ satisfying

$$\gamma_j \cdot \hat{\mu}'_j = \hat{\mu}'_j \cdot \hat{\tau}_j, \tag{6}$$

which is guaranteed by Theorem 6.1.9. Write

$$\hat{\theta}_{|\hat{\mathbb{D}}_j} = \hat{\mu}'_j \cdot \bar{\delta}, \quad \text{for a morphism} \quad \bar{\delta}\colon \hat{\mathbb{D}}_j \to Y_k. \tag{7}$$

Since $\hat{\theta}$ is A-equivariant, we get from (6) and (7) that

$$\gamma_j \bar{\delta} = \hat{\tau}_j^{-1}\ \bar{\delta}; \quad \text{in particular,} \quad \bar{\delta}(\hat{p}_j) = \hat{\tau}_j\, \bar{\delta}(\hat{p}_j), \tag{8}$$

where γ_j acts on $\hat{\Sigma}$ and not on Y_k. But it is easy to see that

$$(Y_k)^{\hat{\tau}_j} := \left\{ gQ_k : \hat{\tau}_j g Q_k = gQ_k \right\} \quad \text{is given by} \quad (Y_k)^{\hat{\tau}_j} = \bigcup_{w \in W} L_j w Q_k,$$

where L_j is the Levi subgroup of P_j containing H. Thus, $\bar{\delta}(\hat{p}_j) = l_j \bar{w}_j Q_k$ for some $\bar{w}_j \in N(H)$ and $l_j \in L_j$. Thus, $U^-_{Q_k} \cdot Q_k$ being an open subset of Y_k, $\bar{\delta}$ lands as a map $\bar{\delta}\colon \hat{\mathbb{D}}_j \to l_j \bar{w}_j U^-_{Q_k} Q_k$, where $U^-_{Q_k}$ is the opposite unipotent radical of Q_k. Define $\delta\colon \hat{\mathbb{D}}_j \to l_j \bar{w}_j U^-_{Q_k} \subset G$ obtained from the isomorphism $U^-_{Q_k} \cdot Q_k/Q_k \subset Y_k \simeq U^-_{Q_k} \subset G$. Let $\delta_o\colon \hat{\mathbb{D}}_j \to U^-_{Q_k}$ be the map $\delta_o := (l_j \bar{w}_j)^{-1}\delta$. Then, by (8) (since $(\text{Ad}\, l_j^{-1})\hat{\tau}_j^{-1} = \hat{\tau}_j^{-1}$),

$$\gamma_j \cdot \delta_o = \left(\bar{w}_j^{-1} \hat{\tau}_j^{-1} \bar{w}_j \right) \delta_o \left(\bar{w}_j^{-1} \hat{\tau}_j \bar{w}_j \right). \tag{9}$$

Now, consider a new section of $\hat{E}_{|\hat{\mathbb{D}}_j}$:

$$\hat{\mu}_j := \hat{\mu}'_j \cdot (l_j \bar{w}_j \delta_o \bar{w}_j^{-1}).$$

Then, by (6), (7) and (9),

$$\gamma_j \cdot \hat{\mu}_j = \hat{\mu}_j \cdot \hat{\tau}_j \quad \text{and} \quad \hat{\theta}_{|\hat{\mathbb{D}}_j} = \hat{\mu}_j \cdot \bar{w}_j Q_k.$$

This proves (4) and (5).

We next show that the relative position of the section θ of the bundle E/Q_k with respect to the quasi-parabolic structure on E at p_j is given by $W_{P_j} w_j W_{Q_k}$, where $w_j \in W$ is the image of $\bar{w}_j$ in W (cf. Definition 6.1.4 for the definition of the relative position). By (5),

$$\begin{aligned}
\hat{\theta}_{|\hat{\mathbb{D}}^*_j} &= \hat{\mu}^*_j \bar{w}_j Q_k \\
&= \hat{\sigma}_{|\hat{\mathbb{D}}^*_j} \hat{\beta}_j \bar{w}_j Q_k, \ \text{by (2) of the proof of Theorem 6.1.12, where} \\
&\qquad \hat{\sigma} \ \text{is an } A\text{-equivariant section of } \hat{E}_{|\hat{\Sigma}^*} \\
&= \hat{\sigma}_{|\hat{\mathbb{D}}^*_j} \beta_j (\hat{t}_j)^{-\bar{\tau}_j} \bar{w}_j Q_k, \quad \text{see (4) and (5) of the proof of Theorem 6.1.12,} \\
&\qquad \text{where } \beta_j \text{ is } \gamma_j\text{-invariant} \\
&= \hat{\sigma}_{|\hat{\mathbb{D}}^*_j} \beta_j \bar{w}_j Q_k. \qquad (10)
\end{aligned}$$

But $\hat{\sigma}_{|\hat{\mathbb{D}}^*_j} \cdot \beta_j$ descends (since $\hat{\sigma}$ is the pull-back of a section σ of $E_{|\Sigma^*}$) and extends to give a section μ_j of $E_{|\mathbb{D}_j}$ (cf. Proposition 5.2.4 and (14) of the proof of Theorem 6.1.12) and hence

$$\theta_{|\mathbb{D}_j} = \mu_j \bar{w}_j Q_k. \qquad (11)$$

Now, from the definition of the relative position as in Definition 6.1.4(d), since $\mu_j(p_j)P_j$ gives the quasi-parabolic structure on E at p_j (cf. Exercise 5.2.E.3, especially the equation (*) therein), we get from (11) that $W_{P_j} w_j W_{Q_k}$ is the relative position of θ at p_j.

We finally compute the degree of the line bundle $\mathscr{S} := \hat{\theta}^*(\hat{\mathscr{L}}) \otimes (\pi^* \theta^* \mathscr{L})^*$ over $\hat{\Sigma}$, where

$$\hat{\mathscr{L}} := \hat{E} \times^{Q_k} \mathbb{C}_{\omega_k} \quad \text{and} \quad \mathscr{L} := E \times^{Q_k} \mathbb{C}_{\omega_k}.$$

By Exercise 6.1.E.14, the section $\theta_{|\Sigma^*}: \Sigma^* \to E/Q_k$ lifts to a *holomorphic* section $\Theta: \Sigma^* \to E$, i.e., $\Theta \bmod Q_k = \theta_{|\Sigma^*}$. Moreover, let $\hat{\Theta}: \hat{\Sigma}^* \to \hat{E}$ be the A-equivariant *holomorphic* section given as $\pi^*\Theta$, which lifts $\hat{\theta}_{|\hat{\Sigma}^*}$. Then $\hat{\Theta}$ provides a trivialization $\hat{\Theta}_{\mathscr{S}}$ of $\mathscr{S}_{|\hat{\Sigma}^*}$.

For any $1 \leq j \leq s$, take a section $\hat{\mu}_j$ of $\hat{E}_{|\hat{\mathbb{D}}_j}$ satisfying (4) and (5) and write

$$\hat{\mu}_{j|\hat{\mathbb{D}}_j^*} = \hat{\Theta}_{|\hat{\mathbb{D}}_j^*} \cdot \hat{\beta}_j, \quad \text{for} \quad \hat{\beta}_j \in G((\hat{t}_j)),$$

$$\text{cf. (2) of the proof of Theorem 6.1.12 taking } \hat{\sigma} = \hat{\Theta}. \tag{12}$$

Let μ_j be a section of $E_{|\mathbb{D}_j}$ given above satisfying (11), again replacing $\hat{\sigma} = \hat{\Theta}$. Then, by definition,

$$(\pi^*\mu_j)_{|\hat{\mathbb{D}}_j^*} = \hat{\Theta}_{|\hat{\mathbb{D}}_j^*} \cdot \hat{\beta}_j \cdot (\hat{t}_j)^{\bar{\tau}_j}. \tag{13}$$

Let $\mathbb{1}_{\omega_k}$ be a nonzero vector of $\mathbb{C}_{\omega_k}$ and let $\hat{s}_o$ be the section $\hat{E} \to \hat{E} \times \mathbb{C}_{\omega_k}$, $x \mapsto (x, \mathbb{1}_{\omega_k})$, of the trivial line bundle $\hat{E} \times \mathbb{C}_{\omega_k} \to \hat{E}$ (which is viewed as the pull-back of the line bundle $\mathscr{L}_{Q_k}(-\omega_k)$ over $\hat{E}/Q_k$ as in Definition 6.1.3(c) via the projection $\hat{q}: \hat{E} \to \hat{E}/Q_k$). Similarly, we define the section $s_o: E \to E \times \mathbb{C}_{\omega_k}$. From the identities (5) and (12) (by considering the sections $\hat{s}_o$ and s_o) we get that the line bundle $(\hat{\theta}^*\hat{\mathscr{L}})_{|\hat{\mathbb{D}}_j}$ has a section

$$\begin{aligned}
\hat{\delta}_j &:= \left[\hat{\mu}_j \cdot \bar{w}_j, \mathbb{1}_{\omega_k}\right], \\
&= \left[\hat{\Theta}_{|\hat{\mathbb{D}}_j^*} \cdot \hat{\beta}_j \cdot \bar{w}_j, \mathbb{1}_{\omega_k}\right] \quad \text{over} \quad \hat{\mathbb{D}}_j^* \\
&= \left[\hat{\Theta}_{|\hat{\mathbb{D}}_j^*}, \hat{\beta}_j \bar{w}_j \cdot \mathbb{1}_{\omega_k}\right],
\end{aligned} \tag{14}$$

since $\hat{\beta}_j \bar{w}_j$ has image in Q_k by the identity (10) taking $\hat{\sigma} = \hat{\Theta}$, where $[\hat{x}, \mathbb{1}_{\omega_k}] \in \hat{\mathscr{L}}$ denotes $(\hat{x}, \mathbb{1}_{\omega_k}) \bmod Q_k$, for $\hat{x} \in \hat{E}$. Similarly, using the identities (11) and (13), the line bundle $(\pi^*\theta^*\mathscr{L})_{|\hat{\mathbb{D}}_j}$ has a section

$$\begin{aligned}
\delta_j &:= \left[(\pi^*\mu_j) \cdot \bar{w}_j, \mathbb{1}_{\omega_k}\right] \\
&= \left[\hat{\Theta}_{|\hat{\mathbb{D}}_j^*} \cdot \hat{\beta}_j \cdot \bar{w}_j, \bar{w}_j^{-1} (\hat{t}_j)^{\bar{\tau}_j} \bar{w}_j \cdot \mathbb{1}_{\omega_k}\right] \quad \text{over} \quad \hat{\mathbb{D}}_j^* \\
&= \left[\hat{\Theta}_{|\hat{\mathbb{D}}_j^*} \cdot \hat{\beta}_j \cdot \bar{w}_j, (\hat{t}_j)^{\omega_k(\bar{w}_j^{-1}\bar{\tau}_j)} \cdot \mathbb{1}_{\omega_k}\right] \\
&= \left[\hat{\Theta}_{|\hat{\mathbb{D}}_j^*}, (\hat{t}_j)^{\omega_k(\bar{w}_j^{-1}\bar{\tau}_j)} \hat{\beta}_j \bar{w}_j \cdot \mathbb{1}_{\omega_k}\right].
\end{aligned} \tag{15}$$

From (14) and (15), we get that the line bundle $\mathscr{S}$ over $\hat{\Sigma}$ has section $\hat{\Theta}_{\mathscr{S}}$ over $\hat{\Sigma}^*$ and sections $(\hat{\mu}_j)_{\mathscr{S}}$ over $\hat{\mathbb{D}}_j$ satisfying the following equation over $\hat{\mathbb{D}}_j^*$:

$$(\hat{\mu}_j)_{\mathscr{S}} = \hat{\Theta}_{\mathscr{S}} \cdot (\hat{t}_j)^{-\omega_k(w_j^{-1}\bar{\tau}_j)}. \tag{16}$$

Since $\hat{\Sigma} \to \Sigma$ is of degree $N = \# A$ and, for each $1 \le j \le s$, there are N/d_j isomorphic copies of $\hat{\mathbb{D}}_j$ over $\mathbb{D}_j$, we get from (16) and Exercise 6.1.E.3,

$$\deg(\hat{\theta}^* \hat{\mathscr{L}}) - N \deg(\theta^* \mathscr{L}) = \deg \mathscr{S}$$

$$= \sum_{j=1}^{s} \frac{N}{d_j} \omega_k(w_j^{-1} \bar{\tau}_j).$$

Thus,

$$\deg(\hat{\theta}^* \hat{\mathscr{L}}) = N \left(\deg(\theta^* \mathscr{L}) + \sum_{j=1}^{s} \omega_k(w_j^{-1} \tau_j) \right), \quad \text{since} \quad \tau_j := \frac{\bar{\tau}_j}{d_j},$$

i.e.,

$$\deg \left(\hat{\theta}^* \hat{\mathscr{L}}_{Q_k}(-\omega_k) \right) = N \operatorname{Pardeg} \left(\theta^* \mathscr{L}_{Q_k}(-\omega_k) \right), \tag{17}$$

where Pardeg denotes the parabolic degree of the G-bundle E with respect to the section θ and the line bundle $\mathscr{L}_{Q_k}(-\omega_k)$ for the parabolic markings $\bar{\tau} = (\tau_1, \dots, \tau_s)$ (cf. Definition 6.1.4(d)). This proves the theorem. □

6.1.E Exercises

In the following, Σ is a smooth irreducible projective curve.

(1) Let G be a connected affine algebraic group and let $n \ge 1$. Show that the affine algebraic group $G\left(\mathbb{C}[[t]]/\langle t^n \rangle\right)^+$ (cf. Exercise 1.3.E.10) is a unipotent (in particular, connected) group with Lie algebra $\mathfrak{g} \otimes (t\mathbb{C}[[t]]/\langle t^n \rangle)$, where $G\left(\mathbb{C}[[t]]/\langle t^n \rangle\right)^+$ is the kernel of the homomorphism $G\left(\mathbb{C}[[t]]/\langle t^n \rangle\right) \to G$ induced by the $\mathbb{C}$-algebra homomorphism $\mathbb{C}[[t]]/\langle t^n \rangle \to \mathbb{C}$, $t \mapsto 0$.

Use the above to show that for any $R \in \mathbf{Alg}$ and $n \ge 0$, the canonical homomorphism

$$\pi_n^R : G\left(R[[t]]/\langle t^{n+1} \rangle\right)^+ \to G\left(R[[t]]/\langle t^n \rangle\right)^+$$

is surjective with kernel isomorphic (as a group) to the $\mathbb{C}$-vector space $R \otimes \left(\mathfrak{g} \otimes_{\mathbb{C}} \frac{t^n \mathbb{C}[[t]]}{t^{n+1}\mathbb{C}[[t]]}\right)$, where $G\left(R[[t]]/\langle t^n \rangle\right)^+$ is the kernel of the homomorphism $G\left(R[[t]]/\langle t^n \rangle\right) \to G(R)$ induced by $t \mapsto 0$.

Hint: By Exercises 1.3.E.10 and 1.3.E.6, $R \rightsquigarrow G\left(R[[t]]/\langle t^n \rangle\right)^+$ is a representable group functor, represented by an affine algebraic group with $\mathbb{C}$-points $G\left(\mathbb{C}[[t]]/\langle t^n \rangle\right)^+$.

(2) Let A be a group and let $\{\pi_n : G_{n+1} \to G_n\}_{n\geq 1}$ be an inverse system of A-groups. Assume that each $\pi_n^A : G_{n+1}^A \to G_n^A$ is surjective, where G_n^A denotes the subgroup of A-equivariants in G_n. Let G be the inverse limit of G_n. Then G is canonically an A-group.

Prove that if $H^1(A, G_n) = 0$, for all n, then so is $H^1(A, G) = 0$.

(3) Let $\mathscr{L}$ be a line bundle over Σ with nowhere vanishing sections σ over $\Sigma \backslash p$ and μ over $\mathbb{D}_p$ (for a fixed $p \in \Sigma$), where $\mathbb{D}_p$ is the formal disc centered at p in Σ with a local parameter z. Write

$$\mu_{|\mathbb{D}_p^*} = \sigma_{|\mathbb{D}_p^*} \cdot \beta(z), \quad \text{for} \quad \beta(z) \in \mathbb{C}((z)),$$

where $\mathbb{D}_p^* := \operatorname{Spec} \mathbb{C}((z))$ is the punctured formal disc at p. Then show that $\deg \mathscr{L} = d$, where d is the unique integer such that $z^d \cdot \beta(z) \in \mathbb{C}[[z]]$ with nonzero constant term.

(4) For any connected reductive algebraic group G, show that the two alternative definitions of semistability/stability (cf. Definition 6.1.4(b)) are equivalent.

Moreover, show that if a G-bundle $E \to \Sigma$ is semistable (resp. stable), then for any standard parabolic subgroup P of G and any section μ of $E/P \to \Sigma$,

$$\deg \mu^* \left(\mathscr{L}_P(-\lambda)\right) \leq 0 \left(\text{resp. } \deg \mu^* \left(\mathscr{L}_P(-\lambda)\right) < 0\right),$$

for any nontrivial character λ of P which is trivial restricted to the connected center of G and is dominant (i.e., $\lambda(\alpha_i^\vee) \geq 0$ for all the simple coroots $\alpha_i^\vee$).

(5) Show that a vector bundle $\mathscr{V}$ over Σ is semistable (resp. stable) if and only if the associated frame bundle $F(\mathscr{V})$ (which is a principal GL_n-bundle over Σ, where $n := \operatorname{rank}(\mathscr{V})$) is semistable (resp. stable) in the sense of Definition 6.1.4(b).

Hint: A rank-r subbundle $\mathscr{W}$ of $\mathscr{V}$ is given by a P_r-subbundle $F(\mathscr{V})_{P_r} \subset F(\mathscr{V})$ (induced by a section μ of $F(\mathscr{V})/P_r \to \Sigma$) by taking the associated vector bundle $F(\mathscr{V})_{P_r} \times^{P_r} \mathbb{C}^r$ and conversely, where P_r is the maximal parabolic subgroup of GL_n stabilizing $\mathbb{C}^r \subset \mathbb{C}^n$ under the standard representation. Now,

$$\deg(\mathscr{W}) = \deg\left(\mu^* \mathscr{L}_{P_r}(-\omega_r)\right)$$

and $n\omega_r - r\omega_n$ is a character of P_r which vanishes on its center, where

$$\omega_r\left(\operatorname{diag}(t_1, \ldots, t_n)\right) := t_1 \ldots t_r.$$

(6) Let $\mathscr{V}$ be a semistable vector bundle over Σ. Then show the following.

(a) For any nonzero $\mathscr{O}_\Sigma$-submodule $\mathscr{F}$ of $\mathscr{V}$,

$$\mu(\mathscr{F}) \leq \mu(\mathscr{V}),$$

where the slope $\mu(\mathscr{F})$ has the same definition as in Definition 6.1.4(a) for any $\mathscr{O}_\Sigma$-module $\mathscr{F}$.

(b) For any nonzero $\mathscr{O}_\Sigma$-module quotient $\mathscr{Q}$ of $\mathscr{V}$,

$$\mu(\mathscr{Q}) \geq \mu(\mathscr{V}).$$

(7) (a) Let $(\Sigma, \vec{p})$ be an s-pointed smooth irreducible projective curve, where $\vec{p} = \{p_1, \ldots, p_s\}$ and let $\mathscr{V}$ be a rank-n vector bundle over Σ. A *parabolic structure* for $\mathscr{V}$ at p_i consists of a partial flag in the fiber:

$$V_i^1 \subsetneq V_i^2 \subsetneq \cdots \subsetneq V_i^{l_i} = V_{p_i},$$

together with a set of markings:

$$1 > \mu_i^1 > \mu_i^2 > \cdots > \mu_i^{l_i} \geq 0,$$

where $V_{p_i} := \mathscr{V}_{p_i}$. Such a $\mathscr{V}$ with a parabolic structure is called a *parabolic vector bundle*.

The *parabolic degree* of $\mathscr{V}$ (with the above parabolic structure) is defined to be

$$\operatorname{pardeg}(\mathscr{V}) := \deg \mathscr{V} + \sum_{1\leq i\leq s} \sum_{1\leq k\leq l_i} \dim(V_i^k / V_i^{k-1})\, \mu_i^k,$$

where we set $V_i^0 = (0)$. The *parabolic slope* of $\mathscr{V}$ (with the above parabolic structure) is defined to be

$$\mu_{\text{par}}(\mathscr{V}) := \operatorname{pardeg}(\mathscr{V}) / \operatorname{rank}(\mathscr{V}).$$

The parabolic structure on $\mathscr{V}$ defines a parabolic structure on any subbundle $\mathscr{W}$ by defining a flag $\{W_i^d\}_{1\leq d\leq m_i}$ in the fiber W_{p_i} by removing repeated terms in the filtration and renumbering them successively:

$$V_i^1 \cap W_{p_i} \subset V_i^2 \cap W_{p_i} \subset \cdots \subset V_i^{l_i} \cap W_{p_i} = W_{p_i}.$$

Further, we define the markings ν_i^d, $1 \leq d \leq m_i$ by setting

$$\nu_i^d := \mu_i^k,$$

where k is the smallest integer with $W_i^d \subset V_i^k$.

Finally, the parabolic bundle $\mathscr{V}$ is defined to be *parabolic semistable* (resp. *parabolic stable*) if for every proper nonzero subbundle $\mathscr{W}$, we have

$$\mu_{\text{par}}(\mathscr{W}) \leq \mu_{\text{par}}(\mathscr{V}) \ \left(\text{resp. } \mu_{\text{par}}(\mathscr{W}) < \mu_{\text{par}}(\mathscr{V})\right).$$

Now, prove that the above notion of parabolic semistability (resp. parabolic stability) corresponds precisely to the notion of parabolic semistability (resp. parabolic stability) for the corresponding frame bundle $F(\mathscr{V})$ as in Definition 6.1.4(d). Write down the precise parabolic subgroups P_i, the sections σ_i of $F(\mathscr{V})_{p_i}/P_i$ and the markings τ_i (cf. Definition 6.1.2) under this correspondence.

(b) Show that a parabolic semistable vector bundle $\mathscr{V}$ has a filtration by parabolic semistable subbundles

$$\mathscr{V} = \mathscr{V}_1 \supsetneq \mathscr{V}_2 \supsetneq \cdots \supsetneq \mathscr{V}_\ell = (0)$$

such that (under the canonical parabolic structure)

(b_1) $\mu_{\text{par}}(\mathscr{V}_i) = \mu_{\text{par}}(\mathscr{V}),\quad$ for all $1 \leq i \leq \ell - 1$, and

(b_2) $\mathscr{V}_i/\mathscr{V}_{i+1}$ is parabolic stable, and

(b_3) $\text{gr}\,\mathscr{V} := \oplus_{i=1}^{\ell-1} \mathscr{V}_i/\mathscr{V}_{i+1}$ is independent (up to parabolic isomorphism) of the above filtration of $\mathscr{V}$ with properties (b_1) and (b_2).

(8) Let $f\colon G \to H$ be a surjective homomorphism between connected reductive groups such that the identity component of $\operatorname{Ker} f$ is a torus and let E be a G-bundle over Σ. Then, show that if $E(H)$ is stable (resp. semistable) then accordingly so is E.

(9) Let $\mathscr{V}$ be a vector bundle over Σ of degree d and rank r. Assume that $(d,r) = 1$. Then, show that $\mathscr{V}$ is semistable if and only if it is stable.

(10) Let $\mathscr{V}$ be a semistable vector bundle over Σ of degree 0. Then, show that any nonzero section of $\mathscr{V}$ is no-where zero.

Hint: A nonzero section gives rise to an injective $\mathscr{O}_\Sigma$-module map from $\mathscr{O}_\Sigma(D)$ to $\mathscr{V}$ for some effective divisor D.

(11) For a semisimple group G, show that any semistable G-bundle over $\mathbb{P}^1$ is trivial.

(12) Let E be a G-bundle over Σ for a connected reductive group G. Then E is semistable (resp. stable; polystable) if and only if $E(G/Z)$ is semistable (resp. stable; polystable), where Z is contained in the center of G.

Prove its analogue for the A-equivariant case.

Observe that Exercise 8 is a weaker version of this exercise.

(13) Let H be a connected affine algebraic group and Σ a smooth projective curve. Then show that any H-bundle over Σ is locally trivial in the Zariski topology.

Hint: Use the result that over a smooth affine curve any U-bundle is trivial, where U is a unipotent group. Moreover, prove that if H is reductive, then any H-bundle over Σ is Zariski locally trivial.

(14) Let H be a connected affine algebraic group and Σ a smooth projective curve. Then show that any holomorphic H-bundle over $\Sigma^* = \Sigma \backslash \{p_1, \dots, p_s\}$ (for $p_j \in \Sigma$ and $s \geq 1$) is *holomorphically* trivial.

Hint: Show that any holomorphic line bundle over Σ^* is holomorphically trivial by using the cohomology sequence corresponding to the sheaf exact sequence induced from $\mathscr{O}_{\text{hol}} \to \mathscr{O}^*_{\text{hol}}, f \mapsto e^{2\pi i f}$:

$$0 \to \mathbb{Z} \to \mathscr{O}_{\text{hol}} \to \mathscr{O}^*_{\text{hol}} \to 0.$$

(15) (a) A vector bundle $\mathscr{V}$ over Σ is polystable (cf. Definition 6.1.4) if and only if it is a direct sum of stable vector bundles all of which have the same slope.

(b) Let $\mathscr{V} = \mathscr{V}_1 \oplus \mathscr{V}_2$, where $\mathscr{V}_i$ are semistable vector bundles over Σ of the same slope μ. Then show that $\mathscr{V}$ is semistable.

Thus, a polystable vector bundle $\mathscr{V}$ is semistable.

Hint: Take any vector subbundle $\mathscr{W} \subset \mathscr{V}$ and consider its projection to $\mathscr{V}_1$. Now apply Exercise 6.1.E.6(a) or the construction $(*)$ in the proof of Lemma 6.3.22 to conclude that $\mu(\mathscr{W}) \leq \mu$.

(16) Following the notation in Definition 6.1.16, show that an A-equivariant vector bundle $\mathscr{V}$ over $\hat{\Sigma}$ is A-polystable if and only if we can write

$$\mathscr{V} = \oplus V_i,$$

where each V_i is an A-stable vector bundle all of which have the same slope.

6.2 Harder–Narasimhan Filtration for G-Bundles

In this section we assume that Σ is a smooth irreducible projective curve over $\mathbb{C}$ and G is a connected reductive group with a fixed Borel subgroup B and maximal torus $H \subset B$ with their Lie algebras $\mathfrak{g}$, $\mathfrak{b}$ and $\mathfrak{h}$, respectively. Let $Z(\mathfrak{g})$ $(\subset \mathfrak{h})$ be the center of $\mathfrak{g}$. By simple roots $\{\alpha_1, \dots, \alpha_\ell\}$ and fundamental weights $\{\omega_1, \ldots, \omega_\ell\}$, we mean the corresponding objects for the semisimple Lie algebra $\mathfrak{g}/Z(\mathfrak{g})$. In particular, $\omega_i, \alpha_i \in (\mathfrak{h}/Z(\mathfrak{g}))^*$.

Definition 6.2.1 Let $\pi : E \to \Sigma$ be a G-bundle. Then, a P-subbundle $E_P \subset E$ for a standard parabolic subgroup P of G (cf. Definition 5.1.1) is called a *Harder–Narasimhan reduction* (also called *HN reduction* for short or *canonical reduction*) if it satisfies the following two conditions.

(a) The associated L-bundle $E_P(L)$, obtained from the P-bundle E_P via the extension of the structure group $P \to P/U \simeq L$, is semistable, where L is the Levi subgroup of P containing H and U is the unipotent radical of P.

(b) For any nontrivial character λ of P such that $\lambda \in \bigoplus_{i=1}^{\ell} \mathbb{Z}_+\alpha_i$, where $\mathbb{Z}_+ := \mathbb{Z}_{\geq 0}$ (in particular, λ is trivial restricted to the identity component of the center of G),

$$\deg\ (E_P(\lambda)) > 0, \text{ where } E_P(\lambda) := E_P \times^P \mathbb{C}_\lambda.$$

By Theorem 6.2.3, such a reduction exists and is unique.

If we realize $E_P \subset E$ via a section μ of the bundle $E/P \to \Sigma$ (cf. Lemma 5.1.2), then the line bundle

$$E_P \times^P \mathbb{C}_\lambda \simeq \mu^* (\mathscr{L}_P(-\lambda)) \quad \text{(cf. Definition 6.1.3(c))}. \tag{1}$$

If E itself is semistable, then clearly it is an HN reduction.

Definition 6.2.2 For any G-bundle $E \to \Sigma$, define the integer

$$d_E = \min \left\{ \deg \mu^* \left(T_v(E/Q) \right) \right\},$$

where Q runs over all the standard parabolic subgroups of G and μ runs over all the sections of $E/Q \to \Sigma$. (Here the relative tangent bundle $T_v(E/Q)$ is as defined in Definition 6.1.3(c).)

Since any $\mu^*(T_v(E/Q))$ is a quotient of the adjoint bundle

$$\operatorname{ad} E := E \times^G \mathfrak{g} \quad (G \text{ acting on } \mathfrak{g} \text{ via the adjoint action}),$$

$\deg \mu^*(T_v(E/Q))$ is bounded from below by using the Riemann–Roch theorem for smooth curves (Hartshorne, 1977, Chap. IV, Theorem 1.3). Thus, d_E is indeed an integer.

Theorem 6.2.3 *Let P be a standard parabolic subgroup and let $E_P \subset E$ be a P-subbundle of a G-bundle $\pi : E \to \Sigma$ given by a section μ of $E/P \to \Sigma$ (via Lemma 5.1.2) satisfying the following conditions:*

(α) *deg* $\mu^* \left(T_v(E/P) \right) = d_E$.

(β) *There does not exist any parabolic* $\tilde{P} \supsetneq P$ *with a section* $\tilde{\mu}$ *of* $E/\tilde{P} \to \Sigma$ *such that*

$$deg\ \tilde{\mu}^* \left(T_v(E/\tilde{P})\right) = d_E.$$

Then E_P *is a HN reduction of* E. *Thus, a HN reduction of* E *exists. Moreover,* E_P *is the unique HN reduction.*

Further,

$$H^0\left(\Sigma, E_P \times^P \mathfrak{g}/\mathfrak{p}\right) = 0. \tag{1}$$

Proof Let $E_P \subset E$ be a reduction satisfying conditions (α) and (β). We first prove that E_P satisfies condition (a) of Definition 6.2.1, i.e., $E_L := E_P(L)$ is semistable.

We choose $B_L := B \cap L$ as the Borel subgroup of L. Assume, for contradiction, that E_L is not semistable, i.e., by Definition 6.1.4(b) there exists a standard parabolic subgroup Q of L and a section σ of $E_L/Q \to \Sigma$ such that

$$\deg \sigma^* \left(T_v(E_L/Q)\right) < 0. \tag{2}$$

Consider the surjective group homomorphism $p_L \colon P \to P/U \simeq L$ and let $P_1 := p_L^{-1}(Q)$. Since $p_L(B) = B_L$, $P_1 \subset P$ is a standard parabolic subgroup of G. Since the homomorphism p_L induces an isomorphism:

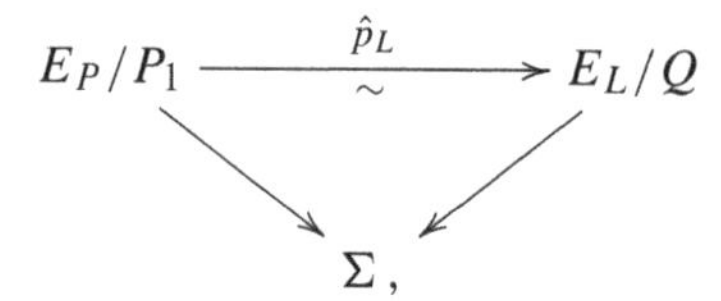

the section σ of E_L/Q induces a section μ_1 of $E_P/P_1 \subset E/P_1$. As earlier, we denote the Lie algebra of any group by the corresponding Gothic character. Then, we have the following exact sequence of P_1-modules:

$$0 \to \mathfrak{p}/\mathfrak{p}_1 \to \mathfrak{g}/\mathfrak{p}_1 \to \mathfrak{g}/\mathfrak{p} \to 0. \tag{3}$$

Since $\mathfrak{p}/\mathfrak{p}_1 \simeq \mathfrak{l}/\mathfrak{q}$ as Q-modules, the above exact sequence gives rise to the following exact sequence of vector bundles over Σ (cf. Definition 6.1.3(c)):

$$0 \to \sigma^* \left(T_v(E_L/Q)\right) \to \mu_1^* \left(T_v(E/P_1)\right) \to \mu^* \left(T_v(E/P)\right) \to 0. \tag{4}$$

Observe that $\pi \circ \mu_1 = \mu$, where $\pi : E/P_1 \to E/P$ is the projection. From (4), we get

$$\begin{aligned} \deg \mu_1^* \left(T_v(E/P_1)\right) &= \deg \sigma^* \left(T_v(E_L/Q)\right) + \deg \mu^* \left(T_v(E/P)\right) \\ &< \deg \mu^* \left(T_v(E/P)\right), \quad \text{by (2)}. \end{aligned}$$

This contradicts the choice (α). Thus, E_L is semistable.

We next show that E_P satisfies condition (b) of Definition 6.2.1.

For any nontrivial character λ of P such that $\lambda = \sum_{i=1}^{\ell} n_i\alpha_i$, with $n_i \in \mathbb{Z}_+$, we need to show that

$$\deg\left(E_P \times^P \mathbb{C}_\lambda\right) > 0. \tag{5}$$

Fix $\alpha_k \notin S_P$ (where S_P is the set of simple roots of L) and let $P_2 \supset P$ be the parabolic subgroup with $S_{P_2} = S_P \cup \{\alpha_k\}$ and let L_2 be its Levi component containing H again. (If $P = G$, there does not exist any nontrivial character λ of P such that $\lambda \in \sum_{i=1}^{\ell} \mathbb{Z}_+\alpha_i$.) Then the image of P under the homomorphism $p_{L_2}\colon P_2 \to L_2$ is a parabolic subgroup Q of L_2 giving rise to an isomorphism (as Q-modules, viewing Q as a subgroup of P under the embedding $L_2 \subset P_2$):

$$\mathfrak{p}_2/\mathfrak{p} \simeq \mathfrak{l}_2/\mathfrak{q}. \tag{6}$$

Similar to the exact sequence (4), we get the exact sequence

$$0 \to \sigma_2^*\left(T_v(E_{L_2}/Q)\right) \to \mu^*\left(T_v(E/P)\right) \to \mu_2^*\left(T_v(E/P_2)\right) \to 0, \tag{7}$$

where μ_2 is the section μ followed by the projection $E/P \to E/P_2$, $E_{L_2} := E_{P_2}(L_2)$ and σ_2 is the section of $E_{L_2}/Q \simeq E_{P_2}/P \subset E/P$ induced by μ. From the exact sequence (7), we get

$$\deg \mu^*\left(T_v(E/P)\right) = \deg \sigma_2^*\left(T_v(E_{L_2}/Q)\right) + \deg \mu_2^*\left(T_v(E/P_2)\right). \tag{8}$$

From the 'maximality' of P with the minimality of $\deg \mu^*\left(T_v(E/P)\right)$ as in (β), we get

$$\deg \sigma_2^*\left(T_v(E_{L_2}/Q)\right) < 0. \tag{9}$$

Now, by Definition 6.1.3(c),

$$T_v(E_{L_2}/Q) \simeq E_{L_2} \times^Q (\mathfrak{l}_2/\mathfrak{q}) \simeq E_{P_2} \times^P (\mathfrak{p}_2/\mathfrak{p}). \tag{10}$$

Clearly, $\wedge^{\text{top}}(\mathfrak{l}_2/\mathfrak{q}) \simeq \wedge^{\text{top}}(\mathfrak{p}_2/\mathfrak{p})$ is a P-module and the character

$$\mathrm{ch}_P\left(\wedge^{\text{top}}(\mathfrak{p}_2/\mathfrak{p})\right) = -\theta_k, \text{ where } \theta_k = m_k\alpha_k + \sum_{\alpha_i \in S_P} m_i^k\alpha_i, \tag{11}$$

for some $m_k \geq 1$ and $m_i^k \in \mathbb{Z}_+$. Combining (9)–(11), we get

$$\deg \sigma_2^*\left(T_v(E_{L_2}/Q)\right) = -\deg\left(E_P \times^P \mathbb{C}_{\theta_k}\right) < 0.$$

Thus,

$$\deg\left(E_P \times^P \mathbb{C}_{d\theta_k}\right) > 0, \text{ for any } d > 0. \tag{12}$$

From (12), we get that for some $N \gg 0$ and some $m_i \in \mathbb{Z}_+$,

$$\deg\left(E_P \times^P \mathbb{C}_\beta\right) > 0, \tag{13}$$

where $\lambda = \sum_{i=1}^{\ell} n_i\alpha_i$ is as in (5) (in particular, λ being a nontrivial character of P, $n_k > 0$ for some $\alpha_k \notin S_P$) and $\beta = N\sum_{\alpha_k \notin S_P} n_k\alpha_k + \sum_{\alpha_i \in S_P} m_i\alpha_i$. Now, $N\lambda$ and β are both characters of P (or equivalently characters of L) and clearly they coincide on $Z(\mathfrak{l})$, where $Z(\mathfrak{l})$ is the center of the Lie algebra $\mathfrak{l}$ of L. Of course, being characters of L, they both vanish on the commutator $[\mathfrak{l}, \mathfrak{l}]$. Hence, $N\lambda = \beta$ on $\mathfrak{l}$ and hence they coincide on L (L being connected). Thus, by (13), we get $\deg\left(E_P \times^P \mathbb{C}_{N\lambda}\right) > 0$, which gives $\deg\left(E_P \times^P \mathbb{C}_\lambda\right) > 0$. This proves (5), proving the first part of the theorem.

We now prove the uniqueness of the HN reduction.

Let $E_P \subset E$ and $E_{P'} \subset E$ be two HN reductions (for standard parabolic subgroups P and P') given by sections μ and μ' of $E/P \to \Sigma$ and $E/P' \to \Sigma$, respectively. The L-bundle E_L obtained from the extension of the structure group via $P \to P/U \simeq L$ is semistable by the definition of HN reduction, where $L \supset H$ (resp. U) is the Levi subgroup (resp. unipotent radical) of P. Consider the P-module filtration:

$$V_0 = 0 \subset V_1 \subset V_2 \subset \ldots \subset V_k = \mathfrak{g}/\mathfrak{p}$$

such that each $A_i := V_i/V_{i-1}$, $1 \le i \le k$, is an irreducible P-module. In particular, U acts trivially on each A_i (cf. Exercise 6.2.E.1).

Similarly, consider the P-module filtration

$$W_0 = 0 \subset W_1 \subset W_2 \subset \ldots \subset W_n = \mathfrak{u}$$

such that each $B_j := W_j/W_{j-1}$ is an irreducible P-module. Let $\mathscr{V}_i$, $\mathscr{A}_i$, $\mathscr{W}_j$ and $\mathscr{B}_j$ be the vector bundles over Σ associated to the P-bundle E_P by the P-modules V_i, A_i, W_j and B_j, respectively. For $1 \le i \le k$ and $1 \le j \le n$,

$$\mathscr{V}_i \subset \mu^*\left(T_v(E/P)\right) \quad \text{and} \quad \mathscr{W}_j \subset \operatorname{ad} E_P, \tag{14}$$

where $\operatorname{ad} E_P$ is the adjoint bundle $E_P \times^P \mathfrak{p} \to \Sigma$. We also let $\mathscr{B}_{n+1}$ be the vector bundle over Σ associated to $E_P \to \Sigma$ via the P-module $\mathfrak{p}/\mathfrak{u}$. Then, since $\mathfrak{l} \simeq \mathfrak{p}/\mathfrak{u}$, it is easy to see that

$$\mathscr{B}_{n+1} \simeq \operatorname{ad}(E_L). \tag{15}$$

Since each A_i and B_j (for $1 \le i \le k$ and $1 \le j \le n$) is an irreducible L-module, E_L is semistable and $\mathscr{B}_{n+1}$ is the associated adjoint bundle, by

Theorem 6.1.7, we get that each of vector bundles $\mathscr{A}_i$ $(1 \leq i \leq k)$ and $\mathscr{B}_j$ $(1 \leq j \leq n+1)$ is semistable (cf. Exercise 6.1.E.5). Clearly, for any $1 \leq i \leq k$ and any $1 \leq j \leq n$,

$$-\operatorname{ch}\left(\wedge^{\text{top}}(A_i)\right) \in \bigoplus_{p=1}^{\ell} \mathbb{Z}_+\alpha_p, \quad \operatorname{ch}\left(\wedge^{\text{top}}(B_j)\right) \in \bigoplus_{p=1}^{\ell} \mathbb{Z}_+\alpha_p, \tag{16}$$

and both of these are clearly nontrivial characters of P. Hence, by Definition 6.2.1(b), for any $1 \leq i \leq k$ and $1 \leq j \leq n$,

$$\deg \mathscr{A}_i < 0 \quad \text{and} \quad \deg \mathscr{B}_j > 0. \tag{17}$$

Moreover, since $\wedge^{\text{top}}(\mathfrak{l})$ is a trivial L-module,

$$\deg \mathscr{B}_{n+1} = 0. \tag{18}$$

In exactly the same way, we consider filtrations $V'_{i'}$ of $\mathfrak{g}/\mathfrak{p}'$ and $W'_{j'}$ of $\mathfrak{u}'$ giving rise to vector bundles $\mathscr{V}'_{i'}$, $\mathscr{A}'_{i'}$, $\mathscr{W}'_{j'}$ and $\mathscr{B}'_{j'}$ over Σ. Analogous to (17) and (18), we get, for all i' and j',

$$\deg \mathscr{A}'_{i'} < 0 \quad \text{and} \quad \deg \mathscr{B}'_{j'} \geq 0. \tag{19}$$

Moreover, the vector bundles $\mathscr{A}'_{i'}$ and $\mathscr{B}'_{j'}$ are semistable.

By the following lemma, there is no nonzero $\mathscr{O}_\Sigma$-linear map from any $\mathscr{B}_j$ to $\mathscr{A}'_{i'}$. Thus, working through the filtration $\mathscr{W}_j$ of $\operatorname{ad} E_P$ and the filtration $\mathscr{V}'_{i'}$ of $\mu'^*\left(T_v(E/P')\right)$, we get that

$$\operatorname{Hom}_{\mathscr{O}_\Sigma}\left(\operatorname{ad} E_P, \mu'^*\left(T_v(E/P')\right)\right) = 0. \tag{20}$$

The exact sequence of P-modules

$$0 \to \mathfrak{p} \to \mathfrak{g} \to \mathfrak{g}/\mathfrak{p}$$

gives rise to the exact sequence of vector bundles over Σ:

$$0 \to \operatorname{ad} E_P \to \operatorname{ad} E \to \mu^*\left(T_v(E/P)\right) \to 0,$$

and a similar sequence for the reduction (P', μ'). Thus, from (20), we get that

$$\operatorname{ad} E_P \subset \operatorname{ad} E_{P'}.$$

Similarly,

$$\operatorname{ad} E_{P'} \subset \operatorname{ad} E_P.$$

Thus,

$$\operatorname{ad} E_P = \operatorname{ad} E_{P'}, \quad \text{as subbundles of } \operatorname{ad} E. \tag{21}$$

We assert that from (21) we get

$$E_P = E_{P'} \text{ as subbundles of } E. \tag{22}$$

Take $x \in \Sigma$ and $e_x \in E_x$ (the fiber over x) such that

$$(E_P)_x = e_x \cdot P \quad \text{and} \quad (E_{P'})_x = e_x \cdot g_x \cdot P', \quad \text{for some} \quad g_x \in G. \tag{23}$$

Then, by definition,

$$(\operatorname{ad} E_P)_x = [e_x, \mathfrak{p}] \quad \text{and} \quad (\operatorname{ad} E_{P'})_x = [e_x g_x, \mathfrak{p}'],$$

where $[e_x, \mathfrak{p}]$ is the set of equivalence classes of (e_x, Y) in $E \times^G \mathfrak{g}$ as Y ranges over $\mathfrak{p}$. Since

$$[e_x, \mathfrak{p}] = [e_x g_x, \mathfrak{p}'] = [e_x, (\operatorname{Ad} g_x) \cdot \mathfrak{p}'], \quad \text{by (21)}, \tag{24}$$

we get

$$\mathfrak{p} = (\operatorname{Ad} g_x) \cdot \mathfrak{p}', \quad \text{equivalently} \quad P = g_x P' g_x^{-1}.$$

But since P and P' are both standard parabolic subgroups, we get (cf. (Borel, 1991, Theorem 11.16 and Corollary 11.17)) that $g_x \in P$ and $P' = P$. Thus, from (23),

$$(E_P)_x = (E_{P'})_x \quad \text{for all} \quad x \in \Sigma,$$

proving that $E_P = E_{P'}$ and hence E_P is unique.

To prove identity (1), from the filtration $\mathscr{V}_i$ of $E_P \times^P \mathfrak{g}/\mathfrak{p}$, it suffices to prove that $H^0(\Sigma, \mathscr{A}_i) = 0$ for all $1 \leq i \leq k$. But, as shown above, $\mathscr{A}_i$ are semistable vector bundles and further $\deg \mathscr{A}_i < 0$ (cf. (17)). Thus, $H^0(\Sigma, \mathscr{A}_i) = 0$ (e.g., by the next Lemma 6.2.4 applied to $\mathscr{E} = \mathscr{O}_\Sigma$ and $\mathscr{F} = \mathscr{A}_i$). This proves identity (1).

This proves the theorem modulo the next lemma. □

Lemma 6.2.4 *Let $\mathscr{E}$ and $\mathscr{F}$ be two semistable vector bundles over Σ such that*

$$\mu(\mathscr{E}) > \mu(\mathscr{F}). \tag{1}$$

Then

$$\operatorname{Hom}_{\mathscr{O}_\Sigma}(\mathscr{E}, \mathscr{F}) = 0.$$

Proof If possible, take a nonzero $f \in \operatorname{Hom}_{\mathscr{O}_\Sigma}(\mathscr{E}, \mathscr{F})$. Then by Exercise 6.1.E.6,

$$\mu(\mathscr{E}) \leq \mu(f(\mathscr{E})) \leq \mu(\mathscr{F}),$$

where $\mu(\mathscr{C}) := \frac{\deg \mathscr{C}}{\operatorname{rank} \mathscr{C}}$, for any $\mathscr{O}_\Sigma$-module $\mathscr{C}$. This is a contradiction to (1). This proves the lemma. □

From the uniqueness of the HN reduction, we get the following.

Corollary 6.2.5 *Let $\hat{\Sigma}$ be a smooth irreducible projective curve with the action of a finite group A and let $\hat{E}$ be an A-equivariant G-bundle over $\hat{\Sigma}$. Then its HN reduction $\hat{E}_P$ remains stable under the A-action, i.e., $A\cdot\hat{E}_P \subset \hat{E}_P$.*

Let $f : G \hookrightarrow G'$ be an embedding between connected reductive algebraic groups. Choose a Borel subgroup B of G (resp. B' of G') and a maximal torus $H \subset B$ of G (resp. $H' \subset B'$ of G') such that

$$B' \cap G = B \quad \text{and} \quad H' \cap G = H.$$

We fix these choices.

Let E be a G-bundle and let $E' := E(G')$ be the associated (principal) G'-bundle. Thus, $E \subset E'$ can be thought of as a G-subbundle of E'. Let $E_P \subset E$ and $E'_{P'} \subset E'$ be the HN reductions to P and P', for standard parabolic subgroups P of G and P' of G'.

Theorem 6.2.6 *With the notation as above, assume the following.*

$$\textit{For any } g \in G' \textit{ such that if } P \subset gP'g^{-1}, \textit{ then } g \in P', \tag{1}$$

and

$$U(P' \cap G) \subset U(P'), \tag{2}$$

where $U(P')$ denotes the unipotent radical of P'. Then

$$E_P = E'_{P'} \cap E \quad \textit{as subsets of} \quad E'.$$

Proof Consider the filtrations $\mathscr{V}_i \subset \mu^*(T_v(E/P))$ and $\mathscr{W}_j \subset \operatorname{ad} E_P$ as in (14) of the proof of Theorem 6.2.3, where the P-subbundle $E_P \subset E$ is given by a section μ of $E/P \to \Sigma$. Similarly, consider the filtrations $\mathscr{V}'_{i'} \subset \mu'^*\left(T_v(E'/P')\right)$ and $\mathscr{W}'_{j'} \subset \operatorname{ad} E'_{P'}$. As in the proof of Theorem 6.2.3 (using Lemma 6.2.4), we conclude that (considered as subsets of $E \times^G \mathfrak{g}'$)

$$\operatorname{ad} E_P \subset \operatorname{ad} E'_{P'}.$$

So far we have not used any of the assumptions (1) and (2). By an argument towards the end of the proof of Theorem 6.2.3, we get (by using assumption (1)): $E_P \subset E'_{P'}$, which gives

$$E_P \subset E'_{P'} \cap E, \tag{3}$$

and

$$P \subset P' \cap G. \tag{4}$$

Let $P_1 := P' \cap G$, which is a standard parabolic subgroup of G. From (3) and (4), we get

$$E_{P_1} := E_P \times^P P_1 = E'_{P'} \cap E. \tag{5}$$

Since the reduction E_P is a HN reduction, clearly E_{P_1} satisfies property (b) of HN reduction as in Definition 6.2.1.

So far, we have not used the assumption (2). Now, by assumption (2),

$$U(P_1) \subset U(P') \cap G.$$

Conversely, $U(P') \cap G$ being a normal unipotent subgroup of P_1, $U(P') \cap G \subset U(P_1)$. Thus,

$$U(P') \cap G = U(P_1). \tag{6}$$

The inclusions $P_1 \hookrightarrow P'$ and $U(P_1) \subset U(P')$ induce the commutative diagram

$$\begin{array}{ccc} P_1 & \longrightarrow & P_1/U(P_1) \\ \downarrow & & \downarrow \\ P' & \longrightarrow & P'/U(P'), \end{array}$$

where the right vertical map is injective by virtue of (6). This gives that

$$E_{P_1}(L_1) \hookrightarrow E'_{P'}(L'),$$

where L_1 (resp. L') is the Levi component of P_1 (resp. P') containing H (resp. H').

Now, since $E'_{P'}(L')$ is semistable (since $E'_{P'}$ is a HN reduction of E') so is its adjoint bundle $\operatorname{ad} E'_{P'}(L')$ (by Theorem 6.1.7). Moreover, $\operatorname{ad} E'_{P'}(L')$ has degree 0. Similarly, $\operatorname{ad} E_{P_1}(L_1)$ has degree 0. Thus, $\operatorname{ad} E_{P_1}(L_1)$ is a semistable vector bundle (by the definition of semistability of vector bundles as in Definition 6.1.4(a)). Thus, by Lemma 6.1.5, $E_{P_1}(L_1)$ is a semistable L_1-bundle. Hence, E_{P_1} satisfies property (a) of HN reduction as well. Thus, E_{P_1} is a HN reduction of E. From the uniqueness of HN reduction (cf. Theorem 6.2.3), we get that

$$P = P_1 \quad \text{and} \quad E_P = E_{P_1}.$$

Combining this with (5), we get

$$E_P = E'_{P'} \cap E, \quad \text{proving the theorem.}$$ □

Corollary 6.2.7 *Let $G \hookrightarrow G'$ be an embedding of connected reductive groups and let E be a G-bundle over Σ. Then, we have the following:*

(a) If $E(G')$ is semistable, then so is E.

(b) If E is semistable and G is not contained in any proper (not necessarily standard) parabolic subgroup of G', then $E(G')$ is semistable.

Proof (a) Let $E_P \subset E$ be the HN reduction of E. Since $E(G')$ is semistable, this is the HN reduction of $E(G')$. Thus, from Theorem 6.2.6, $E_P = E$ proving (a).

(Observe that the conditions (1) and (2) of Theorem 6.2.6 are trivially satisfied since $U(G) = \{1\}$.)

(b) Let $E'_{P'} \subset E(G')$ be the HN reduction. By the assumption in (b), condition (1) of Theorem 6.2.6 is clearly satisfied. By the proof of Theorem 6.2.6 (specifically identity (4), which does not require assumption (2) of Theorem 6.2.6), we get

$$G \subset P' \cap G, \quad \text{which gives} \quad G \subset P'.$$

But since, by assumption, there is no proper parabolic subgroup of G' containing G, we get $P' = G'$ and hence $E(G')$ is semistable. This proves (b). □

Remark 6.2.8 (1) The assumption in the (b)-part of Corollary 6.2.7 that there is no proper parabolic subgroup of G' containing G is, in general, required. Take, e.g., $G = H$, $G' = \mathrm{SL}_2(\mathbb{C})$, where H is the standard maximal torus of $\mathrm{SL}_2(\mathbb{C})$. Take any line bundle $\mathscr{L}$ over Σ of positive degree and let E be the corresponding G-bundle. Then $E' = E(G')$ corresponds to the frame bundle of the rank-2 vector bundle $\mathscr{L} \oplus \mathscr{L}^*$, which clearly is not semistable.

(2) Conditions (1) and (2) in Theorem 6.2.6 are missing in the corresponding theorem (Biswas and Holla, 2004, Theorem 5.1). Their proof has a gap which necessitated imposing conditions (1) and (2).

6.2.E Exercises

(1) Let H be an algebraic group and let V be an irreducible H-module. Show that the unipotent radical $U(H)$ of H acts trivially on V.

Hint: A unipotent group fixes a nonzero vector in any representation.

(2) Let $\mathscr{V} \to \Sigma$ be a vector bundle. Then there is a unique filtration of $\mathscr{V}$ by subbundles:

$$0 = \mathscr{V}_0 \subsetneq \mathscr{V}_1 \subsetneq \cdots \subsetneq \mathscr{V}_n = \mathscr{V},$$

such that each $\mathscr{V}_i/\mathscr{V}_{i-1}$ is a semistable vector bundle and, moreover, for all $2 \leq i \leq n$,

$$\mu(\mathscr{V}_i/\mathscr{V}_{i-1}) < \mu(\mathscr{V}_{i-1}/\mathscr{V}_{i-2}),$$

cf. Definition 6.1.4(a) for the definition of μ.

This filtration is called the Harder–Narasimhan (for short HN) filtration of $\mathscr{V}$.

Let $F(\mathscr{V})$ be the frame bundle of $\mathscr{V}$ and let $F(\mathscr{V})_P$ be the HN reduction of $F(\mathscr{V})$, where P is a standard parabolic subgroup of GL_N, N being the rank of $\mathscr{V}$ (cf. Definition 6.2.1). Then, the filtration of $\mathbb{C}^N$ induced by the parabolic subgroup P gives rise to a filtration of the vector bundle $\mathscr{V} = F(\mathscr{V}) \times^{\mathrm{GL}_N} \mathbb{C}^N$ from the reduction $F(\mathscr{V})_P$. Show that this filtration is the unique HN-filtration of $\mathscr{V}$.

Conversely, show that the HN filtration of $\mathscr{V}$ gives rise to the HN reduction of the GL_N-bundle $F(\mathscr{V})$.

(3) Let $f: G \to H$ be a homomorphism between connected reductive algebraic groups such that the identity component of $\operatorname{Ker} f$ is a torus and let E be a G-bundle over Σ. Then, if $E(H)$ is semistable, so is E. (Compare this exercise with Exercise 6.1.E.8.)

Hint: Express f as the composite $G \to^f f(G) \hookrightarrow^i H$. By Corollary 6.2.7(a), $E(f(G))$ is semistable. Now use Exercise 6.1.E.8.

(4) Following the definition and assumptions as in Definition 6.1.16, show that an A-equivariant G-bundle $\hat{E}$ over $\hat{\Sigma}$ is A-semistable if and only if it is semistable.

Hint: Use Theorem 6.2.3.

6.3 A Topological Construction of Semistable G-Bundles (Result of Narasimhan–Seshadri and its Generalization)

Let G be a connected reductive group and let K be a maximal compact subgroup (which is an $\mathbb{R}$-analytic group unique up to a conjugation). Let Σ be a smooth projective irreducible curve of genus $g \geq 1$. *This will be our tacit assumption through this section unless stated otherwise.*

Choose any base point $\infty \in \Sigma$. Then, as is well known (cf. (Spanier, 1966, Chap. 3, §8.12)), the fundamental group $\pi_1(\Sigma) = \pi_1(\Sigma, \infty)$ is isomorphic with

$$\pi_1(\Sigma) \simeq F(a_1, b_1, a_2, b_2, \ldots, a_g, b_g)/\langle \Pi_{i=1}^{g}[a_i, b_i] \rangle, \qquad (*)$$

where F denotes the free group, $[a_i, b_i]$ is the commutator $a_i b_i a_i^{-1} b_i^{-1}$ and $\langle\ \rangle$ denotes the normal subgroup generated by the enclosed element(s). Moreover, under this isomorphism,

$$\pi_1(\Sigma \setminus p, \infty) \simeq F(a_1, b_1, a_2, b_2, \ldots, a_g, b_g), \text{ for any } p \neq \infty \in \Sigma.$$

Definition 6.3.1 For any group homomorphism $\rho\colon \pi_1(\Sigma) \to G$, we get a holomorphic G-bundle E_ρ over Σ defined by extension of the structure group of the principal $\pi_1(\Sigma)$-bundle $q\colon \tilde{\Sigma} \to \Sigma$ to G via ρ, where $\tilde{\Sigma}$ is the simply-connected cover of Σ, i.e.,

$$E_\rho := \tilde{\Sigma} \times^{\pi_1(\Sigma)} G.$$

By the GAGA principle (Serre 1958, §6.3), E_ρ is an algebraic G-bundle over Σ (cf. Section 1.1).

If $\operatorname{Im} \rho \subset gKg^{-1}$ for some $g \in G$, then ρ is called a *unitary homomorphism* of $\pi_1(\Sigma)$ and E_ρ is called a *unitary G-bundle*. A representation of $\pi_1(\Sigma)$ in a finite-dimensional vector space V is called *unitary* if the corresponding homomorphism $\rho_o\colon \pi_1(\Sigma) \to \mathrm{GL}_V$ is unitary. Equivalently, V is unitary if it admits a positive-definite Hermitian form invariant under $\pi_1(\Sigma)$.

The homomorphism ρ is called *irreducible* if $\operatorname{Im} \rho$ is not contained in any proper (not necessarily standard) parabolic subgroup of G.

A vector bundle $\mathscr{V}$ over Σ is called *unitary* if there exists a finite-dimensional vector space V and a unitary representation $\rho_o\colon \pi_1(\Sigma) \to \mathrm{GL}_V$ such that $\mathscr{V} \simeq E_{\rho_o}(V) := E_{\rho_o} \times^{\mathrm{GL}(V)} V$.

Lemma 6.3.2 *Let $R_G(g)$ be the set of all the homomorphisms from $\pi_1(\Sigma)$ to G and let $R_K(g)$ be its subset consisting of unitary homomorphisms ρ with $\operatorname{Im} \rho \subset K$. Then $R_G(g)$ acquires an affine variety structure and $R_K(g)$ is a compact $\mathbb{R}$-analytic subset.*

Moreover, there exists a 'universal' $\mathbb{C}$-analytic G-bundle $\theta\colon \mathscr{E} \to \Sigma \times R_G(g)$ such that for any $\rho \in R_G(g)$, $\mathscr{E}_{|\Sigma \times \rho} \simeq E_\rho$.

Proof Consider the morphism of varieties:

$$\xi\colon (G \times G)^g \to [G, G],\ \ ((x_1, y_1), \ldots, (x_g, y_g)) \mapsto [x_1, y_1] \ldots [x_g, y_g].$$

Then, by the identification (*) as at the beginning of this section, mapping $a_i \mapsto x_i, b_i \mapsto y_i$, we get

$$R_G(g) = \xi^{-1}(1).$$

Thus, $R_G(g)$ acquires a natural affine variety structure as the reduced scheme corresponding to the scheme-theoretic fiber over 1. Now, $R_K(g) := R_G(g) \cap K^{\times 2g}$ and $K \subset G$ is an $\mathbb{R}$-analytic subgroup. Hence, $R_K(g)$ has a natural $\mathbb{R}$-analytic space structure and $R_K(g)$ being a closed subset of $K^{\times 2g}$ is compact in the analytic topology.

We now construct the family $\mathscr{E}$ over $\Sigma \times R_G(g)$.

Consider the right holomorphic action of $\pi_1(\Sigma)$ on $\tilde{\Sigma} \times R_G(g) \times G$ by

$$(\tilde{x}, \rho, g) \cdot \gamma = (\tilde{x} \cdot \gamma, \rho, \rho(\gamma^{-1})g),$$
$$\text{for } \tilde{x} \in \tilde{\Sigma}, \rho \in R_G(g), g \in G \text{ and } \gamma \in \pi_1(\Sigma).$$

Since the action of $\pi_1(\Sigma)$ on $\tilde{\Sigma}$ is fixed point free and properly discontinuous, so is its action on $\tilde{\Sigma} \times R_G(g) \times G$. Thus, we get a $\mathbb{C}$-analytic space

$$\mathscr{E} = \left(\tilde{\Sigma} \times R_G(g) \times G\right) / \pi_1(\Sigma)$$

together with holomorphic projection

$$\theta : \mathscr{E} \to \Sigma \times R_G(g), [\tilde{x}, \rho, g] \mapsto (q(\tilde{x}), \rho),$$

where $[\tilde{x}, \rho, g]$ denotes the $\pi_1(\Sigma)$-orbit of $(\tilde{x}, \rho, g)$. Then, θ is a $\mathbb{C}$-analytic principal G-bundle under the right action of G on $\mathscr{E}$ via the right multiplication on the G-factor (local triviality of θ is easy to see since $q : \tilde{\Sigma} \to \Sigma$ is locally trivial). By construction,

$$\mathscr{E}_{|\Sigma \times \rho} \simeq E_\rho, \quad \text{for any} \quad \rho \in R_G(g). \qquad \square$$

Definition 6.3.3 Let $\hat{\Sigma}$ be a smooth irreducible projective curve with the faithful action of a finite group A and let $\Sigma := \hat{\Sigma}/A$ be the quotient (smooth) curve. *If the genus g of Σ is 0, we assume that there are at least three ramification points of $\hat{\Sigma} \to \Sigma$.* Let $\hat{q} : \tilde{\hat{\Sigma}} \to \hat{\Sigma}$ be the simply-connected cover of $\hat{\Sigma}$ and let π_1 be the fundamental group of $\hat{\Sigma}$ with respect to a fixed base point in $\hat{\Sigma}$. Then, there exists a subgroup $\pi \subset \mathrm{Aut}_{\mathrm{hol}}(\tilde{\hat{\Sigma}})$ such that π acts discontinuously on $\tilde{\hat{\Sigma}}$ and π_1 is a normal subgroup of π (π_1 acting of course properly discontinuously without fixed points on $\tilde{\hat{\Sigma}}$). Moreover, $\pi/\pi_1 \simeq A$ and they satisfy

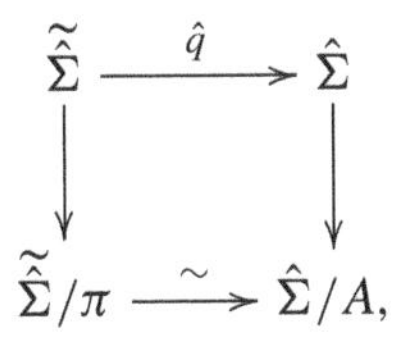

where the two vertical maps are the canonical orbit maps and the bottom horizontal morphism is induced by

$$\widetilde{\hat{\Sigma}}/\pi \simeq \left(\widetilde{\hat{\Sigma}}/\pi_1\right) \Big/ (\pi/\pi_1) \simeq \hat{\Sigma}/A.$$

In fact, π is the subgroup of $\mathrm{Aut}_{\mathrm{hol}}(\widetilde{\hat{\Sigma}})$ consisting of those automorphisms σ of $\widetilde{\hat{\Sigma}}$ which commute with the quotient $\bar{q}\colon \widetilde{\hat{\Sigma}} \to \Sigma$, i.e., $\bar{q} \circ \sigma = \bar{q}$. Then, it is well known (e.g. (Serre, 1992, §6.4) or (Jones and Singerman, 1987, §5.10)) that π has a presentation

$$\pi \simeq F(a_1, b_1, a_2, b_2, \ldots, a_g, b_g, c_1, \ldots, c_s)/M,$$

where F is the free group and M is the normal subgroup generated by

$$\mu := \left(\Pi_{i=1}^{g}[a_i, b_i]\right) \cdot c_1 \cdots c_s,\ c_1^{d_1}, \ldots, c_s^{d_s}.$$

Here d_j is the ramification index of p_j (i.e., the order of the isotropy subgroup of any preimage of p_j in $\hat{\Sigma}$, where $p_1, \ldots, p_s \in \Sigma$ are precisely the ramification points).

Let G be a connected reductive group. For any group homomorphism $\hat{\rho}\colon \pi \to G$ we get an A-equivariant G-bundles $\hat{E}_{\hat{\rho}}$ over $\hat{\Sigma}$ as follows:

$$\hat{E}_{\hat{\rho}} := \widetilde{\hat{\Sigma}} \times^{\pi_1} G \to \hat{\Sigma},$$

where π_1 acts on G via the left multiplication through the representation $\hat{\rho}_{|\pi_1}$. The A-equivariant structure on $\hat{E}_{\hat{\rho}}$ is given by fixing an identification $A \simeq \pi/\pi_1$ and defining

$$\gamma \cdot [z, g] = [z \cdot \gamma^{-1}, \hat{\rho}(\gamma)g], \quad \text{for } \gamma \in \pi, z \in \widetilde{\hat{\Sigma}} \text{ and } g \in G,$$

where $[z, g] \in \hat{E}_{\hat{\rho}}$ denotes the equivalence class of (z, g). This action clearly descends to give an A-equivariant structure on $\hat{E}_{\hat{\rho}}$.

We can clearly extend the definition of unitary (resp. irreducible) homomorphisms $\hat{\rho}\colon \pi \to G$. Similarly, we can define a unitary representation of π. If $\hat{\rho}$ is unitary, we call the corresponding A-equivariant G-bundle $\hat{E}_{\hat{\rho}}$ *A-unitary*. Similarly, an A-equivariant vector bundle $\hat{\mathscr{V}}$ over $\hat{\Sigma}$ is called *A-unitary* if there exists a finite-dimensional vector space V and a unitary representation $\hat{\rho}_o\colon \pi \to \mathrm{GL}_V$ such that $\hat{\mathscr{V}} \simeq \hat{E}_{\hat{\rho}_o}(V)$, as A-equivariant vector bundles.

Proposition 6.3.4 *We follow the notation and assumptions as in the above Definition 6.3.3.*

(a) *For any unitary homomorphism $\hat{\rho}\colon \pi \to G$, the associated A-equivariant G-bundle $\hat{E}_{\hat{\rho}}$ is A-semistable.*
(b) *Further, such on $\hat{E}_{\hat{\rho}}$ is A-stable if and only if $\hat{\rho}$ is irreducible.*

Proof (a) Let $\hat{\rho}$ be a unitary homomorphism. To prove that $\hat{E}_{\hat{\rho}}$ is A-semistable, by Exercise 6.2.E.4 and Lemma 6.1.5, it suffices to show that the adjoint vector bundle ad $\hat{E}_\rho$ over $\hat{\Sigma}$ is semistable, where $\rho := \hat{\rho}_{|\pi_1}$. But ad $\hat{E}_\rho$ is the vector bundle associated to the $\mathrm{SL}_{\mathfrak{g}}$-bundle $\hat{E}_{\mathrm{Ad}\,\rho}$, where Ad ρ is the composite homomorphism

$$\pi_1 \xrightarrow{\rho} G \xrightarrow{\mathrm{Ad}} \mathrm{SL}_{\mathfrak{g}}\,.$$

Since ρ is unitary, so is Ad ρ.

We now show that for a unitary homomorphism $\rho_o\colon \pi_1 \to \mathrm{SL}_V$ (for a finite-dimensional vector space V), the corresponding vector bundle $\mathscr{V} = \hat{E}_{\rho_o}(V) := \hat{E}_{\rho_o} \times^{\mathrm{SL}_V} V$ is semistable.

Let $\mathscr{W}$ be a vector subbundle of $\mathscr{V}$ of rank r. Since deg $\mathscr{V} = 0$, we need to show that deg $\mathscr{W} \leq 0$. Consider the line bundle

$$\wedge^r \mathscr{W} \subset \wedge^r \mathscr{V}.$$

Assume, if possible, that deg $\mathscr{W} > 0$. Then, there exists a degree 0 line bundle $\mathscr{L}$ over $\hat{\Sigma}$ such that

$$H^0(\hat{\Sigma}, \mathscr{L} \otimes \wedge^r \mathscr{W}) \neq 0. \tag{1}$$

Since V is a unitary representation of π_1, so is $\wedge^r V$. Also, the line bundle $\mathscr{L}$ being of degree 0 comes from a unitary character χ (i.e., 1-dimensional unitary representation $\mathbb{C}_\chi$) of π_1 (cf. Exercise 6.3.E.3). Choose a positive-definite π_1-invariant Hermitian form on $\mathbb{C}_\chi \otimes \wedge^r V$ and decompose

$$\mathbb{C}_\chi \otimes \wedge^r V = V_o \oplus V_o^\perp,$$

where $V_o := [\mathbb{C}_\chi \otimes \wedge^r V]^{\pi_1}$ is the subspace of π_1-invariants and $V_o^\perp$ is its ortho-complement. Decompose the vector bundle accordingly as

$$\mathscr{L} \otimes \wedge^r \mathscr{V} = \mathscr{V}_o \oplus \mathscr{V}_o^\perp.$$

Of course, $\mathscr{V}_o$ is a trivial vector bundle. Consider the projections $p_1\colon \mathscr{L} \otimes \wedge^r \mathscr{V} \to \mathscr{V}_o$ and $p_2\colon \mathscr{L} \otimes \wedge^r \mathscr{V} \to \mathscr{V}_o^\perp$ and let i denote the inclusion $i\colon \mathscr{L} \otimes \wedge^r \mathscr{W} \hookrightarrow \mathscr{L} \otimes \wedge^r \mathscr{V}$. By the next Lemma 6.3.6,

$$H^0\left(\hat{\Sigma}, \mathscr{V}_o^\perp\right) = 0 \tag{2}$$

and hence

$$p_2 \circ i = 0, \quad \text{by} \quad (1). \tag{3}$$

Further, since $\deg(\mathscr{L} \otimes \wedge^r \mathscr{W}) > 0$ and $\mathscr{V}_o$ is a trivial vector bundle,

$$p_1 \circ i = 0. \tag{4}$$

Combining (3) and (4), we get $i = 0$, which is a contradiction. Hence, $\deg \mathscr{W} \leq 0$, proving the (a)-part of the proposition.

(b) Assume that $\hat{\rho}\colon \pi \to G$ is an irreducible unitary homomorphism. We need to show that $\hat{E}_{\hat{\rho}}$ is an A-stable G-bundle.

Let Q_k be a standard maximal parabolic subgroup of G and let $\hat{\mu}$ be an A-equivariant section of $\hat{E}_{\hat{\rho}}/Q_k \to \hat{\Sigma}$. Then, following Definition 6.1.16, we need to show that

$$\deg \hat{\mu}^* \left(\hat{\mathscr{L}}_{Q_k}(-\bar{\omega}_k) \right) < 0, \tag{5}$$

where $d > 0$ is chosen so that $\bar{\omega}_k := d\omega_k$ is a character of Q_k and $\hat{\mathscr{L}}_{Q_k}(-\bar{\omega}_k) := \hat{E}_{\hat{\rho}} \times^{Q_k} \mathbb{C}_{\bar{\omega}_k}$ over $\hat{E}_{\hat{\rho}}/Q_k$. Since, by the (a)-part, $\hat{E}_{\hat{\rho}}$ is A-semistable, we get

$$\deg \hat{\mu}^* \left(\hat{\mathscr{L}}_{Q_k}(-\bar{\omega}_k) \right) \leq 0.$$

Assume, if possible, that

$$\deg(\hat{\mu}^* \hat{\mathscr{L}}_{Q_k}(-\bar{\omega}_k)) = 0. \tag{6}$$

In particular, $\hat{\mu}^* \hat{\mathscr{L}}_{Q_k}(-\bar{\omega}_k)$ is a A-unitary line bundle (cf. Exercise 6.3.E.3). Let $V(\bar{\omega}_k)$ be the irreducible representation of G with highest weight $\bar{\omega}_k$. Then, we get an embedding

$$j : G/Q_k \hookrightarrow \mathbb{P}(V(\bar{\omega}_k)), \quad gQ_k \mapsto [gv_+],$$

where v_+ is a nonzero highest-weight vector of $V(\bar{\omega}_k)$ and $[gv_+]$ denotes the line through gv_+. Let τ be the tautological line bundle over $\mathbb{P}(V(\bar{\omega}_k))$ restricted to G/Q_k. Then, since τ is the homogeneous line bundle over G/Q_k corresponding to the character $\bar{\omega}_k$ of Q_k, we get that, as A-equivariant line bundles over $\hat{E}_{\hat{\rho}}/Q_k$:

$$\hat{\mathscr{L}}_{Q_k}(-\bar{\omega}_k) \simeq \hat{E}_{\hat{\rho}} \times^{Q_k} \mathbb{C}v_+ \to \hat{E}_{\hat{\rho}}/Q_k.$$

Thus,

$$\hat{\mu}^* \hat{\mathscr{L}}_{Q_k}(-\bar{\omega}_k) \subset \hat{E}_{\hat{\rho}} \times^G C(G \cdot v_+) \subset \hat{E}_{\hat{\rho}}(V(\bar{\omega}_k)), \tag{7}$$

where $C(G \cdot v_+)$ is the cone $\mathbb{C} \cdot (G \cdot v_+)$ inside $V(\bar{\omega}_k)$. Assuming (6), we get that $\hat{\mu}^*(\hat{\mathscr{L}}_{Q_k}(-\bar{\omega}_k)^*) \otimes \hat{E}_{\hat{\rho}}(V(\bar{\omega}_k))$ is a A-unitary vector bundle and, by (7),

$$H^0\left(\hat{\Sigma}, \hat{\mu}^*\left(\hat{\mathscr{L}}_{Q_k}(-\bar{\omega}_k)^*\right) \otimes \hat{E}_{\hat{\rho}}(V(\bar{\omega}_k))\right)^A \neq 0.$$

Thus, by the following Lemma 6.3.6 and the first inclusion in (7), we get that there exists a $g \in G$ such that the line $\mathbb{C}gv_+$ is stable under π. This gives, from the embedding $j\colon G/Q_k \hookrightarrow \mathbb{P}(V(\bar{\omega}_k))$, that $\operatorname{Im}(\hat{\rho}) \subset gQ_kg^{-1}$. This is a contradiction to the assumption that ρ is irreducible. Thus, (5) is satisfied and hence $\hat{E}_{\hat{\rho}}$ is A-stable.

Conversely, assume that $\hat{\rho}$ is unitary and $\hat{E}_{\hat{\rho}}$ is A-stable. Then we need to show that $\hat{\rho}$ is irreducible. Assume, for contradiction, that $\hat{\rho}$ is not irreducible. Thus, $\operatorname{Im}\hat{\rho} \subset gQ_kg^{-1}$, for some $g \in G$ and a standard maximal parabolic subgroup Q_k of G. Since $\hat{E}_{\hat{\rho}} \simeq \hat{E}_{g^{-1}\hat{\rho}g}$ (cf. Exercise 6.3.E.1), we can assume that

$$\operatorname{Im}\hat{\rho} \subset Q_k.$$

Then

$$\hat{E}_{\hat{\rho}}/Q_k \simeq \hat{E}_{\hat{\rho}} \times^G G/Q_k \simeq \tilde{\hat{\Sigma}} \times^{\pi_1} G/Q_k \supset \tilde{\hat{\Sigma}} \times^{\pi_1} Q_k/Q_k \simeq \hat{\Sigma}.$$

This gives rise to an A-equivariant section $\hat{\mu}$ of $\hat{E}_{\hat{\rho}}/Q_k$ over $\hat{\Sigma}$. It is easy to see that (as A-equivariant line bundles over $\hat{\Sigma}$)

$$\hat{\mu}^*\left(\hat{\mathscr{L}}_{Q_k}(-\bar{\omega}_k)\right) \simeq \tilde{\hat{\Sigma}} \times^{\pi_1} \mathbb{C}_{\bar{\omega}_k}, \tag{8}$$

where π_1 acts on the 1-dimensional space $\mathbb{C}_{\bar{\omega}_k}$ via the character $\bar{\omega}_k$ of Q_k through the homomorphism $\hat{\rho}_{|\pi_1}\colon \pi_1 \to Q_k$. Since $\hat{\rho}$ is unitary, by (8), we have that $\hat{\mu}^*(\hat{\mathscr{L}}_{Q_k}(-\bar{\omega}_k))$ is a A-unitary line bundle. Further, $\hat{E}_{\hat{\rho}}$ being A-stable,

$$\deg \hat{\mu}^*(\hat{\mathscr{L}}_{Q_k}(\bar{\omega}_k)) > 0. \tag{9}$$

Thus, for $N \gg 0$, $H^0(\hat{\Sigma}, \hat{\mu}^*(\hat{\mathscr{L}}_{Q_k}(N\bar{\omega}_k))) \neq 0$ (cf. (Hartshorne, 1977, Chap. IV, Corollary 3.3)). Hence, by (8) and the next lemma for $A = (1)$, π_1 acts trivially on $\mathbb{C}_{N\bar{\omega}_k}$. In particular, by (8) again, $\hat{\mu}^*(\hat{\mathscr{L}}_{Q_k}(N\bar{\omega}_k))$ is (non-equivariantly) a trivial line bundle over $\hat{\Sigma}$, which gives $\deg \hat{\mu}^*(\hat{\mathscr{L}}_{Q_k}(N\bar{\omega}_k)) = 0$ and hence $\deg \hat{\mu}^*(\hat{\mathscr{L}}_{Q_k}(\bar{\omega}_k)) = 0$. This contradicts (9). This contradiction shows that for $\hat{E}_{\hat{\rho}}$ to be A-stable, $\hat{\rho}$ must be irreducible. This proves the proposition modulo the following Lemma 6.3.6. □

Remark 6.3.5 From the above proof we see that any unitary line bundle $\mathscr{L}$ over $\hat{\Sigma}$ has $\deg \mathscr{L} = 0$.

Lemma 6.3.6 *We follow the notation and assumptions as in Definition 6.3.3. Let* $\hat{\rho}_o \colon \pi \to \mathrm{GL}_V$ *be a finite-dimensional complex representation such that* $\operatorname{Im} \hat{\rho}_o$ *leaves a positive-definite Hermitian form* $\{\cdot,\cdot\}$ *on* V *invariant. Then, there is a canonical isomorphism from the group cohomology*

$$H^0(\pi, V) \to H^0(\hat{\Sigma}, \mathscr{V}_{\hat{\rho}_o})^A,$$

where $\mathscr{V}_{\hat{\rho}_o}$ *is the* A*-equivariant vector bundle* $\hat{E}_{\hat{\rho}_o}(V)$ *over* $\hat{\Sigma}$.

Moreover, any section $s \in H^0(\hat{\Sigma}, \mathscr{V}_{\hat{\rho}_o})^A$ *pulled back to* $\widetilde{\hat{\Sigma}}$ *is of the form* $\tilde{s}(\tilde{x}) = (\tilde{x}, v_o)$*, for some fixed* $v_o \in V^\pi$.

Proof Decompose $V = V_o \oplus V_o^\perp$, where $V_o := V^\pi$ is the subspace of π-invariants in V and $V_o^\perp$ is the ortho-complement of V_o in V (which is clearly a π-module). Then

$$H^0(\pi, V) \simeq V_o \quad \text{and} \quad \mathscr{V}_{\hat{\rho}_o} = \hat{E}_{\hat{\rho}_o}(V_o) \oplus \hat{E}_{\hat{\rho}_o}(V_o^\perp). \tag{1}$$

Of course, V_o being a trivial π-module and $\hat{\Sigma}$ being irreducible and projective,

$$H^0(\hat{\Sigma}, \hat{E}_{\hat{\rho}_o}(V_o))^A \simeq V_o. \tag{2}$$

So, to prove the lemma, by (1) and (2), it suffices to show that

$$H^0(\hat{\Sigma}, \hat{E}_{\hat{\rho}_o}(V_o^\perp))^A = 0. \tag{3}$$

Now, by the definition, for any π-module W, the associated A-equivariant vector bundle $\mathscr{W} := \widetilde{\hat{\Sigma}} \times^{\pi_1} W$ has

$$H^0(\hat{\Sigma}, \mathscr{W})^A = \left\{\text{Hol. maps } f \colon \widetilde{\hat{\Sigma}} \to W \text{ satisfying the following identity}\right\} \tag{4}$$

$$f(\tilde{x} \cdot \sigma) = \sigma^{-1} \cdot f(\tilde{x}), \quad \text{for all } \tilde{x} \in \widetilde{\hat{\Sigma}} \text{ and } \sigma \in \pi. \tag{5}$$

If $\mathscr{W}$ is a A-unitary vector bundle, then, for any such f, we get

$$||f(\tilde{x} \cdot \sigma)|| = ||f(\tilde{x})||, \quad \text{for all } \tilde{x} \in \widetilde{\hat{\Sigma}} \text{ and } \sigma \in \pi.$$

Thus, $||f(\tilde{x})||$ descends to a continuous function on Σ; in particular, it attains a maximum α say at $\tilde{x}_o \in \widetilde{\hat{\Sigma}}$. Choosing an appropriate orthonormal basis of W, we can write (for $n = \dim W$)

$$f = (f_1, \ldots, f_n), \quad f(\tilde{x}_o) = (\alpha, 0, \ldots, 0).$$

Now, $f_1 \colon \widetilde{\hat{\Sigma}} \to \mathbb{C}$ being a holomorphic map, $\operatorname{Im} f_1$ is open (unless it is a constant). In particular, if f_1 is non-constant, there exists $\tilde{y} \in \widetilde{\hat{\Sigma}}$ with $f(\tilde{y}) = (\alpha + \epsilon, f_2(\tilde{y}), \ldots, f_n(\tilde{y}))$, for some $\epsilon > 0$. This is a contradiction since $||f(\tilde{y})|| \leq \alpha$. Thus, f_1 is a constant giving

$$f(\tilde{x}) = (\alpha, f_2(\tilde{x}), \ldots, f_n(\tilde{x})), \quad \text{for all} \quad \tilde{x} \in \widetilde{\hat{\Sigma}}.$$

But, since $||f(\tilde{x})|| \leq \alpha$, we get $f_2 = \ldots = f_n \equiv 0$, i.e.,

$$f(\tilde{x}) = (\alpha, 0, \ldots, 0). \tag{6}$$

Thus, by (4) and (5),

$$H^0(\hat{\Sigma}, \mathscr{W})^A \simeq W^\pi.$$

Since $(V_o^\perp)^\pi = 0$, we get (3). This proves the first part of the lemma. The second part follows from (4), (5) and (6). □

As a corollary of the proof of Lemma 6.3.6, we get the following with the same notation and assumptions as in Definition 6.3.3.

Corollary 6.3.7 *Let $\hat{\rho}$, $\hat{\rho}'$ be unitary homomorphisms $\pi \to G$. Then $\hat{E}_{\hat{\rho}} \simeq \hat{E}_{\hat{\rho}'}$ (as A-equivariant holomorphic G-bundles over $\hat{\Sigma}$) if and only if*

$$\hat{\rho}' = g\hat{\rho}g^{-1}, \quad \textit{for some} \quad g \in G.$$

Moreover, if $\hat{\rho}$ and $\hat{\rho}'$ both have images in a maximal compact subgroup K of G, then g (as above) can be taken to lie in K.

Proof If $\hat{\rho}'$ is conjugate of $\hat{\rho}$, then $\hat{E}_{\hat{\rho}} \simeq \hat{E}_{\hat{\rho}'}$ (see Exercise 6.3.E.1).

Conversely, assume that $\hat{E}_{\hat{\rho}} \overset{\varphi}{\simeq} \hat{E}_{\hat{\rho}'}$ as A-equivariant G-bundles. By conjugating $\hat{\rho}'$ by some $g \in G$, we can (and will) assume that $\operatorname{Im} \hat{\rho}$ and $\operatorname{Im} \hat{\rho}'$ both lie in the same maximal compact subgroup K of G. Similar to condition (5) in the proof of Lemma 6.3.6, we get that φ is induced from a holomorphic function $\overline{\varphi} \colon \widetilde{\hat{\Sigma}} \to G$ satisfying

$$\overline{\varphi}(\tilde{x} \cdot \sigma)\hat{\rho}(\sigma^{-1}) = \hat{\rho}'(\sigma^{-1})\overline{\varphi}(\tilde{x}), \quad \text{for} \quad \tilde{x} \in \widetilde{\hat{\Sigma}} \quad \text{and} \quad \sigma \in \pi, \tag{1}$$

in the sense that the map $(\tilde{x}, g) \mapsto (\tilde{x}, \overline{\varphi}(\tilde{x})g)$, for $\tilde{x} \in \widetilde{\hat{\Sigma}}$, $g \in G$ descends to give the isomorphism φ.

Take a faithful representation $i \colon G \hookrightarrow \mathrm{GL}_V$ and realize $W := \operatorname{End} V$ as a G-module under the conjugation:

$$g \cdot A = i(g) A\, i(g)^{-1}, \quad \text{for} \quad g \in G \quad \text{and} \quad A \in \operatorname{End} V.$$

Put the standard positive-definite Hermitian product on W:

$$\{A, B\} = \quad \text{trace} \quad AB^*,$$

where B^* is the adjoint of B taken with respect to a fixed K-invariant positive-definite Hermitian product on V. Clearly $\{,\}$ is K-invariant; in particular, it is invariant under both $\hat{\rho}$ and $\hat{\rho}'$ (this is where we use the assumption that $\operatorname{Im}\hat{\rho}$, $\operatorname{Im}\hat{\rho}' \subset K$).

From (1) we get, for any $\tilde{x} \in \tilde{\hat{\Sigma}}$ and $\sigma \in \pi$,

$$\begin{aligned} ||i(\overline{\varphi}(\tilde{x}\sigma))|| &= ||i(\hat{\rho}'(\sigma^{-1}))i(\overline{\varphi}(\tilde{x}))i(\hat{\rho}(\sigma))|| \\ &= ||i\overline{\varphi}(\tilde{x})||, \quad \text{since } \hat{\rho}(\sigma), \hat{\rho}'(\sigma^{-1}) \in K. \end{aligned}$$

Thus, $||i \circ \overline{\varphi}||$ descends to a continuous function on $\hat{\Sigma}$. By the same argument as in the proof of Lemma 6.3.6, we get that $\overline{\varphi}\colon \tilde{\hat{\Sigma}} \to G$ is a constant function, say $\overline{\varphi}(\tilde{\hat{\Sigma}}) = g_o \in G$. Thus, by (1),

$$\hat{\rho}(\sigma^{-1}) = g_o^{-1}\hat{\rho}'(\sigma^{-1})g_o \text{ , for all } \sigma \in \pi.$$

Hence, $\hat{\rho}$ and $\hat{\rho}'$ are conjugate.

The 'Moreover' assertion follows from (Helgason, 1978, Chap. VI, Theorem 1.1), proving the corollary. □

Lemma 6.3.8 *We follow the notation and assumptions as in Definition 6.3.3. Let $\rho\colon \pi_1 \to G$ be a unitary homomorphism such that the corresponding holomorphic G-bundle $E_\rho := \tilde{\hat{\Sigma}} \times^{\pi_1} G$ over $\hat{\Sigma}$ is holomorphically A-equivariant, where $\pi_1 := \pi_1(\hat{\Sigma})$. Then, ρ extends to a unitary homomorphism $\hat{\rho}\colon \pi \to G$ such that*

$$\hat{E}_{\hat{\rho}} \simeq E_\rho, \text{ as } A\text{-equivariant holomorphic } G\text{-bundles.} \tag{1}$$

Proof Any A-equivariant structure on E_ρ (using the pull-back of E_ρ to $\tilde{\hat{\Sigma}} \times G$) is given by

$$\sigma \cdot [\tilde{x}, g] = [\tilde{x}\sigma^{-1}, \varphi_\sigma(\tilde{x})g], \text{ for } \sigma \in \pi, \tilde{x} \in \tilde{\hat{\Sigma}} \text{ and } g \in G, \tag{2}$$

where $\varphi_\sigma : \tilde{\hat{\Sigma}} \to G$ is a holomorphic map satisfying

(a) $\varphi_{\sigma_1\sigma_2}(\tilde{x}) = \varphi_{\sigma_1}(\tilde{x} \cdot \sigma_2^{-1})\varphi_{\sigma_2}(\tilde{x})$, for $\sigma_1, \sigma_2 \in \pi$ and $\tilde{x} \in \tilde{\hat{\Sigma}}$, and

(b) $\varphi_\sigma(\tilde{x}\mu^{-1})\rho(\mu) = \rho(\sigma\mu\sigma^{-1})\varphi_\sigma(\tilde{x})$, for $\mu \in \pi_1, \sigma \in \pi$ and $\tilde{x} \in \tilde{\hat{\Sigma}}$.

Moreover, since $A = \pi/\pi_1$; in particular, π_1 acts trivially on E_ρ. Thus,

(c) $\varphi_\mu(\tilde{x}) = \rho(\mu)$, for all $\mu \in \pi_1$ and $\tilde{x} \in \tilde{\hat{\Sigma}}$.

Since ρ is unitary, by using the same argument as in the proof of Corollary 6.3.7, we get (using (b) and comparing it with identity (1) of Corollary 6.3.7):

$\varphi_\sigma : \hat{\widetilde{\Sigma}} \to G$ is a constant function with image denoted $\bar{\varphi}_\sigma \in G$. In particular, by (a),

$$\bar{\varphi}_{\sigma_1\sigma_2} = \bar{\varphi}_{\sigma_1}\bar{\varphi}_{\sigma_2}. \tag{3}$$

Further, by (c), we get

$$\bar{\varphi}_{|\pi_1} = \rho.$$

Thus, setting $\hat{\rho} = \bar{\varphi} : \pi \to G$ we get (1) from (2) and (3). Of course, $\hat{\rho}$ is unitary since π_1 is of finite index in π. □

Definition 6.3.9 (A construction of topological G-bundles) We take Σ and G as at the beginning of this section. For any continuous map $c : S^1 \to G$, construct a topological principal G-bundle F_c over Σ as follows. Fix a base point $p \in \Sigma$ and take an open disc D_p in Σ around p. Fix a homotopy equivalence $h : D_p^* \to S^1$, where $D_p^* = D_p \backslash \{p\}$, and let $\overline{c} : D_p^* \to G$ be the composite $c \circ h$. Let $\Sigma^* := \Sigma \backslash \{p\}$. Take the trivial G-bundles

$$D_p \times G \to D_p \quad \text{and} \quad \Sigma^* \times G \to \Sigma^*$$

and 'clutch' them via $\overline{c}$ to get a topological G-bundle F_c over Σ, i.e.,

$$F_c := (D_p \times G) \sqcup (\Sigma^* \times G)/\sim,$$

where

$$(x,g) \in D_p \times G \sim (x, \overline{c}(x)g) \in \Sigma^* \times G, \quad \text{for} \quad x \in D_p^* \quad \text{and} \quad g \in G. \tag{1}$$

The projection $F_c \to \Sigma$ is obtained by the projections to the first factor. It can be seen that the topological G-bundle F_c (up to an isomorphism) does not depend upon the choices of c in its homotopy class, p, D_p and h (cf. Exercise 6.3.E.2). Thus, we get the 'clutching' map

$$\eta : [S^1, G] \to \mathrm{Bun}_G^{\mathrm{top}}(\Sigma), \quad [c] \mapsto F_c,$$

where $[S^1, G]$ is the set of (free) homotopy classes of maps from $S^1 \to G$ and $\mathrm{Bun}_G^{\mathrm{top}}(\Sigma)$ is the set of isomorphism classes of topological principal G-bundles over Σ.

Lemma 6.3.10 *The above map*

$$\eta : [S^1, G] \mapsto \mathrm{Bun}_G^{\mathrm{top}}(\Sigma)$$

is a bijection. Of course, $[S^1, G]$ is bijective with $\pi_1(G)$.

Proof Since D_p is contractible, any G-bundle over D_p is trivial (cf. (Steenrod, 1951, Corollary 11.6)). Further, since Σ^* is homotopic to a 1-complex and the classifying space BG has trivial fundamental group (G being connected), any G-bundle over Σ^* is trivial. From this, we see that η is surjective.

To prove the injectivity of η, take two continuous maps $c_1, c_2 \colon S^1 \to G$ and assume that there exists a G-bundle isomorphism:

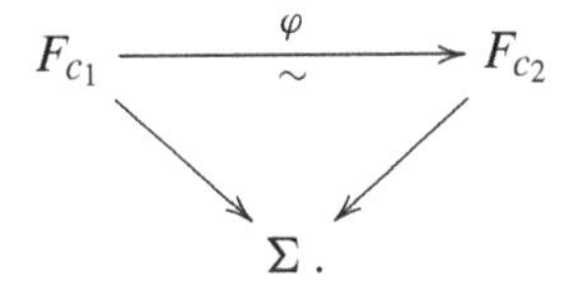

For any $i = 1, 2$, take the section μ_i of $F_{c_i|D_p}$ given by

$$\mu_i(z) = (z, 1) \in D_p \times G, \quad \text{for any } z \in D_p$$

and the section σ_i of $F_{c_i|\Sigma^*}$ given by

$$\sigma_i(z) = (z, 1) \in \Sigma^* \times G, \quad \text{for any } z \in \Sigma^*.$$

Thus, by identity (1) of Definition 6.3.9,

$$\mu_i = \sigma_i \cdot \bar{c}_i \ \text{ over } D_p^*. \tag{1}$$

Also, let μ_2' (resp. σ_2') be the section of $F_{c_2|D_p}$ (resp. $F_{c_2|\Sigma^*}$) given by

$$\mu_2' := \varphi \circ \mu_1 \ \text{ and } \ \sigma_2' := \varphi \circ \sigma_1.$$

Then, we get continuous functions $\alpha \colon D_p \to G$ and $\beta \colon \Sigma^* \to G$ such that

$$\mu_2 = \mu_2' \cdot \alpha \text{ over } D_p \text{ and } \sigma_2 = \sigma_2' \cdot \beta \text{ over } \Sigma^*. \tag{2}$$

From relations (1) and (2), we get

$$\bar{c}_1(z) = \beta(z)\bar{c}_2(z)\alpha(z)^{-1} \ \text{ for } z \in D_p^*. \tag{3}$$

Since α is defined as a continuous function on D_p,

$$\alpha \colon D_p^* \to G \ \text{ is homotopically trivial.} \tag{4}$$

Similarly, since β is defined as a continuous function on Σ^*, $\beta_{|D_p^*} \colon D_p^* \to G$ is also homotopically trivial, as the following argument shows.

Denote $\beta_{|D_p^*} = \bar{\beta}$. By (Spanier, 1966, Chap. III, §8), the induced map $i_* \colon \pi_1(D_p^*, \infty) \to \pi_1(\Sigma^*, \infty)$ from the inclusion $i \colon D_p^* \to \Sigma^*$, where ∞ is any base point in the boundary of D_p, takes a generator

$$\sigma \in \pi_1(D_p^*, \infty) \mapsto [a_1, b_1] \dots [a_g, b_g] \in \pi_1(\Sigma^*, \infty)$$

(cf. the beginning of this Section 6.3 for the description of $\pi_1(\Sigma^*,\infty)$). Thus, the composite map

$$\bar{\beta}_* = \beta_* \circ i_* : \pi_1(D_p^*,\infty) \to \pi_1(G,\beta(\infty))$$

is trivial since $\pi_1(G,\beta(\infty))$ is an abelian group. From this we conclude that

$$\bar{\beta}: D_p^* \to G \text{ is homotopically trivial.} \tag{5}$$

Combining (3)–(5), we conclude that c_1 and c_2 are homotopic, proving the injectivity of η. □

Lemma 6.3.11 *Let genus g of Σ be ≥ 2 and let G be a connected semisimple group. Then, for any $F \in \mathrm{Bun}_G^{\mathrm{top}}(\Sigma)$, there exists an irreducible (unitary) homomorphism $\rho_F : \pi_1(\Sigma) \to K \subset G$ such that $E_{\rho_F} \simeq F$ as topological G-bundles, where E_{ρ_F} is as in Definition 6.3.1.*

Proof Let $\alpha : \tilde{K} \to K$ be the simply-connected cover of K. Fix any $c \in \mathrm{Ker}\,\alpha$. Then, of course, c is a central element of $\tilde{K}$ (hence c belongs to any maximal torus of $\tilde{K}$) and K being semisimple, $\mathrm{Ker}\,\alpha$ is finite. We assert that there exists a group homomorphism

$$\tilde{\rho}_c : \pi_1(\Sigma^*) \to \tilde{K}, \quad \text{such that} \quad \tilde{\rho}_c(\mu) = c,$$

where, as earlier, $\Sigma^* := \Sigma \setminus p$, $\infty \in \Sigma$ is any point lying on the boundary ∂D_p of a small disc D_p around p in Σ, $\pi_1(\Sigma^*)$ denotes $\pi_1(\Sigma^*,\infty)$ and μ is a generator of $\mathrm{Ker}\,\pi_1(\Sigma^*) \to \pi_1(\Sigma,\infty)$ (which is an infinite cyclic group generated by orientation preserving homeomorphism $(S^1,1) \simeq (\partial D_p,\infty) \subset \Sigma^*$).

Recall, from the beginning of this Section 6.3, that we have generators $\{a_i,b_i\}_{1\le i\le g} \subset \pi_1(\Sigma^*)$ such that

$$\pi_1(\Sigma^*) = F(a_1,b_1,a_2,b_2,\ldots,a_g,b_g\} \tag{1}$$

and $\mu = \Pi_{i=1}^g [a_i,b_i]$. Fix a maximal torus $\tilde{T} \subset \tilde{K}$ and take a Weyl group element $w \in W$ such that the map

$$c_w : \tilde{T} \to \tilde{T}, \quad \tilde{t} \mapsto w\tilde{t}w^{-1}\tilde{t}^{-1}$$

has finite kernel. (Such a $w \in W$ exists; e.g., we can take w to be a Coxeter element.) Since c_w has finite kernel, it is surjective. Take $\tilde{t}_o \in \tilde{T}$ such that

$$c_w(\tilde{t}_o) = c. \tag{2}$$

Now, consider the homomorphism (since $g \geq 2$ by assumption)

$\tilde{\rho}_c : \pi_1(\Sigma^*) \to \tilde{K}$, $a_1 \mapsto \dot{w}$, $b_1 \mapsto \tilde{t}_o$, $a_2 \mapsto \tilde{t}'$, $b_2 \mapsto \tilde{t}'$, $a_i,b_i \mapsto 1$ for $i > 2$, where $\tilde{t}' \in \tilde{T}$ is any element such that $\tilde{T}$ is the smallest closed subgroup containing $\tilde{t}'$ and $\dot{w}$ is a representative of w in the normalizer $N_{\tilde{K}}(\tilde{T})$ of $\tilde{T}$ in $\tilde{K}$. From (2), we see that $\tilde{\rho}_c(\mu) = c$. Since $c \in \operatorname{Ker}\alpha$, $\alpha \circ \tilde{\rho}_c : \pi_1(\Sigma^*) \to K$ descends to a homomorphism $\rho_c : \pi_1(\Sigma) \to K \subset G$. We next claim that ρ_c is an irreducible homomorphism, i.e., $\operatorname{Im}\rho_c$ is not contained in any proper parabolic subgroup P of G.

Assume, if possible, that

$$\operatorname{Im}\rho_c \subset P. \tag{3}$$

By the definition of ρ_c,

$$\overline{\operatorname{Im}\rho_c} \supset \{T, \overline{\dot{w}}\}, \quad \text{where } T := \alpha(\tilde{T}) \text{ and } \overline{\dot{w}} := \alpha(\dot{w}).$$

If (3) were true,

$$\{T, \overline{\dot{w}}\} \subset P \cap K.$$

In particular, T is a maximal torus of $P \cap K$. Moreover, P being a proper parabolic subgroup, there exists a nontrivial connected subgroup $Z \subset T$ centralizing $P \cap K$. Thus, $\overline{\dot{w}}$ commutes with Z. From this we see that c_w has infinite kernel contradicting the choice of w. Thus, (3) is not possible, i.e., ρ_c is irreducible.

Take any

$$c \in \operatorname{Ker}\alpha \simeq \pi_1(K) \simeq \pi_1(G)$$

and let $\tilde{\rho}_c : \pi_1(\Sigma^*) \to \tilde{K} \subset \tilde{G}$ be as above, where $\tilde{G}$ is the simply-connected cover of G. This descends to give an irreducible (unitary) homomorphism $\rho_c : \pi_1(\Sigma) \to K \subset G$. Let E_{ρ_c} be the corresponding stable G-bundle over Σ (cf. Proposition 6.3.4 for $A = (1)$). Then

$$E_{\rho_c} \simeq \eta(c), \quad \text{as topological } G\text{-bundles}, \tag{4}$$

where c also denotes the corresponding element of the fundamental group $\pi_1(K) \simeq \pi_1(G)$ (cf. Exercise 6.3.E.4).

Thus, the lemma follows from Lemma 6.3.10. □

Continuing the assumption at the beginning of this Section 6.3, let G be a connected complex reductive group and let K be a maximal compact subgroup. Let Σ be a smooth irreducible projective curve of any genus $g \geq 1$ and let $\rho : \pi_1(\Sigma) \to K \subset G$ be a (unitary) homomorphism. Let E_ρ be the associated holomorphic G-bundle over Σ (cf. Definition 6.3.1). For any finite-dimensional complex representation $V_{\mathbb{C}}$ of G, we denote the associated vector bundle $E_\rho \times^G V_{\mathbb{C}}$ by $E_\rho(V_{\mathbb{C}})$. With this notation, we have the following proposition.

Proposition 6.3.12 *Let V be a finite-dimensional real representation of K. Then the natural map (induced from the sheaf embedding $L(V_\rho) \hookrightarrow E_\rho(V_\mathbb{C})$)*

$$i \colon H^1\left(\Sigma, L(V_\rho)\right) \to H^1\left(\Sigma, E_\rho(V_\mathbb{C})\right)$$

is an isomorphism of real vector spaces, where $L(V_\rho)$ is the local system over Σ obtained from the representation V of $\pi_1(\Sigma)$ through ρ, $V_\mathbb{C} := V \otimes_\mathbb{R} \mathbb{C}$ is the complexification of V which is canonically a G-module (obtained as the complexification of the K-module V) and $H^1\left(\Sigma, E_\rho(V_\mathbb{C})\right)$ is the coherent cohomology of the vector bundle $E_\rho(V_\mathbb{C})$ over Σ.

Further, since by assumption, the genus g of Σ is at least one, there is a natural isomorphism

$$j \colon H^1\left(\pi_1(\Sigma), V_\rho\right) \simeq H^1\left(\Sigma, L(V_\rho)\right). \tag{1}$$

Proof Choose a Hermitian metric on the complex curve Σ (which is automatically Kähler) and a K-invariant positive-definite Hermitian form on $V_\mathbb{C}$. By the Hodge decomposition applied to the local system $L(V_\rho \otimes_\mathbb{R} \mathbb{C})$ over Σ (cf. (Griffiths and Harris, 1978, Chapter 0, §7) where the corresponding result for the trivial local system is proved; the proof applies equally to the local systems), we get (using the Dolbeault isomorphism, cf. (Griffiths and Harris, 1978, Chapter 0, §3))

$$\begin{aligned}
H^1\left(\Sigma, L(V_\rho \otimes_\mathbb{R} \mathbb{C})\right) &\simeq H^1\left(\Sigma, E_\rho(V_\mathbb{C})\right) \oplus H^0\left(\Sigma, \Omega^1(\Sigma) \otimes E_\rho(V_\mathbb{C})\right) \\
&\simeq H^1\left(\Sigma, E_\rho(V_\mathbb{C})\right) \oplus \overline{H^1\left(\Sigma, E_\rho(V^*_\mathbb{C})\right)}, \\
&\simeq H^1\left(\Sigma, E_\rho(V_\mathbb{C})\right) \oplus \overline{H^1\left(\Sigma, E_\rho(V_\mathbb{C})\right)},
\end{aligned} \tag{2}$$

since V has a K-invariant positive-definite form, where $\Omega^1(\Sigma)$ is the sheaf of holomorphic 1-forms on Σ and $\bar{M}$ for a $\mathbb{C}$-vector space M denotes the same space as M wherein the complex multiplication is twisted by the conjugation.

From this it is easy to see that the natural $\mathbb{R}$-linear map $i \colon H^1(\Sigma, L(V_\rho)) \to H^1(\Sigma, E_\rho(V_\mathbb{C}))$ is injective. By the isomorphism (2),

$$\begin{aligned}
\dim_\mathbb{R} H^1(\Sigma, L(V_\rho)) = \dim_\mathbb{C} H^1\left(\Sigma, L(V_\rho \otimes_\mathbb{R} \mathbb{C})\right) &= 2 \dim_\mathbb{C} H^1\left(\Sigma, E_\rho(V_\mathbb{C})\right) \\
&= \dim_\mathbb{R} H^1\left(\Sigma, E_\rho(V_\mathbb{C})\right).
\end{aligned} \tag{3}$$

From the injectivity of i and the equality of the dimensions as in (3), we get that i is an isomorphism.

The isomorphism (1) follows from Cartan and Eilenberg (1956, Chap. XVI, §9) by using the contractibility of the simply-connected cover $\tilde{\Sigma}$ of Σ (since $g \geq 1$). □

Remark 6.3.13 For a complete proof of Proposition 6.3.12, we refer to Narasimhan and Seshadri (1964, Proposition 4.4). Moreover, in the setting of Definition 6.3.3, for a (unitary) homomorphism $\hat{\rho}\colon \pi \to K$, the map i of the above Proposition 6.3.12 is A-equivariant and hence we get an isomorphism:

$$H^1\left(\hat{\Sigma}, L(V_\rho)\right)^A \to H^1\left(\hat{\Sigma}, \hat{E}_{\hat{\rho}}(V_{\mathbb{C}})\right)^A, \text{ where } \rho = \hat{\rho}_{|\pi_1(\hat{\Sigma})}.$$

Corollary 6.3.14 *Let $\rho\colon \pi_1(\Sigma) \to K$ be a (unitary) homomorphism and let $\operatorname{ad}\rho$ be the corresponding adjoint representation of $\pi_1(\Sigma)$ in $\mathfrak{k} := \operatorname{Lie} K$. As earlier, we assume that the genus g of Σ is at least 1. Then*

$$\dim H^1(\pi_1(\Sigma), \operatorname{ad}\rho) = 2\dim H^0(\pi_1(\Sigma), \operatorname{ad}\rho) + 2(\dim K)(g-1). \quad (1)$$

Further, ρ is irreducible if and only if

$$\dim H^0(\pi_1(\Sigma), \operatorname{ad}\rho) = \dim \mathfrak{z}, \quad (2)$$

where $\mathfrak{z}$ is the center of $\mathfrak{k}$.

Thus, ρ is irreducible if and only if

$$\dim H^1(\pi_1(\Sigma), \operatorname{ad}\rho) = 2\left((\dim K)(g-1) + \dim \mathfrak{z}\right). \quad (3)$$

Proof By Proposition 6.3.12,

$$\dim_{\mathbb{R}} H^1(\pi_1(\Sigma), \operatorname{ad}\rho) = 2\dim_{\mathbb{C}} H^1\left(\Sigma, \operatorname{ad} E_\rho\right), \quad (4)$$

where $\operatorname{ad} E_\rho = E_\rho \times^G \mathfrak{g}$.

By the Riemann–Roch theorem (cf. (Fulton, 1998, Example 15.2.1))

$$\dim_{\mathbb{C}} H^0\left(\Sigma, \operatorname{ad} E_\rho\right) = \dim_{\mathbb{C}} H^1\left(\Sigma, \operatorname{ad} E_\rho\right) + (\dim K)(1-g), \quad (5)$$

since the adjoint action of G on $\wedge^{\text{top}}(\mathfrak{g})$ is trivial.

Combining (4) and (5), we get

$$\begin{aligned}\dim_{\mathbb{R}} H^1(\pi_1(\Sigma), \operatorname{ad}\rho) &= 2(\dim K)(g-1) + 2\dim_{\mathbb{C}} H^0\left(\Sigma, \operatorname{ad} E_\rho\right)\\ &= 2(\dim K)(g-1) + 2\dim_{\mathbb{C}} H^0(\pi_1(\Sigma), (\operatorname{ad}\rho)_{\mathbb{C}}),\\ &\qquad \text{by Lemma 6.3.6 for } A = (1)\\ &= 2(\dim K)(g-1) + 2\dim_{\mathbb{R}} H^0(\pi_1(\Sigma), \operatorname{ad}\rho).\end{aligned}$$

This proves (1).

We next prove (2). If $\dim H^0(\pi_1(\Sigma), \operatorname{ad}\rho) > \dim \mathfrak{z}$, then there exists a non-central element $x \in \mathfrak{k}$ fixed by $\pi_1(\Sigma)$. Thus, $\operatorname{Im}\pi_1(\Sigma) \subset Z_K(x)$, where $Z_K(x)$ is the centralizer of x in K, which is a proper Levi subgroup of K (since x is non-central) contained in a parabolic subgroup of G. Thus, ρ is not irreducible.

Conversely, if ρ is not irreducible, then $\operatorname{Im}\rho \subset K \cap P$, where P is a proper parabolic subgroup of G. But $K \cap P$ being compact, $K \cap P \subset L_P$, for some Levi subgroup L_P of P. Recall that for any Levi subgroup L_P of a proper parabolic subgroup P, the centralizer $\mathfrak{z}_{\mathfrak{k}}(K \cap L_P)$ of $K \cap L_P$ in $\mathfrak{k}$ satisfies

$$\mathfrak{z}_{\mathfrak{k}}(K \cap L_P) \supsetneq \mathfrak{z}.$$

Hence

$$\dim H^0(\pi_1(\Sigma), \operatorname{ad}\rho) > \dim \mathfrak{z},$$

contradicting (2). This shows that ρ is irreducible and hence (2) is proved. Combining (1) and (2) we, of course, get (3). □

Let K be a compact connected Lie group. For any integer $g \geq 1$, let F_g be the free group on the symbols $\{a_1, b_1, a_2, b_2, \ldots, a_g, b_g\}$. Define the map

$$\beta\colon K^{2g} \to [K,K], \quad \big((h_1,k_1),(h_2,k_2),\ldots,(h_g,k_g)\big) \mapsto \Pi_{i=1}^{g}[h_i,k_i].$$

Any $\bar{\rho} = \big((h_1,k_1),\ldots,(h_g,k_g)\big) \in K^{2g}$ determines a group homomorphism $\tilde{\rho}\colon F_g \to K$ taking $a_i \mapsto h_i$ and $b_i \mapsto k_i$. If $\bar{\rho} \in \beta^{-1}(e)$, then the homomorphism $\tilde{\rho}$ descends to a group homomorphism $\rho\colon \pi_1(\Sigma) \to K$, where g is the genus of Σ (cf. equation (*) at the beginning of this section).

Proposition 6.3.15 *For any $\bar{\rho} \in \beta^{-1}(e)$,*

$$\operatorname{Ker}((d\beta)_{\bar{\rho}}) \simeq Z^1(\pi_1(\Sigma), \operatorname{ad}\rho), \tag{1}$$

where $Z^1(\pi_1(\Sigma), \operatorname{ad}\rho)$ denotes the space of 1-cocycles of $\pi_1(\Sigma)$ with coefficients in $\operatorname{ad}\rho = \mathfrak{k}$ (in the standard cochain complex as in Serre (1997, Chap. I, §5.1)).

Proof For any $\bar{\sigma} \in K^{2g}$, the tangent space $T_{\bar{\sigma}}(K^{2g})$ is identified with $T_{\bar{e}}(K^{2g}) = \mathfrak{k}^{2g}$ under the right multiplication by $\bar{\sigma}^{-1}$, where $\bar{e} := ((e,e),(e,e), \ldots,(e,e))$ and similarly the tangent space $T_k(K)$ is identified with $T_e(K) = \mathfrak{k}$. For any $\alpha \in F_g$, define the function

$$\Phi_\alpha\colon K^{2g} \to K \quad \text{by} \quad \Phi_\alpha(\bar{\sigma}) = \tilde{\sigma}(\alpha). \tag{2}$$

Then, for $\alpha_1, \alpha_2 \in F_g$, clearly

$$\Phi_{\alpha_1\alpha_2}(\bar{\sigma}) = \Phi_{\alpha_1}(\bar{\sigma})\Phi_{\alpha_2}(\bar{\sigma}), \quad \text{for any} \quad \bar{\sigma} \in K^{2g}. \tag{3}$$

For any $\bar{\rho} \in K^{2g}$ and $v \in T_{\bar{\rho}}(K^{2g})$, define the function

$$\mathfrak{F}_v\colon F_g \to \mathfrak{k} \quad \text{by} \quad \mathfrak{F}_v(\alpha) = (d\Phi_\alpha)_{\bar{\rho}}(v),$$

where we have identified $T_{\Phi_\alpha(\bar{\rho})}(K) \simeq \mathfrak{k}$ as above. Then we claim that

$$\mathfrak{F}_v \in Z^1(F_g, \operatorname{ad} \tilde{\rho}). \tag{4}$$

For $\alpha_1, \alpha_2 \in F_g$, by (3),

$$\begin{aligned}\mathfrak{F}_v(\alpha_1\alpha_2) &= d(\Phi_{\alpha_1} \cdot \Phi_{\alpha_2})_{\bar{\rho}}(v) \\ &= (d\Phi_{\alpha_1})_{\bar{\rho}}(v) + \left(\operatorname{Ad} \Phi_{\alpha_1}(\bar{\rho})\right) \cdot \left((d\Phi_{\alpha_2})_{\bar{\rho}}(v)\right),\end{aligned} \tag{5}$$

where the last equality follows from the following equality for any $\bar{\sigma} \in K^{2g}$:

$$\begin{aligned}&\Phi_{\alpha_1}(\bar{\sigma})\Phi_{\alpha_2}(\bar{\sigma}) \cdot \Phi_{\alpha_2}(\bar{\rho})^{-1}\Phi_{\alpha_1}(\bar{\rho})^{-1} \\ &\quad = \left(\Phi_{\alpha_1}(\bar{\sigma})\Phi_{\alpha_1}(\bar{\rho})^{-1}\right) \cdot \left(\Phi_{\alpha_1}(\bar{\rho}) \cdot \left(\Phi_{\alpha_2}(\bar{\sigma}) \cdot \Phi_{\alpha_2}(\bar{\rho})^{-1}\right) \cdot \Phi_{\alpha_1}(\bar{\rho})^{-1}\right).\end{aligned}$$

Rewritten, the identity (5), of course, is the identity

$$\mathfrak{F}_v(\alpha_1\alpha_2) = \mathfrak{F}_v(\alpha_1) + (\operatorname{Ad} \tilde{\rho}(\alpha_1)) \cdot (\mathfrak{F}_v(\alpha_2)),$$

which proves (4).

Now, by definition,

$$\beta(\bar{\sigma}) = \Phi_{\prod_{i=1}^g [a_i, b_i]}(\bar{\sigma}),$$

and hence, for any $v \in T_{\bar{\rho}}(K^{2g})$,

$$(d\beta)_{\bar{\rho}}(v) = \mathfrak{F}_v\left(\Pi_{i=1}^g [a_i, b_i]\right).$$

Thus,

$$v \in \operatorname{Ker}(d\beta)_{\bar{\rho}} \iff \mathfrak{F}_v\left(\Pi_{i=1}^g [a_i, b_i]\right) = 0. \tag{6}$$

So far, in the proof, we took an arbitrary $\bar{\rho} \in K^{2g}$. But now we take $\bar{\rho} \in \operatorname{Ker}\beta$ so that $\operatorname{ad}\tilde{\rho}$ is a $\pi_1(\Sigma)$-module. In this case, by (6) and (*) at the beginning of the section, we get a linear map

$$\mathfrak{F}\colon \operatorname{Ker}(d\beta)_{\bar{\rho}} \to Z^1(\pi_1(\Sigma), \operatorname{ad}\rho),\ v \mapsto \mathfrak{F}_v. \tag{7}$$

We claim that $\mathfrak{F}$ is an isomorphism.

Take $v \in \operatorname{Ker}\mathfrak{F}$, i.e., $\mathfrak{F}_v \equiv 0$. In particular, $\mathfrak{F}_v(a_i) = \mathfrak{F}_v(b_i) = 0$. By the definition of $\mathfrak{F}_v$, this gives that for all the coordinate projections $\pi_j\colon K^{2g} \to K$, $1 \le j \le 2g$, $(d\pi_j)_{\bar{\rho}}(v) = 0$. This, of courses, forces $v = 0$, i.e., $\mathfrak{F}$ is injective.

To prove that $\mathfrak{F}$ is surjective, take $\delta \in Z^1(\pi_1(\Sigma), \operatorname{ad}\rho)$. Since δ is a cocycle and $\{a_i, b_i\}$ generate $\pi_1(\Sigma)$, δ is completely determined by its values $\delta(a_i), \delta(b_i) \in \operatorname{ad}\rho$. Consider the vector

$$v = \left((\delta(a_1), \delta(b_1)), \ldots, (\delta(a_g), \delta(b_g))\right) \in \mathfrak{k}^{2g}.$$

Then, it is easy to see from the definition of $\mathfrak{F}_v$ that

$$\mathfrak{F}_v(a_i) = \delta(a_i) \quad \text{and} \quad \mathfrak{F}_v(b_i) = \delta(b_i), \quad \text{for all } 1 \leq i \leq g.$$

But since both of δ and $\mathfrak{F}_v$ are cocycles for F_g with coefficients in $\operatorname{ad}\tilde{\rho}$, and they coincide on a_i and b_i, we get that $\delta = \mathfrak{F}_v$. Moreover, since $\delta \in Z^1(\pi_1(\Sigma), \operatorname{ad}\rho)$ and $\delta = \mathfrak{F}_v$ as cocycles for F_g, by (6), we get that $v \in \operatorname{Ker}(d\beta)_{\bar{\rho}}$. This proves the surjectivity of $\mathfrak{F}$ as well. Hence $\mathfrak{F}$ is an isomorphism, proving the proposition. □

Combining Corollary 6.3.14 and Proposition 6.3.15, we get the following result.

Corollary 6.3.16 *Let $\beta\colon K^{2g} \to [K,K]$ be the map given above Proposition 6.3.15. Take $\bar{\rho} \in \beta^{-1}(e)$. Then, $(d\beta)_{\bar{\rho}}$ is of maximal rank (equal to $\dim[K,K]$) if and only if the corresponding representation $\rho\colon \pi_1(\Sigma) \to K$ is irreducible.*

Thus,

$$M_g(K) := \{\bar{\rho} \in \beta^{-1}(e) : \rho \text{ is irreducible}\}$$

is an $\mathbb{R}$-analytic (smooth) manifold of dimension $(2g-1)\dim K + \dim \mathfrak{z}$ with the tangent space at any $\bar{\rho}$ identified with $Z^1(\pi_1(\Sigma), \operatorname{ad}\rho)$.

Proof By Proposition 6.3.15,

$$\begin{aligned}
\operatorname{rank}(d\beta)_{\bar{\rho}} &= 2g\dim K - \dim Z^1(\pi_1(\Sigma), \operatorname{ad}\rho) \\
&= 2g\dim K - \dim H^1(\pi_1(\Sigma), \operatorname{ad}\rho) \\
&\quad + \dim H^0(\pi_1(\Sigma), \operatorname{ad}\rho) - \dim K \\
&= (2g-1)\dim K - \dim H^0(\pi_1(\Sigma), \operatorname{ad}\rho) - 2(\dim K)(g-1), \\
&\qquad \text{by identity (1) of Corollary 6.3.14} \\
&= \dim K - \dim H^0(\pi_1(\Sigma), \operatorname{ad}\rho). \qquad (1)
\end{aligned}$$

By identity (2) of Corollary 6.3.14, ρ is irreducible if and only if

$$\dim H^0(\pi_1(\Sigma), \operatorname{ad}\rho) = \dim \mathfrak{z}.$$

Thus, by (1),

$$\operatorname{rank}(d\beta)_{\bar{\rho}} = \dim[K,K] \text{ if and only if } \rho \text{ is irreducible.}$$ □

An infinitesimal deformation map for a family of fiber bundles is defined in Kodaira and Spencer (1958a, §7). We recall the definition for a family of G-bundles over Σ parameterized by a smooth variety.

Definition 6.3.17 Let $\mathcal{E} \to \Sigma \times T$ be a family of G-bundles over Σ parameterized by a smooth variety T. For $t_o \in T$, any tangent vactor $v \in T_{t_o}(T)$ is given by $\theta_v \colon \operatorname{Spec} \mathbb{C}(\epsilon) \to T$ such that its restriction to $\operatorname{Spec} \mathbb{C}$ (under $\epsilon \mapsto 0$) corresponds to the point t_o (cf. Definition B.7). Thus, pulling $\mathcal{E}$ via $\hat{\theta}_v := \mathrm{Id}_\Sigma \times \theta_v$, we get a G-bundle denoted $\mathcal{E}_v$ over $\Sigma(\epsilon)$, where $\Sigma(\epsilon) := \Sigma \times \operatorname{Spec} \mathbb{C}(\epsilon)$. Clearly,

$$\mathcal{E}_{v|\Sigma} = \mathcal{E}_{t_o}, \text{ where } \mathcal{E}_{t_o} := \mathcal{E}_{|\Sigma \times t_o}. \tag{1}$$

Take an affine Zariski open cover $\{U_i\}_i$ of Σ such that $\mathcal{E}_{v|U_i(\epsilon)}$ are trivial. (This is possible by Ramanathan (1983, Proposition 4.3) for $\mathcal{E}_{t_o}$ and affineness of U_i gives the result for $\mathcal{E}_v$.) Taking sections $s_i^\epsilon \in \Gamma(U_i(\epsilon), \mathcal{E}_v)$, we get transition functions

$$g_{ij}^\epsilon \colon (U_i \cap U_j)(\epsilon) \to G \text{ given by } s_i^\epsilon g_{ij}^\epsilon = s_j^\epsilon.$$

Thus, $\{g_{ij}^\epsilon\}$ satisfy the cocycle condition

$$g_{ij}^\epsilon g_{jk}^\epsilon = g_{ik}^\epsilon : (U_i \cap U_j \cap U_k)(\epsilon) \to G. \tag{2}$$

Moreover, by (1), $g_{ij} := g^\epsilon_{ij|U_i \cap U_j}$ provide transition functions for the bundle $\mathcal{E}_{t_o}$. Let $U_i \cap U_j = \operatorname{Spec}(R_{ij})$ for a $\mathbb{C}$-algebra R_{ij} (observe that $U_i \cap U_j$ is affine by Hartshorne (1977, Chap. II, Exercise 4.3)). Then, we can view $g_{ij}^\epsilon \in G(R_{ij}(\epsilon))$, where $R_{ij}(\epsilon) := R_{ij} \otimes \mathbb{C}(\epsilon)$. Consider the exact sequence of groups (cf. Lemma B.11 and Definition B.15(b)):

$$\mathfrak{g} \otimes R_{ij} \xrightarrow{\iota} G(R_{ij}(\epsilon)) \xrightarrow{\theta_{ij}} G(R_{ij}),$$

where $\mathfrak{g} := \operatorname{Lie} G$ and θ_{ij} is induced by taking $\epsilon \mapsto 0$. By definition, $\theta_{ij}(g_{ij}^\epsilon) = g_{ij}$. Write

$$g_{ij}^\epsilon = \iota(h_{ij}) \cdot g_{ij}, \text{ where } h_{ij} : U_i \cap U_j \to \mathfrak{g}$$

and g_{ij} is thought of as an element of $G(R_{ij}(\epsilon))$ under the embedding $G(R_{ij}) \hookrightarrow G(R_{ij}(\epsilon))$ induced by $R_{ij} \hookrightarrow R_{ij}(\epsilon)$. Thus, by (2) and Definition B.17, we get the cocycle condition

$$h_{ik} = h_{ij} + \operatorname{Ad}(g_{ij})(h_{jk}), \text{ as morphisms } U_i \cap U_j \cap U_k \to \mathfrak{g}.$$

Hence $\{h_{ij}\}$ give rise to an element $\bar{\mathcal{E}}_v$ of $H^1(\Sigma, \operatorname{ad} \mathcal{E}_{t_o})$ in the Čech realization of cohomology (Hartshorne, 1977, Chap. III, §4). It is easy to see that the element $\bar{\mathcal{E}}_v \in H^1(\Sigma, \operatorname{ad} \mathcal{E}_{t_o})$ does not depend upon the choice of the open cover $\{U_i\}$ or the sections s_i^ϵ.

The *Kodaira–Spencer infinitesimal deformation map* of the family $\mathscr{E}$ at t_o is defined by

$$\eta\colon T_{t_o}(T) \to H^1(\Sigma, \operatorname{ad} \mathscr{E}_{t_o}),\ \eta(v) := \bar{\mathscr{E}}_v.$$

Then, η is a $\mathbb{C}$-linear map.

We can extend the above definition for any $\mathbb{R}$-analytic family of holomorphic G-bundles over Σ parameterized by a smooth $\mathbb{R}$-analytic space T to get an $\mathbb{R}$-linear map (as above) $\eta\colon T_{t_o}(T) \to H^1(\Sigma, \operatorname{ad} \mathscr{E}_{t_o})$, for any $t_o \in T$.

Let $R^s_K(g)$ be the set of irreducible homomorphisms from $\pi_1(\Sigma)$ to K. By Corollary 6.3.16, $R^s_K(g)$ is an $\mathbb{R}$-analytic (smooth) manifold. By Lemma 6.3.2, $R_G(g)$ parameterizes a 'universal' $\mathbb{C}$-analytic family $\theta\colon \mathscr{E} \to \Sigma \times R_G(g)$ of holomorphic G-bundles over Σ. Let us consider its restriction $\theta^s_K : \mathscr{E}^s_K \to \Sigma \times R^s_K(g)$ to $\Sigma \times R^s_K(g)$ giving rise to an $\mathbb{R}$-analytic family of holomorphic G-bundles over Σ parameterized by $R^s_K(g)$. The following proposition determines its infinitesimal deformation map $\eta = \eta(\theta^s_K)$.

Proposition 6.3.18 *For any $\rho \in R^s_K(g)$, the infinitesimal deformation map*

$$\eta\colon T_\rho\left(R^s_K(g)\right) \to H^1(\Sigma, \operatorname{ad} E_\rho)$$

coincides with the composition of the maps

$$T_\rho\left(R^s_K(g)\right) \underset{\sim}{\overset{\mathfrak{F}}{\to}} Z^1\left(\pi_1(\Sigma), \operatorname{ad}\rho\right) \overset{q}{\to} H^1\left(\pi_1(\Sigma), \operatorname{ad}\rho\right)$$
$$\underset{\sim}{\overset{j}{\to}} H^1(\Sigma, L(\operatorname{ad}\rho)) \underset{\sim}{\overset{i}{\to}} H^1(\Sigma, \operatorname{ad} E_\rho),$$

where the isomorphism $\mathfrak{F}$ is as defined in the proof of Proposition 6.3.15 (identifying $R^s_K(g)$ canonically with $M_g(K)$), the map q is the standard projection, $H^1(\Sigma, L(\operatorname{ad}\rho))$ denotes the singular cohomology of Σ with coefficients in the local system $L(\operatorname{ad}\rho)$ and the isomorphisms j and i are as in Proposition 6.3.12.

In particular, η is surjective.

Proof We identify $R^s_K(g)$ as the subset $M_g(K)$ of K^{2g}, taking

$$\sigma \mapsto (\sigma(a_1), \sigma(b_1), \ldots, \sigma(a_g), \sigma(b_g)).$$

Take a small enough finite open cover $\{U_k\}$ of Σ such that we can find a holomorphic section s_k of the simply-connected cover $\pi : \tilde{\Sigma} \to \Sigma$ over U_k. Thus, whenever $U_k \cap U_l \neq \emptyset$, we get an element $\alpha_{k,l} \in \pi_1(\Sigma)$ such that

$$s_l = s_k \cdot \alpha_{k,l}.$$

Clearly, $\alpha_{k,l}$ satisfy the cocycle condition

$$\alpha_{k,l} \cdot \alpha_{l,m} = \alpha_{k,m} \quad \text{whenever} \quad U_k \cap U_l \cap U_m \neq \emptyset. \tag{1}$$

Take a tangent vector $v \in T_\rho(R^s_K(g))$. By the definition of $\mathfrak{F}$, $\mathfrak{F}_v(\alpha) = (d\Phi_\alpha)_\rho(v)$, where $\Phi_\alpha \colon R^s_K(g) \to K$ is the function $\Phi_\alpha(\sigma) = \sigma(\alpha)$. Take the cohomology class $\bar{\delta} \in H^1(\pi_1(\Sigma), \operatorname{ad}\rho)$ represented by a cocycle $\delta \in Z^1(\pi_1(\Sigma), \operatorname{ad}\rho)$. Then $j(\bar{\delta})$ is the cohomology class given by the Čech 1-cocycle

$$(U_k, U_l) \mapsto \left[s_{k|U_k \cap U_l}, \delta(\alpha_{k,l})\right] \in H^0\left(U_k \cap U_l, L(\operatorname{ad}\rho) := \tilde{\Sigma} \times^{\pi_1(\Sigma)} \operatorname{ad}\rho\right),$$

and so is the composite $i \circ j$ (cf. (Hartshorne, 1977, Chap. III, Lemma 4.4)). Thus, the composite map $i \circ j \circ q \circ \mathfrak{F}$ takes v to the cohomology class of $\operatorname{ad} E_\rho$ determined by the Čech 1-cocycle

$$(U_k, U_l) \mapsto \left[s_{k|U_k \cap U_l}, \left(d\Phi_{\alpha_{k,l}}\right)_\rho(v)\right], \tag{2}$$

where, as above, $\Phi_{\alpha_{k,l}}$ denotes the function $R^s_K(g) \to K, \sigma \mapsto \sigma(\alpha_{k,l})$ and $(d\Phi_{\alpha_{k,l}})_\rho(v) \in T_e(K) = \mathfrak{k} \subset \mathfrak{g}$ identifying $T_{\rho(\alpha_{k,l})}(K)$ with $T_e(K)$ under the right translation.

By the definition of the infinitesimal deformation map η as above in Definition 6.3.17, it can be seen that the cohomology class of the above Čech 1-cocycle (2) coincides with $\eta(v)$ (cf. Exercise 6.3.E.12). This proves the proposition. □

The following definition is an analogue of Kodaira and Spencer (1958b, Definition 2).

Definition 6.3.19 Let $\mathscr{F} \to \Sigma \times T$ be an $\mathbb{R}$-analytic family of holomorphic G-bundles over Σ (parameterized by an $\mathbb{R}$-analytic space T). Then, this family is said to be ($\mathbb{R}$-analytically) *complete* at $t_o \in T$ if for any $\mathbb{R}$-analytic family $\mathscr{F}' \to \Sigma \times T'$ with $t'_o \in T'$ such that $\mathscr{F}'_{t'_o} \simeq \mathscr{F}_{t_o}$, there exists an open neighborhood $U_{t'_o} \subset T'$ of t'_o and an $\mathbb{R}$-analytic map $f \colon U_{t'_o} \to T$ such that $f(t'_o) = t_o$ and the family $\mathscr{F}'$ restricted to $\Sigma \times U_{t'_o}$ is isomorphic to the pull-back family $(\operatorname{Id} \times f)^*(\mathscr{F})$.

The family $\mathscr{F}$ is called ($\mathbb{R}$-analytically) *complete* if it is complete at each $t \in T$.

We recall the following general result, the proof of which can be extracted from Ramanathan (1983, Remark 8.11) or Biswas and Ramanan (1994, Theorem 3.1) (also see the proof of Kodaira and Spencer (1958b, Theorem); and for vector bundles see the article by Nitsure (2009)). The result hinges upon the fact that $H^2(\Sigma, \operatorname{ad}\mathscr{F}_{t_o}) = 0$, since Σ is a curve.

Theorem 6.3.20 *Let $\mathscr{F} \to \Sigma \times T$ be an $\mathbb{R}$-analytic family of holomorphic G-bundles over Σ parameterized by an $\mathbb{R}$-analytic space T. Let $t_o \in T$ be a smooth point such that the infinitesimal deformation map*

$$T_{t_o}(T) \to H^1(\Sigma, \operatorname{ad} \mathscr{F}_{t_o})$$

is surjective. Then the family $\mathscr{F}$ is complete at t_o.

Conversely, if $\mathscr{F}$ is complete at t_o, then the above deformation map is surjective.

As a corollary of the above theorem and Proposition 6.3.18, we obtain the following result.

Corollary 6.3.21 *Let $\mathscr{F} \to \Sigma \times T$ be an $\mathbb{R}$-analytic family of holomorphic G-bundles over Σ. Then the subset*

$$T_o := \left\{ t \in T : \mathscr{F}_t \simeq E_\rho \text{ for some } \rho \in R^s_K(g) \right\}$$

is an open subset of T.

Proof Take $t_o \in T_o$ so that $\mathscr{F}_{t_o} \simeq E_\rho$ (for some $\rho \in R^s_K(g)$). By Theorem 6.3.20 and Proposition 6.3.18, the family $\theta^s_K : \mathscr{E}^s_K \to \Sigma \times R^s_K(g)$ (cf. the discussion above Proposition 6.3.18) is ($\mathbb{R}$-analytically) complete. Applying its completeness at ρ, we get that there exists an open subset $U_{t_o} \subset T$ containing t_o and an $\mathbb{R}$-analytic map $f : U_{t_o} \to R^s_K(g)$ such that the family $\mathscr{F}_{|\Sigma \times U_{t_o}}$ is isomorphic with the pull-back family $\bar{f}^*(\theta^s_K)$, where $\bar{f} := I_\Sigma \times f$. In particular, for any $t \in U_{t_o}$, $\mathscr{F}_t \simeq E_{\rho'}$ for some $\rho' \in R^s_K(g)$, i.e., $t \in T_o$. Thus, T_o is open in T proving the corollary. □

Lemma 6.3.22 *Let $f : \mathscr{V} \to \mathscr{W}$ be a nonzero $\mathscr{O}_\Sigma$-module homomorphism between two semistable vector bundles over Σ such that at least one of them is stable. Assume further that they both have the same rank and the same degree. Then f is an isomorphism.*

Proof We first recall the following general construction from Narasimhan and Seshadri (1965, §4).

For any nonzero $\mathscr{O}_\Sigma$-module homomorphism $f : \mathscr{E} \to \mathscr{F}$ between any two vector bundles (not necessarily of the same rank) over Σ, since the structure sheaf $\mathscr{O}_\Sigma$ is a sheaf of PIDs, f has the following canonical factorization (obtained from the following commutative diagram):

$$\begin{array}{ccccccccc}
0 & \longrightarrow & \mathscr{E}_1 & \longrightarrow & \mathscr{E} & \xrightarrow[\pi]{} & \mathscr{E}_2 & \longrightarrow & 0 \\
 & & & & \downarrow{\scriptstyle f} & & \downarrow{\scriptstyle f'} & & \\
0 & \longleftarrow & \mathscr{F}_2 & \longleftarrow & \mathscr{F} & \xleftarrow[i]{} & \mathscr{F}_1 & \longleftarrow & 0
\end{array} \qquad (*)$$

where $\mathscr{E}_i, \mathscr{F}_i (i = 1, 2)$ are vector bundles, the above rows are exact and f' is of maximal rank (i.e., $\mathscr{E}_2$ and $\mathscr{F}_1$ are of the same rank say r and the induced map $\wedge^r(\mathscr{E}_2) \to \wedge^r(\mathscr{F}_1)$ is nonzero). *Then, $\mathscr{F}_1$ (resp. $\mathscr{E}_1$) is called the vector subbundle of $\mathscr{F}$ (resp. $\mathscr{E}$) generated by the image (resp. kernel) of f.*

We now come to the proof of the lemma assuming that $\mathscr{W}$ is stable. If f is of maximal rank, since $\deg \mathscr{V} = \deg \mathscr{W}$, f is an isomorphism. This is because a nonzero section of a degree 0 line bundle over Σ is nowhere zero.

So, assume that f is not of maximal rank and consider the decomposition (*) for $f = i \circ f' \circ \pi : \mathscr{V} \xrightarrow{\pi} \mathscr{V}_2 \xrightarrow{f'} \mathscr{W}_1 \xrightarrow{i} \mathscr{W}$. Since $\deg(E \otimes F) = (\deg E)(\operatorname{rank} F) + (\deg F)(\operatorname{rank} E)$ for vector bundles E, F over Σ (as can easily be seen from the Chern character of $E \otimes F$),

$$0 = \deg\left(\mathscr{V}^* \otimes \mathscr{V}\right) = \deg\left(\mathscr{V}^* \otimes \mathscr{V}_1\right) + \deg\left(\mathscr{V}^* \otimes \mathscr{V}_2\right), \tag{1}$$

where $\mathscr{V}_1 := \operatorname{Ker} \pi$. Since $\mathscr{V}$ is semistable,

$$\deg\left(\mathscr{V}^* \otimes \mathscr{V}_1\right) = (\operatorname{rank} V) \cdot \deg(\mathscr{V}_1) - (\operatorname{rank} \mathscr{V}_1) \cdot (\deg \mathscr{V}) \leq 0. \tag{2}$$

Thus, combining (1) and (2), we get

$$\deg(\mathscr{V}^* \otimes \mathscr{V}_2) \geq 0. \tag{3}$$

Since f' is of maximal rank, we get

$$\operatorname{rank} \mathscr{V}_2 = \operatorname{rank} \mathscr{W}_1 \quad \text{and} \quad \deg \mathscr{W}_1 \geq \deg \mathscr{V}_2. \tag{4}$$

Thus, $\mathscr{V}$ and $\mathscr{W}$ having the same rank by assumption,

$$\begin{aligned} \deg\left(\mathscr{W}^* \otimes \mathscr{W}_1\right) &= \deg\left(\mathscr{V}^* \otimes \mathscr{W}_1\right), \quad \text{since } \deg \mathscr{V} = \deg \mathscr{W} \text{ by assumption} \\ &\geq \deg\left(\mathscr{V}^* \otimes \mathscr{V}_2\right), \quad \text{by (4)} \\ &\geq 0, \quad \text{by (3).} \end{aligned} \tag{5}$$

But $\mathscr{W}_1$ is a proper subbundle of the stable bundle $\mathscr{W}$ (since f is assumed to be not of maximal rank). Thus

$$\deg\left(\mathscr{W}^* \otimes \mathscr{W}_1\right) < 0. \tag{6}$$

Then (5) and (6) contradict each other, and hence f must be of maximal rank. But since $\deg \mathscr{V} = \deg \mathscr{W}$, f must be an isomorphism. This proves the lemma when $\mathscr{W}$ is stable.

The case when $\mathscr{V}$ is stable can be handled similarly. □

Definition 6.3.23 Let $\phi : G \to \mathrm{GL}_V$ be a finite-dimensional representation and let $V = \bigoplus_{i=1}^r V_i$ be its decomposition into irreducible components. Let $\phi_i : G \to \mathrm{GL}_{V_i}$ be the restriction of ϕ to V_i.

Let

$$C := \{(z_1\phi_1(g), \ldots, z_r\phi_r(g)) \in \mathrm{GL}_V : z_i \in \mathbb{C}^* \text{ and } g \in G\}.$$

Then C being the image of an algebraic group homomorphism $(\mathbb{C}^*)^r \times G \to \mathrm{GL}_V$, C is closed in GL_V. Let

$$\bar{C} = \text{closure of } C \text{ in } \mathrm{End}\, V.$$

Then, C (and hence $\bar{C}$) is stable under left and right multiplications by $\phi(g)$ for any $g \in G$.

Let E and E' be two G-bundles over Σ. Then their fiber product $F: E \underset{\Sigma}{\times} E' \to \Sigma$ is canonically a $G \times G$-bundle. Consider the vector bundle

$$\mathbf{Hom}\big(E(V), E'(V)\big) = F(\mathrm{End}\, V),$$

where $G \times G$ acts on End V via

$$(g,h) \cdot f = \phi(g) f \phi(h)^{-1}, \text{ for } g,h \in G \text{ and } f \in \mathrm{End}\, V. \tag{1}$$

The subsets End V_i, C, $\bar{C} \subset$ End V are clearly stable under the above action of $G \times G$ (where End V_i is a block of End V through the decomposition $V = \oplus V_i$). Thus, we get fiber subbundles $F(C) \subset F(\bar{C}) \subset F(\mathrm{End}\, V)$ and also the vector subbundle $F(\mathrm{End}\, V_i)$.

Proposition 6.3.24 *With the notation as above, let E be a stable G-bundle and E' a semistable G-bundle of the same topological type (i.e., they are topologically isomorphic). Let*

$$s = (s_1, \ldots, s_r) \in H^0(\Sigma, F(\mathrm{End} V_1 \times \cdots \times \mathrm{End} V_r)) = \bigoplus_{i=1}^{r} H^0(\Sigma, F(\mathrm{End} V_i))$$

be such that $s(\Sigma) \subset F(\bar{C})$. Then, any s_i is either 0 or an isomorphism $E(V_i) \to E'(V_i)$.

Further, if each s_i is nonzero and if $\phi: G \to \mathrm{GL}_V$ is a faithful representation, then there exists $(z_1, \ldots, z_r) \in (\mathbb{C}^*)^r$ such that the section $(z_1 s_1, \ldots, z_r s_r)$ is induced from a G-bundle isomorphism $\bar{s}: E \to E'$.

Before we come to the proof of the proposition, we need the following two lemmas.

As earlier, we fix a maximal compact subgroup K of G and take $K' := [K, K]$ as a maximal compact subgroup of the commutator subgroup $G' := [G, G]$. Fix a maximal abelian subalgebra $\mathfrak{a}'$ of $\mathfrak{k}' :=$ Lie K'. Then $\mathfrak{h}' := \mathfrak{a}' \oplus i\,\mathfrak{a}'$ is a Cartan subalgebra of $\mathfrak{g}' :=$ Lie G' and $\mathfrak{h} := \mathfrak{h}' \oplus z(\mathfrak{g})$ is a Cartan subalgebra of $\mathfrak{g} =$ Lie G, where $z(\mathfrak{g})$ is the center of $\mathfrak{g}$. Let $\Pi = \{\alpha_1, \ldots, \alpha_\ell\} \subset \mathfrak{h}'^*$ be a system of simple roots of $\mathfrak{g}'$. Then they take real values on $i\mathfrak{a}'$. Let (M,ρ) be an irreducible representation of G with highest weight Λ (with respect to the above choice of simple roots). For any subset $\Phi \subset \Pi$, define the $\mathfrak{h}$-module projection

$$p_\Phi : M \to M^\Phi \subset M, \text{ where } M^\Phi := \oplus_{\lambda \in \left(\Lambda - \sum_{\alpha_i \in \Phi} \mathbb{Z}_+ \alpha_i\right)} M_\lambda, \ \mathbb{Z}_+ := \mathbb{Z}_{\geq 0}$$

and M_λ is the λth weight space of M. Recall that any weight λ also takes real values on $i\mathfrak{a}$, where $\mathfrak{a} := \mathfrak{a}' \oplus z(\mathfrak{k})$. Let

$$C_M := \{z\rho(g) : z \in \mathbb{C}^*, g \in G\} \subset \operatorname{End} M$$

and let $\bar{C}_M$ be its closure in $\operatorname{End} M$. With this notation, we have the following lemma (having fixed $\mathfrak{h}'$ as above).

Lemma 6.3.25 *For any $f \in \bar{C}_M$, there exists a system of simple roots $\Pi \subset \mathfrak{h}'^*$ (depending upon f) and a subset $\Phi \subset \Pi$ such that*

$$f = z\rho(g) p_\Phi \rho(g'), \text{ for some } g, g' \in G \text{ and } z \in \mathbb{C}. \tag{1}$$

Conversely, any element f of the form (1) *lies in $\bar{C}_M$.*
In particular,

$$\{Im\, f : f \in \bar{C}_M\} = \{gM^\Phi : g \in G, \Phi \subset \Pi \text{ and } \Pi \text{ ranges over systems of simple roots in } \mathfrak{h}'^*\}.$$

Proof Take a sequence

$$z_n \rho(g_n) \to f, \text{ for } z_n \in \mathbb{C}^* \text{ and } g_n \in G'.$$

(Observe that the center of G acts by a scalar on M due to Schur's lemma and hence we can choose $g_n \in G'$.) Decompose

$$g_n = k_n a_n k_n' \text{ with } k_n, k_n' \in K' \text{ and } a_n \in A',$$

where A' is the real subgroup of G' with Lie algebra $i\mathfrak{a}'$ (cf. (Knapp, 2002, Theorem 7.39)). Replacing g_n by a suitable subsequence, we can assume that $k_n \to k, k_n' \to k'$ and there exist h_n all belonging to the same Weyl chamber inside $i\mathfrak{a}'$ such that $a_n = \operatorname{Exp}(h_n)$. Let Π be the set of simple roots corresponding to this Weyl chamber. Thus, $\alpha_i(h_n) \geq 0$ for all $\alpha_i \in \Pi$ and all n. By passing to a further subsequence (if needed) and reordering the simple roots, let $0 \leq q \leq \ell$ be such that

$$\alpha_i(h_n) \to x_i, \text{ for } 1 \leq i \leq q, \text{ and } \alpha_i(h_n) \to \infty, \text{ for } q < i \leq \ell.$$

Take $h \in i\mathfrak{a}'$ such that

$$\alpha_i(h) = x_i, \text{ for } 1 \le i \le q \\ = 0, \text{ for } q < i \le \ell.$$

Then

$$f = \rho(k)\rho(\operatorname{Exp} h)\left(\lim_{n\to\infty} z_n\rho(a_n \operatorname{Exp}(-h))\right)\rho(k').$$

Thus

$$f = \rho(k)\rho(\operatorname{Exp} h)\lim_{n\to\infty}\left(z_n\rho(\operatorname{Exp}(\bar{h}_n))\right)\rho(k'), \tag{2}$$

where $\bar{h}_n := h_n - h \in i\mathfrak{a}'$ is such that

$$\alpha_i(\bar{h}_n) \to 0 \text{ for all } 1 \le i \le q \quad \text{and } \alpha_i(\bar{h}_n) \to \infty, \text{ for } q < i \le \ell.$$

Let $\Phi := \{\alpha_1, \ldots, \alpha_q\}$. The operator $z_n\rho(\operatorname{Exp}(\bar{h}_n))$ restricted to any weight space M_λ is given by $z_n e^{\lambda(\bar{h}_n)}$. In particular, denoting

$$P_\lambda = \lim_{n\to\infty}\left(z_n\rho(\operatorname{Exp}(\bar{h}_n))_{|M_\lambda}\right), \tag{3}$$

we get

$$P_\lambda = \left(\lim_{n\to\infty} e^{(\lambda-\Lambda)(\bar{h}_n)}\right)P_\Lambda = P_\Lambda \cdot \operatorname{Id}_{M_\lambda}, \text{ if } \lambda \in \Lambda - \sum_{\alpha_i\in\Phi}\mathbb{Z}_+\alpha_i \\ = 0, \text{ otherwise}, \tag{4}$$

where, M_Λ being a 1-dimensional space, we think of P_Λ as a scalar. Combining (2)–(4), we get (1).

For the converse, choose $h_n \in i\mathfrak{a}$ such that

$$\alpha_i(h_n) = 0, \text{ for } \alpha_i \in \Phi \text{ and } \alpha_i(h_n) = n \text{ for } \alpha_i \notin \Phi.$$

Set $z_n = ze^{-\Lambda(h_n)}$. Then the sequence $z_n\rho(\operatorname{Exp}(h_n)) \to zp_\Phi$. By the $G \times G$-invariance of $\bar{C}_M$, we get that any f of the form (1) lies in $\bar{C}_M$. This proves the lemma. □

Definition 6.3.26 Let M and Φ be as above. Then a subspace $M' \subset M$ is said to be of *type* Φ if $M' = \rho(g)M^\Phi$ for some $g \in G$, where M^Φ is defined above Lemma 6.3.25.

Clearly, the set of subspaces of M of type Φ can be viewed as a G-stable subset of the Grassmannian $\operatorname{Gr}(d^\Phi, M)$ of d^Φ-dimensional subspaces of M, where $d^\Phi := \dim M^\Phi$.

For a G-bundle E over Σ, a vector subbundle of $E(M)$ is said to be of *type* Φ, if the fibers are subspaces of M of type Φ.

Lemma 6.3.27 *Let E be a stable (resp. semistable) G-bundle over Σ and let M be an irreducible representation of G. Then, for any proper nonzero subbundle $\mathscr{W}$ of $E(M)$ of type Φ (for a subset $\Phi \subset \Pi$), we have*

$$\mu(\mathscr{W}) < \mu(E(M)) \ \ (\textit{resp. } \mu(\mathscr{W}) \leq \mu(E(M))). \tag{1}$$

Proof Let P be the stabilizer of $M^\Phi \in \mathrm{Gr}(d^\Phi, M)$ in G. Then clearly P is a standard parabolic subgroup of G (with respect to the choice of simple roots Π). This gives an embedding

$$G/P \hookrightarrow \mathrm{Gr}(d^\Phi, M), \ \ gP \mapsto gM^\Phi. \tag{2}$$

The subbundle $\mathscr{W}$ gives rise to a section $\sigma_{\mathscr{W}}$ of $E\big(\mathbb{P}(\wedge^{d^\Phi} M)\big)$. Since $\mathscr{W}$ is of type Φ, $\sigma_{\mathscr{W}}$ lands inside $E(G/P)$ (under the embedding (2)), giving rise to a reduction of the structure group of the G-bundle E to P (cf. Lemma 5.1.2). Let χ_Φ (resp. χ) be the character of the action of P on $\wedge^{d^\Phi}(M^\Phi)$ (resp. $\wedge^r M$, where $r := \dim M$). Let Z be the center of G. Then, since Z acts on M via scalars, the character $\bar{\chi}_\Phi := \chi_\Phi^r \cdot \chi^{-d^\Phi}$ is trivial restricted to Z. Moreover, since the line $\wedge^{d^\Phi}(M^\Phi) \subset \wedge^{d^\Phi}(M)$ is P-stable, the character χ_Φ (and hence $\bar{\chi}_\Phi$) is dominant. If $P \cap G'$ were to act trivially on $\wedge^{d^\Phi}(M^\Phi)$, then the line $\wedge^{d^\Phi}(M^\Phi) \subset \wedge^{d^\Phi}(M)$ would be G-stable and hence M^Φ would be G-stable, i.e., $M^\Phi = 0$ or M. But since $\mathscr{W} \neq 0$ is a proper subbundle of $E(M)$, M^Φ cannot be (0) or M. Thus, $\bar{\chi}_\Phi$ is a nontrivial character of P. Hence, by Exercise 6.1.E.4, if E is stable (resp. semistable)

$$\deg \sigma_{\mathscr{W}}^* (\mathscr{L}_P(-\bar{\chi}_\Phi)) < 0 \ \ \big(\text{resp. } \deg \sigma_{\mathscr{W}}^* (\mathscr{L}_P(-\bar{\chi}_\Phi)) \leq 0\big). \tag{3}$$

But

$$\deg \sigma_{\mathscr{W}}^* (\mathscr{L}_P(-\bar{\chi}_\Phi)) = r \deg \mathscr{W} - d^\Phi \deg E(M). \tag{4}$$

Combining (3) and (4), we get the lemma. □

Remark 6.3.28 By Theorem 6.1.7, if E is semistable, then so is $E(M)$. Thus, inequality (1) of Lemma 6.3.27 in the semistable case follows from the definition of semistable vector bundles.

We return now to the proof of Proposition 6.3.24.

Proof of Proposition 6.3.24 We fix an $1 \leq i \leq r$ and denote V_i by M. Let $s_i \neq 0$ and let $\mathscr{F}_1$ be the vector subbundle of $E'(M)$ generated by the image of s_i (cf. proof of Lemma 6.3.22). Thus, similar to the diagram (*) in the same, we have

$$\begin{array}{ccccccccc} 0 & \longrightarrow & \mathscr{E}_1 & \longrightarrow & E(M) & \xrightarrow[\pi]{} & \mathscr{E}_2 & \longrightarrow & 0 \\ & & & & \Big\downarrow{\scriptstyle s_i} & & \Big\downarrow{\scriptstyle f'} & & \\ 0 & \longleftarrow & \mathscr{F}_2 & \longleftarrow & E'(M) & \xleftarrow[i]{} & \mathscr{F}_1 & \longleftarrow & 0. \end{array} \tag{$\mathcal{D}$}$$

Let $d = \operatorname{rank} \mathscr{F}_1$ and let $\operatorname{Gr}(d, M)$ be the Grassmannian of d-dimensional subspaces of M. Then $\mathscr{F}_1$ can be thought of as a section of $E' \times^G \operatorname{Gr}(d, M)$. Let

$$\mathcal{A} := \{N \in \operatorname{Gr}(d, M) : N = \operatorname{Im} f_i, \text{ for some } f = (f_1, \ldots, f_r) \in \bar{C}\},$$

where $\bar{C}$ is as in Definition 6.3.23. Clearly, $\mathcal{A}$ is stable under the action of G on $\operatorname{Gr}(d, M)$ (since $\bar{C}$ is $G \times G$-stable). Moreover, by Lemma 6.3.25, the stabilizer of any $N \in \mathcal{A}$ is a parabolic subgroup (since so is for M^Φ). Thus, G-orbits in $\mathcal{A}$ are closed in $\operatorname{Gr}(d, M)$ and, by Lemma 6.3.25, $\mathcal{A}$ has finitely many G-orbits all of which are of the form $\{gM^\Phi\}_{g \in G}$, for some M^Φ. In particular, $\mathcal{A}$ is a closed subset of $\operatorname{Gr}(d, M)$, which is a finite disjoint union $\mathcal{A} = \sqcup \mathcal{A}_j$ of closed subsets, with each $\mathcal{A}_j$ (with reduced structure) isomorphic with G/P_j (for a parabolic subgroup P_j).

Since the fibers of $\mathscr{F}_1$ coincide with $\operatorname{Im} s_i$ on a dense open subset U of Σ (which of course is connected), we can think of $\operatorname{Im}(s_{i\,|U}) \in H^0\big(U, E'(\mathcal{A}_{j_o})\big)$ for some fixed j_o. But then $\mathscr{F}_1 \in H^0\big(\Sigma, E'(\mathcal{A}_{j_o})\big)$, $E'(\mathcal{A}_{j_o})$ being closed in $E'(\operatorname{Gr}(d, M))$. Therefore, by Lemma 6.3.27,

$$\mu(\mathscr{F}_1) \leq \mu(E'(M)), \tag{1}$$

since $s_i \neq 0$ and $\mathcal{A}_{j_o}$ consists of subspaces of M of fixed type Φ_{j_o}. Considering the dual of the diagram ($\mathcal{D}$), we get

$$\begin{array}{ccccccccc} 0 & \longrightarrow & \mathscr{F}_2^* & \longrightarrow & E'(M^*) & \longrightarrow & \mathscr{F}_1^* & \longrightarrow & 0 \\ & & & & \Big\downarrow{\scriptstyle s_i^*} & & \Big\downarrow & & \\ 0 & \longleftarrow & \mathscr{E}_1^* & \longleftarrow & E(M^*) & \longleftarrow & \mathscr{E}_2^* & \longleftarrow & 0, \end{array} \tag{$\mathcal{D}_1$}$$

where $s_i^* \colon E'(M^*) \to E(M^*)$ is the dual morphism. Similar to (1), using Lemma 6.3.27 again, since E is a stable G-bundle, we get

$$\mu(\mathscr{E}_2^*) < \mu(E(M^*)), \text{ if } d < \dim M. \tag{2}$$

But

$$\mu(E(M^*)) = -\mu(E(M)) = -\mu(E'(M)),$$

since E and E' are of the same topological type (by assumption). Thus, (2) gives

$$\mu(E'(M)) < \mu(\mathscr{E}_2) \leq \mu(\mathscr{F}_1), \tag{3}$$

where the last inequality follows since f' is of maximal rank. Now, (3) contradicts (1), proving that $d = \dim M$, i.e., $s_i : E(M) \to E'(M)$ is an isomorphism over a nonempty open subset U of Σ. Think of s_i as a section of the degree 0 line bundle $E(\wedge^d M)^* \otimes E'(\wedge^d M)$ over Σ which does not vanish over U and hence it must not vanish anywhere. Thus, s_i is an isomorphism. This proves the first part of the proposition.

We now prove the second part. By the first part, since each s_i is an isomorphism, and since C is closed in GL_V and $s(\Sigma) \subset F(\bar{C})$ (by assumption), we get that $s(\Sigma) \subset F(C)$, i.e., for any $x \in \Sigma$, there exists $\mathfrak{z}(x) = (z_1(x), \dots, z_r(x)) \in (\mathbb{C}^*)^r$ such that

$$(z_1(x)s_1(x), \dots, z_r(x)s_r(x)) \in F_x(\phi(G)),$$

where $\phi(G) \subset \operatorname{End} V$ is stable under $G \times G$-action on $\operatorname{End} V$ given by (1) of Definition 6.3.23. Let H be the closed subgroup of $(\mathbb{C}^*)^r$ defined by

$$H = \left\{(y_1, \dots, y_r) \in (\mathbb{C}^*)^r : (y_1 \operatorname{Id}_{V_1}, \dots, y_r \operatorname{Id}_{V_r}) \in \phi(G)\right\}.$$

Then, for any $\mathfrak{y} = (y_1, \dots, y_r) \in (\mathbb{C}^*)^r$,

$$(z_1(x)y_1s_1(x), \dots, z_r(x)y_rs_r(x)) \in F_x(\phi(G)) \iff \mathfrak{y} \in H.$$

Hence, the function

$$\hat{\mathfrak{z}} \colon \Sigma \to (\mathbb{C}^*)^r/H, \quad x \mapsto (z_1(x), \dots, z_r(x)) \cdot H$$

is a well-defined morphism. But, since Σ is a projective variety and $(\mathbb{C}^*)^r/H$ is affine, the function $\hat{\mathfrak{z}}$ is a constant. Write

$$\hat{\mathfrak{z}}(x) = (z_1, \dots, z_r) \cdot H, \quad \text{for any } x \in \Sigma,$$

where $(z_1, \dots, z_r) \in (\mathbb{C}^*)^r$ is a fixed point. Thus, taking

$$\tilde{s} = (z_1s_1, \dots, z_rs_r),$$

we get that $\tilde{s}(\Sigma) \subset F(\phi(G))$. Finally, since ϕ is an embedding, we get that $\tilde{s}$ is induced from a G-bundle isomorphism $\bar{s} \colon E \to E'$. This proves the proposition. □ □

We have the following general result.

Lemma 6.3.29 *Let X and T be $\mathbb{C}$-analytic spaces with X compact and let $\{\mathscr{W}_i\}_{1\le i\le r}$ be $\mathbb{C}$-analytic vector bundles over $X\times T$. Let $\mathscr{W}=\bigoplus_{i=1}^r \mathscr{W}_i$ and let $\mathscr{C}\subset\mathscr{W}$ be a closed $\mathbb{C}$-analytic subset which is stable under the homothety action of $(\mathbb{C}^*)^r$ on $\mathscr{W}_1\oplus\cdots\oplus\mathscr{W}_r$. Then*

(a) The set $S_{\mathscr{C}} := \bigcup_{t\in T}\left\{s^t\in H^0(X,\mathscr{W}_t) : s^t(X)\subset\mathscr{C}\right\}$ has a natural structure of a $\mathbb{C}$-analytic space such that the projection $S_{\mathscr{C}}\to T$ is holomorphic, where $\mathscr{W}_t := \mathscr{W}_{|X\times t}$.

Moreover, its subset

$$S'_{\mathscr{C}} := \bigcup_{t\in T}\{s^t\in S_{\mathscr{C}} : s^t=(s^t_1,\ldots,s^t_r)$$
$$\text{has each } s^t_i\neq 0, \text{ where } s^t_i\in H^0(X,\mathscr{W}_{i,t})\}$$

is an open subset of $S_{\mathscr{C}}$. In fact, $S'_{\mathscr{C}}$ is the complement of a closed $\mathbb{C}$-analytic subset of $S_{\mathscr{C}}$.

(b) Consider the projectivization $\mathbb{P}(S'_{\mathscr{C}}) := \left\{[s^t] : t\in T \text{ and } s^t\in S'_{\mathscr{C}}\right\}$, where

$$[s^t] := \left([s^t_1],\ldots,[s^t_r]\right) \quad \text{with } [s^t_i]\in\mathbb{P}\left(H^0(X,\mathscr{W}_{i,t})\right).$$

Then $\mathbb{P}(S'_{\mathscr{C}})$ has a natural structure of a $\mathbb{C}$-analytic space such that $S'_{\mathscr{C}}\to\mathbb{P}(S'_{\mathscr{C}})$ is a holomorphic submersion and $\mathbb{P}(S'_{\mathscr{C}})\to T$ is a proper holomorphic map.

In particular, the set

$$\left\{t\in T : \exists\, s^t\in H^0(X,\mathscr{W}_t) \text{ with } s^t\in S'_{\mathscr{C}}\right\}$$

is a closed $\mathbb{C}$-analytic subset of T.

Proof (a) Let $\mathrm{Hol}(X,\mathscr{W})$ be the space of holomorphic maps from X to $\mathscr{W}$ with the topology of uniform convergence. Then $\mathrm{Hol}(X,\mathscr{W})$ has a natural structure of a $\mathbb{C}$-analytic space such that for any $\mathbb{C}$-analytic space Y, any map $Y\to\mathrm{Hol}(X,\mathscr{W})$ is holomorphic if and only if the corresponding map $Y\times X\to\mathscr{W}$ is holomorphic (cf. (Barlet and Magnússon, 2014, Chap. IV, §9.4)[1]; also see (Douady, 1966)). In particular, the evaluation map $\mathrm{Hol}(X,\mathscr{W})\times X\to\mathscr{W}$ is holomorphic. Hence, the subspace $\mathrm{Hol}(X,\mathscr{C})$ is a (closed) $\mathbb{C}$-analytic subspace of $\mathrm{Hol}(X,\mathscr{W})$.

[1] We thank D. Barlet for this reference.

Consider the composite projections $\pi_X : \mathscr{W} \to X \times T \to X$ and $\pi_T : \mathscr{W} \to X \times T \to T$. Define

$$\text{Hol}_o(X, \mathscr{W}) := \{\text{holomorphic } s : X \to \mathscr{W} : \pi_T \circ s \text{ is a constant}\}.$$

Then $\text{Hol}_o(X, \mathscr{W})$ is a (closed) $\mathbb{C}$-analytic subspace of $\text{Hol}(X, \mathscr{W})$, being the inverse image of the set of constant maps under the holomorphic map $\text{Hol}(X, \mathscr{W}) \to \text{Hol}(X, T)$, $s \mapsto \pi_T \circ s$ (cf. Exercise 6.3.E.5).

By definition,

$$S_{\mathscr{C}} = \{s \in \text{Hol}_o(X, \mathscr{C}) : \pi_X \circ s = \text{Id}_X\}. \tag{1}$$

Of course, the map $\hat{\pi}_X : \text{Hol}_o(X, \mathscr{C}) \to \text{Hol}(X, X)$ induced from π_X is holomorphic, since the corresponding map given by $\text{Hol}(X, \mathscr{C}) \times X \to X$, $(s, x) \mapsto \pi_X(s(x))$, is holomorphic. Since $S_{\mathscr{C}} = (\hat{\pi}_X)^{-1}(\text{Id}_X)$, $S_{\mathscr{C}}$ is a $\mathbb{C}$-analytic space.

The projection $S_{\mathscr{C}} \subset \text{Hol}_o(X, \mathscr{W}) \to T$ is given by $s \mapsto \pi_T(s(x))$ for any (fixed) $x \in X$. Hence, it is holomorphic.

Considering the (closed) $\mathbb{C}$-analytic subset $S_{\mathscr{C}(i)}$ of $S_{\mathscr{C}}$ for $\mathscr{C}(i) = \mathscr{C} \cap \left(\mathscr{W}_1 \oplus \cdots \oplus \underline{0} \oplus \cdots \oplus \mathscr{W}_r\right)$, where $\underline{0}$ is the zero vector bundle over $X \times T$ placed in the ith slot, we get that $S_{\mathscr{C}} \backslash S'_{\mathscr{C}} = \bigcup_{i=1}^r S_{\mathscr{C}(i)}$ is a closed subset of $S_{\mathscr{C}}$ and hence $S'_{\mathscr{C}}$ is an open subset of $S_{\mathscr{C}}$. This proves the (a)-part of the lemma.

(b) The standard action of $(\mathbb{C}^*)^r$ on $S'_{\mathscr{C}}$ (by homothety in each factor $H^0(X, \mathscr{W}_{i,t})$) is, of course, fixed-point free. Moreover, it is holomorphic. This follows since

$$(\mathbb{C}^*)^r \times S_{\mathscr{W}} \times X \to \mathscr{W}, \ \left((z_1, \ldots, z_r), s^t, x\right) \mapsto \Sigma z_i s_i^t(x)$$

is holomorphic, where $S_{\mathscr{W}}$ is defined by (1) taking $\mathscr{C} = \mathscr{W}$ and $s^t = (s_1^t, \ldots, s_r^t)$ with $s_i^t \in H^0(X, \mathscr{W}_{i,t})$. Also, since $S'_{\mathscr{C}}$ consists of nonzero sections in each $\mathscr{W}_{i,t}$, the action of $(\mathbb{C}^*)^r$ on $S'_{\mathscr{C}}$ is proper. Hence, the orbit space $S'_{\mathscr{C}}/(\mathbb{C}^*)^r$ is a $\mathbb{C}$-analytic space and the quotient map $S'_{\mathscr{C}} \to S'_{\mathscr{C}}/(\mathbb{C}^*)^r$ is holomorphic submersion (cf. (Cartan, 1957)). In particular, the holomorphic map $S'_{\mathscr{C}} \to T$ which clearly descends to a map $S'_{\mathscr{C}}/(\mathbb{C}^*)^r \to T$ is holomorphic. Introduce a positive-definite continuous Hermitian form on the vector bundle $\mathscr{W}$. Then the subset

$$S'_{\mathscr{C}}(1) = \Big\{s^t = (s_1^t, \ldots, s_r^t) \in S'_{\mathscr{C}} \text{ with}$$
$$t \in T,\ s_i^t \in H^0(X, \mathscr{W}_{i,t}) \text{ and } ||s_i^t|| = 1\Big\},$$

where $||s_i^t|| := \sup_{x \in X} |s_i^t(x)|$, maps surjectively onto $S'_{\mathscr{C}}/(\mathbb{C}^*)^r$.

By Montel's theorem (cf. (Rudin, 1966, Theorem 14.6)), the map $S'_{\mathscr{C}}(1) \to T$ is proper and hence so is the map $S'_{\mathscr{C}}/(\mathbb{C}^*)^r \to T$. Now, the (b)-part of the lemma follows since

$$S'_{\mathscr{C}}/(\mathbb{C}^*)^r \simeq \mathbb{P}(S'_{\mathscr{C}}).$$

The 'In particular' part of the lemma follows from Remmert's theorem asserting that the image of a proper holomorphic map (between $\mathbb{C}$-analytic spaces) is a (closed) $\mathbb{C}$-analytic subspace (cf. (Remmert, 1957)). □

As a consequence of Lemmas 6.3.2, 6.3.29 and Proposition 6.3.24, we get the following.

Proposition 6.3.30 *Let $\mathscr{F} \to \Sigma \times T$ be a $\mathbb{C}$-analytic family of stable G-bundles over Σ (parameterized by a $\mathbb{C}$-analytic space T). Then the subset*

$$T_u := \left\{t \in T : \mathscr{F}_t \simeq E_\rho \text{ for some unitary representation } \rho \text{ of } \pi_1(\Sigma) \text{ in } G\right\}$$

is a closed subset of T.

Proof We can of course assume that T is connected so that each $\mathscr{F}_t$ is of the same topological type.

Recall the definition of the tautological family $\theta: \mathscr{E} \to \Sigma \times R_G(g)$ from Lemma 6.3.2. Consider the fiber product

$$\mathscr{H} := \mathscr{F} \underset{\Sigma}{\times} \mathscr{E} \to \Sigma \times T \times R_G(g) \text{ of}$$
$$\mathscr{F} \to \Sigma \times T \to \Sigma \text{ and } \mathscr{E} \to \Sigma \times R_G(g) \to \Sigma.$$

Then $\mathscr{H}$ is a family of $G \times G$-bundles over Σ parameterized by $T \times R_G(g)$ with fiber $(\mathscr{H})_{t,\rho} = \mathscr{F}_t \times E_\rho$. Choose a faithful representation $\phi: G \to \mathrm{GL}_V$ and consider the $G \times G$-stable subset $\bar{C} \subset \bigoplus_{i=1}^r \mathrm{End}\, V_i \subset \mathrm{End}\, V$ as in Definition 6.3.23. Applying Lemma 6.3.29 for $X = \Sigma$, T replaced by $T \times R_G(g)$, $\mathscr{W}_i = \mathscr{H}(\mathrm{End}\, V_i)$, $\mathscr{C} = \mathscr{H}(\bar{C})$, we get that the subset

$$F := \left\{(t, \rho) \in T \times R_G(g) : \exists s \in H^0\left(\Sigma, \mathscr{W}_{(t,\rho)}\right) \text{ with } s \in S'_{\mathscr{C}}\right\}$$

is a closed $\mathbb{C}$-analytic subset of $T \times R_G(g)$, where $S'_{\mathscr{C}}$ is as defined in Lemma 6.3.29. Hence, $F_K := F \cap (T \times R_K(g))$ is a closed $\mathbb{R}$-analytic subset of $T \times R_K(g)$, where $R_K(g)$ is as in Lemma 6.3.2. But the projection $p_T : T \times R_K(g) \to T$ is proper (since $R_K(g)$ is compact) and hence $p_T(F_K)$ is closed in T. Now, using Propositions 6.3.24 and 6.3.4(a) for $A = (1)$, we get that $T_u = p_T(F_K)$ and hence T_u is closed in T, proving the proposition. □

Lemma 6.3.31 *Let $\mathscr{F} \to \Sigma \times T$ be a $\mathbb{C}$-analytic family of G-bundles over Σ (parameterized by a $\mathbb{C}$-analytic space T). Then the subset*

$$T_s := \{t \in T : \mathscr{F}_t \text{ is a stable } G\text{-bundle}\}$$

is an open subset which is the complement of a (closed) $\mathbb{C}$-analytic subset of T.

Proof Take a standard maximal parabolic subgroup Q_k of G and take the irreducible representation $V_k := V(d\omega_k)$ of G with highest weight $d\omega_k$, where ω_k is the kth fundamental weight required to vanish on the center $\mathfrak{z}(\mathfrak{g})$ of $\mathfrak{g}$ and $d\omega_k$ is a suitable positive multiple of ω_k so that it is a character of the maximal torus of G. Thus, we get an embedding $G/Q_k \hookrightarrow \mathbb{P}(V_k)$, $gQ_k \mapsto [gv_+]$, where v_+ is a highest-weight vector of V_k and $[gv_+]$ is the line through $g \cdot v_+$. Let $J = J_\Sigma$ be the Jacobian of Σ (i.e., the group of isomorphism classes of degree 0 line bundles over Σ) and $\mathcal{P} \to \Sigma \times J$ the *Poincaré line bundle* (cf. (Arbarello et al., 1985, Chap. IV, §2)). Define a $\mathbb{C}$-analytic family of vector bundles $\mathscr{F}_{\mathcal{P}}(V_k) \to \Sigma \times (J \times T)$ by

$$\mathscr{F}_{\mathcal{P}}(V_k)_{(j,t)} := j^* \otimes \mathscr{F}_t(V_k).$$

Consider the closed $\mathbb{C}$-analytic cone $\mathscr{C} \subset \mathscr{F}_{\mathcal{P}}(V_k)$ over $\pi^*(\mathscr{F}(G/Q_k)) \subset \pi^*(\mathbb{P}(\mathscr{F}(V_k))) = \mathbb{P}(\mathscr{F}_{\mathcal{P}}(V_k))$, where $\pi : \Sigma \times J \times T \to \Sigma \times T$ is the projection. Then, by Lemma 6.3.29, the subset

$$Z_k := \left\{(j,t) \in J \times T : \exists \mu \neq 0 \in H^0\left(\Sigma, \mathscr{F}_{\mathcal{P}}(V_k)_{(j,t)}\right),\ \mu(\Sigma) \subset \mathscr{C}\right\}$$

is a (closed) $\mathbb{C}$-analytic subset of $J \times T$. Let $\tilde{Z}_k \subset T$ be the image of Z_k under the projection $J \times T \to T$. Since J is compact, by a theorem of Remmert (cf. (Remmert, 1957)), $\tilde{Z}_k$ is a (closed) $\mathbb{C}$-analytic subset of T.

We next claim that

$$\tilde{Z}_k \subset T \backslash T_s. \tag{1}$$

Take $t \in \tilde{Z}_k$. Thus, there exists $j \in J$ such that there exists nonzero $\mu \in H^0(\Sigma, j^* \otimes \mathscr{F}_t(V_k))$ with $\mu(\Sigma) \subset \mathscr{C}$. Hence, μ gives rise to a section $\bar{\mu}$ of $\mathbb{P}(\mathscr{F}_t(V_k))$ over a nonempty Zariski open subset $U \subset \Sigma$ (where μ is nonzero) with the image contained in $\mathscr{F}_t(G/Q_k)$. Since Σ is a curve and the fibers of $\mathbb{P}(\mathscr{F}_t(V_k))$ are projective varieties, the section $\bar{\mu}$ extends holomorphically to the whole of Σ with the image contained in $\mathscr{F}_t(G/Q_k)$. The section $\bar{\mu}$ of course provides a reduction of the structure group of $\mathscr{F}_t$ to Q_k (cf. Lemma 5.1.2). Let $\tau \to \mathbb{P}(\mathscr{F}_t(V_k))$ be the tautological line bundle, the pull-back of which to $\mathscr{F}_t(G/Q_k)$ can easily be seen to be $\mathscr{L}_{Q_k}(-d\omega_k)$ (following the notation of Definition 6.1.3(c)).

Let $\bar{\mu}^*(\tau)$ be the pull-back line bundle over Σ. Then, the section μ thought of as a bundle morphism $j \to \mathscr{F}_t(V_k)$ has its image contained in the line bundle $\bar{\mu}^*(\tau)$. Thus, $j^* \otimes \bar{\mu}^*(\tau)$ has a nonzero section showing that $\deg \bar{\mu}^*(\tau) \geq 0$. Thus, $\mathscr{F}_t$ is not stable (cf. Definition 6.1.4(b)), i.e., $t \in T\backslash T_s$, proving (1).

Conversely, take $t \in T\backslash T_s$. Thus, there exists a standard maximal parabolic subgroup Q_k such that $\deg \bar{\mu}^* \mathscr{L}_{Q_k}(-d\omega_k) \geq 0$ for a section $\bar{\mu}$ of $\mathscr{F}_t(G/Q_k)$. Thus, there exists a $j \in J$ and a nonzero $\mathscr{O}_\Sigma$-module morphism $j \to \bar{\mu}^* \mathscr{L}_{Q_k}(-d\omega_k)$ over Σ such that the corresponding (nonzero) section

$$\sigma(j^* \otimes \bar{\mu}^* \mathscr{L}_{Q_k}(-d\omega_k)) \in H^0\left(\Sigma, \mathscr{F}_{\mathcal{P}}(V_k)_{(j,t)}\right)$$

has its image contained in $\mathscr{C}$. Hence, $t \in \tilde{Z}_k$, i.e.,

$$T\backslash T_s \subset \bigcup_k \tilde{Z}_k, \tag{2}$$

where $\{k\}$ parameterizes the standard maximal parabolic subgroups Q_k of G. Combining (1) and (2), we get the lemma. □

Lemma 6.3.32 *Let E_0 and E_1 be two holomorphic G-bundles of the same topological type over Σ. Then there exists a holomorphic family $\mathscr{E}$ of G-bundles parameterized by $\mathbb{C}$ such that*

$$\mathscr{E}_0 \simeq E_0 \quad and \quad \mathscr{E}_1 \simeq E_1. \tag{1}$$

Further, if E_0 and E_1 are stable G-bundles, then such a holomorphic family $\mathscr{E}$ satisfying (1) *can be chosen over a nonempty connected open subset T of $\mathbb{C}$ containing $\{0, 1\}$ such that $\mathbb{C}\backslash T$ is a (closed) $\mathbb{C}$-analytic subset of $\mathbb{C}$ and $\mathscr{E}_t$ is stable for each $E \in T$.*

Proof Let $E \to X$ be a C^∞ principal G-bundle over a holomorphic manifold X. Then a connection form ∇ over E (which is a $\mathfrak{g}$-valued C^∞ 1-form on E) induces a unique structure of holomorphic G-bundle on E satisfying Koszul (1960, Proposition 1, §6.4) if and only if the corresponding curvature form Ω satisfies $\Omega^{0,2} = 0$, where $\Omega^{0,2}$ is the component of Ω of type $(0, 2)$ with respect to the holomorphic structure on X (cf. (Koszul, 1960, Proposition 3, §6.4)). Conversely, the structure of a holomorphic G-bundle on E gives rise to a (not necessarily unique) connection form ∇ on E with $\Omega^{0,2} = 0$ such that the corresponding holomorphic structure on E coincides with the original holomorphic structure (cf. (Koszul, 1960, §6.4)).

Taking $X = \Sigma$, since Σ is of complex dimension 1, the condition $\Omega^{0,2} = 0$ is automatically satisfied. Since the holomorphic bundles E_0 and E_1 are of the some topological type, we can assume that they correspond to (different) holomorphic structures on the same underlying C^∞ principal G-bundle E over Σ. Choose connection forms ∇_0 and ∇_1 on E which give rise to the holomorphic structures E_0 and E_1 respectively. Now, consider the C^∞ product G-bundle $E \times \mathbb{C} \to \Sigma \times \mathbb{C}$ and define the connection form ∇ on $E \times \mathbb{C}$ by $\nabla_z = z\nabla_1 + (1-z)\nabla_0$ for $z \in \mathbb{C}$, i.e.,

$$\nabla(w, v) = \nabla_z(w), \quad \text{for} \quad w \in T_e(E) \quad \text{and} \quad v \in T_z(\mathbb{C}).$$

Thus, the connection form $\nabla = z\pi_E^*(\nabla_1) + (1-z)\pi_E^*(\nabla_0)$, where $\pi_E : E \times \mathbb{C} \to E$ is the projection.

From the definition of the curvature: $\Omega = d\nabla + \frac{1}{2}[\nabla, \nabla]$, it is easy to see that $\Omega^{0,2} = 0$ for the above connection form ∇ on $E \times \mathbb{C}$. Thus, we get the structure of a holomorphic bundle on $\mathscr{E} := E \times \mathbb{C}$ such that the holomorphic structure restricted to $E \times 0$ (resp. $E \times 1$) is isomorphic with E_0 (resp. E_1). This proves the first part of the proposition.

The second part follows immediately from the first part and Lemma 6.3.31 by observing that the complement of a (closed) $\mathbb{C}$-analytic subset of $\mathbb{C}$ is automatically connected. □

Definition 6.3.33 A G-bundle E over Σ is said to be of *degree* 0 if for any character χ of G, the line bundle $E \times^G \mathbb{C}_\chi$ has degree 0, where $\mathbb{C}_\chi$ is the 1-dimensional representation of G given by the character χ. (This definition coincides with the definition of degree 0 vector bundles.)

With all these preparations, we are now ready to prove the following celebrated theorem.

Theorem 6.3.34 *Let G be a connected reductive group and Σ a smooth irreducible curve of genus $g \geq 2$. Let $E \to \Sigma$ be a stable G-bundle of degree 0. Then there exists a unique (up to conjugacy by G) irreducible unitary representation $\rho : \pi_1(\Sigma) \to G$ such that (as holomorphic bundles)*

$$E \simeq E_\rho.$$

Proof Let $Z_o := G/G'$, where G' is the commutator $[G, G]$. Then Z_o is a (connected) torus. Let $E(Z_o)$ be the bundle obtained from E by extension of the structure group $G \to Z_o$. We claim that $E(Z_o)$ is topologically trivial since E is of degree 0 (by assumption). To prove this, since Z_o is a torus, it suffices to observe that a degree 0 line bundle over Σ is topologically trivial.

The topological triviality of $E(Z_o)$ allows a topological reduction of the structure group of E to G', i.e., there is a topological G'-bundle E' which is isomorphic topologically with E under the extension of the structure group to G. By Lemma 6.3.11, since G' is semisimple, there exists an irreducible unitary representation $\rho_o\colon \pi_1(\Sigma) \to G'$ such that

$$E_{\rho_o} \simeq E' \text{ as topological } G'\text{-bundles}$$

and hence

$$E_{\rho_o}(G) \simeq E \text{ as topological } G\text{-bundles},$$

where $E_{\rho_o}(G)$ denotes the extension of the structure group G' of E_{ρ_o} to G. Observe that ρ_o clearly remains irreducible considered as a homomorphism $\pi_1(\Sigma) \to G$.

Take a holomorphic family of stable G-bundles $\mathscr{E} \to \Sigma \times T$, such that T is a connected open subset of $\mathbb{C}$ containing $\{0, 1\}$ and $\mathscr{E}_0 \simeq E$, $\mathscr{E}_1 \simeq E_{\rho_o}(G)$ (cf. Lemma 6.3.32). Let

$$T_o := \{t \in T : \mathscr{E}_t \simeq E_\sigma, \text{ for some unitary irreducible representation } \sigma \text{ of } \pi_1(\Sigma) \text{ in } G\}.$$

Then, by Corollary 6.3.21, T_o is an open subset of T. Further, by Proposition 6.3.30, T_o is a closed subset of T. (Observe that if $\mathscr{E}_t \simeq E_\sigma$ for some unitary representation σ of $\pi_1(\Sigma)$, then σ is automatically irreducible by Proposition 6.3.4 for $A = (1)$ since each $\mathscr{E}_t$ is stable). Of course, T_o is nonempty since $1 \in T_o$. Thus, $T_o = T$. The uniqueness of ρ (up to conjugation by G) follows from Corollary 6.3.7 for $A = (1)$. This proves the theorem. □

Recall the definition of polystable bundles from Definition 6.1.4(c). Then we have the following generalization of Theorem 6.3.34.

Theorem 6.3.35 *Let G be a connected reductive group and let E be a holomorphic G-bundle over a smooth irreducible projective curve Σ of genus $g \geq 2$. Then E is polystable of degree 0 if and only if $E \simeq E_\rho$ (as holomorphic G-bundles) for a unitary representation $\rho\colon \pi_1(\Sigma) \to G$.*

Proof Assume first that E is polystable of degree 0. Then E admits a reduction E_L to a Levi subgroup L such that E_L is stable of degree 0 (as an L-bundle). (To prove this, observe that for any character χ of L, there exists a character χ' of G and a character χ'' of L trivial on the center of G such that

$$\chi^N = \chi'_{|L} \cdot \chi'' \quad \text{for some } N \gg 0.)$$

Thus, by Theorem 6.3.34,

$$E_L \simeq E_{\rho_L}, \tag{1}$$

for an irreducible unitary representation $\rho_L\colon \pi_1(\Sigma) \to L$. Let ρ be the same representation thought of as $\pi_1(\Sigma) \to G$. Then, from (1), we get $E \simeq E_\rho$.

Conversely, take a unitary representation $\rho\colon \pi_1(\Sigma) \to K \subset G$. Then, if it is not irreducible, there exists a proper parabolic subgroup P of G and a Levi subgroup L_P of P such that

$$\operatorname{Im}\rho \subset L_P$$

(since $P \cap K$ is contained in a Levi subgroup of P). Continuing this way (inducting on the semisimple rank of G), we find a Levi subgroup L with $\operatorname{Im}\rho \subset L$ and $\rho_L\colon \pi_1(\Sigma) \to L$ is irreducible, where $\rho_L = \rho$. Thus, from Proposition 6.3.4 for $A = (1)$, E_{ρ_L} is a stable L-bundle. Further, since E_{ρ_L} has a discrete structure group (thereby a flat connection), by the Chern–Weil theory,

$$\deg\left(E_{\rho_L} \times^L \mathbb{C}_\chi\right) = 0, \quad \text{for any character } \chi \text{ of } L.$$

In particular, $E_{\rho_L}(G) = E_\rho$ is polystable of degree 0. This proves the theorem. □

Definition 6.3.36 Let Σ be a smooth irreducible projective curve, G a connected semisimple algebraic group and let $E \to \Sigma$ be a holomorphic G-bundle. Then a C^∞-connection ∇ on E is called

(a) *complex connection* if the corresponding holomorphic structure on E (cf. the proof of Lemma 6.3.32) coincides with the original holomorphic structure.

(b) *unitary connection* if there exists a C^∞-reduction $E_K \subset E$ of the structure group of E to a maximal compact subgroup K of G and ∇ is reducible to E_K (i.e., ∇ is obtained as the direct image of a C^∞-connection on E_K).

Observe that the condition of ∇ being unitary is equivalent to the requirement that the holonomy group of ∇ is relatively compact.

(c) *Einstein connection* if the curvature form of ∇ is identically zero.

(d) *Einstein–Hermitian connection* if it satisfies the above properties (a)–(c).

Observe that since E admits a C^∞-reduction $E_K \subset E$ of the structure group (G/K being contractible), E admits a unique complex unitary connection (cf. (Kobayashi and Nomizu, 1969, Theorem 10.1 on p. 178 and Remark on p. 185)).

Moreover, the existence of an Einstein–Hermitian connection on E is equivalent to the unitarity of E (as in Definition 6.3.1) using the Holonomy Theorem (Koszul, 1960, Chap. 4).

Recall the definition of A-unitary G-bundles and A-unitary vector bundles from Definition 6.3.3, the notation of which we will follow.

Lemma 6.3.37 *Let $\hat{E}$ be an A-equivariant G-bundle over $\hat{\Sigma}$, where G is a connected semisimple group. Then $\hat{E}$ is A-unitary if and only if $\hat{E}(\mathfrak{g})$ is A-unitary vector bundle.*

The lemma is clearly false if G were a torus.

Proof Of course, if $\hat{E}$ is A-unitary, then so is $\hat{E}(\mathfrak{g})$. Conversely, assume that $\hat{E}(\mathfrak{g})$ is A-unitary. Then we show that $\hat{E}$ is A-unitary.

The bracket $\mathfrak{g} \otimes \mathfrak{g} \to \mathfrak{g}$, $x \otimes y \mapsto [x, y]$, being G-equivariant, induces an A-equivariant bundle morphism

$$\varphi \colon \hat{E}(\mathfrak{g} \otimes \mathfrak{g}) \to \hat{E}(\mathfrak{g})$$

between A-unitary bundles. By Lemma 6.3.6, the bracket map $\mathfrak{g} \otimes \mathfrak{g} \to \mathfrak{g}$ must be π-equivariant, where the action $\hat{\rho}$ of π on $\mathfrak{g}$ comes from the assumption that $\hat{E}(\mathfrak{g})$ is A-unitary. Thus, the representation $\hat{\rho} \colon \pi \to \mathrm{Aut}(\mathfrak{g})$ has its image inside $G_F := \mathrm{Aut}_{\mathrm{Lie}}(\mathfrak{g})$, where $\mathrm{Aut}_{\mathrm{Lie}}(\mathfrak{g})$ is the group of Lie algebra automorphisms of $\mathfrak{g}$. Thus, $\hat{E}(G_F)$ is A-unitary.

Assume now that G is of adjoint type (i.e., its center is trivial). Then we have the exact sequence of groups

$$1 \to G \to G_F \to F \to 1, \qquad (*)$$

where F is the (finite) group of outer automorphisms of $\mathfrak{g}$ (its finiteness follows from the Whitehead Lemma (Hilton and Stammbach, 1997, Chap. VII, Proposition 6.1)). Since $\hat{E}(G_F)(F)$ admits a canonical A-equivariant section (coming from the embedding $\hat{E} = \hat{E}(G) \hookrightarrow \hat{E}(G_F)$), $\hat{E}(G_F)(F)$ being an A-equivariant principal F-bundle, it is A-equivariantly trivial. Hence, the composite map

$$\pi \xrightarrow{\hat{\rho}} G_F \to F$$

is trivial (use Corollary 6.3.7), i.e., $\hat{\rho}(\pi) \subset G$, which proves that $\hat{E}$ is A-unitary (in the case G is an adjoint group).

We now prove the A-unitarity of $\hat{E}$ when G is an arbitrary connected semisimple group. Consider the exact sequence

$$1 \to Z \to G \to G_{\mathrm{ad}} \to 1,$$

where Z is the center of G and G_{ad} is the corresponding adjoint group. We have already established that $\hat{E}(G_{\mathrm{ad}})$ is A-unitary. Thus, the A-unitarity of $\hat{E}$ follows from the following lemma. □

Lemma 6.3.38 *We follow the notation as in the above Lemma 6.3.37. Let $G \to H$ be a surjective morphism of connected semisimple algebraic groups with finite kernel. Let $\hat{E} = \hat{E}(G)$ be an A-equivariant G-bundle over $\hat{\Sigma}$ such that $\hat{E}(H)$ is A-unitary. Then so is $\hat{E}$.*

Proof Since $\hat{E}(H)$ is A-unitary, in particular unitary, it admits a unique Einstein–Hermitian connection ∇_H (cf. Definition 6.3.36). Moreover, by its uniqueness, ∇_H is A-invariant. Let ∇_G be the connection induced from ∇_H on $\hat{E}$ (using the isomorphism of tangent spaces of G and H). Then it is easy to see that ∇_G is an A-invariant Einstein–Hermitian connection (cf. (Ramanathan and Subramanian, 1988, Lemma 2)). Thus the bundle $\hat{E}$ is given by a representation of the fundamental group

$$\pi_1(\hat{\Sigma}) \to K \text{ for a maximal compact subgroup } K.$$

Moreover, since $\hat{E}$ is an A-equivariant G-bundle, by Lemma 6.3.8 we get that $\hat{E}$ is A-unitary. □

Let $\hat{\Sigma}$ be an irreducible smooth projective curve with faithful action of a finite group A and let G be a connected reductive group. The following equivariant generalization of Theorem 6.1.7 holds by the same proof.

Lemma 6.3.39 *Let $f: G \to G'$ be a homomorphism between connected reductive groups such that $f(Z^o(G)) \subset Z^o(G')$, where $Z^o(G)$ denotes the identity component of the center of G. Then, if $\hat{E} \to \hat{\Sigma}$ is a A-semistable (resp. A-polystable) G-bundle, then so is $\hat{E}(G')$ obtained from $\hat{E}$ by extension of the structure group to G'.*

In particular, for any A-semistable (resp. A-polystable) G-bundle $\hat{E}$, ad $\hat{E}$ *is an A-semistable (resp. A-polystable) vector bundle.*

Lemma 6.3.40 *Let $\hat{\Sigma}, G$ and A be as above but we assume that $\hat{\Sigma}$ has genus $\hat{g} \geq 2$. Let $\hat{E}$ be an A-polystable G-bundle over $\hat{\Sigma}$. Then, $\hat{E}$ is polystable.*

In particular, $\hat{E}$ is A-semistable.

Proof[2] Observe first that $\hat{E}$ is A-polystable (resp. polystable) if and only if $\hat{E}(G/Z)$ is A-polystable (resp. polystable), where Z is the center of G (cf. Exercise 6.1.E.12). Thus, we can assume that G is semisimple. By Lemma 6.3.39, ad $\hat{E}$ is A-polystable vector bundle. By Exercise 6.1.E.16, we can write ad $\hat{E} = \oplus_i \mathscr{V}_i$, where $\mathscr{V}_i$ are A-stable vector subbundles of ad $\hat{E}$ all with the same slope; in particular, $\mathscr{V}_i$ are A-semistable and hence semistable by Exercise 6.2.E.4. Fix any $\mathscr{V} = \mathscr{V}_i$ and let $\mathscr{V}^o$ be the socle of $\mathscr{V}$, which is an

[2] We thank V. Balaji for this proof.

A-equivariant vector subbundle of $\mathscr{V}$ such that the slope $\mu(\mathscr{V}^o) = \mu(\mathscr{V})$ (cf. (Mehta and Ramanathan, 1984, Definition 2.1 and Lemma 2.2)). If $\mathscr{V}$ were not polystable, then by the same reference and Exercise 6.1.E.15, $\mathscr{V}^o \subsetneqq \mathscr{V}$ and since $\mu(\mathscr{V}^o) = \mu(\mathscr{V})$, it contradicts the A-stability of $\mathscr{V}$. Hence, $\mathscr{V}$ is polystable and hence so is ad $\hat{E}$. Clearly, ad $\hat{E}$ is of degree 0. Thus, by Theorem 6.3.35, ad $\hat{E}$ is a unitary vector bundle. Hence, by Lemma 6.3.37, $\hat{E}$ is a unitary G-bundle, and thus is polystable by Theorem 6.3.35.

The 'In particular' part of the lemma follows from Definition 6.1.4(c) and Exercise 6.2.E.4. □

We now come to the following equivariant generalization of Theorem 6.3.35.

Theorem 6.3.41 *Let $\hat{\Sigma}$ be an irreducible smooth projective curve with faithful action of a finite group A such that $\Sigma := \hat{\Sigma}/A$ has genus $g \geq 2$ and G a connected reductive group. Then an A-equivariant G-bundle $\hat{E}$ over $\hat{\Sigma}$ is A-unitary if and only it is A-polystable of degree* 0.

In particular, an A-equivariant G-bundle over $\hat{\Sigma}$ is A-polystable if and only it is polystable.

Proof Assume first that $\hat{E}$ is A-unitary, i.e., there is a unitary homomorphism $\hat{\rho}\colon \pi \to G$ (following the notation of Definition 6.3.3) with

$$\hat{E} \simeq \hat{E}_{\hat{\rho}}, \quad \text{as } A\text{-equivariant } G\text{-bundles.} \tag{1}$$

Then, as in the proof of Theorem 6.3.35, there exists a Levi subgroup L with $\operatorname{Im}\hat{\rho} \subset L$ and $\hat{\rho}_L\colon \pi \to L$ is irreducible, where $\hat{\rho}_L := \hat{\rho}$. Thus, the corresponding bundle $\hat{E}_{\hat{\rho}_L}$ is A-stable by Proposition 6.3.4(b). Moreover, for any character χ of L,

$$\deg\left(\hat{E}_{\hat{\rho}_L} \times^L \mathbb{C}_\chi\right) = 0,$$

since $\hat{E}_{\hat{\rho}_L}$ has discrete structure group. Thus $\hat{E}$ is A-polystable of degree 0 by (1).

Conversely, assume that $\hat{E}$ is A-polystable of degree 0. Then, by Lemma 6.3.40, $\hat{E}$ is polystable (of degree 0). (Observe that $\hat{\Sigma}$ has genus $\hat{g} \geq 2$ by (Hartshorne, 1977, Chap. IV, Example 2.5.4) since $g \geq 2$ by assumption.) Now, by Theorem 6.3.35, as G-bundles,

$$\hat{E} \simeq E_\rho, \quad \text{for a unitary homomorphism } \rho\colon \pi_1(\hat{\Sigma}) \to G.$$

But, since $\hat{E}$ is an A-equivariant G-bundle, by Lemma 6.3.8, we get that ρ lifts to a unitary homomorphism $\hat{\rho}\colon \pi \to G$ such that

$$\hat{E} \simeq \hat{E}_{\hat{\rho}}, \text{ as } A\text{-equivariant } G\text{-bundles.}$$

Thus, $\hat{E}$ is A-unitary, proving the first part of the theorem.

We now prove the 'In particular' part. Of course, by Lemma 6.3.40, if $\hat{E}$ is A-polystable then it is polystable. For the converse part, using Exercise 6.1.E.12, we can assume that G is semisimple. Now, if $\hat{E}$ is polystable, then by Theorem 6.3.35 and Lemma 6.3.8, $\hat{E}$ is A-unitary. Thus, by the first part of the theorem, $\hat{E}$ is A-polystable. This proves the theorem. □

Let $\hat{\Sigma}$ and A be as in Theorem 6.3.41 (in particular, Σ has genus ≥ 2) and G a connected (not necessarily simply-connected) semisimple group. Using Theorem 6.3.41 and Lemma 6.3.37, we get the following generalization of Lemma 6.3.37.

Proposition 6.3.42 *Let $\hat{E}$ be an A-equivariant G-bundle over $\hat{\Sigma}$ and let $\theta\colon G \to \mathrm{GL}_V$ be a representation with finite kernel. Then the vector bundle $\hat{E}(V)$ is A-unitary if and only if $\hat{E}$ is A-unitary.*

Proof Clearly, if $\hat{E}$ is A-unitary, then so is $\hat{E}(V)$.

Conversely, assume that $\hat{E}(V)$ is A-unitary. Then so is $\hat{E}(W)$ for any GL_V-module W. In particular, for $W := V^* \otimes V = \mathrm{End}_V$, $\hat{E}(W)$ is A-unitary. Consider the G-module embedding

$$d\theta : \mathfrak{g} \hookrightarrow \mathrm{End}_V .$$

(Observe that $d\theta$ is injective since θ has finite kernel.) Take a G-submodule M of End_V such that

$$\mathrm{End}_V \simeq \mathfrak{g} \oplus M, \quad \text{as} \quad G\text{-modules.} \tag{1}$$

The bundle $\hat{E}(W)$ breaks up as a direct sum of A-equivariant bundles:

$$\hat{E}(W) = \hat{E}(\mathfrak{g}) \oplus \hat{E}(M), \tag{2}$$

obtained from the decomposition (1). Since $\hat{E}(W)$ is A-unitary (since so is $\hat{E}(V)$), by Theorem 6.3.41 for GL_W, $\hat{E}(W)$ is A-polystable of degree 0. Decompose

$$\hat{E}(W) = \bigoplus_{i=1}^{k} V_i,$$

where each V_i is an A-stable vector bundle of degree 0 (cf. Exercise 6.1.E.16). Let $\pi : \hat{E}(W) \to \hat{E}(\mathfrak{g})$ be the projection obtained from the decomposition (2) and choose the smallest subset $S \subset \{1, \dots, k\}$ such that $\pi_{|V_S} : V_S \to \hat{E}(\mathfrak{g})$ is surjective, where $V_S := \bigoplus_{i \in S} V_i$. We claim that $\pi_{|V_S}$ is an isomorphism.

Let K_S be the kernel of $\pi_{|V_S}$. Then clearly K_S is an A-equivariant vector bundle of degree 0. For any $i \in S$, let $\pi_i : K_S \to V_i$ be the projection on the ith factor. Then either $\pi_i \equiv 0$ or π_i is surjective since $\deg(K_S) = \deg V_i = 0$ and V_i is A-stable (cf. Exercise 6.3.E.11). (Observe that K_S is A-semistable since it is a degree 0 subbundle of an A-semistable vector bundle V_S of degree 0.) We next show that $\pi_i \equiv 0$ for all $i \in S$. For, if not, assume that $\pi_i \neq 0$ for some i and hence it is surjective. Thus, for any $y \in V_i$ we can choose $x \in K_S$ such that $\pi_i(x) = y$. Decompose (obtained from the decomposition $V_S = \bigoplus_{i \in S} V_i$):

$$x = \sum_{j \in S} x_j, \quad \text{with} \quad x_j \in V_j \quad \text{so that} \quad x_i = y.$$

Hence,

$$0 = \pi(x) = \pi(y) + \sum_{\substack{j \neq i \\ j \in S}} \pi(x_j).$$

This gives

$$\pi(V_i) \subset \pi\left(\bigoplus_{\substack{j \in S \\ j \neq i}} V_j\right).$$

This contradicts the minimality of S, proving that $\pi_i \equiv 0$ for all $i \in S$, i.e., $K_S = (0)$. This proves that $\hat{E}(\mathfrak{g}) \simeq \bigoplus_{i \in S} V_i$ and hence $\hat{E}(\mathfrak{g})$ is A-polystable of degree 0. Thus, by Theorem 6.3.41 for $G = \mathrm{GL}_{\mathfrak{g}}$, $\hat{E}(\mathfrak{g})$ is A-unitary. Thus, the proposition follows from Lemma 6.3.37. □

Remark 6.3.43 For any adjoint simple group G not of type $\mathrm{PGL}(n)$, there exist semistable but not stable G-bundles of any topological type. For a proof of a more general result see Ramanathan (1975, Proposition 7.8).

We end the chapter with the following result.

Lemma 6.3.44 *Let G be a connected reductive group and let $\pi : E \to \Sigma$ be a semistable G-bundle. Then, for any $p \in \Sigma$, the restriction map*

$$\mathrm{Aut}(E) \to \mathrm{Aut}(E_p)$$

is injective, where $\mathrm{Aut}(E)$ *denotes the group of automorphisms of the bundle E inducing the identity over the base and $E_p := E_{|p}$.*

Proof Let Z be the center of G. Then we can think of Z as a central subgroup of $\mathrm{Aut}(E)$ by taking the embedding

$$\delta\colon Z \hookrightarrow \mathrm{Aut}(E), \ \ \delta(g)(e) = e \cdot g, \ \text{ for } e \in E, g \in Z.$$

We have the following commutative diagram:

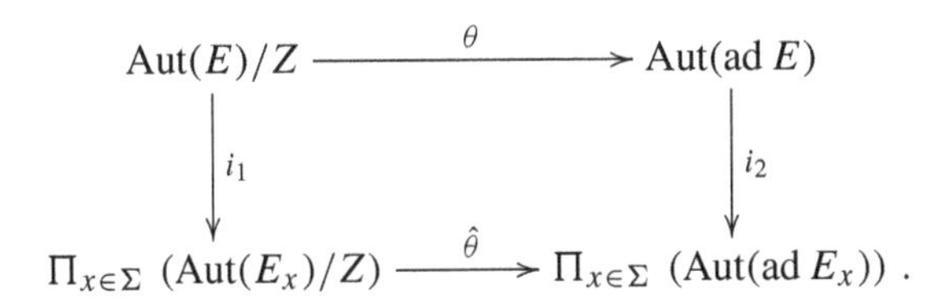

Clearly, i_2 and $\hat{\theta}$ are injective. Moreover, i_1 is also injective since any $\varphi \in \mathrm{Ker}\, i_1$ gives rise to a morphism $\bar{\varphi}\colon \Sigma \to Z$, which must be a constant. Hence, θ is injective. Now, consider the analogue of the above diagram for the fixed point $p \in \Sigma$:

$$\begin{array}{ccc} \mathrm{Aut}(E)/Z & \xrightarrow{\ \theta\ } & \mathrm{Aut}(\mathrm{ad}\, E) \\ \Big\downarrow{\scriptstyle i_1(p)} & & \Big\downarrow{\scriptstyle i_2(p)} \\ \mathrm{Aut}(E_p)/Z & \xrightarrow{\ \hat{\theta}(p)\ } & \mathrm{Aut}(\mathrm{ad}\, E_p)\,. \end{array}$$

Observe that $\mathrm{ad}\, E$ is a degree 0 vector bundle over Σ. We next claim that for any degree 0 semistable vector bundle $\mathscr{V}$ over Σ, the map $\bar{i}_2(p)\colon \mathrm{End}\,\mathscr{V} \to \mathrm{End}\,\mathscr{V}_p$ is injective, where $\mathrm{End}\,\mathscr{V}$ denotes the set of $\mathscr{O}_\Sigma$-module endomorphisms of $\mathscr{V}$. The injectivity of $\bar{i}_2(p)$ follows from the vanishing

$$H^0\left(\Sigma, \mathscr{O}_\Sigma(-p) \otimes \mathbf{End}\mathscr{V}\right) = 0, \tag{1}$$

where **End** is the corresponding sheaf and (1) follows from Lemma 6.2.4.

Taking $\mathscr{V} = \mathrm{ad}\, E$ and using Lemma 6.1.5, we get the injectivity of $i_2(p)$. Hence, from the second commutative diagram, we get the injectivity of $i_1(p)$. The injectivity of $i_1(p)$ implies the injectivity of $\mathrm{Aut}(E) \to \mathrm{Aut}(E_p)$, proving the lemma. □

6.3.E Exercises

In the following Σ is a smooth projective irreducible curve of any genus and G is a connected reductive group.

(1) Let $\rho\colon \pi_1(\Sigma) \to G$ be a homomorphism, where G is any algebraic group. Then, for any $g \in G$, show that the associated G-bundles E_ρ and $E_{g\rho g^{-1}}$ are isomorphic as algebraic G-bundles.

Following the notation as in Definition 6.3.3, prove the same result for $\hat{\rho}\colon \pi \to G$, i.e., $\hat{E}_{\hat{\rho}} \simeq \hat{E}_{g\hat{\rho}g^{-1}}$ as A-equivariant G-bundles.

(2) Following the notation in Definition 6.3.9, show that the topological G-bundle F_c (up to an isomorphism) does not depend upon the choices of c in its homotopy class, p, D_p and h.

Hint: Follow the argument as in Steenrod (1951, §18).

(3) Show that a degree 0 line bundle $\mathscr{L}$ over Σ comes from a unitary character χ (i.e., a 1-dimensional unitary representation $\mathbb{C}_\chi$) of $\pi_1(\Sigma)$.

Moreover, following the notation as in Definition 6.3.3, if $\mathscr{L}$ is an A-equivariant line bundle of degree 0, then show that it is A-unitary.

Hint: The universal bundle $\mathscr{E}_{|\Sigma\times R_K(g)}$ of Lemma 6.3.2 for $G = \mathrm{GL}_1$ (so that $K = S^1$) gives rise to an $\mathbb{R}$-analytic group homomorphism

$$\beta : R_K(g) = (S^1)^{2g} \to \mathrm{Jac}(\Sigma),$$

where $\mathrm{Jac}(\Sigma)$ is the Jacobian variety of Σ consisting of the set of isomorphism classes of degree 0 line bundles over Σ. Show that the above map is injective and hence surjective from the dimensional consideration. For the equivariant version, use the first part together with Lemma 6.3.8.

(4) Prove identity (4) in the proof of Lemma 6.3.11.

(5) Let X, Y be $\mathbb{C}$-analytic spaces such that X is compact. Let $\pi : Y \to T$ be a holomorphic map. Then show that the inverse image of the set of constant maps under the holomorphic map

$$\mathrm{Hol}(X,Y) \to \mathrm{Hol}(X,T), \quad f \mapsto \pi \circ f$$

is a closed $\mathbb{C}$-analytic subspace, where $\mathrm{Hol}(X,Y)$ has a natural $\mathbb{C}$-analytic structure as in the proof of Lemma 6.3.29.

(6) Let $\mathscr{V}$ be a stable vector bundle over Σ. Then show that $\mathscr{V}$ is simple, i.e., $H^0(\Sigma, \mathbf{End}\,\mathscr{V}) = \mathbb{C}$, where $\mathbf{End}\,\mathscr{V}$ denotes the sheaf of $\mathscr{O}_\Sigma$-module endomorphisms of $\mathscr{V}$.

Hint: Let $A := H^0(\Sigma, \mathbf{End}\,\mathscr{V})$ be the endomorphism algebra. Now, use Lemma 6.3.22 and finite-dimensionality of A.

(7) Let E be a G-bundle over Σ. Show that $H^0(\Sigma, \mathrm{ad}\, E) \simeq \mathrm{Lie}(\mathrm{Aut}\, E)$.

Hint:

$$\mathrm{Aut}\, E \simeq \{\varphi : E \to G : \varphi(eg) = g^{-1}\varphi(e)g \;\forall e \in E, g \in G\}$$

and similarly

$$H^0(\Sigma, \operatorname{ad} E) \simeq \{f : E \to \mathfrak{g} : \varphi(eg) = \operatorname{Ad}(g^{-1}) \cdot (\varphi(e)) \, \forall e \in E, g \in G\}.$$

(8) Let $\mathscr{V}_1$ and $\mathscr{V}_2$ be two stable vector bundles over Σ such that $\mu(\mathscr{V}_1) = \mu(\mathscr{V}_2)$. Show that any nonzero $\mathscr{O}_\Sigma$-module map $\varphi\colon \mathscr{V}_1 \to \mathscr{V}_2$ is an isomorphism.

(9) Prove Lemma 6.3.31 with T_s replaced by $T_{ss} := \{t \in T : \mathscr{F}_t$ is a semistable G-bundle$\}$.

Hint: Follow the proof of Lemma 6.3.31.

(10) Give an example of a stable G-bundle E such that $\operatorname{ad} E$ is *not* stable, where G is a simple group.

Hint: For Σ of genus ≥ 2, take a representation $\rho\colon \pi_1(\Sigma) \to \mathrm{SO}_n(\mathbb{R})$ with dense image (cf. Lemma 7.2.9). This representation remains irreducible considered as a homomorphism $\hat{\rho}\colon \pi_1(\Sigma) \to \mathrm{SL}_n(\mathbb{C})$. Thus, $E_{\hat{\rho}}$ is stable $\mathrm{SL}_n(\mathbb{C})$-bundle. However, show that $\operatorname{ad} E_{\hat{\rho}}$ is not stable.

(11) Let $\mathscr{V}$ and $\mathscr{W}$ be two A-equivariant vector bundles over $\hat{\Sigma}$ of degree 0. Assume further that $\mathscr{V}$ (resp. $\mathscr{W}$) is A-stable (resp. A-semistable). Show that any nonzero A-equivariant $\mathscr{O}_{\hat{\Sigma}}$-linear map $f\colon \mathscr{W} \to \mathscr{V}$ is surjective.

Hint: Use the canonical factorization of f as given in the proof of Lemma 6.3.22.

(12) Following Proposition 6.3.18 and its proof, prove that the composite map $i \circ j \circ q \circ \mathfrak{F}$ coincides with the deformation map η.

6.C Comments

Mumford defined the notion of semistable and stable vector bundles over a smooth projective curve Σ as in Definition 6.1.4(a). Its extension to any G-bundles over Σ for a connected reductive group G as in Definition 6.1.4(b) is due to Ramanathan (1975). Definition 6.1.4(c) of polystability for any G-bundle is taken from Ramanan and Ramanathan (1984, Definition 3.16) (though they call it 'quasi-stable'). This extends the earlier definition of polystability for vector bundles, that is why we prefer to call it 'polystable.' The notion of parabolic structure on vector bundles and their semistablity (and stability) as in Exercise 6.1.E.7 is due to Mehta and Seshadri (1980) (also see (Seshadri, 1977) for an announcement of some of the results). Its extension to any G-bundles over Σ is due to Bhosle and Ramanathan (1989), though we have taken our Definition 6.1.4(d) from Teleman and Woodward

(2001, Definition 2.2). It might be mentioned that the paper by Bhosle and Ramanathan (1989) has a serious error in their association of $E(\rho,\tau)$ to a representation ρ in their §2.[3] Lemma 6.1.5 appears in Ramanathan (1996, Corollary 3.18).

A systematic study of A-equivariant vector bundles on $\hat{\Sigma}$ (where $\hat{\Sigma}$ is a smooth projective curve and A is a finite group acting faithfully on $\hat{\Sigma}$) was begun in Narasimhan and Seshadri (1965), wherein many of the results from Narasimhan and Seshadri (1964) were extended to an A-equivariant setting. One of the classical results (Narasimhan and Seshadri, 1965, Corollary 2, §12) (Theorem 6.3.35 for vector bundles) is derived from an analogous unitarity result in the A-equivariant setting (Narasimhan and Seshadri, 1965, Theorem 2, §12). Study of A-equivariant vector bundles on $\hat{\Sigma}$ was continued and expanded in Seshadri (2011).

We have taken Theorem 6.1.9 and its proof from Teleman and Woodward (2001) (though we have provided more complete details). Theorem 6.1.17 is due to Mehta and Seshadri (1980) for vector bundles (also see (Grothendieck, 1956–57), (Seshadri, 2011), (Boden, 1991), (Furuta and Steer, 1992) and (Biswas, 1997)). Theorems 6.1.15 and 6.1.17 for general G are taken from Teleman and Woodward (2001, Theorem 2.3) and Balaji and Seshadri (2015, Proposition 3.1.1, Theorems 5.3.1 and 6.3.5) (Theorem 6.1.17 is also proved in Balaji, Biswas and Nagaraj (2001, Theorem 4.3)). In fact, in Balaji and Seshadri (2015), the restriction $\theta(\tau_j) < 1$ plays no role by using Bruhat–Tits group schemes. Exercise 6.1.E.4 is taken from Ramanathan (1975, Lemma 2.1 and Remark 2.2), Exercise 6.1.E.5 is taken from Ramanathan (1975, Lemma 3.3) and Exercise 6.1.E.7(a) is taken from Bhosle and Ramanathan (1989, §1). Exercise 6.1.E.7(b) is taken from Mehta and Seshadri (1980, Remark 1.16). Exercise 6.1.E.8 is taken from Ramanathan (1975, Proposition 7.1). For Exercise 6.1.E.12 see Ramanathan and Subramanian (1988, Proposition 1) and Ramanathan (1975, Proposition 7.1). Exercise 6.1.E.13 is taken from Kumar, Narasimhan and Ramanathan (1994, Lemma 3.6). Some of these results on parabolic bundles have been extended to G-bundles over an arbitrary smooth projective variety over $\mathbb{C}$ by Balaji, Biswas and Nagaraj (2001).

Harder–Narasimhan (for short HN) filtration of vector bundles over Σ is due to Harder and Narasimhan (1975). Its extension for any G-bundles over Σ was announced by Ramanathan (1979). However, he did not publish its proof. Then, Atiyah and Bott (1982) provided an analogue of the HN filtration (or reduction) for G-bundles over Σ by looking at the original HN filtration of the corresponding adjoint bundle. Behrend (1995) proved the existence and

[3] I thank V. Balaji for pointing this out.

uniqueness of the HN reduction (in any characteristic) of G-bundles over Σ by using a 'complementary polyhedron.' A more bundle-theoretic proof of the existence and uniqueness of the HN reduction of G-bundles over Σ was given by Biswas and Holla (2004) and we have followed their proof in Section 6.2 (Theorem 6.2.3). Identity (1) of Theorem 6.2.3 is taken from Kumar and Narasimhan (1997, Lemma 3.6). Existence and uniqueness of the HN reduction of G-bundles over compact Kähler manifolds was established by Anchouche, Azad and Biswas (2002). The HN reduction of G-bundles over Σ in suitably positive characteristics was also studied by Mehta and Subramanian (2002) and Biswas and Holla (2004). Theorem 6.2.6 and Corollary 6.2.7 are taken from Biswas and Holla (2004), though (as mentioned in Remark 6.2.8) their proof has a gap which required us to put additional hypotheses (1) and (2) in Theorem 6.2.6. Exercise 6.2.E.4 is taken from Balaji, Biswas and Nagaraj (2001, Proposition 4.1).

Several of the results in Section 6.3 (including Lemma 6.3.2, Proposition 6.3.4 in the non-equivariant case, Corollary 6.3.7 in the non-equivariant case, Lemma 6.3.10, Lemma 6.3.11, Corollary 6.3.21, Proposition 6.3.24, Lemma 6.3.25, Lemma 6.3.27 and Lemma 6.3.29 are taken from Ramanathan (1975, 1996). Proposition 6.3.4 and Corollary 6.3.7 in the case of equivariant vector bundles as well as Lemma 6.3.6 in the equivariant case is proved in Seshadri (2011, Proposition 10 (Chap. II), Corollary (Chap. I), Proposition 1 (Chap. I)) (see also (Bhosle and Ramanathan, 1989, Propositions 2.1 and 2.2) for the parabolic analogue of Proposition 6.3.4 and Corollary 6.3.7).

Lemma 6.3.29 is attributed to R.R. Simha in Ramanathan (1975). Lemma 6.3.6 in the non-equivariant case is taken from Narasimhan and Seshadri (1964, Proposition 4.1), though the proof given here is a slight modification of their proof with help from Michael Taylor. Proposition 6.3.12, Corollary 6.3.14, Proposition 6.3.15, Corollary 6.3.16 and Proposition 6.3.18 are taken from Narasimhan and Seshadri (1964). Even though they prove their results for $G = \mathrm{GL}_n$, virtually the same proof works for any G. Proposition 6.3.12 is proved by them, more generally, for any compact, connected, Kähler manifold. Lemma 6.3.22 is taken from Narasimhan and Seshadri (1965, Proposition 4.3). Proposition 6.3.30, Lemmas 6.3.31 and 6.3.32 are due to Ramanathan (1975, §7, §4). Theorem 6.3.34 is a slight variant of Ramanathan (1975, Theorem 7.1). Its extension to Theorem 6.3.35 is straightforward. For $G = \mathrm{GL}_n$, this is a classical result due to Narasimhan and Seshadri (1965, §12, Corollary 2). Exercise 6.3.E.4 is taken from Ramanathan (1975, Proposition 6.1 and Remark 6.2). Exercise 6.3.E.6 is taken from Narasimhan and Seshadri (1965, §4). Exercise 6.3.E.9 is asserted in Ramanathan (1996, proof of Lemma 5.9.1). The analogue of most of the results in Section 6.3 for vector bundles is due to Narasimhan and Seshadri (1964) and Narasimhan and Seshadri (1965).

Corollary 6.3.7 for vector bundles is mentioned in Weil (1938). Theorem 6.3.41 for vector bundles is due to Seshadri (2011, Theorem 4 (Chap. II)). Lemma 6.3.37 is taken from Atiyah and Bott (1982, Lemma 10.12) though part of its proof via Lemma 6.3.38 is taken from Ramanathan and Subramanian (1988, Proposition 1). Proposition 6.3.42 is taken from Balaji, Biswas and Nagaraj (2001, §5).

There is an alternative proof of the Narasimhan–Seshadri theorem for stable vector bundles over Σ using the differential geometry of connections on holomorphic bundles (cf. (Donaldson, 1983)). For its extension to any reductive G and the base Σ replaced by any complex projective manifold, see Ramanathan and Subramanian (1988, Theorem 1).

We have restricted the discussion of parabolic G-bundles to the case when G is a simply-connected simple group. Its generalization to any connected reductive group G (under some restrictions on parabolic weights) can be found in Faltings (1993, §V) and in Balaji and Seshadri (2015, §8.2).

7

Moduli Space of Semistable G-Bundles Over a Smooth Curve

This chapter is devoted to constructing the moduli *space* of vector bundles (more generally parabolic G-bundles) over a smooth irreducible projective curve Σ of any genus $g \geq 0$.

As in Section 1.1 by a variety we mean a reduced (separated) scheme of finite type over $\mathbb{C}$.

We begin by showing that a semistable vector bundle $\mathscr{V}$ over Σ of degree d and rank r with $d > r(2g-1)$ is globally generated and $H^1(\Sigma, \mathscr{V}) = 0$ (cf. Proposition 7.1.3). We recall Grothendieck's Quot Scheme construction in a general setting consisting of a projective scheme X, a very ample line bundle $\mathscr{L}$ and a coherent sheaf E on X together with a (Hilbert) polynomial $P(z) \in \mathbb{Q}[z]$. The quot scheme $Q = Q(E, P)$ has for its closed points the $\mathscr{O}_X$-module quotients of E with Hilbert polynomial $P(z)$ (with respect to $\mathscr{L}$). Moreover, there is a 'tautological' coherent sheaf $\mathscr{U}$ over $X \times Q$ which is a quotient of $E \boxtimes \mathscr{O}_Q$ (cf. Theorem 7.1.6). Take a pair of positive integers (r, d) such that $d > r(2g-1)$ and let $N := d + r(1-g)$. We apply the Grothendieck's Quot Scheme construction as above in the special case when $X = \Sigma$, $E = \mathscr{O}_\Sigma \otimes \mathbb{C}^N$ and $P(z) = N + rhz$, where h is the degree of a fixed very ample line bundle H over Σ. This data gives rise to the quot scheme $Q = Q(E, P(z))$ with a canonical action of GL_N together with the 'tautological' GL_N-equivariant coherent sheaf $\mathscr{U}$ over $\Sigma \times Q$ which is a quotient of $E \boxtimes \mathscr{O}_Q$ and $\mathscr{U}$ is flat over Q. We collect various properties of Q and $\mathscr{U}$ in Proposition 7.1.9. The above results (Proposition 7.1.3, Theorem 7.1.6 and Proposition 7.1.9) together show that the set of isomorphism classes of rank r semistable vector bundles over Σ of fixed degree d is bounded (cf. Corollary 7.1.10). As a consequence of Proposition 7.1.9, we obtain that in any family of vector bundles over Σ parameterized by a noetherian scheme T, the subset T^{ss} (resp. T^s) parameterizing semistable (resp. stable) vector bundles is open (cf. Corollary 7.1.16). Define the GL_N-stable subset $R = R(r,d) \subset Q$ by

$$R := \{q \in Q : \bar{q} \text{ is locally free } \mathcal{O}_\Sigma\text{-module and } H^0(\Sigma, E) \to H^0(\Sigma, \bar{q}) \text{ is an isomorphism}\},$$

where $\bar{q}$ is the sheaf corresponding to the point $q \in Q$. Let R^{ss} (resp. R^s) be the GL_N-stable subset of R consisting of those $q \in R$ such that $\bar{q}$ is semistable (resp. stable) vector bundle over Σ. Then, for d large enough, these are open subsets of R (cf. Theorem 7.1.13). Consider the functor $\mathscr{M}(r,d)$ (resp. $\mathscr{M}^s(r,d)$) of semistable (resp. stable) vector bundles over Σ of rank r and degree d (cf. Definition 7.1.1). The following is the main result of this section (cf. Theorem 7.1.14).

Theorem For Σ as above and any $r \geq 1$ and $d \in \mathbb{Z}$, $\mathscr{M}(r,d)$ has a coarse moduli space $M(r,d)$, which is an irreducible, normal projective variety with rational singularities. In fact, $M(r,d) = R^{ss}//\mathrm{GL}_N$ (the GIT quotient). Moreover, the subfunctor $\mathscr{M}^s(r,d)$ has a coarse moduli space $M^s(r,d)$ which is an open subset of $M(r,d)$. Also, the dimension of $M(r,d)$ is $(g-1)r^2+1$ if $M^s(r,d) \neq \emptyset$. As in Remark 7.1.15, $M^s(r,d) \neq \emptyset$ if $g \geq 2$ and whenever nonempty $M^s(r,d)$ is smooth.

In Section 7.2, we extend the above results to G-bundles over Σ (and more generally to parabolic G-bundles with rational markings), where G is a connected semisimple group. The strategy is to embed $G \hookrightarrow \mathrm{SL}_r$ and reduce the problem about G-bundles to that of SL_r-bundles (i.e., vector bundles of rank r with trivial determinant). The main technical result which allows us to achieve this is Corollary 7.2.4. To handle parabolic G-bundles, we use the bijection between quasi-parabolic G-bundles over Σ and A-equivariant G-bundles of a fixed A-topological type τ over a certain Galois cover $\hat{\Sigma} \to \Sigma$ (with Galois group A) which takes parabolic semistable (resp. stable) G-bundles to A-semistable (resp. A-stable) G-bundles over $\hat{\Sigma}$ (cf. Theorem 6.1.17). *Thus, we consider A-equivariant G-bundles over $\hat{\Sigma}$ of a fixed A-topological type τ in this section.* Our principal result in this section is the following (cf. Theorems 7.2.6, 7.2.8, Corollary 7.2.15, Lemma 7.2.13 and Corollary 7.2.27).

Theorem Let $\hat{\Sigma}$ be a smooth irreducible projective curve with a faithful action of a finite group A and G a connected semisimple group. Fix the topological type τ of A-equivariant G-bundles over $\hat{\Sigma}$. Then, the A-semistable G-bundles over $\hat{\Sigma}$ of topological type τ admit a coarse moduli space $M_\tau^G(\hat{\Sigma})$, which is the good quotient $R_\tau^{ss}(G)//\mathfrak{G}$ (under a connected reductive group $\mathfrak{G}$), where $R_\tau^{ss}(G)$ is a certain analogue of R^{ss} as above (see Corollary 7.2.4 for its precise definition). Further, $M_\tau^G(\hat{\Sigma})$ is an irreducible, normal variety with rational singularity (in particular, Cohen–Macaulay). Moreover, it is a nonempty projective variety if G is simply-connected and the genus g of Σ is at least 2, where $\Sigma := \hat{\Sigma}/A$.

In this case its dimension is given by (cf. Corollary 7.2.27)

$$\dim_{\mathbb{C}} M_\tau^G(\hat{\Sigma}) = (g-1)\dim_{\mathbb{R}} K + \frac{1}{2}\sum_{j=1}^{s} \dim_{\mathbb{R}} O(\tau_j),$$

where K is a maximal compact subgroup of G, $([\tau_1], \dots, [\tau_s])$ is the local type of τ and $O(\tau_j)$ is the K-orbit of τ_j under the adjoint action.

Further, the moduli space $M_\tau^G(\hat{\Sigma})$ has any of its elements uniquely represented by an A-polystable G-bundle (cf. Corollary 7.2.20(a)).

If we take A to be trivial so that $\hat{\Sigma} = \Sigma$, then $M_\tau^G(\Sigma)$ in this case is nothing but the (non-parabolic) moduli space of semistable G-bundles over Σ of topological type τ.

The above theorem together with Theorem 6.3.41 translate into the corresponding results for the parabolic G-bundles (cf. Theorems 7.2.23 and 7.2.29).

Theorem Let G be a connected, simply-connected, simple algebraic group, $(\Sigma, \vec{p})$ a smooth irreducible projective s-pointed ($s \geq 0$) curve of any genus $g \geq 2$ and let $\vec{\tau} = (\tau_1, \dots, \tau_s)$ be a set of markings with τ_j rational points of the fundamental alcove Φ_o. We assume further that $\theta(\tau_j) < 1$ for all j, where θ is the highest root of G. Then, the parabolic semistable G-bundles with markings $\vec{\tau}$ admit a coarse moduli space $M^G_{\mathrm{par},\vec{\tau}}(\Sigma)$, which is a normal irreducible projective variety with rational singularity.

Theorem With the notation as in the above theorem and Definition 7.2.28, for any parabolic stable G-bundle E over Σ with markings $\vec{\tau}$ at $\vec{p}$, there exists an irreducible unitary homomorphism $\hat{\rho}: \pi \to G$ of local type $\vec{\tau}$ (i.e., $\rho(c_j)$ is conjugate of $\mathrm{Exp}(2\pi i \tau_j)$ for all j) such that $E \simeq E_{\hat{\rho}}(\vec{\tau})$ (as parabolic G-bundles). Conversely, for any unitary homomorphism $\hat{\rho}: \pi \to G$ of local type $\vec{\tau}$, the G-bundle $E_{\hat{\rho}}(\vec{\tau})$ over Σ (with some specific sections over $\{p_j\}$) is a parabolic semistable G-bundle with markings $\vec{\tau}$. Further, $E_{\hat{\rho}}(\vec{\tau})$ is parabolic stable if and only if ρ is irreducible (and unitary).

7.1 Moduli Space of Semistable Vector Bundles Over a Smooth Curve

Let Σ be a smooth irreducible projective curve of any genus $g \geq 0$.

Definition 7.1.1 Let r be a positive integer and $d \in \mathbb{Z}$. As in Section 1.1 let $\mathfrak{S}$ be the category of quasi-compact and separated schemes over $\mathbb{C}$ and morphisms between them and let $\mathfrak{S}'$ be the full subcategory consisting of noetherian (separated) schemes. Define the contravariant functor

$$\mathcal{M}(r,d)\colon \mathfrak{S}' \to \mathbf{Set},$$

where $\mathcal{M}(r,d)(S)$ = set of isomorphism classes of families of *semistable* vector bundles over Σ of rank r and degree d parameterized by S. For any morphism of schemes $S \to S'$, we associate the pull-back of the family.

Also, consider the subfunctor

$$\mathcal{M}^s(r,d)\colon \mathfrak{S}' \to \mathbf{Set}$$

defined as above by replacing *semistable* by stable.

Definition 7.1.2 Let $\mathcal{F}\colon \mathfrak{S}' \to \mathbf{Set}$ be a contravariant functor. By a *coarse moduli space* for $\mathcal{F}$ we mean a scheme $M \in \mathfrak{S}$ together with a natural transformation

$$\Phi\colon \mathcal{F} \to \mathrm{Mor}(-,M)$$

such that the following universal property holds.

For any scheme $N \in \mathfrak{S}$ and any natural transformation

$$\Phi'\colon \mathcal{F} \to \mathrm{Mor}(-,N),$$

there exists a unique morphism

$$f_{\Phi'}\colon M \to N$$

such that

$$\Phi' = \Omega_{f_{\Phi'}} \circ \Phi.$$

Here $\mathrm{Mor}(-,N)\colon \mathfrak{S}' \to \mathbf{Set}$ is the functor taking any $T \in \mathfrak{S}'$ to the set $\mathrm{Mor}(T,N)$ of all the morphisms from T to N and for any morphism $f\colon M \to N$,

$$\Omega_f\colon \mathrm{Mor}(-,M) \to \mathrm{Mor}(-,N)$$

is the natural transformation taking any morphism $g\colon T \to M$ to $f \circ g$.

If a coarse moduli space exists, it is clearly unique.

Our main aim in this section is to prove that the functors $\mathcal{M}(r,d)$ and $\mathcal{M}^s(r,d)$ admit coarse moduli spaces $M(r,d)$ and $M^s(r,d)$, respectively, where $M(r,d)$ is an irreducible normal projective variety and $M^s(r,d)$ is an open subset.

Proposition 7.1.3 *Let $\mathscr{V}$ be a semistable vector bundle over Σ of degree d and rank $r \geq 1$. If*

$$d > r(2g-1), \tag{1}$$

then

(a) *$H^1(\Sigma, \mathscr{V}) = 0$, and*
(b) *$\mathscr{V}$ is globally generated, i.e., the natural map $\mathscr{O}_\Sigma \otimes H^0(\Sigma, \mathscr{V}) \to \mathscr{V}$ is surjective.*

In fact, for the validity of (a), we only need $d > r(2g-2)$.

Proof (a) We prove (a) under the assumption $d > r(2g-2)$. Assume to the contrary that $H^1(\Sigma, \mathscr{V}) \neq 0$. Then, by the Serre duality, there exists a nonzero bundle map $\alpha\colon \mathscr{V} \to \omega_\Sigma$, where ω_Σ is the canonical line bundle of Σ. Let $\mathscr{W}$ be the vector subbundle of $\mathscr{V}$ determined by $\operatorname{Ker}\alpha$ (cf. analogue of $\mathscr{E}_1$ in diagram (*) as in the proof of Lemma 6.3.22). Then, from the diagram in the same,

$$\deg \mathscr{W} \geq \deg \mathscr{V} - \deg \omega_\Sigma = d - (2g-2), \tag{2}$$

where the last equality follows from Hartshorne (1977, Chap. IV, Example 1.3.3).

For $r \geq 2$, from (2), we get

$$\mu(\mathscr{W}) = \frac{\deg \mathscr{W}}{r-1} \geq \frac{d-2g+2}{r-1} > \frac{d}{r} = \mu(\mathscr{V}), \quad \text{since } d > r(2g-2).$$

This is a contradiction to the semistability of $\mathscr{V}$. Thus, by contradiction, we get $H^1(\Sigma, \mathscr{V}) = 0$.

If $r = 1$, since $\mathscr{W} = 0$, we again get a contradiction.

(b) For any $x \in \Sigma$, consider the ideal sheaf exact sequence

$$0 \to \mathscr{O}_\Sigma(-x) \to \mathscr{O}_\Sigma \to \mathscr{O}_x \to 0.$$

Tensoring with (the locally free) $\mathscr{V}$, we get the sheaf exact sequence

$$0 \to \mathscr{V}(-x) \to \mathscr{V} \to \mathscr{V}_{|x} \to 0, \tag{3}$$

where

$$\mathscr{V}(-x) := \mathscr{V} \otimes \mathscr{O}_\Sigma(-x).$$

Since $\mathscr{O}_\Sigma(-x)$ is a line bundle and $\mathscr{V}$ is semistable, we get that $\mathscr{V}(-x)$ is semistable (cf. Definition 6.1.4(a)). To show that $\mathscr{V}$ is globally generated, from the long exact cohomology sequence associated to the sheaf sequence (3), it suffices to show that

$$H^1(\Sigma, \mathscr{V}(-x)) = 0. \tag{4}$$

But $\deg \mathscr{V}(-x) = \deg \mathscr{V} - r > r(2g-2)$. Thus, we get (4) from the (a)-part. □

Definition 7.1.4 Let G be an affine algebraic group acting on an algebraic variety X. A *good quotient* of X by G is a pair (Y, ϕ), where Y is a variety and $\phi\colon X \to Y$ is an affine G-invariant surjective morphism satisfying the following conditions:

(a) For any open subset $U \subset Y$, $\phi^*\colon \mathbb{C}[U] \to \mathbb{C}[\phi^{-1}(U)]^G$ is an isomorphism.

(b) For any closed G-stable subset $W \subset X, \phi(W)$ is closed in Y.

(c) For disjoint closed G-stable subsets $W_1, W_2 \subset X$, $\phi(W_1) \cap \phi(W_2) = \emptyset$.

We denote the good quotient by $X//G$. A good quotient is called a *geometric quotient* if all the fibers of ϕ are single G-orbits.

A *categorical quotient* of X by G is a pair (Y, ϕ), where Y is a variety and $\phi\colon X \to Y$ is a G-invariant morphism such that for any scheme Z and G-invariant morphism $\phi'\colon X \to Z$, there exists a unique morphism $\chi\colon Y \to Z$ such that $\chi \circ \phi = \phi'$.

Clearly, a categorical quotient if it exists is unique up to an isomorphism.

For any good quotient $\phi\colon X \to Y$, $\phi^{-1}(U) \to U$ is a good and categorical quotient for any open subset $U \subset Y$ (cf. (Newstead, 2012, Propositions 3.10 and 3.11); even though the proof in the same that ϕ is a categorical quotient is obtained under the assumption that Z is a variety, but the same proof works for any scheme Z).

Lemma 7.1.5 *Let G be a reductive group and let $f\colon X' \to X$ be an affine G-equivariant morphism between G-varieties. If X possesses a good quotient, then so does X'.*

Proof Let $\phi\colon X \to Y$ be a good quotient. For any affine open subset $V \subset Y$, the G-stable open subset $\tilde{V} := f^{-1}(\phi^{-1}(V))$ being affine possesses a good quotient $\phi'_V\colon \tilde{V} \to V'$ (cf. (Newstead, 2012, Theorem 3.5)). In particular, ϕ'_V is a categorical quotient. Thus we get a unique morphism $\psi'_{V'}\colon V' \to Y$ such that restricted to $\tilde{V}$:

$$\psi'_{V'} \circ \phi'_V = \phi \circ f. \tag{1}$$

Now, take a finite cover $\{V_i\}$ of Y by affine open subsets and construct V_i', $\phi_i' := \phi_{V_i}'$, $\psi_i' = \psi_{V_i'}'$ as above. Since $\psi_i'^{-1}(V_i \cap V_j)$ and $\psi_j'^{-1}(V_i \cap V_j)$ are both good quotients of $f^{-1}\phi^{-1}(V_i \cap V_j)$, there exist unique isomorphisms

$$\beta_{ji} : \psi_j'^{-1}(V_i \cap V_j) \xrightarrow{\sim} \psi_i'^{-1}(V_i \cap V_j)$$

such that, restricted to $f^{-1}\phi^{-1}(V_i \cap V_j)$,

$$\beta_{ji} \circ \phi_j' = \phi_i'. \tag{2}$$

From the uniqueness of β_{ji}, they clearly satisfy the cocycle condition. Let Y' be the scheme obtained from glueing the affine varieties $\{V_i'\}$ along the isomorphisms β_{ji} (cf. (Hartshorne, 1977, Chap. II, Exercise 2.12)). Then, by the same, the morphisms ϕ_i' glue to give a morphism $\phi'\colon X' \to Y'$ such that, restricted to $f^{-1}\phi^{-1}(V_i)$, $\phi' = \phi_i'$. Since each V_i' is reduced, so is Y'. We next show that Y' is separated, i.e., the image $\Delta_{Y'}$ of the diagonal map $\Delta\colon Y' \to Y' \times Y'$ is closed in $Y' \times Y'$.

Consider the morphism $\psi'\colon Y' \to Y$ obtained by glueing the maps $\psi_i'\colon V_i' \to Y$. Since Y is a variety, Δ_Y is closed in $Y \times Y$ and hence $D := (\psi' \times \psi')^{-1}\Delta_Y$ is closed in $Y' \times Y'$. Clearly,

$$\Delta_{Y'} \subset D \subset \bigcup_i (V_i' \times V_i'),$$

and $\bigcup_i (V_i' \times V_i')$ is open in $Y' \times Y'$. Now,

$$\Delta_{Y'} \cap (V_i' \times V_i') = \Delta_{V_i'},$$

which is closed in $V_i' \times V_i'$ (since V_i' is a variety). Thus, $\Delta_{Y'}$ is closed in $\bigcup_i (V_i' \times V_i')$ and hence so in D. This proves that $\Delta_{Y'}$ is closed in $Y' \times Y'$, i.e., Y' is separated. Thus, Y' is a variety (cf. (Mumford, 1988, Proposition 4, Chap. I.6)). Further, since each $\phi_i'\colon \tilde{V}_i \to V_i'$ is a good quotient, so is $\phi'\colon X' \to Y'$ (cf. (Newstead, 2012, Proposition 3.10(b))). This proves the lemma. □

Recall the following result due to Grothendieck (1960–61, Théorèm 3.2 and §5). In fact, there is a more general result in the same.

Theorem 7.1.6 *Let X be a projective scheme with a very ample line bundle $\mathscr{L}$ on X. Let E be a coherent sheaf on X and $P(z) \in \mathbb{Q}[z]$ be a polynomial. Consider the contravariant functor*

$$\mathfrak{Q}(E, P)\colon \mathfrak{S}' \to \mathbf{Set}$$

$\mathfrak{Q}(E,P)(S) =$ *the set of all $\mathscr{O}_{X\times S}$-module quotients $\mathscr{F}$ of $E \boxtimes \mathscr{O}_S$ (where two quotients are considered the same if they have the same kernel) such that $\mathscr{F}$*

is flat over S and $\mathscr{F}_{|X\times t}$ has Hilbert polynomial (cf. (Hartshorne, 1977, Chap. III, Exercise 5.2(a))) $P(z)$ with respect to the line bundle $\mathscr{L}$ for any $t \in S$.

For a morphism $f : S' \to S$ in $\mathfrak{S}'$, the corresponding morphism in **Set** *is given by the pull-back $(\mathrm{Id}_X \times f)^*\mathscr{F}$ which is a quotient of $E \boxtimes \mathscr{O}_{S'}$ since the tensor product is right exact.*

Then, $\mathfrak{Q}(E,P)$ is a representable functor (cf. Section 1.1) represented by a projective scheme $Q = Q(E,P)$.

The Zariski tangent space of Q at any (closed) point of $\bar{Q} = \mathfrak{Q}(E,P)(pt)$ corresponding to an $\mathscr{O}_X$-module quotient $F = E/F'$ of E (with Hilbert polynomial $P(z)$) is given by

$$T_F(Q) \simeq \mathrm{Hom}_{\mathscr{O}_X}(F',F). \tag{1}$$

Further, Q is smooth at F if

$$\mathrm{Ext}^1_{\mathscr{O}_X}(F',F) = 0. \tag{2}$$

In particular, taking $S = Q$ and the identity map of Q, we get a 'tautological' (coherent) sheaf $\mathscr{U}$ over $X \times Q$ which is quotient of $E \boxtimes \mathscr{O}_Q$ and $\mathscr{U}$ is flat over Q such that for any closed point $q \in \bar{Q}$,

$$\mathscr{U}_{|X\times q} \simeq \bar{q}, \tag{3}$$

where $\bar{q} \in \mathfrak{Q}(F,P)(pt)$ corresponds to the morphism $pt \to Q$, $pt \mapsto q$.

The scheme $Q = Q(E,P)$ is called the Grothendieck quot scheme (general case).

Definition 7.1.7 Take a pair of positive integers (r,d) such that $d > r(2g-1)$ (cf. Proposition 7.1.3) and let

$$N := d + r(1-g).$$

We apply Theorem 7.1.6 to the special case when $X = \Sigma$, $E = \mathscr{O}_\Sigma \otimes \mathbb{C}^N$ and $P(z) = N + rhz$, where h is the degree of a fixed very ample line bundle H over Σ. Thus, we get the quot scheme $Q = Q(E,P(z))$ and the 'tautological' (coherent) sheaf $\mathscr{U}$ over $\Sigma \times Q$ flat over Q which is a quotient of $E \boxtimes \mathscr{O}_Q$.

Define the subset $R = R(r,d) \subset \bar{Q}$ by

$$\begin{aligned} R := \{q \in \bar{Q} : \bar{q} &\text{ is a locally free } \mathscr{O}_\Sigma\text{-module and} \\ \mathbb{C}^N = H^0(\Sigma,E) &\to H^0(\Sigma,\bar{q}) \text{ is an isomorphism}\}, \end{aligned}$$

where $\bar{q}$ is the sheaf corresponding to the closed point $q \in \bar{Q} = \mathfrak{Q}(E,P)(pt)$. For any $q \in R$, since the Hilbert polynomial $P_{\bar{q}}(z) = N + rhz$ and $\dim H^0(\Sigma,\bar{q}) = N$, we get that

$$H^1(\Sigma,\bar{q}) = 0, \ \ \mathrm{rank}\,\bar{q} = r \ \text{ and } \ \deg\bar{q} = d. \tag{1}$$

Further, for any $q \in R$, $\bar{q}$ is clearly globally generated. Conversely, any vector bunsdle F over Σ of rank r and degree d, which satisfies conditions (a) and (b) of Proposition 7.1.3, is isomorphic with $\bar{q}$ for some $q \in R$.

Let GL_N act on E via its standard action on $\mathbb{C}^N$ (and trivial action on $\mathscr{O}_\Sigma$). Clearly, the action of GL_N on E induces an action of GL_N on Q making $\mathscr{U}$ a GL_N-equivariant sheaf over Q. By the definition of R, $R \subset Q$ is GL_N-stable.

Lemma 7.1.8 *Let $\mathscr{E}$ be a coherent sheaf over $\Sigma \times S$, which is flat over noetherian scheme S. Take $(x,t) \in \Sigma \times S$. Then, $\mathscr{E}_{|\Sigma\times t}$ is locally free at x if and only if $\mathscr{E}$ is locally free at (x,t).*

Proof Let $f\colon R \to S$ be a local homomorphism between noetherian local rings and let M be a finitely generated S-module which is flat over R. Then, by the Nakayama Lemma, M is free over S if and only if $M/\mathfrak{m}M$ is free over $S/\mathfrak{m}\cdot S$, where $\mathfrak{m}$ is the maximal ideal of R. From this the lemma clearly follows. □

Proposition 7.1.9 *With the notation and assumptions as in the above Definition 7.1.7, we have the following:*

(A) *R is an open smooth subset of $\bar{Q}$ of pure dimension $N^2 + r^2(g-1)$ and $\mathscr{U}_{|\Sigma\times R}$ is a rank-r vector bundle.*
(B) *The family $\mathscr{U}_{|\Sigma\times R}$ has local universal property for any family $\mathscr{F}$ of vector bundles over Σ of rank r and degree d satisfying the conditions (a) and (b) of Proposition 7.1.3 parameterized by any noetherian scheme S. This means that for any such family $\mathscr{F}$ and any (closed) point $t_o \in S$, there exists an open neighborhood U_{t_o} of t_o in S and a morphism $f\colon U_{t_o} \to R$ such that*

$$(Id_\Sigma \times f)^*(\mathscr{U}_{|\Sigma\times R}) \simeq \mathscr{F}_{|\Sigma\times U_{t_o}}.$$

(C) *For $q_1, q_2 \in R$, $\bar{q}_1 \simeq \bar{q}_2$ if and only if $q_2 \in \mathrm{GL}_N \cdot q_1$.*
(D) *For any $q \in R$, the stabilizer G_q of q in GL_N satisfies*

$$G_q = \mathrm{Aut}(\bar{q}),$$

where $\mathrm{Aut}(\bar{q})$ is the group of automorphisms of the vector bundle $\bar{q}$ over Σ inducing the identity map at the base. In particular, the center of GL_N acts trivially on R (in fact, on Q).
(E) *R is irreducible.*

Proof (A) The set $V \subset \Sigma \times Q$ of points (x,q) such that $\mathscr{U}$ is locally free at (x,q) is an open subset of $\Sigma \times Q$ (cf. (Hartshorne, 1977, Chap. II, Exercise 5.7(a))). Let

$$R' = \{q \in \bar{Q} : \bar{q} \text{ is locally free } \mathcal{O}_\Sigma\text{-module}\}.$$

Then, by Lemma 7.1.8,

$$R' = \bar{Q} \backslash \pi_Q\left((\Sigma \times Q)\backslash V\right),$$

where π_Q is the projection $\Sigma \times Q \to Q$. But, since π_Q is a proper map, we get that R' is open in $\bar{Q}$. Moreover, by Lemma 7.1.8, $\mathcal{U}_{|\Sigma \times R'}$ is locally free. Since, by assumption, the Hilbert polynomial of $\bar{q} = \mathcal{U}_{|\Sigma \times q}$ (for any $q \in R'$) is $N + rhz$, we get that $\bar{q}$ is of rank r (and degree d) and hence $\mathcal{U}_{|\Sigma \times R'}$ is a rank-r vector bundle. We next prove that R is open in R'.

Consider the projection $\pi_{R'}: \Sigma \times R' \to R'$. The vector bundle $\mathcal{U}_{|\Sigma \times R'}$ which is a quotient of $\mathcal{O}_{\Sigma \times R'} \otimes \mathbb{C}^N$ gives rise to a $\mathcal{O}_{R'}$-module map

$$\text{Ⓗ} : \mathcal{O}_{R'} \otimes \mathbb{C}^N \simeq \pi_{R'_*}\left(\mathcal{O}_{\Sigma \times R'} \otimes \mathbb{C}^N\right) \to \pi_{R'_*}\left(\mathcal{U}_{|\Sigma \times R'}\right).$$

By the semicontinuity theorem (cf. (Hartshorne, 1977, Chap. III, Theorem 12.8)), since $\chi(\Sigma, \bar{q}) = N$, for all $q \in R'$, the subset R'_o of R' consisting of those $q \in R'$ such that $\dim H^0(\Sigma, \bar{q}) = N$, is open in R'.

We now prove that R'_o is smooth. To prove this, by Theorem 7.1.6, it suffices to show that for $q \in R'_o$ (writing $\bar{q} = \left(\mathcal{O}_\Sigma \otimes \mathbb{C}^N\right)/K$ for a vector bundle K over Σ),

$$\mathrm{Ext}^1_{\mathcal{O}_\Sigma}(K, \bar{q}) = 0. \tag{1}$$

By Hartshorne (1977, Chap. III, Propositions 6.3 and 6.7),

$$\mathrm{Ext}^1_{\mathcal{O}_\Sigma}(K, \bar{q}) \simeq H^1(\Sigma, \bar{q} \otimes K^*). \tag{2}$$

Since $q \in R'_o$, we get that $\dim H^0(\Sigma, \bar{q}) = N = \chi(\Sigma, \bar{q})$.

Thus,

$$H^1(\Sigma, \bar{q}) = 0. \tag{3}$$

Considering the long exact cohomology sequence associated to the following exact sequence of vector bundles and using (2) and (3), we get (1).

$$0 \to \bar{q} \otimes \bar{q}^* \to \bar{q} \otimes (\mathcal{O}_\Sigma \otimes \mathbb{C}^N) \to \bar{q} \otimes K^* \to 0. \tag{4}$$

This proves that R'_o is smooth.

Since $\dim H^0(\Sigma, \bar{q}) = N$ for any $q \in R'_o$, by Hartshorne (1977, Chap. III, Corollary 12.9), $\left(\pi_{R'*}(\mathcal{U}_{|\Sigma \times R'})\right)_{|R'_o}$ is a rank-N vector bundle over R'_o. Now, take $q_o \in R$. In particular, $q_o \in R'_o$. The $\mathcal{O}_{R'}$-module map Ⓗ restricted to R'_o is a morphism between vector bundles which is an isomorphism at q_o and hence it remains an isomorphism in an open subset U_{q_o} of q_o in R'_o. This shows that $U_{q_o} \subset R$ and hence R is open in R'_o and hence in $\bar{Q}$.

We next calculate the dimension of R. By Theorem 7.1.6, for any $q \in R$,

$$\begin{aligned}\dim_q R &= \dim \operatorname{Hom}_{\mathscr{O}_\Sigma}(K, \bar{q}) \\ &= \dim H^0(\Sigma, \bar{q} \otimes K^*) \\ &= N^2 + r^2(g-1),\end{aligned}$$

$$\text{since } H^1(\Sigma, \bar{q} \otimes K^*) = 0 \text{ by (1) and (2) and } N := d + r(1-g),$$

where the last equality follows from the Riemann–Roch theorem (cf. (Fulton, 1998, Example 15.2.1)) by using the exact sequence (4). This completes the proof of (A).

(B) Let $\mathscr{F} \to \Sigma \times S$ be a family of vector bundles over Σ of rank r and degree d parameterized by a noetherian scheme S such that $\mathscr{F}_t := \mathscr{F}_{|\Sigma \times t}$ satisfies conditions (a) and (b) of Proposition 7.1.3 for all the (closed) points $t \in S$. Thus, by the Riemann–Roch theorem,

$$\dim H^0(\Sigma, \mathscr{F}_t) = N := d + r(1-g), \quad \text{for all } t \in S.$$

Hence, $(\pi_S)_*(\mathscr{F})$ is a vector bundle of rank N over S (cf. (Kempf, 1978, Theorem 13.1)); in particular, locally trivial in the Zariski topology, i.e., for any $t_o \in S$, there exists an open neighborhood U_{t_o} of t_o in S such that $(\pi_{S_*}\mathscr{F})_{|U_{t_o}}$ is a trivial vector bundle of rank N. From the adjointness of π_S^* (cf. (Hartshorne, 1977, Chap. II, §5)), we get a $\mathscr{O}_{\Sigma\times S}$-module map Ⓗ $: \pi_S^*\pi_{S_*}\mathscr{F} \to \mathscr{F}$ corresponding to the identity map $\pi_{S_*}\mathscr{F} \to \pi_{S_*}\mathscr{F}$. Now, since $(\pi_{S_*}\mathscr{F})_{|U_{t_o}} \simeq \mathscr{O}_{U_{t_o}} \otimes \mathbb{C}^N$, restricting Ⓗ, we get a $\mathscr{O}_{\Sigma\times U_{t_o}}$-module morphism Ⓗ$_{|\Sigma\times U_{t_o}} : \mathscr{O}_{\Sigma\times U_{t_o}} \otimes \mathbb{C}^N \to \mathscr{F}_{|\Sigma\times U_{t_o}}$. Since $\mathscr{F}_t$ is globally generated for all (closed) $t \in S$, it is easy to see that Ⓗ$_{|\Sigma\times U_{t_o}}$ is surjective. Further, the Hilbert polynomial (for any $t \in S$) $P_{\mathscr{F}_t}(z) = N + rhz$. Thus, from the representability of the functor $\mathfrak{Q}(E = \mathscr{O}_\Sigma \otimes \mathbb{C}^N, P(z) = N + rhz)$ (cf. Theorem 7.1.6), we get that there exists a morphism $f: U_{t_o} \to Q(E, P(z))$ such that $\mathscr{F}_{|\Sigma\times U_{t_o}} \simeq f^*(\mathscr{U})$. Since, for any $t \in U_{t_o}$, the restriction of Ⓗ to $\Sigma \times t$ induces an isomorphism in cohomology $H^0(\Sigma, (\pi_{S_*}\mathscr{F})_{|t}) \to H^0(\Sigma, \mathscr{F}_t)$, we get that $\operatorname{Im} f \subset R$. This proves (B).

(C) Since $\mathscr{U}_{|\Sigma\times R} \to \Sigma \times R$ is a GL_N-equivariant vector bundle, for q_1, $q_2 \in R$ with $q_2 \in \mathrm{GL}_N \cdot q_1$, we get

$$\bar{q}_1 \simeq \bar{q}_2.$$

Conversely, take $q_1, q_2 \in R$ such that

$$\psi: \bar{q}_1 \simeq \bar{q}_2.$$

By the definition of R, $\mathscr{O}_\Sigma \otimes \mathbb{C}^N \xrightarrow{f_i} \bar{q}_i$ $(i = 1,2)$ induces isomorphism at H^0. This gives a commutative diagram for some $g \in \mathrm{GL}_N$:

$$\begin{array}{ccc} \mathbb{C}^N = H^0\left(\Sigma, \mathscr{O}_\Sigma \otimes \mathbb{C}^N\right) & \xrightarrow[\sim]{f_{1*}} & H^0(\Sigma, \bar{q}_1) \\ {\scriptstyle g}\downarrow & & \downarrow{\scriptstyle \psi_*} \\ \mathbb{C}^N & \xrightarrow[f_{2*}]{\sim} & H^0(\Sigma, \bar{q}_2). \end{array}$$

Now, it is easy to see that $g \cdot q_1 = q_2$ by considering the surjective evaluations $H^0(\Sigma, \bar{q}_i) \to (\bar{q}_i)_x$ for all $x \in \Sigma$. This proves (C).

(D) We have a group homomorphism

$$\beta_q : G_q \to \mathrm{Aut}(\bar{q}),$$

which is surjective by the proof of the (C)-part. Further, β_q is injective since if $g \in \operatorname{Ker} \beta_q$, then g induces the identity isomorphism:

$$\mathbb{C}^N \simeq H^0(\Sigma, \bar{q}) \xrightarrow{g_*} H^0(\Sigma, \bar{q}) \simeq \mathbb{C}^N,$$

which gives that $g = 1$. This proves the (D)-part.

(E) We finally prove that R is irreducible. Let $\mathscr{F}$ be the family obtained by considering the equivalence classes of extensions

$$0 \to \mathscr{O}_\Sigma \otimes \mathbb{C}^{r-1} \to F \to L_d \to 0, \qquad (*)$$

where L_d runs over the isomorphism classes of degree d line bundles on Σ. For fixed L_d, the equivalence classes of extensions (*) are of course parameterized by

$$\mathrm{Ext}^1_{\mathscr{O}_\Sigma}\left(L_d, \mathscr{O}_\Sigma \otimes \mathbb{C}^{r-1}\right) \simeq H^1(\Sigma, L_d^*)^{\oplus r-1}.$$

For any positive d, $H^0(\Sigma, L_d^*) = 0$ and hence $\dim H^1(\Sigma, L_d^*)$ does not depend upon the choice of L_d. Thus, this family is parameterized by an irreducible variety X. Let

$$X_o = \left\{x \in X : H^1(\Sigma, \mathscr{F}_x) = H^1(\Sigma, \mathscr{F}_x(-p)) = 0\, \forall p \in \Sigma\right\},$$
$$\text{where } \mathscr{F}_x := \mathscr{F}_{|\Sigma \times x}.$$

Observe that the condition $H^1(\Sigma, \mathscr{F}_x(-p)) = 0$ for all $p \in \Sigma$ in the presence of the condition $H^1(\Sigma, \mathscr{F}_x) = 0$ is equivalent to the condition that $\mathscr{F}_x$ is globally generated.

In particular, for any $x \in X_o$, $\mathscr{F}_x$ is globally generated. Conversely, any vector bundle F over Σ of rank r and degree d satisfying conditions (a) and (b) of Proposition 7.1.3 is isomorphic with $\mathscr{F}_x$ for some $x \in X_o$ (cf. (Atiyah, 1957, Theorem 2, §5, Part I)).

Then, X_o is open in X. To prove this, by the Upper Semicontinuity Theorem (Hartshorne, 1977, Chap. III, Theorem 12.8) applied to the projection $\Sigma \times X \to X$ and the vector bundle $\mathscr{F}$ over $\Sigma \times X$, the condition $H^1(\Sigma, \mathscr{F}_x) = 0$ is an open condition in x. To prove the openness of the condition $H^1(\Sigma, \mathscr{F}_x \otimes \mathscr{O}_\Sigma(-p)) = 0$ for all $p \in \Sigma$, apply the Semicontinuity Theorem to the projection $p_{2,3}\colon \Sigma \times X \times \Sigma \to X \times \Sigma$ and the vector bundle $p_{1,2}^*(\mathscr{F}) \otimes p_{1,3}^*(\mathscr{I}_{\Delta(\Sigma)})$ over $\Sigma \times X \times \Sigma$, where $p_{i,j}$ is the projection on the (i, j)th factor, $\Delta(\Sigma)$ is the diagonal of $\Sigma \times \Sigma$ and $\mathscr{I}_{\Delta(\Sigma)}$ is its ideal sheaf in $\Sigma \times \Sigma$.

Thus, by the (B)-part, there exists a finite open over $\{V_j\}_j$ of X_o and morphisms

$$f_j\colon V_j \to R \text{ such that } (\mathrm{Id} \times f_j)^* \left(\mathscr{U}_{|\Sigma\times R}\right) \simeq \mathscr{F}_{|\Sigma\times V_j}.$$

Thus, by the (C)-part,

$$\coprod_j \mathrm{GL}_N \times V_j \to R,\ (g, v_j) \mapsto g \cdot f_j(v_j),\ \text{ for } g \in \mathrm{GL}_N \text{ and } v_j \in V_j,$$

is surjective. Since X_o is irreducible, we get that each V_j is irreducible (in particular, connected) and $\bigcap_j V_j$ is nonempty. From this (using (C) again) it is easy to see that the images of $\mathrm{GL}_N \times V_j$ in R have nonempty intersection hence R is connected and hence R is irreducible (R being smooth by (A)).

This completes the proof of the proposition. □

As a consequence of Propositions 7.1.3, 7.1.9(A) and Theorem 7.1.6, we get the following.

Corollary 7.1.10 *The set of isomorphism classes of rank-r semistable vector bundles over Σ of fixed degree d is bounded in the sense that there exists a family of vector bundles over Σ of rank r and degree d parameterized by a quasi-projective* variety T_o *such that any semistable vector bundle $\mathscr{V}$ over Σ of rank r and degree d occurs in this family.*

Proof We first assume that $d > r(2g-1)$. Then, by Proposition 7.1.3 and the Riemann–Roch theorem (cf. (Fulton, 1998, Example 15.2.1)), for any semistable vector bundle $\mathscr{V}$ of rank r and degree d,

$$\dim H^0(\Sigma, \mathscr{V}) = \chi(\Sigma, \mathscr{V}) = d + r(1-g) =: N, \tag{1}$$

and the Hilbert polynomial of $\mathscr{V}$ with respect to a very ample line bundle H of degree h over Σ (which we fix once for all),

$$P_{\mathscr{V}}(z) = N + rhz. \tag{2}$$

Further, by Proposition 7.1.3, $\mathscr{V}$ is a quotient of the trivial vector bundle $\mathscr{O}_\Sigma \otimes \mathbb{C}^N$. Now, using Theorem 7.1.6, as in Definition 7.1.7, for $X = \Sigma$, $E = \mathscr{O}_\Sigma \otimes \mathbb{C}^N$ and $P(z) = N + rhz$, we get that (corresponding to the identity map of Q) there exists a 'tautological' sheaf $\mathscr{U}$ over $\Sigma \times Q$ which is a quotient of $E \boxtimes \mathscr{O}_Q$ flat over Q satisfying (3) of Theorem 7.1.6, where $Q := Q(\mathscr{O}_\Sigma \otimes \mathbb{C}^N, P(z))$. In particular, any semistable vector bundle $\mathscr{V}$ of rank r and degree $d > r(2g-1)$ occurs in this family restricted to $\Sigma \times R$. Now, restricting $\mathscr{U}_{|\Sigma\times R}$ we get the corollary when $d > r(2g-1)$ by using Proposition 7.1.9(A).

The result clearly extends to an arbitrary d replacing $\mathscr{V}$ by $\mathscr{V} \otimes H^p$ for large enough p so that $d + rhp > r(2g-1)$. □

Definition 7.1.11 Recall (cf. (Newstead, 2012, Chap. 3, §5)) that for a reductive group G acting on a projective variety X with a G-equivariant ample line bundle $\mathcal{L}$ over X, a point $x \in X$ is said to be *semistable* (with respect to the line bundle $\mathcal{L}$) if there exists a G-equivariant section $\sigma \in H^0(X, \mathcal{L}^r)^G$ (for some $r > 0$) such that $\sigma(x) \neq 0$.

A semistable point x is called *stable* if $\dim G \cdot x = \dim G$ and $G \cdot x$ is closed in X^{ss}, where we denote the set of semistable (resp. stable) points of X by X^{ss} (resp. X^s). If we need to emphasize the underlying $\mathcal{L}$, then we denote X^{ss} (resp. X^s) by $X^{ss}(\mathcal{L})$ (resp. $X^s(\mathcal{L})$). Clearly, X^{ss} is a G-stable open (possibly empty) subset of X. Also, X^s is a G-stable open subset of X (cf. (Newstead, 2012, Lemma 3.13)).

Let $Y \subset X$ be a G-stable closed subvariety. Then,

$$Y^{ss}(\mathcal{L}) = Y \cap X^{ss}(\mathcal{L}) \text{ and } Y^s(\mathcal{L}) = Y \cap X^s(\mathcal{L}).$$

Definition 7.1.12 We follow the notation and assumptions from Definition 7.1.7. Let $\check{\mathrm{Gr}} = \check{\mathrm{Gr}}(r, N)$ be the Grassmannian of r-dimensional quotients of $\mathbb{C}^N$. For any $x \in \Sigma$, define the morphism

$$e_x : R \to \check{\mathrm{Gr}}, \quad q \mapsto \bar{q}_x,$$

where R is as in Definition 7.1.7 and $\bar{q}_x$ is the fiber of $\bar{q}$ at x. To see that e_x is indeed a morphism, consider the quotient line bundle $\mathscr{O}_R \otimes \wedge^r(\mathbb{C}^N) \twoheadrightarrow \wedge^r(\mathscr{U})_{|x\times R}$. Dualizing this we get $\wedge^r(\mathscr{U}^*)_{|x\times R}$ as a line subbundle of $\mathscr{O}_R \otimes \wedge^r(\mathbb{C}^{N*})$. This gives rise to a morphism $\tilde{e}_x : R \to \mathbb{P}(\wedge^r(\mathbb{C}^{N^*}))$. Since

$$\check{\mathrm{Gr}} \hookrightarrow \mathbb{P}\left(\wedge^r(\mathbb{C}^{N^*})\right), \quad V \mapsto \wedge^r(V^*),$$

is a closed embedding, we get that e_x is a morphism. Clearly, e_x is GL_N-equivariant under the canonical action of GL_N on $\check{\mathrm{Gr}}$. The action of GL_N on $\check{\mathrm{Gr}}$ clearly descends to an action of PGL_N. For any sequence

$$\vec{x} = (x_1, \dots, x_p) \in \Sigma^p,$$

we get the GL_N-equivariant morphism

$$e_{\vec{x}} : R \to \check{\mathrm{Gr}}^p,\ q \mapsto (\bar{q}_{x_1}, \dots, \bar{q}_{x_p}).$$

Let R^{ss} (resp. R^s) be the GL_N-stable subset of R consisting of those $q \in R$ such that $\bar{q}$ is semistable (resp. stable) vector bundle over Σ. Also, let $\check{\mathrm{Gr}}_p^{ss}$ (resp. $\check{\mathrm{Gr}}_p^{s}$) be the set of semistable (resp. stable) points of $\check{\mathrm{Gr}}^p$ under the diagonal action of SL_N with respect to the ample line bundle $L^{\boxtimes p}$ over $\check{\mathrm{Gr}}^p$, L being any SL_N-equivariant ample line bundle over $\check{\mathrm{Gr}}$. Observe that since $\mathrm{Pic}_{\mathrm{SL}_N}(\check{\mathrm{Gr}}) \simeq \mathbb{Z}$, the choice of SL_N-equivariant ample L does not affect $\check{\mathrm{Gr}}_p^{ss}$ or $\check{\mathrm{Gr}}_p^{s}$.

We recall the following result whose proof can be found in Newstead (2012, Chap. 5, §6) (also see (Seshadri, 1967)).

Theorem 7.1.13 *For any $r \geq 1$ and positive $d > \max\{r(2g-1), r((r+1)g-2), r(r^2g+2g-2)\}$, there exists a set of distinct points $\vec{x} = (x_1, \dots, x_p)$ (for some $p = p(d)$) in Σ such that the above morphism $e_{\vec{x}} \colon R \to \check{Gr}^p$ satisfies:*

(i) *$e_{\vec{x}}$ is an embedding, i.e., an isomorphism onto a locally closed subvariety of $\check{Gr}^p$*
(ii) $R^{ss} = e_{\vec{x}}^{-1}(\check{Gr}_p^{ss})$
(iii) $R^s = e_{\vec{x}}^{-1}(\check{Gr}_p^{s})$
(iv) *The restriction of $e_{\vec{x}}$ to R^{ss}, i.e., $e_{\vec{x}|R^{ss}} : R^{ss} \to \check{Gr}_p^{ss}$ is a proper morphism.*

Observe that since $\check{Gr}_p^{ss}$ and $\check{Gr}_p^{s}$ are open subsets of $\check{Gr}_p$ (cf. Definition 7.1.11), R^{ss} and R^s are open subsets of R by (ii) and (iii) above.

We are now ready to prove that the functor $\mathscr{M}(r,d)$ (resp. $\mathscr{M}^s(r,d)$) of semistable (resp. stable) vector bundles over Σ of rank r and degree d (cf. Definition 7.1.1) has a coarse moduli space (cf. Definition 7.1.2). By definition, $\mathscr{M}(r,d)(pt) :=$ set of isomorphism classes of semistable vector bundles over Σ of rank r and degree d and $\mathscr{M}^s(r,d)(pt)$ is the subset of $\mathscr{M}(r,d)(pt)$ consisting of isomorphism classes of stable vector bundles.

Theorem 7.1.14 *Let Σ be an irreducible smooth projective curve of any genus $g \geq 0$. Let $r \geq 1$ and $d \in \mathbb{Z}$. Then, $\mathscr{M}(r,d)$ has a coarse moduli* *space $M(r,d)$, which is an irreducible, normal projective variety with rational singularities.*

Moreover, the subfunctor $\mathscr{M}^s(r,d)$ has a coarse moduli space $M^s(r,d)$ which is an open subset of $M(r,d)$. The canonical map $\Phi\colon \mathscr{M}(r,d)(pt) \to M(r,d)$ (induced from the coarse moduli property) is surjective. Further, the restriction of Φ to $\mathscr{M}^s(r,d)(pt)$ is a bijection onto $M^s(r,d)$.

Also, the dimension of $M(r,d)$ is $(g-1)r^2+1$ if $M^s(r,d) \neq \emptyset$.

Proof Let $\mathscr{L}$ be a line bundle over Σ. Then, a vector bundle $\mathscr{V}$ over Σ is semistable (resp. stable) if and only if $\mathscr{V} \otimes \mathscr{L}$ is semistable (resp. stable) (cf. Definition 6.1.4(a)). Thus, the functor $\mathscr{M}(r,d)$ (resp. $\mathscr{M}^s(r,d)$) is equivalent with the functor $\mathscr{M}(r,d+rd')$ (resp. $\mathscr{M}^s(r,d+rd')$) for any $d' \in \mathbb{Z}$. In particular, we can (and do) assume that d is large enough satisfying the bound in Theorem 7.1.13.

Thus, by Theorem 7.1.13, R^{ss} can be identified with a SL_N-stable closed subvariety of $\check{\mathrm{Gr}}_p^{ss}$. Take the closure $\bar{R}^{ss}$ of R^{ss} in $\check{\mathrm{Gr}}_p$ with the reduced structure. Then, by Definition 7.1.11 and Theorem 7.1.13,

$$\left(\bar{R}^{ss}\right)^{ss}(L^{\boxtimes p}) = R^{ss} \text{ and } \left(\bar{R}^{ss}\right)^{s}(L^{\boxtimes p}) = R^{s}.$$

Let $\pi\colon R^{ss} \to M(r,d) := R^{ss}//\mathrm{SL}_N$ be the good quotient, which is a projective variety (cf. (Newstead, 2012, Theorem 3.14)). (Observe that the center of GL_N acts trivially on R by Proposition 7.1.9(D).) Moreover, by the same reference of Newstead, there exists an open subset $M^s(r,d)$ of $M(r,d)$ such that

$$\pi^{-1}\left(M^s(r,d)\right) = R^s \quad \text{and} \quad \pi_{|R^s}: R^s \to M^s(r,d) \text{ is a geometric quotient.}$$

We now prove that $M(r,d)$ (resp. $M^s(r,d)$) is the coarse moduli space of the functor $\mathscr{M}(r,d)$ (resp. $\mathscr{M}^s(r,d)$).

Take a family $\mathscr{F}$ of semistable vector bundles over Σ of rank r and degree d parameterized by a noetherian scheme S. Then, by Propositions 7.1.3 and 7.1.9(B), there exists an open cover $\{S_i\}$ of S and a morphism $f_i\colon S_i \to R$ such that

$$\mathscr{F}_{|\Sigma\times S_i} \simeq (\mathrm{Id}_\Sigma \times f_i)^*(\mathscr{U}_{|\Sigma\times R}).$$

In particular, $\mathrm{Im}\, f_i \subset R^{ss}$. Consider the map $\bar{f}_i\colon S_i \to M(r,d)$ obtained from the projection $\pi\colon R^{ss} \to M(r,d)$. By Proposition 7.1.9(C), $\bar{f}_i = \bar{f}_j$ on the intersection $S_i \cap S_j$. Thus, we get a morphism $\bar{f}\colon S \to M(r,d)$. This gives a natural transformation

$$\Phi\colon \mathscr{M}(r,d) \to \mathrm{Mor}(-, M(r,d)).$$

Let N be a scheme and let $\Phi' : \mathscr{M}(r,d) \to \mathrm{Mor}(-,N)$ be any natural transformation. Consider the family $\mathscr{U}_{|\Sigma \times R^{ss}}$. This gives rise to a morphism $\theta_{\mathscr{U}} : R^{ss} \to N$. Since SL_N acts on $\mathscr{U}_{|\Sigma \times R^{ss}}$, the map $\theta_{\mathscr{U}}$ is SL_N-invariant. Thus, a good quotient being a categorical quotient (cf. Definition 7.1.4), the morphism $\theta_{\mathscr{U}}$ descends to a morphism $\bar{\theta}_{\mathscr{U}} : M(r,d) \to N$. Hence, $M(r,d)$ is the coarse moduli space for the functor $\mathscr{M}(r,d)$.

Exactly the same proof, replacing semistable by stable, proves that $M^s(r,d)$ is a coarse moduli space for the functor $\mathscr{M}^s(r,d)$.

Now, $M(r,d)$ is an irreducible, normal variety since so is R by Proposition 7.1.9(A), (E). By Boutot's theorem (cf. (Boutot, 1987)) $M(r,d)$ has rational singularities since so is R.

We next prove the surjectivity of the map $\Phi : \mathscr{M}(r,d)(pt) \to M(r,d)$. Take any $q \in R^{ss}$. Then, $\bar{q} := \mathscr{U}_{|\Sigma \times q}$ is a semistable vector bundle (of rank r and degree d) and hence $\bar{q} \in \mathscr{M}(r,d)(pt)$ with $\Phi(\bar{q}) = \pi(q)$ (by the proof above). Since $\pi : R^{ss} \to M(r,d)$ is surjective, we get the surjectivity of Φ.

To show that $\Phi_{|\mathscr{M}^s(r,d)(pt)}$ is a bijection onto $M^s(r,d)$, using Proposition 7.1.9(C), it suffices to use that $\pi_{|R^s} : R^s \to M^s(r,d)$ is a geometric quotient.

We finally calculate the dimension of $M(r,d)$ when $M^s(r,d) \neq \emptyset$. In this case

$$
\begin{aligned}
\dim M(r,d) &= \dim M^s(r,d) \\
&= \dim R^s - \dim \mathrm{SL}_N, \\
&\quad \text{by Exercise 6.3.E.6 and Proposition 7.1.9(D)} \\
&= N^2 + r^2(g-1) - (N^2 - 1), \quad \text{by Proposition 7.1.9(A)} \\
&= r^2(g-1) + 1.
\end{aligned}
$$

This proves the theorem completely. □

Remark 7.1.15 (a) From the above calculation of $\dim M^s(r,d)$, it is clear that if $g = 0$ and $r \geq 2$, $M^s(r,d) = \emptyset$. The only other possible case in which $M^s(r,d) = \emptyset$ is when $g = 1$ and $(r,d) \neq 1$ (cf. (Narasimhan and Seshadri, 1965, Proposition 9.1 and Theorem 2(A)) for $g \geq 2$; (Le Potier, 1997, Theorem 8.6.2) for $g = 1$).

(b) Whenever $M^s(r,d) \neq \emptyset$, it is a smooth irreducible variety. Of course, it is an irreducible variety being an open subset of irreducible variety $M(r,d)$. To prove its smoothness, by Exercise 6.3.E.6, we get that for any stable vector bundle $\mathscr{V}$ over Σ, $\mathrm{Aut}\,\mathscr{V} \simeq \mathbb{C}^*$. Thus, by Proposition 7.1.9(D), for any $q \in R^s$ its stabilizer $G_q = \mathbb{C}^*$ (under the GL_N-action on R^s). Hence, the action of

SL_N on R^s descends to an action of PGL_N with trivial stabilizers. Moreover, the quotient $M^s(r,d) := \mathrm{PGL}_N \backslash R^s$ is smooth (cf. (Le Potier, 1997, Theorem 8.3.2)).

Observe that if $(r,d) = 1$, then any semistable vector bundle over Σ is stable (cf. Exercise 6.1.E.9). Thus, $R^{ss} = R^s$ and hence $M(r,d) = M^s(r,d)$. In particular, in this case, $M(r,d)$ is a smooth irreducible projective variety. In this case, $\Sigma \times M(r,d)$ carries a universal vector bundle (cf. (Le Potier, 1997, Theorem 8.4.2)).

As a consequence of Proposition 7.1.9(B), we get the following.

Corollary 7.1.16 *(a) Let $\mathscr{F} \to \Sigma \times S$ be a family of vector bundles over Σ of rank r and degree d parameterized by a noetherian scheme S. Let S^{ss} (resp. S^s) be the subset of S consisting of those $t \in S$ such that $\mathscr{F}_t := \mathscr{F}_{|\Sigma \times t}$ is semistable (resp. stable). Then, both of S^{ss} and S^s are open in S.*

(b) Thus, for a family $\mathscr{F} \to \Sigma \times S$ of G-bundles, S^{ss} is open in S, where G is any connected reductive group.

(The corresponding result for S^s (resp S^{ss}) for $\mathbb{C}$-analytic families is proved in Lemma 6.3.31 (resp. Exercise 6.3.E.9).)

Proof (a) Depending upon $\mathscr{F}$, we can choose an ample line bundle L on Σ such that (denoting $\mathscr{F}(L) := \mathscr{F} \otimes (L \boxtimes \mathscr{O}_S)$)

$$R^i \pi_{S_*}(\mathscr{F}(L)) = 0, \quad \text{for all} \quad i > 0, \quad \text{and } \mathscr{F}(L) \text{ is globally generated,}$$

where $\pi : \Sigma \times S \to S$ is the projection (cf. (Hartshorne, 1977, Chap. III, Theorem 8.8)). From this we easily see that $\mathscr{F}_t \otimes L$ is globally generated for all (closed) points $t \in S$ and, moreover, by Kempf (1978, Theorem 13.1),

$$H^i(\Sigma, \mathscr{F}_t \otimes L) = 0, \quad \text{for} \quad i = 1.$$

Thus, $\mathscr{F}_t \otimes L$ satisfies conditions (a) and (b) of Proposition 7.1.3 for all $t \in S$. Hence, the family $\mathscr{F}(L)$ over $\Sigma \times S$ is locally the pull-back of the family $\mathscr{U}_{|\Sigma \times R}$ by Proposition 7.1.9(B). But, since R^{ss} and R^s are open in R by Theorem 7.1.13 (by taking L of large enough degree), we get that S^{ss} and S^s are open in S.

(b) The assertion for any G as in (b) follows immediately from the vector bundle case by using Lemma 6.1.5. □

In the following G is a connected reductive group.

Definition 7.1.17 Let $E \to \Sigma$ be a G-bundle. A reduction E_P of E to a parabolic subgroup P is called *admissible* if for any character χ of P which is trivial restricted to the connected center Z^o of G, the corresponding line bundle

$\mu^*(\mathscr{L}_P(\chi))$ has degree 0, where $\mathscr{L}_P(\chi) := E \times^P \mathbb{C}_{\chi^{-1}}$ and μ is the section of $E/P \to \Sigma$ determined by the reduction P (cf. Lemma 5.1.2). Of course, E is an admissible reduction of itself.

Given an admissible reduction E_P of E, by $\mathrm{gr}(E_P)$ we mean the G-bundle over Σ obtained from the extension of the structure group

$$E_P \to E_P(P/U_P) \to E_P(G),$$

where the first map is obtained from $P \to P/U_P$ (U_P being the unipotent radical of P) and the second map is obtained from identifying P/U_P with a Levi subgroup L_P of P and then the inclusion $L_P \hookrightarrow G$. Since the Levi subgroups L_P are conjugate in P, $\mathrm{gr}(E_P)$ is uniquely defined up to an isomorphism.

Given two G-bundles E' and E'' over Σ, they are said to be *related* if they admit admissible reductions $E'_{P'}$ and $E''_{P''}$ (for some parabolic subgroups P' and P'') such that

$$\mathrm{gr}(E'_{P'}) \simeq \mathrm{gr}(E''_{P''}), \text{ as } G\text{-bundles over } \Sigma.$$

Two G-bundles E' and E'' over Σ are called *equivalent* if they are in the same equivalence class generated by the above relation. In particular, these definitions apply to $G = \mathrm{GL}_r$, i.e., rank-r vector bundles over Σ.

Lemma 7.1.18 *If two G-bundles E' and E'' over Σ are equivalent, then they have the same topological type.*

Proof It suffices to show that for an admissible reduction E_P of a G-bundle E over Σ, $\mathrm{gr}(E_P)$ has the same topological type as E. To prove this, observe that the inclusion $P \hookrightarrow G$ and the composite map

$$P \to P/U_P \xrightarrow{\sim} L_P \hookrightarrow G$$

both induce the same map $\pi_1(P) \to \pi_1(G)$ since L_P is a deformation retract of P. Now, use the topological classification of G-bundles over Σ as in Lemma 6.3.10. □

The following proposition determines the fiber of the map $\Phi: \mathscr{M}(r,d)(pt) \to M(r,d)$ (cf. Theorem 7.1.14). In Section 7.2, the corresponding result in the equivariant setting is proved for any connected semisimple group G and $g \geq 2$ (cf. Corollary 7.2.20). The same proof gives the following result for any genus $g \geq 0$. Alternatively, a direct proof can be found in Le Potier (1997, §7.4).

Proposition 7.1.19 *With the notation and assumptions as in Theorem 7.1.14, the surjective map*

$$\Phi : \mathscr{M}(r,d)(pt) \to M(r,d)$$

has its fibers precisely the equivalence classes of semistable bundles, where $\mathscr{M}(r,d)(pt)$ is the set of isomorphism classes of semistable vector bundles of rank r and degree d.

7.1.E Exercises

In the following Σ is a smooth projective irreducible curve.

(1) Let $\mathbf{Vect}(\Sigma)$ denote the additive category whose objects are algebraic vector bundles (of any rank) over Σ and $\mathscr{O}_\Sigma$-module morphisms between them and let $\mathbf{Vec}_0^{ss}(\Sigma)$ be the full subcategory of semistable vector bundles of degree 0. Then, show that $\mathbf{Vec}_0^{ss}(\Sigma)$ is an abelian category, which is artinian and noetherian. In particular, each object $\mathscr{V}$ in $\mathbf{Vec}_0^{ss}(\Sigma)$ has Jordan–Hölder series with unique $\mathrm{gr}(\mathscr{V})$. Moreover, $\mathrm{gr}(\mathscr{V})$ is a direct sum of stable vector bundles of degree 0.

7.2 Moduli Space of Parabolic Semistable G-Bundles

Let $\hat{\Sigma}$ be an irreducible smooth projective curve of any genus $\hat{g} \geq 0$ equipped with a faithful action of a finite group A and let G be a connected reductive algebraic group over $\mathbb{C}$. We fix an embedding of algebraic groups

$$i : G \hookrightarrow \mathrm{SL}_r \subset \mathrm{GL}_r .$$

We make a slight modification to the construction of the Quot scheme Q from Definition 7.1.7 to take into consideration the action of A on $\hat{\Sigma}$.

Definition 7.2.1 For any G-bundle E over $\hat{\Sigma}$, the corresponding rank-r vector bundle $E(i)$ (obtained from the extension of the structure group via i) is of coarse of degree 0 (since G is embedded in SL_r). Fix an A-stable finite subset $\{y_1, \ldots, y_b\}$ of $\hat{\Sigma}$ and let $\vec{y} = \vec{y}(d')$ be the divisor $d' \sum_{j=1}^{b} y_j$ (for a positive integer d') and let $\mathscr{O}_{\hat{\Sigma}}(\vec{y})$ be the corresponding A-equivariant line bundle over $\hat{\Sigma}$, where we choose d' so that $d := d'b \geq 2\hat{g}$. This guarantees that any semistable vector bundle $\mathscr{V}$ over $\hat{\Sigma}$ of rank r and trivial degree satisfies (cf. Proposition 7.1.3):

(a) $H^1(\hat{\Sigma}, \mathscr{V}(\vec{y})) = 0$, and
(b) $\mathscr{V}(\vec{y})$ is globally generated, where $\mathscr{V}(\vec{y}) := \mathscr{V} \otimes \mathscr{O}_{\hat{\Sigma}}(\vec{y})$.

As in Theorem 7.1.6, let

$$Q = Q\left(E = \left(\mathscr{O}_{\hat{\Sigma}} \otimes \mathbb{C}^N\right) \otimes \mathscr{O}_{\hat{\Sigma}}(-\vec{y}), P(z)\right)$$

be the quot scheme with the tautological coherent sheaf $\mathscr{U}$ over $\hat{\Sigma} \times Q$ flat over Q which is a quotient of $E \boxtimes \mathscr{O}_Q$, where $N := r(d + 1 - \hat{g})$ and $P(z) := r(1 - \hat{g}) + rhz$ (h being the degree of a fixed very ample A-equivariant line bundle H over $\hat{\Sigma}$).

Fix a representation $\mathring{\tau}$ of A in $\mathbb{C}^N$

Then the action of A on $\hat{\Sigma}$ induces an action of A on $\mathscr{O}_{\hat{\Sigma}} \otimes \mathbb{C}^N$ with the $\mathring{\tau}$-action on $\mathbb{C}^N$. Thus, E is an A-equivariant vector bundle over $\hat{\Sigma}$ inducing a canonical action of A on Q and making $\mathscr{U}$ an A-equivariant coherent sheaf (with respect to the diagonal action of A on $\hat{\Sigma} \times Q$). Define

$$R_{\mathring{\tau}} = \Big\{q \in \bar{Q}^A : \bar{q} := \mathscr{U}_{|_{\hat{\Sigma}\times q}} \text{ is a locally free sheaf and } \mathbb{C}^N = H^0\left(\hat{\Sigma}, E(\vec{y})\right) \xrightarrow{\sim} H^0\left(\hat{\Sigma}, \bar{q}(\vec{y})\right) \text{ is an isomorphism}\Big\},$$

where Q^A is the subscheme of A-invariants in Q and $\bar{Q}^A$ is the set of its closed points. Observe that $\bar{q}$, for $q \in R_{\mathring{\tau}}$, (and hence $\bar{q}(\vec{y})$) is an A-equivariant bundle and the above isomorphism is A-equivariant. Further, deg $\bar{q} = 0$ (cf. (1) of Definition 7.1.7). Let $\mathfrak{G}$ be the group of A-equivariant automorphisms of E inducing the identity at the base. Then, clearly

$$\mathfrak{G} = \mathrm{GL}_N^A,$$

the A-invariants under the conjugation action of A on GL_N (induced from the representation $\mathring{\tau}$) and hence $\mathfrak{G}$ is a reductive group. The group $\mathfrak{G}$ keeps Q^A stable, $R_{\mathring{\tau}} \subset \bar{Q}^A$ is a $\mathfrak{G}$-stable smooth open subset and $\mathscr{U}_{|_{\hat{\Sigma}\times R_{\mathring{\tau}}}}$ is an A-equivariant rank-r vector bundle with the action of $\mathfrak{G}$. (To prove that $R_{\mathring{\tau}}$ is a smooth open subset of $\bar{Q}^A$ and $\mathscr{U}_{|_{\hat{\Sigma}\times R_{\mathring{\tau}}}}$ is a rank-r vector bundle, use the fact that $R_{\mathring{\tau}} = R^A$ and then use Proposition 7.1.9(A), where R is as in this proposition.) Let

$$R_{\mathring{\tau}}^{ss} = \{q \in R_{\mathring{\tau}} : \bar{q} \text{ is an } A\text{-semistable vector bundle}\}$$

(cf. Definition 6.1.16 for the definition of A-semistable bundles). By Corollary 7.1.16 and Exercise 6.2.E.4, $R_{\mathring{\tau}}^{ss}$ is open in $R_{\mathring{\tau}}$. We denote the family $\mathscr{U}_{|_{\hat{\Sigma}\times R_{\mathring{\tau}}^{ss}}}$ by $\mathscr{U}_{\mathring{\tau}}^{ss}$. Of course, $R_{\mathring{\tau}}^{ss}$ is $\mathfrak{G}$-stable. Let $R_{\mathring{\tau}}^{ss}(\mathscr{O})$ denote the subset

$$R_{\mathring{\tau}}^{ss}(\mathscr{O}) = \left\{q \in R_{\mathring{\tau}}^{ss} : \wedge^r(\bar{q}) \simeq \mathscr{O}_{\hat{\Sigma}}\right\}.$$

Let

$$R^{ss} := \{q \in \bar{Q} : \bar{q} \text{ is a semistable vector bundle and}$$
$$\mathbb{C}^N = H^0(\hat{\Sigma}, E(\vec{y})) \xrightarrow{\sim} H^0(\hat{\Sigma}, \bar{q}(\vec{y}))\}.$$

Then, by Proposition 7.1.9(A) and Corollary 7.1.16(a), R^{ss} is a smooth variety and $\mathscr{U}_{|\hat{\Sigma}\times R^{ss}}$ is a rank-r vector bundle. Moreover,

$$\det : R^{ss} \to J(\hat{\Sigma}), \quad q \mapsto \wedge^r(\bar{q}),$$

is a morphism, where $J(\hat{\Sigma})$ is the Jacobian variety of $\hat{\Sigma}$, which is an abelian variety consisting of isomorphism classes of degree 0 line bundles over $\hat{\Sigma}$. To show that det is a morphism, the family $(\wedge^r\mathscr{U})_{|\hat{\Sigma}\times R^{ss}}$ gives rise to the morphism det by considering the *Poincaré bundle* over $\hat{\Sigma} \times J(\hat{\Sigma})$ (cf. (Arbarello et al., 1985, Chap. IV, §2)). We further claim that det is a smooth morphism.

First of all det is surjective. Take any $q_o \in R^{ss}$ and let $\det(q_o) = \mathscr{L}_o$. Take $\mathscr{L}_o^{1/r} \in J(\hat{\Sigma})$ (since $J(\hat{\Sigma})$ is a divisible group, this is possible). Now, take any $\mathscr{L} \in J(\hat{\Sigma})$. Then, there exists $q \in R^{ss}$ with $\bar{q} \simeq (\bar{q}_o \otimes \mathscr{L}_o^{-1/r} \otimes \mathscr{L}^{1/r})$ (by the properties (a) and (b)). Clearly, $\det(q) = \mathscr{L}$. This prove the surjectivity of det.

By the generic smoothness (cf. (Hartshorne, 1977, Chap. III, Corollary 10.7)), det is a smooth morphism over a nonempty open subset of $J(\hat{\Sigma})$. Fix $\mathscr{L} \in J(\hat{\Sigma})$ and consider the family parameterized by R^{ss}:

$$\mathscr{U}_{\mathscr{L}} := \mathscr{U}_{|\hat{\Sigma}\times R^{ss}} \otimes \pi^*_{\hat{\Sigma}}(\mathscr{L}) \to \hat{\Sigma} \times R^{ss},$$

where $\pi_{\hat{\Sigma}} : \hat{\Sigma} \times R^{ss} \to \hat{\Sigma}$ is the projection. Then, $\mathscr{U}_{\mathscr{L}}$ satisfies conditions (a) and (b). Thus, by Proposition 7.1.9(B), there exists an open cover$\{V_j\}$ of R^{ss} and a morphism $f_j : V_j \to R^{ss}$ such that

$$\bar{f}_j^*\left(\mathscr{U}_{|\hat{\Sigma}\times R^{ss}}\right) = \mathscr{U}_{\mathscr{L}_{|\hat{\Sigma}\times V_j}}, \quad \text{where } \bar{f}_j = \mathrm{Id}_{\hat{\Sigma}} \times f_j.$$

Let $f : \sqcup V_j \to R^{ss}$ be the morphism defined by $f_{|V_j} = f_j$. It is easy to see that $\overline{f(q)} = \bar{q} \otimes \mathscr{L}$, for any $q \in \sqcup V_j$. Thus, we have the following commutative diagram:

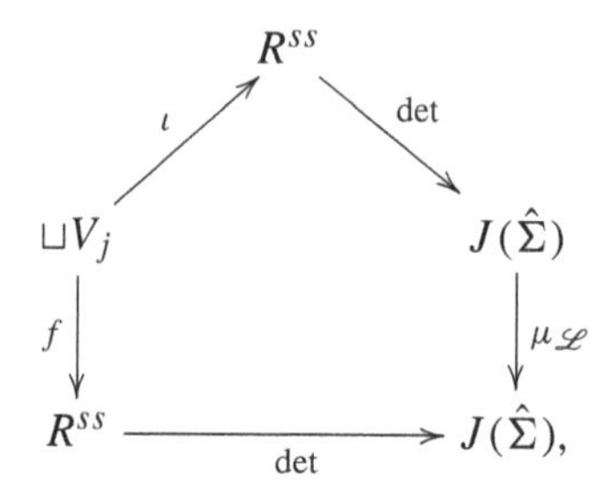

where $\iota_{|V_j} = \mathrm{Id}_{V_j}$ and $\mu_{\mathscr{L}}(\mathscr{L}') = \mathscr{L}' \otimes \mathscr{L}^r$. Since det is a smooth morphism over an open nonempty subset of $J(\hat{\Sigma})$, using the above commutative diagram and Proposition 7.1.9(C), varying $\mathscr{L}$ we get that det is smooth everywhere. In particular, $R^{ss}(\mathscr{O}) := \det^{-1}(\mathscr{O}_{\hat{\Sigma}})$ is a smooth variety and hence

$$R^{ss}_{\mathring{\tau}}(\mathscr{O}) = R^{ss}(\mathscr{O})^A$$

is a closed smooth subvariety of $R^{ss}_{\mathring{\tau}}$.

Definition 7.2.2 Let $i: G \hookrightarrow H$ be an embedding of reductive groups and let $\mathscr{F} \to \hat{\Sigma} \times S$ be a family of A-equivariant H-bundles over $\hat{\Sigma}$ parameterized by a scheme S with the trivial action of A on S. Define the contravariant functor $\Gamma(i, \mathscr{F}): \mathfrak{S}_S \to$ **Set** by $\Gamma(i, \mathscr{F})(f: T \to S) =$ the set of A-equivariant sections σ of $\mathscr{F}_f/G$, where $\mathscr{F}_f := (\mathrm{Id}_{\hat{\Sigma}} \times f)^*(\mathscr{F})$ and $\mathfrak{S}_S$ is the category as in Section 1.1. Clearly, any such σ gives rise to an A-equivariant G-subbundle $\mathscr{F}_f(\sigma)$ of $\mathscr{F}_f$ over $\hat{\Sigma} \times T$ (cf. Lemma 5.1.2). Conversely, any A-equivariant G-subbundle $\mathscr{E}$ of $\mathscr{F}_f$ over $\hat{\Sigma} \times T$ gives rise to an A-equivariant section $\sigma_{\mathscr{E}}$ of $\mathscr{F}_f/G$.

Let τ be an A-equivariant topological G-bundle over $\hat{\Sigma}$. Define a subfunctor $\Gamma^\tau(i, \mathscr{F})$ of $\Gamma(i, \mathscr{F})$ by demanding that for any $f: T \to S$, the G-subbundle $\mathscr{F}_f(\sigma) \subset \mathscr{F}_f$ satisfies that $\mathscr{F}_f(\sigma)_t := \mathscr{F}_f(\sigma)_{|\hat{\Sigma} \times t}$ (for any $t \in T$) is topologically A-equivariant isomorphic with τ.

The non-equivariant analogue of the following result is obtained in Ramanathan (1996, Lemma 4.8.1). (In fact, a more general result is obtained in the same in the non-equivariant setting.) The proof in the equivariant case is similar. Alternatively, the result in the equivariant case can be obtained from the non-equivariant case by using Edixhoven (1992, Proposition 3.1).[1]

Proposition 7.2.3 *With the notation and assumptions as in the above definition, the functor $\Gamma(i, \mathscr{F})$ from $\mathfrak{S}_S$ to* **Set** *is representable (representability is defined in Section 1.1) by a separated of finite type S-scheme $f_o: T_o \to S$.*

In fact, there exists a 'universal' A-equivariant G-subbundle $\mathscr{U}_G$ of $\mathscr{F}_{f_o}$ over $\hat{\Sigma} \times T_o$ (i.e., $\mathscr{U}_G \in \Gamma(i, \mathscr{F})(f_o)$ given by a section s_o) such that under the natural equivalence of the functors $\Gamma(i, \mathscr{F})$ and $\tilde{h}_{T_o/S}$ (cf. Section 1.1 for the definition of $\tilde{h}_{T_o/S}$), if an element $\mathscr{E}$ in $\Gamma(i, \mathscr{F})(f: T \to S)$ (i.e., $\mathscr{E}$ is an A-equivariant G-subbundle of $\mathscr{F}_f$ over $\hat{\Sigma} \times T$) corresponds to a morphism $\psi(f)$:

[1] We thank Balaji for the reference.

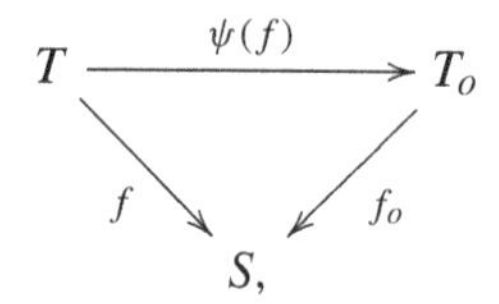

then $\left(\mathrm{Id}_{\hat{\Sigma}} \times \psi(f)\right)^* (\mathscr{U}_G) = \mathscr{E}$.

Moreover, the subfunctor $\Gamma^\tau(i, \mathscr{F})$ *is representable by a S-subscheme* T_o^τ *of* T_o *and the restriction of* $\mathscr{U}_G$ *to* $\hat{\Sigma} \times T_o^\tau$. □

Applying Proposition 7.2.3 to the embedding $i\colon G \hookrightarrow \mathrm{SL}_r \subset \mathrm{GL}_r$ (as at the beginning of this section) and the frame bundle of the family $\mathscr{U}_{\mathring{\tau}}^{ss}$ over $\hat{\Sigma} \times R_{\mathring{\tau}}^{ss}$, where $\mathscr{U}_{\mathring{\tau}}^{ss}$ and $R_{\mathring{\tau}}^{ss}$ are defined in Definition 7.2.1, we obtain the following for any topological type τ of A-equivariant G-bundles over $\hat{\Sigma}$.

Corollary 7.2.4 *The functor* $\Gamma^\tau(i, \mathscr{U}_{\mathring{\tau}}^{ss})$ *from* $\mathfrak{S}_{R_{\mathring{\tau}}^{ss}}$ *to* **Set** *is representable by a separated scheme of finite type* $f_\tau\colon R_\tau^{ss}(G) \to R_{\mathring{\tau}}^{ss}$. *Moreover, there exists a 'universal' A-equivariant G-bundle* $\mathscr{U}_\tau^{ss}(G) \in \Gamma^\tau(i, \mathscr{U}_{\mathring{\tau}}^{ss})(f_\tau)$ *over* $R_\tau^{ss}(G)$ *with the meaning as in the above Proposition 7.2.3.*

From the representability of $\Gamma^\tau(i, \mathscr{U}_{\mathring{\tau}}^{ss})$, *it is easy to see that the fibers of the morphism* $f_\tau\colon R_\tau^{ss}(G) \to R_{\mathring{\tau}}^{ss}$ *are given by (for* $q \in R_{\mathring{\tau}}^{ss}$*):*

$$\begin{aligned} f_\tau^{-1}(q)(\mathbb{C}) &= \textit{the set of A-equivariant G-subbundles of } \hat{q} \textit{ of topological type } \tau, \\ &\qquad \textit{where } \hat{q} \textit{ is the frame bundle of the vector bundle } \bar{q} \\ &= \textit{the set of A-equivariant sections of } \hat{q}/G \textit{ of topological type } \tau. \end{aligned}$$

(By Corollary 6.2.7(a) and Exercise 6.2.E.4, any G-bundle in $f_\tau^{-1}(q)$ *is A-semistable.)*

Further, the action of $\mathfrak{G}$ *on* $R_{\mathring{\tau}}^{ss}$ *and* $\mathscr{U}_{\mathring{\tau}}^{ss}$ *canonically lifts to its action on* $R_\tau^{ss}(G)$ *and the universal A-equivariant G-bundle* $\mathscr{U}_\tau^{ss}(G)$ *parameterized by* $R_\tau^{ss}(G)$ *via* f_τ.

Since G is embedded in SL_r, the morphism $f_\tau\colon R_\tau^{ss}(G) \to R_{\mathring{\tau}}^{ss}$ clearly lands inside $R_{\mathring{\tau}}^{ss}(\mathscr{O})$.

Define the subset $R_\tau^s(G)$ of $R_\tau^{ss}(G)$ by

$$R_\tau^s(G) := \{q \in R_\tau^{ss}(G) : \mathscr{U}_\tau^{ss}(G)_{|\hat{\Sigma}\times q} \text{ is } A\text{-stable}\}.$$

Proposition 7.2.5 *With the notation as in Corollary 7.2.4, the morphism* $f_\tau\colon R_\tau^{ss}(G) \to R_{\mathring{\tau}}^{ss}$ *is an affine morphism.*

Proof Let $\mathscr{U}_{\mathring{\tau}}^{ss} \to \hat{\Sigma} \times R_{\mathring{\tau}}^{ss}$ be the family of A-semistable vector bundles as in Definition 7.2.1 and let $\mathrm{Fr}(\mathscr{U}_{\mathring{\tau}}^{ss})$ be the corresponding family of frame bundles of $\mathscr{U}_{\mathring{\tau}}^{ss}$, which is a family of GL_r-bundles. Define $F = F_{\mathring{\tau}}^{ss}$ to be the total space of

$$\left(\mathrm{Fr}\left(\mathscr{U}_{\mathring{\tau}}^{ss}\right)/G\right)_{|p\times R_{\mathring{\tau}}^{ss}}, \quad \text{where } p \in \hat{\Sigma} \text{ is a fixed point.}$$

Thus, we have a fibration $\pi : F \to R_{\mathring{\tau}}^{ss}$ with fiber GL_r/G. Since GL_r/G is an affine variety (by Matsushima's Theorem), π is an affine morphism. Moreover, the morphism $f_\tau : R_\tau^{ss}(G) \to R_{\mathring{\tau}}^{ss}$ factors through F (via π), i.e., there is a morphism $\theta : R_\tau^{ss}(G) \to F$ making the following triangle commutative:

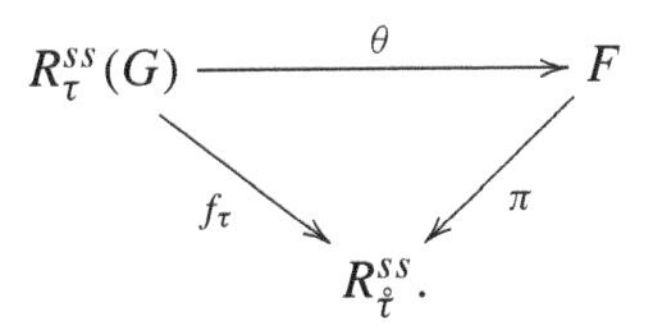

To prove the existence of θ as above, define the natural transformation from the functor $\Gamma^\tau(i, \mathscr{U}_{\mathring{\tau}}^{ss})$ to the functor $\tilde{h}_{F/R_{\mathring{\tau}}^{ss}}$ (cf. Section 1.1) by taking (for any morphism $f : T \to R_{\mathring{\tau}}^{ss}$) and any A-equivariant G-bundle $\mathscr{E} \to \hat{\Sigma} \times T$ obtained as an A-equivariant G-subbundle of $\mathrm{Fr}(\mathscr{U}_{\mathring{\tau}}^{ss})_f$ to the morphism $T \to F, t \mapsto \mathscr{E}_{|p\times t}$, where $\mathscr{E}_{|p\times t}$ can be viewed as an element of the fiber of $\mathrm{Fr}(\mathscr{U}_{\mathring{\tau}}^{ss})/G$ over $p \times f(t)$. Taking $f = f_\tau : R_\tau^{ss}(G) \to R_{\mathring{\tau}}^{ss}$ and the G-bundle $\mathscr{U}_\tau^{ss}(G)$, we get the morphism θ. Clearly, the map θ takes any A-equivariant section σ of $\hat{q}/G$ over $\hat{\Sigma}$ of topological type τ (for any $q \in R_{\mathring{\tau}}^{ss}$) to $\sigma(p)$, where $\hat{q}$ is the frame bundle of $\bar{q}$. We next claim that θ is injective restricted to closed points.

To prove this, it suffices to show that the evaluation map

$$e_p : \Gamma\left(\hat{\Sigma}, \hat{q}/G\right) \to \hat{q}_p/G, \quad \sigma \mapsto \sigma(p),$$

is injective for any A-semistable GL_r-bundle $\hat{q}$, where $\Gamma\left(\hat{\Sigma}, \hat{q}/G\right)$ denotes the space of sections of $\hat{q}/G$ over $\hat{\Sigma}$. Take a GL_r-module W with a vector $w_o \in W$ such that the isotropy of w_o in GL_r is exactly equal to G and $\mathrm{GL}_r \cdot w_o$ is closed in W. Such a choice is possible since GL_r/G is affine (cf. (Popov and Vinberg, 1994, §1)). Thus,

$$\hat{q}/G \hookrightarrow \hat{q}(W) := \hat{q} \times^{\mathrm{GL}_r} W, \quad eG \mapsto [e, w_o], \text{ for } e \in \hat{q},$$

and hence

$$\Gamma\left(\hat{\Sigma}, \hat{q}/G\right) \hookrightarrow \Gamma\left(\hat{\Sigma}, \hat{q}(W)\right). \tag{1}$$

Since $\hat{q}$ is a GL_r-semistable bundle, so is the G-bundle obtained from any section of $\hat{q}/G$ (cf. Corollary 6.2.7). Thus, G being embedded in SL_r, if $\Gamma\left(\hat{\Sigma}, \hat{q}/G\right)$ is nonempty, the vector bundle $\hat{q}(W)$ is semistable of degree 0 (cf. Corollary 6.2.7 and Theorem 6.1.7). Further, by Exercise 6.1.E.10, any nonzero section of a degree 0 semistable vector bundle over $\hat{\Sigma}$ is nowhere zero. From this the injectivity of e_p follows by using (1).

Since $\mathrm{Im}(f_\tau) \subset R^{ss}_{\mathring{\tau}}(\mathscr{O})$, we have $\mathrm{Im}(\theta) \subset F(\mathscr{O}) := \pi^{-1}(R^{ss}_{\mathring{\tau}}(\mathscr{O}))$. We next prove that θ considered as a map $R^{ss}_\tau(G) \to F(\mathscr{O})$ is a proper morphism by using the 'valuative criterion of properness' (cf. (Grothendieck, 1961a, Théorème 7.3.8 and Remarques 7.3.9(i))).

Let $\mathbb{D} = \mathrm{Spec}(\mathbb{C}[[t]])$ be the formal disc and $\mathbb{D}^* = \mathrm{Spec}(\mathbb{C}((t)))$ the formal punctured disc with the trivial action of A. Thus, $\mathbb{D}^*$ is canonically a subscheme of $\mathbb{D}$. We need to show that for any commutative diagram:

$$\begin{array}{ccc} \mathbb{D}^* & \xrightarrow{\hat{\beta}} & R^{ss}_\tau(G) \\ {\scriptstyle j}\downarrow & & \downarrow{\scriptstyle \theta} \\ \mathbb{D} & \xrightarrow{\beta} & F(\mathscr{O}), \end{array}$$

there exists $\gamma\colon \mathbb{D} \to R^{ss}_\tau(G)$ such that

$$\gamma \circ j = \hat{\beta}. \tag{2}$$

By Corollary 7.2.4, the morphism $\hat{\beta}$ corresponds to an A-equivariant section $\sigma_{\hat{\beta}}$ of $\left(\mathrm{Id}_{\hat{\Sigma}} \times (\pi \circ \beta)\right)^* \left(\mathrm{Fr}(\mathscr{U}^{ss}_{\mathring{\tau}})/G\right)$ restricted to $\hat{\Sigma} \times \mathbb{D}^*$. Embed $\mathrm{Fr}(\mathscr{U}^{ss}_{\mathring{\tau}})/G \hookrightarrow (\mathrm{Fr}\,\mathscr{U}^{ss}_{\mathring{\tau}})(W)$, giving rise to the embedding

$$\left(\mathrm{Id}_{\hat{\Sigma}} \times (\pi \circ \beta)\right)^* \left(\mathrm{Fr}(\mathscr{U}^{ss}_{\mathring{\tau}})/G\right) \hookrightarrow \mathscr{V} := \left(\mathrm{Id}_{\hat{\Sigma}} \times (\pi \circ \beta)\right)^* \left(\left(\mathrm{Fr}\,\mathscr{U}^{ss}_{\mathring{\tau}}\right)(W)\right).$$

Extend $\sigma_{\hat{\beta}}$ to a rational section $\bar{\sigma}_{\hat{\beta}}$ of $\mathscr{V}$ over $\hat{\Sigma} \times \mathbb{D}$ with possibly a pole of order $d \geq 1$ along $\hat{\Sigma} \times 0$, where 0 is the unique closed point of $\mathbb{D}$. Thus, $t^d \bar{\sigma}_{\hat{\beta}}$ is a regular section of $\mathscr{V}$ over $\hat{\Sigma} \times \mathbb{D}$ which does not vanish identically over $\hat{\Sigma} \times 0$. But, $\mathscr{V}_{|\hat{\Sigma} \times 0}$ being a semistable vector bundle (cf. Theorem 6.1.7) of degree 0 (by the definition of $R^{ss}_{\mathring{\tau}}(\mathscr{O})$, $\mathscr{V}_{|\hat{\Sigma} \times 0}$ is a vector bundle with $\wedge^r(\mathscr{V}_{|\hat{\Sigma} \times 0}) = \mathscr{O}_{\hat{\Sigma}}$), the section $t^d \bar{\sigma}_{\hat{\beta}}$ does not vanish anywhere over $\hat{\Sigma} \times 0$ (cf. Exercise 6.1.E.10). However, since the morphism $\theta \circ \hat{\beta}$ extends to $\mathbb{D}$, $\bar{\sigma}_{\hat{\beta}|p \times \mathbb{D}}$ is regular. In particular,

$$(t^d \bar{\sigma}_{\hat{\beta}})(p,0) = t^d(0) \cdot \bar{\sigma}_{\hat{\beta}}(p,0) = 0, \quad \text{if } d \geq 1,$$

which is a contradiction. Thus, the section $\sigma_{\hat{\beta}}$ extends to a regular section $\bar{\sigma}_{\hat{\beta}}$ of $\mathscr{V}$ over $\hat{\Sigma} \times \mathbb{D}$. Then, GL_r/G being closed in W and $\hat{\Sigma} \times \mathbb{D}^*$ being dense in $\hat{\Sigma} \times \mathbb{D}$, the section $\bar{\sigma}_{\hat{\beta}}$ is an A-equivariant section of $\left(\mathrm{Id}_{\hat{\Sigma}} \times (\pi \circ \beta)\right)^* \left(\mathrm{Fr}(\mathscr{U}^{ss}_{\mathring{\tau}})/G\right)$ over $\hat{\Sigma} \times \mathbb{D}$. This gives the desired extension γ of $\hat{\beta}$ to $\mathbb{D}$ satisfying (2). Hence, $\theta\colon R^{ss}_\tau(G) \to F(\mathscr{O})$ is a proper morphism. Moreover, as proved earlier, θ is injective on closed points. Hence, θ is a finite (in particular, affine) morphism by a result of Chevalley (cf. (Grothendieck, 1961b, Corollaire 4.4.11)). Further, π is an affine morphism (as observed earlier in the proof), so is the composite $f_\tau = \pi \circ \theta$. This proves the proposition. □

We follow the notation from Definition 7.2.1 except that we now take G to be a connected semisimple group.

Fix an embedding $i\colon G \hookrightarrow \mathrm{SL}_r \subset \mathrm{GL}_r$ and take an A-stable set $\{y_1, \ldots, y_b\} \subset \hat{\Sigma}$ and let $\vec{y} = \vec{y}(d') = d' \sum_{i=1}^{b} y_i$ be the divisor with the corresponding line bundle $\mathcal{O}_{\hat{\Sigma}}(\vec{y})$ where we choose d' so that $d := d'b \geq 2\hat{g}$. Now, take $N = r(d + 1 - \hat{g})$. For any A-semistable G-bundle F over $\hat{\Sigma}$, the associated A-equivariant rank-r vector bundle $F(i)$ is A-semistable (cf. Theorem 6.1.7 and Exercise 6.2.E.4) and it is of degree 0 (since G is semisimple). In particular, for $\mathscr{V} = F(i)$, $\mathscr{V}(\vec{y}) := \mathscr{V} \otimes \mathcal{O}_{\hat{\Sigma}}(\vec{y})$ satisfies conditions (a) and (b) of Definition 7.2.1. Hence, the A-module $H^0(\hat{\Sigma}, F(i)(\vec{y}))$ does not depend upon the choice of A-semistable F over $\hat{\Sigma}$ of a fixed topological type τ (cf. the Atiyah–Bott residue formula (Berline, Getzler and Vergne, 2004, Corollary 6.8)). Let $\mathring{\tau}$ be the representation of A on any such $H^0(\hat{\Sigma}, F(i)(\vec{y}))$.

Theorem 7.2.6 *Let $\hat{\Sigma}$ be a smooth irreducible projective curve with a faithful action of a finite group A and G a connected semisimple group. Fix the topological type τ of A-equivariant G-bundles over $\hat{\Sigma}$. Then, the A-semistable G-bundles over $\hat{\Sigma}$ of topological type τ admit a coarse moduli space $M_\tau^G(\hat{\Sigma})$ (cf. Definition 7.1.2), which is the good quotient $R_\tau^{ss}(G)//\mathfrak{G}$ (under a connected reductive group $\mathfrak{G}$). Further, $M_\tau^G(\hat{\Sigma})$ is a disjoint union of its irreducible components which are all normal with rational singularity (in particular, Cohen–Macaulay).*

It is an irreducible variety by (subsequent) Theorem 7.2.8. Moreover, by (subsequent) Corollary 7.2.15 and Lemma 7.2.13, $M_\tau^G(\hat{\Sigma})$ is, in fact, an (irreducible) projective (nonempty) variety if G is simply connected and the genus g of Σ is at least 2, where $\Sigma := \hat{\Sigma}/A$.

By the uniqueness of the coarse moduli space (cf. Definition 7.1.2), $M_\tau^G(\hat{\Sigma})$ does not depend upon the choice of the embedding $G \hookrightarrow \mathrm{SL}_r$.

If we take A to be trivial so that $\hat{\Sigma} = \Sigma$, then $M_\tau^G(\Sigma)$ in this case is nothing but the (non-parabolic) moduli space of semistable G-bundles over Σ of topological type τ.

Proof In Definition 7.2.1, choose $d > \max\{2\hat{g}-1, (r+1)\hat{g}-2, r^2\hat{g}+2\hat{g}-2\}$ (cf. Theorem 7.1.13). Choose a set of distinct points $\vec{x} = (x_1, \ldots, x_p)$ in $\hat{\Sigma}$ (for some $p = p(d)$) satisfying Theorem 7.1.13. We can assume further that the set $\{x_1, \ldots, x_p\}$ is A-stable. We can define an action of A on $\check{\mathrm{Gr}} = \check{\mathrm{Gr}}(r,N)$, so that the map $e_{\vec{x}} : R \to \check{\mathrm{Gr}}^p$ (as in Definition 7.1.12) is A-equivariant, where R is defined by (cf. Definition 7.1.7)

$$R = \Big\{ q \in \bar{Q} : \bar{q} \text{ is locally free } \mathscr{O}_{\hat{\Sigma}}\text{-module and}$$
$$\mathbb{C}^N = H^0\left(\hat{\Sigma}, E(\vec{y})\right) \xrightarrow{\sim} H^0\left(\hat{\Sigma}, \bar{q}(\vec{y})\right) \Big\}.$$

Now, by the definition,

$$R_{\mathring{\tau}} = R^A, \quad \text{and, by Exercise 6.2.E.4,} \tag{1}$$
$$R^{ss}_{\mathring{\tau}} = (R^{ss})^A, \tag{2}$$

where, as in Definition 7.2.1,

$$R^{ss} := \{q \in R : \bar{q} \text{ is a semistable vector bundle}\}.$$

By the proof of Theorem 7.1.14, there exists a good quotient $R^{ss} \to R^{ss}//\mathrm{SL}_N = R^{ss}//\mathrm{GL}_N$ (since the center of GL_N acts trivially on R as in Proposition 7.1.9(D)). Hence, $\mathfrak{G} := \mathrm{GL}_N^A$ being a reductive subgroup of GL_N, there exists a good quotient $R^{ss} \to R^{ss}//\mathfrak{G}$ (cf. (Bialynicki-Birula and Swiecicka, 1991, Proposition 2.1)). Thus, by Lemma 7.1.5, there exists a good quotient $R^{ss}_{\mathring{\tau}} \to R^{ss}_{\mathring{\tau}}//\mathfrak{G}$. *For later use, observe that* $\mathfrak{G}$*, being a Levi subgroup of* GL_N*, is connected.*

As in Corollary 7.2.4, the action of $\mathfrak{G}$ on $R^{ss}_{\mathring{\tau}}$ and $\mathscr{U}^{ss}_{\mathring{\tau}}$ lifts to its action on $R^{ss}_{\tau}(G)$ and the universal A-equivariant G-bundle $\mathscr{U}^{ss}_{\tau}(G)$ parameterized by $R^{ss}_{\tau}(G)$ via f_τ.

By Proposition 7.2.5, the map $f_\tau : R^{ss}_{\tau}(G) \to R^{ss}_{\mathring{\tau}}$ is an affine morphism. Hence, by Lemma 7.1.5, $R^{ss}_{\tau}(G)$ admits the good quotient $R^{ss}_{\tau}(G)//\mathfrak{G}$. (Even though Lemma 7.1.5 is stated for varieties X', X, the same proof works for schemes X', X of finite type over $\mathbb{C}$.) We claim that

$$M^G_\tau(\hat{\Sigma}) := R^{ss}_{\tau}(G)//\mathfrak{G} \tag{3}$$

is the coarse moduli space for the A-semistable G-bundles over $\hat{\Sigma}$ of topological type τ.

Observe that for $(V_1, s_1), (V_2, s_2) \in R^{ss}_{\tau}(G)$,

$$(V_1, s_1) \simeq (V_2, s_2) \text{ as } A\text{-equivariant } G\text{-bundles if and only if}$$
$$(V_2, s_2) \in \mathfrak{G} \cdot (V_1, s_1), \tag{4}$$

where V_i are A-equivariant GL_r-bundles and s_i are A-equivariant sections of V_i/G. To prove this, we first observe that if (V_i, s_i) lie in the same $\mathfrak{G}$-orbit, then $V_1 \simeq V_2$ as A-equivariant GL_r-bundles. Let $\mathrm{Aut}^A_{\mathrm{GL}_r}(V_1)$ be the group of A-equivariant automorphisms of the GL_r-bundle V_1 inducing the identity map at the base. Identifying V_2 with V_1, we further see that s_1 and s_2 lie in the same orbit of $\mathrm{Aut}^A_{\mathrm{GL}_r}(V_1)$ acting canonically on $\Gamma(\hat{\Sigma}, V_1/G)^A$. Now, it is easy to check that the $\mathrm{Aut}^A_{\mathrm{GL}_r}(V_1)$-orbits in $\Gamma(\hat{\Sigma}, V_1/G)^A$ correspond precisely to the A-equivariant G-reductions of V_1 which are isomorphic as A-equivariant G-bundles. Thus, s_1 and s_2 give isomorphic A-equivariant G-bundles. Conversely, it is easy to see that if $(V_1, s_1), (V_2, s_2) \in R^{ss}_\tau(G)$ are such that $V_1 \simeq V_2$ as A-equivariant GL_r-bundles and A-equivariant reductions s_1, s_2 give isomorphic A-equivariant G-bundles, then (V_i, s_i) lie in the same $\mathfrak{G}$-orbit.

From the above, it is easy to see that for any $x = (V, s) \in R^{ss}_\tau(G)$, the isotropy under the action of $\mathfrak{G}$ is given by

$$\mathfrak{G}_x = \mathrm{Aut}^A(\bar{x}), \tag{5}$$

where $\bar{x}$ is the A-equivariant G-subbundle of V obtained from the section s and $\mathrm{Aut}^A(\bar{x})$ denotes the group of automorphisms of the A-equivariant G-bundle $\bar{x}$.

Take a family $\mathscr{E} \to \hat{\Sigma} \times S$ of A-semistable G-bundles over $\hat{\Sigma}$ of topological type τ (parameterized by a noetherian scheme S). The family $\mathscr{E}$ gives rise to the family $\mathscr{E}(i)$ of vector bundles with trivial determinant obtained from the G-module $\mathbb{C}^r$ through the embedding $i: G \hookrightarrow \mathrm{SL}_r$. As observed above Theorem 7.2.6, the family $\mathscr{E}(i)$ consists of A-equivariant vector bundles $E(i)$ over $\hat{\Sigma}$ with a fixed representation $\mathring{\tau}$ of A on $H^0(\hat{\Sigma}, E(i)(\vec{y}))$ (depending only upon τ). Moreover, by Theorem 6.1.7 and Exercise 6.2.E.4, $E(i)$ is A-semistable vector bundle.

From the local universal property of the family $\mathscr{U}$ over $\hat{\Sigma} \times R$ (actually, the A-equivariant analogue of Proposition 7.1.9(B), which is true by the same proof) there exists an open cover $\{S_i\}$ of S and a morphism $f_i: S_i \to R^A$ (such that as A-equivariant vector bundles)

$$(\mathrm{Id}_{\hat{\Sigma}} \times f_i)^* \left(\mathscr{U}_{|\hat{\Sigma} \times R^A}\right) \simeq \mathscr{E}(i)_{|\hat{\Sigma} \times S_i}. \tag{6}$$

But, since $\mathscr{E}$ is a family of A-semistable G-bundles over $\hat{\Sigma}$,

$$\mathrm{Im}\, f_i \subset R^{ss}_{\mathring{\tau}}.$$

From the representability of the functor $\Gamma^\tau(i, \mathscr{U}^{ss}_{\mathring{\tau}})$ (cf. Corollary 7.2.4), the family $\mathscr{E}$ gives rise to a lift of the morphism $f_i: S_i \to R^{ss}_{\mathring{\tau}}$ to $\bar{f}_i: S_i \to R^{ss}_\tau(G)$ through the map $f_\tau: R^{ss}_\tau(G) \to R^{ss}_{\mathring{\tau}}$ (thus the family $\mathscr{U}^{ss}_\tau(G)$ satisfies the local universal property for any family of A-semistable G-bundles over $\hat{\Sigma}$ of local type τ) and hence a morphism

$$\hat{f}_i: S_i \to M^G_\tau(\hat{\Sigma}) := R^{ss}_\tau(G)//\mathfrak{G}.$$

We claim that over the intersection $S_i \cap S_j$:

$$\hat{f}_{i|_{S_i \cap S_j}} = \hat{f}_{j|_{S_i \cap S_j}}. \tag{7}$$

Take $x \in S_i \cap S_j$ and let $\mathscr{E}_x := \mathscr{E}_{|_{\hat{\Sigma} \times x}}$, $\mathscr{E}_x(i) := \mathscr{E}(i)_{|_{\hat{\Sigma} \times x}}$. By (4), we get

$$\bar{f}_i(x) \in \mathfrak{G} \cdot \bar{f}_j(x), \tag{8}$$

since $\bar{f}_i(x)$ and $\bar{f}_j(x)$ represent isomorphic A-equivariant G-bundles. From this the identity (7) follows. Thus, we get a morphism

$$\hat{f} = \hat{f}_{\mathscr{E}} \colon S \to M_\tau^G(\hat{\Sigma})$$

such that

$$\hat{f}_{|S_i} = \hat{f}_i.$$

This provides the necessary natural transformation

$$\Phi \colon \mathscr{F} \to \operatorname{Mor}\left(-, M_\tau^G(\hat{\Sigma})\right), \mathscr{E} \mapsto \hat{f}_{\mathscr{E}},$$

as in Definition 7.1.2 of coarse moduli space, where $\mathscr{F}$ is the functor of A-semistable G-bundles over $\hat{\Sigma}$ of topological type τ. If N is another scheme with natural transformation $\Phi' \colon \mathscr{F} \to \operatorname{Mor}(-, N)$, then take $\bar{f}_{\Phi'} := \Phi'\left(\mathscr{U}_\tau^{ss}(G)\right) : R_\tau^{ss}(G) \to N$, where the universal A-equivariant G-bundle $\mathscr{U}_\tau^{ss}(G)$ is as in Corollary 7.2.4. Moreover, by (4), $\bar{f}_{\Phi'}$ is $\mathfrak{G}$-invariant (under the trivial action of $\mathfrak{G}$ on N). Thus, by the categorical quotient property (cf. Definition 7.1.4), $\bar{f}_{\Phi'}$ descends to give a morphism $f_{\Phi'} \colon M_\tau^G(\hat{\Sigma}) \to N$. This gives the uniqueness part of the definition of a coarse moduli space. This proves that $M_\tau^G(\hat{\Sigma})$ is a coarse moduli space.

We now prove that $R_\tau^{ss}(G)$ is smooth. In fact, we prove a stronger result.

We freely follow the notation from Definition 7.2.1. Consider the A-equivariant vector bundle $E := \left(\mathscr{O}_{\hat{\Sigma}} \otimes \mathbb{C}^N\right) \otimes \mathscr{O}_{\hat{\Sigma}}(-\vec{y})$ over $\hat{\Sigma}$ and let $\check{\mathrm{G}}\mathrm{r} := \check{\mathrm{G}}\mathrm{r}(r, E)$ be the Grassmannian bundle over $\hat{\Sigma}$ consisting of r-dimensional quotients of E_x over $x \in \hat{\Sigma}$ (cf. (Nitsure, 2005b, §5.15)). Let $\mathcal{T} \to \check{\mathrm{G}}\mathrm{r}$ be the tautological rank-r vector bundle and let $F(\mathcal{T}) \to \check{\mathrm{G}}\mathrm{r}$ be its frame bundle. Let $R'_{\mathring{\tau}}$ be the subset of $\bar{Q}^A$ defined by

$$R'_{\mathring{\tau}} := \left\{q \in \bar{Q}^A : \bar{q} := \mathscr{U}_{|_{\hat{\Sigma} \times q}} \text{ is locally free}\right\}.$$

Then, $R'_{\mathring{\tau}}$ is an open subset of $\bar{Q}^A$ and $\mathscr{U}_{|_{\hat{\Sigma} \times R'_{\mathring{\tau}}}}$ is an A-equivariant rank-r vector bundle (cf. the proof of Proposition 7.1.9(A)). Moreover, we get a bijection

$$R'_{\mathring{\tau}} \simeq \Gamma_A(\hat{\Sigma}, \check{\mathrm{G}}\mathrm{r}), \;\; q \mapsto \bar{q},$$

where $\Gamma_A(\hat{\Sigma}, \check{\mathrm{Gr}})$ denotes the set of A-equivariant regular sections of bundle $\check{\mathrm{Gr}}$ over $\hat{\Sigma}$. Recall the definition of the subset $R_{\mathring{t}}$ of $R'_{\mathring{t}}$ from Definition 7.2.1. By the Semicontinuity Theorem (Hartshorne, 1977, Chap. III, Theorem 12.8) applied to the kernel of $E(\vec{y}) \to \bar{q}(\vec{y})$, $R_{\mathring{t}}$ is an open subset of $R'_{\mathring{t}}$. Let $R'(G)$ be the scheme representing the functor $\Gamma\left(i, \mathscr{U}_{|\hat{\Sigma}\times R'_{\mathring{t}}}\right)$ (cf. Proposition 7.2.3 and Corollary 7.2.4). Then, we get the bijection

$$R'(G) \simeq \Gamma_A(\hat{\Sigma}, F(\mathfrak{T})/G).$$

Replacing $R'_{\mathring{t}}$ by $R_{\mathring{t}}$ in the definition of $R'(G)$, we get an open subscheme $R(G)$ of $R'(G)$. By the A-equivariant analogue of Nitsure (2005b, Theorem 5.23), $R'(G) \simeq \Gamma_A(\hat{\Sigma}, F(\mathfrak{T})/G)$ is an open subscheme of the A-equivariant Hilbert scheme $\mathrm{Hilb}^A_{(F(\mathfrak{T})/G)/\hat{\Sigma}}$ and hence so is $R(G)$ an open subscheme of the A-equivariant Hilbert scheme $\mathrm{Hilb}^A_{(F(\mathfrak{T})/G)/\hat{\Sigma}}$. By the A-equivariant analogue of Illusie (2005, Theorem 8.5.9),[2] an element $f \in \Gamma_A(\hat{\Sigma}, F(\mathfrak{T})/G)$ is a smooth point if

$$H^1\left(\hat{\Sigma}, f^*(T(F(\mathfrak{T})/G))\right)^A = 0, \tag{9}$$

where T denotes the tangent bundle. We now prove that for any $f \in R(G) \subset \Gamma_A(\hat{\Sigma}, F(\mathfrak{T})/G)$, the above condition (9) is satisfied.

Let $\alpha\colon F(\mathfrak{T})/G \to \check{\mathrm{Gr}}$ be the projection. Then it is easy to see that we have the following two exact sequences:

$$0 \to \mathrm{ad}(\bar{q}_G) \to f^*(T(F(\mathfrak{T}))) \to f^*(T(F(\mathfrak{T})/G)) \to 0, \tag{10}$$

and

$$0 \to \bar{q}^* \otimes \bar{q} \to f^*(T(F(\mathfrak{T}))) \to K^*_{\bar{q}} \otimes \bar{q} \to 0, \tag{11}$$

where $\bar{q}_G$ is the G-bundle over $\hat{\Sigma}$ determined by f, $\bar{q}$ is the corresponding rank-r vector bundle $\bar{q}_G \times^G \mathbb{C}^r$ induced from the embedding $G \hookrightarrow \mathrm{SL}_r \subset \mathrm{GL}_r$ and $K_{\bar{q}}$ is the kernel of $E \to \bar{q}$. Further, we have the identification

$$f^*(T(F(\mathfrak{T}))) \simeq E^* \otimes \bar{q} \tag{12}$$

arising from the exact sequence

$$0 \to \bar{q}^* \to E^* \to K^*_{\bar{q}} \to 0$$

[2] I thank N. Nitsure for the reference.

by tensoring with $\bar{q}$. So

$$H^1\left(\hat{\Sigma}, f^*(T(F(\mathfrak{T})))\right) = H^1(\hat{\Sigma}, E^* \otimes \bar{q}) \\ = \mathbb{C}^N \otimes H^1(\hat{\Sigma}, \bar{q}(\vec{y})) \\ = 0, \tag{13}$$

since $\chi(\hat{\Sigma}, \bar{q}(\vec{y})) = N = \dim H^0(\hat{\Sigma}, \bar{q}(\vec{y}))$ by the definition of $R(G)$. Thus, from the long exact cohomology sequence associated to the sequence (10), using the identity (13), we get for any $f \in R(G)$:

$$H^1\left(\hat{\Sigma}, f^*(T(F(\mathfrak{T})/G))\right) = 0; \text{ in particular, } H^1\left(\hat{\Sigma}, f^*(T(F(\mathfrak{T})/G))\right)^A = 0.$$

This proves that $R(G)$ is smooth and hence so is its open subset $R^{ss}(G)$ consisting of semistable G-bundles. But, $R^{ss}_\tau(G)$ is the union of certain irreducible components of $R^{ss}(G)$ (in fact, it is an irreducible component of $R^{ss}(G)$ by Theorem 7.2.8) and hence, in particular, $R^{ss}_\tau(G)$ is smooth.

We now calculate the dimension $\dim_f(R(G))$ of $R(G)$ around any $f \in R^{ss}_\tau(G)$.

By the general result on dimension of Hilbert schemes, we get

$$\begin{aligned}\dim_f(R(G)) &= \dim H^0\left(\hat{\Sigma}, f^*(T(F(\mathfrak{T})/G))\right)^A \\ &= -\dim H^0\left(\hat{\Sigma}, \operatorname{ad}(\bar{q}_G)\right)^A + \dim H^0\left(\hat{\Sigma}, f^*(T(F(\mathfrak{T})))\right)^A \\ &\quad + \dim H^1\left(\hat{\Sigma}, \operatorname{ad}(\bar{q}_G)\right)^A, \text{ by (9) and (13)} \\ &\quad \text{using the cohomology exact sequence associated to (10)} \\ &= \dim H^1\left(\hat{\Sigma}, \operatorname{ad}(\bar{q}_G)\right)^A - \dim H^0\left(\hat{\Sigma}, \operatorname{ad}(\bar{q}_G)\right)^A + \dim \mathfrak{G},\end{aligned} \tag{14}$$

since

$$H^0\left(\hat{\Sigma}, f^*(T(F(\mathfrak{T})))\right)^A \simeq \operatorname{Lie}(\mathrm{GL}_N^A = \mathfrak{G}),$$

by (12) and the definition of $R_{\hat{\tau}}$.

Since $\mathfrak{G}$ is connected, by the defining property of good quotients (cf. Definition 7.1.4), $M^G_\tau(\hat{\Sigma})$ is a disjoint union of its irreducible components. Thus, $M^G_\tau(\hat{\Sigma})$ is normal. Moreover, by Boutot's Theorem (Boutot, 1987), $M^G_\tau(\hat{\Sigma})$ has rational singularity. This proves the theorem. □

For its subsequent application, we isolate the following result proved in the proof of Theorem 7.2.6.

Corollary 7.2.7 *Let $\hat{\Sigma}$, A, τ and G be as in Theorem 7.2.6. Let $\mathscr{E} \to \hat{\Sigma} \times S$ be a family of A-semistable G-bundles over $\hat{\Sigma}$ of topological type τ (parameterized by a noetherian scheme S). Then there is a unique morphism*

$$\hat{f}_{\mathscr{E}} : S \to M_\tau^G(\hat{\Sigma})$$

such that

$$\hat{f}_{\mathscr{E}}(t) = \left[\mathscr{E}_{|_{\hat{\Sigma} \times t}}\right], \quad \textit{for any closed point } t \in S,$$

where $\left[\mathscr{E}_{|_{\hat{\Sigma}\times t}}\right] \in M_\tau^G(\hat{\Sigma})$ denotes the image of (any) $q \in R_\tau^{ss}(G)$ with $\bar{q} \simeq \mathscr{E}_{|_{\hat{\Sigma}\times t}}$ under the canonical map $R_\tau^{ss}(G) \to R_\tau^{ss}(G)//\mathfrak{G} =: M_\tau^G(\hat{\Sigma})$.

It is easy to see that the projection map $R_\tau^{ss}(G) \to M_\tau^G(\hat{\Sigma})$ corresponds to the family $\mathscr{U}_\tau^{ss}(G)$ over $\hat{\Sigma} \times R_\tau^{ss}(G)$.

We remark that the $\mathbb{C}$-analytic analogue of the corollary remains true.

With the notation and assumptions as in Theorem 7.2.6, we have the following:

Theorem 7.2.8 *The coarse moduli space $M_\tau^G(\hat{\Sigma})$ is an irreducible variety. Hence, so is $R_\tau^{ss}(G)$.*

Proof We first prove that $M_\tau^G(\hat{\Sigma})$ is irreducible.

Take two A-semistable G-bundles E_0 and E_1 over $\hat{\Sigma}$ of the same topological type τ. In particular, they have the same underlying A-equivariant C^∞-bundle structure E (of topological type τ). As in the proof of Lemma 6.3.32, choose connections ∇_0 and ∇_1 such that the corresponding holomorphic structures on E coincide with the holomorphic structures E_0 and E_1, respectively. By replacing ∇_i by $\frac{1}{|A|}\sum_{g\in A} g^*\nabla_i$, we can (and will) assume that ∇_0 and ∇_1 are both A-equivariant connections. As in the proof of Lemma 6.3.32, take the connection ∇ on the C^∞-bundle $E \times \mathbb{C} \to \hat{\Sigma} \times \mathbb{C}$ defined by

$$\nabla_z = z\nabla_1 + (1-z)\nabla_0, \quad \text{for} \quad z \in \mathbb{C},$$

which has $\Omega^{0,2} = 0$ proved there. This gives rise to an A-equivariant holomorphic G-bundle structure on $\mathscr{E} := E \times \mathbb{C}$, such that the A-equivariant holomorphic structure restricted to $E \times 0$ (resp. $E \times 1$) is isomorphic with E_0 (resp. E_1). Now, by Exercises 6.3.E.9 and 6.2.E.4, there is an open subset $V \subset \mathbb{C}$ containing $\{0,1\}$ such that $\mathscr{E}_t := \mathscr{E}_{|_{\hat{\Sigma}\times t}}$ is A-semistable G-bundle for all $t \in V$ and V is the complement of a closed analytic subset of $\mathbb{C}$. By the $\mathbb{C}$-analytic analogue of Corollary 7.2.7, we get a $\mathbb{C}$-analytic map

$$\hat{f}_{\mathscr{E}} : V \to M_\tau^G(\hat{\Sigma})$$

with its image containing the classes $[E_0]$ and $[E_1]$. Thus, V being connected, $M_\tau^G(\hat{\Sigma})$ is connected. But, by Theorem 7.2.6, $M_\tau^G(\hat{\Sigma})$ is a disjoint union of its irreducible components. Hence, $M_\tau^G(\hat{\Sigma})$ is irreducible.

As shown in the proof of Theorem 7.2.6, $R_\tau^{ss}(G)$ is a smooth scheme and $\mathfrak{G}$ is connected. Thus, $M_\tau^G(\hat{\Sigma}) := R_\tau^{ss}(G)//\mathfrak{G}$ being irreducible forces $R_\tau^{ss}(G)$ to be irreducible. □

We next prove that $M_\tau^G(\hat{\Sigma})$ is a complete variety. But before we come to its proof we need some preliminary work.

We follow the notation and assumptions as in Definition 6.3.3, but we assume G to be a connected semisimple complex algebraic group. Let $K \subset G$ be a maximal compact subgroup and let $\alpha\colon \tilde{K} \to K$ be the simply-connected cover of K. Since α is surjective, any homomorphism $\hat{\rho}\colon \pi \to K$ lifts to a homomorphism

$$\tilde{\hat{\rho}} : F(a_1, b_1, a_2, b_2, \dots, a_g, b_g, c_1, \dots, c_s) \to \tilde{K}, \text{ i.e., } \alpha \circ \tilde{\hat{\rho}} = \hat{\rho}.$$

Then $\tilde{\hat{\rho}}(\mu) \in \operatorname{Ker}\alpha$, where (as earlier in Definition 6.3.3)

$$\mu := \left(\Pi_{i=1}^g [a_i, b_i]\right) \cdot \left(\Pi_{j=1}^s c_j\right).$$

Lemma 7.2.9 *Assume that $\Sigma := \hat{\Sigma}/A$ has genus $g \geq 2$. For any $c \in \operatorname{Ker}\alpha$, and any elements $x_1, \dots, x_s \in \tilde{K}$, there exists a group homomorphism*

$$\tilde{\hat{\rho}} : F\left(a_1, b_1, a_2, b_2, \dots, a_g, b_g, c_1, \dots, c_s\right) \to \tilde{K}$$

such that $\tilde{\hat{\rho}}(\mu) = c$, $\tilde{\hat{\rho}}(c_j) = x_j$ $(1 \leq j \leq s)$, and $\operatorname{Im}(\tilde{\hat{\rho}})$ is a dense subgroup of $\tilde{K}$ (in the Hausdorff topology).

Proof Since any element of $\tilde{K}$ can be conjugated inside its maximal torus $\tilde{T}$ and any element in $\tilde{T}$ can be written as a commutator $[a,b]$ for a, $b \in \tilde{K}$ (cf. proof of Lemma 6.3.11), any element of $\tilde{K}$ is a commutator. Choose two elements $g, h \in \tilde{K}$ such that the subgroup $\langle g, h\rangle$ generated by them is dense in $\tilde{K}$ (which exists by Kuranishi's Theorem (Kuranishi, 1949)). Now, choose two elements g', $h' \in \tilde{K}$ such that

$$[g', h'] = c \cdot ([g,h](x_1, \dots, x_s))^{-1}.$$

Define

$$\tilde{\hat{\rho}}(a_1) = g', \tilde{\hat{\rho}}(b_1) = h', \tilde{\hat{\rho}}(a_2) = g, \tilde{\hat{\rho}}(b_2) = h,$$
$$\tilde{\hat{\rho}}(a_3) = \tilde{\hat{\rho}}(b_3) = \cdots = \tilde{\hat{\rho}}(a_g) = \tilde{\hat{\rho}}(b_g) = 1,$$
$$\tilde{\hat{\rho}}(c_j) = x_j, \quad \text{for} \quad 1 \leq j \leq s.$$

This uniquely extends to a group homomorphism satisfying the assumptions of the lemma. □

By the presentation of π as in Definition 6.3.3, the above homomorphism $\tilde{\hat{\rho}}$ (with dense image) descends to give rise to a homomorphism

$$\hat{\rho} := \alpha \circ \tilde{\hat{\rho}} : \pi \to K$$

if and only if

$$\alpha(x_j)^{d_j} = 1, \quad \text{for all} \quad 1 \leq j \leq s. \tag{*}$$

Since $\hat{\rho}$ has dense image; in particular, $\hat{\rho}$ is irreducible.

Definition 7.2.10 Let $j: G \to H$ be a homomorphism of connected semisimple groups. Then, by Corollary 7.2.7, Theorem 6.1.7 and Exercise 6.2.E.4, we get a morphism (for any topological type τ of A-equivariant G-bundles over $\hat{\Sigma}$)

$$R_\tau^{ss}(G) \to M_{\tau'}^H(\hat{\Sigma}),$$

where τ' is the topological type of A-equivariant H-bundles obtained by extension of the structure group via j. By the categorical quotient property of $R_\tau^{ss}(G) \to M_\tau^G(\hat{\Sigma})$ (cf. Definition 7.1.4) and identity (4) of the proof of Theorem 7.2.6, we get a morphism

$$\bar{\theta}: M_\tau^G(\hat{\Sigma}) \to M_{\tau'}^H(\hat{\Sigma}).$$

Moreover, if j is an embedding, there is a morphism $\theta: R_\tau^{ss}(G) \to R_{\tau'}^{ss}(H)$ making the following diagram commutative:

$$\begin{array}{ccc} R_\tau^{ss}(G) & \xrightarrow{\theta} & R_{\tau'}^{ss}(H) \\ {\scriptstyle q_1}\downarrow & & \downarrow{\scriptstyle q_2} \\ M_\tau^G(\hat{\Sigma}) & \xrightarrow[\bar{\theta}]{} & M_{\tau'}^H(\hat{\Sigma}), \end{array} \tag{1}$$

where the vertical maps q_i are the canonical (good) quotients. To prove this, take an embedding $i: H \hookrightarrow \mathrm{GL}_r$. Consider the universal A-equivariant G-bundle $\mathscr{U}_\tau^{ss}(G)$ over $\hat{\Sigma} \times R_\tau^{ss}(G)$. Then it is a G-subbundle of $(\mathrm{Id}_{\hat{\Sigma}} \times f_\tau)^*$ $(\mathscr{U}_{\hat{\tau}}^{ss})$ and hence $\mathscr{U}_\tau^{ss}(G) \times^G H$ is a H-subbundle of $(\mathrm{Id}_{\hat{\Sigma}} \times f_\tau)^*(\mathscr{U}_{\hat{\tau}}^{ss})$, where $f_\tau, f_{\tau'}$ are the morphisms as in Corollary 7.2.4. Thus, from the representability of $\Gamma^{\tau'}(i, \mathscr{U}_{\hat{\tau}}^{ss})$, we get the morphism θ:

$$R_\tau^{ss}(G) \xrightarrow{\theta} R_{\tau'}^{ss}(H) \xrightarrow{f_{\tau'}} R_{\hat{\tau}}^{ss}$$

such that $f_{\tau'} \circ \theta = f_\tau$. By the definition of the maps $\theta, \bar{\theta}$, the commutativity of the diagram (1) follows.

Lemma 7.2.11 *Let G be a connected reductive group and assume that $\Sigma := \hat{\Sigma}/A$ has genus ≥ 2. Let $\mathcal{E} \to \hat{\Sigma} \times S$ be a family of A-equivariant G-bundles of degree 0 over $\hat{\Sigma}$ of any topological type τ parameterized by a noetherian scheme S. Assume that there exists $t_o \in S$ such that*

$$\mathcal{E}_{t_o} \simeq \hat{E}_{\hat{\rho}}, \quad \textit{as } A\textit{-equivariant } G\textit{-bundles},$$

for a unitary representation $\hat{\rho}: \pi \to G$ with Zariski dense image, where $\mathcal{E}_{t_o} := \mathcal{E}_{|\hat{\Sigma}\times t_o}$. Then

$$S_s := \{t \in S : \mathcal{E}_t \textit{ is } A\textit{-stable}\} \textit{ contains a Zariski open subset } \mathring{S}_s \textit{ containing } t_o. \tag{1}$$

Moreover, by Theorem 6.3.41,

$$S_u := \{t \in S : \mathcal{E}_t \textit{ is } A\textit{-unitary}\} \textit{ contains } S_s. \tag{2}$$

Proof Choose a surjective homomorphism with finite kernel:

$$\varphi: G \twoheadrightarrow \Pi_{i=1}^d G_i, \quad \text{where } G_i \text{ are simple groups.}$$

This gives rise to surjective projections $\varphi_i : G \to G_i$. Let

$$S_s^i := \{t \in S : \mathcal{E}_t(G_i) \text{ is stable } G_i\text{-bundle}\}.$$

Then $t_o \in S_s^i$ by Proposition 6.3.4. We now prove that S_s^i contains a Zariski open subset $\mathring{S}_s^i$ containing t_o. The set

$$S_s^i(\mathrm{ad}) := \{t \in S : \mathrm{ad}\,(\mathcal{E}_t(G_i)) \text{ is stable vector bundle}\}$$

is an open subset of S by Corollary 7.1.16(a). By Lemma 6.1.5 (actually by its proof), $S_s^i(\mathrm{ad}) \subset S_s^i$. Moreover, since G_i are simple groups and $\hat{\rho}$ has Zariski dense image in G (and hence $\hat{\rho}_{|\pi_1(\hat{\Sigma})}$ also has Zariski dense image in G since $\pi_1(\hat{\Sigma})$ is of finite index in π), $t_o \in S_s^i(\mathrm{ad})$ by Proposition 6.3.4. Thus, abbreviating $\mathring{S}_s^i = S_s^i(\mathrm{ad})$, we get $t_o \in \mathring{S}_s^i \subset S_s^i$. We next have

$$t_o \in \cap_{i=1}^d \mathring{S}_s^i \subset \cap_{i=1}^d S_s^i \subset S_s,$$

where the last inclusion follows from the A-equivariant analogue of Exercise 6.1.E.8 (which is trivial to extend). Taking $\mathring{S}_s := \cap \mathring{S}_s^i$, we get (1). □

The following definition provides an extension of Definition 6.1.10 to any connected semisimple group G from that of simply-connected G.

Definition 7.2.12 Let $\hat{\Sigma}$ be an irreducible smooth projective curve with a faithful action of a finite group A and let $\hat{\Sigma} \to \Sigma := \hat{\Sigma}/A$ be the quotient. Let

$\vec{p} = (p_1, \ldots, p_s)$ be the ramification points of π in Σ and $\vec{d} = (d_1, \ldots, d_s)$ be the signature of Σ (cf. Theorem 6.1.8). *We assume that* $s \geq 3$ *if* $\Sigma = \mathbb{P}^1$. Let G be a connected semisimple group and let E be an A-equivariant G-bundle over $\hat{\Sigma}$. Choose a point $\hat{p}_j \in \hat{\Sigma}$ over p_j and generators $\vec{\gamma} = (\gamma_1, \ldots, \gamma_s)$ of the isotropy groups $(A_{\hat{p}_1}, \ldots, A_{\hat{p}_s})$. Then, $d_j = \sharp A_{\hat{p}_j}$. Let $\hat{\mathbb{D}}_j := \operatorname{Spec}(\mathbb{C}[[\hat{t}_j]])$ be the formal disc with respect to a local parameter $\hat{t}_j$ for $\hat{\Sigma}$ at $\hat{p}_j$. Then $E_{|\hat{\mathbb{D}}_j}$ is trivial as a G-bundle (e.g., by Theorem 5.2.5). Since $\hat{p}_j$ is fixed by $A_{\hat{p}_j}$, $E_{|\hat{\mathbb{D}}_j}$ is an $A_{\hat{p}_j}$-equivariant G-bundle. Thus, by Theorem 6.1.9, we get a homomorphism $\theta_j \colon A_{\hat{p}_j} \to G$ (unique up to conjugation). Let $\tau_j = \theta_j(\gamma_j)$ and let $[\tau_j]$ be its G-conjugacy class. Since $\gamma_j^{d_j} = 1$, τ_j can be chosen to lie in a fixed maximal compact subgroup K of G. We define the *local type of* E to be the conjugacy classes

$$[\vec{\tau}] := ([\tau_1], \ldots, [\tau_s]).$$

The following lemma is easy to prove.

Lemma 7.2.13 *Let G be a connected, simply-connected semisimple group (so that the simply-connected cover $\tilde{K} = K$) and assume that Σ has genus $g \geq 2$. For any topological type τ of A-equivariant G-bundles over $\hat{\Sigma}$, there exist $x_1, \ldots, x_s \in \tilde{K}$ satisfying the condition (*) just before Definition 7.2.10 such that any homomorphism $\hat{\rho} \colon \pi \to K$ with dense image and satisfying $\hat{\rho}(c_j) = x_j$, for all $1 \leq j \leq s$, (guaranteed to exist by Lemma 7.2.9) gives rise to the A-stable G-bundle $\hat{E}_{\hat{\rho}}$ over $\hat{\Sigma}$ (cf. Definition 6.3.3 and Proposition 6.3.4(b)) of topological type τ.*

In fact, $([x_1], \ldots, [x_s])$ is the local type of $\hat{E}_{\hat{\rho}}$ (cf. Exercise 7.2.E.4).

Theorem 7.2.14 *With the notation and assumptions as in Theorem 7.2.6 (in particular, G is connected, semisimple), assume further that $\Sigma := \hat{\Sigma}/A$ has genus $g \geq 2$ and G is simply connected. Then, the coarse moduli space $M_\tau^G(\hat{\Sigma})$ is a nonempty complete variety for any topological type τ.*

By Theorem 7.2.8, it is irreducible.

Proof By Lemma 7.2.13 there exists a representation $\hat{\rho}_o \colon \pi \to K$ with dense image (in particular, irreducible) such that the associated A-equivariant G-bundle $\hat{E}_{\hat{\rho}_o}$ is of topological type τ. Thus, by Proposition 6.3.4(b), $\hat{E}_{\hat{\rho}_o}$ is an A-stable G-bundle over $\hat{\Sigma}$. Let $\mathscr{U}_\tau^{ss}(G) \to \hat{\Sigma} \times R_\tau^{ss}(G)$ be the 'universal' family of A-semistable G-bundles over $\hat{\Sigma}$ (cf. Corollary 7.2.4). Observe that by Theorem 6.1.7 and Exercise 6.2.E.4, any A-semistable G-bundle over $\hat{\Sigma}$ of topological type τ appears in this family. By Lemma 7.2.11, the subset

$$R_\tau^{ss}(G)_u := \{q \in R_\tau^{ss}(G) : \bar{q} \text{ is } A\text{-unitary}\}$$

contains a nonempty (due to the existence of $\hat{E}_{\hat{\rho}_o}$ in $R^{ss}_\tau(G)$) Zariski open subset of $R^{ss}_\tau(G)$, where $\bar{q} := \mathscr{U}^{ss}_\tau(G)_{|_{\hat{\Sigma}\times q}}$.

Let $R^A_G(\hat{\Sigma})$ be the set of all the homomorphisms from π to G and let $R^A_K(\hat{\Sigma})$ be its subset consisting of those homomorphisms with image in K. Then, as in Lemma 6.3.2 (by the some proof) $R^A_G(\hat{\Sigma})$ acquires an affine variety structure and there exists a 'universal' A-equivariant $\mathbb{C}$-analytic G-bundle $\mathscr{E} \to \hat{\Sigma} \times R^A_G(\hat{\Sigma})$ such that for any $\hat{\rho} \in R^A_G(\hat{\Sigma})$, $\mathscr{E}_{|_{\hat{\Sigma}\times\hat{\rho}}} \simeq \hat{E}_{\hat{\rho}}$. Consider the restriction $\mathscr{E}_{|_{R^A_K(\hat{\Sigma})}}$ of the family $\mathscr{E}$ to $R^A_K(\hat{\Sigma})$. Then, $\mathscr{E}_{|_{R^A_K(\hat{\Sigma})}}$ is a family of holomorphic A-semistable G-bundles over $\hat{\Sigma}$ (use Proposition 6.3.4(a)) parameterized by the real analytic space $R^A_K(\hat{\Sigma})$. Let $R^{A,\tau}_K(\hat{\Sigma})$ (resp. $R^{A,\tau}_G(\hat{\Sigma})$) be the closed analytic subset (as evidenced by Lemma 7.2.13) consisting of those $\hat{\rho} \in R^A_K(\hat{\Sigma})$ (resp. $\hat{\rho} \in R^A_G(\hat{\Sigma})$) such that $\hat{E}_{\hat{\rho}}$ is of A-topological type τ. Then, there exists a $\mathbb{C}$-analytic neighborhood U of $R^{A,\tau}_K(\hat{\Sigma})$ in $R^{A,\tau}_G(\hat{\Sigma})$ such that $\mathscr{E}_{|U}$ consists of A-semistable G-bundles over $\hat{\Sigma}$ (use Exercises 6.2.E.4, 6.3.E.9 and Proposition 6.3.4(a)). Thus, by the $\mathbb{C}$-analytic analogue of Corollary 7.2.7, there exists a real analytic map

$$f_{\mathscr{E}} \colon R^{A,\tau}_K(\hat{\Sigma}) \to M^G_\tau(\hat{\Sigma}),$$

such that

$$f_{\mathscr{E}}(\hat{\rho}) = [\hat{E}_{\hat{\rho}}], \quad \text{for any } \hat{\rho} \in R^{A,\tau}_K(\hat{\Sigma}).$$

Consider the surjective morphism (cf. Theorem 7.2.6)

$$\beta \colon R^{ss}_\tau(G) \to M^G_\tau(\hat{\Sigma}) := R^{ss}_\tau(G)//\mathfrak{G}.$$

By Chevalley's theorem (cf. (Hartshorne, 1977, Chap. II, Exercise 3.19)) since $R^{ss}_\tau(G)$ is irreducible (cf. Theorem 7.2.8) and $R^{ss}_\tau(G)_u$ has a nonempty Zariski open subset, $\beta(R^{ss}_\tau(G)_u)$ contains a (nonempty) Zariski open subset V of $M^G_\tau(\hat{\Sigma})$. Moreover, for $q_1, q_2 \in R^{ss}_\tau(G)$ such that $\bar{q}_1 \simeq \bar{q}_2$ as A-equivariant G-bundles, we have $\beta(q_1) = \beta(q_2)$ (cf. identity (4) of the proof of Theorem 7.2.6). Thus,

$$V \subset \beta\left(R^{ss}_\tau(G)_u\right) \subset f_{\mathscr{E}}\left(R^{A,\tau}_K(\hat{\Sigma})\right) \subset M^G_\tau(\hat{\Sigma}).$$

From the above inclusions, we see that ($R^{A,\tau}_K(\hat{\Sigma})$ being compact) $f_{\mathscr{E}}$ is surjective and hence $M^G_\tau(\hat{\Sigma})$ is compact in the analytic topology. Thus, by (Mumford, 1988, Chap. I, §10, Theorem 2), $M^G_\tau(\hat{\Sigma})$ is a complete variety. This proves the theorem. □

Corollary 7.2.15 *If Σ has genus $g \geq 2$ and G is simply connected, then the morphism $\bar{\theta}$ from Definition 7.2.10 is finite if $j \colon G \to H$ is an embedding.*

In particular, if $M^H_{\tau'}(\hat{\Sigma})$ is a projective variety, then so is $M^G_\tau(\hat{\Sigma})$ (by (Stacks, 2019, Tags 087S and 0B3I)).

By a subsequent Theorem 8.5.3, $M^{\mathrm{SL}_V}_{\tau'}(\hat{\Sigma})$ is a projective variety, and hence so is $M^G_\tau(\hat{\Sigma})$ for any connected, simply-connected semisimple G and any A-equivariant topological type τ by taking an embedding $G \hookrightarrow \mathrm{SL}_V$.

Proof Follow the notation from Definition 7.2.10. By Proposition 7.2.5, f_τ and $f_{\tau'}$ are affine morphisms and hence so is θ. Since q_i are good quotients; in particular, for any open $U \subset M^G_\tau(\hat{\Sigma})$, $q_1^{-1}(U) \to U$ is a categorical quotient (cf. Definition 7.1.4). Moreover, q_i are affine morphisms. By the uniqueness of the categorical quotients, from the commutativity of the diagram (1) in Definition 7.2.10, we can easily deduce that $\bar{\theta}$ is an affine morphism. But, by Theorem 7.2.14, $M^G_\tau(\hat{\Sigma})$ is a nonempty complete variety (this is where we used the assumption that G is simply connected). Since an affine morphism from a complete variety to any variety is finite (Stacks, 2019, Tags 01W6 and 01WN), the corollary follows. □

The following definition extends Definition 7.1.17 to the equivariant setting.

Definition 7.2.16 Let $E \to \hat{\Sigma}$ be an A-equivariant G-bundle for a connected reductive group G. Then, an A-equivariant reduction E_P to a parabolic subgroup P is called *A-admissible* if for any character χ of P which is trivial restricted to the center of G, the line bundle $E_P \times^P \mathbb{C}_\chi$ has degree 0.

As earlier, $\hat{\Sigma}$ is a smooth irreducible projective curve with faithful action of a finite group A. Moreover, unless otherwise stated, we assume that $\Sigma := \hat{\Sigma}/A$ has genus ≥ 2 and G is a connected semisimple group.

Lemma 7.2.17 *Let E be an A-semistable G-bundle over $\hat{\Sigma}$. Then, it admits an A-admissible reduction E_P to a standard parabolic subgroup P such that $E_P(P/U)$ is an A-stable P/U-bundle, where U is the unipotent radical of P. Identifying P/U with any Levi subgroup L of P, we get an A-equivariant G-bundle $\mathrm{gr}\, E$:*

$$E_P(P/U) \simeq E_P(L) \hookrightarrow E_P(G) := \mathrm{gr} E.$$

Then, $\mathrm{gr} E$ constructed thus is uniquely determined from E as an A-equivariant G-bundle (up to an isomorphism). By the definition, $\mathrm{gr} E$ is A-polystable; in particular, A-semistable (cf. Lemma 6.3.40).

It is easy to see since L is a deformation retract of P that $\mathrm{gr} E$ has the same topological type as E as A-equivariant G-bundles.

Proof The proof is parallel (in fact, almost identical once we take A-action into consideration) to that of Ramanathan (1996, Proposition 3.12). □

Proposition 7.2.18 *For any A-semistable G-bundle E over $\hat{\Sigma}$, there exists an A-equivariant G-bundle $\mathscr{E} \to \hat{\Sigma} \times \mathbb{A}^1$ such that as A-equivariant G-bundles:*

(a) $\mathscr{E}_{|\hat{\Sigma}\times(\mathbb{A}^1\backslash\{0\})} \simeq p^*(E)$, where $p\colon \hat{\Sigma} \times \mathbb{A}^1\backslash\{0\} \to \hat{\Sigma}$ is the projection.

(b) $\mathscr{E}_{|\hat{\Sigma}\times\{0\}} \simeq gr\, E$.

(c) If E_1 and E_2 are two A-polystable G-bundles of the same topological type τ such that $[E_1] = [E_2] \in M_\tau^G(\hat{\Sigma})$. Then $E_1 \simeq E_2$ as A-equivariant G-bundles.

Thus, any element in $M_\tau^G(\hat{\Sigma})$ (which is an equivalence class of A-semistable G-bundles) contains a unique A-polystable representative.

By Corollary 7.2.7, Lemma 7.2.17 and the (a)- and (b)-parts of this proposition, we see that any element in $M_\tau^G(\hat{\Sigma})$ does contain an A-polystable representative.

This proposition remains valid for $g = 1$ as well in the case $A = (1)$, where g is the genus of Σ (cf. (Laszlo, 1998b, §4)).

Proof By the above Lemma 7.2.17, we can choose an A-admissible reduction E_P of E to a standard parabolic subgroup P such that, as A-equivariant G-bundles,

$$E_P(P/U) \simeq E_P(L) \hookrightarrow E_P(G) \simeq \text{gr}\, E,$$

where U is the unipotent radical of P and L is a Levi component of P.

It is easy to see (cf. (Ramanathan, 1996, Lemma 3.5.12) for more details) that there exists a one-parameter subgroup $\lambda : \mathbb{G}_m (= \mathbb{C}\backslash\{0\}) \to Z(L)$ (for the center $Z(L)$ of L), such that the regular map

$$\mathbb{G}_m \times P \to P, \text{ given by } (t,p) \mapsto \lambda(t)p\lambda(t)^{-1} \text{ for } t \in \mathbb{G}_m \text{ and } p \in P,$$

extends to a regular map

$$\varphi\colon \mathbb{C} \times P \to P \quad \text{satisfying } \varphi(0, lu) = l, \text{ for } l \in L \text{ and } u \in U.$$

To construct the A-equivariant G-bundle $\mathscr{E} \to \hat{\Sigma} \times \mathbb{A}^1$ as in the proposition, consider the A-equivariant trivial group scheme $S \times P \to S$ over $S := \mathbb{C} \times \hat{\Sigma}$ (with the trivial action of A on $\mathbb{C}$). Then, $\mathbb{C} \times E_P \to S$ is canonically an A-equivariant principal homogeneous space under the group scheme $S \times P \to S$ over S. Define the following A-equivariant endomorphism of group scheme over S:

$$\tilde{\varphi}\colon S \times P \to S \times P, \ (t,z,p) \mapsto (t,z,\varphi(t,p)) \text{ for } t \in \mathbb{C}, z \in \hat{\Sigma} \text{ and } p \in P.$$

Now, take the associated A-equivariant principal homogeneous space of $\mathbb{C} \times E_P \to S$ under the extension $\tilde{\varphi}$ which exists since $S \times P \to S$ is an affine algebraic group scheme over S ((Grothendieck, 1971, Expose XI, §4, pp. 11–12)). This, of course, can be viewed as an A-equivariant principal P-bundle $\mathscr{E}(P)$ over S. Let $\mathscr{E}$ be the A-equivariant G-bundle over S obtained from the A-equivariant P-bundle $\mathscr{E}(P)$ via extension of the structure group. It is easy to see that $\mathscr{E}$ satisfies conditions (a) and (b).

To prove the (c)-part, recall from (3) of the proof of Theorem 7.2.6 that the moduli space $M_\tau^G(\hat{\Sigma})$ is by definition the good quotient:

$$M_\tau^G(\hat{\Sigma}) := R_\tau^{ss}(G)//\mathfrak{G},$$

where $R_\tau^{ss}(G)$ is the 'G-quot scheme' and $\mathfrak{G}$ is a Levi subgroup of GL_N (and hence connected). As in the proof of Theorem 7.2.6, let $\mathscr{U}_\tau^{ss}(G)$ be the A-equivariant universal G-bundle over $\hat{\Sigma}$ parameterized by $R_\tau^{ss}(G)$. Let $x \in R_\tau^{ss}(G)$ (resp. y) be an element with the corresponding bundle E_1 (resp. E_2), i.e., $\mathscr{U}_\tau^{ss}(G)_{|\hat{\Sigma}\times x} \simeq E_1$ and similarly for y, and let $\overline{\mathfrak{G}\cdot x}$ (resp. $\overline{\mathfrak{G}\cdot y}$) be the closure of the $\mathfrak{G}$-orbit of x (resp. y) in $R_\tau^{ss}(G)$. Since x and y go to the same element in $M_\tau^G(\hat{\Sigma})$ by assumption, by the good quotient property (cf. Definition 7.1.4),

$$\overline{\mathfrak{G}\cdot x} \cap \overline{\mathfrak{G}\cdot y} \neq \emptyset.$$

Take any z in the above intersection. Then, by Ramanan (1978), there exists an irreducible curve C in $\overline{\mathfrak{G}\cdot x}$ connecting x and z. Since $\mathfrak{G}\cdot x$ is open in $\overline{\mathfrak{G}\cdot x}$, an open subset C_o of the curve C lies in $\mathfrak{G}\cdot x$; in particular, $\mathscr{U}_\tau^{ss}(G)$ restricted to any point of C_o is A-isomorphic with E_1. Thus, by the following Lemma 7.2.19 applied to the family $\mathscr{F} = \mathscr{U}_\tau^{ss}(G)_{|(\hat{\Sigma}\times\overline{\mathfrak{G}\cdot x})}$, the A-equivariant G-bundle $E_z := \mathscr{U}_\tau^{ss}(G)_{|\hat{\Sigma}\times z}$ corresponding to the point z satisfies

$$\operatorname{gr} E_z \simeq E_1, \text{ as } A\text{-equivariant } G\text{-bundles.}$$

Similarly, $\operatorname{gr} E_z \simeq E_2$. Hence we get that

$$E_1 \simeq E_2, \text{ as } A\text{-equivariant } G\text{-bundles.}$$

This proves the (c)-part modulo the next lemma. □

The non-equivariant version of the following lemma is given in Ramanathan (1996, Lemma 3.23). In fact, the same proof in this reference works in the following equivariant setting as well. However, we give a different proof communicated to us by V. Balaji.

Lemma 7.2.19 *Let $\mathscr{F} \to \hat{\Sigma} \times S$ be a family of A-semistable G-bundles of topological type τ parameterized by a scheme S of finite type over $\mathbb{C}$ such that*

$\mathscr{F}_t \simeq E$ for a fixed A-polystable G-bundle E for all t in a dense open subset S_o of S. Then

$$\mathrm{gr}(\mathscr{F}_t) \simeq E, \textit{ for all } t \in S.$$

Proof Take an embedding $i: G \hookrightarrow \mathrm{SL}_r \subset \mathrm{GL}_r$ and consider the family $\mathscr{F}(i)$ obtained from the family $\mathscr{F}$ by extension of the group to GL_r via i. We also let $\mathscr{F}(\mathbb{C}^r)$ be the corresponding family of rank-r vector bundles. By Lemma 6.3.39, $E(\mathbb{C}^r)$ is A-polystable (since G is semisimple by assumption). Hence, by Exercise 6.1.E.16, we can decompose:

$$E(\mathbb{C}^r) = \oplus_j \mathscr{V}_j^{n_j},$$

where $\mathscr{V}_j$ is an A-stable vector subbundle ocuring with multiplicity n_j (and all the $\mathscr{V}_j$ have the same slope). By the Semicontinuity Theorem (Hartshorne, 1977, Chap. III, Theorem 12.8) applied to the family $\{\mathscr{V}_j^* \otimes \mathscr{F}_t(\mathbb{C}^r)\}_{t\in S}$,

$$\dim \mathrm{Hom}_A\left(\mathscr{V}_j, \mathscr{F}_t(\mathbb{C}^r)\right) \geq \dim \mathrm{Hom}_A\left(\mathscr{V}_j, E(\mathbb{C}^r)\right) = n_j, \text{ for any } t \in S.$$

Thus, we get an A-equivariant embedding $\oplus_j \mathscr{V}_j^{n_j} \hookrightarrow \mathscr{F}_t(\mathbb{C}^r)$, which is an A-equivariant isomorphism of vector bundles since both of the sides have the same rank. Hence,

$$\mathscr{F}_t(i) \simeq E(i), \text{ for any } t \in S. \tag{1}$$

In particular, $\mathscr{F}_t(i)$ is A-polystable.

Now, for any $t \in S$, let $(\mathrm{gr}\,\mathscr{F}_t)_{H_t}$ be the unique minimal A-equivariant reduction of $\mathrm{gr}\,\mathscr{F}_t$ to a (not necessarily connected) reductive subgroup $H_t \subset G$ (cf. (Ramanan and Ramanathan, 1984, Theorem 3.19) for the corresponding non-equivariant case, but since the minimal reduction is unique, it is automatically A-equivariant). Clearly, $(\mathrm{gr}\,\mathscr{F}_t)_{H_t}$ is also the minimal A-equivariant reduction of $(\mathrm{gr}\,\mathscr{F}_t)(i)$, thus

$$(\mathrm{gr}\,\mathscr{F}_t)_{H_t} \simeq ((\mathrm{gr}\,\mathscr{F}_t)(i))_{H_t}, \text{ for all } t \in S. \tag{2}$$

From the vector bundle case proved above in (1),

$$E(i) \simeq \mathscr{F}_t(i) \simeq (\mathrm{gr}\,\mathscr{F}_t)\,(i), \text{ for all } t \in S,$$

where the second isomorphism (for any fixed $t \in S$) follows by considering the family $\{\mathscr{G}_s\}_{s\in\mathbb{A}^1}$ joining $\mathscr{F}_t$ and $\mathrm{gr}\,\mathscr{F}_t$ as in Proposition 7.2.18(a) and (b) and then applying the isomorphism (1) to conclude that

$$\mathscr{G}_s(i) \simeq \mathscr{F}_t(i) \simeq E(i), \text{ for } s \in \mathbb{A}^1\backslash\{0\}.$$

Hence, by the above proof for vector bundles applied to the family $\mathscr{G}_s(i)$,

$$\mathscr{G}_0(i) \simeq (\mathrm{gr}\, \mathscr{F}_t)(i) \simeq \mathscr{G}_s(i) \simeq \mathscr{F}_t(i), \quad \text{for any } s \in \mathbb{A}^1.$$

Hence their minimal reductions are all isomorphic:

$$H_t = H \text{ and } ((\mathrm{gr}\, \mathscr{F}_t)(i))_{H_t} \simeq E(i)_H, \quad \text{for all } t \in S,$$

where E_H is the minimal reduction of E and hence $E(i)_H \simeq E_H$ is the minimal reduction of $E(i)$. From this, using (2), we get

$$(\mathrm{gr}\, \mathscr{F}_t)_{H_t=H} \simeq E_H, \quad \text{for all } t \in S,$$

and hence

$$\mathrm{gr}(\mathscr{F}_t) \simeq E, \quad \text{for all } t \in S.$$

This proves the lemma. □

Recall the definition of $R^{ss}_\tau(G)$ and $R^s_\tau(G)$ from Corollary 7.2.4 and let $M^G_\tau(\hat{\Sigma})^s$ be the image of $R^s_\tau(G)$ under the good quotient map $R^{ss}_\tau(G) \to M^G_\tau(\hat{\Sigma}) := R^{ss}_\tau(G)//\mathfrak{G}$ (cf. Theorem 7.2.6).

As a corollary of Proposition 7.2.18, Theorem 6.3.41 and Corollary 6.3.7, we get the following result.

Corollary 7.2.20 *Let the notation and assumptions be as in Lemma 7.2.17; in particular, G is a connected (not necessarily simply-connected) semisimple group and $\Sigma := \hat{\Sigma}/A$ has genus ≥ 2.*

(a) The moduli space $M^G_\tau(\hat{\Sigma})$ has any of its elements uniquely represented by an A-polystable G-bundle.

(b) The real analytic map $\bar{f}_{\mathscr{E}}\colon R^{A,\tau}_K(\hat{\Sigma})/\mathrm{Ad}\, K \to M^G_\tau(\hat{\Sigma})$ induced from $f_{\mathscr{E}}$ (cf. the proof of Theorem 7.2.14) is a homeomorphism.

(c) For any $E \in R^s_\tau(G)$, the orbit $\mathfrak{G} \cdot E$ is closed in $R^{ss}_\tau(G)$. Moreover, the image of $R^{ss}_\tau(G) \setminus R^s_\tau(G)$ in $M^G_\tau(\hat{\Sigma})$ is disjoint from the image of $R^s_\tau(G)$.

Proof (a) By Proposition 7.2.18 and Corollary 7.2.7, for any A-semistable G-bundle E of topological type τ, $[E] = [\mathrm{gr}\, E]$ as elements of $M^G_\tau(\hat{\Sigma})$. Moreover, by Proposition 7.2.18(c), any element of $M^G_\tau(\hat{\Sigma})$ contains a unique A-polystable G-bundle. This proves the (a)-part of the corollary.

(b) By (a) any element in $M^G_\tau(\hat{\Sigma})$ is uniquely represented by an A-polystable G-bundle and any A-polystable bundle is A-unitary and conversely (cf. Theorem 6.3.41). Moreover, for $\hat{\rho}, \hat{\rho}' \in R^{A,\tau}_K(\hat{\Sigma})$, $\hat{E}_{\hat{\rho}} \simeq \hat{E}_{\hat{\rho}'}$ as A-equivariant G-bundles if and only if $\hat{\rho}$ and $\hat{\rho}'$ are conjugate as maps to K (cf. Corollary 6.3.7). Thus, we get that $\bar{f}_{\mathscr{E}}$ is a bijection and hence a homeomorphism (from the compactness of $R_K^{A,\tau}(\hat{\Sigma})/\operatorname{Ad} K$). This proves the (b)-part of the corollary.

(c) For any $E \in R^s_\tau(G)$, the $\mathfrak{G}$-orbit $\mathfrak{G} \cdot E \subset R^{ss}_\tau(G)$ is closed in $R^{ss}_\tau(G)$. For otherwise, there would exist $F \in \overline{\mathfrak{G} \cdot E} \backslash \mathfrak{G} \cdot E$ such that, by Proposition 7.2.18, $E \simeq \operatorname{gr} F$ (as A-equivariant G-bundles). Thus, $\operatorname{gr} F$ is A-stable which forces F to be A-stable, i.e., $F \simeq \operatorname{gr} F \simeq E$. This is a contradiction to the choice of F by identity (4) of the proof of Theorem 7.2.6.

To prove that the image of $R^{ss}_\tau(G) \backslash R^s_\tau(G)$ in $M^G_\tau(\hat{\Sigma})$ is disjoint from the image of $R^s_\tau(G)$, observe that if $[E] = [F]$ for $E \in R^s_\tau(G)$ and $F \in R^{ss}_\tau(G) \setminus R^s_\tau(G)$, then $\operatorname{gr} F \simeq E$ forcing F to be A-stable (as above). A contradiction! This proves the disjointness of the images of $R^{ss}_\tau(G) \setminus R^s_\tau(G)$ and $R^s_\tau(G)$. □

Remark 7.2.21 It is likely that $R^s_\tau(G)$ is Zariski open in $R^{ss}_\tau(G)$. If so, the good quotient $R^{ss}_\tau(G) \to M^G_\tau(\hat{\Sigma})$ would restrict to a geometric quotient $R^s_\tau(G) \to M^G_\tau(\hat{\Sigma})^s$.

The following corollary extends Theorem 7.2.14 and Corollary 7.2.15 for any semisimple G by following exactly the same proof once we use Corollary 7.2.20(b).

Corollary 7.2.22 *Let G be a connected (not necessarily simply-connected) semi-simple group, τ any topological type and $\Sigma := \hat{\Sigma}/A$ has genus ≥ 2. Assume that $M^G_\tau(\hat{\Sigma})$ is nonempty. Then*

(a) $M^G_\tau(\hat{\Sigma})$ is an irreducible complete variety.

(b) For any embedding of connected semisimple groups $G \hookrightarrow H$, $M^G_\tau(\hat{\Sigma}) \to M^H_{\tau'}(\hat{\Sigma})$ is a finite morphism, where τ' is the topological type of the corresponding A-equivariant H-bundles.

Let G be a connected, simple, simply-connected complex algebraic group. Combining Theorems 7.2.6, 7.2.8, Corollary 7.2.20 and Theorem 6.1.17, we obtain the following result for parabolic G-bundles (see Definition 6.1.10).

Theorem 7.2.23 *Let $(\Sigma, \vec{p})$ be a smooth irreducible projective s-pointed ($s \geq 0$) curve of any genus $g \geq 2$ and let $\vec{\tau} = (\tau_1, \dots, \tau_s)$ be a set of markings (cf. Definition 6.1.2) with τ_j rational points (i.e., we can write $\tau_j = \bar{\tau}_j/d_j$ for some positive integers d_j and $\operatorname{Exp}(2\pi i \bar{\tau}_j) = 1$) of the fundamental alcove Φ_o. We assume further that $\theta(\tau_j) < 1$ for all j, where θ is the highest root of G. Then the parabolic semistable G-bundles with markings $\vec{\tau}$ admit a coarse moduli space $M^G_{par,\vec{\tau}}(\Sigma)$, which is a normal irreducible projective variety with rational singularity.*

In fact, it is isomorphic with $M^G_{\vec{\tau}}(\hat{\Sigma})$ for any $\hat{\Sigma}$ and A given by Theorem 6.1.8 for the s-tuple $(d_1, \dots, d_s)$ and for the local type $\vec{\tau}$ (cf. Exercise 7.2.E.4).

(Observe that G being simply connected, any G-bundle over any curve is topologically trivial.)

Thus, $M^G_{par,\vec{\tau}}(\Sigma)$ is homeomorphic with $R_K^{A,\vec{\tau}}(\hat{\Sigma})/\operatorname{Ad} K$.

Remark 7.2.24 The above theorem is proved in terms of semistable parahoric torsors over Σ without the restriction $\theta(\tau_j) < 1$ (cf. (Balaji and Seshadri, 2015, Theorem 8.1.11)).

Let G be a connected, semisimple group and let $K \subset G$ be a maximal compact subgroup. Let $\hat{\Sigma}$ be an irreducible smooth projective curve with a faithful action of a finite group A and let $\Sigma := \hat{\Sigma}/A$ be the quotient of genus $g \geq 0$. If $g = 0$, we assume that there are at least three ramification points. Fix the local type $[\vec{\tau}] = ([\tau_1], \ldots, [\tau_s])$ of A-equivariant G-bundles over $\hat{\Sigma}$ with $\tau_j \in K$ (cf. Definition 7.2.12). By Exercise 7.2.E.4, if G is simply connected (and hence any G-bundle over $\hat{\Sigma}$ is topologically trivial), the topological type of any A-equivariant G-bundle over $\hat{\Sigma}$ is completely determined by the sequence $[\vec{\tau}]$. Let $\vec{d} := (d_1, \ldots, d_s)$ be the signature of Σ (cf. Theorem 6.1.8). Then, $\tau_j^{d_j} = 1$. Observe that $(\operatorname{Ad} G \cdot \tau_j) \cap K = \operatorname{Ad} K \cdot \tau_j$ (cf. (Helgason, 1978, Chap. VI, Theorem 1.1)).

Definition 7.2.25 We follow the notation from Definition 6.3.3. The set of homomorphisms $\hat{\rho} : \pi \to K$ with

$$\hat{\rho}(c_j) \in O(\tau_j) := \operatorname{Ad} K \cdot \tau_j, \quad \text{for } 1 \leq j \leq s, \tag{1}$$

is in bijective correspondence with the fiber over e of the smooth map:

$$\beta_{[\vec{\tau}]} : (K^2)^{\times g} \times \Pi_{j=1}^s O(\tau_j) \to K,$$

$$\Big(\big((k_1, k_1'), \ldots, (k_g, k_g')\big), (l_1, \ldots, l_s)\Big) \mapsto [k_1, k_1'] \ldots [k_g, k_g'] l_1 \ldots l_s.$$

Any such $\bar{\rho} \in \beta_{[\vec{\tau}]}^{-1}(e)$ gives rise to $\hat{\rho} : \pi \to K$ defined by

$$\hat{\rho}(a_i) = k_i,\ \hat{\rho}(b_i) = k_i' \quad (\text{for } 1 \leq i \leq g) \quad \text{and} \quad \hat{\rho}(c_j) = l_j \ \text{ for } 1 \leq j \leq s. \tag{2}$$

As an 'A-equivariant analogue' of Corollary 6.3.16, we get the following. Its proof is similar to that of the same and hence omitted.

Proposition 7.2.26 *With the notation and assumptions as above (in particular, we assume that G is a connected semisimple group) assume further that $g \geq 1$. Then, for any $\bar{\rho} \in \beta_{[\vec{\tau}]}^{-1}(e)$, $(d\beta_{[\vec{\tau}]})_{\bar{\rho}}$ is of maximal rank (equal to* $\dim K$*) if and only if the corresponding representation $\hat{\rho} : \pi \to K$ is irreducible.*

Thus,

$$M_{g,[\vec{\tau}]}(K) := \left\{ \bar{\rho} \in \beta_{[\vec{\tau}]}^{-1}(e) : \hat{\rho} \text{ is irreducible} \right\},$$

which is an open subset of $\beta_{[\vec{\tau}]}^{-1}(e)$, if nonempty is a real analytic (smooth) manifold of dimension (of each of its connected components)

$$(2g-1)\dim_{\mathbb{R}}(K) + \sum_{j=1}^{s} \dim_{\mathbb{R}} O(\tau_j). \qquad (1)$$

Moreover, the tangent space at any $\bar{\rho} \in M_{g,[\vec{\tau}]}(K)$ is identified with $Z^1(\pi, \operatorname{ad}\hat{\rho})$.

Observe that if $g \geq 2$, by Lemma 7.2.9, $M_{g,[\vec{\tau}]}(K)$ is nonempty.

As an immediate consequence of the above Proposition and Corollary 7.2.20, we get the following.

Corollary 7.2.27 *With the notation as in Theorem 7.2.14 (in particular, G is connected, simply-connected semisimple group and $g \geq 2$) for any topological type τ,*

$$\dim_{\mathbb{C}} M_{\tau}^{G}(\hat{\Sigma}) = (g-1)\dim_{\mathbb{R}} K + \frac{1}{2}\sum_{j=1}^{s} \dim_{\mathbb{R}} O(\tau_j).$$

Hence, the same dimension count is available for the coarse moduli space $M_{\text{par},\vec{\tau}}^{G}(\Sigma)$ of parabolic semistable G-bundles over Σ with markings $\vec{\tau}$ as in Theorem 7.2.23 in the case G is a connected, simply-connected, simple group.

Proof Since G is simply connected, the topological type τ of any A-equivariant G-bundle over $\hat{\Sigma}$ is determined by its local type $[\vec{\tau}]$ as observed above Definition 7.2.25. Since $M_{\tau}^{G}(\hat{\Sigma})$ is irreducible (cf. Theorem 7.2.8), we get that $R_K^{A,\tau}(\hat{\Sigma})$ is connected by Corollary 7.2.20(b). Let $R_K^{A,\tau}(\hat{\Sigma})^o$ be the subset of $R_K^{A,\tau}(\hat{\Sigma})$ consisting of irreducible representations $\pi \to K$. Then, $R_K^{A,\tau}(\hat{\Sigma})^o$ is nonempty by Lemma 7.2.13. Moreover, $R_K^{A,\tau}(\hat{\Sigma})$ (resp. $R_K^{A,\tau}(\hat{\Sigma})^o$) coincides with $\beta_{[\vec{\tau}]}^{-1}(e)$ (resp. $M_{g,[\vec{\tau}]}(K)$) following the notation in Proposition 7.2.26. In particular, $R_K^{A,\tau}(\hat{\Sigma})^o$ is an open subset of $R_K^{A,\tau}(\hat{\Sigma})$. Thus, $\bar{f}_{\mathcal{E}}\left(R_K^{A,\tau}(\hat{\Sigma})^o/\operatorname{Ad}K\right)$ is a nonempty open subset of $M_{\tau}^{G}(\hat{\Sigma})$ in the analytic topology. Thus,

$$\begin{aligned}\dim_{\mathbb{C}}\left(M_{\tau}^{G}(\hat{\Sigma})\right) &= \frac{1}{2}\dim_{\mathbb{R}}\left(M_{g,[\vec{\tau}]}(K)/\operatorname{Ad}K\right),\\ &= (g-1)\dim_{\mathbb{R}} K + \frac{1}{2}\sum_{j=1}^{s}\dim_{\mathbb{R}} O(\tau_j), \text{ by Proposition 7.2.26,}\end{aligned}$$

where the last identity follows since $\operatorname{Ad}K$ action on $M_{g,[\vec{\tau}]}(K)$ has only finite isotropies. For, if not, let $\bar{\rho} \in M_{g,[\vec{\tau}]}(K)$ has a nontrivial connected subgroup

H in its isotropy subgroup (under the Ad K action). In particular, a nontrivial torus S is in the isotropy group of $\bar{\rho}$. Thus, the corresponding representation $\hat{\rho}: \pi \to K$ has its image in the centralizer of S, which is a proper Levi subgroup. This is a contradiction to the irreducibility of $\hat{\rho}$.

This proves the corollary. □

Definition 7.2.28 Let $(\Sigma, \vec{p})$ be an s-pointed ($s \geq 1$) smooth projective irreducible curve of genus $g \geq 2$, where $\vec{p} = (p_1, \ldots, p_s)$ and let G be a connected simply-connected simple algebraic group. Let $\vec{\tau} = (\tau_1, \ldots, \tau_s)$ be a set of rational markings as in Definition 6.1.2, i.e., $\tau_j \in \Phi_0$ can be written as

$$\tau_j = \frac{\overline{\tau}_j}{d_j}, \text{ for some positive integers } d_j \text{ and } \operatorname{Exp}(2\pi i \overline{\tau}_j) = 1. \qquad (1)$$

We also assume that $\theta(\tau_j) < 1$, θ being the highest root of $\mathfrak{g}$.

As in Theorem 6.1.8, we fix a Galois cover $\hat{\Sigma} \to \Sigma$ with finite Galois group A associated to $(\Sigma, \vec{p})$ and the sequence $\vec{d} = (d_1, \ldots, d_s)$, ignoring those p_i with $d_i = 1$. Following Definition 6.3.3 (the notation of which we freely follow), we have the commutative diagram:

$$\begin{array}{ccc} \mathbb{H} = \tilde{\hat{\Sigma}} & \xrightarrow{\hat{q}} & \hat{\Sigma} \\ {\scriptstyle f}\downarrow & & \downarrow \\ \tilde{\hat{\Sigma}}/\pi & \xrightarrow{\simeq} & \Sigma = \hat{\Sigma}/A, \end{array}$$

where we have identified the simply-connected cover $\tilde{\hat{\Sigma}}$ of $\hat{\Sigma}$ with the upper half plane $\mathbb{H} := \{x + iy, x, y \in \mathbb{R}, y > 0\}$, since $\hat{\Sigma}$ has genus $\hat{g} \geq 2$ (cf. (Hartshorne, 1977, Chap. IV, Example 2.5.4)).

Let $\hat{\rho}: \pi \to G$ be a homomorphism such that

$$\hat{\rho}(c_j) = g_j \operatorname{Exp}(2\pi i \tau_j) g_j^{-1}, \quad \text{for some } g_j \in G \text{ and all } 1 \leq j \leq s. \qquad (2)$$

By (1), $\operatorname{Exp}(2\pi i \tau_j)^{d_j} = 1$, which is compatible with the relation (2). Any $\hat{\rho}$ satisfying (2) is said to have *local type* $\vec{\tau}$.

Consider the action of $\mathbb{Z}$ on $\mathbb{H}$ generated by $z \mapsto z + 1$. Then

$$\gamma: \mathbb{H} \to D^*, \ z \mapsto \exp(2\pi i z),$$

induces an isomorphism $\overline{\gamma}: \mathbb{H}/\mathbb{Z} \simeq D^*$, where D is the open unit disc in $\mathbb{C}$ centered at 0 and $D^* := D \backslash \{0\}$. Choose a holomorphic isomorphism

$$\beta_j: D \xrightarrow{\sim} D_{p_j} \subset \Sigma, \quad \text{taking} \quad 0 \mapsto p_j,$$

onto a small disc D_{p_j} in Σ centered at p_j and let

$$\beta_j^o \colon D^* \xrightarrow{\sim} D_{p_j}^* \subset \Sigma \backslash \bar{p}$$

be its restriction, where $\bar{p} = \{p_1, \ldots, p_s\}$. Since f is unramified over $\Sigma \backslash \bar{p}$, we get a covering projection $f^o \colon f^{-1}(\Sigma \backslash \bar{p}) \to \Sigma \backslash \bar{p}$ over $\Sigma \backslash \bar{p}$. From the homotopy lifting property of covering projections, we get a lifting

$$\begin{array}{ccc} \mathbb{H} & \xrightarrow{\hat{\beta}_j^o} & f^{-1}(\Sigma \backslash \bar{p}) \\ \gamma \downarrow & & \downarrow f^o \\ D^* & \xrightarrow[\beta_j^o]{} & \Sigma \backslash \bar{p} \end{array} \tag{3}$$

such that $\hat{\beta}_j^o$ is $\mathbb{Z}$-equivariant, where $\mathbb{Z}$ acts on $f^{-1}(\Sigma \backslash \vec{p})$ through the homomorphism $\chi_j \colon \mathbb{Z} \to \pi$, $1 \mapsto c_j$. Moreover, if $\hat{\beta}_j^{o'}$ is another such $\mathbb{Z}$-equivariant lifting, then there exists a $\sigma \in \pi$ commuting with c_j such that

$$\hat{\beta}_j^{o'}(z) = \hat{\beta}_j^o(z) \cdot \sigma, \quad \text{for all } z \in \mathbb{H}.$$

Let $f^{-1}(\Sigma \backslash \bar{p}) \times^{\pi} G \to \Sigma \backslash \bar{p}$ be the holomorphic G-bundle obtained from the homomorphism $\hat{\rho} \colon \pi \to G$. Observe that by virtue of the diagram (3), $\hat{\beta}_j^o$ induces an isomorphism of G-bundles between the restriction of $f^{-1}(\Sigma \backslash \bar{p}) \times^{\pi} G \to \Sigma \backslash \bar{p}$ to D^* and $f_j \colon \mathbb{H} \times^{\mathbb{Z}} G \to D^*$, where the homomorphism $\mathbb{Z} \to G$ is the composite $\hat{\rho} \circ \chi_j$. Define a map $\hat{s}_j \colon \mathbb{H} \to G$ by $\hat{s}_j(z) = g_j \operatorname{Exp}(-2\pi i z \tau_j) g_j^{-1}$. This map descends to give a section s_j of the bundle f_j defined by $s_j(x) = [\hat{x}, \hat{s}_j(\hat{x})]$, where $\hat{x} \in \mathbb{H}$ is any element in the preimage of x under γ. (By condition (2), s_j is indeed a well-defined section of f_j.)

Now, we define a holomorphic G-bundle $E_{\hat{\rho}}(\vec{\tau})$ over Σ by gluing the G-bundle $f^{-1}(\Sigma \backslash \bar{p}) \times^{\pi} G \to \Sigma \backslash \bar{p}$ with the trivial bundles $D_{p_j} \times G \to D_{p_j}$ by identifying the trivial section $(-, 1)$ of $D_{p_j} \times G \to D_{p_j}$ over $D_{p_j}^*$ with the section $\hat{\beta}_j^o \circ s_j$ of $f^{-1}(\Sigma \backslash \bar{p}) \times^{\pi} G \to \Sigma \backslash \bar{p}$ over $D_{p_j}^*$. The bundle $E_{\hat{\rho}}(\vec{\tau})$ clearly has a section over p_j by taking the point $(p_j, 1)$ in $D_{p_j} \times G$.

Theorem 7.2.29 *With the notation and assumptions as in the above definition, for any unitary homomorphism $\hat{\rho} \colon \pi \to G$ of local type $\vec{\tau}$ (i.e., $\hat{\rho}(c_j)$ is conjugate of $Exp(2\pi i \tau_j)$ for all j), the G-bundle $E_{\hat{\rho}}(\vec{\tau})$ over Σ with the above sections over $\{p_j\}$ is a parabolic semistable G-bundle with markings $\vec{\tau}$. Further, $E_{\hat{\rho}}(\vec{\tau})$ is parabolic stable if and only if $\hat{\rho}$ is irreducible (and unitary).*

Moreover, for unitary homomorphisms $\hat{\rho}$ and $\hat{\rho}'$ (of local type $\vec{\tau}$), $E_{\hat{\rho}}(\vec{\tau}) \simeq E_{\hat{\rho}'}(\vec{\tau})$ (as parabolic G-bundles) if and only if

$$\hat{\rho}' = g\hat{\rho}g^{-1}, \quad \textit{for some} \quad g \in G.$$

Conversely, for any parabolic stable G-bundle E over Σ with markings $\vec{\tau}$ at $\vec{p}$, there exists an irreducible unitary homomorphism $\hat{\rho}: \pi \to G$ of local type $\vec{\tau}$ such that $E \simeq E_{\hat{\rho}}(\vec{\tau})$ (as parabolic G-bundles).

Proof The result follows from Theorem 6.3.41, Proposition 6.3.4 and Corollary 6.3.7 by using the correspondence between parabolic bundles over Σ and A-equivariant bundles over $\hat{\Sigma}$ as in Theorem 6.1.17. □

7.2.E Exercises

(1) With the notation as in Definition 6.3.3, for any group homomorphism $\hat{\rho}: \pi \to G$ such that $\hat{\rho}(\pi_1) \subset K$ for a compact subgroup K, there exists a compact subgroup K' such that

$$\operatorname{Im} \hat{\rho} \subset K'.$$

Hint: Embed $G \hookrightarrow \mathrm{GL}_N$ as a closed subgroup. By assumption, there exists a π_1-invariant positive-definite Hermitian form $\{,\}$ on $\mathbb{C}^N$. Now, since $A \simeq \pi/\pi_1$ is finite, we can average $\{,\}$ to get a π-invariant Hermitian form on $\mathbb{C}^N$.

Let G be a connected reductive group with a maximal compact subgroup K and let $\hat{\Sigma}$ and A be as at the beginning of Section 7.2 in the following exercises.

(2) Let $\mathscr{F} \to \hat{\Sigma} \times S$ be a family of A-equivariant G-bundles (for a connected reductive group G) parameterized by a noetherian scheme S. Then, prove that S^{ss} is open in S, where

$$S^{ss} := \{t \in S : \mathscr{F}_{|_{\hat{\Sigma}\times t}} \text{ is an } A\text{-semistable } G\text{-bundle}\}.$$

Hint: Use Corollary 7.1.16(b) and Exercise 6.2.E.4.

(3) Let $\mathscr{F} \to \hat{\Sigma} \times S$ be a real analytic family of holomorphic A-equivariant G-bundles over $\hat{\Sigma}$. Then, the subset

$$\begin{aligned} T_o = \{t \in S : \mathscr{F}_t \simeq \hat{E}_{\hat{\rho}}, &\text{as } A\text{-equivariant } G\text{-bundles,} \\ &\text{for some irreducible representation } \hat{\rho}: \pi \to K\} \end{aligned}$$

is an open subset of S, where $\hat{E}_{\hat{\rho}}$ is as in Definition 6.3.3.

Hint: Follow the proof of Corollary 6.3.21.

(4) Follow the notation as in Definition 7.2.12 and assume that G is a connected, simply-connected semisimple group. Let E_1 and E_2 be two A-equivariant (algebraic) G-bundles over $\hat{\Sigma}$. Then, show that they are isomorphic as topological A-equivariant G-bundles over $\hat{\Sigma}$ if and only if they have the same local type.

Hint: Use Lemma 6.3.10 and topological triviality of G-bundles over open Riemann surfaces together with the result that A-equivariant G-bundles over a space X with free action of A are pull-back of G-bundles over the quotient space X/A.

(5) Let $\hat{\rho}: \pi \to \mathrm{GL}_V$ be a unitary representation and let $\hat{E}_{\hat{\rho}}(V)$ be the associated A-equivariant vector bundle over $\hat{\Sigma}$. Assume that $\wedge^{\mathrm{top}}(\hat{E}_{\hat{\rho}}(V))$ is A-equivariantly trivial line bundle. Show that $\mathrm{Im}\,\hat{\rho} \subset \mathrm{SL}_V$.

Hint: Consider the homomorphism $\bar{\hat{\rho}}: \pi \to \mathrm{GL}_1$, where $\bar{\hat{\rho}} := \det \circ \hat{\rho}$. Then $\hat{E}_{\bar{\hat{\rho}}}$ is A-equivariantly trivial (by assumption). Now, use Corollary 6.3.7 to conclude that $\bar{\hat{\rho}}$ is trivial.

7.C Comments

Most of the results in Section 7.1 (in particular, Theorem 7.1.14 for the degree $d = 0$) are due to Seshadri (1967). But Lemma 7.1.5 is due to Ramanathan (1996, Lemma 5.1) and we have taken Lemma 7.1.8 from Newstead (2012, Lemma 5.4). Our treatment in this section follows closely that of Newstead (2012, Chap. 5). Exercise 7.1.E.1 is taken from Seshadri (1967, Proposition 3.1 and Remark 3.1).

Section 7.2: Most of the results in Section 7.2 for $G = \mathrm{SL}_n$ (equivalently, for vector bundles over $\hat{\Sigma}$ of degree 0) are taken from Seshadri (2011). In particular, Theorem 7.2.6, Corollary 7.2.7, Theorems 7.2.8, 7.2.14, Corollary 7.2.20, Proposition 7.2.26 and Corollary 7.2.27 appear in Seshadri (2011, Lemma 3, Theorems 3 (Chap. I), §5 (Chap. II)). These results, barring Proposition 7.2.26 and Corollary 7.2.27 for any connected semisimple G (proved in this section) as well as Proposition 7.2.5 and Corollary 7.2.15, are taken from Balaji, Biswas and Nagaraj (2001, §5). It may however be mentioned that we have provided more details in some of the proofs in the same. These results also appear in Teleman and Woodward (2001, §§2.6 and 3.2), where some of the details are missing. We also refer to Balaji and Seshadri (2015) for several of these results proved in a slightly more general setting. The proof of smoothness of $R_\tau^{ss}(G)$ (cf. proof of Theorem 7.2.6) is an adaptation from Ramanathan (1996, proof of Lemma 4.13.3). We also refer to Simpson (1994, Lemma 10.7) for its dimension calculation.

The notion of parabolic structure on a vector bundle over a curve Σ and its semistability (and stability) is due to Seshadri (1977) (inspired by Weil (1938)). Mehta and Seshadri proved that the parabolic semistable vector bundles (over any smooth irreducible projective curve of any genus ≥ 2) of any degree d and parabolic degree 0 and any rational weights admit a coarse moduli space which is a normal, projective variety. Moreover, they proved that the parabolic stable bundles come from irreducible unitary representations of the fundamental group π of $\Sigma\backslash$ parabolic points (cf. (Mehta and Seshadri, 1980, Theorem 4.1)). These results were announced in Seshadri (1977).

As mentioned in Section 6.C, the notion of parabolic structures on vector bundles and their parabolic stability/semistability was extended in Bhosle and Ramanathan (1989) for any connected reductive group G. They proved Theorems 7.2.23 and 7.2.29 (cf. (Bhosle and Ramanathan, 1989, Theorems I, II)), though some of the crucial details are missing. In general, the paper Bhosle and Ramanathan (1989) has several unexplained details.

In view of the canonical bijection between A-semistable (resp. A-stable) G-bundles over $\hat{\Sigma}$ and the parabolic semistable (resp. stable) G-bundles over Σ (cf. Theorem 6.1.17), results from one can be translated into the other. With this correspondence in mind, Theorem 7.2.23 is equivalent to Theorems 7.2.6, 7.2.8, 7.2.14 and Corollary 7.2.20. Moreover, if we use this correspondence, many of the results in Section 7.2 have their parabolic counterparts available in Bhosle and Ramanathan (1989), though some of their proofs are unclear. The non-equivariant analogues of Proposition 7.2.3, Lemma 7.2.17 and Proposition 7.2.18 are taken from Ramanathan (1996) (also see (Bhosle and Ramanathan, 1989, Proposition 3.1)). Proposition 7.2.26 and Corollary 7.2.27 appear in Bhosle and Ramanathan (1989, Proposition 2.3 and its proof). For an analogue of Theorem 7.2.23 when the weights are not rational, the reader is referred to Balaji, Biswas and Pandey (2017).

Exercises 7.2.E.2 and 7.2.E.3 are taken from Seshadri (2011, Chap. II, Remark 5(iv) and Proposition 7(ii)) where their analogue for vector bundles is proved.

Faltings has given a construction of the coarse moduli space of parabolic semistable vector bundles (more generally, of Higgs G-bundles for a reductive group G) over smooth projective curves Σ of characteristic 0 without using the Geometric Invariant Theory (GIT) (Faltings, 1993). An improvement on Faltings' construction due to Nori gives a construction of the moduli space of semistable vector bundles on smooth projective curves in an arbitrary characteristic. In particular, Nori proved that a vector bundle $\mathscr{V}$ over Σ is semistable if and only if there exists a vector bundle $\mathscr{W}$ over Σ such that $\mathscr{V} \otimes \mathscr{W}$ is cohomologically trivial (i.e., $H^i(\Sigma, \mathscr{V} \otimes \mathscr{W}) = 0$, for all i) (cf. (Seshadri, 1993, Theorem 6.2)).

8

Identification of the Space of Conformal Blocks with the Space of Generalized Theta Functions

In this chapter Σ denotes a smooth irreducible projective curve of any genus $g \geq 0$ (except in Section 8.5 where we assume $g \geq 2$) and G a connected, simply-connected simple algebraic group over $\mathbb{C}$.

Let $\vec{q} = \{q_1, \ldots, q_s\}$ be a set of points of Σ for $s \geq 1$ and let $\Sigma^* := \Sigma \backslash \vec{q}$. Let $\bar{\Gamma} = \bar{\Gamma}_{\vec{q}}$ be the ind-affine group variety with $\mathbb{C}$-points $\Gamma := \mathrm{Mor}(\Sigma^*, G)$ (cf. Lemma 5.2.10) and let Γ^{an} denote the group Γ with the analytic topology. Then, our first theorem asserts that the group Γ^{an} is path-connected and hence $\bar{\Gamma}$ is irreducible (cf. Theorem 8.1.1). As a consequence of this result we derive that the infinite Grassmannian $\bar{X}_G$ is an irreducible ind-projective variety (cf. Corollary 8.1.4). Moreover, any morphism $f : \bar{\Gamma} \to \mathbb{C}^*$ is constant (cf. Corollary 8.1.5). The same result is also true for $\bar{G}[[t]]$ (cf. Exercise 8.1.E.2).

Let us take a point $p \in \Sigma$ and let $\bar{\Gamma}$ be as defined above for $\vec{q} = \{p\}$. Fix $\lambda_c \in \hat{D}$ and recall the central extension in the category of reduced ind-affine group schemes (cf. Proposition 1.4.12):

$$1 \to \mathbb{G}_m \to \bar{\hat{G}}_{\lambda_c} \xrightarrow{\bar{p}} \bar{G}((t)) \to 1.$$

Then, the main result of Section 8.2 asserts that the above central extension for $\lambda_c = 0_c$ splits over $\bar{\Gamma}$. Moreover, the tangent map to the splitting is the identity map (cf. Theorem 8.2.1). Further, the splitting over $\bar{\Gamma}$ is unique (cf. Corollary 8.2.3).

In Section 8.3, after determining the $\bar{\Gamma}$-equivariant Picard group of $\bar{X}_G$ (cf. Proposition 8.3.2), we prove the main result of this section (cf. Theorem 8.3.4) which identifies the space of conformal blocks over Σ with the space of global sections of an associated line bundle on the moduli stack $\mathbf{Parbun}_G(\Sigma, \vec{P})$ of quasi-parabolic G-bundles over Σ. Specifically, we have the following result, the proof of which crucially uses the analogue of the Borel–Weil theorem for affine Kac–Moody groups proved by Kumar and also Mathieu (cf. Theorem 8.3.3).

Theorem Let $(\Sigma, \vec{p})$ be an s-pointed smooth irreducible projective curve (for $s \geq 1$) and let $\vec{\lambda} = (\lambda_1, \ldots, \lambda_s)$ be a set of weights in D_c attached to $\vec{p}$. Pick a point $p \in \Sigma \backslash \vec{p}$ and let $\Sigma^* = \Sigma \backslash p$. Then, there is a canonical isomorphism up to a scalar multiple

$$H^0\left(\mathbf{Parbun}_G(\Sigma, \vec{P}), \bar{\mathscr{L}}(\vec{\lambda})\right) \simeq \mathscr{V}_{\Sigma}^{\dagger}(\vec{p}, \vec{\lambda}),$$

where the line bundle $\bar{\mathscr{L}}(\vec{\lambda})$ is defined above Theorem 8.3.4. We also determine the Picard group of $\mathbf{Parbun}_G(\Sigma, \vec{P})$ in Theorem 8.3.5.

For any family $\mathscr{E}$ of G-bundles over $\Sigma \times S$ (for connected noetherian scheme S) and any G-module V, we define the theta bundle $\Theta(\mathscr{E}(V))$, which is a line bundle over S (cf. Definition 8.4.1). Even though the moduli space $M^G = M^G(\Sigma)$ of semistable G-bundles over Σ does not parameterize a family of G-bundles, we still have a line bundle $\Theta(V)$ over M^G such that $f_{\mathscr{E}}^*(\Theta(V)) \simeq \Theta(\mathscr{E}(V))$ for any family $\mathscr{E}$ consisting of semistable G-bundles over Σ, where $f_{\mathscr{E}} : S \to M^G$ is the morphism induced from the family $\mathscr{E}$ via the coarse moduli property of M^G (cf. Definition 8.4.4).

Recall the definition of the Dynkin index for any Lie algebra homomorphism between simple Lie algebras from Definition A.1. Also, recall from Proposition 5.2.4 the definition of the 'tautological' G-bundle $\mathbf{U}_G$ over $\Sigma \times \bar{X}_G$. In particular, we get the Θ-bundle $\Theta(\mathbf{U}_G(V))$ over $\bar{X}_G$ for any G-module V. The following is a crucial result from this chapter (cf. Theorem 8.4.7).

Theorem For any finite-dimensional representation V of G, $\Theta(\mathbf{U}_G(V)) \simeq \mathscr{L}(0_{d_V})$, as line bundles over $\bar{X}_G$, where $\mathscr{L}(0_c)$ is defined in the proof of Proposition 8.3.2, d_V is the Dynkin index of V and $\mathbf{U}_G(V) := \mathbf{U}_G \overset{G}{\times} V$.

The following theorem asserts that the space of conformal blocks is canonically identified (up to scalar multiples) with the space of generalized theta functions. Precisely, the theorem is as follows (cf. Theorem 8.4.15):

Theorem Let $p \in \Sigma$ be a base point and let V be a representation of G with Dynkin index d_V. Then, for any $a \geq 0$ there exists a canonical isomorphism

$$H^0(M^G(\Sigma), (\Theta(V))^{\otimes a}) \simeq H^0(\bar{X}_G, \mathscr{L}(0_{d_V})^{\otimes a})^{\Gamma},$$

where $\Gamma := G[\Sigma \backslash p]$. Moreover, there exists a canonical identification (up to scalar multiples)

$$H^0\left(\bar{X}_G, \mathscr{L}(0_{d_V})^{\otimes a}\right)^{\Gamma} \simeq \mathscr{V}_{\Sigma}^{\dagger}(p, 0_{d_V \cdot a}),$$

and hence its dimension is given by the Verlinde formula (Theorem 4.2.19).

By Exercise 8.4.E.3, $M^G(\Sigma)$ is a unirational variety.

Section 8.5 is devoted to proving the parabolic analogue of the above theorem. In this section $(\Sigma, \vec{p})$ is an s-pointed smooth irreducible projective curve of any genus $g \geq 2$ and we fix a base point $p \in \Sigma$ different from any $\vec{p}$. We fix rational parabolic weights $\vec{\tau} = (\tau_1, \ldots, \tau_s)$, $\tau_j \in \Phi_0$ (cf. Definition 6.1.2) satisfying $\theta(\tau_j) < 1$ for the highest root θ. This gives rise to standard parabolic subgroups $\vec{P} = (P_1, \ldots, P_s)$ and a Galois cover $\pi : \hat{\Sigma} \to \Sigma$ (with Galois group A). As in Exercise 7.2.E.4, the set of markings $\vec{\tau}$ equivalently provides the topological type (denoted) τ of A-equivariant G-bundles over $\hat{\Sigma}$. We define the quasi-parabolic determinant line bundle over S associated to any family $(\mathscr{E}, \vec{\sigma})$ of quasi-parabolic G-bundles of type $\vec{P}$ over $(\Sigma, \vec{p})$ parameterized by a noetherian scheme S, a finite-dimensional representation V of G, $d \in \mathbb{Z}$ and characters $\vec{\mu} = (\mu_1, \ldots, \mu_s)$, where μ_j is a character of P_j (cf. Definition 8.5.1). We then define in Definition 8.5.4 the parabolic theta bundle $\Theta_{\text{par},G}(V, \tau, d)$ over the A-equivariant moduli space $M_\tau^G(\hat{\Sigma})$ of topological type τ corresponding to a G-module V and $d \in \mathbb{Z}_{>0}$ satisfying (1) of Definition 8.5.4. This line bundle is ample if V is a faithful G-module. With this notation, we have the following parabolic analogue of Theorem 8.4.15, which is the main result of this section (cf. Theorem 8.5.9).

Theorem For any representation V of G, any central charge $c > 0$ and any $\vec{\lambda} = (\lambda_1, \ldots, \lambda_s)$ with $\lambda_j \in D_c$ such that for all j, $\lambda_j(\theta^\vee) < c$ and λ_j satisfies condition (2) of Theorem 8.5.9, we have a canonical isomorphism up to a scalar multiple:

$$H^0\left(M^G_{\tau(\vec{\lambda})}(\hat{\Sigma}), \Theta_{\text{par},G}(V, \tau(\vec{\lambda}), c)\right) \simeq \mathscr{V}^\dagger_\Sigma(\vec{p}, d_V \vec{\lambda}),$$

where $\vec{\tau}(\vec{\lambda}) = \left(\frac{\kappa_{\mathfrak{g}}^{-1}(\lambda_1)}{c}, \ldots, \frac{\kappa_{\mathfrak{g}}^{-1}(\lambda_s)}{c}\right)$, $\tau(\vec{\lambda})$ is the corresponding A-equivariant topological type ($\kappa_{\mathfrak{g}}$ is the isomorphism $\mathfrak{h} \xrightarrow{\sim} \mathfrak{h}^*$ induced from the invariant bilinear form) and the space of vacua $\mathscr{V}^\dagger_\Sigma$ is taken at the central charge cd_V.

Remark 8.5.10 and Exercise 8.5.E.1 contain the values of c and λ_j for which the above theorem is applicable.

8.1 Connectedness of Γ

Let $\vec{q} = \{q_1, \ldots, q_s\}$ be a set of points of Σ for $s \geq 1$ and let $\Sigma^* := \Sigma \backslash \vec{q}$. Recall from Definition 5.2.9 and Lemma 5.2.10 that $\bar{\Gamma} = \bar{\Gamma}_{\vec{q}}$ (with $\mathbb{C}$-points $\Gamma := \bar{\Gamma}(\mathbb{C}) = \text{Mor}(\Sigma^*, G)$) is an ind-affine group variety (filtered by schemes $\bar{\Gamma}^{(n)}$ of finite type over $\mathbb{C}$) which represents the functor:

$$R \in \textbf{Alg} \rightsquigarrow \Gamma(R) := \text{Mor}(\Sigma^*_R, G), \text{ where } \Sigma^*_R := \Sigma^* \times \text{Spec } R.$$

Let Γ^{an} denote the group Γ with the analytic topology.

Theorem 8.1.1 *The group Γ^{an} is path-connected and hence $\bar{\Gamma}$ is irreducible.*

Proof Take any distinct points $p_1, \ldots, p_n, p_{n+1} \in \Sigma^*$ and set (for any $0 \leq i \leq n+1$)

$$\Gamma_i = \Gamma_{\bar{q} \cup \{p_1, \ldots, p_i\}} = \text{Mor}(\Sigma_i^*, G), \quad \text{where} \quad \Sigma_i^* := \Sigma^* \backslash \{p_1, \ldots, p_i\}.$$

Consider the functor

$$\mathscr{F}^\circ \colon R \rightsquigarrow \Gamma_{n+1}(R)/\Gamma_n(R), \text{ where } \Gamma_i(R) := \text{Mor}\big((\Sigma_i^*)_R, G\big).$$

It is easy to see that

$$\mathscr{F}^\circ(R) \hookrightarrow \mathscr{F}^\circ(R'), \quad \text{for any } \mathbb{C} \text{ algebras } R \subset R'; \tag{1}$$

in particular, it satisfies condition (1) of Lemma B.2.

Let $\mathscr{F}$ be the sheafification of $\mathscr{F}^\circ$ (cf. Lemma B.2), which we denote by $\widehat{\Gamma_{n+1}/\Gamma_n}$. We claim that as $\mathbb{C}$-space functors

$$\widehat{\Gamma_{n+1}/\Gamma_n} \simeq \bar{X}_G, \tag{2}$$

where $\mathscr{X}_G := \mathscr{X}_G(p_{n+1})$ denotes the infinite Grassmannian functor based at p_{n+1} (cf. Definition 1.3.5) represented by $\bar{X}_G$ as in Proposition 1.3.18. Take the open cover $\Sigma = (\Sigma \backslash p_{n+1}) \cup \Sigma_n^*$. Given any $\gamma \in \Gamma_{n+1}(R)$, consider the G-bundle E_γ over $\Sigma_R := \Sigma \times \text{Spec}\, R$ obtained via Lemma 5.2.3 from the Laurent series expansion $(\gamma)_{p_{n+1}}$ of γ at p_{n+1} (cf. Corollary 5.2.11): $\Gamma_{n+1}(R) \to G(R((t)))$. Since E_γ is equipped with a section $\sigma_R(\gamma)$ over $(\Sigma \backslash p_{n+1})_R$ (and also a section $\mu_R(\gamma)$ over the formal disc $\mathbb{D}_R$ at p_{n+1}), by Proposition 5.2.7, the pair $(E_\gamma, \sigma_R(\gamma))$ gives rise to a morphism $\theta_\gamma \colon \text{Spec}\, R \to \bar{X}_G$. This is the map

$$\theta \colon \Gamma_{n+1}(R) \to \mathscr{X}_G(R), \gamma \mapsto \theta_\gamma.$$

Moreover, this map factors through $\Gamma_{n+1}(R)/\Gamma_n(R)$, since (for any $\gamma' \in \Gamma_n(R)$) taking the section $\mu_R(\gamma) \cdot (\gamma')_{p_{n+1}}$ over $\mathbb{D}_R$ in Lemma 5.2.3, we get that $(E_{\gamma\gamma'}, \sigma_R(\gamma\gamma'))$ can be taken to be $(E_\gamma, \sigma_R(\gamma))$. Hence, by Proposition 5.2.7,

$$\theta(\gamma\gamma') = \theta(\gamma)$$

giving rise to the map

$$\bar{\theta} \colon \Gamma_{n+1}(R)/\Gamma_n(R) \to \mathscr{X}_G(R).$$

But, since $\mathscr{X}_G$ is a $\mathbb{C}$-space functor, $\bar{\theta}$ extends to the sheafification (cf. Lemma B.2)

$$\hat{\theta} \colon \widehat{\Gamma_{n+1}/\Gamma_n} \to \mathscr{X}_G.$$

Conversely, we define a map $\hat{\psi}: \mathscr{X}_G \to \widehat{\Gamma_{n+1}/\Gamma_n}$ as follows. Fix $R \in \mathbf{Alg}$. Take an $\alpha = (E_R, \sigma_R) \in \mathscr{X}_G(R) = \text{Mor}\left(\text{Spec}\, R, \bar{X}_G\right)$, where E_R is a G-bundle over Σ_R together with a section σ_R over $(\Sigma \backslash p_{n+1})_R$ (cf. Proposition 5.2.7). By Theorem 5.2.5, there exists an étale cover $f: \text{Spec}\, S \to \text{Spec}\, R$ (in particular, S is an fppf R-algebra by (Stacks, 2019, Tag 021N)) such that $E_S := (\text{Id} \times f)^* E_R$ has a section μ_S over $(\Sigma_n^*)_S$. Define the element $\psi_{\mu_S}(\alpha) \in \Gamma_{n+1}(S)$ by

$$\mu_S = \sigma_S \cdot \psi_{\mu_S}(\alpha), \text{ where } \sigma_S \text{ is the pull-back of } \sigma_R \text{ to } \Sigma_S \text{ via } \text{Id} \times f. \quad (3)$$

If we take any other section μ'_S over $(\Sigma_n^*)_S$, then

$$\mu'_S = \mu_S \cdot \psi'_{\mu'_S}, \quad \text{for} \quad \psi'_{\mu'_S} \in \Gamma_n(S).$$

Thus, from (3), we get

$$\psi_{\mu'_S}(\alpha) = \psi_{\mu_S}(\alpha)\psi'_{\mu'_S}, \quad (4)$$

i.e., the coset $\psi_f(\alpha) := \psi_{\mu_S}(\alpha)\Gamma_n(S)$ does not depend upon the choice of the section μ_S over $(\Sigma_n^*)_S$. This gives rise to a well-defined map

$$\mathscr{X}_G(R) \to \Gamma_{n+1}(S)/\Gamma_n(S), \quad \alpha \mapsto \psi_f(\alpha).$$

From the analogue of equation (4) for S replaced by $S \underset{R}{\otimes} S$, it is easy to see that for any $\alpha \in \mathscr{X}_G(R)$, $\psi_f(\alpha)$ belongs to the equalizer of the two maps

$$\Gamma_{n+1}(S)/\Gamma_n(S) \rightrightarrows \Gamma_{n+1}\left(S \underset{R}{\otimes} S\right) / \Gamma_n\left(S \underset{R}{\otimes} S\right),$$

i.e., following the notation of the proof of Lemma B.2, $\psi_f(\alpha) \in K_R(S)$ for the functor $\mathscr{F}^\circ(R) = \Gamma_{n+1}(R)/\Gamma_n(R)$. Thus, we get a well-defined map

$$\hat{\psi}: \mathscr{X}_G(R) \to (\widehat{\Gamma_{n+1}/\Gamma_n})(R), \ \alpha \mapsto \psi_f(\alpha).$$

From their definition, using Lemma 5.2.3 and Proposition 5.2.7, it is easy to see that $\hat{\theta}$ and $\hat{\psi}$ are inverses of each other. This proves assertion (2). Thus, the functor $\widehat{\Gamma_{n+1}/\Gamma_n}$ is a representable functor represented by $\bar{X}_G$.

Consider the morphism $\pi: \bar{\Gamma}_{n+1} \to \bar{X}_G$, induced from the quotient map

$$\Gamma_{n+1}(R) \to \Gamma_{n+1}(R)/\Gamma_n(R) \hookrightarrow (\widehat{\Gamma_{n+1}/\Gamma_n})(R) \simeq \mathscr{X}_G(R),$$

where the ind-affine variety $\bar{\Gamma}_i$ represents the functor $\Gamma_i(R)$ (cf. Lemma 5.2.10).

Recall from the above proof that $\pi(\gamma)$ (for any $\gamma \in \Gamma_{n+1}(R)$) is the G-bundle E_γ over Σ_R obtained from the Laurent series expansion $(\gamma)_{p_{n+1}}$ of γ at p_{n+1} together with a section $\sigma_R(\gamma)$ over $(\Sigma \backslash p_{n+1})_R$ via Lemma 5.2.3.

For any $R \in \mathbf{Alg}$ and any $\alpha = (E_R, \sigma_R) \in \text{Mor}(\text{Spec } R, \bar{X}_G)$, as observed above, there exists an étale (in particular, fppf) cover $\text{Spec } S \to \text{Spec } R$ with an element of $\pi^{-1}(\alpha)$ over Spec S given by the element $\psi_{\mu_S}(\alpha) \in \Gamma_{n+1}(S)$ (cf. (3)).

Since, by definition, $\Gamma_i(\mathbb{C}) = \Gamma_i$, from (1), we see that

$$\Gamma_{n+1}/\Gamma_n \hookrightarrow (\widehat{\Gamma_{n+1}/\Gamma_n})(\mathbb{C}) \simeq \mathscr{X}_G(\mathbb{C}).$$

Moreover, from the above proof using Theorem 5.2.5 and Proposition 5.2.7, it is easy to see that $\hat{\psi} : \mathscr{X}_G(\mathbb{C}) \xrightarrow{\sim} (\widehat{\Gamma_{n+1}/\Gamma_n})(\mathbb{C})$ lands inside Γ_{n+1}/Γ_n. Thus, we get

$$\Gamma_{n+1}/\Gamma_n = (\widehat{\Gamma_{n+1}/\Gamma_n})(\mathbb{C}) \simeq \mathscr{X}_G(\mathbb{C}). \tag{5}$$

This identification gives rise to an ind-projective variety structure on Γ_{n+1}/Γ_n transported from that of $\bar{X}_G$ (cf. Proposition 1.3.24). With this ind-variety structure, Γ_{n+1}/Γ_n represents the functor $\widehat{\Gamma_{n+1}/\Gamma_n}$. It is easy to see (by considering the corresponding map at R-points) that with this ind-variety structure on Γ_{n+1}/Γ_n, the action map

$$\Gamma_{n+1} \times (\Gamma_{n+1}/\Gamma_n) \to \Gamma_{n+1}/\Gamma_n$$

is a morphism of ind-varieties. Further, for any morphism $\alpha\colon \text{Spec } R \to \bar{X}_G$ (for $R \in \mathbf{Alg}$), from the existence of an element of $\pi^{-1}(\alpha)$ over Spec S for certain étale cover Spec S of Spec R (proved above), it is easy to see that $(\Gamma_{n+1}/\Gamma_n)^{\text{an}}$ has the quotient topology obtained from Γ_{n+1}^{an} (since étale morphism is an open map in the analytic topology (Mumford, 1988, Chap. III.5, Corollary 2, p. 252)).

Take a filtration of $\bar{X}_G$ by projective varieties $(\bar{X}_G^N)_{N\geq 0}$ giving the ind-variety $\bar{X}_G$. For any morphism $\alpha\colon Y \to \bar{X}_G$, for Y an affine variety, there exists an étale cover $\beta\colon \tilde{Y} \to Y$ such that the base change of $\pi : \bar{\Gamma}_{n+1} \to \bar{X}_G$ via $\alpha \circ \beta$ has a section. In particular, taking Y to be an affine open subset of $\bar{X}_G^N$ (for any N), we see that $\pi_{|\bar{X}_G^N}$ has a local section in the analytic topology. Thus, $\pi_{|\bar{X}_G^N}$ is a fiber bundle (cf. (Steenrod, 1951, Corollary 7.4)). In particular, since for any ind-variety $X = (X_n)_{n\geq 0}$, any compact subset of X^{an} lies in some X_N as in Exercise 8.1.E.1,

$$\Gamma_{n+1}^{\text{an}} \to (\Gamma_{n+1}/\Gamma_n)^{\text{an}}$$

is a Serre fibration. This gives rise to an exact sequence (cf. (Spanier, 1966, Chap. 7, §2, Theorem 10))

$$\pi_1(X_G^{\text{an}}) \to \pi_0(\Gamma_n^{\text{an}}) \to \pi_0(\Gamma_{n+1}^{\text{an}}) \to \pi_0(X_G^{\text{an}}), \tag{6}$$

where $X_G = \bar{X}_G(\mathbb{C})$. But

$$\pi_1(X_G^{\mathrm{an}}) = \pi_0(X_G^{\mathrm{an}}) = 0, \tag{7}$$

from the Bruhat decomposition (cf. (Kumar, 2002, Proposition 7.4.16) and Proposition 1.3.24). Thus, we get

$$\pi_0(\Gamma_n^{\mathrm{an}}) \simeq \pi_0(\Gamma_{n+1}^{\mathrm{an}}). \tag{8}$$

Now we are ready to prove the theorem. Take

$$\gamma \in \Gamma := \mathrm{Mor}(\Sigma^*, G) = G(\mathbb{C}[\Sigma^*]) \subset G(L),$$

where L is the quotient field of $\mathbb{C}[\Sigma^*]$. Since G is simply connected and, of course, split over L (since $L \supset \mathbb{C}$), by Steinberg's result, $G(L)$ is generated by root subgroups $U_\alpha(L)$ over L (cf. (Steinberg, 2016, Chapter 6)). Moreover, U_α being a unipotent group and $L \supset \mathbb{C}$, $U_\alpha(L) \simeq \mathfrak{u}_\alpha(L)$ under the exponential map. Thus, we can write

$$\gamma = \mathrm{Exp}(f_1 x_1) \ldots \mathrm{Exp}(f_d x_d), \quad \text{for some root vectors } x_i \in \mathfrak{g} \text{ and } f_i \in L.$$

Thus, there exists a finite set $\{p_1, \ldots, p_{n+1}\} \subset \Sigma^*$ given by the poles of f_i such that $\gamma \in \Gamma_{n+1}$. Consider the curve

$$\hat{\gamma}\colon [0,1] \to \Gamma_{n+1}^{\mathrm{an}}, \quad t \mapsto \mathrm{Exp}(tf_1 x_1) \ldots \mathrm{Exp}(tf_d x_d) \ \text{ joining } e \text{ to } \gamma.$$

Since

$$\pi_0(\Gamma^{\mathrm{an}}) \simeq \pi_0(\Gamma_{n+1}^{\mathrm{an}}), \quad \text{by (8)},$$

we get that e and γ lie in the same path component of Γ^{an}, thus Γ^{an} is path-connected. Using Kumar (2002, Lemma 4.2.5) we get that Γ is irreducible. □

The following lemma is false in general if the base field is of characteristic $p > 0$.

Lemma 8.1.2 *Let $f\colon X \to Y$ be a morphism between ind-varieties $X = (X_n)_{n\geq 0}$ and $Y = (Y_m)_{m\geq 0}$. Assume that X is connected in the Zariski topology and*

$$(df)_x\colon T_x(X) \to T_{f(x)}(Y)$$

is the zero map for all $x \in X$, where the (Zariski) tangent space is as in Definition B.7. Then f is constant.

In particular, using (1) *below in the proof, if X is a connected ind-projective variety and Y is an ind-affine variety, then any morphism $f\colon X \to Y$ is constant.*

Proof Take an irreducible component X'_n of X_n. Then, there exists m such that $f(X'_n) \subset Y_m$. From the assumption we get $(d(f_{|X'_n}))_x \equiv 0$, for any $x \in X'_n$. Thus, the base field being $\mathbb{C}$ (in particular, of characteristic 0), f is constant on the smooth locus of X'_n and hence on the whole of X'_n.

Fix $x_o \in X_0$ and let X^o_n be the connected component of X_n in the analytic topology containing x_o. Then, it is both open and closed in X_n in either Zariski or analytic topology and it is a union of certain irreducible components of X_n. Let

$$X^o := \bigcup_{n \geq 0} X^o_n.$$

Then, X^o is both open and closed in X in the Zariski topology (since, for any $n \geq N$, $X^o_n \cap X_N$ being both open and closed in X_N in the Zariski topology is a union of irreducible components of X_N). Thus,

$$X^o = X \quad \text{(since } X \text{ is connected by assumption).} \tag{1}$$

Since $f_{|X'_n}$ is constant, so is $f_{|X^o_n}$. Thus, f is constant on X by (1), proving the lemma. □

As an immediate corollary of the above lemma, we get the following.

Corollary 8.1.3 *Let $f: G \to H$ be an ind-group morphism between ind-group varieties. Assume that G is connected and the derivative $\dot{f}$: $\mathrm{Lie}G \to \mathrm{Lie}H$ is zero, where $\mathrm{Lie}G$ is the Lie algebra of G (cf. Corollary B.21). Then, f is the constant map.*

As a corollary of Theorem 8.1.1 and Proposition 5.2.7, we get the following.

Corollary 8.1.4 *The infinite Grassmannian $\bar{X}_G$ is an irreducible ind-projective variety.*

Proof Let $\Sigma = \mathbb{P}^1, \bar{q} = \{0, \infty\}$ and the ind-variety $\bar{\Gamma}$ with $\bar{\Gamma}(\mathbb{C}) = \mathrm{Mor}(\Sigma \backslash \bar{q}, G)$. Consider the natural transformation between the functors

$$\Gamma(R) = G(R[t,t^{-1}]) \to G\left(R((t))\right)/G\left(R[[t]]\right).$$

This gives rise to the morphism between the corresponding ind-varieties (cf. Lemma 5.2.10 and Proposition 1.3.18):

$$\pi : \bar{\Gamma} \to \bar{X}_G.$$

Using Proposition 5.2.7 and its proof for $\Sigma = \mathbb{P}^1, \Sigma^* = \mathbb{P}^1 \backslash 0$, it is easy to see that π is surjective on $\mathbb{C}$-points (since any G-bundle on $\mathbb{P}^1$ has a trivialization on $\mathbb{P}^1 \backslash \infty$ by Theorem 5.2.5). Since $\bar{\Gamma}$ is irreducible (by Theorem 8.1.1) and both of $\bar{\Gamma}$ and $\bar{X}_G$ are *ind-varieties*, we get the irreducibility of $\bar{X}_G$. Observe that in the proof of Theorem 8.1.1 we used the connectedness and simply-connectedness of $X_G^{\rm an}$; in particular, this corollary builds upon the connectedness of $X_G^{\rm an}$ to prove the stronger result. □

As a corollary of Theorem 8.1.1 and Lemma 8.1.2, we get the following.

Corollary 8.1.5 *Let $\bar{\Gamma}$ be as in Theorem 8.1.1. Then any morphism $f : \bar{\Gamma} \to \mathbb{C}^*$ is constant.*

In particular, any morphism $\bar{G}[t] \to \mathbb{C}^*$ is constant. The same result is also true for $\bar{G}[[t]]$ (cf. Exercise 8.1.E.2). For the notation $\bar{G}[t]$ and $\bar{G}[[t]]$, see Lemma 1.3.2.

Proof Take any unipotent subgroup $(e) \neq U \subset G$. For any $g \in G$, it is easy to see (cf. Definition 5.2.9 and the proof of Lemma 5.2.10) that the functor $R \rightsquigarrow \operatorname{Mor}(\Sigma_R^*, gUg^{-1})$ is represented by a closed ind-subvariety (denoted) $\bar{\Gamma}_{g,U}$ of $\bar{\Gamma}$, where $\bar{\Gamma}_{g,U}$ has $\mathbb{C}$-points $\Gamma_{g,U} := \operatorname{Mor}(\Sigma^*, gUg^{-1})$. Moreover, U being a unipotent group,

$$\operatorname{Exp}: \mathfrak{u} \to U, \quad \text{where} \quad \mathfrak{u} = \operatorname{Lie} U,$$

is an isomorphism of varieties. Thus, $\Gamma_{g,U} \simeq \mathbb{C}[\Sigma^*] \otimes \mathfrak{u}$. Since, there are no nonconstant morphisms from an affine space to $\mathbb{C}^*$,

$$f_{|\bar{\Gamma}_{g,U}} = \text{constant}.$$

In particular,

$$(df_{|\bar{\Gamma}_{g,U}})_x : T_x(\bar{\Gamma}_{g,U}) \to T_{f(x)}(\mathbb{C}^*), \text{ for any } x \in \Gamma_{g,U} \text{ is the zero map.}$$

It is easy to see (cf. Lemma B.14) that

$$\operatorname{Lie}(\bar{\Gamma}_{g,U}) = \mathbb{C}[\Sigma^*] \otimes (\operatorname{Ad} g \cdot \mathfrak{u}).$$

Similarly,

$$\operatorname{Lie}(\bar{\Gamma}) = \mathbb{C}[\Sigma^*] \otimes \mathfrak{g}.$$

Moreover, for any nontrivial U, being a G-stable subspace, $\sum_{g \in G} (\operatorname{Ad} g \cdot \mathfrak{u}) = \mathfrak{g}$. Thus,

$$(df)_e : T_e(\bar{\Gamma}) \to T_{f(e)}(\mathbb{C}^*) \text{ is the zero map.}$$

Replacing f by $f(\gamma \cdot)$ (for any $\gamma \in \Gamma$), we get that

$$(df)_\gamma : T_\gamma(\bar{\Gamma}) \to T_{f(\gamma)}(\mathbb{C}^*) \text{ is the zero map.}$$

Now, $\bar{\Gamma}$ being irreducible (by Theorem 8.1.1), using Lemma 8.1.2, we get the first part of the corollary.

Of course, $\bar{\Gamma}$ for $\Sigma^* = \mathbb{P}^1\backslash\infty$ is nothing but $\bar{G}[t]$. □

8.1.E Exercises

(1) Let $X = (X_n)_{n\geq 0}$ be an ind-variety and let $K \subset X^{\mathrm{an}}$ be a compact subset in the analytic topology X^{an} on X. Show that $K \subset X_N$ for some N.

(2) Show that any morphism $f : \bar{G}[[t]] \to \mathbb{C}^*$ is constant (cf. Corollary 8.1.5).

Hint: Let $\bar{G}[[t]]^+$ be the kernel of $\bar{G}[[t]] \to G, t \mapsto 0$.

Show that the exponential map induces an isomorphism for any $\mathbb{C}$-algebra R: $\mathfrak{g} \otimes tR[[t]] \simeq G(R[[t]])^+ := \mathrm{Ker}\,(G(R[[t]]) \to G(R))$. Thus, $G(R[[t]])^+$ is represented by an affine scheme corresponding to the polynomial ring $\mathbb{C}[\mathbf{X}]$ in countably infinite many variables. Since $\mathbb{C}[\mathbf{X}]$ has no invertible elements except constants, any morphism $\bar{G}[[t]]^+ \to \mathbb{C}^*$ is a constant. Also, any morphism $G \to \mathbb{C}^*$ is constant by Lemma 1.4.9.

8.2 Splitting of the Loop Group Central Extension Over $\bar{\Gamma}$

Let Σ and G be as at the beginning of this chapter and let $p \in \Sigma$ be a base point. Let $\Sigma^* := \Sigma\backslash p$ and let $\Gamma := \mathrm{Mor}(\Sigma^*, G)$. Then $\Gamma = \bar{\Gamma}(\mathbb{C})$, for an irreducible ind-affine group variety $\bar{\Gamma}$ (cf. Definition 5.2.9, Lemma 5.2.10 and Theorem 8.1.1). Moreover, there is a morphism (cf. Corollary 5.2.11) $\bar{\Gamma} \to \bar{G}((t))$ obtained by taking the Laurent series expansion at p (with respect to a parameter t at p).

Fix $\lambda_c \in \hat{D}$ and recall the central extension in the category of reduced ind-affine group schemes (cf. Proposition 1.4.12 and Remark 1.3.26(b)):

$$1 \to \mathbb{G}_m \to \bar{\hat{G}}_{\lambda_c} \xrightarrow{\bar{p}} L_G = \bar{G}((t)) \to 1. \qquad (*_{\lambda_c})$$

Theorem 8.2.1 *The above central extension* $(*_{0_c})$ *for* $\lambda_c = 0_c$ *splits over* $\bar{\Gamma}$. *Moreover, the tangent map to the splitting is the identity map.*

Proof Take $R \in \mathbf{Alg}$ and consider the action of the R-Lie algebra

$$\mathfrak{g}(\Sigma^*_R) := \mathfrak{g}\underset{\mathbb{C}}{\otimes}\mathbb{C}[\Sigma^*]\underset{\mathbb{C}}{\otimes}R \subset \hat{\mathfrak{g}}(R)$$

on $\mathscr{H}(0_c)_R := \mathscr{H}(0_c) \underset{\mathbb{C}}{\otimes} R$ (cf. Definition 1.4.2). By the Residue Theorem (cf. (Hartshorne, 1977, Chap. III, Theorem 7.14.2)) $\mathfrak{g}(\Sigma_R^*)$ is indeed a Lie subalgebra of $\hat{\mathfrak{g}}(R)$. Let

$$V_R := \mathfrak{g}(\Sigma_R^*) \cdot \mathscr{H}(0_c)_R \subset \mathscr{H}(0_c)_R.$$

It is easy to see that

$$V_R = V \underset{\mathbb{C}}{\otimes} R, \quad \text{where} \quad V := V_{\mathbb{C}}. \tag{1}$$

Let $\hat{\rho}_o \colon \hat{\bar{\Gamma}} \to \hat{\mathscr{G}}_{0_c} \to \mathbf{GL}_{\mathscr{H}(0_c)}$ be the restriction of the representation as in Definition 1.4.5, where $\hat{\bar{\Gamma}}$ is the ind-affine group variety obtained as the pull-back of $\bar{p}$ over $\bar{\Gamma}$ for $\lambda_c = 0_c$ and $\hat{\Gamma}$ is the corresponding $\mathbb{C}$-group functor.

We claim that

$$\hat{\Gamma}(R) \cdot V_R \subset V_R. \tag{2}$$

By (1) of Corollary 1.4.7, for any $\hat{\gamma} \in \hat{\Gamma}(R)$, $x \in \mathfrak{g}(\Sigma_R^*)$ and $v \in \mathscr{H}(0_c)_R$ (with the notation in the same),

$$\hat{\rho}_o(\hat{\gamma})\bar{\rho}_R(x)v = \bar{\rho}_R(\mathbf{Ad}_C(p(\hat{\gamma})) \cdot x)\hat{\rho}_o(\hat{\gamma})v.$$

This gives

$$\hat{\rho}_o(\hat{\gamma}) \cdot V_R \subset V_R,$$

$$\text{since } \mathbf{Ad}_C(p(\hat{\gamma})) \cdot \mathfrak{g}(\Sigma_R^*) \subset \mathfrak{g}(\Sigma_R^*) \text{ (by the Residue Theorem).}$$

This proves (2). We next claim that

$$\text{the action of } \hat{\Gamma}(R) \text{ on } \mathscr{H}(0_c)_R / V_R \text{ is via } R^* \cdot I. \tag{3}$$

The homomorphism of $\mathbb{C}$-group functors $\rho \colon \mathscr{L}_G \to \mathbf{PGL}_{\mathscr{H}(0_c)}$ (cf. Theorem 1.4.4) by virtue of (2) descends to a morphism of $\mathbb{C}$-group functors

$$\bar{\rho} \colon \bar{\Gamma} \to \mathbf{PGL}_{\mathscr{H}(0_c)/V}.$$

Since $\mathscr{H}(0_c)/V$ is finite dimensional (cf. Lemma 2.1.4), $\mathbf{PGL}_{\mathscr{H}(0_c)/V}$ is represented by the (finite-dimensional) projective linear group $\mathrm{PGL}_{\mathbb{C}}(\mathscr{H}(0_c)/V)$.

Thus, $\bar{\rho}$ gives rise to a morphism of ind-group varieties

$$\bar{\rho}_{\mathbb{C}} \colon \bar{\Gamma} \to \mathrm{PGL}_{\mathbb{C}}(\mathscr{H}(0_c)/V).$$

It is easy to see (using Lemma 5.2.10 and Definition B.15(b)) that

$$\mathrm{Lie}(\bar{\Gamma}(R)) = \mathfrak{g}(\Sigma_R^*).$$

Thus, the Lie algebra homomorphism induced from $\bar{\rho}_{\mathbb{C}}$ (cf. Exercise B.E.3 and Theorem 1.4.4)

$$\dot{\bar{\rho}}_{\mathbb{C}}\colon \operatorname{Lie}\bar{\Gamma} \to \frac{\operatorname{End}_{\mathbb{C}}(\mathscr{H}(0_c)/V)}{\mathbb{C}\cdot I}$$

is trivial. But, $\bar{\Gamma}$ being an irreducible ind-group variety (cf. Theorem 8.1.1), by Corollary 8.1.3, $\bar{\rho}_{\mathbb{C}}$ is the constant map. Hence, the map induced from $\hat{\rho}_o$ (by virtue of (2)):

$$\tilde{\rho}_{\mathbb{C}}\colon \hat{\bar{\Gamma}} \to \operatorname{GL}_{\mathbb{C}}(\mathscr{H}(0_c)/V)$$

has its image contained in $\mathbb{C}^*\cdot I$. This proves the assertion (3).

Take a nonzero vector $v_+ \in \mathscr{H}(0_c)\backslash V$ (cf. Corollary 3.5.11) and take a decomposition

$$\mathscr{H}(0_c) = \mathbb{C}v_+ \oplus V', \quad \text{where} \quad V' \supset V.$$

Then, by (2) and (3), for any $\hat{\gamma} \in \hat{\Gamma}(R)$, $\hat{\rho}_o(\hat{\gamma})v_+ \in R^*v_+ \oplus V'_R$, where $V'_R := V' \underset{\mathbb{C}}{\otimes} R$, and

$$\hat{\rho}_o(\hat{\gamma})(V'_R) \subset V'_R, \text{ i.e., } \hat{\rho}_o(\hat{\Gamma}(R)) \subset \mathbf{GL}''_{\mathscr{H}(0_c)}(R),$$

where $\mathbf{GL}''_{\mathscr{H}(0_c)}(R) := \{T \in \operatorname{GL}_R(\mathscr{H}(0_c)_R) : Tv_+ \in R^*v_+ \oplus V'_R$ and $T(V'_R) \subset V'_R\}$. It is clear that

$$\pi\left(\mathbf{GL}'_{\mathscr{H}(0_c)}\right) = \pi\left(\mathbf{GL}''_{\mathscr{H}(0_c)}\right),$$

where $\pi\colon \mathbf{GL}_{\mathscr{H}(0_c)} \to \mathbf{PGL}_{\mathscr{H}(0_c)}$ is the canonical homomorphism and $\mathbf{GL}'$ is defined in Lemma 1.4.8. Thus, the morphism $\rho\colon \mathscr{L}_G \to \mathbf{PGL}_{\mathscr{H}(0_c)}$ has its image under $\bar{\Gamma}$ in $\pi(\mathbf{GL}'_{\mathscr{H}(0_c)})$. Hence, Lemma 1.4.8 gives the splitting of $(*_{0_c})$ over $\bar{\Gamma}$.

Moreover, the assertion that the tangent map to the splitting is the identity map follows easily since $[\operatorname{Lie}(\bar{\Gamma}), \operatorname{Lie}(\bar{\Gamma})] = \operatorname{Lie}(\bar{\Gamma})$. □

Remark 8.2.2 The above Theorem 8.2.1 is false for $\Gamma = \operatorname{Mor}(\Sigma^*, G)$ if Σ^* is obtained from Σ by removing more than one point.

As an immediate consequence of Corollary 8.1.5, we get the following.

Corollary 8.2.3 *Following the notation of Proposition 1.4.12, for any $\lambda_c \in \hat{D}$, the morphism $\bar{p}\colon \hat{\bar{G}}_{\lambda_c} \to \bar{G}((t))$ admits a unique (up to a scalar multiple of $z \in \mathbb{C}^*$) regular section over $\bar{G}[t^{-1}]^- \times \bar{G}[[t]]$.*

In particular, the (group) splittings of $\bar{p}$ over $\bar{G}[t^{-1}]$ and $\bar{G}[[t]]$ (cf. Theorem 1.4.11) are unique.

Similarly, following the notation of Theorem 8.2.1, splitting of $\bar{p}\colon \hat{\bar{G}}_{0_c} \to \bar{G}((t))$ over $\bar{\Gamma}$ is unique. In fact, any regular section of $\bar{p}$ over $\bar{\Gamma}$ is unique up to a scalar multiple of $z \in \mathbb{C}^*$.

8.3 Identification of Conformal Blocks with Sections of Line Bundles over Moduli Stack

Definition 8.3.1 Let G be a connected, simply-connected, simple algebraic group over $\mathbb{C}$, $c > 0$ a level and let $(\Sigma, \vec{p} = (p_1, \dots, p_s))$ be an s-pointed smooth irreducible projective curve (for $s \geq 1$). Let $\vec{\lambda} = (\lambda_1, \dots, \lambda_s)$ with $\lambda_i \in D_c$ be a set of weights attached to the points $\vec{p}$. Fix a point $p_o \in \Sigma \backslash \vec{p}$. Recall from Corollary 2.2.3(b) that the space of conformal blocks

$$\mathscr{V}_\Sigma^\dagger(\vec{p}, \vec{\lambda}) \simeq \left[\mathscr{H}(0_c)^* \otimes V(\vec{\lambda})^*\right]^{\mathfrak{g}[\Sigma^*]}, \quad \text{where } \Sigma^* = \Sigma\backslash p_o \text{ and } \mathfrak{g} = \operatorname{Lie} G.$$

Let P_i be the standard parabolic subgroup of G such that its Levi subgroup L_i has for its simple roots

$$\Delta_i = \left\{\alpha_j : \lambda_i(\alpha_j^\vee) = 0\right\},$$

where $\{\alpha_1, \dots, \alpha_\ell\}$ are the simple roots of $\mathfrak{g}$ (cf. Section 1.2). Recall from Theorem 5.2.16 that the moduli stack of quasi-parabolic G-bundles of type $\vec{P} = (P_1, \dots, P_s)$ over $(\Sigma, \vec{p})$ is given by

$$\mathbf{Parbun}_G\left(\Sigma, \vec{P}\right) \simeq \left[\bar{\Gamma}\backslash \left(\bar{X}_G \times \Pi_{i=1}^s G/P_i\right)\right], \tag{1}$$

where $\bar{\Gamma}$ has $\mathbb{C}$-points $G[\Sigma^*]$, $\bar{\Gamma}$ acts on $\bar{X}_G$ through the ind-group morphism $\bar{\Gamma} \to \bar{G}((t))$ (see the beginning of Section 8.2) and $\bar{\Gamma}$ acts on G/P_i via the left multiplication by its evaluation at p_i.

By Exercise C.E.8,

$$\operatorname{Pic}\left[\bar{\Gamma}\backslash \left(\bar{X}_G \times \Pi_{i=1}^s G/P_i\right)\right] \simeq \operatorname{Pic}_{\bar{\Gamma}}\left(\bar{X}_G \times \Pi_{i=1}^s G/P_i\right), \tag{2}$$

where $\operatorname{Pic}_{\bar{\Gamma}}$ is the abelian group of isomorphism classes of $\bar{\Gamma}$-equivariant line bundles. The weight λ_i gives rise to the G-equivariant line bundle $\mathscr{L}_{P_i}(\lambda_i)$ associated to the principal P_i-bundle $G \to G/P_i$ via the 1-dimensional representation $\mathbb{C}_{\lambda_i^{-1}}$ of P_i given by the character λ_i^{-1}. Hence, $\mathscr{L}_{P_i}(\lambda_i)$ is $\bar{\Gamma}$-equivariant. Moreover, $\mathscr{L}_{P_i}(\lambda_i)$ is an ample line bundle.

We next determine $\operatorname{Pic}_{\bar{\Gamma}}(\bar{X}_G)$.

Proposition 8.3.2 *$\mathrm{Pic}_{\bar{\Gamma}}(\bar{X}_G) \simeq \mathbb{Z}$, where $c \in \mathbb{Z}$ corresponds to the $\bar{\Gamma}$-equivariant line bundle $\mathscr{L}(0_c)$ constructed below in the proof.*

Thus, by the uniformization theorem (Theorem 5.2.14) and Exercise C.E.8,

$$\mathrm{Pic}\,(\mathbf{Bun}_G(\Sigma)) \simeq \mathbb{Z}, \tag{1}$$

where $\mathbf{Bun}_G(\Sigma)$ is the moduli stack of G-bundles over Σ.

Proof For any integer $c \geq 1$ and $\lambda_c \in \hat{D}$ recall the central extension $p: \hat{\mathscr{G}}_{\lambda_c} \to \mathscr{L}_G$ and the homomorphism of group functors $\hat{\rho}: \hat{\mathscr{G}}_{\lambda_c} \to \mathbf{GL}_{\mathscr{H}(\lambda_c)}$ from Definition 1.4.5. For any countable-dimensional vector space V over $\mathbb{C}$, recall the definition of the projective space $\mathbb{P}_V$, which is an ind-projective variety (cf. (Kumar, 2002, Example 4.1.3(4)). The corresponding $\mathbb{C}$-space functor is given by (cf. (Eisenbud and Harris, 2000, Exercise VI-18)) $R \rightsquigarrow$ set of R-module direct summands of $V \otimes R$ of rank 1.

Let v_+ be a highest weight vector of $\mathscr{H}(\lambda_c)$ (which is unique up to a scalar multiple). Define the morphism between $\mathbb{C}$-space functors (for any $R \in \mathbf{Alg}$)

$$\hat{\rho}_+(R): \hat{\mathscr{G}}_{\lambda_c}(R) \to \mathscr{H}(\lambda_c) \otimes R,\ g \mapsto g \cdot v_+.$$

This gives rise to the morphism

$$\bar{\rho}_+(R): \hat{\mathscr{G}}_{\lambda_c}(R) \to \mathbb{P}_{\mathscr{H}(\lambda_c)}(R),\ g \mapsto [g \cdot v_+],$$

where $[gv_+]$ denotes the line $R \cdot (gv_+) \subset \mathscr{H}(\lambda_c) \otimes R$. Taking $\lambda_c = 0_c$, by identity (1) of the proof of Theorem 1.4.11 (since G fixes v_+), the map $\bar{\rho}_+(R)$ factors to give the map

$$\rho_+(R): \hat{\mathscr{G}}_{0_c}(R)/p^{-1}(G(R[[t]])) \to \mathbb{P}_{\mathscr{H}(\lambda_c)}(R).$$

Let $\hat{\mathscr{X}}_G$ be the sheafification of the functor (cf. Lemma B.2)

$$R \rightsquigarrow \hat{\mathscr{G}}_{0_c}(R)/p^{-1}(G(R[[t]])).$$

Then ρ_+ extends to a morphism of $\mathbb{C}$-space functors (cf. Lemma B.2) (still denoted by)

$$\rho_+: \hat{\mathscr{X}}_G(R) \to \mathbb{P}_{\mathscr{H}(0_c)}(R).$$

Further, it is easy to see (using Proposition 1.4.12) that the sheafification of the functor

$$R \rightsquigarrow G(R((t)))/G(R[[t]]) \supset \hat{\mathscr{G}}_{0_c}(R)/p^{-1}(G(R[[t]]))$$

which is represented by $\bar{X}_G$ (cf. Proposition 1.3.18) coincides with $\hat{\mathscr{X}}_G$. Thus, we get a morphism of ind-varieties (cf. Proposition 1.3.24)

$$\tilde{\rho}_+: \bar{X}_G \to \mathbb{P}_{\mathscr{H}(0_c)}. \tag{2}$$

Recall from Kumar (2002, Example 4.2.7(c)) the definition of the tautological line bundle θ over $\mathbb{P}_{\mathscr{H}(0_c)}$, such that the fiber over a line $[v] \in \mathbb{P}_{\mathscr{H}(0_c)}$ is the line $\mathbb{C}v$. Now, define the line bundle $\mathscr{L}(0_c)$ over $\bar{X}_G$ as the pull-back of θ^* via $\tilde{\rho}_+$. In the notation of Kumar (2002, §7.2.1) $\mathscr{L}(0_c)$ is the line bundle $\mathscr{L}(c\omega_0)$, where $\omega_0 \in \hat{\mathfrak{h}}^*$ is the zeroth fundamental weight (cf. (1) of Definition 1.2.2 for the notation $\hat{\mathfrak{h}}$): $\omega_{0|\mathfrak{h}} = 0$ and $\omega_0(C) = 1$. By Kumar (2002, Lemma 7.2.2),

$$\mathscr{L}(0_c) \simeq \mathscr{L}(0_1)^{\otimes c} \quad \text{for any integer } c \geq 1.$$

Define

$$\mathscr{L}(0_c) = \mathscr{L}(0_{-c})^* \quad \text{for} \quad c < 0 \quad \text{and} \quad \mathscr{L}(0_0) = \text{trivial line bundle.}$$

By Kumar (2002, Proposition 13.2.19), the map

$$\mathbb{Z} \xrightarrow{\sim} \operatorname{Pic}(\bar{X}_G), \; c \mapsto \mathscr{L}(0_c), \tag{3}$$

is an isomorphism. We next show that for any $c > 0$, $\mathscr{L}(0_c)$ is $\bar{\Gamma}$-equivariant (and hence so is $\mathscr{L}(0_c)$ for any $c \in \mathbb{Z}$). By the morphism of $\mathbb{C}$-space functors

$$\hat{\rho}_+(R) \colon \hat{\mathscr{G}}_{\lambda_c}(R) \to \mathscr{H}(\lambda_c) \otimes R,$$

we see that the line bundle $\mathscr{L}(0_c)$ is $\hat{\bar{G}}_{0_c}$-equivariant and hence it is $\bar{\Gamma}$-equivariant because of the splitting of $\hat{\bar{G}}_{0_c} \to \bar{G}((t))$ over $\bar{\Gamma}$ (cf. Theorem 8.2.1).

Finally, we show that any line bundle $\mathscr{L}$ over any $\bar{\Gamma}$-stable ind-subvariety Y of $\bar{X}_G$ can not admit two different $\bar{\Gamma}$-equivariant structures (where $\bar{\Gamma}$ acts on $\bar{X}_G$ through the morphism $\bar{\Gamma} \to \bar{G}((t))$ as at the beginning of Section 8.2): for, if possible, take (for a closed point $\gamma \in \bar{\Gamma}$ and $x \in Y$) γ'_x and $\gamma''_x \colon \mathscr{L}_x \to \mathscr{L}_{\gamma \cdot x}$, where $\mathscr{L}_x$ is the fiber over x. Define the morphism $f \colon \bar{\Gamma} \times Y \to \mathbb{G}_m$ by

$$\gamma'_x(v_x) = f(\gamma, x)\gamma''_x(v_x), \quad \text{for any} \quad v_x \in \mathscr{L}_x.$$

By Corollary 8.1.5, f only depends upon the variable $x \in Y$. Thus, $f(\gamma, x) = f(1, x) = 1$, for all $\gamma \in \bar{\Gamma}$ and $x \in Y$. Thus, $\mathscr{L}$ has a unique $\bar{\Gamma}$-equivariant structure if there exists one. Thus, by the isomorphism (3), the proposition follows. □

We recall the following analogue of the Borel–Weil theorem for affine Lie algebras, due to Kumar and also due to Mathieu (cf. (Kumar, 2002, Corollary 8.3.12 and Exercise 8.3.E.2)). Observe that we have used Proposition 1.3.24 in the following.

Theorem 8.3.3 *For any $c > 0$, the map (cf. (2) of the proof of Proposition 8.3.2)*

$$\tilde{\rho}_+ : \bar{X}_G \to \mathbb{P}_{\mathscr{H}(0_c)}$$

induces an isomorphism

$$H^0\left(\mathbb{P}_{\mathscr{H}(0_c)}, \theta^*\right) \simeq \mathscr{H}(0_c)^* \xrightarrow{\sim} H^0\left(\bar{X}_G, \mathscr{L}(0_c)\right).$$

Of course, for $c = 0$, $H^0(\bar{X}_G, \mathscr{L}(0_c)) \simeq \mathbb{C}$ (cf. Lemma 8.1.2), since $\bar{X}_G$ is a connected ind-projective variety,

We are now ready to prove the main result of this section. We follow the notation as in Definition 8.3.1. In particular, G is a connected, simply-connected, simple algebraic group over $\mathbb{C}$, $(\Sigma, \vec{p} = (p_1, \ldots, p_s))$ is an s-pointed smooth irreducible projective curve (for $s \geq 1$) and $\vec{\lambda} = (\lambda_1, \ldots, \lambda_s)$ is a set of weights in D_c attached to $\vec{p}$, where $c > 0$ is the central charge. Pick a point $p_o \in \Sigma \backslash \vec{p}$ and let $\Sigma^* = \Sigma \backslash p_o$. Consider the line bundle

$$\mathscr{L}(\vec{\lambda}) := \mathscr{L}(0_c) \boxtimes \mathscr{L}_{P_1}(\lambda_1) \boxtimes \cdots \boxtimes \mathscr{L}_{P_s}(\lambda_s)$$
$$\text{over } \bar{X}_G \times G/P_1 \times \cdots \times G/P_s.$$

By Proposition 8.3.2 and the discussion just above the proposition, $\mathscr{L}(\vec{\lambda})$ is a $\bar{\Gamma}$-equivariant line bundle. Hence, by the isomorphism (2) of Definition 8.3.1, $\mathscr{L}(\vec{\lambda})$ descends to give a line bundle $\bar{\mathscr{L}}(\vec{\lambda})$ over the stack $\left[\bar{\Gamma} \backslash (\bar{X}_G \times \Pi_{j=1}^s G/P_j)\right]$.

Theorem 8.3.4 *With the notation and assumptions as above,*

$$H^0\left(\left[\bar{\Gamma} \backslash \left(\bar{X}_G \times \Pi_{j=1}^s G/P_j\right)\right], \bar{\mathscr{L}}(\vec{\lambda})\right) \simeq \mathscr{V}_\Sigma^\dagger(\vec{p}, \vec{\lambda}). \quad (1)$$

Moreover, the isomorphism is canonical up to a scalar multiple.

Observe that by identity (1) of Definition 8.3.1,

$$\mathbf{Parbun}_G\left(\Sigma, \vec{P}\right) \simeq \left[\bar{\Gamma} \backslash \left(\bar{X}_G \times \Pi_{j=1}^s G/P_j\right)\right].$$

Proof Recall that $\mathbf{Parbun}_G(\Sigma, \vec{P})$ (and hence $[\bar{\Gamma} \backslash (\bar{X}_G \times \Pi_{j=1}^s G/P_j)]$) is an algebraic stack (cf. Theorem 5.1.5). By Proposition C.23,

$$H^0\left(\left[\bar{\Gamma} \backslash \left(\bar{X}_G \times \Pi_{j=1}^s G/P_j\right)\right], \bar{\mathscr{L}}(\vec{\lambda})\right) \simeq H^0\left(\bar{X}_G \times \Pi_{j=1}^s G/P_j, \mathscr{L}(\vec{\lambda})\right)^{\bar{\Gamma}_{\text{funct}}}$$

$$\simeq H^0\left(\bar{X}_G \times \Pi_{j=1}^s G/P_j, \mathscr{L}(\vec{\lambda})\right)^{\operatorname{Lie}\bar{\Gamma}}, \qquad \text{by Lemma B.23}$$

since $\bar{\Gamma}$ is irreducible by Theorem 8.1.1

$$\simeq \left[H^0\left(\bar{X}_G, \mathscr{L}(0_c)\right) \otimes H^0\left(G/P_1, \mathscr{L}_{P_1}(\lambda_1)\right) \otimes \cdots \otimes H^0\left(G/P_s, \mathscr{L}_{P_s}(\lambda_s)\right)\right]^{\operatorname{Lie}\bar{\Gamma}},$$

since $H^0(G/P_j, \mathscr{L}_{P_j}(\lambda_j))$ is finite dimensional

$$\simeq \left[\mathscr{H}(0_c)^* \otimes V(\lambda_1)^* \otimes \cdots \otimes V(\lambda_s)^*\right]^{\operatorname{Lie}\bar{\Gamma}},$$

by Theorem 8.3.3 and the classical Borel–Weil theorem

$$= \mathscr{V}_\Sigma^\dagger(\vec{p}, \vec{\lambda}), \quad \text{by Corollary 2.2.3(b) and (9) of Definition 2.1.1.}$$

This proves (1).

The uniqueness of the isomorphism (1) up to scalar multiples follows from the uniqueness of the isomorphisms (up to scalar multiples)

$$H^0\left(\bar{X}_G, \mathscr{L}(0_c)\right) \simeq \mathscr{H}(0_c)^* \quad \text{and} \quad H^0\left(G/P_j, \mathscr{L}_{P_j}(\lambda_j)\right) \simeq V(\lambda_j)^*.$$

□

Let $X(H)$ denote the character group of any group H. For any $d \in \mathbb{Z}$ and $\lambda_j \in X(P_j)$, same as the definition of $\mathscr{L}(\vec{\lambda})$, we can define the line bundle

$$\mathscr{L}(d, \vec{\lambda}) := \mathscr{L}(0_1)^{\otimes d} \boxtimes \mathscr{L}_{P_1}(\lambda_1) \boxtimes \cdots \boxtimes \mathscr{L}_{P_s}(\lambda_s)$$
$$\text{over } \bar{X}_G \times G/P_1 \times \cdots \times G/P_s.$$

Moreover, as observed above Theorem 8.3.4, $\mathscr{L}(d, \vec{\lambda})$ is a $\bar{\Gamma}$-equivariant line bundle. Hence, by isomorphism (2) of Definition 8.3.1, $\mathscr{L}(d, \vec{\lambda})$ descends to give a line bundle $\bar{\mathscr{L}}(d, \vec{\lambda})$ over the stack $\left[\bar{\Gamma} \backslash \left(\bar{X}_G \times \Pi_{j=1}^s G/P_j\right)\right]$.

Theorem 8.3.5 *With the notation as above,*

$$\operatorname{Pic}\left(\left[\bar{\Gamma} \backslash \left(\bar{X}_G \times \Pi_{j=1}^s G/P_j\right)\right]\right)$$
$$\simeq \mathbb{Z} \times \Pi_{j=1}^s X(P_j), \ (d, \lambda_1, \ldots, \lambda_s) \mapsto \bar{\mathscr{L}}(d, \vec{\lambda}),$$

for $d \in \mathbb{Z}$ and $\lambda_j \in X(P_j)$.

Proof Since

$$\operatorname{Pic}\left(\bar{X}_G \times \Pi_{j=1}^s G/P_j\right) = \operatorname{Pic}(\bar{X}_G) \times \operatorname{Pic}(G/P_j),$$

by Hartshorne (1977, Chap. III, Exercise 12.6(b))

$$\simeq \mathbb{Z} \times \Pi_{j=1}^s X(P_j), \ \text{under } (d, \lambda_1, \ldots, \lambda_s) \mapsto \mathscr{L}(d, \vec{\lambda})$$

by (3) of Proposition 8.3.2 and since $\text{Pic}(G/P_j) \simeq X(P_j)$. (Recall that $H^1(G/P_j, \mathcal{O}_{G/P_j}) = 0$.) Moreover, a line bundle $\mathscr{L}$ over any $\bar{\Gamma}$-ind-variety can admit at most one $\bar{\Gamma}$-equivariant structure (cf. the last part of the proof of Proposition 8.3.2). This proves the theorem. □

8.3.E Exercises

(1) Complete an alternative construction of the line bundle $\mathscr{L}(0_c)$ (for any $c \in \mathbb{Z}$) over $\bar{X}_G$ as follows.

Consider the central extension

$$1 \to \mathbb{G}_m \to \hat{\bar{G}}_{0_1} \xrightarrow{\bar{p}} \bar{G}((t)) \to 1,$$

which splits over $\bar{G}[[t]]$ (cf. Proposition 1.4.12 and Theorem 1.4.11). Thus,

$$\bar{p}^{-1}(\bar{G}[[t]]) \simeq \bar{G}[[t]] \times \mathbb{G}_m.$$

Now, define the homogeneous line bundle over $\hat{\bar{G}}_{0_1}/\bar{p}^{-1}(\bar{G}[[t]])$ given by the character of

$$\bar{p}^{-1}(\bar{G}[[t]]) = \bar{G}[[t]] \times \mathbb{G}_m \to \mathbb{G}_m, (g, z) \mapsto z^{-c}.$$

Show that this line bundle is isomorphic with $\mathscr{L}(0_c)$.

8.4 Identification of Conformal Blocks with Sections of Line Bundles over Non-parabolic Moduli Space

In this section Σ is a smooth irreducible projective curve of any genus $g \geq 0$ with a base point $p \in \Sigma$ and G is a connected, simply-connected simple algebraic group over $\mathbb{C}$. We denote $\Sigma^* = \Sigma \backslash p$.

Definition 8.4.1 (Θ-bundle) Let S be an affine noetherian scheme and let $\mathscr{V} \to \Sigma \times S$ be a vector bundle. Then, there exists a complex of vector bundles $\mathscr{V}_i$ on S (with $\mathscr{V}_i = 0$ for $i \geq 2$ since Σ is 1-dimensional):

$$\mathscr{V}_0 \to \mathscr{V}_1 \to 0 \to \dots \tag{1}$$

such that for any affine base change $f: S' \to S$ (for affine noetherian S'), the ith direct image (under the projection $\Sigma \times S' \to S'$) of the pull-back $(\text{Id}_\Sigma \times f)^*\mathscr{V}$ is given by the ith cohomology of the pull-back of the above complex (1) to S' (cf. (Mumford, 1985, §5); also see (Faltings, 1993, §1)). Define the *determinant line bundle* $\operatorname{Det} \mathscr{V}$ over S to be the product

$$\operatorname{Det} \mathscr{V} := \Lambda^{\mathrm{top}}(\mathscr{V}_1) \otimes \Lambda^{\mathrm{top}}(\mathscr{V}_0)^* \tag{2}$$

up to an isomorphism. This does not depend upon the choice of the complex (1) since the complex (1) is uniquely determined up to a unique quasi-isomorphism. In particular, $\operatorname{Det} \mathscr{V}$ over affine S patches to give its extension over any (not necessarily affine) noetherian scheme S. For an axiomatic development of Det in a more general context we refer to Lang (1988, Chapter 6, §1), where $\operatorname{Det} \mathscr{V}$ is defined as the dual of the above definition (see (Knudsen and Mumford, 1976) for more details). We prefer our convention since it gives rise to ample line bundles in our situations.

From the above base change property of $\{\mathscr{V}_i\}$, we get the following base change property of Det: for any morphism $f : S' \to S$,

$$\operatorname{Det}\left((\mathrm{Id}_\Sigma \times f)^* \mathscr{V}\right) = f^*(\operatorname{Det} \mathscr{V}). \tag{3}$$

In particular, for any closed point $t \in S$,

$$(\operatorname{Det} \mathscr{V})_t = \Lambda^{\mathrm{top}}\left(H^0(\Sigma, \mathscr{V}_t)^*\right) \otimes \Lambda^{\mathrm{top}}(H^1(\Sigma, \mathscr{V}_t)), \ \text{ where } \ \mathscr{V}_t := \mathscr{V}_{|\Sigma \times t}. \tag{4}$$

Let $\mathscr{L}$ be a line bundle on S and let $\pi_S : \Sigma \times S \to S$ be the projection onto the second factor. Then, for the family $\mathscr{V} \otimes \pi_S^* \mathscr{L} \to \Sigma \times S$, we have (for connected noetherian S):

$$\operatorname{Det}\left(\mathscr{V} \otimes \pi_S^* \mathscr{L}\right) = \operatorname{Det} \mathscr{V} \otimes \mathscr{L}^{-\chi(\mathscr{V})}, \tag{5}$$

where $\chi(\mathscr{V}) := \dim H^0(\Sigma, \mathscr{V}_t) - \dim H^1(\Sigma, \mathscr{V}_t)$ is the Euler characteristic of $\mathscr{V}$ restricted to $\Sigma \times t$, for any $t \in S$. (Since S is connected, $\chi(\mathscr{V})$ does not depend upon the choice of $t \in S$ as can be seen from the base change property of $(\mathscr{V}_i)$.)

We now define the Θ-bundle $\Theta(\mathscr{V})$ of a family $\mathscr{V} \to \Sigma \times S$ of rank r and degree d vector bundles over Σ such that $r|d$ by

$$\Theta(\mathscr{V}) := \operatorname{Det} \mathscr{V} \otimes \left(\det \mathscr{V}_{|p \times S}\right)^{\frac{\chi(\mathscr{V})}{r}}, \tag{6}$$

where $\det \mathscr{V}_{|p \times S}$ denotes the line bundle $\Lambda^{\mathrm{top}} \mathscr{V}_{|p \times S}$ and p is the fixed point of Σ. Observe that by the Riemann–Roch theorem (since $\deg \mathscr{V}_t = d$)

$$\chi(\mathscr{V}) = d + r(1-g), \ \text{ thus } \ r|\chi(\mathscr{V}).$$

Observe that, $\Theta(\mathscr{V})$ *does* depend upon the choice of the base point $p \in \Sigma$ in general. Consider, e.g., $\mathscr{V} = \mathscr{O}(D)$, where $S = \Sigma$ and D is the diagonal of $\Sigma \times \Sigma$.

By the definition of the Θ-bundle and (5), we get that

$$\Theta(\mathscr{V}) = \Theta\left(\mathscr{V} \otimes \pi_S^* \mathscr{L}\right), \quad \text{for any line bundle } \mathscr{L} \text{ over } S. \tag{7}$$

Observe that $\deg(\mathscr{V} \otimes \pi_S^* \mathscr{L}) = \deg \mathscr{V}$.

Clearly, Θ satisfies the base change property as in (3).

If $\mathscr{E} \to \Sigma \times S$ is a family of G-bundles and V is a representation of G, then we define

$$\mathrm{Det}(\mathscr{E}(V)) := \mathrm{Det}\left(\mathscr{E} \times^G V\right). \tag{8}$$

Since G is assumed to be simple,

$$\det \mathscr{E}(V) \simeq \mathscr{O}_{\Sigma \times S}. \tag{9}$$

Thus,

$$\Theta(\mathscr{E}(V)) = \mathrm{Det}(\mathscr{E}(V)). \tag{10}$$

In this case since $d = 0$ (by (9)), the condition $r|d$ is automatically satisfied. Moreover, in this case, by (10), $\Theta(\mathscr{E}(V))$ does not depend upon the choice of the base point $p \in \Sigma$.

By (3), we immediately see that $\Theta(\mathscr{E}(V))$ is functorial under the pull-back of families of G-bundles.

Let $M(r,d)$ be the moduli space of semistable vector bundles over Σ of rank $r \geq 1$ and any degree $d \in \mathbb{Z}$ (cf. Theorem 7.1.14). Recall the following theorem which follows immediately from the surjectivity of the morphism i as in Drezet and Narasimhan (1989, Corollaire 4.3). In this reference, there is a standing assumption on the genus $g \geq 2$ and if $g = 2$ then $r \neq 2$. But the surjectivity of i holds good for Σ of any genus $g \geq 0$ and any r (by the same proof).[1] For the ampleness of Θ, see Kumar, Narasimhan and Ramanathan (1994, Lemma 7.5 and Remark 7.6) (also see (Narasimhan and Ramadas, 1993, Theorem 1)).

Theorem 8.4.2 *Let $r|d$ and any $g \geq 0$. Then, there exists an ample line bundle $\Theta = \Theta_{r,d}$ over $M(r,d)$ such that for any family of semistable vector bundles $\mathscr{V} \to \Sigma \times S$ of rank r and degree d*

$$f_{\mathscr{V}}^* \Theta \simeq \Theta(\mathscr{V}), \tag{1}$$

where $f_{\mathscr{V}}\colon S \to M(r,d)$ is the induced map obtained from the coarse moduli property of the functor $\mathscr{M}(r,d)$ (cf. Theorem 7.1.14).

For the following remark, see Drezet and Narasimhan (1989, §0.2.1).

[1] I thank J.-M. Drezet for pointing this out.

Remark 8.4.3 Under the notation of Theorem 8.4.2, assume further that the genus $g \geq 2$ and if $g = 2$ then $r \neq 2$. Then, taking any line bundle $\mathscr{L}_o$ over Σ of degree

$$\deg \mathscr{L}_o = (g-1) - \frac{d}{r}, \tag{1}$$

$$\Theta \simeq \mathscr{O}(D_{\mathscr{L}_o}), \quad \text{for the divisor } D_{\mathscr{L}_o}, \tag{2}$$

which is the closure in $M(r,d)$ of the set of stable vector bundles $\mathscr{W}$ over Σ of rank r and degree d such that $H^0(\Sigma, \mathscr{W} \otimes \mathscr{L}_o) \neq 0$.

For any $r|d$ and $h \in \mathbb{Z}$, consider the isomorphism $\beta\colon M(r,d) \to M(r,d+rh)$, $[V] \mapsto [V \otimes \mathscr{O}_\Sigma(hp)]$. By Exercise 8.4.E.2, $\beta^*(\Theta_{r,d+rh}) = \Theta_{r,d}$. In the sequel, we will identify $\Theta_{r,d+rh}$ with $\Theta_{r,d}$.

Definition 8.4.4 Let V be a representation of G of dimension r. Then, for any semistable G-bundle E over Σ, the associated vector bundle $E(V) := E \times^G V$ (of rank r and degree 0) is semistable (cf. Theorem 6.1.7 and Exercise 6.1.E.5). Thus, given a family $\mathscr{E} \to \Sigma \times S$ of semistable G-bundles over Σ parameterized by a noetherian scheme S, we have the induced morphism from the coarse moduli property of $M(r,d)$ (cf. Theorem 7.1.14 and Corollary 7.2.7):

$$\varphi^V_{\mathscr{E}}\colon S \to M(r,0), t \mapsto [\mathscr{E}(V)_t],$$

where $[\mathscr{E}(V)_t]$ denotes the equivalence class of $\mathscr{E}(V)_t := \mathscr{E}(V)_{|\Sigma \times t}$.

Let $M^G = M^G(\Sigma)$ be the moduli space of semistable G-bundles over Σ (cf. Theorem 7.2.6). Then, by identity (3) of the proof of Theorem 7.2.6, taking $\hat{\Sigma} = \Sigma$ (and thus $A = (e)$ and τ = trivial),

$$M^G := R^{ss}(G)//\operatorname{GL}_N.$$

Moreover, $R^{ss}(G)$ parameterizes a family $\mathscr{U}^{ss}(G)$ of semistable G-bundles over Σ (cf. Corollary 7.2.4). In particular, we get a morphism

$$\varphi^V_{\mathscr{U}^{ss}(G)}\colon R^{ss}(G) \to M(r,0).$$

By the categorical quotient property of $R^{ss}(G) \to R^{ss}(G)//\operatorname{GL}_N$ (cf. Definition 7.1.4) and since $\mathscr{U}^{ss}(G)_{|\Sigma\times x} \simeq \mathscr{U}^{ss}(G)_{|\Sigma\times\gamma\cdot x}$, for any $x \in R^{ss}(G)$ and $\gamma \in \operatorname{GL}_N$ (cf. identity (4) of the proof of Theorem 7.2.6), we get a morphism

$$\varphi^V\colon M^G \to M(r,0), \ [E] \mapsto [E(V)], \tag{1}$$

for $E \in R^{ss}(G)$, where $[E]$ denotes its image in M^G. Finally, define the Θ-line bundle (for any representation V of G of dimension $r \geq 1$): $\Theta(V)$ over M^G via

$$\Theta(V) := (\varphi^V)^*\Theta. \tag{2}$$

For any family $\mathscr{E}$ of semistable G-bundles over $\Sigma \times S$ (for connected noetherian scheme S) and any G-module V of dimension $r \geq 1$, we get the following from (1) of Theorem 8.4.2:

$$f_{\mathscr{E}}^*(\Theta(V)) \simeq \Theta(\mathscr{E}(V)), \tag{3}$$

where $f_{\mathscr{E}}\colon S \to M^G$ is the morphism induced from the family $\mathscr{E}$ via the coarse moduli property of M^G (cf. Corollary 7.2.7).

Definition 8.4.5 (Picard group of $\bar{X}_G$) Consider the Lie algebra homomorphism $\dot{\gamma}_\theta\colon sl_2 \to \mathfrak{g}$, $X = \left(\begin{smallmatrix} 0 & 1 \\ 0 & 0 \end{smallmatrix}\right) \mapsto x_\theta$, $Y = \left(\begin{smallmatrix} 0 & 0 \\ 1 & 0 \end{smallmatrix}\right) \mapsto y_\theta$ and $H = \left(\begin{smallmatrix} 1 & 0 \\ 0 & -1 \end{smallmatrix}\right) \mapsto \theta^\vee$, where θ is the highest root of $\mathfrak{g}$ and $x_\theta \in \mathfrak{g}_\theta$, $y_\theta \in \mathfrak{g}_{-\theta}$ are such that $\langle x_\theta, y_\theta \rangle = 1$. This gives rise to an algebraic group homomorphism

$$\gamma_\theta\colon \mathrm{SL}_2 \to G$$

and hence a homomorphism of the corresponding loop group functors

$$\hat{\gamma}_\theta^R\colon \mathrm{SL}_2(R((t))) \to G(R((t))), \quad \text{for any } \mathbb{C}\text{-algebra } R.$$

Define further a homomorphism

$$\gamma_o^R\colon \mathrm{SL}_2(R) \to \mathrm{SL}_2(R((t))), \quad \begin{pmatrix} a & b \\ c & d \end{pmatrix} \mapsto \begin{pmatrix} d & ct^{-1} \\ bt & a \end{pmatrix}. \tag{1}$$

On composition, we get a homomorphism

$$\hat{\gamma}^R := \hat{\gamma}_\theta^R \circ \gamma_o^R : \mathrm{SL}_2(R) \to G(R((t))). \tag{2}$$

By the definition of γ_o^R, it is easy to see that $\hat{\gamma}^R$ induces a morphism (in fact, an embedding)

$$\hat{\beta} : \mathbb{P}^1 := \mathrm{SL}_2 / B_o \to \bar{X}_G, \tag{3}$$

where B_o is the subgroup of SL_2 consisting of upper-triangular matrices of determinant 1.

Recall the definition of the $\bar{\Gamma}$-equivariant line bundle $\mathscr{L}(0_c)$ over $\bar{X}_G$ from the proof of Proposition 8.3.2. Then we have the following result:

Proposition 8.4.6 *(a)* $\mathbb{Z} \simeq \mathrm{Pic}(\bar{X}_G), c \mapsto \mathscr{L}(0_c)$
(b) $H^2(X_G^{\mathrm{an}}, \mathbb{Z}) \underset{\sim}{\leftarrow} \mathrm{Pic}(\bar{X}_G) \underset{\sim}{\rightarrow} \mathrm{Pic}(\mathbb{P}^1) \underset{\sim}{\rightarrow} H^2(\mathbb{P}^1, \mathbb{Z})$,

where $X_G := \bar{X}_G(\mathbb{C})$, the second isomorphism of (b) is induced from $\hat{\beta}$ and the first and the third isomorphisms of (b) are given by the first Chern class c_1.

Proof For the (a)-part see (3) of the proof of Proposition 8.3.2. For the (b)-part see Kumar (2002, Proof of Proposition 13.2.19) along with Lemma 1.2.3. □

Recall the definition of the Dynkin index for any Lie algebra homomorphism between simple Lie algebras from Definition A.1. Also, recall from Proposition 5.2.4 the definition of the 'tautological' G-bundle $\mathbf{U} = \mathbf{U}_G$ over $\Sigma \times \bar{X}_G$. In particular, we get the Θ-bundle $\Theta(\mathbf{U}_G(V))$ over $\bar{X}_G$ for any G-module V, where $\mathbf{U}_G(V) := \mathbf{U}_G \times^G V$.

Theorem 8.4.7 *For any finite-dimensional (not necessarily irreducible) representation V of G, $\Theta(\mathbf{U}_G(V)) \simeq \mathscr{L}(0_{d_V})$, as line bundles over $\bar{X}_G$, where d_V is the Dynkin index of V.*

Proof If V is a trivial representation of G, the theorem follows trivially (use Exercise 8.4.E.1). So, we assume that V is a nontrivial G-module.

Step I (Reduction to the case $G = \mathrm{SL}_V$): The representation V of G induces the morphism $\phi\colon \bar{X}_G \to \bar{X}_{\mathrm{SL}_V}$. Moreover, by Propositions 5.2.4 and 5.2.7, ϕ satisfies

$$(\mathrm{Id}_\Sigma \times \phi)^* \left(\mathbf{U}_{\mathrm{SL}_V}(V)\right) \simeq \mathbf{U}_G(V).$$

Thus, from the functoriality of the Θ-bundle (cf. Definition 8.4.1),

$$\Theta\left(\mathbf{U}_G(V)\right) = \phi^* \left(\Theta\left(\mathbf{U}_{\mathrm{SL}_V}(V)\right)\right), \tag{1}$$

where $\phi^*\colon \operatorname{Pic} \bar{X}_{\mathrm{SL}_V} \to \operatorname{Pic} \bar{X}_G$ is the induced map.

We next show that, under the identification of Proposition 8.4.6(a), for any $d \in \mathbb{Z}$,

$$\phi^*(d) = d d_V. \tag{2}$$

By the definition as given in the proof of Proposition 8.3.2,

$$\mathscr{L}_{\mathrm{SL}_V}(0_1) := \left(\tilde{\rho}_+^{\mathrm{SL}_V}\right)^* \left(\theta^*_{\mathrm{SL}_V}\right), \tag{3}$$

where $\mathscr{H}_{\mathrm{SL}_V}(0_1)$ is the basic representation with the highest weight 0_1 of the affine Lie algebra $\widehat{sl}_V$, $\tilde{\rho}_+^{\mathrm{SL}_V}\colon \bar{X}_{\mathrm{SL}_V} \to \mathbb{P}_{\mathscr{H}_{\mathrm{SL}_V}(0_1)}$ is the morphism defined by (2) of the proof of Proposition 8.3.2 and θ_{SL_V} is the tautological line bundle over $\mathbb{P}_{\mathscr{H}_{\mathrm{SL}_V}(0_1)}$. By the analogue of (3) for SL_V replaced by G, we get

$$\mathscr{L}_G(0_1) := \left(\tilde{\rho}_+^G\right)^* \left(\theta^*_G\right). \tag{4}$$

Since the highest weight 0_1 of $\widehat{sl}_V$ restricts to the highest weight 0_{d_V} of $\hat{\mathfrak{g}}$ (cf. Lemma A.12), we get the following commutative diagram:

$$\begin{array}{ccc} \bar{X}_G & \xrightarrow{\tilde{\rho}_+^G} & \mathbb{P}_{\mathscr{H}_G(0_{d_V})} \\ \phi \downarrow & & \downarrow \\ \bar{X}_{\mathrm{SL}_V} & \xrightarrow[\tilde{\rho}_+^{\mathrm{SL}_V}]{} & \mathbb{P}_{\mathscr{H}_{\mathrm{SL}_V}(0_1)}, \end{array}$$

where the right vertical map is obtained by taking the projective space of the $\hat{\mathfrak{g}}$-submodule generated by the highest weight subspace of $\mathscr{H}_{\mathrm{SL}_V}(0_1)$. Combining (3) and (4), we get (2).

Thus, we are reduced to proving the theorem for $G = \mathrm{SL}_V$ and its standard module V since its Dynkin index is 1 by definition.

Step II (Reduction to the case $G = \mathrm{SL}_2$): For any $n \geq 2$, embed $i\colon \mathrm{SL}_2 \subset \mathrm{SL}_n$ in the left top 2×2 corner. Parallel to the identity (1), we have (for $V = \mathbb{C}^n$ as a representation of SL_2 via i)

$$\Theta\left(\mathbf{U}_{\mathrm{SL}_2}(V)\right) = \hat{i}^*\left(\Theta\left(\mathbf{U}_{\mathrm{SL}_V}(V)\right)\right), \tag{5}$$

where $\hat{i}\colon \bar{X}_{\mathrm{SL}_2} \to \bar{X}_{\mathrm{SL}_n}$ is the morphism induced from i and $\hat{i}^*$ is the pull-back map between their Picard groups.

We next claim that

$$\Theta\left(\mathbf{U}_{\mathrm{SL}_2}(V)\right) = \Theta\left(\mathbf{U}_{\mathrm{SL}_2}(\mathbb{C}^2)\right). \tag{6}$$

To prove this, decompose $V = \mathbb{C}^2 \oplus \mathbb{C}^{n-2}$ with the trivial action of SL_2 on $\mathbb{C}^{n-2}$. By Exercise 8.4.E.1,

$$\begin{aligned} \Theta\left(\mathbf{U}_{\mathrm{SL}_2}(V)\right) &\simeq \Theta\left(\mathbf{U}_{\mathrm{SL}_2}(\mathbb{C}^2)\right) \otimes \Theta\left(\mathbf{U}_{\mathrm{SL}_2}(\mathbb{C}^{n-2})\right) \\ &= \Theta\left(\mathbf{U}_{\mathrm{SL}_2}(\mathbb{C}^2)\right). \end{aligned}$$

This proves the claim (6). Thus, we are reduced to proving the theorem for $G = \mathrm{SL}_2$ and its standard module $\mathbb{C}^2$.

Step III (Proof of the theorem for $G = \mathrm{SL}_2$): We prove that under the identification of Proposition 8.4.6(b),

$$\Theta\left(\mathbf{U}_{\mathrm{SL}_2}(\mathbb{C}^2)\right) = [\mathbb{P}^1], \tag{7}$$

where $[\mathbb{P}^1]$ denotes the positive cohomology generator of $H^2(\mathbb{P}^1, \mathbb{Z})$. This will of course complete the proof of the theorem.

First of all, the tautological rank-2 vector bundle $\mathbf{U} = \mathbf{U}_{\mathrm{SL}_2}(\mathbb{C}^2)$ can alternatively be described as follows. For any $\bar{g} = gG[[t]] \in X_G$ (for $G = \mathrm{SL}_2$), define the presheaf $\varphi(\bar{g})$ of $\mathscr{O}_\Sigma$-modules by declaring (for any Zariski open subset $W \subset \Sigma$)

$$\varphi(\bar{g})(W) = \begin{cases} H^0(W, \epsilon^2), \text{ if } p \notin W, \quad \text{and} \\ \{\sigma \in H^0(W \backslash p, \epsilon^2) : (\sigma)_p \in g(E \otimes \mathbb{C}[[t]])\}, \quad \text{if } p \in W, \end{cases}$$

where $E := \mathbb{C}^2$, ϵ^2 is the trivial bundle $\Sigma \times E \to \Sigma$ and $(\sigma)_p$ denotes the Laurent series expansion of the rational function σ at p viewed as an element of $E \otimes \mathbb{C}((t))$ with respect to a local parameter t at p. Then, by the characterizing property of the bundle $\mathbf{U}$ as in Proposition 5.2.4, it is easy to see that

$$\mathbf{U}_{|\Sigma \times \bar{g}} \simeq \bar{\varphi}(\bar{g}), \tag{8}$$

where $\bar{\varphi}(\bar{g})$ is the sheaf associated to the presheaf $\varphi(\bar{g})$. If we take $\bar{g} \in \mathbb{P}^1$ under the embedding $\hat{\beta}$ (cf. (3) of Definition 8.4.5), then (from the definition of γ_o as in (1) of Definition 8.4.5)

$$\bar{g}(E \otimes \mathbb{C}[[t]]) \subset t^{-1}\mathbb{C}[[t]] \oplus \mathbb{C}[[t]].$$

In fact, one can easily see that

$$\bar{g}(E \otimes \mathbb{C}[[t]]) = \{t^{-1} f_1 \oplus f_2 : f_1, f_2 \in \mathbb{C}[[t]] \text{ and } af_1(0) - cf_2(0) = 0\}, \tag{9}$$

for $g \in \left(\begin{smallmatrix} a & b \\ c & d \end{smallmatrix}\right) \in \mathrm{SL}_2$.

Consider first the morphism of $\mathscr{O}_{\Sigma \times \mathbb{P}^1}$-modules over $\Sigma \times \mathbb{P}^1$:

$$\tilde{\psi} : (\mathscr{O}_\Sigma(p) \oplus \mathscr{O}_\Sigma) \boxtimes \left(\mathscr{O}_{\mathbb{P}^1} \oplus \mathscr{O}_{\mathbb{P}^1}\right) \to \mathscr{O}_p \boxtimes \mathscr{O}_{\mathbb{P}^1},$$
$$\tilde{\psi}((t^{-1} f_1 \oplus f_2) \boxtimes (f_1' \oplus f_2')) = f_1' f_1(p) - f_2' f_2(p),$$

for $f_1, f_2 \in \mathscr{O}_\Sigma$ and $f_1', f_2' \in \mathscr{O}_{\mathbb{P}^1}$ and let ψ be the restriction of $\tilde{\psi}$ to $(\mathscr{O}_\Sigma(p) \oplus \mathscr{O}_\Sigma) \boxtimes \theta_{\mathbb{P}^1}$, where $\theta_{\mathbb{P}^1}$ is the tautological line bundle over $\mathbb{P}^1$. It is easy to see that ψ is surjective. This gives rise to the surjective morphism

$$\bar{\psi} : (\mathscr{O}_\Sigma(p) \oplus \mathscr{O}_\Sigma) \boxtimes \mathscr{O}_{\mathbb{P}^1} \to \mathscr{O}_p \boxtimes \theta^*_{\mathbb{P}^1}.$$

From the identifications (8) and (9), it is easy to see that

$$\mathbf{U}_o = \operatorname{Ker} \bar{\psi}, \quad \text{where} \quad \mathbf{U}_o := \mathbf{U}_{|_{\Sigma \times \mathbb{P}^1}}. \tag{10}$$

We now calculate the total Chern class $c(\mathbf{U}_o)$ of $\mathbf{U}_o$. By (10), denoting the positive generator of $H^2(\Sigma, \mathbb{Z})$ by $[\Sigma]$,

$$\begin{aligned} c(\mathbf{U}_o) &= c\left(\mathscr{O}_\Sigma(p)\right) c\left(\mathscr{O}_p \boxtimes \theta^*_{\mathbb{P}^1}\right)^{-1} \\ &= (1+[\Sigma]) \cdot (1+[\mathbb{P}^1])^{-1}(1-[\Sigma]+[\mathbb{P}^1]), \text{ since } \mathscr{O}_p = \mathscr{O}/\mathscr{O}(-p) \\ &= (1+[\Sigma]-[\mathbb{P}^1]-[\Sigma][\mathbb{P}^1])(1-[\Sigma]+[\mathbb{P}^1]) \\ &= 1+[\Sigma][\mathbb{P}^1]. \end{aligned}$$

Thus,

$$c_2(\mathbf{U}_o) = [\Sigma][\mathbb{P}^1], \;\; c_1(\mathbf{U}_o) = 0. \tag{11}$$

Let T_{π_2} be the relative tangent bundle along the fibers of the projection $\pi_2 \colon \Sigma \times \mathbb{P}^1 \to \mathbb{P}^1$. By the Grothendieck–Riemann–Roch theorem (Fulton, 1998, Example 15.2.8) applied to the proper morphism π_2, we get

$$\begin{aligned} \operatorname{ch}(R\pi_{2*}\mathbf{U}_o) &= \pi_{2*}(\operatorname{ch}(\mathbf{U}_o) \cdot \operatorname{td}(T_{\pi_2})) \\ &= \pi_{2*}\left[(2-c_2(\mathbf{U}_o))\left(1+\frac{1}{2}c_1(T_{\pi_2})\right)\right], \;\; \text{since } \; c_1(\mathbf{U}_o) = 0, \end{aligned}$$

where ch denotes the Chern character and td denotes the Todd class. Hence,

$$\begin{aligned} c_1(R\pi_{2*}\mathbf{U}_o) &= \pi_{2*}(-c_2(\mathbf{U}_o)) \\ &= \pi_{2*}(-[\Sigma][\mathbb{P}^1]), \quad \text{by (11)} \\ &= -[\mathbb{P}^1]. \end{aligned} \tag{12}$$

This completes the proof of (7) and hence completes the proof of the theorem. □

Remark 8.4.8 A more topological proof of Theorem 8.4.7 is given in Kumar, Narasimhan and Ramanathan (1994, §5).

Definition 8.4.9 Let $\bar{X}_G^{ss} = \{x \in \bar{X}_G : \mathbf{U}_{G|\Sigma \times x}$ is semistable$\}$. Then, by Corollary 7.1.16(b), $\bar{X}_G^{ss} \subset \bar{X}_G$ is a (nonempty) Zariski open subset. Moreover, the family $\mathbf{U}_G \to \Sigma \times \bar{X}_G$ restricted to $\Sigma \times \bar{X}_G^{ss}$ gives rise to a morphism (obtained from the coarse moduli property of M^G):

$$f_{\mathbf{U}_G} \colon \bar{X}_G^{ss} \to M^G.$$

Let V be a representation of G. Then, by the identification (3) of Definition 8.4.4 and Theorem 8.4.7, we get that, as line bundles over $\bar{X}_G^{ss}$,

$$\Theta\left(\mathbf{U}_G(V)_{|\Sigma \times \bar{X}_G^{ss}}\right) = f^*_{\mathbf{U}_G}(\Theta(V)) = \mathscr{L}(0_{d_V})_{|\bar{X}_G^{ss}}. \tag{1}$$

Lemma 8.4.10 *For any representation V of G and $a \in \mathbb{Z}_+$, the morphism $f_{\mathbf{U}_G}: \bar{X}_G^{ss} \to M^G$ induces an isomorphism (via pull-back)*

$$f_{\mathbf{U}_G}^*: H^0\left(M^G, (\Theta(V))^{\otimes a}\right) \xrightarrow{\sim} H^0\left(\bar{X}_G^{ss}, \mathscr{L}(0_{d_V})^{\otimes a}\right)^{\Gamma},$$

where $\bar{\Gamma}$ is the ind-affine group variety as at the beginning of Section 8.1 with $\bar{\Gamma}(\mathbb{C}) = G[\Sigma^*]$. Then $\bar{\Gamma}$ acts on $\bar{X}_G^{ss}$ via its action on $\bar{X}_G$ since $\mathbf{U}_G$ is $\bar{\Gamma}$-equivariant (cf. Lemma 5.2.12) and $\mathscr{L}(0_{d_V})$ is a $\bar{\Gamma}$-equivariant line bundle (cf. Proposition 8.3.2). (For the action of Γ on the cohomology $H^0\left(\bar{X}_G^{ss}, \mathscr{L}(0_{d_V})^{\otimes a}\right)$, see Section 1.1.)

Proof From the $\bar{\Gamma}$-equivariance of $\mathbf{U}_G$, the morphism $f_{\mathbf{U}_G}: \bar{X}_G^{ss} \to M^G$ is $\bar{\Gamma}$-invariant (with the trivial action of $\bar{\Gamma}$ on M^G). Thus, $f_{\mathbf{U}_G}^*((\Theta(V))^{\otimes a})$ canonically acquires the pull-back structure of a $\bar{\Gamma}$-equivariant line bundle over $\bar{X}_G^{ss}$. However, as proved in the proof of Proposition 8.3.2, any line bundle over $\bar{X}_G^{ss}$ cannot admit two different $\bar{\Gamma}$-equivariant structures. Thus, the $\bar{\Gamma}$-equivariant structure on $\mathscr{L}(0_{d_V})^{\otimes a}_{|\bar{X}_G^{ss}}$ coincides with the pull-back structure via $f_{\mathbf{U}_G}$. Thus,

$$f_{\mathbf{U}_G}^*\left(H^0(M^G, (\Theta(V))^{\otimes a})\right) \subset H^0\left(\bar{X}_G^{ss}, \mathscr{L}(0_{d_V})^{\otimes a}\right)^{\Gamma}. \tag{1}$$

Moreover, since any G-bundle over Σ occurs in the family $\mathbf{U}_G$ by virtue of Theorem 5.2.5 and Proposition 5.2.7, the map $f_{\mathbf{U}_G}$ is surjective. Hence, $f_{\mathbf{U}_G}^*$ is injective.

We next prove that $f_{\mathbf{U}_G}^*$ is surjective: take $\sigma \in H^0\left(\bar{X}_G^{ss}, \mathscr{L}(0_{d_V})^{\otimes a}\right)^{\Gamma}$. We first show that it descends to M^G as a set-theoretic section of $(\Theta(V))^{\otimes a}$. Take $x_1, x_2 \in \bar{X}_G^{ss}$ such that $f_{\mathbf{U}_G}(x_1) = f_{\mathbf{U}_G}(x_2)$. Thus, the corresponding G-bundles have the same equivalence classes, i.e.,

$$\left[\mathbf{U}_{G|\Sigma\times x_1}\right] = \left[\mathbf{U}_{G|\Sigma\times x_2}\right]. \tag{2}$$

By Proposition 7.2.18 for $A = (1)$, there exists a G-bundle $\mathscr{E}^1 \to \Sigma \times \mathbb{A}^1$ such that

(a) $\mathscr{E}^1_{|\Sigma\times(\mathbb{A}^1\setminus\{0\})} \simeq p^*(E_1)$, where $E_1 := \mathbf{U}_{G|\Sigma\times x_1}$ and $p: \Sigma \times (\mathbb{A}^1\setminus\{0\}) \to \Sigma$ is the projection, and
(b) $\mathscr{E}^1_{|\Sigma\times\{0\}} \simeq \mathrm{gr}(E_1)$.

Similarly, there is a bundle $\mathscr{E}^2 \to \Sigma \times \mathbb{A}^1$ corresponding to $E_2 := \mathbf{U}_{G|\Sigma\times x_2}$ and

$$\mathrm{gr}(E_1) = \mathrm{gr}(E_2) \quad \text{by (2) and Proposition 7.2.18(c).} \tag{3}$$

By Theorem 5.2.5, there exists an étale cover $\pi : \tilde{\mathbb{A}}^1 \to \mathbb{A}^1$ such that the pull-back $\tilde{\mathscr{E}}^1$ of $\mathscr{E}^1$ to $\Sigma \times \tilde{\mathbb{A}}^1$ is trivial over $\Sigma^* \times \tilde{\mathbb{A}}^1$. Choosing a trivialization over $\Sigma^* \times \tilde{\mathbb{A}}^1$, by Proposition 5.2.7, we get a morphism $\alpha : \tilde{\mathbb{A}}^1 \to \bar{X}_G^{ss}$ (since $\mathrm{gr}(E_1)$ being polystable is semistable) such that $(\mathrm{Id}_\Sigma \times \alpha)^*(\mathbf{U}_G) \simeq \tilde{\mathscr{E}}^1$. Thus, by Corollary 5.2.15, properties (a) and (b) of $\mathscr{E}^1$, and (3), we see that

$$\overline{\Gamma \cdot x_1} \cap \overline{\Gamma \cdot x_2} \neq \emptyset. \tag{4}$$

Since σ is Γ-invariant, this shows that $\sigma(x_1) = \sigma(x_2)$ proving the claim that σ descends to M^G as a set-theoretic section $\bar{\sigma}$ of $(\Theta(V))^{\otimes a}$. Now, by Kumar (2002, Proposition A.12), $\bar{\sigma}$ is automatically a regular section. This proves the Lemma. □

We now aim to prove that any Γ-invariant section of $\mathscr{L}(0_{d_V})^{\otimes a}$ over $\bar{X}_G^{ss}$ extends to $\bar{X}_G$. But, before we come to its proof, we need the following Lemmas 8.4.11 and 8.4.12 and Proposition 8.4.13.

Lemma 8.4.11 *Let X be an irreducible normal variety and let $V \subset X$ be a nonempty open subset. Let $\mathscr{L}$ be a line bundle over X. Then, any element of $R_V := \bigoplus_{n \geq 0} H^0(V, \mathscr{L}^n)$ which is integral over $R := \bigoplus_{n \geq 0} H^0(X, \mathscr{L}^n)$ belongs to R.*

Proof Since R_V and R are graded, it suffices to prove the lemma only for homogeneous elements of R_V. Let $b \in H^0(V, \mathscr{L}^{n_o})$ be integral over R, i.e., b satisfies a relation

$$b^m + a_1 b^{m-1} + \cdots + a_m = 0, \quad \text{for some } a_i \in R. \tag{1}$$

Let $D \subset X \backslash V$ be a prime divisor and let b have a pole of order $d \geq 0$ along D. Then the order of the pole of b^m along D is, of course, md and that of $a_i b^{m-i}$ is $\leq d(m-1)$ for each $i \geq 1$. But, since $b^m + a_1 b^{m-1} + \cdots + a_{m-1} b$ is regular by (1), we are forced to have $d = 0$, i.e., b is regular along D. Hence, $b \in R$. □

Lemma 8.4.12 *Let $f : X \to Y$ be a proper morphism between schemes of finite type over $\mathbb{C}$, where Y is projective and $\mathscr{L}$ an ample line bundle over Y. Then the ring $R_f := \bigoplus_{n \geq 0} H^0(X, f^*\mathscr{L}^n)$ is integral over the ring $R := \bigoplus_{n \geq 0} H^0(Y, \mathscr{L}^n)$.*

Proof First of all, by the projection formula (Hartshorne, 1977, Chap II, Exercise 5.1(d)),

$$H^0(X, f^*\mathscr{L}^n) \simeq H^0(Y, \mathscr{L}^n \otimes f_*\mathscr{O}_X). \tag{1}$$

Since $f_*\mathscr{O}_X$ is a coherent $\mathscr{O}_Y$-module (cf. (Hartshorne, 1977, Chap III, Remark 8.8.1)), we can write (cf. (Hartshorne, 1977, Chap II, Corollary 5.18))

$$0 \to \mathscr{K} \to \mathscr{L}_1 \oplus \cdots \oplus \mathscr{L}_m \to f_*\mathscr{O}_X \to 0$$

for some line bundles $\mathscr{L}_i$ over Y. Since $\mathscr{L}$ is ample over Y, there exists $n_o \geq 1$ such that for $n \geq n_o$, $H^1(Y, \mathscr{L}^n \otimes \mathscr{K}) = 0$ (cf. (Hartshorne, 1977, Chap III, Proposition 5.3)). Thus, the induced map

$$H^0\left(Y, \mathscr{L}^n \otimes (\oplus_i \mathscr{L}_i)\right) \twoheadrightarrow H^0(Y, \mathscr{L}^n \otimes f_*\mathscr{O}_X) \tag{2}$$

is surjective for $n \geq n_o$. We now prove that $\bigoplus_{n\geq 0} H^0(Y, \mathscr{L}^n \otimes \mathscr{L}_i)$ is finitely generated over R for any line bundle $\mathscr{L}_i$ over Y.

Consider the sheaf exact sequence

$$0 \to \mathscr{I}_{\Delta(Y)} \to \mathscr{O}_{Y\times Y} \to \mathscr{O}_Y \to 0,$$

where $\Delta(Y) \subset Y \times Y$ is the diagonal and $\mathscr{I}_{\Delta(Y)}$ is the ideal sheaf of $\Delta(Y)$ in $Y \times Y$. There exists $l_o, m_o \geq 1$ such that for $n \geq l_o$ and $m \geq m_o$,

$$H^1\left(Y \times Y, \mathscr{I}_{\Delta(Y)} \otimes \left(\mathscr{L}^n \boxtimes \left(\mathscr{L}^m \otimes \mathscr{L}_i\right)\right)\right) = 0.$$

Thus, for $n \geq l_o$,

$$H^0(Y, \mathscr{L}^n) \otimes H^0\left(Y, \mathscr{L}^{m_o} \otimes \mathscr{L}_i\right) \twoheadrightarrow H^0\left(Y, \mathscr{L}^{n+m_o} \otimes \mathscr{L}_i\right) \text{ is surjective.}$$

This (using (2)) proves that the ring $\bigoplus_{n\geq 0} H^0(Y, \mathscr{L}^n \otimes f_*\mathscr{O}_X)$ is finitely generated over R, and hence integral (cf. (Atiyah and Macdonald, 1969, Proposition 5.1)). This proves the lemma by (1). □

Let S be an irreducible normal variety parameterizing a family $\mathscr{E} \to \Sigma \times S$ of G-bundles over Σ. Let $S^{ss} \subset S$ be the (open) subset consisting of those $t \in S$ such that $\mathscr{E}_{|\Sigma\times t}$ is semistable (cf. Corollary 7.1.16(b)). Consider the induced morphism

$$f_{\mathscr{E}} : S^{ss} \to M^G.$$

Let V be a representation of G of dimension r. As in Definition 8.4.4 (specifically (1) of the same), this gives rise to the morphism

$$\varphi^V : M^G \to M(r,0),\ [E] \mapsto [E(V)],\ \text{ where } E(V) := E \times^G V.$$

As in the isomorphism (3) of Definition 8.4.4, we have (for any $a \geq 0$):

$$\Theta(\mathscr{E}(V))^{\otimes a}_{|S^{ss}} \simeq f^*_{\mathscr{E}}(\Theta(V))^{\otimes a}.$$

We fix such an isomorphism.

Let $\mathring{S}^{ss} := \{t \in S : \mathscr{E}(V)_{|\Sigma\times t}$ is a semistable vector bundles$\} \supset S^{ss}$ (cf. Theorem 6.1.7).

Proposition 8.4.13 *Assume that $S^{ss} \neq \emptyset$ and take any $a \geq 0$.*

(a) For any $\sigma_o \in H^0(M(r,0), \Theta^{\otimes a})$, the pull-back section

$$\mathring{f}^*_{\mathscr{E}(V)}(\sigma_o) \in H^0(\mathring{S}^{ss}, (\Theta(\mathscr{E}(V)))^{\otimes a})$$

extends to an element of $H^0(S, (\Theta(\mathscr{E}(V)))^{\otimes a})$ if the genus $g \geq 1$, where $\mathring{f}_{\mathscr{E}(V)}\colon \mathring{S}^{ss} \to M(r,0)$ is the morphism induced from the family $\mathscr{E}(V)$.

(b) Assume $g \geq 1$. For any section $\sigma \in H^0(M^G, (\Theta(V))^{\otimes a})$, the pull-back section $f^*_{\mathscr{E}}(\sigma)$ over S^{ss} extends to an element of $H^0(S, (\Theta(\mathscr{E}(V)))^{\otimes a})$.

Proof (a): Let $\mathring{\mathscr{V}} \to \Sigma \times \mathring{S}$ be a family of rank-r and degree-δ vector bundles over Σ parameterized by a noetherian scheme $\mathring{S}$. Fix a very ample line bundle H over Σ of the form $\mathscr{O}_\Sigma(hp)$, where $p \in \Sigma$ is the base point (cf. (Hartshorne, 1977, Chap. IV, Corollary 3.2)). Then, we can find a positive integer $m_{\mathring{S}}$ such that for $m \geq m_{\mathring{S}}$ we have (cf. (Hartshorne, 1977, Chap. III, Theorem 8.8) and (Kempf, 1978, Theorem 13.1)):

(A) $R^i\pi_{\mathring{S}*}(\mathring{\mathscr{V}}(m)) = 0$, for $i \geq 1$.

(B) The natural map $\pi^*_{\mathring{S}}\pi_{\mathring{S}*}(\mathring{\mathscr{V}}(m)) \to \mathring{\mathscr{V}}(m)$ is surjective.

(C) $\pi_{\mathring{S}*}(\mathring{\mathscr{V}}(m))$ is a vector bundle over $\mathring{S}$ of rank $N := \delta + r(hm + 1 - g)$, since $\deg H = h$, where $\pi_{\mathring{S}}$ is the projection on the $\mathring{S}$-factor and $\mathring{\mathscr{V}}(m) := \mathring{\mathscr{V}} \otimes \left(H^m \boxtimes \mathscr{O}_{\mathring{S}}\right)$.

Let $\gamma\colon \mathscr{F}(\mathring{\mathscr{V}}(m)) \to \mathring{S}$ be the frame bundle of $\pi_{\mathring{S}*}(\mathring{\mathscr{V}}(m))$. Now, take large enough m such that $m \geq m_{\mathring{S}}$ and $d = \delta + mhr$ satisfies the bound in Theorem 7.1.13. (Observe that d is the degree of $\mathring{\mathscr{V}}(m)_t$ for any $t \in \mathring{S}$.) Then, by Theorem 7.1.6 applied to the special case of Definition 7.1.7 for the pair (r, $d = \delta + mhr$), we get a GL_N-equivariant morphism $\mathring{\phi}\colon \mathscr{F}(\mathring{\mathscr{V}}(m)) \to R$ (where R is as in Definition 7.1.7) such that

$$(\mathrm{Id}_\Sigma \times \mathring{\phi})^* \left(\mathscr{U}_{|\Sigma \times R}\right) \simeq (\mathrm{Id}_\Sigma \times \gamma)^* \left(\mathring{\mathscr{V}}(m)\right), \tag{1}$$

where $\mathscr{U}$ is the tautological family (cf. Definition 7.1.7 and Proposition 7.1.9(A)).

In the above construction take $\mathring{S} = S$ with the family $\mathring{\mathscr{V}} = \mathscr{E}(V)$ of rank-r vector bundles and degree $\delta = 0$. Consider the following commutative diagram:

$$\begin{array}{ccccc}
\mathring{S}^{ss} & \hookrightarrow & S & \xleftarrow{\gamma} & \mathscr{F}(\mathring{\mathscr{V}}(m)) \\
\downarrow{\scriptstyle \mathring{f}_{\mathscr{E}(V)}} & & & & \downarrow{\scriptstyle \mathring{\phi}} \\
M(r,0) \simeq M(r,mhr) := R^{ss}/\!/\mathrm{SL}_N & \xleftarrow[\pi]{} & R^{ss} & \hookrightarrow & R,
\end{array}$$

where R^{ss} is as defined in Definition 7.1.12 and the identification $M(r,0) \xrightarrow{\sim} M(r,mhr)$ is induced by $[V] \mapsto [V \otimes H^m]$ (cf. the proof of Theorem 7.1.14).

Now, R is smooth by Proposition 7.1.9(A). Further, by the (following) Lemma 8.4.14,

$$\operatorname{codim}_R(R\backslash R^{ss}) \geq 2, \quad \text{if genus } g \geq 1. \tag{2}$$

We are ready to prove (a) now. Since $\mathscr{U}_{|\Sigma\times R}$ is a GL_N-equivariant vector bundle (cf. Definition 7.1.7), $\Theta(\mathscr{U})$ over R acquires a canonical GL_N-equivariant line bundle structure. We retain its SL_N-equivariant structure but we change the action of the center Z of GL_N on $\Theta(\mathscr{U})$ by requiring it to act trivially. (By Proposition 7.1.9(D), Z acts trivially on R.) It is easy to see that the original action of $Z \cap \mathrm{SL}_N$ on $\Theta(\mathscr{U})$ is indeed trivial. Equipped with this new action of GL_N on $\Theta(\mathscr{U})$ over R, to distinguish, we denote it by $\Theta'(\mathscr{U})$. Observe next that any line bundle $\mathcal{L}$ over R^{ss} (and R) can have at most one SL_N-equivariant structure. To prove this follow the last part of the proof of Proposition 8.3.2 and the result that any morphism $\mathrm{SL}_N(\mathbb{C}) \to \mathbb{C}^*$ is constant (cf. Lemma 1.4.9). From the uniqueness of SL_N-structure on any line bundle $\mathcal{L}$ over R^{ss}, it is easy to see that the pull-back GL_N-equivariant structure on $\Theta(\mathscr{U})$, coming from the morphism π, coincides with that of $\Theta'(\mathscr{U})_{|R^{ss}}$.

Let $\pi^*\sigma_o \in H^0(R^{ss}, (\Theta'(\mathscr{U}))^{\otimes a})$ be the pull-back section, which is, of course, GL_N-invariant. This extends as a section $\overline{\pi^*\sigma_o}$ of the GL_N-equivariant line bundle $\Theta'(\mathscr{U})^{\otimes a}$ over R by (2) (which continues to be GL_N-invariant since R^{ss} is dense open in R by Proposition 7.1.9(E)). Pull $\overline{\pi^*\sigma_o}$ via the GL_N-equivariant morphism $\mathring{\phi}$ and push via the GL_N-principal bundle γ to get a section in $H^0(S, (\Theta(\mathscr{E}(V)(m)))^{\otimes a})$ by (1). (It is easy to see using (1) that $\gamma^*(\Theta(\mathscr{E}(V)(m))) \simeq \mathring{\phi}^*(\Theta'(\mathscr{U}))$ as GL_N-equivariant line bundles since there is a unique SL_N-equivariant structure on any line bundle if it exists.) From the commutativity of the above diagram, we get that this section extends the section $\mathring{f}^*_{\mathscr{E}(V)}(\sigma_o)$ over $\mathring{S}^{ss}$ to S. This proves part (a).

(b): Consider the diagram obtained from the morphism $f_{\mathscr{E}}$ and φ^V defined above the proposition:

$$\begin{array}{ccccc}
(\Theta(\mathscr{E}(V)))^{\otimes a}_{|S^{ss}} & \longrightarrow & (\Theta(V))^{\otimes a} & \longrightarrow & \Theta^{\otimes a} \\
\downarrow & & \downarrow & & \downarrow \\
S^{ss} & \xrightarrow{f_{\mathscr{E}}} & M^G & \xrightarrow{\varphi^V} & M(r,0).
\end{array}$$

By Lemma 8.4.11, it suffices to prove that $f^*_{\mathscr{E}}\sigma \in H^0(S^{ss}, (\Theta(\mathscr{E}(V)))^{\otimes a})$ is integral over $R := \bigoplus_{n\geq 0} H^0(S, (\Theta(\mathscr{E}(V)))^{\otimes n})$. By Theorem 7.1.14, $M(r,d)$ is a projective variety and by Theorem 7.2.14, M^G is a complete variety if

$g \geq 2$. Further, M^G is a projective variety if $g = 1$ (cf. Theorem 9.2.1). Since σ is integral over $\mathring{R} := \bigoplus_{n \geq 0} H^0(M(r,0), \Theta^{\otimes n})$ by Lemma 8.4.12, we get a relation

$$\sigma^p + b_1 \sigma^{p-1} + \cdots + b_p = 0, \quad \text{for some } b_i \in \mathring{R}. \tag{3}$$

By the (a)-part, we can extend each $\mathring{f}^*_{\mathscr{E}(V)}(b_i)$ to a section over S. Thus, by (3), we get

$$(f^*_{\mathscr{E}}\sigma)^p + \left(\mathring{f}^*_{\mathscr{E}(V)} b_1\right)_{|S^{ss}} (f^*_{\mathscr{E}}\sigma)^{p-1} + \cdots + \left(\mathring{f}^*_{\mathscr{E}(V)} b_p\right)_{|S^{ss}} = 0.$$

This proves the (b)-part of the proposition. □

I thank V. Balaji for communicating to us the following proof essentially due to Atiyah and Bott (1982, §7).

Lemma 8.4.14 *With the notation as in Definition 7.1.7, for $R = R(r,d)$ where $d > r(2g-1)$, we have*

$$\operatorname{codim}_R(R \backslash R^{ss}) \geq 2, \ \text{ if } g \geq 2 \text{ or } (g = 1, r \geq 2 \text{ and } r | d),$$

where g is the genus of Σ.

Proof Let $R^{un} := R \backslash R^{ss}$ and let $R' \subset R^{un}$ be an open subset consisting of those $q \in R^{un}$ such that the HN filtration $\{\bar{q}_i\}_{0 \leq i \leq n}$ of $\bar{q}$ (cf. Exercise 6.2.E.2) has a fixed collection of slopes $\{\mu(\bar{q}_i/\bar{q}_{i-1})\}_{1 \leq i \leq n}$. The Kodaira–Spencer infinitesimal deformation maps (cf. Definition 6.3.17) at $q \in R'$ give the following commutative diagram:

$$\begin{array}{ccccccccc}
0 & \longrightarrow & T_q(R') & \longrightarrow & T_q(R) & \longrightarrow & T_q(R)/T_q(R') & \longrightarrow & 0 \\
 & & \downarrow & & \downarrow \pi & & \downarrow \pi'' & & \\
0 & \longrightarrow & H^1(\Sigma, \mathbf{End}'(\bar{q})) & \longrightarrow & H^1(\Sigma, \mathbf{End}(\bar{q})) & \longrightarrow & H^1(\Sigma, \mathbf{End}''(\bar{q})) & \longrightarrow & 0,
\end{array}$$

where $\mathbf{End}(\bar{q})$ is the sheaf of endomorphisms of $\bar{q}$, $\mathbf{End}'(\bar{q})$ denotes the subsheaf consisting of those endomorphisms which preserve the HN filtration of $\bar{q}$ and $\mathbf{End}''(\bar{q}) := \mathbf{End}(\bar{q})/\mathbf{End}'(\bar{q})$. The top horizontal sequence is exact by definition and the exactness of the bottom horizontal sequence follows from

$$H^0\left(\Sigma, \mathbf{End}''(\bar{q})\right) = 0. \tag{1}$$

To prove (1), use Lemma 6.2.4. The map $\pi : T_q(R) \to H^1(\Sigma, \mathbf{End}(\bar{q}))$ is surjective, as it follows from the description of the tangent space given by (1) of Theorem 7.1.6 together with the cohomology sequence associated to the sheaf sequence (4) and the cohomology vanishing (3) both from the proof of Proposition 7.1.9. Thus, the map π'' is surjective. Now, by (1), since $\mathrm{gr}(\mathbf{End}''(\bar{q})) = \bigoplus_{1\leq i<j\leq n} \mathbf{Hom}_{\mathcal{O}_\Sigma}(\mathscr{V}_i, \mathscr{V}_j)$ (where $\mathscr{V}_i := \bar{q}_i/\bar{q}_{i-1}$),

$$\begin{aligned}\dim H^1(\Sigma, \mathbf{End}''(\bar{q})) &= (g-1)\,\mathrm{rank}(\mathbf{End}''\,\bar{q}) - \mathrm{degree}(\mathbf{End}''(\bar{q}))\\ &= (g-1)\sum_{1\leq i<j\leq n} r_i r_j + \sum_{1\leq i<j\leq n} (r_j d_i - d_j r_i), \quad (2)\end{aligned}$$

where $r_i := \mathrm{rank}\,\mathscr{V}_i$, $d_i := \deg \mathscr{V}_i$. By the HN filtration property of $\{\bar{q}_i\}_{0\leq i\leq n}$,

$$\mu(\mathscr{V}_i) = \frac{d_i}{r_i} > \frac{d_j}{r_j} = \mu\left(\mathscr{V}_j\right) \quad \text{for any} \quad i<j.$$

Since $\bar{q} \in R^{un}$, we have $n \geq 2$. Thus, by (2) and the surjectivity of π'':

$$\dim\left(T_q(R)/T_q(R')\right) \geq \dim H^1\left(\Sigma, \mathbf{End}''(\bar{q})\right) \geq 2, \text{ if } g\geq 2 \text{ or } g=1 \text{ and } n \geq 3.$$

If $g = 1$ and $n = 2$, we get (by (2)):

$$\begin{aligned}\dim H^1\left(\Sigma, \mathbf{End}''(\bar{q})\right) &= r_2 d_1 - r_1 d_2 > 0\\ &= r d_1 - r_1 d > 0.\end{aligned}$$

But, by assumption, if $g = 1$, then $r|d$. Thus, $r|\dim H^1\left(\Sigma, \mathbf{End}''(\bar{q}\right)$; in particular, $\dim H^1(\Sigma, \mathbf{End}''(\bar{q})) \geq 2$, proving the lemma in the case $g = 1$, $r \geq 2$ and $r|d$ as well. □

We finally come to the main result of this Section 8.4.

Theorem 8.4.15 *Let the triple (G, Σ, p) be as at the beginning of this section and let V be a representation of G (with Dynkin index d_V, cf. Definition A.1). Then, for any $a \geq 0$ there exists a canonical isomorphism:*

$$H^0(M^G, (\Theta(V))^{\otimes a}) \simeq H^0(\bar{X}_G, \mathscr{L}(0_{d_V})^{\otimes a})^{\bar{\Gamma}}, \qquad (1)$$

where the ind-group variety $\bar{\Gamma}$ (with $\bar{\Gamma}(\mathbb{C}) = G[\Sigma^*]$) acts on $\bar{X}_G$ via the morphism $\bar{\Gamma} \to \bar{G}((t))$ (see the beginning of Section 8.2), $\mathscr{L}(0_d)$ is the $\bar{\Gamma}$-equivariant line bundle as in Proposition 8.3.2 and the Θ-bundle $\Theta(V)$ over the moduli space M^G of semistable G-bundles over Σ is defined by (2) of Definition 8.4.4.

Moreover, there exists a canonical identification (up to scalar multiples)

$$H^0\left(\bar{X}_G, \mathscr{L}(0_{d_V})^{\otimes a}\right)^{\bar{\Gamma}} \simeq \left[\mathscr{H}(0_{d_V\cdot a})^*\right]^{\mathrm{Lie}\,\bar{\Gamma}} = \mathscr{V}_\Sigma^\dagger(p, 0_{d_V\cdot a}), \qquad (2)$$

and hence its dimension is given by the Verlinde formula (Theorem 4.2.19). In particular, $H^0(\bar{X}_G, \mathscr{L}(0_{d_V})^{\otimes a})^{\Gamma}$ is finite dimensional.

Proof We first assume that the genus $g \geq 1$. By Lemma 8.4.10, the pull-back map induces an isomorphism

$$f^*_{\mathbf{U}_G} : H^0\left(M^G, (\Theta(V))^{\otimes a}\right) \xrightarrow{\sim} H^0\left(\bar{X}^{ss}_G, \mathscr{L}(0_{d_V})^{\otimes a}\right)^{\Gamma} \tag{3}$$

induced from the $\bar{\Gamma}$-equivariant family $\mathbf{U}_G$ of G-bundles over Σ parameterized by $\bar{X}_G$. Applying Proposition 8.4.13(b) to the family $\mathbf{U}_G$ over $\Sigma \times \bar{X}_G$ (rather its restriction to $\Sigma \times \bar{X}^n_G$), we get that for any $\sigma \in H^0(M^G, (\Theta(V))^{\otimes a})$, $f^*_{\mathbf{U}_G}(\sigma)$ extends to a section over $\bar{X}^n_G$ for each $n \geq 0$, where $\bar{X}^{\circ}_G \subset \bar{X}^1_G \subset \ldots$ is a filtration of $\bar{X}_G$ by Schubert subvarieties $\bar{X}^n_G$ (which are irreducible and normal varieties by Kumar (2002, Theorem 8.2.2(b)); here we have used Proposition 1.3.24). Such a filtration exists by using Kumar (2002, Lemma 1.3.20). Since $\bar{X}^{ss}_G \subset \bar{X}_G$ is a (nonempty) open subset (cf. Corollary 7.1.16(b)) and each $\bar{X}^n_G$ is irreducible, the unique extensions of $f^*_{\mathbf{U}_G}(\sigma)$ to each $\bar{X}^n_G$ patch up to give the extension $\overline{f^*_{\mathbf{U}_G}(\sigma)}$ of $f^*_{\mathbf{U}_G}(\sigma)$ to the whole of $\bar{X}_G$. Moreover, $f^*_{\mathbf{U}_G}(\sigma)$ being Γ-invariant, so is $\overline{f^*_{\mathbf{U}_G}(\sigma)}$. This proves the first isomorphism (1).

The first identification of the second isomorphism (2) follows from the analogue of the Borel–Weil theorem for $\bar{X}_G$ (cf. Theorem 8.3.3) and Lemma B.23 (since $\bar{\Gamma}$ is an irreducible ind-group variety by Theorem 8.1.1).

The second identification of (2) is, of course, the definition of the space of vacua. This proves the theorem for $g \geq 1$.

Assume now that $g = 0$. In this case the isomorphisms (2) are still true (by the same proof). Thus, by Exercise 2.3.E.2(a),

$$\dim H^0\left(\bar{X}_G, \mathscr{L}(0_{d_V})^{\otimes a}\right)^{\Gamma} = 1. \tag{4}$$

Since $\bar{X}^{ss}_G$ is open in $\bar{X}_G$, the restriction map

$$H^0\left(\bar{X}_G, \mathscr{L}(0_{d_V})^{\otimes a}\right)^{\Gamma} \hookrightarrow H^0\left(\bar{X}^{ss}_G, \mathscr{L}(0_{d_V})^{\otimes a}\right)^{\Gamma} \tag{5}$$

is injective. For $g = 0$, M^G is a one-point space (cf. Exercise 6.1.E.11). Thus, by the isomorphism (3) (which is true in any genus),

$$\dim H^0\left(\bar{X}^{ss}_G, \mathscr{L}(0_{d_V})^{\otimes a}\right)^{\Gamma} = 1.$$

Combining (3)–(5), we get (1) of the theorem in the case $g = 0$ as well. This completes the proof of the theorem. □

8.4.E Exercises

Let Σ be as at the beginning of this section.

(1) Let $\mathscr{V}$ and $\mathscr{W}$ be two vector bundles over $\Sigma \times S$, where S is a connected noetherian scheme. Then there is a canonical isomorphism

$$\mathrm{Det}(\mathscr{V} \oplus \mathscr{W}) \simeq \mathrm{Det}(\mathscr{V}) \otimes \mathrm{Det}(\mathscr{W}).$$

Moreover, if $\mathscr{V}$ is a trivial vector bundle then $\mathrm{Det}(\mathscr{V})$ is the trivial line bundle over S:

$$\mathrm{Det}(\mathscr{V}) \simeq \mathscr{O}_S.$$

(2) Let $p \in \Sigma$ be the base point as at the beginning of this section. Show that for any family of vector bundles $\mathscr{V} \to \Sigma \times S$ (for a connected noetherian scheme S),

$$\Theta(\mathscr{V}) = \Theta\left(\mathscr{V} \otimes \pi_\Sigma^* \left(\mathscr{O}(n \cdot p)\right)\right), \quad \text{for any } n \in \mathbb{Z},$$

where $\pi_\Sigma : \Sigma \times S \to \Sigma$ is the projection.

Hint: Use the exact sequence of the families:

$$0 \to \mathscr{V} \otimes \pi_\Sigma^*(\mathscr{O}(-p)) \to \mathscr{V} \to \mathscr{V}_{|p \times S} \to 0.$$

(3) Let G be a connected simply-connected semisimple group. Show that $M^G = M^G(\Sigma)$ is a unirational variety.

Hint: Use the morphism $f_{\mathbf{U_G}} : \bar{X}_G^{ss} \to M^G$ as in Definition 8.4.9.

8.5 Identification of Conformal Blocks with Sections of Line Bundles over Parabolic Moduli Space

In this section $(\Sigma, \vec{p} = (p_1, \dots, p_s))$ is an s-pointed (for any $s \geq 1$) smooth irreducible projective curve of any genus $g \geq 2$. We also fix a base point $p \in \Sigma$ different from any p_i. We take G to be a connected, simply-connected simple algebraic group over $\mathbb{C}$. We fix parabolic weights $\vec{\tau} = (\tau_1, \dots, \tau_s)$, $\tau_j \in \Phi_0$ (cf. Definition 6.1.2) such that $\mathrm{Exp}(2\pi i \tau_j)$ is of finite order and

$$\theta(\tau_j) < 1, \quad \text{for all} \quad 1 \leq j \leq s, \tag{1}$$

where θ is the highest root. The parabolic weights $\vec{\tau}$ provide standard parabolic subgroups of G:

$$\vec{P} = \vec{P}(\vec{\tau}) = (P_1, \dots, P_S),$$

where the Levi subgroup of P_j has for its simple roots $\{\alpha_i : \alpha_i(\tau_j) = 0\}$. Following Definition 6.1.10, we fix positive integers $\vec{d} = (d_1, \ldots, d_s)$ such that $\operatorname{Exp}(2\pi i d_j \tau_j) = 1$ (i.e., $d_j\tau_j \in$ coroot lattice), a Galois cover $\pi : \hat{\Sigma} \to \Sigma$ associated to $(\Sigma, \vec{p})$ and the sequence $\vec{d}$ with Galois group A.

As in Exercise 7.2.E.4, the set of markings $\vec{\tau}$ equivalently provides the topological type (denoted) τ of A-equivariant G-bundles over $\hat{\Sigma}$ (observe that G being simply connected, any G-bundle over $\hat{\Sigma}$ is non-equivariantly topologically trivial).

By virtue of Theorem 6.1.17, for any ind-scheme S, there is a natural correspondence between A-equivariant G-bundles $\hat{\mathscr{F}}$ over $\hat{\Sigma} \times S$ of topological type τ and quasi-parabolic G-bundles $\mathscr{F}$ over $\Sigma \times S$ of type $\vec{P}$. Under this correspondence A-semistable (resp. A-stable) bundles over $\hat{\Sigma}$ correspond to parabolic semistable (resp. stable) G-bundles with markings $\vec{\tau}$. *We will often make this identification:* $\hat{\mathscr{F}} \leftrightarrow \mathscr{F}$.

Definition 8.5.1 (Quasi-parabolic determinant bundle) Let $(\mathscr{E}, \vec{\sigma})$ be a family of quasi-parabolic G-bundles of type $\vec{P}$ over $(\Sigma, \vec{p})$ parameterized by a connected noetherian scheme S, i.e., a G-bundle $\mathscr{E} \to \Sigma \times S$ together with sections σ_j of $\mathscr{E}/P_j$ over $p_j \times S$ (cf. Definition 5.1.4). Let V be a finite-dimensional representation of G. For any $d \in \mathbb{Z}$ and any characters $\vec{\mu} = (\mu_1, \ldots, \mu_s)$, where μ_j is a character of P_j, define a line bundle over S called the *quasi-parabolic determinant bundle*:

$$\operatorname{Det}_{\text{par}}(\mathscr{E}(V), d, \vec{\mu}) := \operatorname{Det}(\mathscr{E}(V))^{\otimes d} \otimes \left(\bigotimes_{j=1}^{s} \sigma_j^* \left(\mathscr{E} \times^{P_j} \mathbb{C}_{\mu_j^{-1}} \right) \right),$$

where $\mathbb{C}_{\mu_j^{-1}}$ denotes the 1-dimensional representation of P_j corresponding to the character μ_j^{-1} and $\operatorname{Det}(\mathscr{E}(V))$ is as defined in Definition 8.4.1. Observe that G being simple (in particular, semisimple)

$$\operatorname{Det}(\mathscr{E}(V)) = \Theta(\mathscr{E}(V))$$

(cf. (10) of Definition 8.4.1).

By Corollary 7.2.4, the equivariant quot scheme $R_\tau^{ss}(G)$ parameterizes a family $\hat{\mathscr{U}}_\tau^{ss}(G)$ of A-semistable G-bundles over $\hat{\Sigma}$ (denoted by $\mathscr{U}_\tau^{ss}(G)$ there). Thus, as mentioned at the beginning of this section, $R_\tau^{ss}(G)$ parameterizes the corresponding family $\mathscr{U}_\tau^{ss}(G)$ of quasi-parabolic G-bundles over Σ of type $\vec{P}$. In particular, for a representation V of G, $d \in \mathbb{Z}$, and any $\vec{\mu}$ as above, we have the corresponding quasi-parabolic determinant bundle over $R_\tau^{ss}(G)$:

$$\operatorname{Det}_{\text{par}}\left(\mathscr{U}_\tau^{ss}(G)(V), d, \vec{\mu}\right).$$

Definition 8.5.2 (Quasi-parabolic family parameterized by $\bar{X}_G \times \Pi_{j=1}^s (G/P_j)$) Let $\mathbf{U}_G \to \Sigma \times \bar{X}_G$ be the tautological bundle as in Proposition 5.2.4 and let $\mathbf{U}_G^{\vec{P}} \to \Sigma \times \bar{X}_G \times \Pi_{j=1}^s (G/P_j)$ be its pull-back via the projection $\Sigma \times \bar{X}_G \times \Pi_{j=1}^s (G/P_j) \to \Sigma \times \bar{X}_G$, where $\bar{X}_G$ is based at p. We realize $\mathbf{U}_G^{\vec{P}}$ as a quasi-parabolic family of G-bundles by defining the sections $\hat{\sigma}_i$ of $\mathbf{U}_G^{\vec{P}}/P_i$ over $p_i \times \bar{X}_G \times \Pi_{j=1}^s (G/P_j)$ by

$$\hat{\sigma}_i(p_i, x, g_1 P_1, \ldots, g_s P_s) = \sigma_{\bar{X}_G}(p_i, x) g_i P_i, \quad \text{for } x \in \bar{X}_G \text{ and } g_j \in G,$$

where the section $\sigma_{\bar{X}_G}$ is as in Proposition 5.2.4. Thus, following Definition 8.5.1, we get the quasi-parabolic determinant bundle (for any representation V of G, $d \in \mathbb{Z}$ and weights $\vec{\mu}$):

$$\mathrm{Det}_{\mathrm{par}}\left(\mathbf{U}_G^{\vec{P}}(V), d, \vec{\mu}\right) \quad \text{over} \quad \bar{X}_G \times \Pi_{j=1}^s (G/P_j).$$

It is easy to see that

$$\mathrm{Det}_{\mathrm{par}}\left(\mathbf{U}_G^{\vec{P}}(V), d, \vec{\mu}\right) = (\mathrm{Det}\, \mathbf{U}_G(V))^{\otimes d} \boxtimes \mathop{\boxtimes}_{j=1}^{s} \mathscr{L}_{P_j}(\mu_j), \tag{1}$$

where $\mathscr{L}_{P_j}(\mu_j)$ is the line bundle over G/P_j as in Definition 8.3.1. Recall that, by Theorem 8.4.7,

$$\mathrm{Det}\, \mathbf{U}_G(V) \simeq \mathscr{L}(0_{d_V}). \tag{2}$$

Let $\kappa : \mathfrak{h} \xrightarrow{\sim} \mathfrak{h}^*$ be the isomorphism induced from the invariant form normalized so that $\langle \theta, \theta \rangle = 2$ for the highest root θ. Let $\vec{\tau}^* := (\tau_1^*, \ldots, \tau_s^*)$, where $\tau_j^* := \kappa(\tau_j)$. Clearly, κ extends to an isomorphism $\mathfrak{g} \to \mathfrak{g}^*$.

We have the following parabolic analogue of Theorem 8.4.2. For $G = \mathrm{SL}_2$ see Narasimhan and Ramadas (1993, Theorem 1) and for any SL_N see Pauly (1996, Théorème 3.3).

Theorem 8.5.3 *Let $G = \mathrm{SL}_V$ for a finite-dimensional vector space V and let $(\Sigma, \vec{p})$, $\vec{\tau}$, $\hat{\Sigma}$, A and τ be as at the beginning of this section (for $G = \mathrm{SL}_V$). Let $d \in \mathbb{Z}_{>0}$ be so that $d\vec{\tau}^*$ is integral, i.e., each $d\tau_j^*$ is an integral weight. Then, there exists an ample line bundle (called the parabolic theta bundle) $\mathring{\Theta}_{par,\, \mathrm{SL}_V}(\tau, d)$ over the equivariant moduli space $M_\tau^{\mathrm{SL}_V}(\hat{\Sigma})$ (cf. Theorem 7.2.6) such that for any family $\hat{\mathscr{F}} \to \hat{\Sigma} \times S$ of A-semistable SL_V-bundles of topological type τ parameterized by a noetherian scheme S, we have*

$$\hat{f}^*_{\hat{\mathscr{F}}}\left(\mathring{\Theta}_{\mathrm{par},\, \mathrm{SL}_V}(\tau, d)\right) \simeq \mathrm{Det}_{\mathrm{par}}\left(\hat{\mathscr{F}}(V), d, d\vec{\tau}^*\right), \tag{1}$$

where $\hat{f}_{\hat{\mathscr{F}}}: S \to M_\tau^{\mathrm{SL}_V}(\hat{\Sigma})$ *is the morphism induced from the coarse moduli property of* $M_\tau^{\mathrm{SL}_V}(\hat{\Sigma})$ *(cf. Corollary 7.2.7) and the quasi-parabolic family* $\mathring{\hat{\mathscr{F}}}$ *corresponds to* $\hat{\mathscr{F}}$ *as at the beginning of this section.*

Since $M_\tau^{\mathrm{SL}_V}(\hat{\Sigma})$ *admits an ample line bundle and* $M_\tau^{\mathrm{SL}_V}(\hat{\Sigma})$ *is complete by Theorem 7.2.14, by (Hartshorne, 1977, Chap. II, Remark 5.16.1 and Theorem 7.6),* $M_\tau^{\mathrm{SL}_V}(\hat{\Sigma})$ *is a projective variety.*

Definition 8.5.4 Let G, $(\Sigma, \vec{p})$, $\vec{\tau}$, $\hat{\Sigma}$, A and τ be as at the beginning of this section. Take a representation V of G. This gives rise to a morphism (cf. Definition 7.2.10):

$$\bar{\rho}: M_\tau^G(\hat{\Sigma}) \to M_{\mathring{\tau}}^{\mathrm{SL}_V}(\hat{\Sigma}),$$

induced from the representation $\rho: G \to \mathrm{SL}_V$, where $\mathring{\tau}$ with $\mathring{\tau}_j \in \Phi_0(sl_V)$ is the topological type of A-equivariant bundle obtained from any A-semistable G-bundle over $\hat{\Sigma}$ of topological type τ by extension of the structure group via ρ. Thus, $\mathrm{Exp}(2\pi i \mathring{\tau}_j)$ is the unique conjugate of $\rho(\mathrm{Exp}(2\pi i \tau_j)) = \mathrm{Exp}(2\pi i \dot{\rho}(\tau_j))$ in SL_V.

Assume that $\vec{\mathring{\tau}} = (\mathring{\tau}_1, \dots, \mathring{\tau}_s)$ satisfies $\mathring{\theta}(\mathring{\tau}_j) < 1$ for all j, where $\mathring{\theta}$ is the highest root of sl_V.

Let $d \in \mathbb{Z}_{>0}$ be such that $d\kappa_{sl_V}(\mathring{\tau}_j)$ is an integral weight for sl_V for all j. (1)

Define the *parabolic theta bundle* over $M_\tau^G(\hat{\Sigma})$ corresponding to the G-module V by

$$\Theta_{\mathrm{par},G}(V,\tau,d) := \bar{\rho}^* \left(\mathring{\Theta}_{\mathrm{par},\,\mathrm{SL}_V}(\mathring{\tau},d) \right). \tag{2}$$

If ρ is an embedding, i.e., V is a faithful representation of G and Σ has genus $g \geq 2$ (which is our blanket assumption in this section), then $\bar{\rho}$ is a finite morphism (cf. Corollary 7.2.15). Thus, $\mathring{\Theta}_{\mathrm{par},\,\mathrm{SL}_V}(\mathring{\tau},d)$ being an ample line bundle, so is $\Theta_{\mathrm{par},G}(V,\tau,d)$ (for faithful G-module V and $g \geq 2$).

Lemma 8.5.5 *For any family* $\hat{\mathscr{F}} \to \hat{\Sigma} \times S$ *of A-semistable G-bundles of topological type* τ *parameterized by a noetherian scheme S, a representation* ρ *of G in V* (ρ *is not necessarily an embedding) such that* $\vec{\tau}$ *and* $\vec{\mathring{\tau}}$ *satisfy the condition* (1) *at the beginning of this section,*

$$\dot{\rho}(\tau_j) \text{ is } \mathrm{SL}_V\text{-conjugate of } \mathring{\tau}_j \text{ for each } j, \tag{1}$$

and $d \in \mathbb{Z}_{>0}$ *such that d satisfies condition* (1) *of Definition 8.5.4, we have*

$$\hat{f}^*_{\hat{\mathscr{F}}} \left(\Theta_{\mathrm{par},G}(V,\tau,d) \right) \simeq \mathrm{Det}_{\mathrm{par}} \left(\mathscr{F}(V), d, d d_V \vec{\tau}^* \right), \tag{2}$$

where $\hat{f}_{\hat{\mathscr{F}}}: S \to M_\tau^G(\hat{\Sigma})$ *is the morphism induced from the family* $\hat{\mathscr{F}}$ *(cf. Corollary 7.2.7) and* d_V *is the Dynkin index of V (cf. Definition A.1). (Observe that by* (6) *below,* $dd_V\vec{\tau}^*$ *is integral for G.)*

Proof Consider the commutative diagram

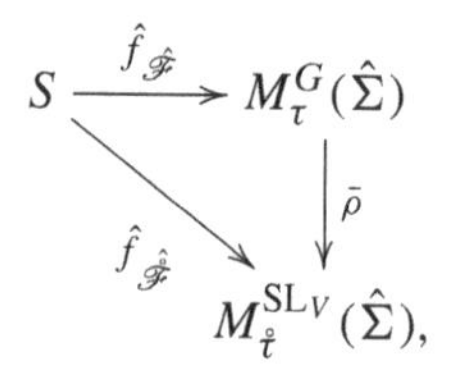

where $\hat{\mathring{\mathscr{F}}} := \hat{\mathscr{F}} \times^G \mathrm{SL}_V$. By Theorem 8.5.3 applied to the family $\hat{\mathring{\mathscr{F}}}$:

$$\hat{f}^*_{\hat{\mathring{\mathscr{F}}}}\left(\mathring{\Theta}_{\mathrm{par},\,\mathrm{SL}_V}(\mathring{\tau},d)\right) \simeq \mathrm{Det}_{\mathrm{par}}\left(\mathscr{F}(V),d,d\vec{\mathring{\tau}}^*\right). \tag{3}$$

Since $\dot{\rho}(\tau_j) = \mathrm{Ad}\,g_j(\mathring{\tau}_j)$, for some $g_j \in \mathrm{SL}_V$ (by assumption), Kempf's parabolic subalgebra $\mathfrak{p}(\dot{\rho}(\tau_j)) = g_j\mathfrak{p}(\mathring{\tau}_j)g_j^{-1}$ (cf. Section 6.1). Thus, from (2) of Definition 8.5.4 and the above commutative diagram together with (3), we need to prove that

$$(\dot{\rho}(\vec{\tau})^*)_{|\mathfrak{g}} = d_V\vec{\tau}^*. \tag{4}$$

For the trivial representation V of G, (4) is trivially true since $d_V = 0$. So, let us assume that V is nontrivial. Thus, $\dot{\rho}: \mathfrak{g} \to sl_V$ is injective. Let $\langle\,,\rangle$ (resp. $\langle\,,\rangle_V$) be the normalized invariant form on $\mathfrak{g}$ (resp. sl_V). Since $\langle\,,\rangle$ is nondegenerate and $d_V \neq 0$, we have

$$\dot{\rho}(\mathfrak{g}) \oplus \dot{\rho}(\mathfrak{g})^\perp = sl_V. \tag{5}$$

Take $y \in \mathfrak{g}$ and $z \in \dot{\rho}(\mathfrak{g})^\perp$. Then, for any $1 \le j \le s$,

$$\begin{aligned}\dot{\rho}(\tau_j)^*(\dot{\rho}(y)+z) &= \langle\dot{\rho}(\tau_j),\dot{\rho}(y)\rangle_V,\\ &= d_V\langle\tau_j,y\rangle, \quad \text{by Definition A.1.}\end{aligned} \tag{6}$$

This gives

$$(\dot{\rho}(\tau_j)^*)_{|\mathfrak{g}} = d_V\tau_j^*, \quad \text{proving (4).}$$

Thus, the lemma is proved. □

Remark 8.5.6 Condition (1) of Lemma 8.5.5 can be rephrased as follows (cf. Exercise 8.5.E.3).

Choose a basis $\mathcal{B}_j$ of V such that $\dot{\rho}(\tau_j)$ is diagonal and dominant in this basis. Then condition (1) is equivalent to the requirement that

$$\mathring{\theta}_j(\dot{\rho}(\tau_j)) < 1, \quad \text{for all } j,$$

where $\mathring{\theta}_j$ is the highest root of SL_V in the basis $\mathcal{B}_j$. To prove this, use the fact that if two dominant elements $h_1, h_2 \in \mathfrak{h}$ are such that $h_1 \in (\mathrm{Ad}\, G) \cdot h_2$, then $h_1 = h_2$.

Let $(\Sigma, \vec{p})$, G, $\vec{\tau}$, $\vec{P}$, $\hat{\Sigma}$, A be as at the beginning of this section. Let $\bar{X}_G^{\vec{P}} := \bar{X}_G \times \Pi_{j=1}^s (G/P_j)$ and let $\mathbf{U}_G^{\vec{P}} \to \Sigma \times \bar{X}_G^{\vec{P}}$ be the quasi-parabolic family of G-bundles as in Definition 8.5.2. This gives rise to the corresponding family $\hat{\mathbf{U}}_G^{\vec{P}}$ of A-equivariant G-bundles over $\hat{\Sigma} \times \bar{X}_G^{\vec{P}}$ (of topological type τ). Let $(\bar{X}_G^{\vec{P}})^{ss}$ be the open subset of $\bar{X}_G^{\vec{P}}$ consisting of those $x \in \bar{X}_G^{\vec{P}}$ such that $(\hat{\mathbf{U}}_G^{\vec{P}})_{|\hat{\Sigma}\times x}$ is A-semistable (cf. Corollary 7.1.16(b) and Exercise 6.2.E.4). This gives rise to a morphism via Corollary 7.2.7,

$$\hat{f}_{\hat{\mathbf{U}}_G^{\vec{P}}} : \left(\bar{X}_G^{\vec{P}}\right)^{ss} \to M_\tau^G(\hat{\Sigma}).$$

We follow the notation as in Definition 8.5.4; in particular, $\vec{\tau}$ and $\vec{\bar{\tau}}$ satisfy condition (1) at the beginning of this section (for a G-module V). By Lemma 8.5.5, and (1) and (2) of Definition 8.5.2, we get (for $\vec{\tau}$ satisfying condition (1) of Lemma 8.5.5 and d satisfying condition (1) of Definition 8.5.4):

$$\hat{f}^*_{\hat{\mathbf{U}}_G^{\vec{P}}} \left(\Theta_{\mathrm{par},G}(V,\tau,d)\right) \simeq \mathscr{L}(0_{dd_V}) \boxtimes \mathop{\boxtimes}_{j=1}^{s} \mathscr{L}_{P_j}\left(dd_V \tau_j^*\right).$$

Lemma 8.5.7 *For any representation V of G, any $d \in \mathbb{Z}_{>0}$ such that d satisfies condition* (1) *of Definition 8.5.4 and $\vec{\tau}$ satisfying condition* (1) *of Lemma 8.5.5,*

$$\hat{f}^*_{\hat{\mathbf{U}}_G^{\vec{P}}} : H^0\left(M_\tau^G(\hat{\Sigma}), \Theta_{\mathrm{par},G}(V,\tau,d)\right)$$
$$\xrightarrow{\sim} H^0\left(\left(\bar{X}_G^{\vec{P}}\right)^{ss}, \mathscr{L}(0_{dd_V}) \boxtimes \mathop{\boxtimes}_{j=1}^{s} \mathscr{L}_{P_j}(dd_V \tau_j^*)\right)^{\Gamma}$$

is an isomorphism, where the $\bar{\Gamma}$-equivariant structure on $\mathscr{L} := \mathscr{L}(0_{dd_V}) \boxtimes \mathop{\boxtimes}_{j=1}^{s} \mathscr{L}_{P_j}(dd_V \tau_j^*)$ is given above Theorem 8.3.4.

Proof The proof is similar to that of Lemma 8.4.10. By the proof of Lemma 5.2.12, it is easy to see that $\mathbf{U}_G^{\vec{P}}$ is a $\bar{\Gamma}$-equivariant quasi-parabolic bundle over $\Sigma \times \bar{X}_G^{\vec{P}}$. Thus, $\hat{\mathbf{U}}_G^{\vec{P}}$ is also $\bar{\Gamma}$-equivariant bundle, where the action of $\bar{\Gamma}$ commutes with that of A. Hence, $\hat{f}_{\hat{\mathbf{U}}_G^{\vec{P}}}$ is $\bar{\Gamma}$-invariant (with the trivial action of $\bar{\Gamma}$ on $M_\tau^G(\hat{\Sigma})$). This gives rise to a $\bar{\Gamma}$-equivariant structure on $\mathscr{L}$. But, by the same proof, as given in the proof of Proposition 8.3.2, $\mathscr{L}$ has a unique $\bar{\Gamma}$-equivariant structure. Thus, the image of $\hat{f}^*_{\hat{\mathbf{U}}_G^{\vec{P}}}$ does land inside

$H^0\big((\bar{X}_G^{\vec{P}})^{ss}, \mathscr{L}\big)^\Gamma$. Moreover, $\hat{f}_{\hat{\mathbf{U}}_G^{\vec{P}}}$ is surjective. To prove this, it suffices to show (by Theorem 6.1.17) that any quasi-parabolic G-bundle over Σ of type $\vec{P}$ occurs in the family $\mathbf{U}_G^{\vec{P}}$, which follows from corollary 5.2.17. Thus, $\hat{f}^*_{\hat{\mathbf{U}}_G^{\vec{P}}}$ is injective.

We now prove that $\hat{f}^*_{\hat{\mathbf{U}}_G^{\vec{P}}}$ is surjective. Take $\sigma \in H^0\big((\bar{X}_G^{\vec{P}})^{ss}, \mathscr{L}\big)^\Gamma$ and take $x_1, x_2 \in (\bar{X}_G^{\vec{P}})^{ss}$ such that

$$\hat{f}_{\hat{\mathbf{U}}_G^{\vec{P}}}(x_1) = \hat{f}_{\hat{\mathbf{U}}_G^{\vec{P}}}(x_2).$$

Thus, the corresponding A-equivariant G-bundles over $\hat{\Sigma}$ have the same equivalence classes, i.e.,

$$\Big[(\hat{\mathbf{U}}_G^{\vec{P}})_{|\hat{\Sigma}\times x_1}\Big] = \Big[(\hat{\mathbf{U}}_G^{\vec{P}})_{|\hat{\Sigma}\times x_2}\Big], \quad \text{as elements of } M_\tau^G(\hat{\Sigma}). \tag{1}$$

By Proposition 7.2.18, there exists an A-equivariant G-bundle $\hat{\mathscr{E}}^1 \to \hat{\Sigma} \times \mathbb{A}^1$ of topological type τ such that

(a) $\hat{\mathscr{E}}^1_{|\hat{\Sigma}\times\mathbb{A}^1\backslash\{0\}} \simeq p^*(\hat{E}^1)$, where $\hat{E}^i := (\hat{\mathbf{U}}_G^{\vec{P}})_{|\hat{\Sigma}\times x_i}$ and $p\colon \hat{\Sigma} \times \mathbb{A}^1\backslash\{0\} \to \hat{\Sigma}$ is the projection, and
(b) $\hat{\mathscr{E}}^1_{|\hat{\Sigma}\times\{0\}} \simeq \mathrm{gr}(\hat{E}^1)$.

Similarly, there is an A-equivariant G-bundle $\hat{\mathscr{E}}^2 \to \hat{\Sigma} \times \mathbb{A}^1$ of topological type τ corresponding to $\hat{E}^2$ and

$$\mathrm{gr}(\hat{E}^1) = \mathrm{gr}(\hat{E}^2), \quad \text{by (1)}. \tag{2}$$

Let $\mathscr{E}^1$ be the corresponding quasi-parabolic G-bundle over $\Sigma \times \mathbb{A}^1$ of type $\vec{P}$. By Theorem 5.2.5, Choose an étale cover $\pi : \widetilde{\mathbb{A}^1} \to \mathbb{A}^1$ such that the pullback $\widetilde{\mathscr{E}^1}$ of $\mathscr{E}^1$ to $\Sigma \times \widetilde{\mathbb{A}^1}$ is trivial over $\Sigma^* \times \widetilde{\mathbb{A}^1}$ as a G-bundle, where $\Sigma^* = \Sigma\backslash p$. Choose a section s of $\widetilde{\mathscr{E}^1}$ over $\Sigma^* \times \widetilde{\mathbb{A}^1}$, giving rise to a morphism (via Proposition 5.2.7) $\alpha\colon \widetilde{\mathbb{A}^1} \to \bar{X}_G$ such that

$$(\mathrm{Id} \times \alpha)^*(\mathbf{U}_G) = \widetilde{\mathscr{E}^1} \quad \text{as } G\text{-bundles}.$$

Let σ_j be the section of $\widetilde{\mathscr{E}^1}/P_j$ over $p_j \times \widetilde{\mathbb{A}^1}$ guaranteed by the quasi-parabolic structure of $\widetilde{\mathscr{E}^1}$. Define the map

$$\tilde{\alpha}\colon \widetilde{\mathbb{A}^1} \to \bar{X}_G \times \Pi_{j=1}^s(G/P_j), \quad \tilde{\alpha}(t) = (\alpha(t), g_1(t)P_1, \ldots, g_s(t)P_s),$$

where $g_j(t)P_j \in G/P_j$ is defined by

$$\sigma_j(p_j,t) = s(p_j,t)g_j(t)P_j.$$

It is easy to see that (as quasi-parabolic G-bundles of type $\vec{P}$)

$$(\mathrm{Id}_\Sigma \times \tilde{\alpha})^* \left(\mathbf{U}_G^{\vec{P}}\right) \simeq \widetilde{\mathscr{E}}^1.$$

Thus, by Corollary 5.2.17, properties (a) and (b) of $\hat{\mathscr{E}}^1$ and (2), we get

$$\overline{\Gamma \cdot x_1} \cap \overline{\Gamma \cdot x_2} \neq \emptyset.$$

Since σ is Γ-invariant, this shows that $\sigma(x_1) = \sigma(x_2)$, i.e., σ descends to a set-theoretic section $\bar{\sigma}$ of $\Theta_{\mathrm{par},G}(V,\tau,d)$ over $M_\tau^G(\hat{\Sigma})$. But then $\bar{\sigma}$ is algebraic by Kumar (2002, Proposition A.12). This completes the proof of the lemma. □

We continue to follow the notation and assumptions as in Definition 8.5.4. Let S be an irreducible normal variety parameterizing a family $\hat{\mathscr{E}} \to \hat{\Sigma} \times S$ of A-equivariant G-bundles over $\hat{\Sigma}$ of topological type τ. Let $S^{ss} \subset S$ be the open subset consisting of those $t \in S$ such that $\hat{\mathscr{E}}_{|\hat{\Sigma}\times t}$ is A-semistable (cf. Corollary 7.1.16(b) and Exercise 6.2.E.4). Consider the induced morphism (cf. Corollary 7.2.7)

$$\hat{f}_{\hat{\mathscr{E}}} \colon S^{ss} \to M_\tau^G(\hat{\Sigma}).$$

Let ρ be a representation of G in V. This gives rise to a morphism by extension of the structure group (cf. Definition 7.2.10)

$$\bar{\rho} \colon M_\tau^G(\hat{\Sigma}) \to M_{\mathring{\tau}}^{\mathrm{SL}_V}(\hat{\Sigma}),$$

where $\mathring{\tau}$ is the A-equivariant topological type of the extended bundles. For any $d \in \mathbb{Z}_{>0}$ satisfying condition (1) of Definition 8.5.4, by definition (cf. (2) of Definition 8.5.4):

$$\Theta_{\mathrm{par},G}(V,\tau,d) = \bar{\rho}^* \left(\mathring{\Theta}_{\mathrm{par},\,\mathrm{SL}_V}(\mathring{\tau},d)\right).$$

Let

$$\mathring{S}^{ss} := \{t \in S : \hat{\mathscr{E}}(V)_{|\hat{\Sigma}\times t} \quad \text{is an } A\text{-semistable vector bundle}\} \supset S^{ss}$$

(by Theorem 6.1.7).

We have the following A-equivariant analogue of Proposition 8.4.13.

Proposition 8.5.8 *Assume that $S^{ss} \neq \emptyset$.*

(a) For any $\sigma_o \in H^0\big(M_{\mathring{\tau}}^{\mathrm{SL}_V}(\hat{\Sigma}), \mathring{\Theta}_{\mathrm{par},\,\mathrm{SL}_V}(\mathring{\tau},d)\big)$, the pull-back section

$$\hat{\mathring{f}}^*_{\hat{\mathscr{E}}(V)}(\sigma_o) \in H^0\left(\mathring{S}^{ss}, \mathrm{Det}_{\mathrm{par}}\left(\mathscr{E}(V), d, d\vec{\mathring{\tau}}^*\right)\right)$$

extends to an element of $H^0\left(S, \mathrm{Det}_{\mathrm{par}}\left(\mathscr{E}(V), d, d\vec{\mathring{\tau}}^*\right)\right)$, *where* $\hat{\mathring{f}}_{\hat{\mathscr{E}}(V)}: \mathring{S}^{ss} \to M_{\mathring{\tau}}^{\mathrm{SL}_V}(\hat{\Sigma})$ *is the induced map.*

(b) Assume that $\dot{\rho}(\tau_j) \in (\mathrm{Ad}\ \mathrm{SL}_V)\mathring{\tau}_j$ *for all* j. *Then, for any section* $\sigma \in H^0\big(M_\tau^G(\hat{\Sigma}), \Theta_{\mathrm{par},G}(V,\tau,d)\big)$, *the pull-back section*

$$\hat{f}^*_{\hat{\mathscr{E}}}(\sigma) \in H^0\left(S^{ss}, \mathrm{Det}_{\mathrm{par}}\left(\mathscr{E}(V), d, dd_V\vec{\tau}^*\right)\right)$$

(cf. Lemma 8.5.5) extends to a section in $H^0\left(S, \mathrm{Det}_{\mathrm{par}}\left(\mathscr{E}(V), d, dd_V\vec{\tau}^*\right)\right)$.

Proof (a) Consider the family $\mathring{\mathscr{V}} := \hat{\mathscr{E}}(V) \to \hat{\Sigma} \times S$ of rank r and degree 0 A-equivariant vector bundles over $\hat{\Sigma}$, where $r := \dim V$. Recall the quot scheme construction from Definition 7.2.1. Let $\pi^{-1}(p) = \{y_1, \ldots, y_k\}$ and denote $\vec{y} = \{y_1, \ldots, y_k\}$, where $\pi: \hat{\Sigma} \to \Sigma$ is the projection. Let $\mathscr{O}_{\hat{\Sigma}}(\vec{y}) = \mathscr{O}_{\hat{\Sigma}}\big(\sum_{i=1}^k y_i\big)$ be the corresponding A-equivariant line bundle. Take h large enough so that $hk > 2\hat{g}$ and hence $H := \mathscr{O}_{\hat{\Sigma}}(h\vec{y})$ is a very ample line bundle over $\hat{\Sigma}$ (cf. (Hartshorne, 1977, Chap. IV, Corollary 3.2)). As in the proof of Proposition 8.4.13(a), we can find a positive integer m_S such that for $m \geq m_S$, we have

(A) $R^i\pi_{S*}(\mathring{\mathscr{V}}(m)) = 0$, for $i \geq 1$. Equivalently, $H^i(\hat{\Sigma} \times t, \mathring{\mathscr{V}}(m)) = 0$ for $i > 0$ and $t \in S$.
(B) The natural map $\pi_S^*\pi_{S_*}(\mathring{\mathscr{V}}(m)) \to \mathring{\mathscr{V}}(m)$ is surjective; in particular, $(\mathring{\mathscr{V}}(m))_{|\hat{\Sigma}\times t}$ is globally generated for any $t \in S$.
(C) $\pi_{S*}(\mathring{\mathscr{V}}(m))$ is a vector bundle over S of rank $N := r(hkm + 1 - \hat{g})$, where π_S is the projection on the S-factor and

$$\mathring{\mathscr{V}}(m) := \mathring{\mathscr{V}} \otimes \left(H^m \boxtimes \mathscr{O}_S\right).$$

The A-equivariant structure on $\mathring{\mathscr{V}}(m)$ gives rise to a natural A-equivariant structure on the vector bundle $\pi_{S*}(\mathring{\mathscr{V}}(m))$ compatible with the natural A-module structure on $H^0(\hat{\Sigma} \times t, \mathring{\mathscr{V}}(m))$ for any $t \in S$. Since $H^i(\hat{\Sigma} \times t, \mathring{\mathscr{V}}(m)) = 0$ for $i > 0$, the A-modules $\{H^0(\hat{\Sigma} \times t, \mathring{\mathscr{V}}(m))\}_{t\in S}$ of dimension N are all isomorphic (cf. (Berline, Getzler and Vergne, 2004, Corollary 6.8)). We put an A-module structure denoted $\tilde{\mathring{\tau}}$ on $\mathbb{C}^N$ isomorphic to these.

Let $\gamma: \mathscr{F}(\mathring{\mathscr{V}}(m)) \to S$ be the A-equivariant frame bundle of $\pi_{S_*}(\mathring{\mathscr{V}}(m))$, i.e., for any $t \in S$, $\gamma^{-1}(t)$ consists of all the A-equivariant linear isomorphisms $\mathbb{C}^N \to H^0\big(\hat{\Sigma} \times t, \mathring{\mathscr{V}}(m)\big)$. Thus, γ is a principal $\mathfrak{G}$-bundle where $\mathfrak{G} := \mathrm{GL}_N^A$ under the conjugation action of A.

Now, take large enough $m \geq m_S$ (so that $d := hkmr$ satisfies the bound in Theorem 7.1.13 for $\hat{\Sigma}$; observe that d is the degree of $\mathring{\mathscr{V}}(m)_t$ for any $t \in S$). By Theorem 7.1.6, we get a $\mathfrak{G}$-equivariant morphism

$$\mathring{\phi}\colon \mathscr{F}(\mathring{\mathscr{V}}(m)) \to R_{\mathring{\tilde{t}}}(\mathscr{O}),$$

such that

$$\left(\mathrm{Id}_{\hat{\Sigma}} \times \mathring{\phi}\right)^* \left(\hat{\mathscr{U}}_{|\hat{\Sigma}\times R_{\mathring{\tilde{t}}}(\mathscr{O})}\right) \simeq \left(\mathrm{Id}_{\hat{\Sigma}} \times \gamma\right)^* (\mathring{\mathscr{V}}), \tag{1}$$

where $R_{\mathring{\tilde{t}}}$ and the tautological A-equivariant bundle $\hat{\mathscr{U}}$ over $R_{\mathring{\tilde{t}}}$ are as defined in Definition 7.2.1 (denoted $\mathscr{U}$ in the same) and $R_{\mathring{\tilde{t}}}(\mathscr{O}) := \{q \in R_{\mathring{\tilde{t}}} : \wedge^r(\bar{q}) \simeq \mathscr{O}_{\hat{\Sigma}}\}$.

Consider the following commutative diagram:

$$\begin{array}{ccccc}
\mathring{S}^{ss} & \hookrightarrow & S & \xleftarrow{\ \gamma\ } & \mathscr{F}(\mathring{\mathscr{V}}(m)) \\
\Big\downarrow \hat{\mathring{f}}_{\hat{\mathscr{E}}(V)} & & & & \Big\downarrow \mathring{\phi} \\
M^{\mathrm{SL}_V}_{\mathring{\tilde{t}}}(\hat{\Sigma}) & \xleftarrow[\pi]{} & R^{ss}_{\mathring{\tilde{t}}}(\mathscr{O}) & \hookrightarrow & R_{\mathring{\tilde{t}}}(\mathscr{O}),
\end{array}$$

where $R^{ss}_{\mathring{\tilde{t}}}(\mathscr{O}) := \{q \in R_{\mathring{\tilde{t}}}(\mathscr{O}) : \bar{q}$ is A-semistable$\}$. Now, $R_{\mathring{\tilde{t}}}(\mathscr{O})$ is smooth (cf. Exercise 8.5.E.4). Further,

$$\operatorname{codim}_{R_{\mathring{\tilde{t}}}(\mathscr{O})}\left(R_{\mathring{\tilde{t}}}(\mathscr{O})\backslash R^{ss}_{\mathring{\tilde{t}}}(\mathscr{O})\right) \geq 2. \tag{2}$$

We prove this by essentially following the same proof as that of the corresponding Lemma 8.4.14. Specifically, in Lemma 8.4.14, replace R by $R_{\mathring{\tilde{t}}}$ and R^{ss} by $R^{ss}_{\mathring{\tilde{t}}}$. Then, it suffices to prove (following the notation of Lemma 8.4.14) that

$$\dim\left(H^1(\hat{\Sigma},\mathscr{V})^A\right) \geq 2, \ \text{ where } \mathscr{V} := \mathbf{End}''(\bar{q}). \tag{3}$$

First of all, by the Leray spectral sequence for affine morphisms (cf. (Hartshorne, 1977, Chap. III, Exercise 8.2)) applied to $\pi : \hat{\Sigma} \to \Sigma$, we get

$$H^i(\hat{\Sigma},\mathscr{V}) \simeq H^i(\Sigma,\pi_*\mathscr{V}), \ \text{ for any } i \geq 0.$$

Moreover, since A is a finite group and we are working over $\mathbb{C}$, we get

$$H^i(\Sigma,(\pi_*\mathscr{V})^A) \simeq H^i(\Sigma,\pi_*\mathscr{V})^A.$$

Thus, combining the above two isomorphisms, we get

$$\begin{aligned}
\dim\left(H^1(\hat{\Sigma},\mathscr{V})^A\right) &= \dim\left(H^1(\Sigma,(\pi_*\mathscr{V})^A)\right) \\
&= -\chi\left(\Sigma,(\pi_*\mathscr{V})^A\right), \ \text{by (1) of the proof of Lemma 8.4.14}
\end{aligned}$$

$$
\begin{aligned}
&= (g-1)\operatorname{rank}((\pi_*\mathscr{V})^A) - \deg((\pi_*\mathscr{V})^A),\\
&\quad \text{by the Riemann–Roch theorem for curves}\\
&= (g-1)\operatorname{rank}(\mathscr{V}) - \frac{\deg(\mathscr{V})}{|A|} + k, \quad \text{for some rational}\\
&\quad k \geq 0 \text{ arising from the `parabolic weights' by}\\
&\quad \text{Seshadri (2011, Chap. I, discussion above Proposition 3)}\\
&\geq 2, \text{ by identity (2) of the proof of Lemma 8.4.14,}\\
&\quad \text{since } g \geq 2 \text{ by assumption.}
\end{aligned}
$$

This proves (3) and hence (2).

Let

$$\pi^*\sigma_o \in H^0\left(R^{ss}_{\tilde{\mathring{t}}}(\mathscr{O}), \mathcal{L}\right), \quad \text{where } \mathcal{L} := \operatorname{Det}_{\text{par}}\left(\mathscr{U}, d, d\vec{\mathring{t}}^*\right)_{|R^{ss}_{\tilde{\mathring{t}}}},$$

be the pull-back section (cf. Theorem 8.5.3), which is clearly $\mathfrak{G}$-invariant under the pull-back $\mathfrak{G}$-equivariant structure on $\mathcal{L}_{|R^{ss}_{\tilde{\mathring{t}}}(\mathscr{O})}$. By Theorem 7.2.8, $R^{ss}_{\tilde{\mathring{t}}}(\mathscr{O})$ is irreducible. Let $R_{\tilde{\mathring{t}}}(\mathscr{O})^o$ be the irreducible component of $R_{\tilde{\mathring{t}}}(\mathscr{O})$ containing $R^{ss}_{\tilde{\mathring{t}}}(\mathscr{O})$. Now, extend $\pi^*\sigma_o$ to $R_{\tilde{\mathring{t}}}(\mathscr{O})^o$ (which is possible by (2)) and extend it to the other components by defining it to be 0. We denote the extended section over $R_{\tilde{\mathring{t}}}(\mathscr{O})$ thus obtained by $\overline{\pi^*\sigma_o}$. By Exercise 8.5.E.2, the $\mathfrak{G}$-equivariant structure on the line bundle $\mathcal{L}_{|R^{ss}_{\tilde{\mathring{t}}}(\mathscr{O})}$ extends uniquely to the line bundle $\mathcal{L}_{|R_{\tilde{\mathring{t}}}(\mathscr{O})^o}$. Moreover, since $\mathfrak{G}$ acts on the family $\mathscr{U}$ of parabolic bundles, the line bundle $\mathcal{L}$ over $R_{\tilde{\mathring{t}}}(\mathscr{O})$ acquires a canonical structure of a $\mathfrak{G}$-equivariant line bundle. Since $\mathfrak{G}$ is connected (cf. proof of Theorem 7.2.6), $R_{\tilde{\mathring{t}}}(\mathscr{O})' := R_{\tilde{\mathring{t}}}(\mathscr{O})\backslash R_{\tilde{\mathring{t}}}(\mathscr{O})^o$ is $\mathfrak{G}$-stable. We endow $\mathcal{L}_{|R_{\tilde{\mathring{t}}}(\mathscr{O})'}$ with this $\mathfrak{G}$-equivariant structure and keep the $\mathfrak{G}$-equivariant structure on $\mathcal{L}_{|R_{\tilde{\mathring{t}}}(\mathscr{O})^o}$ as above. With this $\mathfrak{G}$-equivariant structure on $\mathcal{L}$ over $R_{\tilde{\mathring{t}}}(\mathscr{O})$, clearly $\overline{\pi^*\sigma_o}$ is $\mathfrak{G}$-invariant. Pull $\overline{\pi^*\sigma_o}$ via the $\mathfrak{G}$-equivariant morphism $\mathring{\phi}$ and push via the $\mathfrak{G}$-principal bundle γ to get a section $\hat{\sigma}_o$ in $H^0(S, \operatorname{Det}_{\text{par}}(\mathscr{E}(V), d, d\vec{\mathring{t}}^*))$ (use (1)). Using (1) again, we get

$$\mathring{\phi}\left(\gamma^{-1}(\mathring{S}^{ss})\right) \subset R^{ss}_{\tilde{\mathring{t}}}(\mathscr{O}).$$

Thus, $\mathfrak{G}$ being connected and S being irreducible (by assumption) $\operatorname{Im}\mathring{\phi} \subset R_{\tilde{\mathring{t}}}(\mathscr{O})^o$. Thus, $\hat{\sigma}_o$ extends the section $\hat{\mathring{f}}^*_{\mathscr{E}(V)}(\sigma_o)$ from $\mathring{S}^{ss}$ to S. This proves the (a)-part of the proposition.

The proof of the (b)-part is identical to that of Proposition 8.4.13(b) by observing (cf. Theorem 7.2.14) that $M^G_\tau(\hat{\Sigma})$ is a complete variety when $g \geq 2$. $\square$

Recall the definition of $D_c = D_c(\mathfrak{g})$ from (1) of Definition 2.1.1. Finally, we come to the main result of this section, which is the parabolic analogue of Theorem 8.4.15.

Theorem 8.5.9 *With the notation and assumptions as in Lemma 8.5.7, there is a canonical isomorphism*

$$H^0\left(M_\tau^G(\hat{\Sigma}), \Theta_{\mathrm{par},G}(V,\tau,d)\right) \simeq H^0\left(\bar{X}_G^{\vec{P}}, \mathscr{L}(0_{dd_V}) \boxtimes \mathop{\boxtimes}_{j=1}^{s} \mathscr{L}_{P_j}(dd_V\tau_j^*)\right)^\Gamma. \tag{1}$$

Thus, for any representation ρ of G in V, any central charge $c > 0$ and any $\vec{\lambda} = (\lambda_1, \ldots, \lambda_s)$ with $\lambda_j \in D_c$ such that for all j, $\lambda_j(\theta^\vee) < c$ and under some basis of V,

$$\kappa_{sl_V}(\dot{\rho}(h_j)) \text{ is a dominant integral weight of } \mathrm{SL}_V \text{ with } \mathring{\theta}(\dot{\rho}(h_j)) < c, \quad \text{where } h_j := \kappa_{\mathfrak{g}}^{-1}\lambda_j, \tag{2}$$

we have

$$H^0\left(M_{\tau(\vec{\lambda})}^G(\hat{\Sigma}), \Theta_{\mathrm{par},G}(V,\tau(\vec{\lambda}),c)\right) \simeq \mathscr{V}_\Sigma^\dagger(\vec{p}, d_V\vec{\lambda}), \tag{3}$$

where $\vec{\tau}(\vec{\lambda}) = (\frac{h_1}{c}, \ldots, \frac{h_s}{c})$, $\tau(\vec{\lambda})$ is the corresponding A-equivariant topological type ($\kappa_{\mathfrak{g}}$ is the isomorphism $\mathfrak{h} \xrightarrow{\sim} \mathfrak{h}^*$ as above Theorem 8.5.3) and the space of vacua $\mathscr{V}_\Sigma^\dagger$ is taken at the central charge cd_V.

Moreover, the isomorphism (3) is canonical up to a scalar multiple.

If V is an irreducible representation of G with highest weight μ, then the condition (2) can be equivalently rephrased as:

$$\kappa_{sl_V}(\dot{\rho}(h_j)) \text{ is an integral weight of } \mathrm{SL}_V \text{ and } \mu(h_j) - (w_o\mu)(h_j) < c\,\forall j, \tag{4}$$

under one (and hence any) weight basis of V (i.e., basis consisting of weight vectors of V), where w_o is the longest element of the Weyl group of G.

Proof Observe that $g \geq 2$ by the blanket assumption in this section. The proof of (1) is identical to the proof of (1) of Theorem 8.4.15 by using Proposition 8.5.8(b) in the place of Proposition 8.4.13(b) and taking the filtration of $\bar{X}_G^{\vec{P}}$:

$$\bar{X}_G^0 \times \Pi_{j=1}^s(G/P_j) \subset \bar{X}_G^1 \times \Pi_{j=1}^s(G/P_j) \subset \cdots,$$

where $\bar{X}_G^n$ is the filtration of $\bar{X}_G$ as in the proof of Theorem 8.4.15. By the proof of Lemma 8.5.7, $(\bar{X}_G^{\vec{P}})^{ss}$ is nonempty.

We now prove (3). Since $\lambda_j(\theta^\vee) < c$ by assumption, $\vec{\tau}(\vec{\lambda})$ is a set of parabolic weights of G satisfying (1) at the beginning of this section. Now, applying (1) to $\vec{\tau} = \vec{\tau}(\vec{\lambda})$ and using Theorem 8.3.3, the classical Borel–Weil theorem for G/P_j, Remark 8.5.6 and Lemma B.23, we get (3).

To prove (4), take a basis $\{e_\mu, \dots, e_{w_o\mu}\}$ of V consisting of weight vectors such that e_μ (resp. $e_{w_o\mu}$) is of weight μ (resp. $w_o\mu$). Since h_j is (clearly) a dominant element of $\mathfrak{h}$, i.e., $\alpha_i(h_j) \geq 0$ for all the simple roots α_i of $\mathfrak{g}$, we get that

$$\mu(h_j) \geq \nu(h_j) \text{ and } (w_o\mu)(h_j) \leq \nu(h_j)$$

for all the weights ν of V. Thus, a basis $\mathcal{B}_j$ of V which makes $\dot{\rho}(h_j)$ dominant for SL_V is a permutation of $\{e_\mu, \dots, e_{w_o\mu}\}$, where e_μ (resp. $e_{w_o\mu}$) continues to be the first (resp. last) element of the basis. Hence,

$$\langle \mathring{\theta}, \kappa_{sl_V}(\dot{\rho}(h_j))\rangle = \mu(h_j) - (w_o\mu)(h_j),$$

where $\mathring{\theta}$ is the highest root of SL_V. From this (4) follows. □

Remark 8.5.10 If $\langle \lambda_j, \Lambda\rangle \subset \mathbb{Z}$ for the weight lattice $\Lambda \subset \mathfrak{h}^*$ of G, then $\kappa_{sl_V}(\dot{\rho}(\kappa_{\mathfrak{g}}^{-1}\lambda_j))$ is an integral weight of SL_V. This follows since $\kappa_{\mathfrak{g}}^{-1}(\lambda_j)$ belongs to the coroot lattice of G. Hence, its image $\dot{\rho}(\kappa_{\mathfrak{g}}^{-1}(\lambda_j))$ lies in the coroot lattice of SL_V with respect to any weight basis of V. Thus, $\kappa_{sl_V}(\dot{\rho}(\kappa_{\mathfrak{g}}^{-1}(\lambda_j)))$ is an integral weight of SL_V.

If $G = \mathrm{SL}_n$, taking its standard representation in $V = \mathbb{C}^n$, we get that for any $\vec{\lambda} = (\lambda_1, \dots, \lambda_s)$ with $\lambda_j \in D_c$ satisfying $\lambda_j(\theta^\vee) < c$, we have (since $d_V = 1$ by definition):

$$H^0\left(M^{\mathrm{SL}_n}_{\tau(\vec{\lambda})}(\hat{\Sigma}), \mathring{\Theta}_{\mathrm{par},\,\mathrm{SL}_n}\left(\tau(\vec{\lambda}), c\right)\right) \simeq \mathscr{V}^\dagger_\Sigma(\vec{p}, \vec{\lambda}),$$

where $\vec{\tau}(\vec{\lambda}) = \left(\kappa_{sl_n}^{-1}(\frac{\lambda_1}{c}), \dots, \kappa_{sl_n}^{-1}(\frac{\lambda_s}{c})\right)$.

For any integral weight λ_j, for G of type B_ℓ ($\ell \geq 3$) and D_ℓ ($\ell \geq 4$) (resp. C_ℓ, $\ell \geq 2$), $\langle \lambda_j, \Lambda\rangle \subset \mathbb{Z}$ if $\lambda_j \in 4.\Lambda$ (resp. $\lambda_j \in 2.\Lambda$). Moreover, $d_V = 2$ (resp. $d_V = 1$) for the defining representation V of B_ℓ or D_ℓ (resp. C_ℓ) (cf. Corollary A.9). Thus, $\mathscr{V}^\dagger_\Sigma(\vec{p}, \vec{\lambda})$ at central charge c can be represented as the global sections of some line bundle over $M^G_\tau(\hat{\Sigma})$ (for some τ depending upon $\vec{\lambda}$) if each λ_j satisfies

(1) $\lambda_j \in D_c$ and $\lambda_j(\theta^\vee) < c$ for G of type A_ℓ ($\ell \geq 1$).
(2) c is even, $\lambda_j \in 8\Lambda$, $\lambda_j(\theta^\vee) < c$ and $\langle \lambda_j, \omega_1\rangle < \frac{c}{2}$, for G of type B_ℓ ($\ell \geq 3$) and D_ℓ ($\ell \geq 4$), where ω_1 is the first fundamental weight.
(3) $\lambda_j \in 2\Lambda$ and $\lambda_j(\theta^\vee) < c$ for G of type C_ℓ ($\ell \geq 2$).

To prove these, use (4) of Theorem 8.5.9 and the tables (Bourbaki, 2002, Planche I–IV).

Remark 8.5.11 Since we are taking $G = \mathrm{SL}_V$ in Theorem 8.5.3, $\det \mathring{\mathscr{F}}(V)$ is trivial.

Observe that for $G = \mathrm{SL}_V$ any standard parabolic subgroup of the loop group $G((t))$ can be conjugated to a subgroup of $G[[t]]$ by an outer automorphism. Hence various results in this section which required $\mathring{\theta}(\mathring{\tau}_j) < 1$ remain true without this restriction.

8.5.E Exercises

(1) Show that the space of vacua $\mathscr{V}_\Sigma^\dagger(\vec{p}, \vec{\lambda})$ at central charge c for an s-pointed smooth irreducible projective curve $(\Sigma, \vec{p})$ of genus ≥ 2 is represented as the space of global sections of some line bundle over $M_\tau^G(\hat{\Sigma})$ (for some τ depending upon $\vec{\lambda}$) if each λ_j satisfies (following the indexing convention as in Bourbaki (2002, Planche V–IX)):

(a) c is multiple of 6, $\lambda_j \in 18\Lambda$, $\langle \omega_1 + \omega_6, \lambda_j \rangle < c$ and $\lambda_j(\theta^\vee) < c$ for G of type E_6.
(b) c is multiple of 12, $\lambda_j \in 24\Lambda$, $\langle \omega_7, \lambda_j \rangle < \frac{c}{2}$ and $\lambda_j(\theta^\vee) < c$ for G of type E_7.
(c) c is multiple of 60, $\lambda_j \in 60\Lambda$ and $\lambda_j(\theta^\vee) < \frac{c}{2}$ for G of type E_8.
(d) c is multiple of 6, $\lambda_j \in 12\Lambda$, $\langle \omega_4, \lambda_j \rangle < \frac{c}{2}$ and $\lambda_j(\theta^\vee) < c$ for G of type F_4.
(e) c is multiple of 2, $\lambda_j \in 6\Lambda$, $\langle \omega_1, \lambda_j \rangle < \frac{c}{2}$ and $\lambda_j(\theta^\vee) < c$ for G of type G_2.

Hint: Use Remark 8.5.10, Theorem 8.5.9 and the chart of minimum d_V from Corollary A.9.

(2) Let X be an irreducible normal H-variety for an algebraic group H and $\mathcal{L}$ a line bundle over X. Let $U \subset X$ be an open H-stable subset such that $\operatorname{codim}_X(X \backslash U) \geq 2$. If $\mathcal{L}_{|U}$ has a structure of an H-equivariant line bundle, then show that this structure on $\mathcal{L}_{|U}$ extends uniquely to the structure of an H-equivariant line bundle over X, i.e., the restriction map $\operatorname{Pic}_H(X) \to \operatorname{Pic}_H(U)$ is an isomorphism.

(3) Prove the assertion in Remark 8.5.6.

(4) With the notation and assumptions as in Definition 7.2.1, prove that the morphism (defined in the same for R^{ss})

$$\det \colon R \to J(\hat{\Sigma}), \; q \mapsto \wedge^r(\bar{q})$$

is a smooth morphism.

Hint: Let $\mathscr{O}_{\hat{\Sigma} \times \hat{\Sigma}}(-\Delta)$ be the ideal sheaf of the diagonal $\Delta \subset \hat{\Sigma} \times \hat{\Sigma}$. Also, let $\mathfrak{U}$ be the family of vector bundles over $\hat{\Sigma}$ parameterized by

$R \times J(\hat{\Sigma})$ defined by $(q, \mathscr{L}) \rightsquigarrow \mathscr{U}_{|\hat{\Sigma} \times q} \otimes \mathscr{L}$, for $q \in R$ and $\mathscr{L} \in J(\hat{\Sigma})$, where $\mathscr{U}_{|\hat{\Sigma} \times R}$ is the tautological vector bundle as in Theorem 7.1.6. Now, let $Y = \hat{\Sigma} \times \hat{\Sigma} \times R \times J(\hat{\Sigma})$ and define the vector bundle $\mathcal{F}$ over Y by taking the tensor product

$$\mathcal{F} := \pi_{1,2}^* \left(\mathscr{O}_{\hat{\Sigma} \times \hat{\Sigma}}(-\Delta) \right) \otimes \pi_{1,3,4}^* (\mathcal{U}),$$

where π is the projection on the assigned components. Then, by the Semi-continuity Theorem (Hartshorne, 1977, Chap. III, Theorem 12.8), the set

$$Z = \{(z, q, \mathscr{L}) \in \hat{\Sigma} \times R \times J(\hat{\Sigma}) : H^1(\hat{\Sigma}, \mathcal{F}_{|\hat{\Sigma} \times z \times q \times \mathscr{L}}) \neq 0\}$$

is closed in $\hat{\Sigma} \times R \times J(\hat{\Sigma})$. Let $p_{2,3} \colon \hat{\Sigma} \times R \times J(\hat{\Sigma}) \to R \times J(\hat{\Sigma})$ be the projection. Then show that $V := \left(R \times J(\hat{\Sigma})\right) \backslash p_{2,3}(Z)$ is an open subset of $R \times J(\hat{\Sigma})$ containing $R \times \{\mathscr{L}_o\}$, where $\mathscr{L}_o$ is the trivial line bundle over $\hat{\Sigma}$. Show that, for any $(q, \mathscr{L}) \in V$, the vector bundle $\bar{q} \otimes \mathscr{L}$ over $\hat{\Sigma}$ satisfies conditions (a) and (b) of Proposition 7.1.3.

Take an increasing filtration $(R_n)_{n \geq 1}$ of R by open relatively compact subsets of R (in the analytic topology) with $\cup R_n = R$. Then, there exists a neighborhood J_n of $\mathscr{L}_o \in J(\hat{\Sigma})$ (again in the analytic topology) such that $R_n \times J_n \subset V$. Fix any $\mathscr{L} \in J_n$ and consider the restriction of the vector bundle $\mathcal{U}$ to $\hat{\Sigma} \times R_n \times \mathscr{L}$. Since the vector bundle $\bar{q} \otimes \mathscr{L}$ over $\hat{\Sigma}$ satisfies conditions (a) and (b) of Proposition 7.1.3 for any $(q, \mathscr{L}) \in V$, we get an open cover $\{R_n^j\}_j$ of R_n and a morphism $f_j \colon R_n^j \to R$ such that $\bar{f}_j^* (\mathscr{U}_{|\hat{\Sigma} \times R}) = \mathcal{U}_{|\hat{\Sigma} \times R_n^j \times \{\mathscr{L}\}}$, where $\bar{f}_j = \mathrm{Id}_{\hat{\Sigma}} \times f_j$. By the generic smoothness (cf. (Hartshorne, 1977, Chap. III, Corollary 10.7)), det is a smooth morphism over a nonempty open subset of $J(\hat{\Sigma})$. Further, by a characterization of smoothness in terms of the tangent map as in Hartshorne (1977, Chap. III, Proposition 10.4), it suffices to show the smoothness of det restricted to R_n. The proof in this case follows by an argument similar to that given in Definition 7.2.1 by varying $\mathscr{L}$ in J_n and observing that $\cup_k (J_n)^k = J(\hat{\Sigma})$.

8.C Comments

Theorem 8.1.1 is due to Drinfeld (his proof is produced in Laszlo and Sorger (1997)). The homotopy type of Γ^{an} (where Γ^{an} is as in Theorem 8.1.1) is determined in Teleman (1998, §II). The corresponding result for $\Sigma = \mathbb{P}^1$ was obtained in Garland and Raghunathan (1975) (and also by

Quillen (Mitchell, 1988)). Corollary 8.1.5 is taken from Kumar, Narasimhan and Ramanathan (1994, Corollary 2.6). A different proof of Corollary 8.1.4 as well as reducedness and irreducibility of $\bar{\Gamma}$ for $G = \mathrm{SL}_N$ is given in Beauville and Laszlo (1994).

Theorem 8.2.1 is proved in Kumar (1997a, §§2.6, 2.7). However, we have taken its proof from Sorger (1999a). This result for $G = \mathrm{SL}_N$ is obtained in Beauville and Laszlo (1994).

Proposition 8.3.2 is due to Laszlo and Sorger (1997) for classical G and G_2. For any semisimple G it was obtained in Teleman (1998, §V) (also see Sorger (1999a) and Faltings (2003, Theorem 17) for any simply-connected G). Theorem 8.3.4 for classical G and G_2 is due to Laszlo and Sorger (1997). The identification of $H^0(\mathbf{Bun}_G(\Sigma), \mathscr{L}(0_c))$ with the space of conformal blocks $\mathscr{V}_\Sigma^\dagger(p, 0_c)$ (i.e., the non-parabolic case of Theorem 8.3.4) for $G = \mathrm{SL}_N$ is obtained in Beauville and Laszlo (1994) and in Faltings (1994) for the classical groups and G_2. The Picard group $\mathrm{Pic}(\mathbf{Parbun}_G(\Sigma, \vec{P}))$ (as in Theorem 8.3.5) is determined in Laszlo and Sorger (1997, Proposition 8.7).

Proposition 8.4.6 is taken from Kumar, Narasimhan and Ramanathan (1994, Proposition 2.3). The Dynkin index of a representation is introduced in the theory of G-bundles over a curve by Kumar, Narasimhan and Ramanathan (1994). Theorem 8.4.7 is due to Kumar, Narasimhan and Ramanathan (1994, Theorem 5.4). The proof presented here generally follows the proof in this reference, though less topological (and more geometric) with some input from Laszlo and Sorger (1997, §6). Lemmas 8.4.10, 8.4.11 are taken respectively from Kumar, Narasimhan and Ramanathan (1994, , Proposition 6.4, Lemma 7.1). Lemma 8.4.12 is taken from Kumar (1997a, Lemma 7.2). Proposition 8.4.13 in its present form is taken from Kumar (1997a, Proposition 7.3 and Lemma 7.5). The proof of Lemma 8.4.14 presented here is essentially due to Atiyah and Bott (1982, §7). The first instance of the dimension of the space $H^0(M^{\mathrm{SL}_V}(\Sigma), \Theta)$ (i.e., the Verlinde formula in this case) is given by Beauville, Narasimhan and Ramanan (1989). Theorem 8.4.15 is independently due to Faltings (1994) and Kumar, Narasimhan and Ramanathan (1994, §6) for general G (also see Kumar (1997a)); and also due to Beauville and Laszlo (1994) for $G = \mathrm{SL}_N$ as well as GL_N. Exercise 8.4.E.3 is taken from Kumar, Narasimhan and Ramanathan (1994, Corollary 6.3). Our treatment in Section 8.4 generally follows Kumar, Narasimhan and Ramanathan (1994) and Kumar (1997a).

In relation to the definition of $\mathrm{Det}_{\mathrm{par}}(\mathscr{E}(V), d, \vec{\mu})$ (as in Definition 8.5.1) we refer to the paper Biswas and Raghavendra (1993, Proposition 4.11), where they identify the parabolic determinant of cohomology in terms of a determinant bundle on the A-equivariant bundles and some modification at the ramification locus. Lemma 8.5.7 and Proposition 8.5.8 are parabolic analogues of the corresponding non-parabolic results in Section 8.4. Theorem 8.5.9 for $G = \mathrm{SL}_N$ is due to Pauly (1996). To my knowledge its generalization to general G is not available in the literature. In particular, Remark 8.5.10 and Exercise 8.5.E.1 are new for general G. Exercise 8.5.E.2, though well known, is stated in Kumar and Narasimhan (1997, Lemma 3.9).

In Teleman (1998) the cohomology of certain evaluation vector bundles twisted by powers of the fundamental line bundle over the moduli stack $\mathbf{Bun}_G(\Sigma)$ is determined.

9

Picard Group of Moduli Space of G-Bundles

In this chapter $\Sigma = \Sigma_g$ denotes a smooth irreducible projective curve of any genus $g \geq 1$ (except in Section 9.1, where we take $g \geq 2$) with a base point p and G a connected simply-connected simple algebraic group.

Let $M^G = M^G(\Sigma)$ be the moduli space of semistable G-bundles over Σ and let $\mathrm{Pic}(M^G)$ be its Picard group. The following is the main result of Section 9.1 (cf. Theorem 9.1.3).

Theorem With the notation as above, $\mathrm{Pic}(M^G) \approx \mathbb{Z}$. In fact, there is a natural embedding

$$\bar{f}^*\colon \mathrm{Pic}(M^G) \hookrightarrow \mathrm{Pic}_{\bar{\Gamma}}(\bar{X}_G) \simeq \mathbb{Z}.$$

Moreover, for any G-module V,

$$\bar{f}^*(\Theta(V)) = \mathscr{L}(0_{d_V}),$$

where d_V is the Dynkin index of V and $\Theta(V)$ is the theta bundle on the moduli space M^G as in Definition 8.4.4. (For the notation $\bar{\Gamma}, \bar{X}_G$, see the beginning of Section 9.1 and the line bundle $\mathscr{L}(0_c)$ is defined in Proposition 8.3.2.)

The above result and a result of Grauert–Riemenschneider allow us to prove the following second main result of this Section (cf. Theorem 9.1.6).

Theorem The dualizing sheaf ω of the moduli space M^G is locally free. In particular, M^G is a Gorenstein variety.

Further, for any G-module V, $H^i(M^G, \Theta(V)) = 0$, for all $i > 0$. In particular,

$$\chi(M^G, \Theta(V)) = \dim H^0(M^G, \Theta(V)),$$

where χ is the Euler–Poincaré characteristic.

Section 9.2 is devoted to recalling the identification of $M^G(\Sigma)$ (for Σ of genus 1) with a weighted projective space (cf. Theorems 9.2.1 and 9.2.3).

As in Corollary A.9, let $m(\mathfrak{g})$ be the least common multiple of the coefficients of the coroot $\theta^\vee$ written in terms of the simple coroots $\alpha_i^\vee$, where θ is the highest root of $\mathfrak{g}$. Let ω_i be any fundamental weight such that the Dynkin index $d_{V(\omega_i)} = m(\mathfrak{g})$ (cf. Corollary A.9 for a complete list of such ω_i and $m(\mathfrak{g})$).

The main aim of Section 9.3 is to explicitly determine $\operatorname{Pic}(M^G)$ for an arbitrary G. The following is our main result (cf. Theorem 9.3.8).

Theorem For any Σ with genus $g \geq 1$, the Picard group $\operatorname{Pic}(M^G)$ is freely generated by the Θ-bundle $\Theta(V(\omega_i))$, where ω_i is any fundamental weight such that $d_{V(\omega_i)} = m(\mathfrak{g})$.

In particular, for $g \geq 2$, $\operatorname{Im}(\bar{f}^*)$ is freely generated by $\mathscr{L}_{0_{m(\mathfrak{g})}}$.

We further prove that $M^G(\Sigma)$ is locally factorial if and only if G is of type A_ℓ ($\ell \geq 1$) or C_ℓ ($\ell \geq 2$) (cf. Theorem 9.3.9 and Remark 9.3.10).

We now briefly outline the idea of the proof of the above displayed theorem. Recall that the underlying real analytic space $\bar{R}_K(g)$ of $M^G(\Sigma)$ admits a description as the space of representations of the fundamental group $\pi_1(\Sigma)$ into a fixed compact form K of G modulo the conjugation action (cf. Theorem 9.3.3). In particular, $\bar{R}_K(g)$ depends only upon g and G (and not on the specific choice of the projective curve Σ). Moreover, this description gives rise to a standard embedding $i_g : \bar{R}_K(g) \hookrightarrow \bar{R}_K(g+1)$. Let V be any G-module. We first show that the first Chern class of the theta bundle $\Theta(V) = \Theta_V(\Sigma, G)$ does not depend upon the choice of the smooth projective curve Σ, as long as g is fixed (cf. Proposition 9.3.4). We denote it by $c(\Theta_V(g, G))$. We next show that the first Chern class of $\Theta_V(g+1, G)$ restricts to the first Chern class of $\Theta_V(g, G)$ under the embedding i_g (cf. Proposition 9.3.7). This result is proved by first reducing the case of general G to SL_n and then reducing the case of SL_n to SL_2. The corresponding result for SL_2 is obtained by showing that the inclusion $\bar{R}_{\mathrm{SU}_2}(g) \hookrightarrow \bar{R}_{\mathrm{SU}_2}(g+1)$ induces isomorphism in cohomology $i_g^* : H^2(\bar{R}_{\mathrm{SU}_2}(g+1), \mathbb{Z}) \to H^2(\bar{R}_{\mathrm{SU}_2}(g), \mathbb{Z})$ (cf Proposition 9.3.6). The last result for H^2 with rational coefficients is fairly well known (and follows easily by observing that the symplectic form on $\bar{R}_K(g+1)$ restricts to the symplectic form on $\bar{R}_K(g)$) but the result with integral coefficients is more delicate. Its proof involves the calculation of the determinant bundle of the Poincaré bundle on $\Sigma \times J(\Sigma)$, $J(\Sigma)$ being the Jacobian of Σ which consists of the isomorphism classes of degree 0 line bundles on Σ. By virtue of the above-mentioned two propositions (Propositions 9.3.4 and 9.3.7), to prove our main result determining $\mathrm{Pic}(M^G)$ for any $g \geq 1$, it suffices to consider the case of genus $g = 1$. In the genus $g = 1$ case, the result follows from the description of M^G as a weighted projective space as in Section 9.2.

9.1 Picard Group of Moduli Space of G-Bundles – Its Isomorphism with $\mathbb{Z}$

In this section Σ denotes a smooth irreducible projective curve of any genus $g \geq 2$ with a base point p and G a connected simply-connected simple algebraic group over $\mathbb{C}$.

Let $M^G = M^G(\Sigma)$ be the moduli *space* of semistable G-bundles over Σ (cf. Theorem 7.2.6, where we take $A = (1)$ and hence $\hat{\Sigma} = \Sigma$) and let $\bar{X}_G$ be the infinite Grassmannian with the tautological bundle $\mathbf{U_G}$ over $\Sigma \times \bar{X}_G$ (cf. Proposition 5.2.4). As in Definition 8.4.9, let

$$\bar{X}_G^{ss} := \{x \in \bar{X}_G : (\mathbf{U}_G)_{\Sigma \times x} \text{ is a semistable } G\text{-bundle}\}.$$

Then $\bar{X}_G^{ss}$ is open in $\bar{X}_G$ (cf. Corollary 7.1.16(b)). Further, the family $\mathbf{U_G}$ restricted to $\Sigma \times \bar{X}_G^{ss}$ gives rise to a morphism (cf. Corollary 7.2.7):

$$f = f_{\mathbf{U_G}} \colon \bar{X}_G^{ss} \to M^G.$$

Let $\bar{\Gamma}$ be the ind-affine irreducible variety (cf. Definition 5.2.9, Lemma 5.2.10 and Theorem 8.1.1) with $\mathbb{C}$-points $\bar{\Gamma}(\mathbb{C}) = \Gamma := \mathrm{Mor}(\Sigma^*, G)$, where $\Sigma^* := \Sigma \backslash p$. Then $\bar{\Gamma}$ acts on $\bar{X}_G$ through the morphism $\bar{\Gamma} \to \bar{G}((t))$ as at the beginning of Section 8.2 and $\mathbf{U_G}$ is $\bar{\Gamma}$-equivariant (cf. Lemma 5.2.12). Thus, $\bar{\Gamma}$ keeps $\bar{X}_G^{ss}$ stable.

Lemma 9.1.1 *The $\bar{\Gamma}$-invariant morphism $f \colon \bar{X}_G^{ss} \to M^G$ (with the trivial action of $\bar{\Gamma}$ on M^G) induces an injective map*

$$f^* \colon \mathrm{Pic}(M^G) \hookrightarrow \mathrm{Pic}_{\bar{\Gamma}}(\bar{X}_G^{ss}).$$

In fact, the induced map $\mathrm{Pic}(M^G) \hookrightarrow \mathrm{Pic}(\bar{X}_G^{ss})$ *itself is injective.*

Proof Let $\mathfrak{L} \in \mathrm{Pic}(M^G)$ be in the kernel of f^*. Thus, in particular, $f^*(\mathfrak{L})$ admits a nowhere-vanishing regular section σ on the whole of $\bar{X}_G^{ss}$. Fix $m \in M^G$ and a trivialization for $\mathfrak{L}_{|m}$. This canonically induces a trivialization for the bundle $f^*(\mathfrak{L})_{|f^{-1}(m)}$. In particular, the section $\sigma_{|f^{-1}(m)}$ can be viewed as a morphism $\sigma_m \colon f^{-1}(m) \to \mathbb{C}^*$. But $f^{-1}(m)$ is a certain union of Γ-orbits say $f^{-1}(m) = \cup_{i \in I} \Gamma x_i$, for $x_i \in \bar{X}_G^{ss}$ and, moreover, $\overline{\Gamma x_i} \cap \overline{\Gamma x_j} \neq \emptyset$, for any $i, j \in I$ (cf. (4) of the proof of Lemma 8.4.10). Fixing $i \in I$, we get a morphism

$\sigma_{m,i} : \bar{\Gamma} \to \mathbb{C}^*$, defined as $\sigma_{m,i}(\gamma) = \sigma_m(\gamma x_i)$, for $\gamma \in \bar{\Gamma}$. Now, by Corollary 8.1.5, $\sigma_{m,i}$ is a constant map for any $i \in I$, and hence $\sigma_m : f^{-1}(m) \to \mathbb{C}^*$ itself is a constant map. Thus, the section σ descends to a set-theoretic section $\hat{\sigma}$ of the line bundle $\mathfrak{L}$, which is regular by Kumar (2002, Proposition A.12). Of course, the section $\hat{\sigma}$ does not vanish anywhere on M^G (since σ was chosen to be nowhere-vanishing on $\bar{X}_G^{ss}$). This proves that $\mathfrak{L}$ is a trivial line bundle on M^G, thereby proving the lemma. □

We state the following crucial 'lifting' result.

Proposition 9.1.2 *The restriction map* $\mathrm{Pic}_{\bar{\Gamma}}(\bar{X}_G) \to \mathrm{Pic}_{\bar{\Gamma}}(\bar{X}_G^{ss})$ *is an isomorphism.*

We postpone the proof of the proposition to §9.1.13, but we derive some of its consequences. As an immediate consequence of the above proposition, Lemma 9.1.1 and Proposition 8.3.2, we get the first part of the following main result of this section. For the second part, use Definition 8.4.4 and Theorem 8.4.7.

For the definition of the $\bar{\Gamma}$-equivariant line bundle $\mathscr{L}(0_c)$, see Proposition 8.3.2.

Theorem 9.1.3 *The injective map f^* (as in Lemma 9.1.1) together with the above isomorphism gives an embedding*

$$\bar{f}^* : \mathrm{Pic}\,(M^G) \hookrightarrow \mathrm{Pic}_{\bar{\Gamma}}(\bar{X}_G) \simeq \mathbb{Z}. \tag{1}$$

In particular, $\mathrm{Pic}\,(M^G) \approx \mathbb{Z}$.

Moreover, for any G-module V,

$$\bar{f}^*(\Theta(V)) = \mathscr{L}(0_{d_V}), \quad \textit{where } d_V \textit{ is the Dynkin index of } V. \tag{2}$$

Proof In view of Propositions 9.1.2 and 8.3.2 and Lemma 9.1.1, (1) follows.

To prove (2), using Theorem 8.4.7, it suffices to observe that the forgetful map $\mathrm{Pic}_{\bar{\Gamma}}(\bar{X}_G^{ss}) \to \mathrm{Pic}(\bar{X}_G^{ss})$ is injective, which follows from the last part of the proof of Proposition 8.3.2.

Recall that M^G is a projective variety (cf. Corollary 7.2.15) of positive dimension (cf. Corollary 7.2.27). Thus, $\mathrm{Pic}\,(M^G)$ is forced to be nonzero and hence, by (1), $\mathrm{Pic}\,(M^G) \simeq \mathbb{Z}$. □

In Section 9.3, we will explicitly determine the image of $\bar{f}^*$.

We recall the following well-known result.

Lemma 9.1.4 *Let Y be a Cohen–Macaulay variety and let $U \subset Y$ be an open subset such that $\mathrm{codim}_Y(Y \backslash U) \geq 2$. Now, let $\mathcal{S}_1$ and $\mathcal{S}_2$ be two reflexive sheaves on Y such that ${\mathcal{S}_1}_{|U} \approx {\mathcal{S}_2}_{|U}$. Then, the sheaf $\mathcal{S}_1$ is isomorphic with $\mathcal{S}_2$ on the whole of Y.*

Proof (due to N. Mohan Kumar) We recall the following two facts from commutative algebra.

Fact 1: If M, N are modules of a noetherian local ring with depth $M, N > 1$, and $0 \to M \to N \to K \to 0$ is an exact sequence, then depth $K > 0$.

Fact 2: If M is reflexive, then for any localization $M_{\mathfrak{p}}$ of M at a prime ideal $\mathfrak{p}$, depth $M_{\mathfrak{p}} > 1$, unless the dimension of the local ring itself is less than 2 (i.e., M satisfies the 'Serre condition' S_2).

Let $i : U \hookrightarrow Y$ be the inclusion. Then, from the above facts (and the assumptions of the lemma), one can check that $i_* i^* \mathcal{S}_j = \mathcal{S}_j$ (for $j = 1, 2$). Thus, any homomorphism $i^*\mathcal{S}_1 \to i^*\mathcal{S}_2$ on U gives rise to a homomorphism $\mathcal{S}_1 \to \mathcal{S}_2$, i.e., Hom $(\mathcal{S}_1, \mathcal{S}_2) \to$ Hom $(i^*\mathcal{S}_1, i^*\mathcal{S}_2)$ is surjective. Injectivity is clear using reflexivity. This proves the lemma. □

Definition 9.1.5 Fix an embedding $i : G \hookrightarrow \mathrm{SL}_r \subset \mathrm{GL}_r$. For any $d \geq 2g$, define $R^d(G)$ and the tautological G-bundle $\mathscr{U}^d(G)$ over $\Sigma \times R^d(G)$ via Proposition 7.2.3 from the smooth subscheme $R^d = R_{\mathring{\tau}}$ of the quot scheme Q^d with respect to the line bundle $\mathscr{O}_\Sigma(dp)$ (for $\mathring{\tau}$ = trivial and $A = (1)$ so that $\hat{\Sigma} = \Sigma$) and the frame bundle of the tautological rank-r vector bundle $\mathscr{U}^d$ parameterized by R^d (cf. Definition 7.2.1). Then GL_{N_d} acts on R^d as well as $\mathscr{U}^d$, where $N_d := r(d + 1 - g)$. Moreover, the GL_{N_d} action lifts to an action on $R^d(G)$ and $\mathscr{U}^d(G)$ via the canonical morphism $R^d(G) \to R^d$ (cf. the analogue of Corollary 7.2.4 for $R^{ss}_\tau(G)$ replaced by $R^d(G)$). By the same proof as that of the smoothness of $R^{ss}_\tau(G)$ as in the proof of Theorem 7.2.6 (in fact, as a particular case of that proof), one gets that $R^d(G)$ is smooth. Let

$$(R^d)^{ss}(G) := \{E \in R^d(G) : \bar{E} \text{ is a semistable } G\text{-bundle}\},$$

where $\bar{E} := \mathscr{U}^d(G)_{|\Sigma \times E}$. Then $(R^d)^{ss}(G)$ is open in $R^d(G)$ by Corollary 7.1.16(b), and, of course, it is GL_{N_d}-stable. The family $\mathscr{U}^d(G)$ restricted to $(R^d)^{ss}(G)$ is denoted by $(\mathscr{U}^d)^{ss}(G)$.

Let

$$(R^d)^s(G) := \{E \in R^d(G) : \bar{E} \text{ is a stable } G\text{-bundle}\}.$$

Then $(R^d)^s(G)$ is an open subset of $(R^d)^{ss}(G)$ (Ramanathan, 1996, Proposition 5.8)). Now, let

$$\mathring{R}^d(G) := \{E \in (R^d)^s(G) : \mathrm{Aut}(\bar{E}) \simeq Z\},$$

where Z is the center of G.

Recall from the proof of Theorem 7.2.6 that, for $d > \max\{2g - 1, (r + 1)g - 2, r^2 g + 2g - 2\}$, the good quotient

$$\pi : (R^d)^{ss}(G) \to (R^d)^{ss}(G)//\operatorname{GL}_{N_d} \text{ gives the moduli space } M^G.$$

Recall the definition of the dualizing sheaf ω_X of a projective scheme X from Hartshorne (1977, Chap. III, §7). If X is Cohen–Macaulay and equi-dimensional, ω_X coincides with the top nonzero homology sheaf of the dualizing complex.

Theorem 9.1.6 *The dualizing sheaf ω of the moduli space M^G is locally free. In particular, M^G is a Gorenstein variety.*

Moreover, $\bar{f}^*(\omega) = \mathscr{L}(0_{-2h})$, where h is the dual Coxeter number of the Lie algebra $\mathfrak{g}$ (cf. Example A.7).

Further, for any line bundle $\mathfrak{L}$ on M^G such that $\bar{f}^*(\mathfrak{L}) = \mathscr{L}(0_d)$ for some $d > -2h$,

$$H^i(M^G, \mathfrak{L}) = 0, \quad \text{for all } i > 0. \tag{1}$$

In particular, for any finite-dimensional representation V of G,

$$H^i(M^G, \Theta(V)) = 0, \quad \text{for all } \quad i > 0, \tag{2}$$

where $\Theta(V)$ is the theta bundle on the moduli space M^G (cf. Definition 8.4.4).

Proof Let

$$\mathring{M}^G := \{[E] \in M^G : \bar{E} \text{ is a stable } G\text{-bundle and } \operatorname{Aut} \bar{E} = Z\},$$

i.e., $\mathring{M}^G = \pi(\mathring{R}^d(G))$.

On the set of stable bundles in the moduli space there are no identifications, i.e., if E_1 and E_2 are two stable G-bundles on Σ such that $[E_1] = [E_2]$, then E_1 is isomorphic with E_2 (cf. Proposition 7.2.18(c)).

[1] We next prove that $\mathring{M}^G$ contains an open subset $\mathring{M}^G_s$ of M^G lying in the smooth locus of M^G such that

$$\operatorname{codim}_{M^G}(M^G \backslash \mathring{M}^G_s) \geq 2 \text{ (unless the curve } \Sigma \text{ has genus 2 and } G = \operatorname{SL}_2, \text{ in which case it is 1).} \tag{3}$$

Take d large enough. For any $E \in (R^d)^{ss}(G)$, by identity (5) of the proof of Theorem 7.2.6, the isotropy (under the GL_{N_d}-action on $R^d(G)$) $I_E \simeq \operatorname{Aut}(\bar{E})$. Since Z is contained in the automorphism group for any G-bundle E, I_E is the smallest possible isotropy group for any $E \in \mathring{R}^d(G)$, for the action of GL_{N_d} on $(R^d)^{ss}(G)$. Clearly, $\mathring{R}^d(G)$ is a GL_{N_d}-stable subset of $(R^d)^s(G)$. Moreover, for any $E \in (R^d)^s(G)$, the orbit $\operatorname{GL}_{N_d} \cdot E$ is closed in $(R^d)^{ss}(G)$, $\pi((R^d)^s(G))$ is open in M^G, $\pi^{-1}\left(\pi((R^d)^s(G))\right) = (R^d)^s(G)$

[1] As suggested by G. Faltings in a private communication, $M^G \backslash \mathring{M}^G$ is a constructible set.

and $\pi_{|(R^d)^s(G)}\colon (R^d)^s(G) \to \pi((R^d)^s(G))$ is a geometric quotient (cf. Corollary 7.2.20(c)). By Faltings (1993, Theorem II.6), $\mathring{M}^G$ contains an open subset $\mathring{M}^G_s$ of M^G satisfying (3). Hence, $\mathring{R}^d(G)_s \to \mathring{M}^G_s$ is a principal bundle, where $\mathring{R}^d(G)_s := \pi^{-1}(\mathring{M}^G_s)$ (use (5) of the proof of Theorem 7.2.6). Since $(R^d)^{ss}(G)$ is smooth (cf. Proof of Theorem 7.2.6) and hence so is $\mathring{R}^d(G)_s$, we get that $\mathring{M}^G_s$ is smooth.

Further, for any $\bar{E} \in \mathring{M}^G_s$, the tangent space $T_{\bar{E}}(\mathring{M}^G_s)$ can be identified with $H^1(\Sigma, \mathrm{ad}\,\bar{E})$ (cf. (Ramanathan, 1975, proof of Theorem 4.3)). (Observe that $H^0(\Sigma, \mathrm{ad}\,\bar{E}) = 0$, by Exercise 6.3.E.7.) Thus, the fiber of the canonical bundle of $\mathring{M}^G_s$ at $\bar{E}$ can be identified with $\wedge^{\mathrm{top}}(H^1(\Sigma, \mathrm{ad}\,\bar{E})^*)$, where $\wedge^{\mathrm{top}}$ is the top exterior power. This gives, from the definition of the Θ-bundle as in Definition 8.4.4, that

$$\Theta(\mathrm{ad})^*_{|\mathring{M}^G_s} = \omega_{|\mathring{M}^G_s}.$$

But, since $\Theta(\mathrm{ad})^*$ is a line bundle on the whole of M^G and since any line bundle is a reflexive sheaf (cf. (Hartshorne, 1977, Chap. II, Exercise 5.1)), $\Theta(\mathrm{ad})^*$ is a reflexive sheaf on M^G. Since the dualizing sheaf ω of a normal variety is always reflexive; the moduli space M^G is Cohen–Macaulay and normal (cf. Theorem 7.2.6); and by (3) $\mathrm{codim}_{M^G}(M^G \backslash \mathring{M}^G_s) \geq 2$ (unless the curve Σ is of genus 2 and $G = \mathrm{SL}_2$); we obtain from Lemma 9.1.4:

$$\omega \approx \Theta(\mathrm{ad})^*, \quad \text{on the whole of } M^G. \tag{4}$$

We next prove the validity of (4) for $G = \mathrm{SL}_2$ and $g = 2$. In this case $M^G \simeq \mathbb{P}^3$ (cf. (Narasimhan and Ramanan, 1969, §7, Theorem 2)). Thus, ω is a line bundle. Moreover, by Theorem 9.1.3, $\bar{f}^*(\Theta(V)) = \mathscr{L}(0_1)$, where V is the standard 2-dimensional representation of SL_2. Thus, by Proposition 8.3.2 and Theorem 9.1.3,

$$\bar{f}^*\colon \mathrm{Pic}(M^G) \simeq \mathrm{Pic}_{\bar{\Gamma}}(\bar{X}_G).$$

In particular, $\bar{f}^*(\mathscr{O}(1)) = \mathscr{L}(0_1)$. But, since $\omega_{\mathbb{P}^3} \simeq \mathscr{O}(-4)$, $\bar{f}^*(\omega_{\mathbb{P}^3}) = \mathscr{L}(0_{-4})$. Also, $\bar{f}^*(\Theta(\mathrm{ad})) = \mathscr{L}(0_4)$, by Example A.7 and Theorem 9.1.3. Hence, from the injectivity of $\bar{f}^*$, $\omega_{\mathbb{P}^3} \simeq \Theta(\mathrm{ad})^*$, proving (4) in this case as well.

Of course, (4) gives that M^G is a Gorenstein variety (by definition) for any G and any $g \geq 2$.

Now, the assertion that $\bar{f}^*(\omega) = \mathscr{L}(0_{-2h})$ follows from Theorem 9.1.3 and Example A.7 by using (4).

Finally, we come to the proof of the cohomology vanishing (1). By Serre duality (Hartshorne, 1977, Chap. III, Corollary 7.7) (denoting $\dim M^G = n$),

$$H^i(M^G, \mathfrak{L})^* \approx H^{n-i}(M^G, \mathfrak{L}^* \otimes \omega)$$
$$= H^{n-i}(M^G, \mathfrak{L}^* \otimes \Theta(\mathrm{ad})^*), \text{ by (4).} \tag{5}$$

But $\bar{f}^*(\mathfrak{L}^* \otimes \Theta(\mathrm{ad})^*) = \mathscr{L}(0_{-d-2h})$. Now, since $\mathrm{Pic}\,(M^G) \approx \mathbb{Z}$ (by Theorem 9.1.3), we get that the line bundle $\mathring{\mathfrak{L}} := \mathfrak{L} \otimes \Theta(\mathrm{ad})$ is ample on M^G (using the assumption $d > -2h$). (Observe that by Theorem 8.4.15 and Corollary 3.5.11, $H^0(M^G, \Theta(V)) \neq 0$ for any nontrivial G-module V.) Since the moduli space M^G has rational singularity (cf. Theorem 7.2.6), the vanishing of $H^i(M^G, \mathfrak{L})$ (for $i > 0$) follows from (5) and a result of Grauert–Riemenschneider (1970) (cf. also Kumar (2002, Theorem A.33)). To prove this, take a proper desingularization $\beta\colon \hat{M}^G \to M^G$. Since M^G has rational singularity, for any ample line bundle $\mathring{\mathfrak{L}}$ over M^G,

$$H^i(M^G, \mathring{\mathfrak{L}}^*) \simeq H^i(\hat{M}^G, \beta^*(\mathring{\mathfrak{L}}^*)).$$

Moreover, taking large enough N such that $\mathring{\mathfrak{L}}^N$ is very ample providing an embedding $i\colon M^G \hookrightarrow \mathbb{P}^d$ with $i^*(\mathscr{O}(1)) \simeq \mathring{\mathfrak{L}}^N$, we get that $\beta^*(\mathring{\mathfrak{L}}^N) \simeq \beta^* i^*(\mathscr{O}(1))$. Now, apply the Grauert–Riemenschneider theorem in the form as in Kumar (2002, Theorem A.33).

To prove (2), observe by Theorem 9.1.3 that $\bar{f}^*(\Theta(V)) \simeq \mathscr{L}(0_{d_V})$ and $d_V \geq 0$.

This completes the proof of the theorem. □

Corollary 9.1.7 *For any finite-dimensional representation V of G,*

$$\chi(M^G, \Theta(V)) = \dim H^0(M^G, \Theta(V)),$$

where χ is the Euler–Poincaré characteristic:

$$\chi(M^G, \Theta(V)) = \sum_i (-1)^i \dim H^i(M^G, \Theta(V)).$$

Fix an embedding $i\colon G \hookrightarrow \mathrm{SL}_r$, for some r. In particular, any principal G-bundle E on Σ gives rise to a vector bundle $E(i)$ of rank r on Σ (associated to the standard representation of SL_r).

Definition 9.1.8 For any integer $d \geq 1$, define the subset of $\bar{X}_G$:

$$\bar{X}_G^d = \{x \in \bar{X}_G : H^1(\Sigma, \bar{x}(i) \otimes \mathscr{O}_\Sigma(dp - y)) = 0, \text{ for all } y \in \Sigma\},$$

where $\bar{x} := (\mathbf{U}_G)_{|\Sigma \times x}$ and $p \in \Sigma$ is the fixed base point. Then, it is easy to see (e.g., by using the Serre duality)

$$\bar{X}_G^1 \subset \bar{X}_G^2 \subset \cdots.$$

Lemma 9.1.9 *Each $\bar{X}_G^d$ is open in $\bar{X}_G$. Moreover, $\bar{X}_G^{ss} \subset \bar{X}_G^{2g}$, where*

$$\bar{X}_G^{ss} = \{x \in \bar{X}_G : \bar{x} \text{ is a semistable } G\text{-bundle}\},$$

where g is the genus of the curve Σ.

Further,

$$\cup_{d \geq 1} \bar{X}_G^d = \bar{X}_G. \tag{1}$$

Proof Let $\mathbf{U}_G(i)$ be the rank-r vector bundle associated to the tautological bundle $\mathbf{U}_G$ over $\Sigma \times \bar{X}_G$ via the representation i. Define a vector bundle $\tilde{\mathbf{U}}_G^d(i)$ over $\Sigma \times \Sigma \times \bar{X}_G$ by

$$\tilde{\mathbf{U}}_G^d(i) := \pi_{1,2}^* \left(\mathscr{O}_{\Sigma \times \Sigma}(-\Delta(\Sigma))\right) \otimes \pi_1^* \left(\mathscr{O}_\Sigma(dp)\right) \otimes \pi_{1,3}^* \left(\mathbf{U}_G(i)\right),$$

where $\pi_{i,j}$ is the projection from $\Sigma \times \Sigma \times \bar{X}_G$ to the (i,j)-factor and $\Delta(\Sigma)$ is the diagonal of $\Sigma \times \Sigma$. Then,

$$\tilde{\mathbf{U}}_G^d(i)_{|\Sigma \times y \times \bar{X}_G} = \pi_1^*(\mathscr{O}_\Sigma(dp - y)) \otimes \mathbf{U}_G(i), \quad \text{for any } y \in \Sigma.$$

Applying the Upper Semicontinuity Theorem (Hartshorne, 1977, Chap. III, Theorem 12.8) to the morphism $\pi_{2,3}$ and the vector bundle $\tilde{\mathbf{U}}_G^d(i)$, we get that the set

$$S_d := \{(y,x) \in \Sigma \times \bar{X}_G : H^1(\Sigma, \bar{x}(i) \otimes \mathscr{O}_\Sigma(dp - y)) \neq 0\}$$

is a closed subset of $\Sigma \times \bar{X}_G$. Thus, $\pi_2(S_d)$ is a closed subset of $\bar{X}_G$, where $\pi_2 \colon \Sigma \times \bar{X}_G \to \bar{X}_G$ is the projection on the second factor, which is a proper map. It is easy to see that $\bar{X}_G^d = \bar{X}_G \setminus \pi_2(S_d)$. This proves that $\bar{X}_G^d$ is open in $\bar{X}_G$.

For $x \in \bar{X}_G^{ss}, \bar{x}(i)$ is a semistable vector bundle (cf. Theorem 6.1.7), and hence by the proof of Proposition 7.1.3(a), if

$$H^1(\Sigma, \bar{x}(i) \otimes \mathscr{O}_\Sigma(dp - y)) \neq 0, \quad \text{for some } y \in \Sigma,$$

then $d \leq 2g - 1$. Thus, $\bar{X}_G^{ss} \subset \bar{X}_G^{2g}$.

Take any $d \geq 1$. To prove (1), take $x \in \bar{X}_G$. Consider the projection $p_2 \colon \Sigma \times \Sigma \to \Sigma$ to the first factor and apply the Semicontinuity Theorem to the bundle $\tilde{\mathbf{U}}_G^d(i)_{|\Sigma \times \Sigma \times x}$ over $\Sigma \times \Sigma$, to get

$$C_d := \{y \in \Sigma : H^1(\Sigma, \bar{x}(i) \otimes \mathscr{O}_\Sigma(dp - y)) = 0\}$$

is an open subset of Σ. But $\mathscr{O}_\Sigma(dp)$ being ample (cf. (Hartshorne, 1977, Chap. IV, Corollary 3.2(b)), for any fixed $y \in \Sigma$,

$$H^1(\Sigma, \bar{x}(i) \otimes \mathscr{O}_\Sigma(dp - y)) = 0, \quad \text{for large enough } d = d_y.$$

Since C_d is open for any d, we can take large enough d (independent of y) such that $C_d = \Sigma$. This proves (1), completing the proof of the lemma. □

Recall that we have chosen an embedding $G \hookrightarrow \mathrm{SL}_r$. Clearly, any $\bar{X}_G^d$ is stable under the action of $\bar{\Gamma}$. For any $d \geq 2g$, by Lemma 9.1.9, $\bar{X}_G^d$ contains $\bar{X}_G^{ss}$ as a $\bar{\Gamma}$-stable open subset. Also, for any $d \geq 2g$, recall the definition of GL_{N_d}-variety $R^d(G)$ from Definition 9.1.5, where $N_d := r(d+1-g)$. For any $d \geq 2g$, by Proposition 7.1.3 and Theorem 6.1.7, any semistable G-bundle over Σ appears in the family $\mathscr{U}^d(G)$ parameterized by $R^d(G)$ and $(R^d)^{ss}(G)$ is a GL_{N_d}-stable open subset.

Proposition 9.1.10 *For any $d \geq 2g$, as quotient stacks we have the equivalence*

$$\left[\bar{\Gamma}\backslash\bar{X}_G^d\right] \simeq \left[\mathrm{GL}_{N_d}\backslash R^d(G)\right]. \tag{1}$$

Moreover, the above equivalence restricts to an equivalence

$$\left[\bar{\Gamma}\backslash\bar{X}_G^{ss}\right] \simeq \left[\mathrm{GL}_{N_d}\backslash(R^d)^{ss}(G)\right]. \tag{2}$$

Proof Observe first that for any vector bundle $\mathcal{W}$ over Σ, the following two conditions are equivalent:

(a) $H^1(\Sigma, \mathcal{W} \otimes \mathscr{O}_\Sigma(-y)) = 0$ for all $y \in \Sigma$.
(b) $H^1(\Sigma, \mathcal{W}) = 0$ and $\mathcal{W}$ is globally generated.

Let $\mathbf{Bun}_G^d(\Sigma)$ be the stack whose objects over a scheme S are G-bundles $\mathcal{E}$ over $\Sigma \times S$ such that for any $s \in S$, $\mathcal{W}_s := \mathcal{E}_s(i) \otimes \mathscr{O}_\Sigma(dp)$ satisfies the equivalent conditions (a) and (b), where $\mathcal{E}_s(i) := \mathcal{E}(i)_{|\Sigma\times s}$. Then, following the proof of the Uniformization Theorem 5.2.14, there is an equivalence of stacks:

$$\mathbf{Bun}_G^d(\Sigma) \simeq \left[\bar{\Gamma}\backslash\bar{X}_G^d\right]. \tag{3}$$

We next show that there is an equivalence of stacks:

$$\mathbf{Bun}_G^d(\Sigma) \simeq \left[\mathrm{GL}_{N_d}\backslash R^d(G)\right]: \tag{4}$$

For any scheme S, let $\mathcal{E}$ be a G-bundle over $\Sigma \times S$ as above, i.e., $\mathcal{W}_s := \mathcal{E}(i)_{|\Sigma\times s} \otimes \mathscr{O}_\Sigma(dp)$ satisfies the equivalent conditions (a) and (b) for all $s \in S$. Let $\mathcal{W} := \mathcal{E}(i) \otimes \pi_\Sigma^*(\mathscr{O}_\Sigma(dp))$ be the vector bundle over $\Sigma \times S$ (where $\pi_\Sigma : \Sigma \times S \to \Sigma$ is the projection) and let $\mathscr{F}(\mathcal{W})$ be the frame bundle of $\pi_{S*}(\mathcal{W})$, which is a principal GL_{N_d}-bundle over S since $H^0(\Sigma, \mathcal{W}_s)$ has dimension N_d for any $s \in S$ because of the condition (b) satisfied by $\mathcal{W}_s$ (cf. (Kempf, 1978, Theorem 13.1)). We define a GL_{N_d}-equivariant morphism $\varphi_{\mathcal{E}} : \mathscr{F}(\mathcal{W}) \to R^d(G)$ as follows. An element in $\mathscr{F}(\mathcal{W})$ over $s \in S$ corresponds to an $\mathscr{O}_\Sigma$-linear isomorphism

$$\mathscr{O}_\Sigma \otimes \mathbb{C}^{N_d} \xrightarrow{\sim} \mathscr{O}_\Sigma \otimes H^0(\Sigma, \mathcal{W}_s) \xrightarrow{\text{ev}} \mathcal{W}_s.$$

This gives rise to a surjective bundle map over Σ:

$$\left(\mathscr{O}_\Sigma \otimes \mathbb{C}^{N_d}\right) \otimes \mathscr{O}_\Sigma(-dp) \to \mathcal{E}_s(i),$$

since $\mathcal{E}_s(i) \otimes \mathscr{O}_\Sigma(dp)$ is globally generated by assumption. Moreover, $\mathcal{E}_s$ sits inside $\mathcal{E}_s(\bar{i})$ as a G-subbundle, where $\mathcal{E}_s(\bar{i})$ is the SL_r-bundle obtained from $\mathcal{E}_s$ via the embedding $G \hookrightarrow \mathrm{SL}_r$. This is our morphism $\varphi_{\mathcal{E}} : \mathscr{F}(\mathcal{W}) \to R^d(G)$, which is clearly GL_{N_d}-equivariant. Thus, $\mathcal{E} \rightsquigarrow (\mathscr{F}(\mathcal{W}), \varphi_{\mathcal{E}})$ provides a morphism between the stacks: $\mathbf{Bun}_G^d(\Sigma) \to \left[\mathrm{GL}_{N_d} \backslash R^d(G)\right]$. It is easy to see that it is an equivalence of stacks, proving (4).

Combining (3) and (4), we of course get the equivalence (1) of the proposition.

Further, it is easy to see that the equivalence (1) restricts to the equivalence (2). □

For a standard parabolic subgroup P of G with Levi subgroup L containing the maximal torus H, recall that the character group

$$X(P) = \oplus_{\alpha_i \notin S_P} \mathbb{Z}\omega_i,$$

where $S_P \subset \{\alpha_1, \ldots, \alpha_\ell\}$ is the set of simple roots of L (cf. Definition 6.1.3). Let $E \to \Sigma$ be a G-bundle with a reduction E_P of structure group to P. Consider the function

$$\mu_{E_P} : X(P) \to \mathbb{Z},\ \lambda \mapsto \deg(E_P(\lambda)),$$

where $E_P(\lambda)$ is as in Definition 6.2.1. Extend this function (still denoted) $\mu_{E_P} : X(H) \to \mathbb{Z}$ by taking $\lambda \mapsto 0$ for $\lambda \in \oplus_{\alpha_i \in S_P} \mathbb{Z}\omega_i$. Then, μ_{E_P} is called the *type of the reduction* E_P.

We record the following lemma without proof; the proof being similar to that of Bruguieres (1985, Theorem 4, p. 90) once we use (1) of Theorem 6.2.3.

Lemma 9.1.11 *Let $\mathcal{E}$ be a family of G-bundles over Σ parameterized by a smooth variety S. Assume that at each point $t \in S$ the infinitesimal deformation map (cf. Definition 6.3.17):*

$$T_t(S) \to H^1(\Sigma, \mathrm{ad}(\mathcal{E}_t))$$

is surjective, where $\mathcal{E}_t = \mathcal{E}_{|\Sigma \times t}$ and $T_t(S)$ is the tangent space of S at t. For $\mu \in Hom(X(H), \mathbb{Z})$, let S_μ be the subset of S consisting of those points $t \in S$ such that the canonical reduction (cf. Definition 6.2.1) of $\mathcal{E}_t$ is of type μ. Then,

S_μ is nonempty only for finitely many μ. Moreover, S_μ is locally closed and smooth, and the normal space at $t \in S_\mu$ is given by $H^1(\Sigma, \mathcal{E}_{t,\mathfrak{s}})$, where $\mathcal{E}_{t,\mathfrak{s}}$ is the vector bundle associated to the canonical reduction $\mathcal{E}_{t,P}$ of $\mathcal{E}_t$ by the representation of P on $\mathfrak{s} := \mathfrak{g}/\mathfrak{p}$.

Proposition 9.1.12 *For any $d \geq 2g$, the codimension of $R^d(G)\backslash(R^d)^{ss}(G)$ in $R^d(G)$ is at least 2.*

Proof The family $\mathscr{U}^d(G)$ parameterized by $R^d(G)$ satisfies the hypothesis of the above Lemma 9.1.11, i.e., $R^d(G)$ is smooth (cf. Definition 9.1.5) and the infinitesimal deformation map at any $t \in R^d(G)$ is surjective. To prove the latter, following Definition 9.1.5 and the proof of Theorem 7.2.6 (specifically (6) and the following discussion), we use Proposition 7.2.3 for $\Gamma(i, \mathscr{U}^d)$, where $i\colon G \hookrightarrow \mathrm{SL}_r \subset \mathrm{GL}_r$ and $\mathscr{U}^d$ is the family of vector bundles over Σ parameterized by R^d as in Definition 9.1.5. Then, we get that the family $\mathscr{U}^d(G)$ of G-bundles over Σ parameterized by $R^d(G)$ has the local universal property, i.e., it is complete. Thus, by Theorem 6.3.20, the corresponding deformation map at any $t \in R^d(G)$ is surjective.

So, it suffices to prove that for $t \in R^d(G)\backslash(R^d)^{ss}(G)$, we have $\dim H^1(\Sigma, \mathscr{U}^d(G)_{t,\mathfrak{s}}) \geq 2$.

By (1) of Theorem 6.2.3, $H^0(\Sigma, \mathscr{U}^d(G)_{t,\mathfrak{s}}) = 0$. Thus, by the Riemann–Roch theorem for smooth curves (cf. (Fulton, 1998, Example 15.2.1)),

$$\dim H^1(\Sigma, \mathscr{U}^d(G)_{t,\mathfrak{s}}) = -\deg \mathscr{U}^d(G)_{t,\mathfrak{s}} + \dim(\mathfrak{s})(g-1), \tag{1}$$

where recall that g is the genus of Σ. Further, since $t \in R^d(G)\backslash(R^d)^{ss}(G)$, we have $\mathfrak{g} \neq \mathfrak{p}$. By identity (17) of the proof of Theorem 6.2.3, $\deg \mathscr{U}^d(G)_{t,\mathfrak{s}} < 0$. This gives (using (1) and the assumption $g \geq 2$) that $\dim H^1(\Sigma, \mathscr{U}^d(G)_{t,\mathfrak{s}}) \geq 2$, proving the proposition. □

We are now ready to prove Proposition 9.1.2.

§9.1.13. *Proof of Proposition 9.1.2.* For any $d \geq 2g$, we have the following commutative diagram:

$$\begin{array}{cccc}
\mathrm{Pic}_{\mathrm{GL}_{N_d}}\left(R^d(G)\right) & = \mathrm{Pic}\left(\left[\mathrm{GL}_{N_d}\backslash R^d(G)\right]\right) & \simeq \mathrm{Pic}\left(\left[\bar{\Gamma}\backslash \bar{X}^d_G\right]\right) & = \mathrm{Pic}_{\bar{\Gamma}}\left(\bar{X}^d_G\right) \\
\bar{\gamma}_1 \downarrow & \gamma_1 \downarrow & \gamma_2 \downarrow & \bar{\gamma}_2 \downarrow \\
\mathrm{Pic}_{\mathrm{GL}_{N_d}}\left((R^d)^{ss}(G)\right) & = \mathrm{Pic}\left(\left[\mathrm{GL}_{N_d}\backslash (R^d)^{ss}(G)\right]\right) & \simeq \mathrm{Pic}\left(\left[\bar{\Gamma}\backslash \bar{X}^{ss}_G\right]\right) & = \mathrm{Pic}_{\bar{\Gamma}}\left(\bar{X}^{ss}_G\right),
\end{array}$$

where the vertical maps $\gamma_i, \bar{\gamma}_i$ are induced from the inclusions, all the equalities follow from Exercise C.E.8 (observe that all the above stacks are algebraic stacks by Lemma C.19 and Proposition 9.1.10), the two horizontal isomorphisms $\simeq$ follow from Proposition 9.1.10 and the map $\bar{\gamma}_1$ is an isomorphism by Proposition 9.1.12 and Exercise 8.5.E.2. Thus, from the above commutative diagram, we get that $\bar{\gamma}_2$ is an isomorphism. This gives, in particular, that for any $d \geq 2g$,

$$\mathrm{Pic}_{\bar{\Gamma}}\left(\bar{X}_G^{d+1}\right) \simeq \mathrm{Pic}_{\bar{\Gamma}}\left(\bar{X}_G^{d}\right), \quad \text{under the restriction map.} \tag{2}$$

Thus,

$$\mathrm{Pic}_{\bar{\Gamma}}\left(\bar{X}_G^{ss}\right) \simeq \varprojlim_{d} \mathrm{Pic}_{\bar{\Gamma}}\left(\bar{X}_G^{d}\right), \tag{3}$$

where the inverse limit is induced from the inclusions: $\bar{X}_G^1 \subset \bar{X}_G^2 \subset \bar{X}_G^3 \subset \cdots$. But by Exercise 9.1.E.2, under the restriction map,

$$\mathrm{Pic}_{\bar{\Gamma}}\left(\bar{X}_G\right) \simeq \varprojlim_{d} \mathrm{Pic}_{\bar{\Gamma}}\left(\bar{X}_G^{d}\right). \tag{4}$$

Combining (3) and (4), we get that the restriction map

$$\mathrm{Pic}_{\bar{\Gamma}}\left(\bar{X}_G\right) \to \mathrm{Pic}_{\bar{\Gamma}}\left(\bar{X}_G^{ss}\right)$$

is an isomorphism. This completes the proof of Proposition 9.1.2. □

9.1.E Exercises

(1) Show that any $\bar{\Gamma}$-invariant morphism $f\colon \bar{X}_G^{ss} \to \mathbb{C}$ is a constant. Thus, so is any $\bar{\Gamma}$-invariant morphism $f\colon \bar{X}_G^{d} \to \mathbb{C}$ (for any $d \geq 2g$).

Hint: By Proposition C.23, f gives rise to a morphism $\bar{f}\colon [\bar{\Gamma}\backslash \bar{X}_G^{ss}] \to \mathbb{C}$ and hence using Proposition 9.1.10 a GL_{N_d}-invariant morphism $\hat{f}\colon (R^d)^{ss}(G) \to \mathbb{C}$. Since M^G is the good quotient $(R^d)^{ss}(G)//\mathrm{GL}_{N_d}$ for large enough d (cf. Theorem 7.2.6), the morphism $\hat{f}$ gives rise to a morphism $\mathring{f}\colon M^G \to \mathbb{C}$ (cf. Definition 7.1.4). But, M^G being an irreducible projective variety, $\mathring{f}$ is a constant.

(2) Show that the restriction map

$$\mathrm{Pic}_{\bar{\Gamma}}\left(\bar{X}_G\right) \to \varprojlim_{d} \mathrm{Pic}_{\bar{\Gamma}}\left(\bar{X}_G^{d}\right) \quad \text{is an isomorphism.}$$

Thus,

$$\mathrm{Pic}_{\bar{\Gamma}}\left(\bar{X}_G\right) \to \mathrm{Pic}_{\bar{\Gamma}}\left(\bar{X}_G^{d}\right) \quad \text{is an isomorphism for any } d \geq 2g.$$

Hint: Use Exercise 1 together with the isomorphism (2) in §9.1.13.

(3) Show that the ample line bundle $\Theta = \Theta_{r,0}$ as in Theorem 8.4.2 restricted to $M^{\mathrm{SL}_r} \hookrightarrow M(r,0)$ is the ample *generator* of $\mathrm{Pic}(M^{\mathrm{SL}_r})$.

Hint: Use Theorem 9.1.3.

The result remains true for $g = 1$ as well by Tu (1993, Theorem 8).

9.2 Moduli of G-Bundles over Elliptic Curves – An Explicit Determination

In this section Σ denotes a smooth irreducible projective curve of genus 1, i.e., Σ is an elliptic curve with a base point p and G is a connected simply-connected simple algebraic group over $\mathbb{C}$. We identify $M^G = M^G(\Sigma)$ with a weighted projective space and show that the generator of $\mathrm{Pic}(M^G)$ is $\Theta(V(\omega_i))$ for a fundamental weight ω_i.

We recall the following theorem due independently to Laszlo (1998b, Theorem 4.16) and Friedman, Morgan and Witten (1998, §2).

Theorem 9.2.1 *Let Σ be as above (of genus 1). Then, there is a natural variety isomorphism between the moduli space M^G and $(\Sigma \otimes_{\mathbb{Z}} Q^\vee)/W$, where $Q^\vee$ is the coroot lattice of G and W is its Weyl group acting canonically on $Q^\vee$ (and acting trivially on Σ).*

In particular, M^G is a projective variety.

Definition 9.2.2 Let $N = (n_0, \ldots, n_\ell)$ be a $\ell + 1$-tuple of positive integers. Consider the polynomial ring $\mathbb{C}[z_0, \ldots, z_\ell]$ graded by $\deg z_i = n_i$. The scheme $\mathrm{Proj}(\mathbb{C}[z_0, \ldots, z_\ell])$ is said to be the *weighted projective space* of type N and we denote it by $\mathbb{P}(N)$.

Consider the standard (nonweighted) projective space $\mathbb{P}^\ell := \mathrm{Proj}(\mathbb{C}[w_0, \ldots, w_\ell])$, where each $\deg w_i = 1$. Then, the graded algebra homomorphism $\mathbb{C}[z_0, \ldots, z_\ell] \to \mathbb{C}[w_0, \ldots, w_\ell], z_i \mapsto w_i^{n_i}$, induces a morphism $\delta \colon \mathbb{P}^\ell \to \mathbb{P}(N)$.

The following theorem is due to Looijenga (1976). His proof had a gap; a complete proof of a more general result is outlined in Bernshtein and Shvartsman (1978).

Theorem 9.2.3 *Let Σ be an elliptic curve. Then the variety $(\Sigma \otimes_{\mathbb{Z}} Q^\vee)/W$ is the weighted projective space of type $(1, a_1^\vee, a_2^\vee, \ldots, a_\ell^\vee)$, where $a_i^\vee$ are the coefficients of the coroot $\theta^\vee$ written in terms of the simple coroots $\{\alpha_i^\vee\}$ (and, as earlier, ℓ is the rank of G).*

Hence, by Theorem 9.2.1, M^G is isomorphic with the weighted projective space of type $(1, a_1^\vee, a_2^\vee, \ldots, a_\ell^\vee)$.

The following table lists the weighted projective space isomorphic to M^G corresponding to any G. In this table the entries beyond 1 are precisely the numbers $(a_1^\vee, a_2^\vee, \ldots, a_\ell^\vee)$ following the convention as in Bourbaki (2002, Planche I–IX).

Type of G	Type of the weighted projective space
A_ℓ $(\ell \geq 1)$, C_ℓ $(\ell \geq 2)$	$(1,1,1,\ldots,1)$
B_ℓ $(\ell \geq 3)$	$(1,1,2,\ldots,2,1)$
D_ℓ $(\ell \geq 4)$	$(1,1,2,\ldots,2,1,1)$
G_2	$(1,1,2)$
F_4	$(1,2,3,2,1)$
E_6	$(1,1,2,2,3,2,1)$
E_7	$(1,2,2,3,4,3,2,1)$
E_8	$(1,2,3,4,6,5,4,3,2)$.

We recall the following result from the theory of weighted projective spaces (see, e.g., (Beltrametti and Robbiano, 1986, Lemma 3B.2.c and Theorem 7.1.c)).

Theorem 9.2.4 *Let $N = (n_0, \ldots, n_\ell)$ and assume $gcd\{n_0, \ldots, n_\ell\} = 1$. Then we have the following.*

(a) $\mathrm{Pic}(\mathbb{P}(N)) \simeq \mathbb{Z}$. In fact, the morphism δ of Definition 9.2.2 induces an injective map $\delta^* : \mathrm{Pic}(\mathbb{P}(N)) \to \mathrm{Pic}(\mathbb{P}^\ell)$.

Moreover, the ample generator of $\mathrm{Pic}(\mathbb{P}(N))$ maps to $\mathscr{O}_{\mathbb{P}^\ell}(s)$ under δ^*, where s is the least common multiple of $\{n_0, \ldots, n_\ell\}$. We denote this ample generator by $\mathscr{O}_{\mathbb{P}(N)}(s)$.

(b) For any $d \geq 0$,

$$H^0(\mathbb{P}(N), \mathscr{O}_{\mathbb{P}(N)}(s)^{\otimes d}) = \mathbb{C}[z_0, \ldots, z_\ell]_{ds},$$

where $\mathbb{C}[z_0, \ldots, z_\ell]_{ds}$ denotes the subspace of $\mathbb{C}[z_0, \ldots, z_\ell]$ consisting of homogeneous elements of degree ds.

9.3 Picard Group of Moduli Space of G-Bundles – Explicit Determination

In this section Σ denotes a smooth irreducible projective curve of any genus $g \geq 1$ with a base point p and G a connected simply-connected simple algebraic group over $\mathbb{C}$.

Recall Theorem 9.1.3 which asserts that for $g \geq 2$:

$$\bar{f}^*\colon \operatorname{Pic}(M^G) \hookrightarrow \operatorname{Pic}_{\bar{\Gamma}}(\bar{X}_G) \simeq \mathbb{Z}.$$

Moreover, by Theorems 9.2.3 and 9.2.4, $\operatorname{Pic}(M^G) \simeq \mathbb{Z}$, for $g = 1$ as well. The following is the main result of this section.

Theorem 9.3.1

$$\operatorname{Pic}(M^G) = \langle \Theta(V), V \in \mathcal{R}(G) \rangle,$$

where the notation $\langle \quad \rangle$ denotes the group generated by the elements in the bracket and $\mathcal{R}(G)$ is the set of isomorphism classes of irreducible G-modules.

Before we can prove the theorem, we need various results proved below.

Lemma 9.3.2

$$c\colon \operatorname{Pic}(M^G) \simeq H^2(M^G, \mathbb{Z}),$$

where c maps any line bundle $\mathfrak{L}$ to its first Chern class $c_1(\mathfrak{L})$ and M^G is taken with its analytic topology.

In particular,

$$H^2(M^G, \mathbb{Z}) \simeq \mathbb{Z}.$$

The first Chern class of the ample generator of $\operatorname{Pic}(M^G)$ *is called the positive generator of $H^2(M^G, \mathbb{Z})$.*

Proof Consider the following exact sequence of abelian groups:

$$0 \to \mathbb{Z} \to \mathbb{C} \xrightarrow{f} \mathbb{C}^* \to 0,$$

where $f(x) = e^{2\pi i x}$. This gives rise to the following exact sequence of sheaves on M^G endowed with the analytic topology:

$$0 \to \bar{\mathbb{Z}} \to \bar{\mathcal{O}}_{M^G} \to \bar{\mathcal{O}}^*_{M^G} \to 0,$$

where $\bar{\mathcal{O}}_{M^G}$ is the sheaf of holomorphic functions on M^G, $\bar{\mathcal{O}}^*_{M^G}$ is the sheaf of invertible elements of $\bar{\mathcal{O}}_{M^G}$ and $\bar{\mathbb{Z}}$ is the constant sheaf corresponding to the abelian group $\mathbb{Z}$.

The above sequence, of course, induces the following long exact sequence in cohomology:

$$\begin{aligned}\cdots \to H^1(M^G, \bar{\mathcal{O}}_{M^G}) \to H^1(M^G, \bar{\mathcal{O}}^*_{M^G}) &\xrightarrow{\bar{c}} H^2(M^G, \mathbb{Z}) \\ &\to H^2(M^G, \bar{\mathcal{O}}_{M^G}) \to \cdots .\end{aligned}$$

First of all,

$$\mathrm{Pic}(M^G) \simeq H^1(M^G, \mathscr{O}^*_{M^G}), \tag{1}$$

where $\mathscr{O}_{M^G}$ is the sheaf of algebraic functions on M^G and $\mathscr{O}^*_{M^G}$ is the subsheaf of invertible elements of $\mathscr{O}_{M^G}$.

Moreover, by GAGA (cf. (Serre, 1956)) , M^G being a projective variety,

$$H^1(M^G, \mathscr{O}^*_{M^G}) \simeq H^1(M^G, \bar{\mathscr{O}}^*_{M^G}), \tag{2}$$

and also, for any $p \geq 0$,

$$H^p(M^G, \mathscr{O}_{M^G}) \simeq H^p(M^G, \bar{\mathscr{O}}_{M^G}). \tag{3}$$

By Theorem 9.1.6 for $g \geq 2$ and by Theorem 9.2.3 for $g = 1$ (using Dolgachev (1982, §1.4)), $H^i(M^G, \mathscr{O}_{M^G}) = 0$ for $i > 0$. Hence, under the identification (1) by (2) and (3) and the above long exact cohomology sequence,

$$\mathrm{Pic}(M^G) \underset{\sim}{\overset{c}{\to}} H^2(M^G, \mathbb{Z}),$$

where c is the map $\bar{c}$ under the above identifications. Moreover, as it is well known, c is the first Chern class map. This proves the lemma. □

As at the beginning of Section 6.3,

$$\pi_1(\Sigma) = F(a_1, \ldots, a_g, b_1, \ldots, b_g)/\langle [a_1, b_1] \cdots [a_g, b_g] \rangle, \tag{*}$$

where F denotes the free group and $\langle\, \rangle$ denotes the normal subgroup generated by the enclosed element.

Let us fix a maximal compact subgroup K of G. Consider the real analytic map

$$\xi : (K \times K)^g \to K, \ ((x_1, y_1), \ldots, (x_g, y_g)) \mapsto [x_1, y_1] \ldots [x_g, y_g].$$

Recall the definition of the compact real analytic space $R_K(g) = R_K(\Sigma)$ from Lemma 6.3.2. Then, by the identification (*), mapping $a_i \mapsto x_i, b_i \mapsto y_i$, we get

$$R_K(g) \simeq \xi^{-1}(1).$$

Recall the following result for $A = (1)$ from Corollary 7.2.20(b) for $g \geq 2$. The result holds for $g = 1$ as well (cf., e.g., (Laszlo, 1998b, §4.20)).

Theorem 9.3.3 *The real analytic map*

$$\bar{f}(\Sigma) = \bar{f}_{\mathscr{E}}(\Sigma) : R_K(g)/\mathrm{Ad}K \to M^G = M^G(\Sigma)$$

induced from the family $\mathcal{E}$ (cf. the proof of Theorem 7.2.14) is a homeomorphism, where $\operatorname{Ad}K$ *acts on* $R_K(g)$ *via conjugation on each factor and* Σ *is a smooth irreducible projective curve of genus* $g \geq 1$.

In the sequel, we will often make this identification.

Proposition 9.3.4 *For any* $V \in \mathcal{R}(G)$, $c(\Theta(V))$, *under the above identification* $\bar{f}(\Sigma)$, *does not depend on the choice of the projective variety structure on* Σ *(for any fixed* g*).*

Proof Let $\rho: G \to \mathrm{SL}_V$ be the given representation. By taking a K-invariant Hermitian form on V and an orthonormal basis, we get $\rho(K) \subset \mathrm{SU}_n$, where $n := \dim V$. For any principal G-bundle E on Σ, let E_{SL_V} be the principal SL_V-bundle over Σ obtained by the extension of the structure group via ρ. Then, if E is semistable, so is E_{SL_V}, giving rise to a variety morphism $\hat{\rho}: M^G \to M^{\mathrm{SL}_V}$ (cf. Definition 7.2.10). Hence, we get the commutative diagram

$$\begin{array}{ccc} \bar{R}_K(g) & \xrightarrow{\bar{\rho}} & \bar{R}_{\mathrm{SU}_n}(g) \\ {\scriptstyle \bar{f}(\Sigma)}\downarrow & & \downarrow{\scriptstyle \bar{f}(\Sigma)} \\ M^G & \xrightarrow{\hat{\rho}} & M^{\mathrm{SL}_V}, \end{array} \qquad (\mathrm{D}_1)$$

where $\bar{R}_K(g) := R_K(g)/\operatorname{Ad}K$ and $\bar{\rho}$ is induced from the commutative diagram

$$\begin{array}{ccc} K^{2g} & \xrightarrow{\xi} & K \\ {\scriptstyle \rho^{2g}}\downarrow & & \downarrow{\scriptstyle \rho} \\ \mathrm{SU}_n^{2g} & \xrightarrow{\xi} & \mathrm{SU}_n . \end{array}$$

The diagram (D_1) induces the following commutative diagram in cohomology:

$$\begin{array}{ccc} H^2(M^{\mathrm{SL}_V},\mathbb{Z}) & \xrightarrow{\hat{\rho}^*} & H^2(M^G,\mathbb{Z}) \\ {\scriptstyle \bar{f}(\Sigma)^*}\downarrow{\scriptstyle \simeq} & & {\scriptstyle \bar{f}(\Sigma)^*}\downarrow{\scriptstyle \simeq} \\ H^2(\bar{R}_{\mathrm{SU}_n}(g),\mathbb{Z}) & \xrightarrow{\bar{\rho}^*} & H^2(\bar{R}_K(g),\mathbb{Z}). \end{array} \qquad (\mathrm{D}_2)$$

By the construction of the Θ-bundle as in Definition 8.4.4, $\hat{\rho}^*(\Theta) = \Theta(V)$, where $\hat{\rho}^*$ also denotes the pull-back of line bundles and Θ is the line bundle over M^{SL_V} as in Theorem 8.4.2.

Thus, using the functoriality of the Chern class, we get

$$\hat{\rho}^*(c(\Theta)) = c(\Theta(V)). \tag{1}$$

By Lemma 9.3.2 and Exercise 9.1.E.3, $c(\Theta)$ is the unique positive generator of $H^2(M^{\mathrm{SL}_V}, \mathbb{Z})$ and thus it is independent of the choice of the projective variety structure on Σ under the identification $\bar{f}(\Sigma)$. (Since the Teichmüller space is connected (Harris and Morrison, 1998, Chap. 2, §C), the positive generator of $H^2(M^{\mathrm{SL}_V}, \mathbb{Z})$ can not flip in $H^2(\bar{R}_{\mathrm{SU}_n}(g), \mathbb{Z})$ under a different projective variety structure on Σ.) Consequently, by (1) and the above commutative diagram (D_2), $c(\Theta(V))$ is independent of the choice of the projective variety structure on Σ. □

From now on we will denote the cohomology class $c(\Theta(V))$ *in* $H^2(\bar{R}_K(g), \mathbb{Z})$*, under the identification* $\bar{f}(\Sigma)^*$*, by* $c(\Theta_V(g, G))$*.*

Definition 9.3.5 Consider the embedding

$$i_g = i_g(K) \colon \bar{R}_K(g) \hookrightarrow \bar{R}_K(g+1)$$

induced by the inclusion of $K^{2g} \hookrightarrow K^{2g+2}$ via $(k_1, \ldots, k_{2g}) \mapsto (k_1, \ldots, k_{2g}, 1, 1)$.

By virtue of the map i_g, we will identify $\bar{R}_K(g)$ as a subspace of $\bar{R}_K(g+1)$. In particular, we get the following induced sequence of maps in the second cohomology:

$$H^2(\bar{R}_K(1), \mathbb{Z}) \overset{i_1^*}{\leftarrow} H^2(\bar{R}_K(2), \mathbb{Z}) \overset{i_2^*}{\leftarrow} H^2(\bar{R}_K(3), \mathbb{Z}) \overset{i_3^*}{\leftarrow} \cdots.$$

Proposition 9.3.6 *For* $K = \mathrm{SU}_2$*, the maps* $i_g^* \colon H^2(\bar{R}_K(g+1), \mathbb{Z}) \to H^2(\bar{R}_K(g), \mathbb{Z})$ *are isomorphisms for any* $g \geq 1$*.*

In fact, i_g^* *takes the positive generator of* $H^2(\bar{R}_{\mathrm{SU}_2}(g+1), \mathbb{Z})$ *to the positive generator of* $H^2(\bar{R}_{\mathrm{SU}_2}(g), \mathbb{Z})$*, where (as earlier) by* positive generator of $H^2(\bar{R}_K(g), \mathbb{Z})$*, we mean the image of the positive generator of* $H^2(M^G(\Sigma_g), \mathbb{Z})$ *under* $\bar{f}(\Sigma_g)^*$ *for any smooth irreducible projective curve* Σ_g *of genus* g*.*

We shall prove this proposition later at the end of this section.

Proposition 9.3.7 *For any* $V \in \mathcal{R}(G)$ *and any* $g \geq 1$*,* $i_g^*(c(\Theta_V(g+1, G))) = c(\Theta_V(g, G))$*.*

Proof We first claim that it suffices to prove the proposition for $G = \mathrm{SL}_n$ and the standard n-dimensional representation V of SL_n.

Let $\rho\colon G \to \mathrm{SL}_V$ be the given representation. Choosing a K-invariant Hermitian form on V, we get that $\rho(K) \subset \mathrm{SU}_V$. Consider the following commutative diagram:

$$\begin{array}{ccc} \bar{R}_K(g) & \overset{i_g}{\hookrightarrow} & \bar{R}_K(g+1) \\ \bar{\rho}\downarrow & & \downarrow\bar{\rho} \\ \bar{R}_{\mathrm{SU}_V}(g) & \overset{i_g}{\hookrightarrow} & \bar{R}_{\mathrm{SU}_V}(g+1), \end{array}$$

where $\bar{\rho}$ is the map defined in the proof of Proposition 9.3.4. It induces the commutative diagram:

$$\begin{array}{ccc} H^2(\bar{R}_K(g),\mathbb{Z}) & \overset{i_g^*}{\leftarrow} & H^2(\bar{R}_K(g+1),\mathbb{Z}) \\ \bar{\rho}^*\uparrow & & \bar{\rho}^*\uparrow \\ H^2(\bar{R}_{\mathrm{SU}_V}(g),\mathbb{Z}) & \overset{i_g^*}{\leftarrow} & H^2(\bar{R}_{\mathrm{SU}_V}(g+1),\mathbb{Z}). \end{array}$$

Therefore, using the commutativity of the above diagram and (1) of Proposition 9.3.4, assuming that $i_g^*(c(\Theta_V(g+1, \mathrm{SL}_V))) = c(\Theta_V(g, \mathrm{SL}_V))$, we get $i_g^*(c(\Theta_V(g+1,G))) = c(\Theta_V(g,G))$. Hence, Proposition 9.3.7 is established for any G provided we assume its validity for $G = \mathrm{SL}_V$ and its standard representation in V.

We further reduce the proposition from SL_n to SL_2. As in the proof of Proposition 9.3.4, consider the mappings

$$\bar{\rho}\colon \bar{R}_{\mathrm{SU}_2}(g) \to \bar{R}_{\mathrm{SU}_n}(g) \text{ and } \hat{\rho}\colon M^{\mathrm{SL}_2} \to M^{\mathrm{SL}_n}$$

induced by the inclusions

$$\mathrm{SU}_2 \to \mathrm{SU}_n \quad \text{and } \mathrm{SL}_2 \to \mathrm{SL}_n\,,$$

given by $m \mapsto \mathrm{diag}(m, 1, \dots, 1)$.

The maps $\bar{\rho}$ and $\hat{\rho}$ induce the commutative diagram

$$\begin{array}{ccc} H^2(M^{\mathrm{SL}_n},\mathbb{Z}) & \xrightarrow{\hat{\rho}^*} & H^2(M^{\mathrm{SL}_2},\mathbb{Z}) \\ {\scriptstyle \bar{f}(\Sigma)^*}\Big\downarrow{\scriptstyle \simeq} & & {\scriptstyle \simeq}\Big\downarrow{\scriptstyle \bar{f}(\Sigma)^*} \\ H^2(\bar{R}_{\mathrm{SU}_n}(g),\mathbb{Z}) & \xrightarrow{\bar{\rho}^*} & H^2(\bar{R}_{\mathrm{SU}_2}(g),\mathbb{Z}). \end{array}$$

By the defining property of the Θ-bundle (cf. Theorem 8.4.2 and Exercise 8.4.E.1), $\hat{\rho}^*(\Theta_n) = \Theta_2$, where Θ_n denotes the theta bundle for SL_n. Thus, using the functoriality of the Chern class, we get

$$\hat{\rho}^*(c(\Theta_n)) = c(\Theta_2). \tag{1}$$

Since $c(\Theta_n)$ is the unique positive generator of $H^2(M^{\mathrm{SL}_n},\mathbb{Z})$ for any n (cf. Exercise 9.1.E.3 and Lemma 9.3.2), we see that $\hat{\rho}^*$ is surjective and hence an isomorphism by Lemma 9.3.2. Consider the following commutative diagram:

$$\begin{array}{ccc} H^2(\bar{R}_{\mathrm{SU}_n}(g),\mathbb{Z}) & \overset{i_g^*}{\leftarrow} & H^2(\bar{R}_{\mathrm{SU}_n}(g+1),\mathbb{Z}) \\ \bar{\rho}^* \downarrow & & \bar{\rho}^* \downarrow \\ H^2(\bar{R}_{\mathrm{SU}_2}(g),\mathbb{Z}) & \overset{i_g^*}{\leftarrow} & H^2(\bar{R}_{\mathrm{SU}_2}(g+1),\mathbb{Z}). \end{array}$$

Assume the validity of the proposition for $G = \mathrm{SL}_2$ and the standard representation V_2, i.e.,

$$i_g^*(c(\Theta_{V_2}(g+1,\,\mathrm{SL}_2))) = c(\Theta_{V_2}(g,\,\mathrm{SL}_2)). \tag{2}$$

Then, using the commutativity of the above diagram and (1) together with the fact that $\bar{\rho}^*$ is an isomorphism (since so is $\hat{\rho}^*$), we get that $i_g^*(c(\Theta_V(g+1,\,\mathrm{SL}_n))) = c(\Theta_V(g,\,\mathrm{SL}_n))$. Finally, (2) follows from Proposition 9.3.6 and Exercise 9.1.E.3. Hence the proposition is established for any G (once we prove Proposition 9.3.6). □

As in Corollary A.9, let $m(\mathfrak{g})$ be the least common multiple of the numbers $\{1,a_1^\vee,a_2^\vee,\ldots,a_\ell^\vee\}$, where $a_i^\vee$ are the coefficients of the coroot $\theta^\vee$ written in terms of the simple coroots $\alpha_i^\vee$. Let ω_i be any fundamental weight such that the Dynkin index $d_{V(\omega_i)} = m(\mathfrak{g})$ (cf. Corollary A.9 for a complete list of such ω_i).

Assuming Proposition 9.3.6, we now prove the following.

Proof of Theorem 9.3.1 We first prove the theorem for $g = 1$. Let ω_i be any fundamental weight such that the Dynkin index $d_{V(\omega_i)} = m(\mathfrak{g})$. Using Theorems 9.2.3 and 9.2.4, we get that

$$\Theta(V(\omega_i)) = \mathcal{O}_{\mathbb{P}(1,a_1^\vee,a_2^\vee,\ldots,a_\ell^\vee)}(m(\mathfrak{g}))^{\otimes p} \tag{*}$$

for some positive integer p. Of course, for any $c \geq 1$,

$$D_c = \{(n_1,\ldots,n_\ell) \in (\mathbb{Z}_{\geq 0})^\ell : \sum_{i=1}^{\ell} n_i a_i^\vee \leq c\},$$

where $(n_1,\ldots,n_\ell)$ denotes the weight $\sum_{j=1}^{\ell} n_j\omega_j$.

Using Theorems 4.2.19 and 8.4.15 for $g = 1$, we see that

$$\dim H^0(M^G,\Theta(V(\omega_i))) = |D_{m(\mathfrak{g})}|.$$

On the other hand, by Theorems 9.2.3 and 9.2.4(b),

$$\dim H^0(M^G, \mathscr{O}_{\mathbb{P}(1,a_1^\vee,a_2^\vee,\ldots,a_\ell^\vee)}(m(\mathfrak{g}))^{\otimes p}) = \dim(\mathbb{C}[z_0,\ldots,z_\ell]_{pm(\mathfrak{g})}) = |D_{pm(\mathfrak{g})}|.$$

Hence, in the equation (*), $p = 1$ and $\Theta(V(\omega_i))$ is the (ample) generator of $\mathrm{Pic}(M^G)$. This proves Theorem 9.3.1 for $g = 1$.

We now come to the proof of the theorem for $g \geq 2$.

Denote the subgroup $\langle \Theta(V), V \in \mathcal{R}(G)\rangle$ of $\mathrm{Pic}(M^G)$ by $\mathrm{Pic}^\Theta(M^G)$.

Set $H^2_\Theta(\bar{R}_K(g)) := c(\mathrm{Pic}^\Theta(M^G))$. By virtue of Proposition 9.3.4, this is well defined, i.e., $H^2_\Theta(\bar{R}_K(g))$ does not depend upon the choice of the projective variety structure on Σ. Moreover, by Proposition 9.3.7, $i_g^*(H^2_\Theta(\bar{R}_K(g+1))) = H^2_\Theta(\bar{R}_K(g))$.

Thus, we get the following commutative diagram, where the upward arrows are inclusions and the maps in the bottom horizontal sequence are induced from the maps i_g^*.

$$\begin{array}{ccccccc} H^2(\bar{R}_K(1)) & \overset{i_1^*}{\leftarrow} & H^2(\bar{R}_K(2)) & \overset{i_2^*}{\leftarrow} & H^2(\bar{R}_K(3)) & \overset{i_3^*}{\leftarrow} & \cdots \\ \uparrow & & \uparrow & & \uparrow & & \\ H^2_\Theta(\bar{R}_K(1)) & \twoheadleftarrow & H^2_\Theta(\bar{R}_K(2)) & \twoheadleftarrow & H^2_\Theta(\bar{R}_K(3)) & \twoheadleftarrow & \cdots . \end{array}$$

By this theorem for $g = 1$ proved above and Lemma 9.3.2, $H^2(\bar{R}_K(1)) = H^2_\Theta(\bar{R}_K(1))$. Then, i_1^* is surjective and hence an isomorphism (by using Lemma 9.3.2 again). Thus, by the commutativity of the above diagram, the inclusion $H^2_\Theta(\bar{R}_K(2)) \hookrightarrow H^2(\bar{R}_K(2))$ is an isomorphism. Arguing the same way, we get that $H^2(\bar{R}_K(g)) = H^2_\Theta(\bar{R}_K(g))$ for all g. This completes the proof of the theorem by virtue of the isomorphism c of Lemma 9.3.2 modulo the proof of Proposition 9.3.6. □

Using Theorem 9.3.1, we get the following.

Theorem 9.3.8 *For any Σ with $g \geq 1$ and G as at the beginning of this section, the Picard group $\mathrm{Pic}(M^G)$ is freely generated by the Θ-bundle $\Theta(V(\omega_i))$, where ω_i is any fundamental weight such that $d_{V(\omega_i)} = m(\mathfrak{g})$.*

In particular, for $g \geq 2$,

$$\mathrm{Im}(\bar{f}^*) \textit{ is freely generated by } \mathscr{L}_{0_{m(\mathfrak{g})}}, \tag{1}$$

where $\bar{f}^*$ is as in Theorem 9.1.3.

Proof By Theorem 9.3.1,

$$\mathrm{Pic}(M^G) = \langle \Theta(V), V \in \mathcal{R}(G)\rangle.$$

Thus, if $g \geq 2$, by Theorems 9.1.3 and A.10,

$$\mathrm{Im}(\bar{f}^*) = \langle \mathscr{L}_{0_{d_V}}, V \in \mathcal{R}(G) \rangle = \langle \mathscr{L}_{0_{m(\mathfrak{g})}} \rangle.$$

This proves (1) (in the case $g \geq 2$).

Since $\bar{f}^*$ is injective, by the above description of $\mathrm{Im}(\bar{f}^*)$, $\Theta(V(\omega_i))$ freely generates $\mathrm{Pic}(M^G)$ in the case $g \geq 2$.

For $g = 1$, as in the proof of Theorem 9.3.1, $\Theta(V(\omega_i))$ generates $\mathrm{Pic}(M^G)$. This completes the proof of the theorem. □

Recall that an irreducible variety X is called *locally factorial variety* if all its local rings are unique factorization domains (UFD). Since any UFD is integrally closed, such an X is automatically normal (cf. (Eisenbud, 1995, Proposition 4.10)).

Theorem 9.3.9 *Let Σ and G be as at the beginning of this section (in particular $g \geq 1$). Then, the moduli space $M^G = M^G(\Sigma)$ is not locally factorial for any G of type B_ℓ ($\ell \geq 3$); D_ℓ ($\ell \geq 4$); G_2; F_4; E_6; E_7; E_8.*

Proof Let $\mathrm{Cl}(X)$ denote the class of Weil divisors modulo linear equivalence. Then there is a canonical homomorphism $\eta = \eta_X \colon \mathrm{Pic}\, X \to \mathrm{Cl}(X)$. By Hartshorne (1977, Chap. II, Proposition 6.11), if an irreducible variety X is locally factorial then the above homomorphism η is an isomorphism.

We now show that in all the cases of G as in the theorem, M^G is not locally factorial by showing that η is not surjective.

As in the proof of Theorem 9.1.6, let $\mathring{M}_s^G \subset M^G$ be a smooth open subset consisting of stable G-bundles E with $\mathrm{Aut}\, E = Z$, where Z is the center of G. Then, the morphism $f \colon \bar{X}_G^{ss} \to M^G$ (see the beginning of Section 9.1) restricts to a principal $\bar{\Gamma}/Z$-bundle $\mathring{f} \colon \mathring{\bar{X}}_G^{ss} \to \mathring{M}_s^G$, where the ind-affine group variety $\bar{\Gamma}$ is as at the beginning of Section 9.1 with $\mathbb{C}$-points $\Gamma := \mathrm{Mor}(\Sigma^*, G)$ (cf. Exercise 5.2.E.5 and the proof of Theorem 9.1.6). Consider the commutative diagram

$$\begin{array}{ccc} \mathrm{Pic}(M^G) & \xrightarrow{\eta} & \mathrm{Cl}(M^G) \\ {\scriptstyle\gamma_1}\downarrow & & \downarrow{\scriptstyle\gamma_2}\,\wr \\ \mathrm{Pic}(\mathring{M}_s^G) & \underset{\mathring{\eta}}{\xrightarrow{\sim}} & \mathrm{Cl}(\mathring{M}_s^G), \end{array}$$

where the vertical maps γ_i are the canonical restriction maps and $\eta := \eta_{M^G}, \mathring{\eta} := \mathring{\eta}_{\mathring{M}_s^G}$.

Assume now that $g \geq 2$. Then, by (3) in the proof of Theorem 9.1.6, for all $G \neq \mathrm{SL}_2$, $\mathrm{codim}_{M^G}(M^G \setminus \mathring{M}_s^G) \geq 2$. Thus, γ_2 is an isomorphism (cf. (Fulton, 1998, Proposition 1.8)). Moreover, since $\mathring{M}_s^G$ is smooth, $\mathring{\eta}$ is an isomorphism (cf. (Hartshorne, 1977, Chap. II, Remark 6.11.1A)). Now, the line bundle $\mathscr{L}(0_1)_{|\mathring{X}_G^{ss}}$ is $\bar{\Gamma}$-equivariant. Moreover, since Z acts trivially on the representation space $\mathscr{H}(0_1)$, it acts trivially on $\mathscr{L}(0_1)$. (Observe that Z is central in $\hat{G}_{0_c}$ for any $c > 0$ by Kumar (2002, Theorem 13.2.8).) Thus, $\mathscr{L}(0_1)_{|\mathring{X}_G^{ss}}$ is $\bar{\Gamma}/Z$-equivariant line bundle over the $\bar{\Gamma}/Z$-principal bundle $\mathring{f}$. Hence, $\mathscr{L}(0_1)_{|\mathring{X}_G^{ss}}$ descends to a line bundle $\bar{\mathscr{L}}(0_1)$ over $\mathring{M}_s^G$ (cf. Lemma C.17). But, by Theorem 9.3.8 and Corollary A.9, $\bar{\mathscr{L}}(0_1)$ is *not* in the image of γ_1 for any G listed in the theorem. Thus, γ_1 is not surjective and hence so is η. This proves the theorem for $g \geq 2$.

For $g = 1$, the result follows from Theorem 9.2.3 together with Beltrametti and Robbiano (1986, Theorem 7.1.d). □

Remark 9.3.10 It is known (cf. (Drezet and Narasimhan, 1989) for A_ℓ-types and (Laszlo and Sorger, 1997, §9) for A_ℓ, C_ℓ-types) that M^G is locally factorial if G is of type A_ℓ ($\ell \geq 1$) or C_ℓ ($\ell \geq 2$). Thus, local factoriality of M^G is equivalent to G being *special* in the sense of Serre, i.e., all the G-bundles are locally trivial in the Zariski topology.

We now begin the preparation to prove Proposition 9.3.6.

Definition 9.3.11 Take $G = \mathrm{SL}_2$ and $K = \mathrm{SU}_2$. Let $J(\Sigma)$ be the Jacobian of $\Sigma = \Sigma_g$. Recall that the underlying set of the variety $J(\Sigma)$ consists of all the isomorphism classes of degree 0 line bundles on Σ. Consider the morphism $\xi = \xi_\Sigma : J(\Sigma) \to M^G$, taking $\mathfrak{L} \mapsto [\mathfrak{L} \oplus \mathfrak{L}^{-1}]$ (by Exercise 6.1.E.15(b), $\mathfrak{L} \oplus \mathfrak{L}^{-1}$ is indeed a semistable vector bundle). Since $J(\Sigma)$ is a complete variety, its image is a closed subset of M^G. We denote this subset by M^G_{dec} (decomposable semistable bundles of rank-2 with trivial determinant). Since $J(\Sigma)$ is reduced, the scheme theoretic image M^G_{dec} acquires the reduced subscheme structure (cf. (Hartshorne, 1977, Chap. II, Exercise 3.11(d))). By Exercise 6.1.E.15(a), $\mathfrak{L} \oplus \mathfrak{L}^{-1}$ is polystable. Thus, by Proposition 7.2.18 for $A = (1)$, $\xi^{-1}(\xi(\mathfrak{L})) = \{\mathfrak{L}, \mathfrak{L}^{-1}\}$. (Observe that if $\mathfrak{L} \oplus \mathfrak{L}^{-1} \simeq \mathfrak{L}' \oplus \mathfrak{L}'^{-1}$, for $\mathfrak{L}, \mathfrak{L}' \in J(\Sigma)$, then using Lemma 6.3.22, we see that $\mathfrak{L} \simeq \mathfrak{L}'$ or $\mathfrak{L}^{-1} \simeq \mathfrak{L}'$.) The Jacobian $J(\Sigma)$ admits the involution τ taking $\mathfrak{L} \mapsto \mathfrak{L}^{-1}$.

Let D be the maximal torus of SU_2 consisting of the diagonal matrices. Similar to the identification $\bar{f}(\Sigma)$ as in Theorem 9.3.3, setting $\bar{J}_g := D^{2g}$ and using Exercise 6.3.E.3 together with Corollary 6.3.7, there is an isomorphism of $\mathbb{R}$-analytic spaces $\bar{f}_D(\Sigma) : \bar{J}_g \to J(\Sigma)$ making the following diagram commutative:

$$\begin{array}{ccc} \bar{J}_g & \xrightarrow[\sim]{\bar{f}_D(\Sigma)} & J(\Sigma) \\ {\scriptstyle f_g}\downarrow & & \downarrow{\scriptstyle \xi} \\ \bar{R}_K(g) & \xrightarrow[\sim]{\bar{f}(\Sigma)} & M^G, \end{array} \tag{E}$$

where $f_g((t_1,t_2),\ldots,(t_{2g-1},t_{2g})) = [(t_1,t_2),\ldots,(t_{2g-1},t_{2g})]$, for $t_i \in D$.

Recall the definition of the map $i_g\colon \bar{R}_K(g) \to \bar{R}_K(g+1)$ from Definition 9.3.5 and let $r_g\colon \bar{J}_g \to \bar{J}_{g+1}$ be the map $(t_1,\ldots,t_{2g}) \mapsto (t_1,\ldots,t_{2g},1,1)$. Then we have the following commutative diagram:

$$\begin{array}{ccc} \bar{J}_g & \xrightarrow{f_g} & \bar{R}_K(g) \\ {\scriptstyle r_g}\downarrow & & \downarrow{\scriptstyle i_g} \\ \bar{J}_{g+1} & \xrightarrow{f_{g+1}} & \bar{R}_K(g+1). \end{array} \tag{F}$$

Set $y_g := f_g^*(x_g)$ where $f_g^*\colon H^2(\bar{R}_K(g),\mathbb{Z}) \to H^2(\bar{J}_g,\mathbb{Z})$ is the map in cohomology induced from f_g and $x_g \in H^2(\bar{R}_K(g),\mathbb{Z})$ is the positive generator.

Lemma 9.3.12 *$y_g \neq 0$ and $r_g^*(y_{g+1}) = y_g$ as elements of $H^2(\bar{J}_g,\mathbb{Z})$.*

Proof There exists a tautological line bundle $\mathcal{P}$, called the Poincaré bundle on $\Sigma \times J(\Sigma)$ such that, for each $\mathfrak{L} \in J(\Sigma)$, $\mathcal{P}$ restricts to the line bundle $\mathfrak{L}$ on $\Sigma \times \mathfrak{L}$, and $\mathcal{P}$ restricted to $p \times J(\Sigma)$ is trivial for the fixed base point $p \in \Sigma$ (cf. (Arbarello et al., 1985, Chap. IV, §2)).

Let $\mathcal{F}$ be the rank-2 vector bundle $\mathcal{P} \oplus \hat{\tau}^*(\mathcal{P})$ over the base space $\Sigma \times J(\Sigma)$, and think of $\mathcal{F}$ as a family of rank-2 bundles on Σ parameterized by $J(\Sigma)$, where $\hat{\tau}\colon \Sigma \times J(\Sigma) \to \Sigma \times J(\Sigma)$ is the involution $I \times \tau$.

By definition, we have $x_g = c(\Theta(V_2))$ for the standard representation V_2 of SL_2 under the identification $\bar{f}(\Sigma)$, where, as earlier, c denotes the first Chern class. Using the functoriality of Chern class and Theorem 8.4.2,

$$\xi^*(x_g) = c(\operatorname{Det} \mathcal{F}), \tag{1}$$

where $\operatorname{Det}\mathcal{F}$ denotes the determinant line bundle over $J(\Sigma)$ associated to the family $\mathcal{F}$ (cf. Definition 8.4.1). Recall that the fiber of $\operatorname{Det}\mathcal{F}$ at any $\mathfrak{L} \in J(\Sigma)$ is given by the expression

$$\begin{aligned}
\operatorname{Det}\mathcal{F}_{|\mathfrak{L}} &= \wedge^{\text{top}}\big(H^0(\Sigma,\mathfrak{L}\oplus\mathfrak{L}^{-1})^*\big)\otimes\wedge^{\text{top}}\big(H^1(\Sigma,\mathfrak{L}\oplus\mathfrak{L}^{-1})\big)\\
&= \wedge^{\text{top}}\big(H^0(\Sigma,\mathfrak{L})^*\oplus H^0(\Sigma,\mathfrak{L}^{-1})^*\big)\otimes\wedge^{\text{top}}\big(H^1(\Sigma,\mathfrak{L})\oplus H^1(\Sigma,\mathfrak{L}^{-1})\big)\\
&= \wedge^{\text{top}}\big(H^0(\Sigma,\mathfrak{L})^*\big)\otimes\wedge^{\text{top}}\big(H^0(\Sigma,\mathfrak{L}^{-1})^*\big)\\
&\quad\otimes\wedge^{\text{top}}\big(H^1(\Sigma,\mathfrak{L})\big)\otimes\wedge^{\text{top}}(H^1(\Sigma,\mathfrak{L}^{-1}))\\
&= \big(\operatorname{Det}\mathcal{P}\big)_{|\mathfrak{L}}\otimes\big(\tau^*(\operatorname{Det}\mathcal{P})\big)_{|\mathfrak{L}}. \qquad (2)
\end{aligned}$$

Applying the Grothendieck–Riemann–Roch theorem (cf. (Fulton, 1998, Example 15.2.8)) for the projection $\Sigma\times J(\Sigma)\xrightarrow{\pi} J(\Sigma)$ gives

$$\operatorname{ch}(R\pi_*\mathcal{P}) = \pi_*(\operatorname{ch}\mathcal{P}\cdot\operatorname{Td}T_\pi), \qquad (3)$$

where ch is the Chern character and $\operatorname{Td}T_\pi$ denotes the Todd genus of the relative tangent bundle of $\Sigma\times J(\Sigma)$ along the fibers of π. By the definition of $\operatorname{Det}\mathcal{P}$ and $R\pi_*\mathcal{P}$,

$$c(\operatorname{Det}\mathcal{P}) = -\operatorname{ch}(R\pi_*\mathcal{P})_{[2]}, \qquad (4)$$

where, for a cohomology class y, $y_{[n]}$ denotes the component of y in H^n. Since $\mathcal{P}$ restricted to $p\times J(\Sigma)$ is trivial and for any $\mathfrak{L}\in J(\Sigma)$, $\mathcal{P}$ restricts to the line bundle $\mathfrak{L}$ on $\Sigma\times\mathfrak{L}$ (with the trivial Chern class), we get

$$c(\mathcal{P})\in H^1(\Sigma)\otimes H^1(J(\Sigma)). \qquad (5)$$

Thus, using (3) and (4),

$$\begin{aligned}
-c(\operatorname{Det}\mathcal{P}) &= \pi_*\left((\operatorname{ch}\mathcal{P}\cdot\operatorname{Td}T_\pi)_{[4]}\right)\\
&= \pi_*\left(\frac{c(\mathcal{P})^2}{2}+\frac{c(\mathcal{P})\cdot c(T_\pi)}{2}\right)\\
&= \pi_*\left(c(\mathcal{P})^2\right)/2. \qquad (6)
\end{aligned}$$

(For the expression of the Todd class of a vector bundle, see Fulton (1998, Example 3.2.4).) The last equality follows from (5), since the cup product $c(\mathcal{P})\cdot c(T_\pi)$ vanishes, $c(T_\pi)$ being in $H^2(\Sigma)\otimes H^0(J(\Sigma))$.

Recall the presentation of $\pi_1(\Sigma)$ given just above Theorem 9.3.3. Then, $H_1(\Sigma,\mathbb{Z}) = \oplus_{i=1}^g\mathbb{Z}a_i\oplus\oplus_{i=1}^g\mathbb{Z}b_i$ for $\Sigma=\Sigma_g$. Moreover, the $\mathbb{Z}$-module dual basis $\{a_i^*,b_i^*\}_{i=1}^g$ of $H^1(\Sigma,\mathbb{Z}) = \operatorname{Hom}_{\mathbb{Z}}(H_1(\Sigma,\mathbb{Z}),\mathbb{Z})$ satisfies $a_i^*\cdot a_j^* = 0 = b_i^*\cdot b_j^*$, $a_i^*\cdot b_j^* = \delta_{ij}[\Sigma]$, where $[\Sigma]$ denotes the positive generator of $H^2(\Sigma,\mathbb{Z})$ (cf. (Hatcher, 2001, Example 3.7)).

Having fixed a base point p in Σ as earlier, define the morphism (cf. (Hartshorne, 1977, Chap. IV, Remark 4.10.9))

$$\psi\colon \Sigma \to J(\Sigma),\ x \mapsto \mathscr{O}(x-p).$$

Of course, $J(\Sigma)$ is canonically identified as $H^1(\Sigma,\mathscr{O}_\Sigma)/H^1(\Sigma,\mathbb{Z})$. Thus, as an $\mathbb{R}$-analytic space, we can identify

$$\begin{aligned} J(\Sigma) &\simeq H^1(\Sigma,\mathbb{R})/H^1(\Sigma,\mathbb{Z}) \simeq H^1(\Sigma,\mathbb{Z})\otimes_{\mathbb{Z}}(\mathbb{R}/\mathbb{Z}) \\ &\simeq \operatorname{Hom}_{\mathbb{Z}}\big(H_1(\Sigma,\mathbb{Z}),\mathbb{R}/\mathbb{Z}\big) = \bar{J}_g \end{aligned} \tag{7}$$

obtained from the $\mathbb{R}$-vector space isomorphism (cf. Proposition 6.3.12 for a more general result)

$$H^1(\Sigma,\mathbb{R}) \simeq H^1(\Sigma,\mathscr{O}_\Sigma),$$

induced from the inclusion $\mathbb{R} \subset \mathscr{O}_\Sigma$, where the last equality in (7) follows by using the basis $\{a_1,b_1,\dots,a_g,b_g\}$ of $H_1(\Sigma,\mathbb{Z})$. This identification $\bar{J}_g \simeq J(\Sigma)$ is our identification $\bar{f}_D(\Sigma)$ used in Diagram (E). The induced map, under the identification (7),

$$\psi_*\colon H_1(\Sigma,\mathbb{Z}) \to H_1(J(\Sigma),\mathbb{Z}) \simeq H^1(\Sigma,\mathbb{Z})$$

is the Poincaré duality isomorphism. To see this, identify

$$\operatorname{Hom}_{\mathbb{Z}}\big(H_1(\Sigma,\mathbb{Z}),\mathbb{R}/\mathbb{Z}\big) \simeq \operatorname{Hom}_{\mathbb{Z}}\big(H^1(\Sigma,\mathbb{Z}),\mathbb{R}/\mathbb{Z}\big) \tag{8}$$

using the Poincaré duality isomorphism: $H_1(\Sigma,\mathbb{Z}) \simeq H^1(\Sigma,\mathbb{Z})$. Then, under the identifications (7) and (8), the map

$$\psi\colon \Sigma \to \operatorname{Hom}_{\mathbb{Z}}\big(H^1(\Sigma,\mathbb{Z}),\mathbb{R}/\mathbb{Z}\big)$$

can be described as

$$\psi(x)([\omega]) = e^{2\pi i \int_p^x \omega},$$

for any closed 1-form ω on Σ representing the cohomology class $[\omega] \in H^1(\Sigma,\mathbb{Z})$ (cf. (Milne, 1986, Theorem 2.5)), where $\int_p^x \omega$ denotes the integral of ω along any path in Σ from p to x.

Since

$$\psi_*\colon H_1(\Sigma,\mathbb{Z}) \to H_1(J(\Sigma),\mathbb{Z}) \simeq H^1(\Sigma,\mathbb{Z})$$

is the Poincaré duality isomorphism, it is easy to see that the induced cohomology map

$$\psi^*\colon H^1(J(\Sigma),\mathbb{Z}) \simeq H_1(\Sigma,\mathbb{Z}) \to H^1(\Sigma,\mathbb{Z})$$

is given by

$$\psi^*(a_i) = b_i^*, \ \psi^*(b_i) = -a_i^* \quad \text{for all } 1 \le i \le g. \tag{9}$$

In particular, ψ^* is an isomorphism. Moreover, the isomorphism does not depend on the choice of p.

Consider the map

$$\Sigma \times \Sigma \overset{I \times \psi}{\to} \Sigma \times J(\Sigma).$$

Let $\mathcal{P}' := (I \times \psi)^*(\mathcal{P})$. Then, by Hartshorne (1977, Chap. III, Exercise 12.4), $\mathcal{P}'$ is the unique line bundle over $\Sigma \times \Sigma$ satisfying the following properties:

$$\mathcal{P}'|_{\Sigma \times x} = \mathcal{O}(x - p) \ \text{ and } \ \mathcal{P}'|_{p \times \Sigma} \text{ is trivial.}$$

Consider the following line bundle over $\Sigma \times \Sigma$:

$$\mathcal{O}_{\Sigma \times \Sigma}(\Delta) \otimes (\mathcal{O}(-p) \boxtimes 1) \otimes (1 \boxtimes \mathcal{O}(-p)),$$

where Δ denotes the diagonal in $\Sigma \times \Sigma$. Clearly, this bundle also satisfies the restriction properties mentioned above and hence it must be isomorphic with $\mathcal{P}'$. Consequently,

$$c(\mathcal{P}') = c(\mathcal{O}_{\Sigma \times \Sigma}(\Delta)) + c(\mathcal{O}(-p) \boxtimes 1) + c(1 \boxtimes \mathcal{O}(-p)).$$

Using the definition of $\mathcal{P}'$ and the functoriality of the Chern classes,

$$c(\mathcal{P}') = c((I \times \psi)^*(\mathcal{P})) = (I \times \psi)^* c(\mathcal{P}). \tag{10}$$

By (5), $c(\mathcal{P}) \in H^1(\Sigma) \otimes H^1(J(\Sigma))$, and hence $c(\mathcal{P}') \in H^1(\Sigma) \otimes H^1(\Sigma)$. Moreover,

$$c(\mathcal{O}(-p) \boxtimes 1) + c(1 \boxtimes \mathcal{O}(-p)) \in \left(H^2(\Sigma) \otimes H^0(\Sigma)\right) \oplus \left(H^0(\Sigma) \otimes H^2(\Sigma)\right).$$

Thus, $c(\mathcal{P}')$ is the component of $c(\mathcal{O}_{\Sigma \times \Sigma}(\Delta))$ in $H^1(\Sigma) \otimes H^1(\Sigma)$. Hence, by Milnor and Stasheff (1974, Theorem 11.11),

$$c(\mathcal{P}') = -\sum_{i=1}^{g} a_i^* \otimes b_i^* + \sum_{i=1}^{g} b_i^* \otimes a_i^*.$$

Therefore, by (10),

$$c(\mathcal{P}) = -\sum_{i=1}^{g} a_i^* \otimes \psi^{*-1}(b_i^*) + \sum_{i=1}^{g} b_i^* \otimes \psi^{*-1}(a_i^*),$$

and thus, by (6),

$$\begin{aligned} c(\operatorname{Det}\mathcal{P}) &= -\frac{1}{2}\pi_*(c(\mathcal{P})^2) \\ &= -\frac{1}{2}\pi_*\Big(\big(-\sum_{i=1}^{g} a_i^* \otimes \psi^{*-1}(b_i^*) + \sum_{i=1}^{g} b_i^* \otimes \psi^{*-1}(a_i^*)\big)^2\Big) \\ &= -\frac{1}{2}\pi_*\left(\sum_{i=1}^{g} a_i^* \cdot b_i^* \otimes \psi^{*-1}(b_i^*) \cdot \psi^{*-1}(a_i^*)\right. \\ &\qquad \left. + \sum_{i=1}^{g} b_i^* \cdot a_i^* \otimes \psi^{*-1}(a_i^*) \cdot \psi^{*-1}(b_i^*)\right) \\ &= -\sum_{i=1}^{g} \psi^{*-1}(b_i^*) \cdot \psi^{*-1}(a_i^*) \quad \in H^2(J(\Sigma), \mathbb{Z}). \end{aligned} \tag{11}$$

Now, the involution τ of $J(\Sigma)$ induces the map $-I$ on $H^1(J(\Sigma), \mathbb{Z})$ (since, under the identification $\bar{f}_D(\Sigma)\colon \bar{J}_g \to J(\Sigma)$, τ corresponds to the map $x \mapsto x^{-1}$ for $x \in \bar{J}_g$). Therefore,

$$\tau^*(c(\operatorname{Det}\mathcal{P})) = c(\operatorname{Det}\mathcal{P}).$$

Hence, by the identities (1), (2) and (11),

$$\begin{aligned} \xi^*(x_g) &= c(\operatorname{Det}\mathcal{F}) \\ &= 2c(\operatorname{Det}\mathcal{P}) \\ &= 2\sum_{i=1}^{g} \psi^{*-1}(a_i^*) \cdot \psi^{*-1}(b_i^*), \end{aligned} \tag{12}$$

which is clearly a nonvanishing class in $H^2(J(\Sigma), \mathbb{Z})$. Moreover, for any $g \geq 2$, under the identification (7), the map $r_{g-1}\colon \bar{J}_{g-1} \to \bar{J}_g$ corresponds to the map $H_1(\Sigma_g, \mathbb{Z}) \to H_1(\Sigma_{g-1}, \mathbb{Z}), a_i \mapsto a_i, b_i \mapsto b_i$ for $1 \leq i \leq g-1, a_g \mapsto 0, b_g \mapsto 0$. Thus, by (9) and (12), under the identification $\bar{f}_D(\Sigma)$, $\xi^*(x_g)$ restricts, via r_{g-1}^*, to the class $\xi^*(x_{g-1})$ for any $g \geq 2$. But, by the commutative diagram (E), under the identifications $\bar{f}_D(\Sigma)$ and $\bar{f}(\Sigma)$, $\xi^*(x_g) = y_g$. This proves Lemma 9.3.12. □

Proof of Proposition 9.3.6 Let $i_g^*(x_{g+1}) = d_g x_g$, for some $d_g \in \mathbb{Z}$. By the above Lemma 9.3.12 and the commutative diagram (F), we see that

$$f_g^*(d_g x_g) = f_g^* i_g^*(x_{g+1}) = r_g^*(f_{g+1}^*(x_{g+1})), \text{ i.e., } d_g y_g = y_g.$$

Since the cohomology of $\bar{J}_g$ is torsion free and y_g is a nonvanishing class, we get $d_g = 1$. This concludes the proof of Proposition 9.3.6. □

9.C Comments

Lemmas 9.1.1 and 9.1.9; Theorems 9.1.3 and 9.1.6; and Proposition 9.1.12 are taken from Kumar and Narasimhan (1997).

Theorems 9.3.1, 9.3.8 and Lemma 9.3.2 are taken from Boysal and Kumar (2005, Theorems 1.3, 2.4 and Lemma 1.4). We have followed their proof closely in Section 9.3 (including that of Propositions 9.3.4, 9.3.6, 9.3.7 and Lemma 9.3.12). Theorem 9.3.9 is obtained in Laszlo and Sorger (1997) for B_ℓ, D_ℓ types; in Beauville, Laszlo and Sorger (1998, Proposition 13.2) for G of types $B_\ell(\ell \geq 3), D_\ell(\ell \geq 4), F_4$ and G_2; and for other exceptional groups in Sorger (1999a). It is also proved in Beauville, Laszlo and Sorger (1998, Propositions 13.5, 13.6) that if G is semisimple but not simply connected, then for any $\delta \in \pi_1(G)$, the moduli space $M_\delta^G(\Sigma)$ of semistable G-bundles corresponding to the component δ is *not* locally factorial but always Gorenstein.

Let G be a connected semisimple group, Σ a smooth irreducible projective curve of any genus $g \geq 1$ and $\delta \in \pi_1(G)$. In this case the Picard group $\mathrm{Pic}(M_\delta^G(\Sigma))$ as well as $\mathrm{Pic}(\mathbf{Bun}_G^\delta(\Sigma))$ is determined in Beauville, Laszlo and Sorger (1998) for non-simply-connected classical G as well as G_2, where $\mathbf{Bun}_G^\delta(\Sigma)$ is the moduli stack of G-bundles of topological type δ over Σ. There is a discussion of the Pfaffian bundle in Laszlo and Sorger (1997).

Choose three integers r, s, d such that $r \geq 2, s|r|ds$ and let G be the group SL_r/μ_s (μ_s being the subgroup of the center of SL_r consisting of the sth roots of unity). Let (Σ, p) be a pointed smooth irreducible projective curve of genus $g \geq 1$ and let $\mathbf{Bun}_G(d)$ be the moduli stack of G-bundles over Σ of degree $\exp\left(\frac{2\pi i d}{r}\right)$ and let $\mathbf{Bun}_{\mathrm{SL}_r}(d)$ be the moduli stack of rank-r vector bundles over Σ with determinant $\mathcal{O}_\Sigma(dp)$. Then, as proved in Laszlo (1997), the natural morphism $\mathrm{Pic}(\mathbf{Bun}_G(d)) \to \mathrm{Pic}(\mathbf{Bun}_{\mathrm{SL}_r}(d)) \simeq \mathbb{Z}$ is surjective with kernel isomorphic to $H^1_{et}(\Sigma, \mathbb{Z}/d\mathbb{Z}) \simeq (\mathbb{Z}/d\mathbb{Z})^{2g}$.

As proved in Mehta and Ramadas (1996), for a generic smooth irreducible projective curve Σ, the moduli space $M_{\mathrm{par}}(2,d)$ (resp. $(M_{\mathrm{par}}(2,\mathcal{L}))$ of parabolic semistable bundles of rank 2 over Σ of fixed degree d (resp. fixed determinant $\mathcal{L}$) is Frobenius split for any prime $p \geq 5$. Further, Sun and Zhou (2018) proved that $M_{\mathrm{par}}(r,\mathcal{L})$ is of globally F-regular type for any positive integer r. They also proved these results when Σ has a single node. These results give rise to the higher cohomology vanishing of ample line bundles, Factorization Theorem for the generalized θ-functions (without going through the conformal blocks) and the Verlinde formula over $M_{\mathrm{par}}(r,d)$ (cf. (Narasimhan and Ramadas, 1993), (Mehta and Ramadas, 1996), (Sun, 2000, 2003), (Su and Zhou, 2018)). The paper Narasimhan and Ramadas (1993) does not require char. p methods.

Appendix A

Dynkin Index

Introduction. For a Lie algebra homomorphism $f : \mathfrak{g}_1 \to \mathfrak{g}_2$ between simple Lie algebras, the *Dynkin index* of f is defined to be the unique complex number d_f satisfying

$$\langle f(x), f(y)\rangle = d_f \langle x, y\rangle, \quad \text{for all } x, y \in \mathfrak{g}_1,$$

where $\langle \cdot, \cdot \rangle$ is the normalized invariant bilinear form on $\mathfrak{g}_i$. For a finite-dimensional representation V of a simple Lie algebra $\mathfrak{g}$, its Dynkin index is defined to be d_{f_V}, where $f_V : \mathfrak{g} \to sl_V$ is the induced Lie algebra homomorphism. Let $V(\lambda)$ be an irreducible finite-dimensional representation of $\mathfrak{g}$ with highest weight λ. Then, by Lemma A.2,

$$d_{V(\lambda)} = \left(\|\lambda + \rho\|^2 - \|\rho\|^2\right) \frac{\dim_{\mathbb{C}} V(\lambda)}{\dim_{\mathbb{C}} \mathfrak{g}}.$$

We give two other expressions of d_V for any finite-dimensional representation V of $\mathfrak{g}$, one in terms of the formal character of V (cf. Lemma A.3); and the other in terms of the decomposition of V as a module for $sl_2 = sl_2(\theta)$ passing through the highest root space of $\mathfrak{g}$ (cf. Lemma A.6). Using Lemma A.3, we determine the Dynkin index of the adjoint representation of $\mathfrak{g}$ and show it to be equal to 2 times the dual Coxeter number of $\mathfrak{g}$ (cf. Example A.7). In Proposition A.8, we determine explicitly the Dynkin index of all the fundamental representations of each simple Lie algebra $\mathfrak{g}$. As a consequence of this calculation and a formula for the Dynkin index of the tensor product $V \otimes W$ (cf. Corollary A.4), one easily obtains that Dynkin index of any representation is a non-negative integer divisible by $m(\mathfrak{g})$ (cf. Theorem A.10), where $m(\mathfrak{g})$ is the least common multiple of the coefficients of $\theta^\vee$ written in terms of the simple coroots and θ is the highest root of $\mathfrak{g}$ (cf. Corollary A.9 for the values of $m(\mathfrak{g})$).

Let G be the connected, simply-connected complex algebraic group with Lie algebra $\mathfrak{g}$. The Lie subalgebra $sl_2(\theta) \subset \mathfrak{g}$ gives rise to an algebraic group homomorphism $\gamma_\theta \colon \mathrm{SL}_2 \to G$. Then, by a result of Bott–Samelson, the induced map in singular homology with integral coefficients

$$\gamma_{\theta *} \colon H_3(\mathrm{SL}_2, \mathbb{Z}) \to H_3(G, \mathbb{Z})$$

is an isomorphism (cf. Theorem A.14). We give two different proofs of this theorem. The first proof uses the topology of the infinite Grassmannian $\bar{X}_G$; specifically, identifying the second singular homology of $\bar{X}_G^{an}$ with the second homology of a certain copy of $\mathbb{P}^1$ induced from a morphism of SL_2 in the loop group $G((t))$. The second proof is based on the study of minimal geodesics in the maximal compact subgroup K of G and certain results from Morse Theory.

We make a 'functorial' choice of a generator ω_G of $H^3(G, \mathbb{Z}) \simeq \mathbb{Z}$ for any simple, simply-connected algebraic group G (cf. Definition A.16). Using the above result of Bott–Samelson, we get the following topological characterization of the Dynkin index (cf. Theorem A.17).

Let $f \colon \mathfrak{g}_1 \to \mathfrak{g}_2$ be a Lie algebra homomorphism between simple Lie algebras and let $\bar{f} \colon G_1 \to G_2$ be the associated algebraic group homomorphism, where G_i is the connected, simply-connected complex algebraic group with Lie algebra $\mathfrak{g}_i$ ($i = 1, 2$). Then, the induced map $\bar{f}^* \colon H^3(G_2, \mathbb{Z}) \to H^3(G_1, \mathbb{Z})$ is the multiplication by the Dynkin index d_f. In particular, d_f is a non-negative integer.

Definition A.1 Let $f \colon \mathfrak{g}_1 \to \mathfrak{g}_2$ be a Lie algebra homomorphism between (finite-dimensional) simple Lie algebras over $\mathbb{C}$. Then, there exists a unique number $d_f \in \mathbb{C}$, called the *Dynkin index* of f, satisfying

$$\langle f(x), f(y) \rangle = d_f \langle x, y \rangle, \quad \text{for all } x, y \in \mathfrak{g}_1,$$

where $\langle \cdot, \cdot \rangle$ is the (nondegenerate) invariant (symmetric) bilinear form on $\mathfrak{g}_i$ normalized so that $\langle \theta_i, \theta_i \rangle = 2$ for the highest root θ_i of $\mathfrak{g}_i$, $1 \le i \le 2$. (The existence and uniqueness of d_f is easy to see.) Let G_i be the connected, simply-connected complex algebraic group with Lie algebra $\mathfrak{g}_i$ and choose maximal compact subgroup $K_i \subset G_i$ such that $f(\mathfrak{k}_1) \subset \mathfrak{k}_2$, where $\mathfrak{k}_i$ is the Lie algebra of K_i. Taking $x = y \in \mathfrak{k}_1$ (since $\langle x, x \rangle < 0$ and also $\langle f(x), f(x) \rangle < 0$ for such an x if f is nonzero), we get

$$d_f \in \mathbb{R}_{\ge 0}.$$

If $g \colon \mathfrak{g}_2 \to \mathfrak{g}_3$ is another Lie algebra homomorphism (where $\mathfrak{g}_3$ is a simple Lie algebra as well), then

$$d_{g \circ f} = d_f \cdot d_g. \tag{1}$$

For a finite-dimensional representation V of $\mathfrak{g}_1$ given by a Lie algebra homomorphism $f_V : \mathfrak{g}_1 \to sl_V$, where sl_V is the Lie algebra of traceless endomorphisms of V, we define the Dynkin index d_V of V by

$$d_V = d_{f_V}. \tag{2}$$

Clearly, for two representations V_1 and V_2 of $\mathfrak{g}_1$,

$$d_{V_1 \oplus V_2} = d_{V_1} + d_{V_2}. \tag{3}$$

Lemma A.2 *Let $\mathfrak{g}$ be a (finite-dimensional) simple Lie algebra and let $V(\lambda)$ be an irreducible finite-dimensional representation of $\mathfrak{g}$ with highest weight λ. Then,*

$$d_{V(\lambda)} = \left(\|\lambda + \rho\|^2 - \|\rho\|^2 \right) \frac{\dim_{\mathbb{C}} V(\lambda)}{\dim_{\mathbb{C}} \mathfrak{g}},$$

where $\|\mu\|^2$ denotes $\langle \mu, \mu \rangle$ and 2ρ is the sum of all the positive roots.

In particular, $d_{V(\lambda)}$ is a strictly positive real number for any $\lambda \neq 0$. In fact, we will see later in this appendix that $d_{V(\lambda)}$ is a positive integer.

Proof It is easy to see that the invariant normalized bilinear form in sl_V, for any finite-dimensional vector space V, is given by

$$\langle A, B \rangle = \operatorname{trace}(AB), \quad \text{for } A, B \in sl_V. \tag{1}$$

Let $f = f_V : \mathfrak{g} \to sl_V$ be the Lie algebra homomorphism induced from the representation $V = V(\lambda)$. Thus, by definition,

$$d_V \langle x, y \rangle = \operatorname{trace}\big(f(x) f(y)\big), \quad \text{for all } x, y \in \mathfrak{g}. \tag{2}$$

Choose any basis $\{e_i\}$ of $\mathfrak{g}$ and let $\{e^i\}$ be the dual basis with respect to the normalized form $\langle\ ,\ \rangle$ on $\mathfrak{g}$. Consider the Casimir element

$$\Omega = \sum e_i e^i \in U(\mathfrak{g}).$$

Then, Ω acts on V via

$$\Omega_V = \sum_i f(e_i) f(e^i) \in \operatorname{End}(V). \tag{3}$$

But, V being irreducible of highest weight λ,

$$\Omega_V = \left(\|\lambda + \rho\|^2 - \|\rho\|^2 \right) \mathrm{I}_V, \tag{4}$$

where I_V is the identity operator of V (cf. (Goodman and Wallach, 2009, Lemma 3.3.8)).

Combining (2)–(4), we get

$$\begin{aligned}\left(\dim_{\mathbb{C}} \mathfrak{g}\right) d_V &= \sum_i \operatorname{trace}\left(f(e_i)\, f(e^i)\right)\\ &= \operatorname{trace}(\Omega_V)\\ &= \left(\|\lambda+\rho\|^2 - \|\rho\|^2\right) \dim_{\mathbb{C}} V.\end{aligned}$$

This proves the lemma. □

Lemma A.3 *Let $\mathfrak{g}$ be as in Lemma A.2 and let V be a finite-dimensional representation of $\mathfrak{g}$ with its formal character ch V given by*

$$\operatorname{ch} V = \sum_{\lambda \in \mathfrak{h}^*} n_\lambda\, e^\lambda, \quad n_\lambda \in \mathbb{Z}_{\geq 0}, \tag{1}$$

where $\mathfrak{h} \subset \mathfrak{g}$ is a Cartan subalgebra. Then

$$d_V = \frac{1}{2} \sum_\lambda n_\lambda \lambda(\theta^\vee)^2, \tag{2}$$

where $\theta^\vee \in \mathfrak{h}$ is the coroot associated to the highest root θ of $\mathfrak{g}$.

Proof Let $f = f_V : \mathfrak{g} \to sl_V$ be the Lie algebra homomorphism induced from the representation V. Then, by identity (2) of Lemma A.2,

$$d_V \langle x, y\rangle = \operatorname{trace}\big(f(x)\, f(y)\big), \quad \text{for all } x, y \in \mathfrak{g}.$$

Taking $x = y = \theta^\vee$ in the above, we get

$$\begin{aligned}2d_V &= \operatorname{trace}\big(f(\theta^\vee)^2\big)\\ &= \sum_\lambda n_\lambda \lambda(\theta^\vee)^2, \quad \text{by (1)}.\end{aligned}$$

This proves (2). □

As a consequence of the above lemma, we get the following.

Corollary A.4 *Let $\mathfrak{g}$ be as in Lemma A.2 and let V and W be two finite-dimensional representations of $\mathfrak{g}$. Then*

$$d_{V \otimes W} = d_V \dim W + d_W \dim V.$$

Proof Let $\mathrm{ch}\, V = \sum_{\lambda \in \mathfrak{h}^*} n_\lambda e^\lambda$ and $\mathrm{ch}\, W = \sum_{\lambda \in \mathfrak{h}^*} m_\lambda e^\lambda$, for some $n_\lambda, m_\lambda \in \mathbb{Z}_{\geq 0}$. Then, of course,

$$\mathrm{ch}(V \otimes W) = \sum_{\lambda, \mu} n_\lambda m_\mu e^{\lambda + \mu}.$$

Hence, by Lemma A.3,

$$\begin{aligned} d_{V \otimes W} &= \frac{1}{2} \sum_{\lambda, \mu} n_\lambda m_\mu \left((\lambda + \mu)(\theta^\vee) \right)^2 \\ &= \frac{1}{2} \sum_{\lambda, \mu} n_\lambda m_\mu \big(\lambda(\theta^\vee)^2 + \mu(\theta^\vee)^2 + 2\lambda(\theta^\vee)\mu(\theta^\vee) \big) \\ &= d_V \Big(\sum_\mu m_\mu \Big) + d_W \Big(\sum_\lambda n_\lambda \Big) + \Big(\sum_\lambda n_\lambda \, \lambda(\theta^\vee) \Big) \Big(\sum_\mu m_\mu \mu(\theta^\vee) \Big) \\ &= d_V (\dim W) + d_W (\dim V) + \Big(\sum_\lambda n_\lambda \lambda(\theta^\vee) \Big) \Big(\sum_\mu m_\mu \mu(\theta^\vee) \Big). \end{aligned}$$

But $\sum_\lambda n_\lambda \lambda(\theta^\vee) = \mathrm{tr}\,(f_V(\theta^\vee)) = 0$, thus the last term of the above sum is zero. This proves the corollary. □

Definition A.5 Take any nonzero $x_\theta \in \mathfrak{g}_\theta$ and let $y_\theta \in \mathfrak{g}_{-\theta}$ be such that $\langle x_\theta, y_\theta \rangle = 1$. Following Definition 8.4.5, define the Lie algebra embedding $\phi = \dot{\gamma}_\theta : sl_2 \to \mathfrak{g}$ by

$$\phi(X) = x_\theta, \quad \phi(Y) = y_\theta, \quad \phi(H) = \theta^\vee.$$

Then, from the definition of d_ϕ, it is easy to see that

$$d_\phi = 1. \tag{1}$$

We use this embedding ϕ to give another expression for d_V.

Lemma A.6 *Let $\mathfrak{g}$ be as in Lemma A.2 and let V be a finite-dimensional representation of $\mathfrak{g}$. Decompose V as an sl_2-module via ϕ:*

$$V = \oplus_{i \geq 1} m_i V_i,$$

where V_i is the irreducible sl_2-module of dimension i and m_i is its multiplicity in V. Then

$$d_V = \sum_{i \geq 2} m_i \binom{i+1}{3}. \tag{1}$$

Proof By virtue of identity (1) of Definition A.5, and identities (1) and (3) of Definition A.1, it suffices to prove that for the sl_2-module V_i,

$$d_{V_i} = \binom{i+1}{3}, \quad \text{where we interpret } \binom{2}{3} = 0. \tag{2}$$

But this follows immediately from Lemma A.2. □

In the following, we calculate d_{ad} for the adjoint representation ad of $\mathfrak{g}$.

Example A.7 We have the root space decomposition

$$\mathfrak{g} = \mathfrak{h} \oplus \oplus_{\alpha \in \Delta} \mathfrak{g}_\alpha,$$

where Δ is the set of all the roots of $\mathfrak{g}$ and $\mathfrak{g}_\alpha$ is the root space corresponding to the root α. Thus,

$$\text{ch(ad)} = \dim \mathfrak{h}\, e^0 + \sum_{\alpha \in \Delta} e^\alpha.$$

Hence, by Lemma A.3,

$$\begin{aligned} d_{\text{ad}} &= \frac{1}{2} \sum_{\alpha \in \Delta} \alpha(\theta^\vee)^2 \\ &= \sum_{\alpha \in \Delta^+} \alpha(\theta^\vee)^2, \quad \text{where } \Delta^+ \text{ is the set of positive roots} \\ &= 4 + \sum_{\alpha \in \Delta^+ \backslash \{\theta\}} \alpha(\theta^\vee)^2. \end{aligned}$$

But, by Bourbaki (2002, p. 294), for any $\alpha \in \Delta^+ \backslash \{\theta\}$, $\alpha(\theta^\vee) \in \{0, 1\}$. In particular, $\alpha(\theta^\vee)^2 = \alpha(\theta^\vee)$. Thus,

$$\begin{aligned} d_{\text{ad}} &= 4 + \Big(\sum_{\alpha \in \Delta^+ \backslash \{\theta\}} \alpha \Big)(\theta^\vee) \\ &= 4 + (2\rho - \theta)(\theta^\vee) \\ &= 2 + 2\rho(\theta^\vee) \\ d_{\text{ad}} &= 2(1 + \rho(\theta^\vee)). \end{aligned} \tag{1}$$

The number $1 + \rho(\theta^\vee)$ is called the *dual Coxeter number* of $\mathfrak{g}$. Its values are given as below.

Type of $\mathfrak{g}$	Dual Coxeter number
A_ℓ $(\ell \geq 1)$	$\ell+1$
B_ℓ $(\ell \geq 3)$	$2\ell-1$
C_ℓ $(\ell \geq 2)$	$\ell+1$
D_ℓ $(\ell \geq 4)$	$2\ell-2$
G_2	4
F_4	9
E_6	12
E_7	18
E_8	30

We give below the values of $d_i = d_{V(\omega_i)}$ for all the fundamental weights $\{\omega_1, \ldots, \omega_\ell\}$ of any simple Lie algebra $\mathfrak{g}$. These values d_i are obtained by using Lemma A.2. To calculate $\|\omega_i+\rho\|^2 - \|\rho\|^2$ and read off $\dim \mathfrak{g}$ we can use, e.g., (Bourbaki, 2002, Plates I–IX). To calculate $\dim V(\omega_i)$ for classical $\mathfrak{g}$ (i.e., $\mathfrak{g}$ of type A_ℓ, B_ℓ, C_ℓ or D_ℓ), we can use, e.g., (Bröcker and tom Dieck, 1985, Chap. VI, §§5, 6). To read off $\dim V(\omega_i)$ for exceptional $\mathfrak{g}$ (i.e., $\mathfrak{g}$ of type G_2, F_4, E_6, E_7 and E_8), we can use the tables (Bremner, Moody and Patera, 1985).

In the following, we interpret $\binom{n}{m} = 0$ for $n > 0$ and $m < 0$. We follow the indexing convention as in Bourbaki (2002, Plates I–IX).

Proposition A.8 *The values of d_i for any simple Lie algebra $\mathfrak{g}$ (by their type) are given as follows.*

(1) A_ℓ $(\ell \geq 1) : d_i = \binom{\ell-1}{i-1}$ *for* $1 \leq i \leq \ell$
(2) B_ℓ $(\ell \geq 3) : d_i = 2\binom{2\ell-1}{i-1}$ *for* $1 \leq i \leq \ell-1$ *and* $d_\ell = 2^{\ell-2}$
(3) C_ℓ $(\ell \geq 2) : d_i = \binom{2\ell-2}{i-1} - \binom{2\ell-2}{i-3}$ *for* $1 \leq i \leq \ell$
(4) D_ℓ $(\ell \geq 4) : d_i = 2\binom{2\ell-2}{i-1}$ *for* $1 \leq i \leq \ell-2$ *and* $d_{\ell-1} = d_\ell = 2^{\ell-3}$
(5) G_2 *: 2, 8 respectively*
(6) F_4 *: 18, 882, 126, 6, respectively*
(7) E_6 *: 6, 24, 150, 1800, 150, 6, respectively*
(8) E_7 *: 36, 360, 4680, 297000, 17160, 648, 12, respectively*

(9) E_8 : *1500, 85500, 5292000, 8345660400, 141605100, 1778400, 14700, 60, respectively.*

Express $\theta^\vee$ as a sum of simple coroots $\theta^\vee = \sum m_i \alpha_i^\vee$ and let $m(\mathfrak{g})$ be the least common multiple of $\{m_i\}_{1 \le i \le \ell}$.

The following corollary follows immediately from the above proposition. (For the C_ℓ case use Exercise A.E.1.)

Corollary A.9 *For any simple Lie algebra $\mathfrak{g}$, the following is the complete list of fundamental representations $V(\omega_i)$ such that $m(\mathfrak{g}) = d_{V(\omega_i)}$.*

Type of $\mathfrak{g}$	$m(\mathfrak{g})$	*list of* ω_i *such that* $d_{V(\omega_i)} = m(\mathfrak{g})$
A_ℓ $(\ell \ge 1)$	*1*	ω_1, ω_ℓ
B_ℓ $(\ell \ge 3)$	2	ω_1 *(and also* ω_3 *if* $\ell = 3$*)*
C_ℓ $(\ell \ge 2)$	*1*	ω_1
D_ℓ $(\ell \ge 4)$	2	ω_1 *(and also* ω_3 *and* ω_4 *if* $\ell = 4$*)*
G_2	2	ω_1
F_4	6	ω_4
E_6	6	ω_1, ω_6
E_7	*12*	ω_7
E_8	*60*	ω_8

Theorem A.10 *Let $\mathfrak{g}$ be a simple Lie algebra. For any finite-dimensional representation V of $\mathfrak{g}$, the Dynkin index d_V is a non-negative integer divisible by $m(\mathfrak{g})$.*

(In the last corollary it is shown that, for any $\mathfrak{g}$, there exists a fundamental representation $V(\omega_i)$ of $\mathfrak{g}$ such that $d_{V(\omega_i)} = m(\mathfrak{g})$. In fact, the full list of such ω_i is given there. See also Exercise A.E.2.)

Proof From Proposition A.8, we see that $d_{V(\omega_i)} \in \mathbb{Z}$ for all the fundamental representations $V(\omega_i)$. Next, recall that the representation ring $R(\mathfrak{g}) := \bigoplus_{\lambda \in D} \mathbb{Z}[V(\lambda)]$ of $\mathfrak{g}$ is freely generated by the fundamental representations $\{V(\omega_i)\}_{1 \le i \le \ell}$ as a commutative algebra over $\mathbb{Z}$, where the multiplication in $R(\mathfrak{g})$ comes from the tensor product of representations (cf. (Bröcker and tom Dieck, 1985, Chap. VI, §2)). In particular, any representation V of $\mathfrak{g}$ can be written as a $\mathbb{Z}$-linear combination of products of fundamental representations. Thus, by Corollary A.4 and (3) of Definition A.1, we get that $d_V \in \mathbb{Z}$. Moreover, by Definition A.1, $d_V \in \mathbb{R}_{\ge 0}$. Hence, $d_V \in \mathbb{Z}_{\ge 0}$. This proves the first part of the theorem.

To prove that $m(\mathfrak{g})$ divides d_V, by the preceding argument, it suffices to show that $m(\mathfrak{g})$ divides $d_{V(\omega_i)}$ for each fundamental weight ω_i. But this follows case-by-case from Proposition A.8. □

Remark A.11 (a) A uniform geometric proof for the assertion $d_V \in \mathbb{Z}$ is given below (cf. Theorem A.17).

(b) The proof given here, for the assertion that $m(\mathfrak{g})$ divides d_V, is case-by-case. I do not know any uniform proof for this assertion, which of course would be desirable.

(c) Observe that the set of fundamental weights ω_i such that $d_{V(\omega_i)} = m(\mathfrak{g})$ forms a single orbit under the action of the Dynkin automorphisms of $\mathfrak{g}$, except only in the case of B_3 where ω_1 and ω_3 are not in the same orbit under the Dynkin automorphisms. Of course, the Dynkin automorphism group is trivial in this case.

Lemma A.12 *Let $f: \mathfrak{g}_1 \to \mathfrak{g}_2$ be a Lie algebra homomorphism between simple Lie algebras. For any $\eta \in \mathbb{C}$, consider the extension of f to $\hat{f}_\eta: \hat{\mathfrak{g}}_1 \to \hat{\mathfrak{g}}_2$ defined by $\hat{f}_\eta(x[P]) = f(x)[P]$, for $x \in \mathfrak{g}_1$ and $P \in K$, and $\hat{f}_\eta(C_1) = \eta C_2$, where $\hat{\mathfrak{g}}_i$, K and C_i are as in Definition 1.2.1.*

Then $\hat{f}_\eta$ is a Lie algebra homomorphism if and only if $\eta = d_f$.

Proof For $P, P' \in K, x, x' \in \mathfrak{g}_1$ and $\lambda, \lambda' \in \mathbb{C}$,

$$\hat{f}_\eta\left(\left[x[P] + \lambda C_1, x'[P'] + \lambda' C_1\right]\right) = f([x,x'])[PP'] + \operatorname{Res}_{t=0}(P'dP)\langle x, x'\rangle \eta C_2,$$

whereas

$$\begin{aligned}\left[\hat{f}_\eta(x[P]+\lambda C_1), \hat{f}_\eta(x'[P']+\lambda' C_1)\right] &= \left[f(x)[P]+\lambda\eta C_2, f(x')[P']+\lambda'\eta C_2\right] \\ &= [f(x), f(x')][PP'] \\ &\quad + \operatorname{Res}_{t=0}(P'dP)\langle f(x), f(x')\rangle C_2.\end{aligned}$$

Comparing the above two expressions, the lemma follows immediately. □

Let G be the connected, simply-connected complex algebraic group with $\mathfrak{g}$ as its Lie algebra. Recall that the second homotopy group of any Lie group is trivial (cf. (Bröcker and tom Dieck, 1985, Chap. V, Proposition 7.5)). Thus, by the Hurewicz Theorem (Spanier, 1966, Chap. 7, §5, Theorem 5), the third homotopy group

$$\pi_3(G) \simeq H_3(G, \mathbb{Z}),$$

where $H_3(G, \mathbb{Z})$ is the third singular homology group of G with integral coefficients. Moreover, from the homotopy long exact sequence (cf. (Spanier, 1966,

Chap. 7, §2, Theorem 10)) associated to the path fibration $\Omega_e(G) \to PG \xrightarrow{\pi} G$, we get that

$$\pi_3(G) \simeq \pi_2(\Omega_e(G)),$$

where PG is the path space consisting of all the continuous maps γ from the unit interval $[0,1] \to G$ with $\gamma(0) = e$, $\pi(\gamma) = \gamma(1)$, and $\Omega_e(G) := \pi^{-1}(e)$ is the based loop space of G.

We recall the following classical result due to Bott (1956, Theorems A, B). Observe that for any maximal compact subgroup K of G, K is homotopically equivalent to G and hence $\Omega_e(K)$ is homotopically equivalent to $\Omega_e(G)$.

Theorem A.13 $\pi_2(\Omega_e(G)) \simeq H_2(\Omega_e(G), \mathbb{Z}) \simeq \mathbb{Z}$.

Thus, $\pi_3(G) \simeq H_3(G, \mathbb{Z}) = \mathbb{Z}$.

(Of course, $H_1(G, \mathbb{Z}) = H_2(G, \mathbb{Z}) = 0$.*)*

In particular, by the Universal Coefficient Theorem (cf. (Spanier, 1966, Chap. 5, §5, Theorem 3)), the third singular cohomology

$$H^3(G, \mathbb{Z}) \simeq \mathbb{Z}, \qquad H^3(G, \mathbb{C}) \simeq H^3(G, \mathbb{Z}) \otimes_{\mathbb{Z}} \mathbb{C} \tag{1}$$

and thus the natural map

$$H^3(G, \mathbb{Z}) \to H^3(G, \mathbb{C}) \quad \textit{is injective.} \tag{2}$$

Recall the Lie algebra homomorphism $\phi = \dot{\gamma}_\theta : sl_2 \to \mathfrak{g}$ defined in Definition A.5. This gives rise to an algebraic group homomorphism $\gamma_\theta : \mathrm{SL}_2 \to G$ such that its derivative $d\gamma_\theta = \phi$.

Theorem A.14 *The map* $\gamma_\theta : \mathrm{SL}_2 \to G$ *induces isomorphism in homology* $\gamma_{\theta *} : H_3(\mathrm{SL}_2, \mathbb{Z}) \to H_3(G, \mathbb{Z})$*; thus, also an isomorphism in cohomology* $\gamma_\theta^* : H^3(G, \mathbb{Z}) \to H^3(\mathrm{SL}_2, \mathbb{Z})$.

Proof The map γ_θ of course induces a canonical morphism of ind-varieties

$$\tilde{\gamma}_\theta : \bar{X}_{\mathrm{SL}_2} \to \bar{X}_G,$$

where $\bar{X}_G$ is the infinite Grassmannian endowed with the projective ind-variety structure as in Proposition 1.3.18 (cf. Proposition 1.3.24).

Following Definition 8.4.5, define the group homomorphism of ind-groups:

$$\gamma_o : \mathrm{SL}_2 \to \mathrm{SL}_2((t)), \quad \begin{pmatrix} a & b \\ c & d \end{pmatrix} \mapsto \begin{pmatrix} d & ct^{-1} \\ bt & a \end{pmatrix}. \tag{1}$$

By the definition of γ_o, $\gamma_o(B_o) \subset \mathrm{SL}_2[[t]]$, where B_o is the Borel subgroup of SL_2 consisting of upper-triangular matrices of determinant 1. Hence, we get a morphism of ind-varieties:

$$\tilde{\gamma}_o\colon \mathrm{SL}_2/B_o \to \bar{X}_{\mathrm{SL}_2}.$$

Let

$$\hat{\beta}\colon \mathrm{SL}_2/B_o \to \bar{X}_G \text{ be the composition } \tilde{\gamma}_\theta \circ \tilde{\gamma}_o.$$

By Proposition 8.4.6, the morphism $\hat{\beta}$ induces an isomorphism

$$\hat{\beta}^*\colon H^2\left(\bar{X}_G^{an}, \mathbb{Z}\right) \xrightarrow{\sim} H^2\left(\mathrm{SL}_2/B_o, \mathbb{Z}\right)$$

and hence an isomorphism (since $\bar{X}_G^{an}$ is simply connected)

$$\hat{\beta}_*\colon H_2\left(\mathrm{SL}_2/B_o, \mathbb{Z}\right) \xrightarrow{\sim} H_2\left(\bar{X}_G^{an}, \mathbb{Z}\right).$$

In particular, we have an isomorphism (for $G = \mathrm{SL}_2$):

$$\tilde{\gamma}_{o*}\colon H_2\left(\mathrm{SL}_2/B_o, \mathbb{Z}\right) \xrightarrow{\sim} H_2\left(\bar{X}_{\mathrm{SL}_2}^{an}, \mathbb{Z}\right).$$

Since $\hat{\beta}_* = \tilde{\gamma}_{\theta*} \circ \tilde{\gamma}_{o*}$, we get that

$$\tilde{\gamma}_{\theta*}\colon H_2\left(\bar{X}_{\mathrm{SL}_2}^{an}, \mathbb{Z}\right) \xrightarrow{\sim} H_2\left(\bar{X}_G^{an}, \mathbb{Z}\right) \text{ is an isomorphism.}$$

Finally, by a result of Garland and Raghunathan (1975) and also Quillen (Mitchell, 1988), $\bar{X}_G^{an}$ is homotopically equivalent canonically to the continuous (based) loop space $\Omega_e(G)$. Thus, from the isomorphism $\tilde{\gamma}_{\theta*}$, the induced map $\Omega(\gamma_\theta)\colon \Omega_e(\mathrm{SL}_2) \to \Omega_e(G)$ (induced from the homomorphism γ_θ) induces an isomorphism in second homology $H_2(\Omega_e(\mathrm{SL}_2), \mathbb{Z}) \xrightarrow{\sim} H_2(\Omega_e(G), \mathbb{Z})$. By Theorem A.13,

$$H_2(\Omega_e(G), \mathbb{Z}) \xrightarrow{\sim} \pi_2(\Omega_e(G)) \xrightarrow{\sim} \pi_3(G) \xrightarrow{\sim} H_3(G, \mathbb{Z}),$$

where the second isomorphism is obtained above Theorem A.13. This proves the theorem. □

We give another proof of Theorem A.14.

Another proof of Theorem A.14 Choose a maximal compact subgroup K of G such that $\gamma_\theta(\mathrm{SU}_2) \subset K$. The map γ_θ of course induces a continuous map at the loop space level

$$\Omega(\gamma_\theta)\colon \Omega_e(\mathrm{SU}_2) \to \Omega_e(K).$$

By the definition of $\phi\colon sl_2 \to \mathfrak{g}$ as in Definition A.5, $\phi(H) = \theta^\vee$. Since $\theta^\vee/2$ is not in the coroot lattice of G, it is easy to see (using $\mathrm{Exp}\,(\pi i H) = \begin{pmatrix} -1 & 0 \\ 0 & -1 \end{pmatrix}$) that γ_θ is injective and hence so is $\Omega(\gamma_\theta)$. Under a K-biinvariant

Riemannian metric on K, the geodesics in K starting at e are precisely of the form $t \mapsto \operatorname{Exp}(tx)$, as x ranges over the Lie algebra $\mathfrak{k}$ of K (cf. (Milnor, 1969, Lemma 21.2)). From now on we equip SU_2 and K with biinvariant Riemannian metrics.

We next show that the space of minimal geodesics $\Omega^{\min}(K; e, k_o)$ from e to the point $k_o := \operatorname{Exp}(\pi i \theta^\vee) \in K$ (under the compact open topology) is homeomorphic with the homogeneous space $Z_K(k_o)/Z_K(\theta^\vee)$, where $Z_K(k_o)$ is the centralizer of k_o in K under the conjugation action, i.e., $Z_K(k_o) := \{k \in K : k k_o k^{-1} = k_o\}$
and $Z_K(\theta^\vee)$ is the centralizer of $\theta^\vee$ in K under the adjoint action, i.e., $Z_K(\theta^\vee) := \{k \in K : (\operatorname{Ad} k)\, \theta^\vee = \theta^\vee\}$.

To prove this, take $x \in \mathfrak{k}$ such that $\operatorname{Exp} x = k_o$. We can write, for some $k \in K$,

$$\operatorname{Ad}(k^{-1}) \cdot x = 2\pi i (x_o + y_o), \tag{2}$$

where x_o is in the fundamental alcove

$$\Phi_o := \left\{h \in \mathfrak{h} : \alpha_i(h) \in \mathbb{R}_{\geq 0} \text{ for all the simple roots } \alpha_i \text{ and } \theta(h) \leq 1\right\}$$

and y_o is an element in the coroot lattice $Q^\vee := \oplus_i \mathbb{Z}\alpha_i^\vee$ (cf. Definition 4.2.5). Thus,

$$k^{-1} k_o k = k^{-1} (\operatorname{Exp} x)\, k = \operatorname{Exp}(2\pi i\, x_o). \tag{3}$$

In particular, $\operatorname{Exp}(2\pi i\, x_o)$ and $\operatorname{Exp}(\pi i \theta^\vee)$ are conjugate in K. Since $\Phi_o \to K/\operatorname{Ad} K$, $x_o \mapsto [\operatorname{Exp}(2\pi i\, x_o)]$, is a bijection (cf. (Helgason, 1978, Chap. VII, Theorem 7.9)), and since $\theta^\vee/2 \in \Phi_o$, we get that

$$x_o = \theta^\vee/2. \tag{4}$$

Take $w \in W$ such that $y_o^+ := w y_o \in Q_+^\vee$, where

$$Q_+^\vee := \{y \in Q^\vee : \alpha_i(y) \in \mathbb{R}_{\geq 0}\ \forall \alpha_i\}.$$

The length of the geodesic $\gamma(t) = \operatorname{Exp}(tx)$ on the interval [0,1] being $\|x\|$, we get by (2) and (4),

$$\|x\| = \|\operatorname{Ad}(k^{-1}) \cdot x\| = 2\pi\, \|x_o + y_o\| = 2\pi \left\|\frac{\theta^\vee}{2} + y_o\right\| = 2\pi \left\|\frac{w\theta^\vee}{2} + y_o^+\right\|. \tag{5}$$

Observe next that for any nonzero $y \in Q_+^\vee$, $y - \theta^\vee \in \oplus_i \mathbb{Z}_{\geq 0}\alpha_i^\vee$. (As observed by Jiuzu Hong, this can be seen by considering the Langlands dual Lie algebra $\mathfrak{g}^\vee$ with the root system $\{\beta^\vee : \beta \in \Delta(\mathfrak{g})\}$, where $\Delta(\mathfrak{g})$ is the set of roots of $\mathfrak{g}$. Let $V^\vee(y)$ be the irreducible $\mathfrak{g}^\vee$-module with highest weight y. Since $y \in Q_+^\vee$, the zero weight space $V^\vee(y)_0$ is nonzero. Hence, $\alpha_{i_o}^\vee$ is a weight of $V^\vee(y)$, for some simple coroot $\alpha_{i_o}^\vee$. Thus, so is its any W-translate. In particular, the highest coroot $\theta(\mathfrak{g}^\vee)$ or the highest short coroot $\theta^\vee$ is a weight of $V^\vee(y)$. Thus, in either case, $y - \theta^\vee \in \oplus_i \mathbb{Z}_{\geq 0}\alpha_i^\vee$.)

Thus, for any nonzero $y \in Q_+^\vee$ such that $y \neq \theta^\vee$,

$$\langle y - \theta^\vee, y\rangle = \|y - \theta^\vee\|^2 + \langle y - \theta^\vee, \theta^\vee\rangle > 0. \tag{6}$$

Thus, if $y_o^+ \neq 0, \theta^\vee$, by (6),

$$\begin{aligned}\left\|\frac{w\theta^\vee}{2} + y_o^+\right\|^2 &= \|\theta^\vee/2\|^2 + \langle y_o^+ + w\theta^\vee, y_o^+\rangle \\ &\geq \|\theta^\vee/2\|^2 + \langle y_o^+ - \theta^\vee, y_o^+\rangle > \|\theta^\vee/2\|^2. \end{aligned} \tag{7}$$

Thus, from (5), we get (if $y_o^+ \neq 0, \theta^\vee$):

$$\|x\| > \|\pi\theta^\vee\|.$$

If $y_o^+ = \theta^\vee$, then

$$\left\|\frac{\theta^\vee}{2} + y_o\right\| = \left\|\frac{\theta^\vee}{2} + w^{-1}\theta^\vee\right\| > \|\theta^\vee/2\|, \text{ unless } \langle\theta^\vee, w^{-1}\theta^\vee\rangle = -\|\theta^\vee\|^2.$$

But, if $\langle\theta^\vee, w^{-1}\theta^\vee\rangle = -\|\theta^\vee\|^2$, then, considering $\|\theta^\vee + w^{-1}\theta^\vee\|$, we get $\theta^\vee = -w^{-1}\theta^\vee$, i.e., $\frac{\theta^\vee}{2} + y_o = -\theta^\vee/2$, which is again in the K-orbit of $\theta^\vee/2$.

Thus, γ is a minimal geodesic if and only if

$$x = \mathrm{Ad}(k)(\pi i\theta^\vee), \qquad \text{for some } k \in K,$$

and, moreover, for such an x,

$$\mathrm{Exp}\, x = k_o \Longleftrightarrow k \in Z_K(k_o).$$

This proves the assertion that $\Omega^{\min}(K; e, k_o)$ is canonically homeomorphic with $Z_K(k_o)/Z_K(\theta^\vee)$. In particular, $\Omega^{\min}\big(\mathrm{SU}_2; I, -I\big)$ is canonically homeomorphic with $\mathrm{SU}_2/D \simeq S^2$, where $D \subset \mathrm{SU}_2$ is its diagonal subgroup and $I = \begin{pmatrix} 1 & 0 \\ 0 & 1\end{pmatrix}$.

By a general result, $Z_K(k_o)$ is connected (cf. (Humphreys, 1995, Theorem 2.11)) and hence $\Omega^{\min}(K; e, k_o)$ is connected. We next show that $Z_K(k_o)/Z_K(\theta^\vee)$ is a real 2-dimensional manifold. The complexified Lie algebra $\mathrm{Lie}(Z_K(\theta^\vee)) \otimes \mathbb{C}$ of $Z_K(\theta^\vee)$ is clearly

$$\mathrm{Lie}\big(Z_K(\theta^\vee)\big) \otimes \mathbb{C} = \mathfrak{h} \oplus \bigoplus_{\alpha \in \Delta : \alpha(\theta^\vee) = 0} \mathfrak{g}_\alpha.$$

Since $Z_K(k_o) \supset T$, T being the maximal torus of K with complexified Lie algebra $\mathfrak{h}$, we get that

$$\text{Lie}\big(Z_K(k_o)\big) \otimes \mathbb{C} = \mathfrak{h} \oplus \bigoplus_{\alpha \in S} \mathfrak{g}_\alpha,$$

where $S \subset \Delta$ is a certain subset of roots. Since $-I$ is central in SU_2, it is easy to see that $\{\pm\theta\} \subset S$. If possible, take $\alpha \in S\backslash\{\pm\theta\}$ such that $\alpha(\theta^\vee) \neq 0$. Thus, for any $x_\alpha \in \mathfrak{g}_\alpha$,

$$\text{Exp}(x_\alpha) k_o \, \text{Exp}(-x_\alpha) = k_o.$$

Applying this to $\theta^\vee$ via the adjoint action, we get

$$\text{Ad}\Big(\text{Exp}\big(\text{Ad}(\text{Exp}\, x_\alpha) \cdot (\pi i \theta^\vee)\big)\Big) \cdot \theta^\vee = \theta^\vee.$$

This gives

$$\text{Ad}\Big(\text{Exp}\big(\pi i \theta^\vee - \alpha(\pi i \theta^\vee)\, x_\alpha\big)\Big)\theta^\vee = \theta^\vee. \tag{8}$$

But

$$\text{Ad}\Big(\text{Exp}\big(\pi i \theta^\vee - \alpha(\pi i \theta^\vee)\, x_\alpha\big)\Big)\theta^\vee = \theta^\vee + \alpha(\theta^\vee)\big(e^{\alpha(\pi i \theta^\vee)} - 1\big)\, x_\alpha. \tag{9}$$

Now, since $\alpha \neq \pm\theta$, $\alpha(\theta^\vee) \in \{0, \pm 1\}$ by Bourbaki (2002, p. 294). But, by assumption, $\alpha(\theta^\vee) \neq 0$, thus $\alpha(\theta^\vee) \in \{\pm 1\}$, which is a contradiction to identities (8) and (9). This proves that $Z_K(k_o)/Z_K(\theta^\vee)$ is a 2-dimensional (connected) manifold. Thus, the injective map

$$Z_{\text{SU}_2}(-I)/Z_{\text{SU}_2}(H) \to Z_K(k_o)/Z_K(\theta^\vee), \quad \text{where } H = \begin{pmatrix} 1 & 0 \\ 0 & -1 \end{pmatrix},$$

induced from the inclusion $\gamma_\theta : \text{SU}_2 \to K$ is a (surjective) diffeomorphism. From this we obtain that the map γ_θ induces a homeomorphism

$$\Omega^{\min}(\text{SU}_2 ; I, -I) \xrightarrow{\sim} \Omega^{\min}(K; e, k_o). \tag{10}$$

Further, any non-minimal geodesic from e to k_o in K (in particular, I to $-I$ in SU_2) has index ≥ 4. To prove this, observe that by the argument as above, any *non-minimal* geodesic γ from e to k_o in K is of the form $\gamma(t) = \text{Exp}\, tx$, where $x = \text{Ad}\, k(\pi i \theta^\vee + 2\pi i y_o)$, for some $k \in Z_K(k_o)$ and y_o is a nonzero element of $Q_+^\vee$. Thus, $\alpha_i(y_o) \in \mathbb{Z}_{\geq 0}$ for all the simple roots α_i. In particular, $\theta(y_o) \in \mathbb{Z}_{\geq 0}$ and, in fact, $\theta(y_o) \neq 0$ (for, otherwise, $\alpha_i(y_o) = 0$ for all the simple roots α_i and hence $y_o = 0$, which is a contradiction since $y_o \neq 0$). Thus, $\frac{\theta^\vee}{2} + y_o$ lies in the hyperplane $\theta = d$, for some integer $d \geq 2$. In particular, the straight line segment $t(\frac{\theta^\vee}{2} + y_o)$, $0 \leq t \leq 1$, crosses at least two planes of the diagram of K. Thus, the index of such a nonminimal geodesic is at least 4 (cf. (Milnor, 1969, proof of Theorem 23.5)).

Thus, by Milnor (1969, Theorem 22.1), the following vertical maps are isomorphisms:

$$\begin{array}{ccc} \pi_2\big(\Omega^{\min}(\mathrm{SU}_2\,; I, -I)\big) & \xrightarrow{\sim} & \pi_2\big(\Omega^{\min}(K; e, k_o)\big) \\ \wr\big\downarrow & & \big\downarrow\wr \\ \pi_2\big(\Omega\,(\mathrm{SU}_2\,; I, -I)\big) & \longrightarrow & \pi_2\big(\Omega\,(K; e, k_o)\big), \end{array}$$

where $\Omega\,(K; e, k_o)$ denotes the space of continuous paths in K from e to k_o endowed with the compact open topology, and the horizontal maps are induced from the inclusion γ_θ. By the isomorphism (10), the top horizontal map is an isomorphism and hence so is the bottom horizontal map.

But the map $\Omega\,(K; e, k_o) \hookrightarrow \Omega_e(K)$ obtained by adjoining a fixed path in K from k_o to e at the end to any path in $\Omega\,(K; e, k_o)$ is a homotopy equivalence. Thus, the inclusion γ_θ induces an isomorphism

$$\begin{array}{ccc} \pi_2\big(\Omega_e(\mathrm{SU}_2)\big) & \xrightarrow{\sim} & \pi_2\big(\Omega_e(K)\big) \\ \wr\big\downarrow & & \wr\big\downarrow \\ \pi_3(\mathrm{SU}_2) & \xrightarrow{\sim} & \pi_3(K) \\ \wr\big\downarrow & & \wr\big\downarrow \\ H_3(\mathrm{SU}_2\,, \mathbb{Z}) & \xrightarrow{\sim} & H_3(K, \mathbb{Z}). \end{array}$$

This proves the theorem. □

Let $\mathfrak{g}$ be a simple Lie algebra and let G be the connected, simply-connected complex algebraic group with Lie algebra $\mathfrak{g}$. Choose a maximal compact subgroup $K \subset G$ and let $\mathfrak{k}$ be its Lie algebra. Then, $\mathfrak{k} \otimes_{\mathbb{R}} \mathbb{C} = \mathfrak{g}$. Let $\wedge(\mathfrak{g}^*)^{\mathfrak{g}}$ be the space of $\mathfrak{g}$-invariants in the exterior algebra $\wedge(\mathfrak{g}^*)$ under the coadjoint action. We can view $\wedge(\mathfrak{g}^*)^{\mathfrak{g}}$ as the space of K-biinvariant complex-valued smooth forms on K, i.e., the space of smooth forms on K which are invariant under the action of K on itself via left as well as right multiplication.

We recall the following well-known result due to Chevalley and Eilenberg (1948, Theorem 12.1).

Theorem A.15 *Identifying the singular cohomology $H^*(K, \mathbb{C})$ with the deRham cohomology $H^*_{dR}(K, \mathbb{C})$, the inclusion of $\wedge(\mathfrak{g}^*)^{\mathfrak{g}}$ into the space of all the complex valued smooth forms on K induces an isomorphism (of algebras)*

$$\wedge(\mathfrak{g}^*)^{\mathfrak{g}} \simeq H^*(K, \mathbb{C}) \simeq H^*(G, \mathbb{C}). \tag{1}$$

In particular, $H^3(G,\mathbb{C})$ is generated as a vector space over $\mathbb{C}$ by the element $C_{\mathfrak{g}} \in \wedge^3(\mathfrak{g}^*)^{\mathfrak{g}}$ defined by

$$C_{\mathfrak{g}}(x,y,z) = \langle x, [y,z] \rangle,$$

where $\langle \cdot, \cdot \rangle$ is the normalized invariant form on $\mathfrak{g}$ as in Definition A.1. (Observe that $H^3(G,\mathbb{C})$ is 1-dimensional by the isomorphism (1) *of Theorem A.13.)*

Recall the algebraic group homomorphism $\gamma_\theta \colon \mathrm{SL}_2 \to G$ and the induced isomorphism $\gamma_\theta^* \colon H^3(G,\mathbb{Z}) \simeq H^3(\mathrm{SL}_2,\mathbb{Z})$ from Theorem A.14. The isomorphism (1) *induces the following commutative diagram:*

$$\begin{array}{ccc} H^3(G,\mathbb{Z}) & \xrightarrow[\sim]{\gamma_\theta^*} & H^3(\mathrm{SL}_2,\mathbb{Z}) \\ j\downarrow & & \downarrow j \\ H^3(G,\mathbb{C}) & \xrightarrow[\sim]{\gamma_{\theta\mathbb{C}}^*} & H^3(\mathrm{SL}_2,\mathbb{C}) \\ \wr\downarrow & & \wr\downarrow \\ \wedge^3(\mathfrak{g}^*)^{\mathfrak{g}} & \xrightarrow[\phi^*]{} & \wedge^3(sl_2^*)^{sl_2}, \end{array} \tag{2}$$

where j is the standard map induced from the inclusion $\mathbb{Z} \subset \mathbb{C}$ and ϕ^* is induced from the Lie algebra homomorphism ϕ (cf. Definition A.5). It is easy to see that

$$\phi^*(C_{\mathfrak{g}}) = C_{sl_2}. \tag{3}$$

Definition A.16 Let ω_G be the unique generator of $H^3(G,\mathbb{Z}) \simeq \mathbb{Z}$ such that

$$j(\omega_G) = \alpha_G C_{\mathfrak{g}}, \text{ under isomorphism (1) of Theorem A.15, for some } \alpha_G > 0.$$

(Observe that $C_{\mathfrak{g}} \in H^3(G,\mathbb{R})$ under isomorphism (1) of Theorem A.15 since $\langle \mathfrak{k}, \mathfrak{k} \rangle \subset \mathbb{R}$ for $\mathfrak{k}$. Further, any choice of K does not change the class $C_{\mathfrak{g}}$ because of the G-conjugacy of maximal compact subgroups of G.)

The following theorem provides a topological characterization of the Dynkin index.

Theorem A.17 *Let $f \colon \mathfrak{g}_1 \to \mathfrak{g}_2$ be a Lie algebra homomorphism between simple Lie algebras and let $\bar{f} \colon G_1 \to G_2$ be the associated algebraic group homomorphism (i.e., $d\bar{f} = f$), where G_i is the connected, simply-connected complex algebraic group with Lie algebra $\mathfrak{g}_i$ $(i = 1,2)$. Then, under the above identification of $H^3(G_i,\mathbb{Z})$ with $\mathbb{Z}$, the induced map $\bar{f}^* \colon H^3(G_2,\mathbb{Z}) \to H^3(G_1,\mathbb{Z})$ is the multiplication by the Dynkin index d_f, i.e.,*

$$\bar{f}^*(\omega_{G_2}) = d_f\,\omega_{G_1}. \tag{1}$$

In particular, d_f is a non-negative integer.

Proof Consider the following commutative diagram induced from the isomorphism (1) of Theorem A.15:

$$\begin{array}{ccc} H^3(G_2,\mathbb{Z}) & \xrightarrow{\bar{f}^*} & H^3(G_1,\mathbb{Z}) \\ \downarrow j & & \downarrow j \\ H^3(G_2,\mathbb{C}) & \xrightarrow{\bar{f}^*_{\mathbb{C}}} & H^3(G_1,\mathbb{C}) \\ \wr\downarrow & & \wr\downarrow \\ \wedge^3(\mathfrak{g}_2^*)^{\mathfrak{g}_2} & \xrightarrow[f^*]{} & \wedge^3(\mathfrak{g}_1^*)^{\mathfrak{g}_1}. \end{array}$$

We first show that

$$f^*\big(C_{\mathfrak{g}_2}\big) = d_f\,C_{\mathfrak{g}_1}. \tag{2}$$

$$\begin{aligned} f^*\big(C_{\mathfrak{g}_2}\big)(x,y,z) &= C_{\mathfrak{g}_2}\big(f(x),f(y),f(z)\big), \quad \text{for } x,y,z \in \mathfrak{g}_1 \\ &= \langle f(x),[f(y),f(z)]\rangle \\ &= \langle f(x),f[y,z]\rangle \\ &= d_f\,\langle x,[y,z]\rangle, \quad \text{by the definition of } d_f \\ &= d_f\,C_{\mathfrak{g}_1}(x,y,z). \end{aligned}$$

This proves (2).

Now,

$$\begin{aligned} j\,\bar{f}^*(\omega_{G_2}) &= \bar{f}^*_{\mathbb{C}}\big(j(\omega_{G_2})\big) \\ &= \alpha_{G_2}\,f^*\big(C_{\mathfrak{g}_2}\big), \quad \text{by Definition A.16} \qquad (3) \\ &= \alpha_{G_2}\,d_f\,C_{\mathfrak{g}_1}, \quad \text{by (2).} \qquad (4) \end{aligned}$$

We next show that

$$\alpha_{G_1} = \alpha_{G_2}, \tag{5}$$

i.e., α_G does not depend upon the choice of G. By (3) applied to the homomorphism $(\gamma_\theta)_{G_2}\colon \mathrm{SL}_2 \to G_2$ and identity (3) of Theorem A.15, we get

$$j\,(\gamma_\theta)^*_{G_2}(\omega_{G_2}) = \alpha_{G_2}\,C_{sl_2}. \tag{6}$$

But since $(\gamma_\theta)^*_{G_2}$ is an isomorphism over $\mathbb{Z}$ (cf. diagram (2) of Theorem A.15), we get from (6),

$$j(\omega_{\mathrm{SL}_2}) = \pm\alpha_{G_2} C_{sl_2}.$$

Similarly,

$$j(\omega_{\mathrm{SL}_2}) = \pm\alpha_{G_1} C_{sl_2}.$$

This gives $\alpha_{G_1} = \pm\alpha_{G_2}$. But, $\alpha_{G_i} > 0$, by definition. Hence, $\alpha_{G_1} = \alpha_{G_2}$, proving (5). Combining (4) and (5), we get

$$j\,\bar{f}^*(\omega_{G_2}) = \alpha_{G_1} d_f\, C_{\mathfrak{g}_1} = d_f\, j(\omega_{G_1}).$$

Thus, j being injective,

$$\bar{f}^*\big(\omega_{G_2}\big) = d_f\, \omega_{G_1},$$

proving identity (1).

This proves, in particular, that d_f is an integer. But as in Definition A.1, $d_f \in \mathbb{R}_{\geq 0}$. Thus, $d_f \in \mathbb{Z}_{\geq 0}$. This proves the theorem. □

A.E Exercises

(1) For any positive integer n and non-negative integer m such that $n \geq 2(m+2)$, show that $\binom{n}{m+2} - \binom{n}{m}$ is never equal to 1.

(2) Show that for dominant weights λ and μ of any simple Lie algebra $\mathfrak{g}$ with $\mu \neq 0$,

$$d_{V(\lambda+\mu)} > d_{V(\lambda)}.$$

(*Hint:* Use Lemma A.2 and the Weyl dimension formula together with the fact that $D \subset \bigoplus_{i=1}^{\ell} \mathbb{R}_{\geq 0}\alpha_i$, where D is as in Definition 1.2.6.)

A.C Comments

The definition of the Dynkin index as in Definition A.1, Lemma A.2 and Corollary A.4 are due to Dynkin (1957, §2). Lemmas A.3, A.6 and Example A.7 are taken from Kumar, Narasimhan and Ramanathan (1994, §5). The Dynkin index $d_{V(\omega_i)}$ of the fundamental representations $V(\omega_i)$ was originally determined by Dynkin (1957, §2). However, some of his values are incorrect. We have taken the (correct) values of these as in Proposition A.8 from

Kumar and Narasimhan (1997, §4). Theorem A.10 is taken from Kumar and Narasimhan (1997, §4).

Theorem A.14 is due to Bott and Samelson (1958, Chap. III, Proposition 10.2(A)). (To obtain this theorem from the same, we need to identify $H_3(G, \mathbb{Z}) \simeq \pi_3(G) \simeq \pi_2(\Omega_e(G)) \simeq \pi_2(\Omega_e(K))$ as in this chapter.) The first proof of Theorem A.14 is based on the topology of infinite Grassmannian $\bar{X}_G^{an}$ (specifically its second singular homology), whereas the second proof is influenced by the proof of the Bott periodicity (cf. (Milnor, 1969, §23)). Theorem A.17 is due to Kumar, Narasimhan and Ramanathan (1994, Corollary 5.6), though the proof given here is different. (Observe that in this latter reference, the result is proved when $\mathfrak{g}_2 = sl(V)$, but using identity (1) of Definition A.1, this gives the general result.) The integrality of d_f (as in Theorem A.17) is originally due to Dynkin (1957, §2). His proof requires some case-by-case consideration and as he wrote 'using rather complicated arguments'.

Appendix B

$\mathbb{C}$-Space and $\mathbb{C}$-Group Functors

Introduction. We collect here the definition of $\mathbb{C}$-space functor and $\mathbb{C}$-group functor (and its Lie algebra) and their basic properties. A $\mathbb{C}$-space functor generalizes the notion of ind-schemes.

Given any covariant functor $\mathscr{F}^o$ from the category **Alg** of $\mathbb{C}$-algebras to the category **Set** of sets which is a separated presheaf, we can uniquely sheafify it to get a $\mathbb{C}$-space functor $\mathscr{F}$ (cf. Lemma B.2). Example B.4 contains some important examples of $\mathbb{C}$-group functors: $\mathbf{GL}_V$, $\mathbf{End}_V$ and $\mathbf{PGL}_V$, where V is a (not necessarily finite-dimensional) vector space over $\mathbb{C}$. We define the fiber product of $\mathbb{C}$-space functors and open subfunctor of a $\mathbb{C}$-space functor in Definition B.5. The notion of the tangent space of a $\mathbb{C}$-space functor $\mathscr{F}$ is defined in Definition B.7. The tangent space at an R-point of $\mathscr{F}$ (for $R \in \mathbf{Alg}$) is an R-module under a mild restriction on $\mathscr{F}$ satisfying the condition (E) as defined in Definition B.7 (cf. Lemma B.8). The functors $\mathbf{GL}_V$ and $\mathbf{PGL}_V$ do satisfy the condition (E) and we determine their tangent spaces (cf. Example B.12 and Lemma B.13).

For a $\mathbb{C}$-group functor $\mathscr{G}$ satisfying the condition (E), the tangent space at the identity is canonically a $\mathbb{C}$-space functor which is functorial with respect to the homomorphisms of $\mathbb{C}$-group functors satisfying the condition (E) (cf. Lemma B.14). We define the adjoint representation of $\mathbb{C}$-group functors satisfying the condition (E) and also a bracket $[\, ,\,]$ in the tangent space at the identity of a $\mathbb{C}$-group functor satisfying the condition (L), where the condition (L) is defined in Definition B.15 (cf. Definition B.17). The bracket is functorial with respect to the homomorphisms of $\mathbb{C}$-group functors satisfying the condition (L) (cf. Lemma B.19). We further show that this bracket makes the tangent space at the identity into a Lie algebra (cf. Corollary B.21).

Let Γ be an irreducible ind-group variety and let $X = (X_n)_{n \geq 0}$ be an ind-variety with an action by Γ. Let $\mathscr{V}$ be a Γ-equivariant vector bundle over X

and let $\sigma \in H^0(X, \mathscr{V})$. Then we prove that σ is $\Gamma(\mathbb{C})$-invariant if and only if σ is annihilated by Lie Γ (cf. Lemma B.23).

Definition B.1 A $\mathbb{C}$-space functor (resp. $\mathbb{C}$-group functor) is a covariant functor

$$\mathscr{F}: \mathbf{Alg} \to \mathbf{Set} \quad (\text{resp. } \mathbf{Group})$$

which is a sheaf for the fppf (faithfully flat of finite presentation–Fidèlement Plat de Présentation Finie) topology, i.e., for any $R \in \mathbf{Alg}$ and any faithfully flat finitely presented R-algebra R', the diagram

$$\mathscr{F}(R) \to \mathscr{F}(R') \rightrightarrows \mathscr{F}(R' \underset{R}{\otimes} R') \tag{1}$$

is exact (in particular, $\mathscr{F}(R) \to \mathscr{F}(R')$ is one-to-one), where **Set** (resp. **Group**) is the category of sets (resp. groups), **Alg** is the category of commutative algebras with identity over $\mathbb{C}$ and the two maps $\mathscr{F}(R') \rightrightarrows \mathscr{F}(R' \underset{R}{\otimes} R')$ are induced by the algebra homomorphisms

$$R' \to R' \otimes_R R', \;\; r' \mapsto r' \otimes 1 \text{ and } r' \mapsto 1 \otimes r'.$$

From now on we shall abbreviate faithfully flat finitely presented R-algebra by fppf R-algebra.

By a $\mathbb{C}$-*functor morphism* $\varphi: \mathscr{F} \to \mathscr{F}'$ between two $\mathbb{C}$-space functors, we mean a natural transformation between them, i.e., a set map $\varphi_R: \mathscr{F}(R) \to \mathscr{F}'(R)$ for any $R \in \mathbf{Alg}$ such that the following diagram is commutative for any algebra homomorphism $R \to S$:

$$\begin{array}{ccc} \mathscr{F}(R) & \xrightarrow{\varphi_R} & \mathscr{F}'(R) \\ \downarrow & & \downarrow \\ \mathscr{F}(S) & \xrightarrow{\varphi_S} & \mathscr{F}'(S). \end{array}$$

A $\mathbb{C}$-space functor $\mathscr{F}'$ is called a *subfunctor* of $\mathscr{F}$ if

$$\mathscr{F}'(R) \subset \mathscr{F}(R), \quad \text{for any } R \in \mathbf{Alg}.$$

Direct limits exist in the category of $\mathbb{C}$-space ($\mathbb{C}$-group) functors.

For any ind-scheme $X = (X_n)_{n \geq 0}$ over $\mathbb{C}$, the functor h_X: $R \rightsquigarrow \operatorname{Mor}(\operatorname{Spec} R, X)$ (cf. Section 1.1) is a $\mathbb{C}$-space functor by virtue of the Faithfully Flat Descent (cf. (Grothendieck, 1971, VIII 5.1, 1.1 and 1.2) or

(Stacks, 2019, Tag 023T)). This allows us to realize the category of ind-schemes over $\mathbb{C}$ as a full subcategory of the category of $\mathbb{C}$-space functors. A $\mathbb{C}$-space functor isomorphic to h_X (for an ind-scheme X) is called a *representable functor* represented by X.

Lemma B.2 *Let $\mathscr{F}^o$: **Alg** $\to$ **Set** be a covariant functor. Assume that $\mathscr{F}^o$ is a separated presheaf, i.e.,*

$$\mathscr{F}^o(R) \to \mathscr{F}^o(R') \text{ is one-to-one} \tag{1}$$

*for any $R \in$ **Alg** and any fppf R-algebra R'.*

Then, there exists a $\mathbb{C}$-space functor $\mathscr{F}$ containing $\mathscr{F}^o$ (i.e., $\mathscr{F}^o(R) \subset \mathscr{F}(R)$ for any R) such that for any $\mathbb{C}$-space functor $\mathscr{G}$ and a natural transformation $\theta^o \colon \mathscr{F}^o \to \mathscr{G}$, there exists a unique natural transformation $\theta \colon \mathscr{F} \to \mathscr{G}$ extending θ^o.

Moreover, such a $\mathscr{F}$ is unique up to a unique isomorphism extending the identity map of $\mathscr{F}^o$.

We call such a $\mathscr{F}$ the fppf-sheafification of $\mathscr{F}^o$.

If $\mathscr{F}^o$ is a $\mathbb{C}$-group functor, then its fppf-sheafification $\mathscr{F}$ is a $\mathbb{C}$-group functor.

Proof It is clear that once $\mathscr{F}$ exists, it is unique. We now prove the existence.

Define, for any $R \in$ **Alg**,

$$\mathscr{F}(R) = \varinjlim_{R'} K_R(R'), \tag{2}$$

where R' ranges over fppf R-algebras R', $K_R(R')$ is the kernel (or equalizer) of

$$\mathscr{F}^o(R') \rightrightarrows \mathscr{F}^o\left(R' \underset{R}{\otimes} R'\right),$$

and $R' \leq R''$ means that there exists a R-algebra homomorphism $R' \to R''$.

To make good sense of the limit (2), observe first that since R' ranges only over finitely presented R-algebras, the limit is being taken over a *set*. Secondly, for $R' \leq R''$, i.e., there exists a R-algebra homomorphism $f \colon R' \to R''$, the induced map $\mathscr{F}^o(f) \colon K_R(R') \to K_R(R'')$ is independent of the choice of f. To prove this, take another R-algebra homomorphism $g \colon R' \to R''$. Then, we assert that

$$\mathscr{F}^o(f)_{|K_R(R')} = \mathscr{F}^o(g)_{|K_R(R')} \colon K_R(R') \to K_R(R''). \tag{3}$$

Consider the R-algebra homomorphism

$$\theta \colon R' \underset{R}{\otimes} R' \to R'' \underset{R}{\otimes} R'', \quad x \otimes y \mapsto f(x) \otimes g(y).$$

This induces the map

$$\mathscr{F}^o(\theta)\colon \mathscr{F}^o\left(R'\underset{R}{\otimes}R'\right) \to \mathscr{F}^o\left(R''\underset{R}{\otimes}R''\right).$$

From this we see that the two maps

$$\mathscr{F}^o(R') \rightrightarrows \mathscr{F}^o\left(R''\underset{R}{\otimes}R''\right) \text{ restricted to } K_R(R') \text{ coincide,} \tag{4}$$

where the above two maps are induced from the R-algebra homomorphisms

$$R' \rightrightarrows R''\underset{R}{\otimes}R'',\ x \mapsto f(x)\otimes 1;\ x \mapsto 1\otimes g(x).$$

Moreover, since

$$\mathscr{F}^o(i_1)_{|K_R(R'')} = \mathscr{F}^o(i_2)_{|K_R(R'')},$$

and

$$\operatorname{Im}\left(\mathscr{F}^o(f)_{|K_R(R')}\right),\quad \operatorname{Im}\left(\mathscr{F}^o(g)_{|K_R(R')}\right) \subset K_R(R''),$$

where

$$i_1\colon R'' \to R''\underset{R}{\otimes}R'',\ x\mapsto x\otimes 1 \text{ and } i_2\colon R'' \to R''\underset{R}{\otimes}R'',\ x \mapsto 1\otimes x,$$

we get that

$$\mathscr{F}^o(i_1)\circ\mathscr{F}^o(g)|_{K_R(R')} = \mathscr{F}^o(i_1)\circ\mathscr{F}^o(f)_{|K_R(R')}, \quad \text{by (4).} \tag{5}$$

Since i_1 is an fppf extension (cf. (Matsumura, 1989, §7)), $\mathscr{F}^o(i_1)$ is injective. Thus, from (5), we get (3).

Observe that by condition (1), $\mathscr{F}^o(R) \subset K_R(R')$ for any R'. Thus, $\mathscr{F}^o(R) \subset \mathscr{F}(R)$.

To prove that $\mathscr{F}$ is indeed a sheaf for the fppf topology, take $R \in \mathbf{Alg}$ and R' a fppf R-algebra. We need to show the exactness of the following diagram:

$$\mathscr{F}(R) \to \mathscr{F}(R') \rightrightarrows \mathscr{F}\left(R'\underset{R}{\otimes}R'\right). \tag{6}$$

Recall that if $R \to R'$ and $R' \to R''$ are faithfully flat extensions, then so is $R \to R''$ (cf. (Matsumura, 1989, §7)). Similarly, if $R \to R'$ is finitely presented and $R' \to R''$ is finitely presented, then so is $R \to R''$ (cf. (Stacks, 2019, Tag 00F4)).

Take any fppf R-algebra S. Then, it is easy to show the exactness of the following:

$$K_R\left(R'\underset{R}{\otimes}S\right) \to K_{R'}\left(R'\underset{R}{\otimes}S\right) \rightrightarrows K_{R'\underset{R}{\otimes}R'}\left(R'\underset{R}{\otimes}R'\underset{R}{\otimes}S\right).$$

Taking limits of the above exact sequence, we get the exactness of (6). This proves the lemma. □

Remark B.3 In fact, we can sheafify any functor $\mathscr{F}^o$ (which is not necessarily a separated presheaf) to get a $\mathbb{C}$-space functor $\mathscr{F}$ though the injectivity $\mathscr{F}^o(R) \to \mathscr{F}(R)$ is lost.

Example B.4 (1) Let V be a (not necessarily finite-dimensional) vector space over $\mathbb{C}$. Then, the functor $\mathbf{GL}_V$ is a $\mathbb{C}$-group functor defined by

$$R \in \mathbf{Alg} \rightsquigarrow \mathbf{GL}_V(R) = \text{ group of } R\text{-linear automorphisms of } V_R := V \otimes_{\mathbb{C}} R.$$

It is easy to see that it is a sheaf for the fppf topology. In particular, it satisfies condition (1) of Lemma B.2 by using the fact that any faithfully flat homomorphism $R \to S$ is injective (cf. (Matsumura, 1989, Theorem 7.5)).

Similarly, we define the $\mathbb{C}$-space functor $\mathbf{End}_V$ (consisting of R-modules) $R \in \mathbf{Alg} \rightsquigarrow \mathbf{End}_V(R) = R$-module of R-linear endomorphisms of V_R.

(2) Let V be as in the above example (1). Then, the functor $\mathbf{PGL}_V$ is a $\mathbb{C}$-group functor defined as the fppf-sheafification (cf. Lemma B.2) of the functor

$$R \in \mathbf{Alg} \rightsquigarrow \mathbf{GL}_V(R)/R^*, \tag{1}$$

where R^* is the group of units of R viewed as a central subgroup of $\mathrm{GL}_R(V_R)$ under $r \mapsto r\,\mathrm{I}$ (for $r \in R^*$); I being the identity map. (Observe that the above functor satisfies condition (1) of Lemma B.2.)

The map $\mathbf{GL}_V(R) \to \mathbf{GL}_V(R)/R^*$ induces a morphism of $\mathbb{C}$-group functors $\mathbf{GL}_V \to \mathbf{PGL}_V$ (use Lemma B.2).

Definition B.5 (Fiber product) (a) Let $\mathscr{F}_1, \mathscr{F}_2, \mathscr{F}$ be $\mathbb{C}$-space functors with $\mathbb{C}$-functor morphisms

$$\varphi_1 \colon \mathscr{F}_1 \to \mathscr{F} \text{ and } \varphi_2 \colon \mathscr{F}_2 \to \mathscr{F}.$$

Then, their *fiber product* $\mathscr{F}_1 \times_{\mathscr{F}} \mathscr{F}_2$ is the functor

$$R \rightsquigarrow \mathscr{F}_1(R) \times_{\mathscr{F}(R)} \mathscr{F}_2(R) := \{(x, y) \in \mathscr{F}_1(R) \times \mathscr{F}_2(R) : \varphi_1(x) = \varphi_2(y)\}.$$

Then, it is easy to see that the fiber product $\mathscr{F}_1 \times_{\mathscr{F}} \mathscr{F}_2$ is indeed a $\mathbb{C}$-space functor, i.e., it is a sheaf for the fppf topology.

If $\mathscr{F}_1, \mathscr{F}_2, \mathscr{F}$ are $\mathbb{C}$-group functors and φ_1, φ_2 are homomorphisms of $\mathbb{C}$-group functors, then their fiber product $\mathscr{F}_1 \times_{\mathscr{F}} \mathscr{F}_2$ is a $\mathbb{C}$-group functor.

(b) Let $\mathscr{F}$ be a $\mathbb{C}$-space functor. A subfunctor $\mathscr{F}'$ of $\mathscr{F}$ is called an *open subfunctor* if for any $R_o \in \mathbf{Alg}$ and any $\psi \in \mathscr{F}(R_o)$ (which gives rise to a $\mathbb{C}$-functor morphism $\bar{\psi}: h_{\operatorname{Spec} R_o} \to \mathscr{F}$) the canonical morphism

$$\mathscr{F}' \times_{\mathscr{F}} h_{\operatorname{Spec} R_o} \to h_{\operatorname{Spec} R_o}$$

corresponds to an open immersion $V \hookrightarrow \operatorname{Spec} R_o$, where h_X for any ind-scheme X is defined in Section 1.1.

Let $\{\mathscr{F}_i\}_{i\in I}$ be a collection of open subfunctors of $\mathscr{F}$. Then, it is called an *open covering* of $\mathscr{F}$ if for any $R_o \in \mathbf{Alg}$ and any $\psi \in \mathscr{F}(R_o)$, we have an open covering:

$$\operatorname{Spec} R_o = \cup_i V_i,$$

where $V_i \subset \operatorname{Spec} R_o$ corresponds to $\mathscr{F}_i$ as above.

For any ind-scheme X over $\mathbb{C}$, the $\mathbb{C}$-space functor h_X has its open subfunctors given precisely by h_V for an open subset V of X. Further, a collection $\{h_{V_i}\}$ of open subfunctors of h_X is an open covering of h_X if and only if $\cup_i V_i = X$ (cf. Exercise B.E.10).

Definition B.6 For morphisms $f: R' \to R$ and $g: R'' \to R$ in **Alg**, define the *fiber product algebra* $R' \times_R R''$ by

$$R' \times_R R'' = \left\{(r',r'') \in R' \times R'' : f(r') = g(r'')\right\}.$$

Then, $R' \times_R R''$ is a $\mathbb{C}$-subalgebra of the direct product algebra $R' \times R''$. We have the Cartesian diagram:

$$\begin{array}{ccc} R' \times_R R'' & \xrightarrow{\pi'} & R' \\ {\scriptstyle \pi''}\downarrow & & \downarrow{\scriptstyle f} \\ R'' & \xrightarrow[g]{} & R, \end{array}$$

where π' and π'' are the two projections.

For any algebra $R \in \mathbf{Alg}$, let $R(\epsilon)$ be the algebra $R[\epsilon]/\langle\epsilon^2\rangle$ and let

$$\theta'_R: R(\epsilon') \to R \quad \text{and} \quad \theta''_R: R(\epsilon'') \to R$$

be the augmentation homomorphisms $\epsilon' \mapsto 0$, $\epsilon'' \mapsto 0$. Then, their fiber product can be seen to be the algebra $R[\epsilon',\epsilon'']/\langle\epsilon',\epsilon''\rangle^2$ with map π'_o (resp. π''_o) given by $\epsilon' \mapsto \epsilon'$ and $\epsilon'' \mapsto 0$ (resp. $\epsilon' \mapsto 0, \epsilon'' \mapsto \epsilon''$) (cf. Exercise B.E.1).

Let $\mathscr{F}: \mathbf{Alg} \to \mathbf{Set}$ be a $\mathbb{C}$-space functor. The algebra homomorphisms π'_o and π''_o give rise to the map to the fiber product

$$\mu_R: \mathscr{F}\left(R[\epsilon',\epsilon'']/\langle\epsilon',\epsilon''\rangle^2\right) \to \mathscr{F}(R(\epsilon')) \times_{\mathscr{F}(R)} \mathscr{F}(R(\epsilon'')).$$

Definition B.7 (a) Let $x \in \mathscr{F}(R)$. Then, the *tangent space* $T_x(\mathscr{F})_R$ of the $\mathbb{C}$-space functor $\mathscr{F}$ at x is by definition the fiber of the canonical map (induced from the augmentation) over x:

$$\mathscr{F}(R(\epsilon)) \to \mathscr{F}(R), \quad \epsilon \mapsto 0.$$

(b) We say that the $\mathbb{C}$-space functor $\mathscr{F}$ satisfies the *condition* (E_R) if the above map μ_R is a bijection. If μ_R is a bijection for all $R \in \mathbf{Alg}$, we say that $\mathscr{F}$ satisfies the *condition* (E).

The $\mathbb{C}$-space functor h_X represented by an ind-scheme X satisfies the condition (E) (cf. Exercise B.E.4).

If μ_R is a bijection for all the finite-dimensional $\mathbb{C}$-algebras R, then we say that $\mathscr{F}$ satisfies the *condition (E) finitely*.

Lemma B.8 *Let $\mathscr{F}$ be a $\mathbb{C}$-space functor and let $R \in$ **Alg**. Assume that $\mathscr{F}$ satisfies the condition (E_R). Then, for any $x \in \mathscr{F}(R)$, the tangent space $T_x(\mathscr{F})_R$ is naturally an R-module.*

Proof The scalar multiplication by any $z \in R$ on $T_x(\mathscr{F})_R$ is induced from the algebra homomorphism m_z:

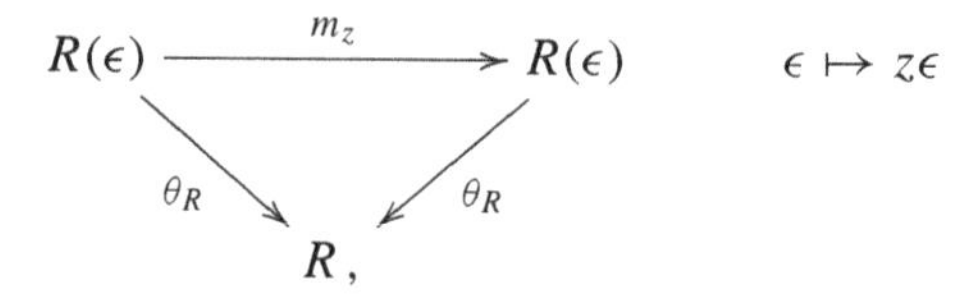

where θ_R is the augmentation map. (In fact, for the scalar multiplication by R on $T_x(\mathscr{F})_R$, we do not require $\mathscr{F}$ to satisfy the condition (E_R).)

To define the additive structure, take $v, w \in T_x(\mathscr{F})_R$ and think of $v \in \mathscr{F}(R(\epsilon'))$,
$w \in \mathscr{F}(R(\epsilon''))$ over x. Define

$$v + w := \delta_R \circ \mu_R^{-1}(v, w),$$

where

$$\delta_R \colon \mathscr{F}\left(R[\epsilon', \epsilon'']/\langle \epsilon', \epsilon'' \rangle^2\right) \to \mathscr{F}(R(\epsilon))$$

is the map induced from $\epsilon' \mapsto \epsilon$, $\epsilon'' \mapsto \epsilon$. It is easy to see that the above structures make $T_x(\mathscr{F})_R$ into an R-module. □

Remark B.9 If $\mathscr{F}$ is a representable functor represented by the ind-scheme X (so that $X(\mathbb{C}) = \mathscr{F}(\mathbb{C})$) (where $X(\mathbb{C}) := \text{Mor}(\text{Spec}\,\mathbb{C}, X)$ are the closed points of X), then the tangent space $T_x(\mathscr{F})_\mathbb{C}$ (For any $x \in X(\mathbb{C})$) coincides with the classical definition of the Zariski tangent space of the ind-scheme X at x (cf. Exercise B.E.2).

Let $\mathcal{G}$ be a $\mathbb{C}$-group functor and let $1 \in \mathcal{G}(\mathbb{C})$ be the identity. Then, for any $R \in \mathbf{Alg}$, the induced homomorphism from the unique $\mathbb{C}$-algebra homomorphism $\mathbb{C} \to R$ takes 1 to the identity (still denoted by 1) of $\mathcal{G}(R)$.

We recall the following simple result (which is classically used to prove that any H-space has an abelian fundamental group).

Lemma B.10 *Let M be a group and let $\eta\colon M \times M \to M$ be a binary map which is a group homomorphism (under the product group structure on $M \times M$). If M contains an element e such that $\eta(e,x) = \eta(x,e) = x$, for all $x \in M$, then M is a commutative group and $\eta(x,y) = xy$ for all $x, y \in M$.*

Proof Take any $x, y \in M$. We have $x = \eta(x,e) = \eta((1,e)(x,1)) = \eta(1,e)\eta(x,1) = \eta(x,1)$. Similarly, $\eta(1,y) = y$. Thus, $\eta(x,y) = \eta(x,1)\eta(1,y) = xy$. Also, $\eta(x,y) = \eta(1,y)\eta(x,1) = yx$. Hence, $xy = yx$. This proves the lemma. □

Lemma B.11 *Let $\mathcal{G}$ be a $\mathbb{C}$-group functor satisfying the condition (E_R) for an $R \in$ **Alg**. Then, the kernel of the group homomorphism*

$$\mathcal{G}(\theta_R)\colon \mathcal{G}(R(\epsilon)) \to \mathcal{G}(R)$$

coincides with the tangent space $T_1(\mathcal{G})_R$ as a group. In particular, $\operatorname{Ker}\mathcal{G}(\theta_R)$ is an abelian group, which, in addition, has a canonical R-module structure.

Proof Denote the product in $T_1(\mathcal{G})_R$ coming as the subgroup of $\mathcal{G}(R(\epsilon))$ by $\cdot$ and the product in $T_1(\mathcal{G})_R$ from its R-module structure by $+$. We apply Lemma B.10 for $M = T_1(\mathcal{G})_R$ and $\eta(x,y) = x + y$. We first prove that η is a group homomorphism under the product $\cdot$ on M:

Take $x_1, y_1, x_2, y_2 \in M$. Then, we need to prove that

$$(x_1 + y_1) \cdot (x_2 + y_2) = x_1 \cdot x_2 + y_1 \cdot y_2. \tag{1}$$

Since μ_R is a group isomorphism,

$$\left(\mu_R^{-1}(x_1,y_1)\right) \cdot \left(\mu_R^{-1}(x_2,y_2)\right) = \mu_R^{-1}(x_1 \cdot x_2, y_1 \cdot y_2). \tag{2}$$

Thus, δ_R being a group homomorphism, from (2) and the definition of $+$ (as in the proof of Lemma B.8), we get

$$(x_1 + y_1) \cdot (x_2 + y_2) = x_1 \cdot x_2 + y_1 \cdot y_2.$$

This proves (1).

Taking $e = 1$, it is easy to see that

$$1 + x = x + 1 = x. \tag{3}$$

Equations (1) and (3), together with Lemma B.10, imply the lemma. □

Example B.12 Let V be any vector space over $\mathbb{C}$. Then, as in Example B.4, $\mathbf{GL}_V$ is a $\mathbb{C}$-group functor. It is easy to see that this satisfies the condition (E).

As shown below, the tangent space $T_1(\mathbf{GL}_V)_R$ coincides with $\mathrm{End}_R(V_R)$ for any $R \in \mathbf{Alg}$ (where $V_R := V \otimes_{\mathbb{C}} R$). Thus, the tangent space functor $T_1(\mathbf{GL}_V)$ is identified with the functor $\mathbf{End}_V$ (cf. Example B.4).

Define the group homomorphism

$$\theta\colon \mathrm{End}_R(V_R) \to \mathrm{Aut}_{R(\epsilon)}(V_{R(\epsilon)}),\ \theta(f)(v) = v + f(v)\epsilon,$$
$$\text{for } v \in V_R \subset V_{R(\epsilon)} = V_R \oplus V_R\epsilon.$$

Then it is easy to see that $\mathrm{Im}\,\theta = T_1(\mathbf{GL}_V)_R$ and, moreover, θ is a R-linear bijection onto its image with respect to the R-linear structure on $T_1(\mathbf{GL}_V)_R$.

We thank B. Conrad for communicating the following simpler (than our original) proof.

Lemma B.13 *The functor* $\mathbf{PGL}_V$ *satisfies the condition (E) for any (not necessarily finite dimensional)* $\mathbb{C}$*-vector space* V.

Further, by Exercise B.E.3,

$$T_1\left(\mathbf{PGL}_V\right)_R \simeq \mathrm{End}_R(V_R)/R \cdot I_{V_R}. \tag{1}$$

Proof Take any $R \in \mathbf{Alg}$. To prove the bijectivity of μ_R for $\mathscr{F} = \mathbf{PGL}_V$ (cf. Definition B.6), it suffices to prove that for any fppf R-algebra S, the following map for $\mathbf{PGL}_V$ (cf. the proof of Lemma B.2 for the notation)

$$\mu_R(S)\colon K_{R_{\epsilon',\epsilon''}}\left(R_{\epsilon',\epsilon''}\underset{R}{\otimes}S\right) \to K_{R(\epsilon')}\left(R(\epsilon')\underset{R}{\otimes}S\right)\underset{K_R(S)}{\times}K_{R(\epsilon'')}\left(R(\epsilon'')\underset{R}{\otimes}S\right)$$

is bijective, where $R_{\epsilon',\epsilon''} := R[\epsilon',\epsilon'']/\langle\epsilon',\epsilon''\rangle^2$. By Example B.12, since $\mathbf{GL}_V$ is a $\mathbb{C}$-group functor satisfying the condition (E), the natural map

$$\mu_S\colon \mathbf{GL}_V\left(R_{\epsilon',\epsilon''}\underset{R}{\otimes}S\right) \simeq \mathbf{GL}_V\left(R(\epsilon')\underset{R}{\otimes}S\right)\underset{\mathbf{GL}_V(S)}{\times}\mathbf{GL}_V\left(R(\epsilon'')\underset{R}{\otimes}S\right)$$

induced from the ring homomorphisms

$$R_{\epsilon',\epsilon''} \to R(\epsilon'),\ \epsilon'' \mapsto 0 \quad\text{and}\quad R_{\epsilon',\epsilon''} \to R(\epsilon''),\ \epsilon' \mapsto 0,$$

is an isomorphism. Let $\mathrm{PGL}_V(R) := \mathrm{GL}_R(V_R)/R^* \cdot \mathrm{Id}$. From the above isomorphism μ_S for $\mathbf{GL}_V$ (and also the same isomorphism for $\mathbf{GL}_1$), it is easy to see that the natural map

$$\psi_S\colon \mathrm{PGL}_V\left(R_{\epsilon',\epsilon''}\underset{R}{\otimes}S\right) \to \mathrm{PGL}_V\left(R(\epsilon')\underset{R}{\otimes}S\right)\underset{\mathrm{PGL}_V(S)}{\times}\mathrm{PGL}_V\left(R(\epsilon'')\underset{R}{\otimes}S\right)$$

is an isomorphism (for any fppf R-algebra S). Consider the following commutative diagram, where ψ_S and $\psi_{S\otimes S}$ are isomorphisms:

$$\begin{array}{ccc}
K_{R_{\epsilon',\epsilon''}}\left(R_{\epsilon',\epsilon''}\underset{R}{\otimes}S\right) & \xrightarrow{\mu_R(S)} & K_{R(\epsilon')}\left(R(\epsilon')\underset{R}{\otimes}S\right)\underset{K_R(S)}{\times}K_{R(\epsilon'')}\left(R(\epsilon'')\underset{R}{\otimes}S\right) \\
\big\downarrow & & \big\downarrow \\
\mathrm{PGL}_V\left(R_{\epsilon',\epsilon''}\underset{R}{\otimes}S\right) & \xrightarrow[\simeq]{\psi_S} & \mathrm{PGL}_V\left(R(\epsilon')\underset{R}{\otimes}S\right)\underset{\mathrm{PGL}_V(S)}{\times}\mathrm{PGL}_V\left(R(\epsilon'')\underset{R}{\otimes}S\right) \\
i_1\Big\downarrow\Big\downarrow i_2 & & \bar{i}_1\Big\downarrow\Big\downarrow \bar{i}_2 \\
\mathrm{PGL}_V\left(R_{\epsilon',\epsilon''}\underset{R}{\otimes}\left(S\underset{R}{\otimes}S\right)\right) & \xrightarrow[\psi_{S\otimes S}]{\simeq} & \mathrm{PGL}_V\left(R(\epsilon')\underset{R}{\otimes}\left(S\underset{R}{\otimes}S\right)\right)\underset{\mathrm{PGL}_V\left(S\underset{R}{\otimes}S\right)}{\times}\mathrm{PGL}_V\left(R(\epsilon'')\underset{R}{\otimes}\left(S\underset{R}{\otimes}S\right)\right).
\end{array} \qquad (\mathscr{D})$$

From this we readily see that $\mu_R(S)$ is an isomorphism. This proves the lemma. □

Lemma B.14 *Let $\mathscr{G}$ be a $\mathbb{C}$-group functor satisfying the condition (E). Then, the tangent space at 1 is a $\mathbb{C}$-space functor*

$$R \in \mathbf{Alg} \rightsquigarrow T_1(\mathscr{G})_R.$$

Moreover, $T_1(\mathscr{G})_R$ is naturally an R-module and for any $\mathbb{C}$-algebra homomorphism $\varphi\colon R \to S$, the induced map $T_1(\mathscr{G})_R \to T_1(\mathscr{G})_S$ is an R-module map, where the S-module $T_1(\mathscr{G})_S$ is thought of as an R-module via φ.

We denote this functor as $\mathfrak{G}$, i.e., $\mathfrak{G}(R) = T_1(\mathscr{G})_R$.

Further, for any homomorphism $\mathsf{f}\colon \mathscr{G} \to \mathscr{H}$ of $\mathbb{C}$-group functors satisfying the condition (E), we get an induced morphism $\dot{\mathsf{f}}\colon \mathfrak{G} \to \mathfrak{H}$ such that $\dot{\mathsf{f}}_R\colon \mathfrak{G}(R) \to \mathfrak{H}(R)$ is an R-module homomorphism.

Proof By Lemma B.8, $\mathfrak{G}(R)$ is an R-module. Moreover, it is easy to see that for a $\mathbb{C}$-algebra homomorphism $\varphi\colon R \to S$, the induced map $\mathfrak{G}(R) \to \mathfrak{G}(S)$ is an R-module map. We next prove that $\mathfrak{G}$ is a sheaf for the fppf topology.

Take any $R \in \mathbf{Alg}$ and fppf R-algebra S and consider the commutative diagram

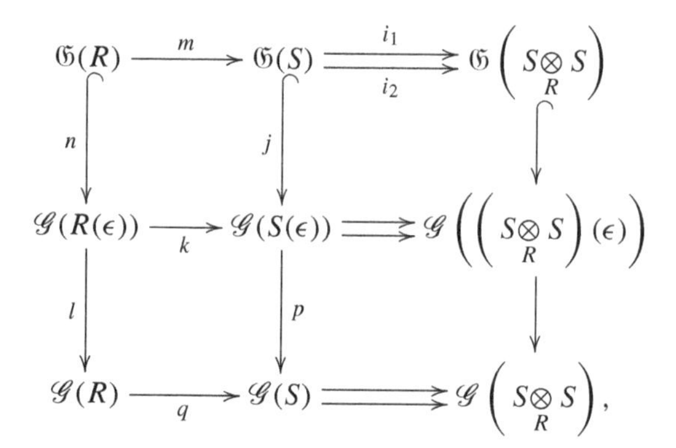

where we have canonically identified $\big(S \otimes_R S\big)(\epsilon) \xrightarrow{\sim} S(\epsilon) \underset{R(\epsilon)}{\otimes} S(\epsilon)$, $s_1 \otimes s_2 + (t_1 \otimes t_2)\epsilon \mapsto s_1 \otimes s_2 + t_1 \otimes (t_2\epsilon)$, for $s_1, s_2, t_1, t_2 \in S$.

From the above diagram, we see that for any $x \in \mathfrak{G}(S)$ which goes to the same element of $\mathfrak{G}\big(S \otimes_R S\big)$ under the two homomorphisms i_1 and i_2, there exists an element $\hat{x} \in \mathscr{G}(R(\epsilon))$ such that

$$j(x) = k(\hat{x}). \tag{1}$$

Further, $q \circ l(\hat{x}) = p \circ k(\hat{x}) = p \circ j(x) = 1$, where the second equality follows from (1). Moreover, q being injective, $l(\hat{x}) = 1$. Hence, $\hat{x} = n(y)$, for some $y \in \mathfrak{G}(R)$. Now, $m(y)$ and x both go to the same element in $\mathscr{G}(S(\epsilon))$ under j (since $j \circ m(y) = k \circ n(y) = k(\hat{x}) = j(x)$). But, j being injective, $m(y) = x$. This proves that $\mathfrak{G}$ is a sheaf for the fppf topology.

The above argument can be summarized[1] by saying that the Weil restriction $\mathscr{G}_\epsilon$ of $\mathscr{G}$ induced from the homomorphism $\mathbb{C} \to \mathbb{C}(\epsilon)$ (i.e., $\mathscr{G}_\epsilon(R) = \mathscr{G}(R(\epsilon))$) is a $\mathbb{C}$-group functor. Moreover, the kernel $\mathfrak{G}$ of the morphism $l\colon \mathscr{G}_\epsilon \to \mathscr{G}$ between $\mathbb{C}$-group functors is a sheaf for the fppf topology and hence $\mathfrak{G}$ is a $\mathbb{C}$-space functor.

The assertion about the induced morphism $\dot{\mathfrak{f}}\colon \mathfrak{G} \to \mathfrak{H}$ follows easily from the definitions. This proves the lemma. □

Definition B.15 (a) Let $\mathscr{G}$ be any $\mathbb{C}$-group functor satisfying the condition (E) and let $\varphi\colon R \to S$ be a $\mathbb{C}$-algebra homomorphism. Then, for any $x \in$

[1] We thank B. Conrad for this observation.

$\mathfrak{G}(R)$ and $\alpha \in S$ such that $\alpha^2 = 0$, we define $e^{\alpha x} \in \mathscr{G}(S)$ as the image of x under

$$\mathfrak{G}(R) \to \mathscr{G}(R(\epsilon)) \to \mathscr{G}(S),$$

where the second map is induced from the homomorphism $\varphi_\alpha \colon R(\epsilon) \to S$ such that $\varphi_{\alpha|R} = \varphi$ and $\varphi_\alpha(\epsilon) = \alpha$.

(b) A $\mathbb{C}$-group functor $\mathscr{G}$ satisfying the condition (E) is said to satisfy the *condition* (L_R) for $R \in \mathbf{Alg}$ if the $\mathbb{C}$-module map $\mathfrak{G}(\mathbb{C}) \to \mathfrak{G}(R)$ (cf. Lemma B.14) induces an R-module isomorphism:

$$\mathfrak{G}(\mathbb{C}) \underset{\mathbb{C}}{\otimes} R \xrightarrow{\sim} \mathfrak{G}(R).$$

If $\mathscr{G}$ satisfies the condition (L_R) for all $R \in \mathbf{Alg}$, then we say that $\mathscr{G}$ satisfies the *condition* (L). If $\mathscr{G}$ satisfies the conditions (E_R) and (L_R) for all finite-dimensional $\mathbb{C}$-algebras R, then we say that $\mathscr{G}$ satisfies the *condition* (L) *finitely*.

By Exercise (B.E.5), if $\mathscr{G}$ is representable represented by an ind-group scheme $G = (G_n)_{n \geq 0}$ with each G_n a scheme of finite type over $\mathbb{C}$, then $\mathscr{G}$ satisfies the condition (L). (By Exercise (B.E.4), it satisfies the condition (E).) If G is any ind-group scheme, then $\mathscr{G}$ satisfies the condition (L) finitely. (For the definition of ind-group schemes see Section 1.1.)

Lemma B.16 *With the notation as in the above Definition B.15,*

(1) $e^{(\alpha+\beta)x} = e^{\alpha x} \cdot e^{\beta x}$, *for* $\alpha, \beta \in S$ *such that* $\alpha, \beta, \alpha+\beta$ *are of square zero.*
(2) $e^{\alpha(x+y)} = e^{\alpha x} \cdot e^{\alpha y}$, *for* $x, y \in \mathfrak{G}(R)$.
(3) For any homomorphism $\mathfrak{f} \colon \mathscr{G} \to \mathscr{H}$ *between* $\mathbb{C}$-*group functors satisfying the condition* (E),

$$\mathfrak{f}(e^{\alpha x}) = e^{\alpha \dot{\mathfrak{f}}(x)}, \quad \text{for } x \in \mathfrak{G}(R) \text{ and } \alpha \in S \text{ with } \alpha^2 = 0,$$

where $\dot{\mathfrak{f}} \colon \mathfrak{G} \to \mathfrak{H}$ *is the induced morphism (cf. Lemma B.14).*

Proof To prove (1), consider

$$\begin{array}{c} \mathfrak{G}(R) \hookrightarrow \mathscr{G}(R(\epsilon)) \xrightarrow{r_R} \mathscr{G}\left(\dfrac{R[\epsilon',\epsilon'']}{\langle\epsilon',\epsilon''\rangle^2}\right) \overset{\mu_R}{\simeq} \mathscr{G}(R(\epsilon')) \times_{\mathscr{G}(R)} \mathscr{G}(R(\epsilon'')) \\ \Big\downarrow \theta_{\alpha,\beta} \\ \mathscr{G}(S), \end{array}$$

where r_R is induced from the R-algebra homomorphism

$$R(\epsilon) \to R[\epsilon', \epsilon'']/\langle \epsilon', \epsilon'' \rangle^2, \qquad \epsilon \mapsto \epsilon' + \epsilon'',$$

the property (E_R) for $\mathscr{G}$ gives the isomorphism μ_R and $\theta_{\alpha,\beta}$ is induced from the R-algebra homomorphism

$$R[\epsilon', \epsilon'']/\langle \epsilon', \epsilon'' \rangle^2 \to S, \quad \epsilon' \mapsto \alpha, \epsilon'' \mapsto \beta.$$

The proof of (2) and (3) is straightforward. □

Definition B.17 (Adjoint representation) (a) Let $\mathscr{G}$ be a $\mathbb{C}$-group functor satisfying the condition (E). Then, for any $R \in \mathbf{Alg}$, we have a representation $\mathrm{Ad} = \mathrm{Ad}_R : \mathscr{G}(R) \to \mathrm{Aut}_R(\mathfrak{G}(R))$ defined as follows.

For any $g \in \mathscr{G}(R(\epsilon))$ and $x \in \mathfrak{G}(R)$, let

$$(\mathrm{Ad}\, g)x := g \cdot x \cdot g^{-1}.$$

Since $\mathfrak{G}(R)$ is a normal subgroup of $\mathscr{G}(R(\epsilon))$, $(\mathrm{Ad}\, g)x \in \mathfrak{G}(R)$. Clearly, $\mathrm{Ad}\, g : \mathfrak{G}(R) \to \mathfrak{G}(R)$ is additive. Moreover, since $\mathfrak{G}(R)$ is abelian (cf. Lemma B.11), the action of $\mathscr{G}(R(\epsilon))$ factors through the quotient $\mathscr{G}(R)$. By the definition of the R-module structure on $\mathfrak{G}(R)$ (cf. the proof of Lemma B.8), for any $z \in R$, the action of z on $\mathfrak{G}(R)$ is induced from the R-algebra homomorphism m_z taking $\epsilon \mapsto z\epsilon$:

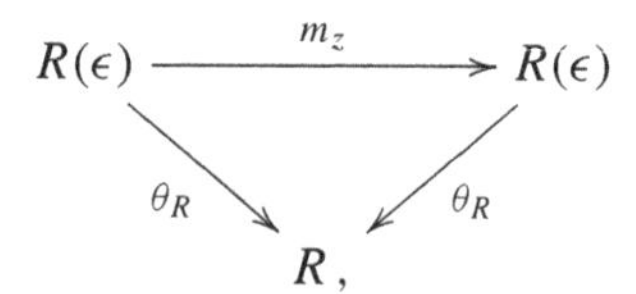

where θ_R is the augmentation map. In particular, the induced action of m_z on $\mathscr{G}(R)$ is trivial. Thus, for $g \in \mathscr{G}(R)$,

$$\begin{aligned}(\mathrm{Ad}\, g)(zx) &= g \cdot zx \cdot g^{-1} \\ &= z(g \cdot x \cdot g^{-1}).\end{aligned}$$

Thus, $\mathrm{Ad}\, g : \mathfrak{G}(R) \to \mathfrak{G}(R)$ is an R-module isomorphism.

Ad is functorial, i.e., for a morphism $\varphi : \mathscr{G} \to \mathscr{H}$ of $\mathbb{C}$-group functors satisfying the condition (E),

$$\dot{\varphi}\,((\mathrm{Ad}\, g)x) = (\mathrm{Ad}\, \varphi(g))\, \dot{\varphi}(x), \quad \text{for } g \in \mathscr{G}(R) \text{ and } x \in \mathfrak{G}(R). \tag{1}$$

(b) Let $\mathscr{G}$ be any group functor satisfying the condition (E) and let $\rho : \mathscr{G} \to \mathbf{GL}_V$ be a homomorphism of $\mathbb{C}$-group functors for a $\mathbb{C}$-vector space V. In this case V (or ρ) is called a *representation* of $\mathscr{G}$. This induces the morphism (cf. Example B.12)

$$\dot{\rho} \colon \mathfrak{G} \to \mathbf{End}_V .$$

(c) Assume further that $\mathscr{G}$ satisfies the condition (L) (so that $\mathfrak{G}(R) = \mathfrak{G}(\mathbb{C}) \underset{\mathbb{C}}{\otimes} R$). Thus,

$$\mathrm{Aut}_R(\mathfrak{G}(R)) = \mathbf{GL}_{\mathfrak{G}(\mathbb{C})}(R) \quad \text{(cf. Example B.4).}$$

In this case we get the homomorphism of group functors

$$\mathbf{Ad} \colon \mathscr{G} \to \mathbf{GL}_{\mathfrak{G}(\mathbb{C})}.$$

This induces the morphism

$$\mathbf{ad} \colon \mathfrak{G} \to \mathbf{End}_{\mathfrak{G}(\mathbb{C})}.$$

In this case, i.e., $\mathscr{G}$ satisfies the condition (L), define a R-bilinear bracket in $\mathfrak{G}(R)$ by

$$[x, y] := (\mathrm{ad}\ x)(y), \quad \text{for} \quad x, y \in \mathfrak{G}(R) = \mathfrak{G}(\mathbb{C}) \underset{\mathbb{C}}{\otimes} R. \tag{2}$$

If $\mathscr{G}$ satisfies the condition (L) only finitely, just as above we can define the bracket in $\mathfrak{G}(R)$ for any finite-dimensional $\mathbb{C}$-algebra R.

Lemma B.18 *Let $\mathscr{G}$ be a $\mathbb{C}$-group functor satisfying the condition (E) and let $\rho \colon \mathscr{G} \to \mathbf{GL}_V$ be a representation in a complex vector space V. Then, for any $x \in \mathfrak{G}(R)$ and $v \in V_R := V \underset{\mathbb{C}}{\otimes} R$,*

$$\rho(e^{\epsilon x})v = v + \epsilon \dot{\rho}(x)(v), \ \textit{as elements of}\ V_{R(\epsilon)} := V_R \oplus \epsilon V_R. \tag{1}$$

Proof Consider the following commutative diagram:

$$\begin{array}{ccccc} \mathfrak{G}(R) & \longrightarrow & \mathscr{G}(R(\epsilon)) & \longrightarrow & \mathscr{G}(R) \\ \downarrow{\scriptstyle \dot{\rho}} & & \downarrow{\scriptstyle \rho} & & \downarrow{\scriptstyle \rho} \\ \mathrm{End}_R(V_R) & \xrightarrow{\theta} & \mathrm{Aut}_{R(\epsilon)}(V_{R(\epsilon)}) & \longrightarrow & \mathrm{Aut}_R(V_R). \end{array}$$

From the commutativity of the above diagram and the definition of the map θ (cf. Example B.12), identity (1) follows, proving the lemma. □

Lemma B.19 *Let $\varphi \colon \mathscr{G} \to \mathscr{H}$ be a homomorphism of $\mathbb{C}$-group functors, where $\mathscr{G}$ and $\mathscr{H}$ satisfy the condition (L) (in particular, they satisfy the condition (E)). Then, the induced morphism $\dot{\varphi} \colon \mathfrak{G} \to \mathfrak{H}$ (cf. Lemma B.14)* *respects the bracket, i.e., for any* $R \in \mathbf{Alg}$, *the* R*-linear map* $\dot{\varphi}_R : \mathfrak{G}(R) \to \mathfrak{H}(R)$ *satisfies*

$$\dot{\varphi}_R[x,y] = [\dot{\varphi}_R x, \dot{\varphi}_R y], \ \textit{for any } x, y \in \mathfrak{G}(R). \tag{1}$$

If $\mathscr{G}$ *and* $\mathscr{H}$ *satisfy the condition* (L) *only finitely, then the above equation (1) is true for finite-dimensional* $\mathbb{C}$*-algebras* R.

Proof Consider the adjoint representations

$$\mathbf{Ad}_{\mathscr{G}} : \mathscr{G} \to \mathbf{GL}_{\mathfrak{G}(\mathbb{C})} \text{ and } \mathbf{Ad}_{\mathscr{H}} : \mathscr{H} \to \mathbf{GL}_{\mathfrak{H}(\mathbb{C})}$$

as in Example B.17(c). By Lemma B.18, for $x, y \in \mathfrak{G}(R) = \mathfrak{G}(\mathbb{C}) \otimes R$, as elements of $\mathfrak{G}(\mathbb{C}) \otimes_{\mathbb{C}} R(\epsilon)$,

$$\mathrm{Ad}_{\mathscr{G}}(e^{\epsilon x})(y) = y + \epsilon(\mathrm{ad}\, x)(y). \tag{2}$$

Applying $\dot{\varphi}_{R(\epsilon)}$ to the above equation (2), we get

$$\dot{\varphi}_{R(\epsilon)}\left(\mathrm{Ad}_{\mathscr{G}}(e^{\epsilon x})(y)\right) = \dot{\varphi}_R y + \epsilon \dot{\varphi}_R\left((\mathrm{ad}\, x)(y)\right). \tag{3}$$

Now,

$$\begin{aligned}\dot{\varphi}_{R(\epsilon)}\left(\mathrm{Ad}_{\mathscr{G}}(e^{\epsilon x})(y)\right) &= \varphi_{R(\epsilon)}\left(e^{\epsilon x} \cdot y \cdot e^{-\epsilon x}\right)\\ &= e^{\epsilon(\dot{\varphi}_R x)} \cdot (\dot{\varphi}_R y) \cdot e^{-\epsilon(\dot{\varphi}_R x)}, \text{ by Lemma B.16(3)}\\ &= \mathrm{Ad}_{\mathscr{H}}\left(e^{\epsilon \dot{\varphi}_R x}\right)(\dot{\varphi}_R y)\\ &= \dot{\varphi}_R(y) + \epsilon\,(\mathrm{ad}\, \dot{\varphi}_R x)\,(\dot{\varphi}_R y), \text{ by Lemma B.18.} \end{aligned} \tag{4}$$

Equating the right-hand side of (3) and (4), we get

$$\dot{\varphi}_R\left((\mathrm{ad}\, x)(y)\right) = (\mathrm{ad}\, \dot{\varphi}_R x)\,(\dot{\varphi}_R y). \tag{5}$$

By the definition of the bracket, the above equation (5) gives

$$\dot{\varphi}_R[x,y] = [\dot{\varphi}_R x, \dot{\varphi}_R y].$$

This proves the lemma. □

Lemma B.20 *Let* $\mathscr{G}$ *be a* $\mathbb{C}$*-group functor satisfying the condition* (L) *and let* $S \in \mathbf{Alg}$ *with elements* $\alpha, \beta \in S$ *of vanishing square. Then, for any homomorphism* $\varphi : R \to S$ *and elements* $x, y \in \mathfrak{G}(R)$, *we have*

$$e^{\alpha x} e^{\beta y} e^{-\alpha x} e^{-\beta y} = e^{\alpha\beta[x,y]}, \ \textit{as elements of } \mathscr{G}(S). \tag{1}$$

If $\mathscr{G}$ *satisfies the condition* (L) *only finitely, then the above equation (1) is satisfied if* R, S *are finite-dimensional* $\mathbb{C}$*-algebras.*

Proof It clearly suffices to prove the lemma for

$$S = R(\epsilon, \epsilon') := R(\epsilon)(\epsilon') \quad \text{and} \quad \alpha = \epsilon,\ \beta = \epsilon'.$$

Now, as elements of $\mathscr{G}(R(\epsilon, \epsilon'))$,

$$\begin{aligned} e^{\epsilon x} e^{\epsilon' y} e^{-\epsilon x} &= e^{\epsilon'(y + \epsilon[x,y])}, \text{ by Lemma B.18 applied to } \rho = \mathbf{Ad} \\ &= e^{\epsilon' \epsilon [x,y]} \cdot e^{\epsilon' y}, \text{ by Lemma B.16(2).} \end{aligned}$$

Thus,

$$e^{\epsilon x} e^{\epsilon' y} e^{-\epsilon x} e^{-\epsilon' y} = e^{\epsilon' \epsilon [x,y]},$$

proving the lemma. □

As a corollary of the above Lemmas B.18 and B.20, we get the following.

Corollary B.21 *Let $\mathscr{G}$ be a $\mathbb{C}$-group functor satisfying the condition (L). Recall the definition of the R-bilinear bracket $[\,,\,]$ in $\mathfrak{G}(R)$ from Definition B.17(c). Let $\rho: \mathscr{G} \to \mathbf{GL}_V$ be a representation in a complex vector space V. Then, for $x, y \in \mathfrak{G}(R)$, under the standard bracket in* $\operatorname{End}_R(V_R)$,

$$\dot{\rho}[x,y] = [\dot{\rho}(x), \dot{\rho}(y)] \in \operatorname{End}_R(V_R). \tag{1}$$

In particular, the bracket $[\,,\,]$ in $\mathfrak{G}(R)$ makes it into an R-Lie algebra.

If $\mathscr{G}$ satisfies the condition (L) only finitely, then the above equation (1) is satisfied and the bracket in $\mathfrak{G}(R)$ makes it into an R-Lie algebra for any finite-dimensional $\mathbb{C}$-algebra R.

In particular, the bracket defined in $T_1(\mathbf{GL}_V)_R \simeq$ $\operatorname{End}_R(V_R)$ (cf. Example B.12) via Definition B.17(c) coincides with the standard bracket in $\operatorname{End}_R(V_R)$ *for any finite-dimensional $\mathbb{C}$-algebra R. (Observe that by Exercise B.E.6,* $\mathbf{GL}_V$ *satisfies the condition (L) finitely.)*

We also denote $\mathfrak{G}$ by $\operatorname{Lie} \mathscr{G}$.

Proof As elements of $\mathbf{End}_V(R(\epsilon')(\epsilon))$,

$$\begin{aligned} &\operatorname{Id} + \epsilon\epsilon' \dot{\rho}[x,y] \\ &= \rho\left(e^{\epsilon\epsilon'[x,y]}\right), \quad \text{by Lemma B.18} \\ &= \rho(e^{\epsilon x}) \rho(e^{\epsilon' y}) \rho(e^{-\epsilon x}) \rho(e^{-\epsilon' y}), \text{ by Lemma B.20} \\ &= (\operatorname{Id} + \epsilon \dot{\rho}(x))(\operatorname{Id} + \epsilon' \dot{\rho}(y))(\operatorname{Id} - \epsilon \dot{\rho}(x))(\operatorname{Id} - \epsilon' \dot{\rho}(y)), \text{ by Lemma B.18} \\ &= \operatorname{Id} + \epsilon\epsilon' \left(\dot{\rho}(x)\dot{\rho}(y) - \dot{\rho}(y)\dot{\rho}(x)\right). \end{aligned}$$

Thus, we get (in $\operatorname{End}_R(V_R)$):

$$\dot{\rho}[x,y] = [\dot{\rho}(x), \dot{\rho}(y)], \quad \text{proving (1).}$$

Now, taking for ρ the Adjoint representation in $V = \mathfrak{G}(\mathbb{C})$, we get (for any $z \in \mathfrak{G}(R)$)

$$[[x,y],z] = [x,[y,z]] - [y,[x,z]],$$

which is exactly the Jacobi identity.

Further, from Lemma B.20, we get in $\mathscr{G}(R(\epsilon,\epsilon'))$:

$$\begin{aligned} e^{-\epsilon\epsilon'[x,y]} &= e^{\epsilon' y} e^{\epsilon x} e^{-\epsilon' y} e^{-\epsilon x} \\ &= e^{\epsilon\epsilon'[y,x]}. \end{aligned}$$

This gives $-[x,y] = [y,x]$. Thus, $[,]$ is an R-Lie algebra bracket in $\mathfrak{G}(R)$. This proves the corollary. □

Definition B.22 Let Γ be an ind-group scheme and let $X = (X_n)_{n\geq 0}$ be an ind-finite type scheme with an action of Γ. Let $\mathscr{V}$ be a Γ-equivariant vector bundle over X (cf. Section 1.1). Let $R \in$ **Alg**. Then,

$$\mathscr{V}_R := \mathscr{V} \times \operatorname{Spec} R \to X_R := X \times \operatorname{Spec} R$$

is a vector bundle. Any element $\gamma \in \Gamma(R)$ acts as a vector bundle automorphism of $\mathscr{V}_R$ given by

$$\gamma \cdot (v,y) = (\gamma(y)v, y), \text{ for } \gamma \in \Gamma(R) := \operatorname{Mor}(\operatorname{Spec} R, \Gamma), v \in \mathscr{V} \text{ and } y \in \operatorname{Spec} R. \quad (1)$$

Thus, the abstract group $\Gamma(R)$ acts on the cohomology

$$H^*(X_R, \mathscr{V}_R) := \varprojlim_n H^*((X_n)_R, (\mathscr{V}_R)_{|(X_n)_R}).$$

Take now R to be a finite-dimensional $\mathbb{C}$-algebra. By the degenerate spectral sequence and the projection formula (cf. (Hartshorne, 1977, Chap. III, Exercises 8.2 and 8.3)):

$$H^p((X_n)_R, \mathscr{V}_R) \simeq H^p(X_n, \mathscr{V}) \otimes R.$$

Hence,

$$\begin{aligned} H^p(X_R, \mathscr{V}_R) &= \varprojlim_n H^p((X_n)_R, \mathscr{V}_R) \\ &\simeq \varprojlim_n \left(H^p(X_n, \mathscr{V}) \otimes R\right) \\ &= H^p(X, \mathscr{V}) \otimes R, \quad \text{since } R \text{ is finite dimensional.} \end{aligned}$$

Thus, for finite-dimensional $\mathbb{C}$-algebras R, we get a functorial group homomorphism

$$\rho_R : \Gamma(R) \to \mathbf{GL}_{H^p(X,\mathscr{V})}(R).$$

It is easy to see that for any $\gamma \in \Gamma(R)$, $\rho_R(\gamma)$ is R-linear. Hence, by Corollary B.21, for any finite-dimensional $\mathbb{C}$-algebra R, we get a Lie algebra homomorphism

$$\dot{\rho}_R \colon \operatorname{Lie}\Gamma \otimes R \to \operatorname{End}_R\left(H^p(X,\mathscr{V}) \otimes R\right).$$

(Observe that by Definition B.15(b), Γ satisfies the condition (L) finitely.) Moreover, $\dot{\rho}_R$ satisfies (cf. (1) of Lemma B.18)

$$\rho_{R(\epsilon)}(e^{\epsilon x})\sigma = \sigma + \epsilon\dot{\rho}_R(x)\sigma, \quad \text{for } \sigma \in H^p(X,\mathscr{V}) \otimes R \text{ and } x \in \operatorname{Lie}\Gamma \otimes R. \tag{2}$$

Lemma B.23 *Let Γ be an irreducible ind-group variety and let $X = (X_n)_{n\geq 0}$ be an ind-variety with an action by Γ (see Section 1.1). Let $\mathscr{V}$ be a Γ-equivariant vector bundle over X and let $\sigma \in H^0(X,\mathscr{V})$. Then, the following are equivalent:*

(a) *σ is $\Gamma(\mathbb{C})$-invariant.*
(b) *σ is annihilated by Lie Γ under the action of Lie Γ on $H^0(X,\mathscr{V})$ defined above in Definition B.22.*

Moreover, any σ satisfying the above equivalent conditions is functorially Γ-invariant (cf. Definition C.22).

Proof (a) $\Rightarrow$ (b) Observe first that σ being $\Gamma(\mathbb{C})$-invariant, σ_R is $\Gamma(R)$-invariant for any finite-dimensional $\mathbb{C}$-algebra R, where σ_R is the section of $\mathscr{V} \times \operatorname{Spec} R \to X \times \operatorname{Spec} R$, defined by $\sigma_R(x,y) = (\sigma x, y)$, for $x \in X$ and $y \in \operatorname{Spec} R$, and the action of $\Gamma(R)$ on $H^0(X_R,\mathscr{V}_R)$ is as in Definition B.22. To prove this, take any $\gamma \in \Gamma(R)$. Then, by (1) of Definition B.22, for any $x \in X(\mathbb{C})$ and $y \in \operatorname{Spec} R$,

$$\begin{aligned}\sigma_R(\gamma\cdot(x,y)) &= (\sigma(\gamma(y)\cdot x), y)\\ &= \gamma\cdot\sigma_R(x,y), \quad \text{since } \sigma \text{ is } \Gamma(\mathbb{C})\text{-invariant.}\end{aligned}$$

Thus, by (2) of Definition B.22 applied to $R = \mathbb{C}$, (b) follows.

(b) $\Rightarrow$ (a) Fix $x_o \in X(\mathbb{C})$ and define the morphism $\sigma_{x_o} : \Gamma \to \mathscr{V}_{x_o}$, $\gamma \mapsto \gamma\cdot\sigma(\gamma^{-1}x_o)$. Clearly, σ_{x_o} is a morphism. Further, for any $\gamma_o \in \Gamma(\mathbb{C})$, we have the following commutative diagram:

$$\begin{array}{ccc} \Gamma & \xrightarrow{\sigma_{x_o}} & \mathscr{V}_{x_o} \\ L_{\gamma_o} \downarrow \wr & & \gamma_o \downarrow \wr \\ \Gamma & \xrightarrow[\sigma_{\gamma_o \cdot x_o}]{} & \mathscr{V}_{\gamma_o \cdot x_o}, \end{array}$$

where L_{γ_o} is the left multiplication by γ_o. From the assumption that σ is annihilated by Lie Γ and the above commutative diagram, we see that

$$\left(\dot{\sigma}_{x_o}\right)_\gamma \equiv 0, \quad \text{for any} \quad \gamma \in \Gamma.$$

Thus, by Lemma 8.1.2, since Γ is irreducible, we get that σ is $\Gamma(\mathbb{C})$-invariant.

To prove that σ is functorially Γ-invariant, observe that Γ and X being ind-varieties; in particular, $\Gamma \times X$ has a filtration by (reduced) varieties $\Gamma_n \times X_n$. Moreover, the varieties are 'determined' by their $\mathbb{C}$-points (cf. (Hartshorne, 1977, Chap. II, Proposition 2.6)). From this it is easy to conclude that σ is functorially Γ-invariant. □

B.E Exercises

In the following exercises $R \in \mathbf{Alg}$.

(1) Show that $R[\epsilon', \epsilon'']/\langle \epsilon', \epsilon'' \rangle^2$ is the fiber product algebra of the augmentation homomorphisms:

$$\theta' \colon R[\epsilon']/\langle \epsilon'^2 \rangle \to R \text{ and } \theta'' \colon R[\epsilon'']/\langle \epsilon''^2 \rangle \to R, \; \epsilon' \mapsto 0, \epsilon'' \mapsto 0$$

(cf. Definition B.6).

(2) Let $\mathscr{F}$ be a representable functor represented by the ind-scheme $X = (X_n)_{n \geq 0}$. Then, show that for any $x \in X(\mathbb{C})$, the tangent space $T_x(\mathscr{F})_{\mathbb{C}}$ coincides with the Zariski tangent space $T_x(X) := \varinjlim_n T_x(X_n)$ of the ind-scheme X at x as a $\mathbb{C}$-vector space.

(3) Let V be any (not necessarily finite-dimensional) vector space over $\mathbb{C}$. Then show that for the $\mathbb{C}$-group functor $\mathbf{PGL}_V$ (cf. Example B.4(2)), the tangent space

$$T_1\,(\mathbf{PGL}_V)_R \simeq \mathrm{End}_R(V_R)/R \cdot I_{V_R}.$$

Hint: Following the notation as in the proof of Lemma B.2, for any fppf R-algebra S, show that the fiber of $K_{R(\epsilon)}(R(\epsilon) \otimes_R S) \to K_R(S)$ over 1 coincides with $\mathrm{End}_R(V_R)/R \cdot I_{V_R}$. In particular, it does not depend upon the choice of S and hence remains the same in the limit.

(4) Let $\mathscr{F}$ be a representable functor represented by an ind-scheme X. Then, show that $\mathscr{F}$ satisfies the condition (E) (cf. Definition B.7).

(5) Let $\mathscr{F}$ be a representable functor represented by an ind-scheme X. Assume further that $X = (X_n)_{n\geq 0}$ is filtered by schemes of finite type over $\mathbb{C}$. Then, show that for any $x \in X(\mathbb{C})$ and $R \in \mathbf{Alg}$,

$$T_x(\mathscr{F})_R \simeq T_x(\mathscr{F})_{\mathbb{C}} \underset{\mathbb{C}}{\otimes} R,$$

where x is also thought of as an element of $\mathscr{F}(R)$ under the map $\mathscr{F}(\mathbb{C}) \to \mathscr{F}(R)$.

In particular, let $\mathscr{G}$ be a representable $\mathbb{C}$-group functor represented by an ind-group scheme $G = (G_n)_{n\geq 0}$, where each G_n is a scheme of finite type over $\mathbb{C}$. Then, $\mathscr{G}$ satisfies the condition (L) (cf. Definition B.15). Observe that $\mathscr{G}$ satisfies the condition (E) by Exercise 4.

Prove further that if G is any ind-group scheme, then $\mathscr{G}$ satisfies the condition (L) finitely.

(6) For a complex vector space V, show that the $\mathbb{C}$-group functor $\mathbf{GL}_V$ (which satisfies the condition (E) as in Example B.12) satisfies the condition (L) if and only if V is finite dimensional. But, show that $\mathbf{GL}_V$ always satisfies the condition (L) finitely.

Also, the functor $\mathbf{PGL}_V$ satisfies the condition (L) finitely (by Lemma B.13 it satisfies the condition (E)).

Hint: Use Example B.12 and Exercise 3.

(7) Let $\mathscr{F}_1, \mathscr{F}_2, \mathscr{F}$ be $\mathbb{C}$-space functors with $\mathbb{C}$-functor morphisms

$$\varphi_1 \colon \mathscr{F}_1 \to \mathscr{F} \text{ and } \varphi_2 \colon \mathscr{F}_2 \to \mathscr{F}.$$

(a) Show that if $\mathscr{F}_1$ and $\mathscr{F}_2$ satisfy the condition (E) and the map μ_R (as in Definition B.6) is injective for $\mathscr{F}$ (for any $\mathbb{C}$-algebra R), then $\mathscr{F}_1 \times_{\mathscr{F}} \mathscr{F}_2$ also satisfies the condition (E).

(b) If $\mathscr{F}_1, \mathscr{F}_2, \mathscr{F}$ are $\mathbb{C}$-group functors satisfying the condition (E) and φ_1, φ_2 are homomorphisms of $\mathbb{C}$-group functors. Then, show that for any $\mathbb{C}$-algebra R,

$$T_e(\mathscr{F}_1 \times_{\mathscr{F}} \mathscr{F}_2)_R = (T_e(\mathscr{F}_1)_R) \times_{T_e(\mathscr{F})_R} (T_e(\mathscr{F}_2)_R).$$

Conclude thus that if $\mathscr{F}_1, \mathscr{F}_2, \mathscr{F}$ satisfy the condition (L) (resp. the condition (L) finitely), then $\mathscr{F}_1 \times_{\mathscr{F}} \mathscr{F}_2$ also satisfies the condition (L) (resp. the condition (L) finitely). In this case the Lie bracket in $T_e(\mathscr{F}_1 \times_{\mathscr{F}} \mathscr{F}_2)_R$ coincides with the (fiber) product Lie bracket in $(T_e(\mathscr{F}_1)_R) \times_{T_e(\mathscr{F})_R}$
$(T_e(\mathscr{F}_2)_R)$ for any R (resp. any finite-dimensional R).

(8) Let $\varphi\colon \mathscr{F} \to \mathscr{G}$ be a morphism of $\mathbb{C}$-space functors. Let $\mathscr{G}' \subset \mathscr{G}$ be an open subfunctor. Then show that

$$\varphi^{-1}(\mathscr{G}')(R) := \varphi_R^{-1}(\mathscr{G}'(R))$$

is an open subfunctor of $\mathscr{F}$ provided it is a $\mathbb{C}$-space functor (i.e., it satisfies the sheaf property).

Prove further that $\varphi^{-1}(\mathscr{G}')$ is a $\mathbb{C}$-space functor if for any fppf R-algebra R',

$$\mathscr{F}(R) \cap \left(\varphi^{-1}(\mathscr{G}')(R')\right) = \varphi^{-1}(\mathscr{G}')(R). \tag{1}$$

Moreover, if a collection $\{\mathscr{G}_i\}$ of open subfunctors is an open cover of $\mathscr{G}$ and if each $\varphi^{-1}(\mathscr{G}_i)$ is a $\mathbb{C}$-space functor, then show that $\{\varphi^{-1}(\mathscr{G}_i)\}$ is an open cover of $\mathscr{F}$.

(9) Let $\varphi\colon \mathscr{F} \to \mathscr{G}$ be a morphism of $\mathbb{C}$-space functors. Let $\mathscr{G}' \subset \mathscr{G}$ be a subfunctor (in particular, $\mathscr{G}'$ is a $\mathbb{C}$-space functor). Assume that for any $R \in \mathbf{Alg}$ and any $x \in \mathscr{F}(R)$, there exists a fppf R-algebra $S = S_x$ (possibly depending upon x) such that

$$\varphi(x_S) \in \mathscr{G}'(S), \quad \text{where } x_S \text{ is the image of } x \text{ in } \mathscr{F}(S).$$

Then show that $\operatorname{Im}\varphi \subset \mathscr{G}'$.

(10) For any ind-scheme X over $\mathbb{C}$, prove that the $\mathbb{C}$-space functor h_X has its open subfunctors given precisely by h_V for an open subset V of X. Further, prove that a collection $\{h_{V_i}\}$ of open subfunctors of h_X is an open covering of h_X if and only if $\cup_i V_i = X$.

B.C Comments

The material in this chapter is fairly standard. We refer to, e.g., Demazure and Gabriel (1980, Chap. I, §1 and Chap. II, §4), Eisenbud and Harris (2000, Chap. VI) and Conrad, Gabber and Prasad (2015, Appendix A.7). For the sheafification of a functor $\mathscr{F}^o\colon \mathbf{Alg} \to \mathbf{Set}$ as in Lemma B.2, also see Milne (2013, Chap. I, Aside 7.19). Lemma B.23 is taken from Beauville and Laszlo (1994, Proposition 7.4). Exercise B.E.10 is taken from Eisenbud and Harris (2000, Exercises VI-8 and VI-11).

Appendix C

Algebraic Stacks

Introduction. We give a brief introduction to *Algebraic stacks*. Rather than defining and discussing stacks in full generality, for clarity, we restrict ourselves to the situation we need to use in this book. The base field is taken to be, as before, $\mathbb{C}$.

We recall the standard definition of étale, smooth and fppf coverings of schemes in Definition C.2. We define a category (resp. groupoid fibration) over the category $\mathfrak{S}$ of quasi-compact and separated schemes over $\mathbb{C}$ in Definition C.3 (resp. Definition C.5) and then a stack in Definition C.10. A stack is a groupoid fibration $\mathscr{T}$ over $\mathfrak{S}$ which satisfies that isomorphisms are a sheaf for $\mathscr{T}$ and, moreover, every descent datum is effective (cf. Definitions C.9, C.8). We define the fiber product (resp. representability) of stacks in Definition C.12 (resp. Definition C.13). Then, we define the class of algebraic stacks and its subclasses: Deligne–Mumford stacks and smooth stacks in Definition C.14. We can speak of the dimension of an algebraic stack (cf. Proposition C.15), which, in general, can be a negative integer.

One of the main examples of stack we discuss is the quotient stack $[\Gamma\backslash X]$, where Γ is an ind-group scheme acting on an ind-scheme X (more generally, a $\mathbb{C}$-group functor acting on a $\mathbb{C}$-space functor) from the left (cf. Example C.18(b)). If X is a scheme of finite type over $\mathbb{C}$ and Γ is an affine algebraic group, $[\Gamma\backslash X]$ is of dimension $\dim X - \dim \Gamma$. Moreover, it is a smooth stack if and only if X is smooth (cf. Lemma C.19). In particular, taking $X = \operatorname{Spec} \mathbb{C}$, we get that the stack $\mathbf{Bun}_G$ of G-bundles (cf. Example C.4(d)) is a smooth stack. Taking $G = \mathrm{GL}_r$, we get that the stack $\mathbf{Vect}^r$ of rank-r vector bundles is a smooth stack. We define quasi-coherent sheaves (in particular, vector bundles) over an algebraic stack $\mathscr{T}$ and the space of global sections of vector bundles over $\mathscr{T}$ (cf. Definitions C.20 and C.21). We define open, closed, quasi-compact locally-closed substacks of a stack as well as union of stacks in Definition C.24.

We have the following result on the global sections of vector bundles over stacks (cf. Proposition C.23).

Let Γ be an ind-group scheme acting on an ind-scheme X such that $[\Gamma \backslash X]$ is an algebraic stack and let $\mathscr{V}$ be a Γ-equivariant vector bundle over X. Let $\bar{\mathscr{V}}$ be the induced vector bundle over the quotient stack $[\Gamma \backslash X]$. Then, the pull-back

$$\mathbf{f}_o^*\colon H^0\left([\Gamma \backslash X], \bar{\mathscr{V}}\right) \to H^0(X, \mathscr{V})$$

is an isomorphism onto $H^0(X, \mathscr{V})^\Gamma$, where $\mathbf{f}_o\colon \mathfrak{S}_X \mapsto [\Gamma \backslash X]$ is the standard presentation.

We follow the standard definition of categories and functors between them; see, e.g., Spanier (1966, Chapter 1, §§1, 2).

As in Section 1.1, let $\mathfrak{S}$ be the category of quasi-compact and separated schemes over $\mathbb{C}$ and morphisms between them. *Unless otherwise stated, by a scheme S we will mean a scheme $S \in \mathfrak{S}$.* For a fixed scheme $S \in \mathfrak{S}$, recall the definition of the category $\mathfrak{S}_S$ of S-schemes from Section 1.1.

Example C.1 If $S = \operatorname{Spec} \mathbb{C} = \{p\}$, clearly $\mathfrak{S}_p$ is just the category $\mathfrak{S}$ itself.

Recall that a morphism $f\colon S \to S'$ between schemes is said to be of *locally finite presentation* if there exists an affine open cover $\{S_i = \operatorname{Spec} R_i\}_{i \in I}$ of S such that for each i, $f(S_i) \subset S_i'$ for an affine open subset $S_i' = \operatorname{Spec} R_i'$ of S' and R_i is finitely presented R_i'-algebra. Also, recall the definitions of smooth and étale morphisms from Mumford and Oda (2015, Definition 5.3.1).

Let $f'\colon S' \to S$ and $f''\colon S'' \to S$ be two morphisms in $\mathfrak{S}$. Recall that their *fiber product* $S' \times_S S''$ is, by definition, the scheme theoretic inverse image of the diagonal $\Delta(S) \subset S \times S$ under the morphism $(f', f'')\colon S' \times S'' \to S \times S$. Observe that if S', S'' and S are quasi-compact and separated schemes then so is the fiber product $S' \times_S S''$ (cf. (Grothendieck, 1960, Chap. I, Proposition 5.5.1 and Corollaire 5.5.3) to prove that $S' \times_S S''$ is separated. Its quasi-compactness is easy to see). Moreover, if S' is separated, then so is any morphism $S' \to S$ for any (not necessarily separated) scheme S (cf. (Grothendieck, 1960)).

Clearly, there are projection maps $\pi'\colon S' \times_S S'' \to S'$, $\pi''\colon S' \times_S S'' \to S''$. It satisfies the following universal property. For any commutative diagram in $\mathfrak{S}$:

$$\begin{array}{ccc} T & \xrightarrow{h''} & S'' \\ {\scriptstyle h'}\downarrow & & \downarrow{\scriptstyle f''} \\ S' & \xrightarrow[f']{} & S, \end{array} \tag{D}$$

there exists a unique morphism $\theta : T \to S' \times_S S''$ such that $\pi' \circ \theta = h'$ and $\pi'' \circ \theta = h''$. If θ is an isomorphism, the diagram (D) is called a *Cartesian diagram.*

Definition C.2 (Étale, smooth and fppf coverings) For any $S \in \mathfrak{S}$, we define an *étale covering* (resp. *smooth covering*, resp. *fppf covering*) of S to be a collection of étale (resp. smooth, resp. flat of locally finite presentation) morphisms $\{\alpha_i : S_i \to S\}_{i \in I}$ such that the disjoint union $\coprod_{i \in I} S_i \to S$ is surjective, where I is an indexing set.

For any covering $\{\alpha_i : S_i \to S\}_{i \in I}$ of S, we define $S_{ij} = S_i \underset{S}{\times} S_j$ (for any i, $j \in I$) and similarly S_{ijk}.

Recall that a property P of morphisms of schemes is said to be *invariant under base change* if, for any Cartesian diagram (D), f'' has that property, then so does h'. Examples of such properties are: smoothness and hence étale maps (cf. (Mumford and Oda, 2015, Proposition 5.3.3)); flatness (cf. (Hartshorne, 1977, Chap. III, Proposition 9.2)); open immersion (clearly); separated maps (cf. (Grothendieck, 1960, Chap. I, Corollaire 5.5.3)); quasi-compact morphisms (cf. (Grothendieck, 1960, §6.6)); proper maps (cf. (Stacks, 2019, Tag 01W4)); closed embeddings (cf. (Hartshorne, 1977, Chap. II, Exercise 3.11(a))); surjective maps (clearly); morphisms locally of finite presentation (cf. (Vistoli, 2005, Proposition 1.2)); finite morphisms. Moreover, all these properties descend through faithfully flat quasi-compact morphisms (in particular, ètale surjective base change) (cf. (Vistoli, 2005, Propositions 1.13 and 1.15)).

Since the property of being étale (resp. smooth; flat of locally finite presentation) morphism is preserved under base change, for any fixed $j \in I$, $\{S_{ij} \to S_j\}_{i \in I}$ is an étale (resp. smooth; fppf) cover of S_j if $\{S_i \to S\}_{i \in I}$ is an étale (resp. smooth; fppf) cover of S.

Definition C.3 (Category over $\mathfrak{S}$) A *category* $\mathscr{T}$ *over* $\mathfrak{S}$ is a category $\mathscr{T}$ together with a covariant functor $\pi : \mathscr{T} \to \mathfrak{S}$ called the *projection*. An object $T \in \mathscr{T}$ is said to *lie over* $S \in \mathfrak{S}$ (or is a *lifting* of S) if $\pi(T) = S$. A similar definition applies for a morphism in $\mathscr{T}$ to lie over a morphism in $\mathfrak{S}$.

For a scheme $S \in \mathfrak{S}$, the *fiber category* $\mathscr{T}_S$ of $\mathscr{T}$ over S is, by definition, the subcategory of $\mathscr{T}$ consisting of all the objects lying over S and all the morphisms in $\mathscr{T}$ lying over the identity morphism of S.

Let $\mathscr{T}_1$ and $\mathscr{T}_2$ be two categories over $\mathfrak{S}$. Then, by a *functor of categories over* $\mathfrak{S}$, we mean a covariant functor $\mathbf{f}\colon \mathscr{T}_1 \to \mathscr{T}_2$ which commutes with the projections to $\mathfrak{S}$.

Let $\mathbf{g}\colon \mathscr{T}_1 \to \mathscr{T}_2$ be another functor over $\mathfrak{S}$. Then, by a *natural transformation* φ from $\mathbf{f}$ to $\mathbf{g}$ over $\mathfrak{S}$, we mean a functor which associates to any object

$E \in \mathscr{T}_1$ a morphism $\varphi_E \in \mathrm{Mor}_{\mathscr{T}_2}(\mathbf{f}(E), \mathbf{g}(E))$ such that the following two conditions are satisfied.

(a) For any morphism $f: E_1 \to E_2$ in $\mathscr{T}_1$, the following diagram is commutative:

$$\begin{array}{ccc} \mathbf{f}(E_1) & \xrightarrow{\mathbf{f}(f)} & \mathbf{f}(E_2) \\ \Big\downarrow{\scriptstyle \varphi_{E_1}} & & \Big\downarrow{\scriptstyle \varphi_{E_2}} \\ \mathbf{g}(E_1) & \xrightarrow[\mathbf{g}(f)]{} & \mathbf{g}(E_2), \end{array}$$

and

(b) For any object $E \in \mathscr{T}_1$, the morphism $\varphi_E: \mathbf{f}(E) \to \mathbf{g}(E)$ induces the identity morphism in $\mathfrak{S}$.

A natural transformation φ from $\mathbf{f}$ to $\mathbf{g}$ over $\mathfrak{S}$ is called a *natural equivalence* over $\mathfrak{S}$, if each of φ_E (for any object $E \in \mathscr{T}_1$) is an isomorphism.

A functor $\mathbf{f}: \mathscr{T}_1 \to \mathscr{T}_2$ over $\mathfrak{S}$ is said to be an *equivalence of categories* over $\mathfrak{S}$ if these exists a functor $\mathbf{f}': \mathscr{T}_2 \to \mathscr{T}_1$ over $\mathfrak{S}$ such that there exists a natural equivalence over $\mathfrak{S}$ between $\mathbf{f}' \circ \mathbf{f}: \mathscr{T}_1 \to \mathscr{T}_1$ and the identity functor $\mathrm{Id}_{\mathscr{T}_1}: \mathscr{T}_1 \to \mathscr{T}_1$ and similarly between $\mathbf{f} \circ \mathbf{f}': \mathscr{T}_2 \to \mathscr{T}_2$ and $\mathrm{Id}_{\mathscr{T}_2}: \mathscr{T}_2 \to \mathscr{T}_2$.

See Exercise (C.E.2) for an equivalent characterization for a functor $\mathbf{f}: \mathscr{T}_1 \to \mathscr{T}_2$ over $\mathfrak{S}$ to be equivalence of categories over $\mathfrak{S}$.

Example C.4 (a) For any scheme $S \in \mathfrak{S}$, the category $\mathfrak{S}_S$ is a category over $\mathfrak{S}$ by taking any object $(f: T \to S) \mapsto T$ and a morphism

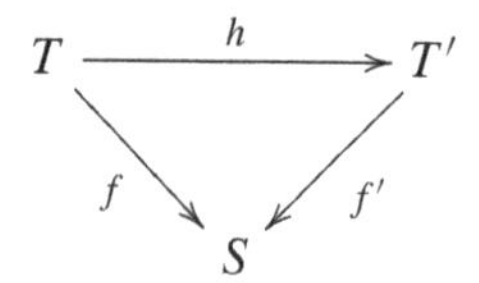

in the category $\mathfrak{S}_S$ to the morphism $h: T \to T'$.

For a category $\mathscr{T}$ over $\mathfrak{S}$, a scheme $S \in \mathfrak{S}$ and a covariant functor $\mathbf{f}: \mathfrak{S}_S \to \mathscr{T}$ over $\mathfrak{S}$, we can associate an object $E \in \mathscr{T}$ lying over S by defining $E = \mathbf{f}(\mathrm{Id}: S \to S)$.

(b) For any schemes $S, T \in \mathfrak{S}$, any covariant functor $\mathbf{f}: \mathfrak{S}_S \to \mathfrak{S}_T$ over $\mathfrak{S}$ gives rise to a morphism $f = f(\mathbf{f}): S \to T$ by taking

$$f := \mathbf{f}(\mathrm{Id} : S \to S).$$

Conversely, any morphism $f: S \to T$ gives rise to a covariant functor of categories $\mathbf{f}: \mathfrak{S}_S \to \mathfrak{S}_T$ over $\mathfrak{S}$ by defining

$$\mathbf{f}(\pi': S' \to S) = f \circ \pi': S' \to T$$

and

$$\mathbf{f}\left(\begin{array}{ccc} S' & \xrightarrow{\ h\ } & S'' \\ & {\scriptstyle \pi'}\searrow \quad \swarrow{\scriptstyle \pi''} & \\ & S & \end{array}\right) = \left(\begin{array}{ccc} S' & \xrightarrow{\ h\ } & S'' \\ & {\scriptstyle f\circ\pi'}\searrow \quad \swarrow{\scriptstyle f\circ\pi''} & \\ & T & \end{array}\right).$$

Clearly, the induced morphism $f(\mathbf{f}) = f$.

(c) Fix a positive integer r. Let $\mathbf{Vect}^r$ be the category of rank-r vector bundles over the category $\mathfrak{S}$. Its objects are locally free sheaves of rank r over schemes. Let $\pi_1 : E_1 \to S_1$ and $\pi_2 : E_2 \to S_2$ be two vector bundles of rank r over the schemes S_1 and S_2. By a *morphism between these two vector bundles*, we mean morphisms $f, \bar{f}$ making the following diagram commutative:

$$\begin{array}{ccc} E_1 & \xrightarrow{\ f\ } & E_2 \\ {\scriptstyle \pi_1}\downarrow & & \downarrow{\scriptstyle \pi_2} \\ S_1 & \xrightarrow[\ \bar{f}\ ]{} & S_2 \end{array} \qquad (*)$$

and further satisfying that f arises from an isomorphism of E_1 with $\bar{f}^*(E_2)$ as locally free sheaves over S_1. Then, $\mathbf{Vect}^r$ is a category over $\mathfrak{S}$ taking a vector bundle $\pi : E \to S$ to the base scheme S and a morphism (*) in $\mathbf{Vect}^r$ to the morphism $\bar{f} : S_1 \to S_2$.

For any scheme $S \in \mathfrak{S}$, the fiber category $\mathbf{Vect}^r(S)$ clearly consists of all the rank-r vector bundles $\pi : E \to S$ as its objects and the morphisms are the vector bundle isomorphisms f:

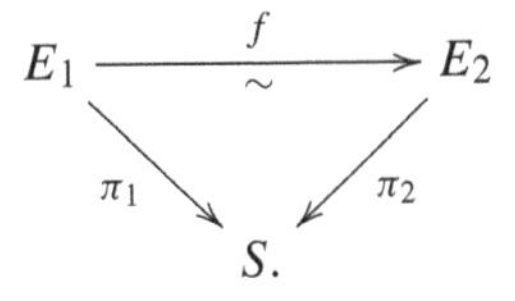

For any scheme $S \in \mathfrak{S}$, as in Example (a), any covariant functor $\mathbf{f}: \mathfrak{S}_S \to \mathbf{Vect}^r$ over $\mathfrak{S}$ gives rise to an associated rank-r vector bundle $\pi : E \to S$ as $\mathbf{f}(\mathrm{Id}: S \to S)$. Conversely, a vector bundle $\pi : E \to S$ of rank r gives rise to a covariant functor $\mathbf{f}_\pi : \mathfrak{S}_S \to \mathbf{Vect}^r$ over $\mathfrak{S}$ by taking

$$\mathbf{f}_\pi(f : T \to S) = f^*E := T \underset{S}{\times} E,$$

and for a morphism h:

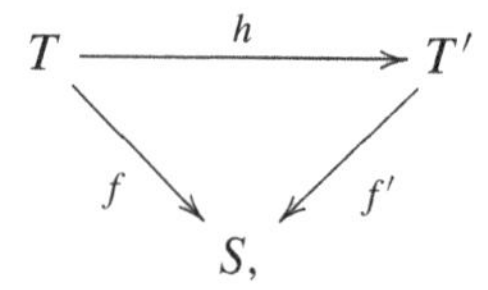

$$\mathbf{f}_\pi(h)\colon f^*(E) \to f'^*(E),\ \ \mathbf{f}_\pi(h)(t,e) = (h(t),e),\ \text{ for } (t,e) \in T \underset{S}{\times} E.$$

Clearly, the associated vector bundle of the functor $\mathbf{f}_\pi$ is π.

(d) Let G be any (not necessarily reductive) affine algebraic group over $\mathbb{C}$. By a principal G-bundle (for short G-bundle) over a scheme $S \in \mathfrak{S}$, we mean a scheme E on which G acts algebraically from the right and a G-equivariant morphism $\pi : E \to S$ (with G acting trivially on S), such that π is *locally iso-trivial* (i.e., locally trivial in the étale topology on S).

Now, define the category $\mathbf{Bun}_G$ over $\mathfrak{S}$ with the objects in $\mathbf{Bun}_G$ being the G-bundles E over schemes S. Let $\pi_1\colon E_1 \to S_1$ and $\pi_2\colon E_2 \to S_2$ be two G-bundles. By a morphism between these two G-bundles (also called a *G-bundle morphism*), we mean a G-equivariant morphism $f\colon E_1 \to E_2$ and a morphism $\bar{f}\colon S_1 \to S_2$ making the following diagram commutative:

$$\begin{array}{ccc} E_1 & \xrightarrow{f} & E_2 \\ {\scriptstyle \pi_1}\downarrow & & \downarrow{\scriptstyle \pi_2} \\ S_1 & \xrightarrow[\bar{f}]{} & S_2. \end{array} \tag{*}$$

Then, $\mathbf{Bun}_G$ is a category over $\mathfrak{S}$ taking a G-bundle $\pi : E \to S$ to the base scheme S and a morphism (*) in $\mathbf{Bun}_G$ to the morphism $\bar{f}\colon S_1 \to S_2$.

For any scheme $S \in \mathfrak{S}$, the fiber category $\mathbf{Bun}_G(S)$ consists of G-bundles $\pi : E \to S$ as its objects and the morphisms are the G-bundle morphisms f:

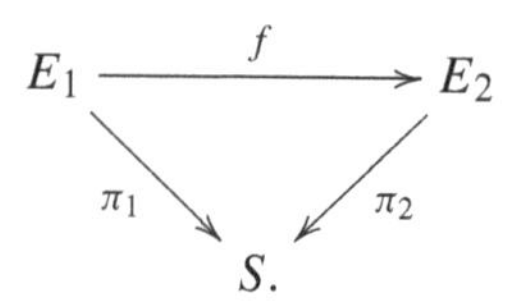

As in Example (c), for any scheme $S \in \mathfrak{S}$, any covariant functor $\mathbf{f}\colon \mathfrak{S}_S \to \mathbf{Bun}_G$ over $\mathfrak{S}$ gives rise to an associated G-bundle $\pi : E \to S$ by taking $\mathbf{f}(\mathrm{Id}\colon S \to S)$. Conversely, any G-bundle $\pi : E \to S$ gives rise to a functor $\mathbf{f}_\pi : \mathfrak{S}_S \to \mathbf{Bun}_G$ over $\mathfrak{S}$ such that the associated G-bundle of the morphism $\mathbf{f}_\pi$ is π.

The functor $\mathbf{f}_r : \mathbf{Vect}^r \to \mathbf{Bun}_{\mathrm{GL}_r}$ over $\mathfrak{S}$, which associates to any rank-r vector bundle E over S its frame bundle $\mathbf{f}_r(E)$ defined by

$$\mathbf{f}_r(E)(\theta : T \to S) = \text{ set of } \mathcal{O}_T\text{-module isomorphisms } \mathcal{O}_T^r \simeq \theta^*(E),$$

is an equivalence of categories over $\mathfrak{S}$.

Recall that a *groupoid category* by definition is a category where all the morphisms are isomorphisms.

Definition C.5 A category $\mathcal{T}$ over $\mathfrak{S}$ is called a *groupoid fibration* over $\mathfrak{S}$ (or a *category fibered in groupoids* over $\mathfrak{S}$) if for any morphism $f : S_1 \to S_2$ in $\mathfrak{S}$ and $T_2 \in \mathcal{T}$ lying over S_2, their exists a morphism $f_{T_2} : T_1 \to T_2$ lying over f in $\mathcal{T}$ (in particular, $T_1 \in \mathcal{T}$ lies over S_1) and the lifting is unique up to a unique isomorphism, i.e., for any other morphism $f'_{T_2} : T'_1 \to T_2$ lying over f, there is a *unique* isomorphism $\alpha : T'_1 \to T_1$ lying over Id_{S_1} such that $f_{T_2} \circ \alpha = f'_{T_2}$.

By Exercise C.E.1, any f_{T_2} lying over an isomorphism $f : S_1 \to S_2$ is an isomorphism in $\mathcal{T}$. In particular, the fiber category $\mathcal{T}_S$ of $\mathcal{T}$ over any $S \in \mathfrak{S}$ (defined in Definition C.3) is a groupoid.

Let $f : S_1 \to S_2$ be a morphism in $\mathfrak{S}$ and let T'_2 and T''_2 be two objects in $\mathcal{T}$ lying over S_2 with a morphism (and hence an isomorphism) $\alpha : T'_2 \to T''_2$ in the fiber category $\mathcal{T}_{S_2}$. Let $f_{T'_2} : T'_1 \to T'_2$ and $f_{T''_2} : T''_1 \to T''_2$ be two lifts of f. (Observe that T'_1 and $f_{T'_2}$ are not unique and similarly for T''_1 and $f_{T''_2}$, but we fix one.) Then, by the definition of groupoid fibration, there exists a unique morphism (and hence an isomorphism) $f^*\alpha$ in the fiber category $\mathcal{T}_{S_1}$, making the following diagram commutative:

$$\begin{array}{ccc} T'_1 & \xrightarrow{f_{T'_2}} & T'_2 \\ {\scriptstyle f^*\alpha}\downarrow & & \downarrow{\scriptstyle \alpha} \\ T''_1 & \xrightarrow[f_{T''_2}]{} & T''_2. \end{array}$$

Let $g : S_2 \to S_3$ be another morphism in $\mathfrak{S}$ and let $T \in \mathcal{T}$ lie over S_3. Fix some lifts

$$T_{g\circ f} \xrightarrow{(g\circ f)_T} T \qquad \text{and} \qquad T_g \xrightarrow{g_T} T$$

of $g \circ f$ and g, respectively. Then, there exists a unique lift $T_{g\circ f} \xrightarrow{\hat{f}_{T_g}} T_g$ of f making the following diagram commutative:

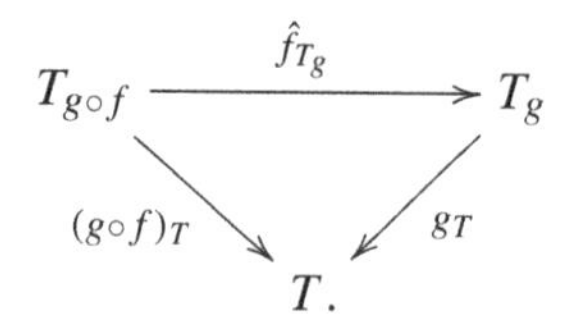

Example C.6 It is easy to see that the categories $\mathfrak{S}_S$ (for any scheme $S \in \mathfrak{S}$) and $\mathbf{Bun}_G$ (for any affine algebraic group G), defined in Example C.4, are groupoid fibrations over $\mathfrak{S}$. To prove it for $\mathfrak{S}_S$, for any morphism $f: S_1 \to S_2$ in $\mathfrak{S}$ and $T_2 = (\pi : S_2 \to S)$, we take $T_1 = (\pi \circ f : S_1 \to S)$ and $f_{T_2} = f$.

To prove it for $\mathbf{Bun}_G$, for $f: S_1 \to S_2$ in $\mathfrak{S}$ and $T_2 = (\pi_2 : E_2 \to S_2)$ in $\mathbf{Bun}_G$, we take $T_1 = \big(f^*E_2 := S_1 \times_{S_2} E_2 \to S_1\big)$ and the lifting of f is taken to be induced from the standard projection map $\pi_2 : S_1 \times_{S_2} E_2 \to E_2$.

In particular, $\mathbf{Vect}^r$ is a groupoid fibration over $\mathfrak{S}$.

Definition C.7 (Choice of a specific lift) Let $\mathscr{T}$ be a category fibered in groupoids over $\mathfrak{S}$. From now on we consider those categories $\mathscr{T}$ fibered in groupoids over $\mathfrak{S}$ such that for any morphism $f: S_1 \to S_2$ in $\mathfrak{S}$ and any object $E_2 \in \mathscr{T}$ over S_2, there is a preassigned object denoted $f^*(E_2)$ over S_1 (depending upon f) and a lift (to be denoted) $f_{E_2}: f^*(E_2) \to E_2$ of f in $\mathscr{T}$. We call such a $\mathscr{T}$ as a *category fibered in groupoids over* $\mathfrak{S}$ *with preassigned pull-backs*. A recipe for such pull-backs will be clear in all the examples of fibered categories that of interest to us. For example, the categories $\mathfrak{S}_S$ (for any $S \in \mathfrak{S}$) and $\mathbf{Bun}_G$ are assigned pull-backs as in Example C.6. When the reference to f is clear, we will simply denote $f^*(E_2)$ by $E_{2|S_1}$. For another morphism $g: S_2 \to S_3$ in $\mathfrak{S}$ and an object $E_3 \in \mathscr{T}$ over S_3, we get the chosen lifts $g_{E_3}: g^*(E_3) \to E_3$ of g and $(g\circ f)_{E_3}: (g\circ f)^*(E_3) \to E_3$ of $g\circ f$. Then, by Definition C.5, there exists a unique lift $\hat{f}_{g^*(E_3)}$ of f making the following diagram commutative:

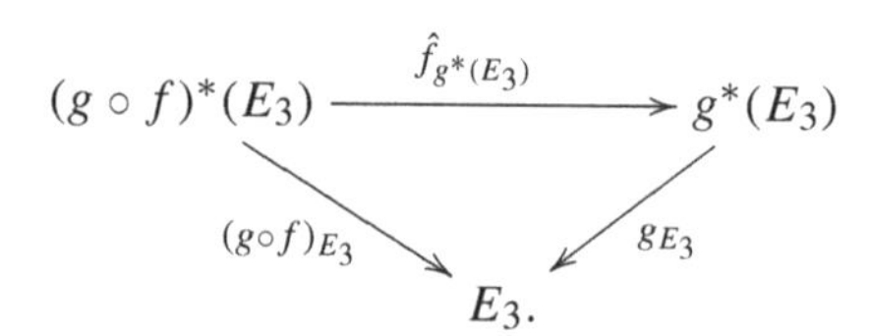

Observe that by our pre-assigned choice, we do not necessarily have $(g \circ f)^*(E_3) = f^*(g^*(E_3))$ and, even if they are equal, $f_{g^*(E_3)}$ may not be equal to $\hat{f}_{g^*(E_3)}$.

As above, let $\mathscr{T}$ be a category fibered in groupoids over $\mathfrak{S}$ with pre-assigned pull-backs.

Definition C.8 (Descent datum) A *descent datum* for $\mathscr{T}$ over a scheme $S \in \mathfrak{S}$ consists of the following:

(a) A fppf cover $\{S_i\}_{i\in I}$ of S.

(b) For each $i \in I$, an object $E_i \in \mathscr{T}$ over S_i.

(c) For each $i, j \in I$, an isomorphism $\alpha_{ij}\colon E_{i|S_{ij}} \to E_{j|S_{ij}}$ in the fiber over S_{ij}.

(d) (Cocycle condition) For each $i, j, k \in I$, the isomorphisms α_{ij} satisfy

$$\alpha_{ik\,|S_{ijk}} = \alpha_{jk\,|S_{ijk}} \circ \alpha_{ij\,|S_{ijk}}. \tag{1}$$

Here S_{ij}, S_{ijk} are defined in Definition C.2 and $\alpha_{ij\,|S_{ijk}}\colon E_{i\,|S_{ijk}} \to E_{j\,|S_{ijk}}$ is the unique morphism in the fiber over S_{ijk} making the following diagram commutative (as guaranteed by the definition of groupoid fibration in Definition C.5):

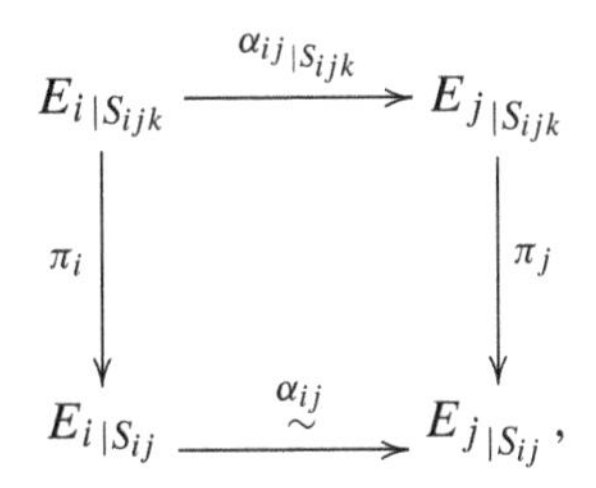

where the morphism π_i is the unique morphism $\hat{f}_{g^*(E_i)}$ lifting $S_{ijk} \to S_{ij}$ (as in Definition C.7 for $S_{ijk} \to S_{ij} \to S_i$ and the object E_i over S_i) making the following diagram commutative:

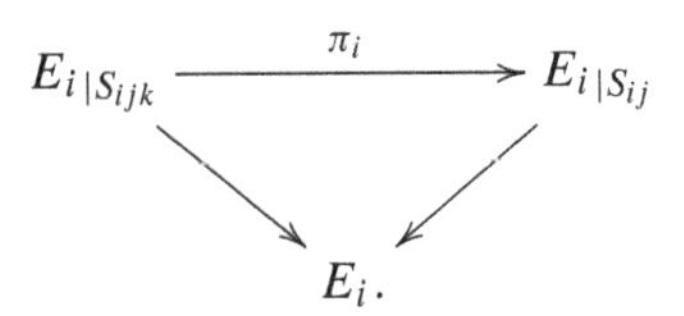

The map π_j is defined similarly.

The above descent datum is said to be *effective* if there exists an object $E \in \mathscr{T}$ over S together with isomorphisms (for all $i \in I$)

$$\alpha_i\colon E_{|S_i} \to E_i$$

in the fiber over S_i such that (in the fiber over S_{ij})

$$\alpha_{ij} = \alpha_{j\,|S_{ij}} \circ \left(\alpha_{i\,|S_{ij}}\right)^{-1},$$

where we first define the (unique) morphism θ_i lifting the map $S_{ij} \to S_i$ (as the map $\hat{f}_{g^*(E)}$ defined in Definition C.7 for $S_{ij} \to S_i \to S$ and the object E over S) making the following diagram commutative:

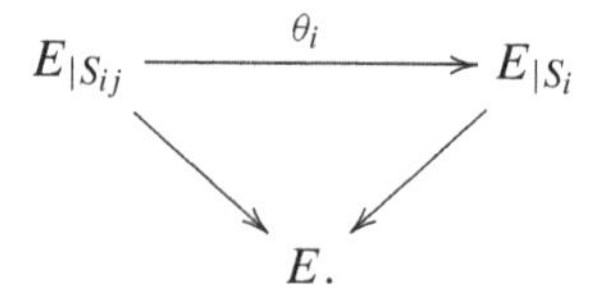

Now, we define $\alpha_{i\,|S_{ij}} : E_{|S_{ij}} \to E_{i\,|S_{ij}}$ in the fiber over S_{ij} to be the unique morphism making the following diagram commutative:

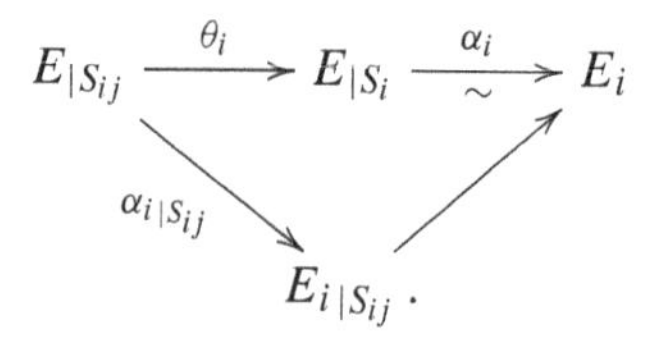

Definition C.9 Let $\mathscr{T}$ be a category fibered in groupoids over $\mathfrak{S}$ with pre-assigned pull-backs. Then, the *isomorphisms are said to be a sheaf for* $\mathscr{T}$ if the following 'glueing' property holds:

For any scheme $S \in \mathfrak{S}$, any fppf covering $\{S_i\}_{i \in I}$ of S, any E, E' objects over S and any collection of isomorphisms $\beta_i : E_{|S_i} \to E'_{|S_i}$ in the fiber over S_i such that

$$\beta_{i\,|S_{ij}} = \beta_{j\,|S_{ij}} : E_{|S_{ij}} \to E'_{|S_{ij}}$$

in the fiber over S_{ij}, where $\beta_{i_{|S_{ij}}}$ is the unique morphism in the fiber over S_{ij} making the following diagram commutative (θ_i is defined in Definition C.8):

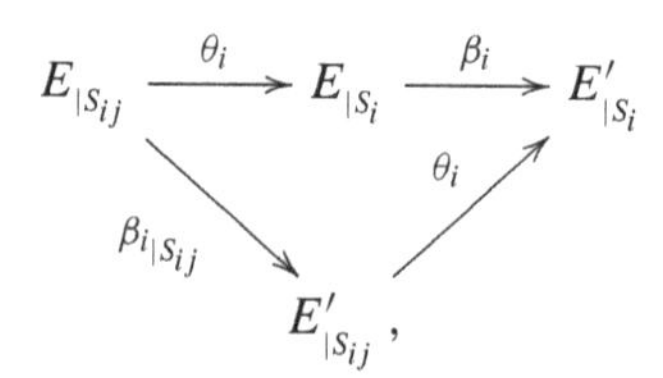

there is a unique isomorphism $\beta : E \to E'$ in the fiber over S such that $\beta_{|S_i} = \beta_i$, where $\beta_{|S_i}$ is the unique morphism in the fiber over S_i making the following diagram commutative (cf. Definition C.5):

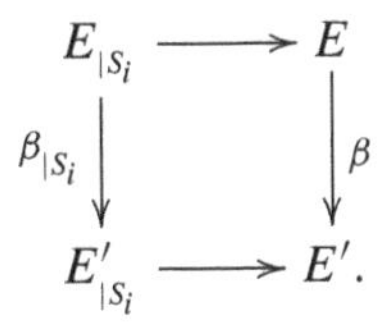

Finally, we come to the following crucial definition.

Definition C.10 A *stack* is a category $\mathscr{T}$ fibered in groupoids over $\mathfrak{S}$ with pre-assigned pull-backs such that the following two properties are satisfied:

(a) isomorphisms are a sheaf for $\mathscr{T}$, and

(b) every (fppf) descent datum is effective.

By a *morphism* $\mathbf{f}\colon \mathscr{T}_1 \to \mathscr{T}_2$ of stacks, we mean a covariant functor $\mathbf{f}\colon \mathscr{T}_1 \to \mathscr{T}_2$ of categories over $\mathfrak{S}$ (cf. Definition C.3).

Observe that for any descent datum in a stack $\mathscr{T}$, its effectivity gives rise to $(E,\alpha_i)_{i\in I}$ as in Definition C.8. Now, by the 'sheaf' property of $\mathscr{T}$, we get that for any other possibility $(E',\alpha'_i)_{i\in I}$, there exists a unique isomorphism (in the fiber over S) $\beta\colon E \to E'$ such that $\alpha'_i \circ (\beta_{|S_i}) = \alpha_i$, for all $i \in I$.

Let $\hat{\mathfrak{S}}$ be the category of schemes which can be written as an arbitrary disjoint union of quasi-compact and separated schemes over $\mathbb{C}$, and all the morphisms between them.

Example C.11 (1) A $\mathbb{C}$-space functor $\mathscr{F}$ can canonically be thought of as a stack $\bar{\mathscr{F}}$ as follows. Its objects are the elements of the set of $\mathbb{C}$-functor morphisms $\mathrm{Mor}(S,\mathscr{F}) = \mathscr{F}(S)$, where S varies over $\mathfrak{S}$ and we think of the scheme S as a $\mathbb{C}$-space functor (as in Definition B.1). The morphisms from $\varphi \in \mathrm{Mor}(S,\mathscr{F})$ to $\varphi' \in \mathrm{Mor}(S',\mathscr{F})$ consist of morphisms $f\colon S \to S'$ making the following diagram commutative:

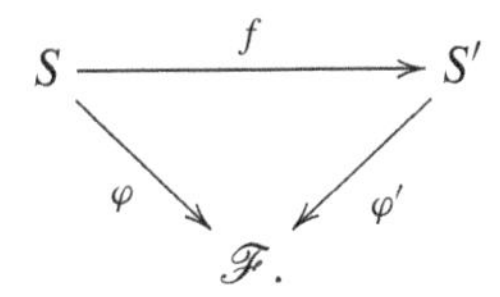

Then, $\bar{\mathscr{F}}$ is a category over $\mathfrak{S}$ by taking $\varphi \in \mathrm{Mor}(S,\mathscr{F})$ to S and the morphism $f\colon S \to S'$ (from $\varphi \in \mathrm{Mor}(S,\mathscr{F})$ to $\varphi' \in \mathrm{Mor}(S',\mathscr{F})$) to f itself. Then, it is easy to see that $\bar{\mathscr{F}}$ is a stack.

In particular, the category $\mathfrak{S}_S$ over $\mathfrak{S}$, for any $S \in \mathfrak{S}$, defined in Example C.4(a) is a stack. The same definition works to define the category $\mathfrak{S}_S$ over $\mathfrak{S}$ for any $S \in \hat{\mathfrak{S}}$. Then, the category $\mathfrak{S}_S$ over $\mathfrak{S}$ is a stack for any

$S \in \hat{\mathfrak{S}}$. More generally, for an ind-scheme S, $\mathfrak{S}_S$ (defined the same way as $\mathfrak{S}_S$ for $S \in \mathfrak{S}$) is a stack.

(2) Let G be an affine algebraic group. Then, $\mathbf{Bun}_G$ (defined in Example C.4(d)) is a stack (cf. (Laumon and Moret-Bailly, 1999, §3.4.2), also see Wang (2011, Theorem 1.0.1) for a more general result). In particular, taking $G = \mathrm{GL}_r$, we get that $\mathbf{Vect}^r$ (defined in Example C.4(c)) is a stack. In fact, $\mathbf{Bun}_G$ is an algebraic stack (cf. (Laumon and Moret-Bailly, 1999, §4.6.1), also see (Wang, 2011, Theorem 1.0.1) for a more general result). (For the definition of algebraic stack, see Definition C.14.)

(3) Let Γ be an (affine) ind-group scheme over $\mathbb{C}$ acting on an ind-scheme X from the left. Then, we have the quotient stack $[\Gamma\backslash X]$ defined in Example C.18(b). This is one of our most important examples. If Γ is an affine algebraic group and $X = \operatorname{Spec}\mathbb{C}$, then $[\Gamma\backslash X]$ is the category $\mathbf{Bun}_\Gamma$ over $\mathfrak{S}$.

Definition C.12 (Fiber product of stacks) Let $\mathbf{f}\colon \mathscr{T}_1 \to \mathscr{T}$ and $\mathbf{g}\colon \mathscr{T}_2 \to \mathscr{T}$ be morphisms of stacks. Their fiber product is defined to be the category $\mathscr{T}_1 \underset{\mathscr{T}}{\times} \mathscr{T}_2$ over $\mathfrak{S}$ as follows.

Its objects are the triples (E_1, E_2, α), where E_1 (resp. E_2) is an object of $\mathscr{T}_1$ (resp. $\mathscr{T}_2$) both lying over the same scheme S in $\mathfrak{S}$, α is a morphism $\mathbf{f}(E_1) \to \mathbf{g}(E_2)$ in the fiber $\mathscr{T}_S$ of $\mathscr{T}$. Its morphisms from (E_1, E_2, α) to (E_1', E_2', α') are given by the pairs (β_1, β_2), where the morphisms $\beta_1 : E_1 \to E_1'$ and $\beta_2 : E_2 \to E_2'$ in $\mathscr{T}_1$ and $\mathscr{T}_2$, respectively lie over the same morphism $S \to S'$ in $\mathfrak{S}$, and they satisfy

$$\mathbf{g}(\beta_2) \circ \alpha = \alpha' \circ \mathbf{f}(\beta_1) : \mathbf{f}(E_1) \to \mathbf{g}(E_2').$$

Clearly, the projections

$$\pi_1 : \mathscr{T}_1 \underset{\mathscr{T}}{\times} \mathscr{T}_2 \to \mathscr{T}_1, \quad (E_1, E_2, \alpha) \mapsto E_1; \quad (\beta_1, \beta_2) \mapsto \beta_1$$

$$\pi_2 : \mathscr{T}_1 \underset{\mathscr{T}}{\times} \mathscr{T}_2 \to \mathscr{T}_2, \quad (E_1, E_2, \alpha) \mapsto E_2; \quad (\beta_1, \beta_2) \mapsto \beta_2,$$

provide covariant functors over $\mathfrak{S}$.

By Exercise C.E.3, $\mathscr{T}_1 \underset{\mathscr{T}}{\times} \mathscr{T}_2$ is a stack.

These functors can be organized into the following (in general non-commutative) diagram:

$$\begin{array}{ccc} \mathscr{T}_1 \underset{\mathscr{T}}{\times} \mathscr{T}_2 & \xrightarrow{\pi_2} & \mathscr{T}_2 \\ \pi_1 \downarrow & & \downarrow \mathbf{g} \\ \mathscr{T}_1 & \xrightarrow[\mathbf{f}]{} & \mathscr{T}. \end{array}$$

It is easy to see from the definition that the functor $\mathbf{g} \circ \pi_2$ is naturally equivalent to the functor $\mathbf{f} \circ \pi_1$ over $\mathfrak{S}$ (see Definition C.3). Further, for any diagram of covariant functors over $\mathfrak{S}$ (where $\mathscr{F}$ is any category over $\mathfrak{S}$):

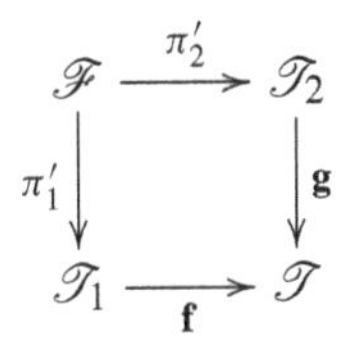

such that $\mathbf{f} \circ \pi_1'$ is naturally equivalent to the functor $\mathbf{g} \circ \pi_2'$ over $\mathfrak{S}$, there exists a unique covariant functor $\theta \colon \mathscr{F} \to \mathscr{T}_1 \times_{\mathscr{T}} \mathscr{T}_2$ over $\mathfrak{S}$ up to a natural equivalence over $\mathfrak{S}$ such that $\pi_1 \circ \theta$ is naturally equivalent to π_1' over $\mathfrak{S}$ and $\pi_2 \circ \theta$ is naturally equivalent to π_2' over $\mathfrak{S}$.

Now, we can define the representability of stacks and representability of morphisms between them.

Definition C.13 A stack $\mathscr{T}$ over $\mathfrak{S}$ is *representable* by a scheme $S \in \hat{\mathfrak{S}}$ if there is an equivalence of categories over $\mathfrak{S}$ between $\mathfrak{S}_S$ and $\mathscr{T}$. (By Example C.11(1), $\mathfrak{S}_S$ is indeed a stack.) Observe that by Example C.4(b), such an S is unique up to an isomorphism of schemes.

A morphism $\mathbf{f} \colon \mathscr{T}_1 \to \mathscr{T}_2$ of stacks is called *representable* if for any scheme $S \in \mathfrak{S}$ and any morphism $\mathbf{g}_S \colon \mathfrak{S}_S \to \mathscr{T}_2$ of stacks, the fiber product $\mathfrak{S}_S \times_{\mathscr{T}_2} \mathscr{T}_1$ is representable by a scheme say $S_{\mathbf{f}} \in \hat{\mathfrak{S}}$.

Let P be a property of morphisms of schemes which is invariant under base change (cf. Definition C.2) and descends through étale surjective base change. A *representable morphism* $\mathbf{f}$: $\mathscr{T}_1 \to \mathscr{T}_2$ *of stacks has property* P if for any morphism of stacks $\mathfrak{S}_S \to \mathscr{T}_2$ (for any scheme $S \in \mathfrak{S}$), the corresponding morphism $S_{\mathbf{f}} \to S$ induced from the projection $\hat{\pi}_1$ (cf. Example C.4(b)):

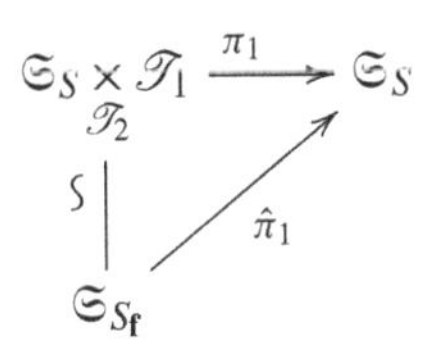

has property P, where $\hat{\pi}_1$ is the morphism making the above triangle commutative. In this case, we say that π_1 *has property* P.

We finally come to the definition of Deligne–Mumford and algebraic stacks.

Definition C.14 Let $\mathscr{T}$ be a stack such that the diagonal $\Delta \colon \mathscr{T} \to \mathscr{T} \times \mathscr{T}$ is representable (by a scheme in $\hat{\mathfrak{S}}$) and it is separated and quasi-compact. Then, such a stack $\mathscr{T}$ is called a *Deligne–Mumford stack* (resp. *algebraic* *stack*) if there exists a scheme $S_o \in \hat{\mathfrak{S}}$ and a representable morphism $\mathbf{f}_o : \mathfrak{S}_{S_o} \to \mathscr{T}$ of stacks such that it is an étale (resp. smooth) surjection (cf. (Laumon and Moret-Bailly, 1999, Définition 4.1)). In this case, such a morphism $\mathbf{f}_o : \mathfrak{S}_{S_o} \to \mathscr{T}$ is called an *étale presentation* (resp. *smooth presentation*) of $\mathscr{T}$. By abuse of notation we say that S_o is an étale (resp. smooth) presentation of the stack $\mathscr{T}$. If an étale (or smooth) presentation S_o of $\mathscr{T}$ can be chosen so that S_o is a smooth $\mathbb{C}$-scheme, than we call $\mathscr{T}$ to be a *smooth stack*.

Observe that if for some smooth presentation S_o of $\mathscr{T}$, S_o is smooth, then for any other smooth presentation S'_o, S'_o is smooth as well (cf. (Mumford and Oda, 2015, Proposition 5.3.5)).

It may be mentioned that in a more conventional definition of algebraic stack one only requires the diagonal Δ to be representable by an algebraic space, but in our definition we require it to be in the category $\hat{\mathfrak{S}}$.

A smooth presentation $\mathbf{f}_o : \mathfrak{S}_{S_o} \to \mathscr{T}$ is said to have *equi-dimensional fibers* if for any field $K \supset \mathbb{C}$ and any morphism $\varphi : \mathfrak{S}_{\mathrm{Spec}\, K} \to \mathscr{T}$, the dimension of the fiber product $\mathfrak{S}_{\mathrm{Spec}\, K} \times_{\mathscr{T}} \mathfrak{S}_{S_o} \to \mathfrak{S}_{\mathrm{Spec}\, K}$ over $\mathrm{Spec}\, K$ does not depend upon the isomorphism φ and K.

Proposition C.15 *Let $\mathbf{f}_o : \mathfrak{S}_{S_o} \to \mathscr{T}$ be a smooth presentation of an algebraic stack $\mathscr{T}$ with equi-dimensional fibers such that $S_o \in \hat{\mathfrak{S}}$ is a disjoint union of schemes of finite type over $\mathbb{C}$ which are all equi-dimensional of the same dimension. (Whenever it exists, we call such a presentation $\mathfrak{S}_{S_o}$ of $\mathscr{T}$ an* equi-dimensional smooth presentation *of $\mathscr{T}$.) Then, for any scheme $S \in \mathfrak{S}$ of finite type over $\mathbb{C}$ and any stack morphism $\mathfrak{S}_S \to \mathscr{T}$, the relative dimension of the induced smooth surjective morphism $S_{\mathbf{f}_o} \to S$ is independent of S, where $S_{\mathbf{f}_o}$ is defined by $\mathfrak{S}_{S_{\mathbf{f}_o}} \simeq \mathfrak{S}_S \times_{\mathscr{T}} \mathfrak{S}_{S_o}$. Let us denote this dimension by $\mu(\mathbf{f}_o)$.*

Further,

$$d(\mathscr{T}) := \dim S_o - \mu(\mathbf{f}_o)$$

is independent of the choice of equi-dimensional smooth presentation S_o of the stack $\mathscr{T}$.

The number $d(\mathscr{T})$ (which can be a negative integer) is called the dimension of the algebraic stack $\mathscr{T}$. (Observe that $d(\mathscr{T})$ makes sense if we can choose an equi-dimensional smooth presentation S_o.)

From now on, we denote $d(\mathscr{T})$ by $\dim(\mathscr{T})$, *whenever it makes sense.*

Proof Observe first that since $\mathbf{f}_o : \mathfrak{S}_{S_o} \to \mathscr{T}$ has equi-dimensional fibers (by assumption), for any morphism $\mathfrak{S}_S \to \mathscr{T}$ (where S is a scheme of finite type over $\mathbb{C}$), the induced smooth morphism $S_{\mathbf{f}_o} \to S$ has a constant relative dimension, which does not depend upon the choice of S (cf. (Hartshorne, 1977, Chap. III, Theorem 10.2)).

To check that $\dim S_o - \mu(\mathbf{f}_o)$ is independent of the choice of equi-dimensional smooth presentation $\mathbf{f}_o \colon \mathfrak{S}_{S_o} \to \mathscr{T}$, take another such presentation $\mathbf{f}'_o \colon \mathfrak{S}_{S'_o} \to \mathscr{T}$ for an equi-dimensional $S'_o \in \hat{\mathfrak{S}}$. These two presentations give rise to morphisms:

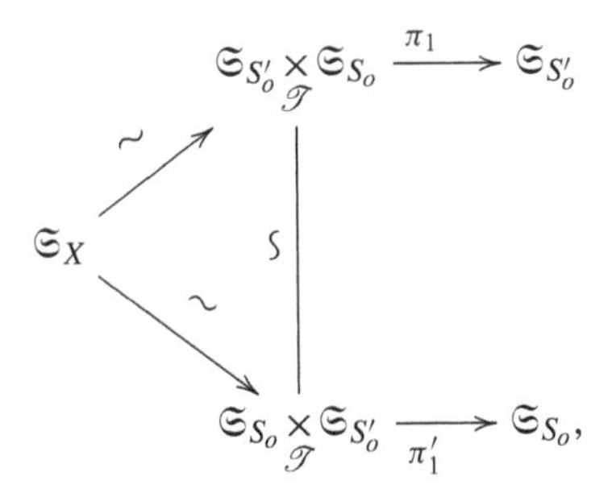

where X is the scheme $(S'_o)_{\mathbf{f}_o} \simeq (S_o)_{\mathbf{f}'_o} \in \hat{\mathfrak{S}}$. The morphisms π_1 and π'_1 give rise to smooth and surjective morphisms: $X \xrightarrow{\hat{\pi}_1} S'_o$ and $X \xrightarrow{\hat{\pi}'_1} S_o$. From this we get that

$$\dim X = \dim S'_o + \mu(\mathbf{f}_o) = \dim S_o + \mu(\mathbf{f}'_o).$$

Thus, we get

$$\dim S_o - \mu(\mathbf{f}_o) = \dim S'_o - \mu(\mathbf{f}'_o),$$

proving the proposition. □

The following definition extends the definition of principal G-bundles (for affine algebraic groups) as in Example C.4(d) to Γ-torsors for $\mathbb{C}$-group functors Γ.

Definition C.16 Let Γ be a $\mathbb{C}$-group functor (cf. Definition B.1) and let $S \in \mathfrak{S}$ be a scheme. By a Γ-*torsor* over S, we mean a $\mathbb{C}$-space functor $\mathbf{E}$ with the left action of Γ together with a Γ-invariant morphism $\pi \colon \mathbf{E} \to S$ (with the trivial action of Γ on S), (where we identify S with the functor h_S as in Definition B.1) such that under an fppf cover S' of S, the pull-back $\mathbf{E}' := S' \times_S \mathbf{E}$ is isomorphic with $\Gamma \times S'$ where the induced action of Γ on $\mathbf{E}'$ corresponds to the left multiplication of Γ on the Γ-factor.

By a *morphism* between two Γ-torsors π and π', we mean a pair $(\varphi, \bar{\varphi})$, where $\varphi \colon \mathbf{E} \to \mathbf{E}'$ is a Γ-equivariant morphism of $\mathbb{C}$-space functors and $\bar{\varphi} \colon S \to S'$ is a morphism of schemes making the following diagram commutative:

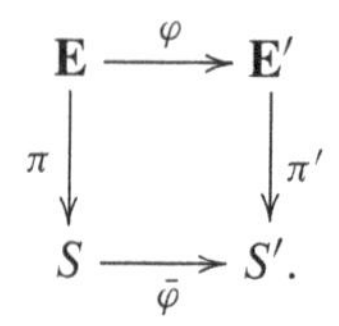

This definition clearly extends to the situation where S is replaced by a $\mathbb{C}$-space functor $\mathbf{B}$. Thus, a Γ-invariant morphism $\pi : \mathbf{E} \to \mathbf{B}$ between $\mathbb{C}$-space functors (where Γ and $\mathbf{E}$ are as above) is called a Γ-*torsor* if for any scheme S and a morphism $f : S \to \mathbf{B}$, the pull-back $f^*(\mathbf{E}) \to S$ is a Γ-torsor in the above sense.

Theorem C.17 *Let $\pi : \mathbf{E} \to \mathbf{B}$ be a Γ-torsor between $\mathbb{C}$-space functors and let $\theta : \mathbf{F} \to \mathbf{E}$ be a Γ-equivariant G-torsor, where $\mathbf{F}$ is a $\mathbb{C}$-space functor with left action of Γ and G is an affine algebraic group (of finite type over $\mathbb{C}$). Then θ descends to give a G-torsor (denoted $\mathbf{F}_\Gamma$) over $\mathbf{B}$.*

Moreover, any Γ-equivariant section σ of θ descends to give a section $\bar{\sigma}$ of the G-torsor $\mathbf{F}_\Gamma$ over $\mathbf{B}$.

Even though we do not need, the result continues to be true with G replaced by any $\mathbb{C}$-group functor.

Proof [1] We first show that $\Gamma(S)$ acts freely on $\mathbf{E}(S)$ for any $\mathbb{C}$-algebra S. Let

$$\gamma_S \cdot x_S = x_S, \quad \text{for } \gamma_S \in \Gamma(S) \text{ and } x_S \in \mathbf{E}(S). \tag{1}$$

Then, by the definition of Γ-torsor, for some fppf extension $S \to S'$,

$$\mathbf{E} \times_{\mathbf{B}} \bar{S}' \simeq \Gamma \times \bar{S}', \quad \Gamma\text{-equivariantly}, \tag{2}$$

where $\bar{S}$ denotes Spec S and the map $\bar{S}' \to \mathbf{B}$ is obtained from $\pi(x_S) = \pi(\gamma_S \cdot x_S)$. Taking the image of (1) under $\mathbf{E}(S) \to \mathbf{E}(S')$ and $\Gamma(S) \to \Gamma(S')$, we get $\gamma_{S'} \cdot x_{S'} = x_{S'}$. (Here and later in the proof, we denote $x_{S'}$ (resp. $\gamma_{S'}$) as the image of x_S in $\mathbf{E}(S')$ (resp. γ_S in $\Gamma(S')$). Thus, from (2), we get $\gamma_{S'} = 1$. But, $\Gamma(S) \to \Gamma(S')$ is injective (Γ being a $\mathbb{C}$-space functor), thus $\gamma_S = 1$, proving that $\Gamma(S)$ acts freely on $\mathbf{E}(S)$.

Define the coset space

$$\mathbf{E}^o_\Gamma(S) = \Gamma(S) \backslash \mathbf{E}(S).$$

The Γ-invariant morphism $\pi : \mathbf{E} \to \mathbf{B}$ clearly induces a map $\pi^o_\Gamma : \mathbf{E}^o_\Gamma \to \mathbf{B}$.

[1] I thank B. Conrad and X. Zhu for their independent (though somewhat similar) proof presented here.

We claim that π_Γ^o is injective (i.e., $\pi_\Gamma^o(S)\colon \mathbf{E}_\Gamma^o(S) \to \mathbf{B}(S)$ is injective for any $\mathbb{C}$-algebra S). Take $x_S, x'_S \in \mathbf{E}(S)$ such that

$$\pi(x_S) = \pi_\Gamma^o(\Gamma(S)x_S) = \pi_\Gamma^o(\Gamma(S)x'_S) = \pi(x'_S).$$

Take S' satisfying (2). Then, by (2), $x_{S'} = \gamma_{S'} \cdot x'_{S'}$, for a unique $\gamma_{S'} \in \Gamma(S')$ (since $\Gamma(S')$ acts freely on $\mathbf{E}(S')$). From the uniqueness of $\gamma_{S'}$ and the sheaf property of Γ and $\mathbf{E}$, we easily see that $x_S = \gamma_S \cdot x'_S$, for a $\gamma_S \in \Gamma(S)$, proving the injectivity of π_Γ^o.

The injectivity of π_Γ^o and the injectivity $\mathbf{B}(S) \to \mathbf{B}(S')$ for any fppf extension $S \to S'$ immediately gives the injectivity

$$\mathbf{E}_\Gamma^o(S) \hookrightarrow \mathbf{E}_\Gamma^o(S'). \tag{3}$$

In fact, we see that for any $\mathbb{C}$-space functor $\mathbf{H}$ with the left action of Γ such that the action of $\Gamma(S)$ on $\mathbf{H}(S)$ is free for any $\mathbb{C}$-algebra S, we get the injectivity (for any fppf extension $S \to S'$)

$$\mathbf{H}_\Gamma^o(S) \hookrightarrow \mathbf{H}_\Gamma^o(S'). \tag{4}$$

To prove this, take $x_S, x'_S \in \mathbf{H}(S)$ such that $x_{S'} = \gamma_{S'} \cdot x'_{S'}$ for a unique $\gamma_{S'} \in \Gamma(S')$. From the uniqueness of $\gamma_{S'}$, it is easy to see that there exists $\gamma_S \in \Gamma(S)$ such that $x_S = \gamma_S \cdot x'_S$, which proves (4).

Let $\mathbf{E}_\Gamma$ (resp. $\mathbf{H}_\Gamma$) be the sheafification of $\mathbf{E}_\Gamma^o$ (resp. $\mathbf{H}_\Gamma^o$) (cf. Lemma B.2).

It is easy to see that the action of $\Gamma(S)$ on $\mathbf{F}(S)$ is free for any $\mathbb{C}$-algebra S, since so is the action of Γ on $\mathbf{E}$. Thus, we get the sheafification $\mathbf{F}_\Gamma$ of $\mathbf{F}_\Gamma^o$.

We next prove that $\pi_\Gamma^o\colon \mathbf{E}_\Gamma^o \to \mathbf{B}$ induces an isomorphism $\pi_\Gamma\colon \mathbf{E}_\Gamma \to \mathbf{B}$. Since π_Γ^o is injective (proved above), then so is π_Γ. So, we need to prove that $\pi_\Gamma(S)\colon \mathbf{E}_\Gamma(S) \to \mathbf{B}(S)$ is surjecive for any $\mathbb{C}$-algebra S. Take $b_S \in \mathbf{B}(S)$. Then, π being a Γ-torsor, there exists a fppf extension $S \to S'$ together with $x_{S'} \in \mathbf{E}(S')$ such that $\pi(x_{S'}) = b_{S'}$ (use (2)). From this, using the construction of the sheafification as in the proof of Lemma B.2, we get the surjectivity of $\pi_\Gamma(S)$.

The Γ-equivariant morphism $\theta\colon \mathbf{F} \to \mathbf{E}$ clearly gives rise to a morphism

$$\theta_\Gamma\colon \mathbf{F}_\Gamma \to \mathbf{E}_\Gamma.$$

We claim that it is a G-torsor. Let $\mathbf{G}$ denote the corresponding $\mathbb{C}$-group functor. Since $\Gamma \times \mathbf{G}$ acts on $\mathbf{F}$ (obtained from the commuting actions of Γ and $\mathbf{G}$ on $\mathbf{F}$) and $\theta\colon \mathbf{F} \to \mathbf{E}$ (resp. $\pi\colon \mathbf{E} \to \mathbf{B}$) is a G-torsor (resp. Γ-torsor), we get that $\pi \circ \theta\colon \mathbf{F} \to \mathbf{B}$ is a $\Gamma \times \mathbf{G}$-torsor. To prove this, take a scheme $\bar{S}$ and a morphism $f\colon \bar{S} \to \mathbf{B}$. Then, as in (2) since π is a Γ-torsor, we can find a fppf cover $\bar{S}' \to \bar{S}$ such that $\mathbf{E} \times_{\mathbf{B}} \bar{S}' \simeq \Gamma \times \bar{S}'$, giving rise to a morphism

$f'_{\mathbf{E}} \colon \bar{S}' \to \mathbf{E}$ and its lift (since θ is a G-torsor) $f'_{\mathbf{F}} \colon \bar{S}' \to \mathbf{F}$ (i.e., $\theta \circ f'_{\mathbf{F}} = f'_{\mathbf{E}}$). Now, define a $\Gamma \times \mathbf{G}$-equivariant morphism

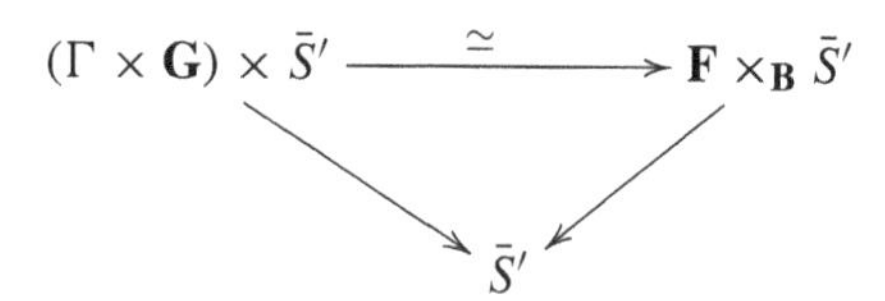

taking $((\gamma, g), z) \mapsto (\gamma \cdot (g \cdot f'_{\mathbf{F}}(z)), z)$, for $z \in \bar{S}', \gamma \in \Gamma$ and $g \in \mathbf{G}$. Then, the above morphism is an isomorphism. From this identification, we get

$$G \times \bar{S}' \simeq \mathbf{F}_\Gamma \times_{\mathbf{B}} \bar{S}'.$$

Thus, $\theta_\Gamma : \mathbf{F}_\Gamma \to \mathbf{E}_\Gamma \simeq \mathbf{B}$ is a G-torsor providing the descent of θ.

Clearly, any Γ-equivariant section σ of θ descends to give a section $\bar{\sigma}$ of $\mathbf{F}_\Gamma \to \mathbf{E}_\Gamma$. This proves the theorem. □

Example C.18 (a) For any scheme $S \in \hat{\mathfrak{S}}$, the stack $\mathfrak{S}_S$ is clearly a Deligne–Mumford stack. It is a smooth stack if and only if S is a smooth $\mathbb{C}$-scheme. Moreover, by Proposition C.15,

$$\dim \mathfrak{S}_S = \dim S,$$

if S is a disjoint union of schemes of finite type over $\mathbb{C}$ which are all equidimensional of the same dimension.

(b) Let Γ be a $\mathbb{C}$-group functor acting on a $\mathbb{C}$-space functor $\mathbf{X}$ from the left. (One important such an example for us to keep in mind is an ind-group Γ acting on an ind-scheme X regarded as $\mathbb{C}$-group functor and $\mathbb{C}$-space functor, respectively.) We define the quotient stack $[\Gamma \backslash \mathbf{X}]$ as follows.

Its objects are pairs (π, φ), where $\pi : \mathbf{E} \to S$ is a Γ-torsor as in Definition C.16 (with left action of Γ on a $\mathbb{C}$-space functor $\mathbf{E}$) over any base scheme $S \in \mathfrak{S}$ and $\varphi \colon \mathbf{E} \to \mathbf{X}$ is a Γ-equivariant morphism of $\mathbb{C}$-space functors. The morphisms from (π, φ) to (π', φ'), for $\pi' \colon \mathbf{E}' \to S'$, consist of Γ-torsor morphisms $(f, \bar{f})$ making the following diagram commutative:

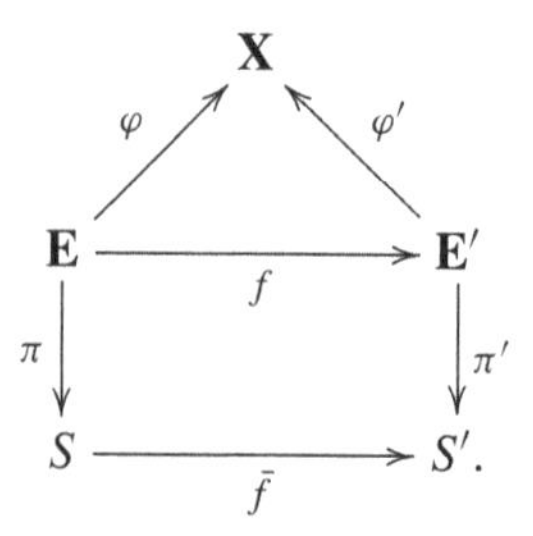

It is a category over $\mathfrak{S}$ under the functor which takes $(\pi, \varphi) \mapsto S$ and the morphism $(f, \bar{f})$ to $\bar{f}$. Then $[\Gamma\backslash\mathbf{X}]$ is a stack (cf. (Laumon and Moret-Bailly, 1999, 3.4.2)). Moreover, the stack $[\Gamma\backslash\mathbf{X}]$ is indeed the quotient of $\mathbf{X}$ by Γ in the category of stacks, i.e., any Γ-invariant morphism from $\mathbf{X}$ to any stack $\mathscr{T}$ factors through $[\Gamma\backslash\mathbf{X}]$ in a unique way.

Even though we defined the category $\mathfrak{S}_S$ over $\mathfrak{S}$ for any $S \in \mathfrak{S}$ or more generally for $S \in \hat{\mathfrak{S}}$ (cf. Example C.11(1)), the same definition defines the category $\mathfrak{S}_S$ over $\mathfrak{S}$ for any ind-scheme S and makes $\mathfrak{S}_S$ into a stack.

Lemma C.19 *Let an ind-group Γ act on an ind-scheme X from the left. Then the stack morphism $\mathbf{f}_o\colon \mathfrak{S}_X \to [\Gamma\backslash X]$ defined below in the proof is a Γ-torsor in the sense that for any $S \in \mathfrak{S}$ and any stack morphism $\mathfrak{S}_S \to [\Gamma\backslash X]$, the fiber product $\mathfrak{S}_S \times_{[\Gamma\backslash X]} \mathfrak{S}_X$ is isomorphic with $\bar{\mathbf{E}}_S$ over $\mathfrak{S}_S$ for a $\mathbb{C}$-space functor $\mathbf{E}_S$, where $\mathbf{E}_S \to S$ is a Γ-torsor. Here $\bar{\mathbf{E}}_S$ is the stack associated to the $\mathbb{C}$-space functor $\mathbf{E}_S$ as in Example C.11(1).*

In particular, if $X \in \mathfrak{S}$ and Γ is an affine algebraic group, then X provides a smooth presentation of $[\Gamma\backslash X]$. Hence, in this case, $[\Gamma\backslash X]$ is an algebraic stack.

Further, if X is equi-dimensional scheme of finite type over $\mathbb{C}$ (and Γ an affine algebraic group), then

$$\dim[\Gamma\backslash X] = \dim X - \dim \Gamma.$$

Moreover, $[\Gamma\backslash X]$ is a smooth stack if and only if X is smooth.

Proof Consider the morphism $\mathbf{f}_o\colon \mathfrak{S}_X \to [\Gamma\backslash X]$ by taking an object $(f\colon Y \to X)$ in $\mathfrak{S}_X$ (so that $Y \in \mathfrak{S}$) to the pair (projection $\pi_Y\colon \Gamma \times Y \to Y, \bar{f}$), where $\bar{f}\colon \Gamma \times Y \to X$ is the Γ-equivariant map $(g, y) \mapsto g \cdot f(y)$. Further, a morphism θ:

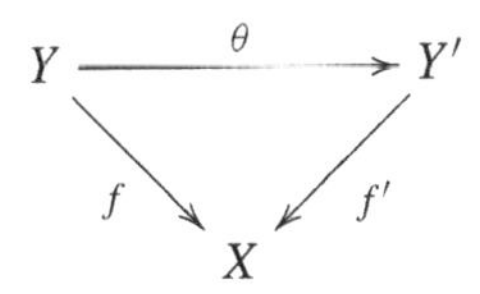

in $\mathfrak{S}_X$ maps to the morphism

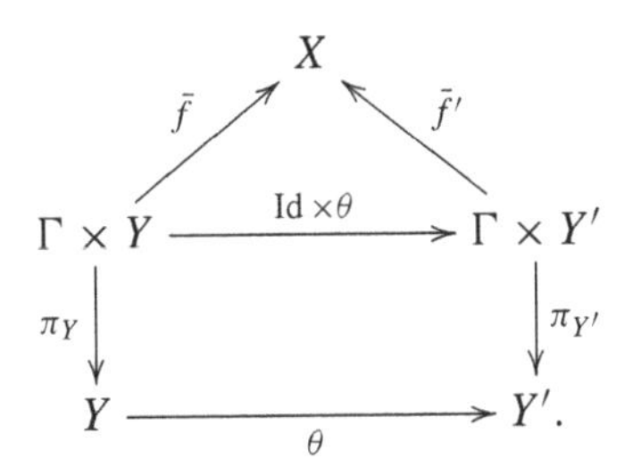

Let $\mathbf{g}_S\colon \mathfrak{S}_S \to [\Gamma\backslash X]$ be a morphism for a scheme S and let $\mathbf{g}_S(\mathrm{Id}\colon S \to S)$ be the Γ-torsor $\pi\colon \mathbf{E}_S \to S$ together with the Γ-equivariant morphism $\xi\colon \mathbf{E}_S \to X$ (where $\mathbf{E}_S$ is a $\mathbb{C}$-space functor). Define a covariant functor Φ over $\mathfrak{S}$ from $\bar{\mathbf{E}}_S$ to the fiber product $\mathfrak{S}_S \times_{[\Gamma\backslash X]} \mathfrak{S}_X$ which takes any morphism $h\colon Y \to \mathbf{E}_S$ to the triple $(\pi \circ h\colon Y \to S,$ $\xi \circ h\colon Y \to X, \alpha)$, where α is given as follows.

Let $\mathbf{g}_S(\pi \circ h)$ be given by the Γ-torsor $p\colon \mathbf{F} \to Y$ (where $\mathbf{F}$ is a $\mathbb{C}$-space functor) together with the Γ-equivariant morphism $\beta\colon \mathbf{F} \to X$. Then, the morphism $\pi \circ h$ in the category $\mathfrak{S}_S$:

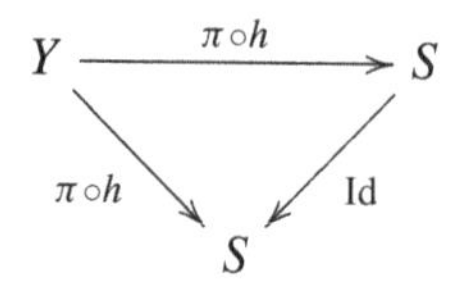

induces a Γ-torsor morphism δ making the following diagram commutative:

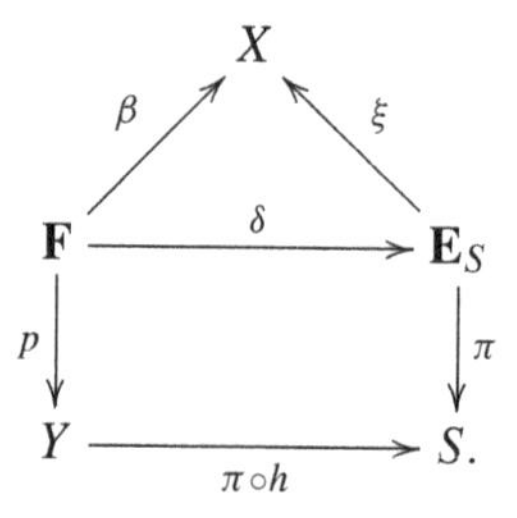

Now, $\mathbf{f}_o(\xi \circ h)$ is the trivial Γ-torsor $\pi_Y\colon \Gamma \times Y \to Y$ together with the Γ-equivariant map $\overline{\xi \circ h}\colon \Gamma \times Y \to X$. Further, in the following triangle:

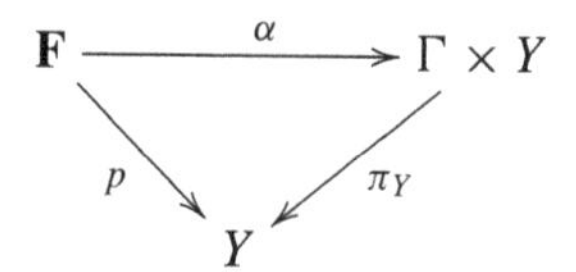

α is the Γ-torsor morphism defined by $\alpha(a) = (g(a), p(a))$, where $g(a) \in \Gamma$ is the unique element such that $\delta(a) = g(a) \cdot h(p(a))$. Then, α makes the following diagram commutative:

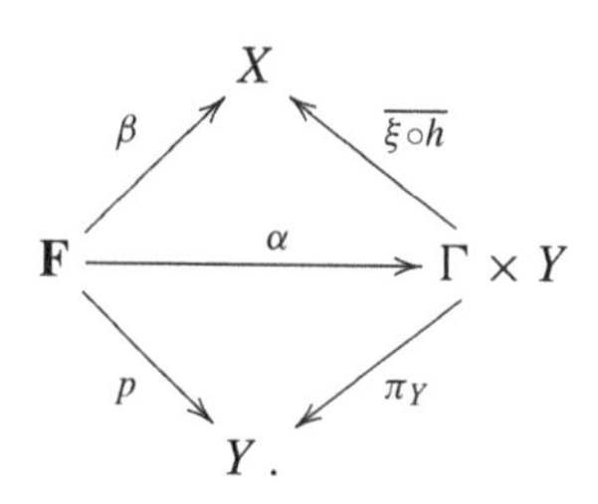

The functor Φ takes any morphism

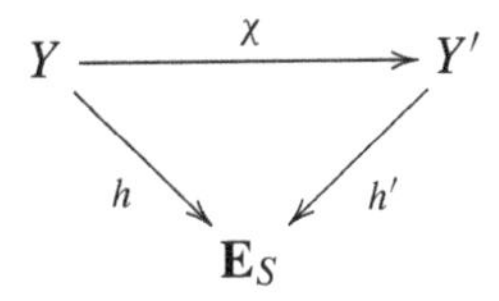

to the morphism in $\mathfrak{S}_S \underset{[\Gamma\backslash X]}{\times} \mathfrak{S}_X$ given by

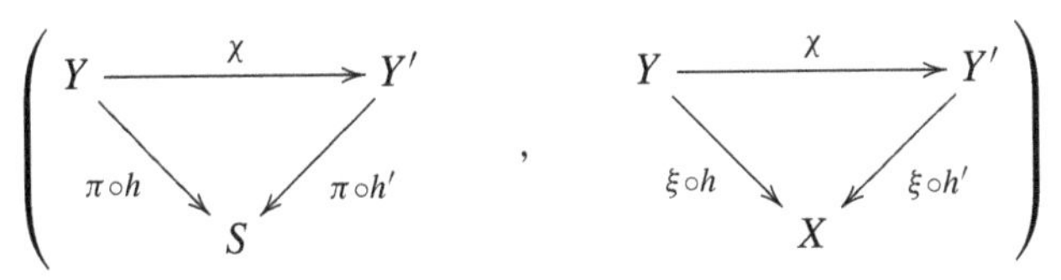

(cf. Definition C.12). We leave it to the reader to prove that the functor $\Phi\colon \bar{\mathbf{E}}_S \to \mathfrak{S}_S \times_{[\Gamma\backslash X]} \mathfrak{S}_X$ defined above is an equivalence of categories over $\mathfrak{S}$.

The morphism $\pi : \mathbf{E}_S \to S$ induces the morphism of stacks $\hat{\pi} : \bar{\mathbf{E}}_S \to \mathfrak{S}_S$ making the following diagram commutative:

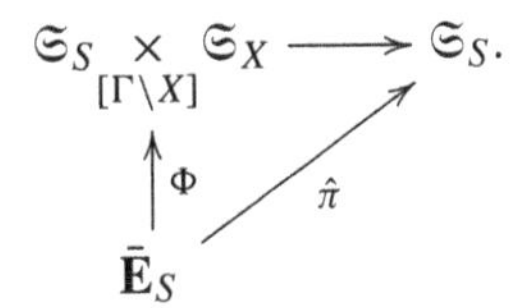

Since $\pi : \mathbf{E}_S \to S$ is a Γ-torsor, the stack morphism $\mathfrak{S}_X \to [\Gamma\backslash X]$ is a Γ-torsor.

Now, assume that $X \in \mathfrak{S}$ and Γ is an affine algebraic group. For any affine algebraic group Γ, any Γ-torsor $\mathbf{E}$ (for $\mathbf{E}$ a $\mathbb{C}$-space functor) over any scheme $S \in \mathfrak{S}$ is automatically a scheme. This follows from the effective descent of relatively affine schemes. Then, $\mathbf{f}_o\colon \mathfrak{S}_X \to [\Gamma\backslash X]$ is a smooth presentation of $[\Gamma\backslash X]$ and hence $[\Gamma\backslash X]$ is an algebraic stack. Moreover, if X is equidimensional scheme of finite type over $\mathbb{C}$, by the definition of $\mu(\mathbf{f}_o)$ (cf. Proposition C.15),

$$\mu(\mathbf{f}_o) = \dim \Gamma.$$

Thus, the dimension of the algebraic stack $[\Gamma\backslash X]$ is given by

$$\begin{aligned} \dim([\Gamma\backslash X]) &= \dim X - \mu(\mathbf{f}_o) \\ &= \dim X - \dim(\Gamma). \end{aligned} \tag{1}$$

In particular, if $\dim X < \dim \Gamma$, we get that

$$\dim([\Gamma\backslash X]) < 0. \tag{2}$$

Observe that $[\Gamma\backslash X]$ is a smooth stack if and only if X is smooth.

When $X = \operatorname{Spec} \mathbb{C}$, then the stack $[\Gamma\backslash X]$ is of special significance. It is the stack $\mathbf{Bun}_\Gamma$ defined in Example C.4(d). From the above, we get that $\mathbf{Bun}_\Gamma$ is a smooth stack of dimension

$$\dim(\mathbf{Bun}_\Gamma) = -\dim(\Gamma). \tag{3}$$

□

Definition C.20 Let **Qcoh** be the groupoid fibration over $\mathfrak{S}$ whose objects are quasi-coherent sheaves $\mathfrak{F}_S$ of $\mathcal{O}_S$-modules over any $S \in \mathfrak{S}$. For any quasi-coherent sheaves $\mathfrak{F}_{S_1}$ and $\mathfrak{F}_{S_2}$ over S_1 and S_2, respectively, by a morphism in $\operatorname{Mor}(\mathfrak{F}_{S_1}, \mathfrak{F}_{S_2})$ we mean a pair $(\varphi, \overline{\varphi})$ consisting of a morphism $\overline{\varphi}\colon S_1 \to S_2$ and an isomorphism $\varphi\colon \mathfrak{F}_{S_1} \xrightarrow{\sim} \overline{\varphi}^*(\mathfrak{F}_{S_2})$ of quasi-coherent sheaves over S_1 (cf. (Hartshorne, 1977, Chap. II, Proposition 5.8(a))). Then, clearly, **Qcoh** is a groupoid fibration over $\mathfrak{S}$ by defining

$$\mathfrak{F}_S \mapsto S \quad \text{and} \quad (\varphi, \overline{\varphi}) \mapsto \overline{\varphi}.$$

In fact, by the descent theory (Grothendieck, 1971, VIII, 5.1, 1.1 et 1.2) or (Stacks, 2019, Tag 023T), **Qcoh** is a stack.

Let $\mathscr{T}$ be an algebraic stack. By a *quasi-coherent sheaf* over $\mathscr{T}$, we mean a morphism of stacks $\mathfrak{F}\colon \mathscr{T} \to$ **Qcoh**. For a scheme $S \in \mathfrak{S}$ and a morphism of stacks $\mathfrak{f}\colon \mathfrak{S}_S \to \mathscr{T}$, we denote the quasi-coherent sheaf $\mathfrak{F} \circ \mathfrak{f}(S \xrightarrow{\text{Id}} S)$ over S by $\mathfrak{f}^*\mathfrak{F}$.

More explicitly, a quasi-coherent sheaf $\mathfrak{F}$ over $\mathscr{T}$ is given by associating a quasi-coherent sheaf $\mathfrak{F}_T$ over S (for any object $T \in \mathscr{T}$ over $S \in \mathfrak{S}$) and for any morphism $\alpha\colon T \to T'$ in $\mathscr{T}$ lying over a morphism $f\colon S \to S'$, an isomorphism of $\mathcal{O}_S$-modules $\theta_\alpha\colon \mathfrak{F}_T \simeq f^*(\mathfrak{F}_{T'})$ satisfying the following cocycle condition for morphisms $\alpha\colon T \to T'$ and $\beta\colon T' \to T''$ lying over $f\colon S \to S'$ and $g\colon S' \to S''$, respectively:

$$f^*(\theta_\beta) \circ \theta_\alpha = \theta_{\beta\circ\alpha},$$

where $\theta_\beta\colon \mathfrak{F}_{T'} \simeq g^*(\mathfrak{F}_{T''})$ gives rise to the canonical isomorpism

$$f^*(\theta_\beta)\colon f^*(\mathfrak{F}_{T'}) \simeq f^*(g^*(\mathfrak{F}_{T''})) \simeq (g \circ f)^*(\mathfrak{F}_{T''}).$$

A quasi-coherent sheaf $\mathfrak{F}$ over $\mathscr{T}$ is called a *rank-r vector bundle* over $\mathscr{T}$ if, for any object $T \in \mathscr{T}$ over $S \in \mathfrak{S}$, the quasi-coherent sheaf $\mathfrak{F}_T$ over S is a locally free $\mathcal{O}_S$-module of rank r. Of course, if $r = 1$, $\mathfrak{F}$ is called a *line bundle* over $\mathscr{T}$.

Let $\mathbf{f}_o\colon \mathfrak{S}_{S_o} \to \mathscr{T}$ be a smooth presentation of an algebraic stack $\mathscr{T}$. Then a quasi-coherent sheaf $\mathfrak{F}$ on $\mathscr{T}$ corresponds to a quasi-coherent sheaf $\mathfrak{F}_{S_o}$ over S_o together with an isomorphism $g\colon \pi_1^*(\mathfrak{F}_{S_o}) \to \pi_2^*(\mathfrak{F}_{S_o})$ as quasi-coherent sheaves over $\mathfrak{S}_{S_o} \underset{\mathscr{T}}{\times} \mathfrak{S}_{S_o}$ (which is isomorphic with $(S_o)_{\mathbf{f}_o}$) such that under the two standard projections $\pi_i\colon \mathfrak{S}_{S_o} \underset{\mathscr{T}}{\times} \mathfrak{S}_{S_o} \to \mathfrak{S}_{S_o}$:

$$\mathfrak{S}_{S_o} \underset{\mathscr{T}}{\times} \mathfrak{S}_{S_o} \overset{\pi_1}{\underset{\pi_2}{\rightrightarrows}} \mathfrak{S}_{S_o},$$

the isomorphism g satisfies the following cocycle condition on the 3-fold stack-theoretic fiber product $\mathfrak{S}_{S_o} \times_{\mathscr{T}} \mathfrak{S}_{S_o} \times_{\mathscr{T}} \mathfrak{S}_{S_o}$:

$$(\pi_{2,3}^* g) \circ (\pi_{1,2}^* g) = \pi_{1,3}^*(g),$$

where $\pi_{i,j}$ are the projections $\mathfrak{S}_{S_o} \times_{\mathscr{T}} \mathfrak{S}_{S_o} \times_{\mathscr{T}} \mathfrak{S}_{S_o} \to \mathfrak{S}_{S_o} \underset{\mathscr{T}}{\times} \mathfrak{S}_{S_o}$ to the (i,j)th factors (cf. Exercise C.E.5).

Given two quasi-coherent sheaves $\mathfrak{F}_1$ and $\mathfrak{F}_2$ over $\mathscr{T}$, by a *morphism* $\varphi\colon \mathfrak{F}_1 \to \mathfrak{F}_2$ of quasi-coherent sheaves, we mean a natural transformation φ from $\mathfrak{F}_1$ to $\mathfrak{F}_2$ over $\mathfrak{S}$ (cf. Definition C.3). This gives rise to the category $\mathbf{Qcoh}(\mathscr{T})$ of quasi-coherent sheaves over $\mathscr{T}$, whose objects are the quasi-coherent sheaves over $\mathscr{T}$ and the morphisms are the morphisms of quasi-coherent sheaves. Of course, similarly we can define the subcategory $\mathbf{Vect}^r(\mathscr{T})$ of rank-r vector bundles over $\mathscr{T}$.

For any morphism $\mathfrak{f}\colon \mathscr{T}_1 \to \mathscr{T}_2$ of algebraic stacks, we can clearly define the pull-back map

$$\mathfrak{f}^*\colon \mathbf{Qcoh}(\mathscr{T}_2) \to \mathbf{Qcoh}(\mathscr{T}_1)$$

taking $\mathfrak{F} \in \mathbf{Qcoh}(\mathscr{T}_2)$ to $\mathfrak{F} \circ \mathfrak{f}$. Similarly, we can define the pull-back map $\mathbf{Vect}^r(\mathscr{T}_2) \to \mathbf{Vect}^r(\mathscr{T}_1)$.

Definition C.21 (Sections of vector bundles over stacks) Let $\mathscr{T}$ be an algebraic stack and let $\mathscr{V}$ be a vector bundle over $\mathscr{T}$ (cf. Definition C.20). Define the *space of global sections* $H^0(\mathscr{T},\mathscr{V})$ of $\mathscr{T}$ with coefficients in $\mathscr{V}$ as the complex vector space consisting of correspondences which assigns to any object T of $\mathscr{T}$ lying over S an element $\sigma_T \in H^0(S,\mathscr{V}_T)$ which satisfy the following compatibility condition.

For any morphism $\alpha \in \mathrm{Mor}\left(T,T'\right)$ in $\mathscr{T}$ over $f\colon S \to S'$, we have:

$$\theta_\alpha(\sigma_T) = f^*(\sigma_{T'}),$$

where the isomorphism $\theta_\alpha\colon \mathscr{V}_T \simeq f^*(\mathscr{V}_{T'})$ is as in Definition C.20.

Definition C.22 Let Γ be an ind-group scheme acting on an ind-scheme X from the left and let $\mathscr{V}$ be a Γ-equivariant vector bundle over X. A section $\sigma \in H^0(X, \mathscr{V})$ (cf. Section 1.1 for the definition of $H^0(X, \mathscr{V})$) is called *functorially* Γ*-invariant* if

$$\phi(\mu^*(\sigma)) = \pi_X^*(\sigma),$$

where $\mu\colon \Gamma \times X \to X$ is the morphism induced from the action, $\pi_X : \Gamma \times X \to X$ is the projection and $\phi\colon \mu^*(\mathscr{V}) \simeq \pi_X^*(\mathscr{V})$ is the isomorphism of vector bundles induced from the Γ-equivariant structure on $\mathscr{V}$ (cf. Section 1.1). We denote by $H^0(X, \mathscr{V})^{\Gamma_{\text{funct}}}$ the subspace of functorially Γ-invariant sections.

Proposition C.23 *Let Γ be an ind-group scheme acting on an ind-scheme X from the left such that $[\Gamma \backslash X]$ is an algebraic stack and let $\mathscr{V}$ be a Γ-equivariant vector bundle over X. Let $\bar{\mathscr{V}}$ be the induced vector bundle over the quotient stack $[\Gamma \backslash X]$ (cf. Exercise C.E.8). Then, the pull-back*

$$\mathbf{f}_o^*\colon H^0\left([\Gamma \backslash X], \bar{\mathscr{V}}\right) \to H^0(X, \mathscr{V})$$

is an isomorphism onto $H^0(X, \mathscr{V})^{\Gamma_{\text{funct}}}$, where $\mathbf{f}_o\colon \mathfrak{S}_X \to [\Gamma \backslash X]$ is the standard morphism as in Lemma C.19.

Proof Since $\mathbf{f}_o \circ \mu = \mathbf{f}_o \circ \pi_X$, we get $\mu^* \circ \mathbf{f}_o^* = \pi_X^* \circ \mathbf{f}_o^*$, from which we see that the image of $H^0\left([\Gamma \backslash X], \bar{\mathscr{V}}\right)$ lies in $H^0(X, \mathscr{V})^{\Gamma_{\text{funct}}}$.

Conversely, take a section $\sigma \in H^0(X, \mathscr{V})^{\Gamma_{\text{funct}}}$. For any scheme S, take an object $(\pi, \varphi) \in [\Gamma \backslash X](S)$, where $\pi : \mathbf{E} \to S$ is a Γ-torsor (with left action of Γ on a $\mathbb{C}$-space functor $\mathbf{E}$) and $\varphi\colon \mathbf{E} \to X$ is a Γ-equivariant morphism. Thus, $\varphi^*\sigma$ is a functorially Γ-invariant section of $\varphi^*\mathscr{V}$ over $\mathbf{E}$. To prove that σ lies in the image of $\mathbf{f}_o^*$, it suffices to show that $\varphi^*\sigma$ is the pull-back via π of a unique section σ_π of $\bar{\mathscr{V}}_S$, where $\bar{\mathscr{V}}_S$ denotes the vector bundle over S induced from $\bar{\mathscr{V}}$ via $(\pi, \varphi) \in [\Gamma \backslash X](S)$. By the definition of $\bar{\mathscr{V}}_S$,

$$\pi^*(\bar{\mathscr{V}}_S) \simeq \varphi^*(\mathscr{V}), \text{ as } \Gamma\text{-equivariant vector bundles over } \mathbf{E}.$$

Since π is locally trivial in the fppf topology, by going to an fppf cover $f\colon S' \to S$, we can assume that $f^*\mathbf{E}$ is the trivial Γ-torsor over S':

$$\begin{array}{ccccc} \Gamma \times S' \simeq f^*\mathbf{E} & \xrightarrow{f'} & \mathbf{E} & \xrightarrow{\varphi} & X \\ \downarrow{\scriptstyle \pi'} & & \downarrow{\scriptstyle \pi} & & \\ S' & \xrightarrow[f]{} & S. & & \end{array}$$

The section $\hat{\sigma} := f'^*\varphi^*\sigma$ of $\pi'^* f^* \bar{\mathscr{V}}_S$ over $f^*\mathbf{E}$ being functorially Γ-invariant, descends to a unique section of $f^*\bar{\mathscr{V}}_S$ over S'. To prove this, observe that $\hat{\sigma}$ being functorially Γ-invariant means that for any morphism $\theta\colon Y \to \Gamma$ (where $Y \in \mathfrak{S}$), the section $(\theta \times I_{S'})^*(\hat{\sigma})$ satisfies the descent condition with respect to the projection $Y \times S' \to S'$. Therefore, the section $\hat{\sigma}$ descends to a unique section of $f^*(\bar{\mathscr{V}}_S)$ over S' (independent of Y). From the faithfully descend property of sections, we get that $\varphi^*\sigma$ itself descends to a unique section σ_π of $\bar{\mathscr{V}}_S$, proving the proposition. □

Definition C.24 Recall that a *subcategory* $\mathscr{C}'$ of a category $\mathscr{C}$ consists of a subcollection of objects and a subcollection of morphisms of $\mathscr{C}$ such that $\mathscr{C}'$ is itself a category under the same composition. A subcategory is called *strictly full* if it is full and closed under isomorphisms.

Let $\mathscr{T}$ be a stack. By a *substack* of $\mathscr{T}$, we mean a strictly full subcategory $\mathscr{T}'$ of $\mathscr{T}$ which is itself a stack (in particular, $\mathscr{T}'$ is a groupoid fibration over $\mathfrak{S}$).

The substack $\mathscr{T}'$ of $\mathscr{T}$ is called *open* (resp. *closed*, resp. *quasi-compact locally closed*) if the inclusion morphism $\mathfrak{i}\colon \mathscr{T}' \to \mathscr{T}$ of stacks is representable and it satisfies the property of being open immersion (resp. closed embedding, resp. quasi-compact locally closed embedding) as defined in Definition C.13.

Let $(\mathscr{T}_i)_{i\in I}$ be a family of groupoid fibrations over $\mathfrak{S}$ with projections $\pi_i\colon \mathscr{T}_i \to \mathfrak{S}$. Then, by the *disjoint union* $\mathscr{T} = \coprod_{i\in I} \mathscr{T}_i$, we mean the groupoid fibration whose objects over $S \in \mathfrak{S}$ are $T = (T_i)_{i\in I}$, where T_i is an object of $\mathscr{T}_i$ over $S_i \in \mathfrak{S}$ such that $(S_i)_{i\in I}$ provides a disjoint open cover of S. Let $T' = (T_i')_{i\in I}$ be another object of $\mathscr{T}$ over $S' \in \mathfrak{S}$, where T_i' is an object of $\mathscr{T}_i$ over $S_i' \in \mathfrak{S}$ such that $(S_i')_{i\in I}$ provides a disjoint open cover of S'. Then, the set of morphisms $\mathrm{Mor}\big((T_i),(T_i')\big)$ is, by definition, a family of morphisms $\alpha = (\alpha_i)_{i\in I}$, where $\alpha_i \in \mathrm{Mor}_{\mathscr{T}_i}(T_i, T_i')$ lies over a morphism $f_i\colon S_i \to S_i'$. The composition of morphisms in $\mathscr{T}$ is the component-wise composition. The covariant functor $\mathscr{T} \to \mathfrak{S}$ takes $T \mapsto \coprod_{i\in I} S_i$ and the morphism (α_i) goes to the morphism $f = \coprod_{i\in I} f_i\colon S = \coprod_{i\in I} S_i \to S' = \coprod_{i\in I} S_i'$. Observe that under $f\colon S \to S'$ we have $f^{-1}(S_i') = S_i$, and the morphism $\alpha_i\colon T_i \to T_i'$ is over the restriction f_i of f to S_i with $\alpha = (\alpha_i)$.

It can be seen that if each $\mathscr{T}_i$ is a stack, then so is $\mathscr{T}$ (cf. Exercise C.E.10).

Let $\mathscr{T}$ be a stack and let $(\mathscr{T}_i)_{i\in I}$ be a family of quasi-compact locally-closed substacks of $\mathscr{T}$. Then, $\mathscr{T}$ is said to be *union* (resp. *disjoint union*) of $(\mathscr{T}_i)_{i\in I}$ if the morphism of stacks (induced by the inclusions $\mathscr{T}_i \to \mathscr{T}$):

$$\theta\colon \coprod_{i\in I} \mathscr{T}_i \to \mathscr{T},$$

which is automatically representable by a scheme in $\hat{\mathfrak{S}}$ (since the morphisms $\mathscr{T}_i \to \mathscr{T}$ are representable by definition), satisfies the property that θ is surjective (resp. universally bijective).

C.E Exercises

(1) Show that in a groupoid fibration $\mathscr{T}$ over $\mathfrak{S}$ (cf. Definition C.5), any morphism $\hat{f}: T_1 \to T_2$ (for $T_1, T_2 \in \mathscr{T}$) lying over an isomorphism $f: S_1 \to S_2$ in $\mathfrak{S}$ is itself an isomorphism in the category $\mathscr{T}$.

(2) Let $\mathbf{f}: \mathscr{T}_1 \to \mathscr{T}_2$ be a functor over $\mathfrak{S}$ between two small categories over $\mathfrak{S}$. Then $\mathbf{f}$ is an equivalence of categories over $\mathfrak{S}$ (cf. Definition C.3) if and only if $\mathbf{f}$ induces a surjection from the objects of $\mathscr{T}_1$ to the objects of $\mathscr{T}_2$ up to an isomorphism of objects in $\mathscr{T}_2$, and for objects $E_1, E_2 \in \mathscr{T}_1$.

$$\mathrm{Mor}_{\mathscr{T}_1}(E_1, E_2) \to \mathrm{Mor}_{\mathscr{T}_2}(\mathbf{f}(E_1), \mathbf{f}(E_2))$$

is a bijection.

(This exercise uses the Axiom of Choice.)

(3) With the notation and assumptions as in Definition C.12, show that the fiber product $\mathscr{T}_1 \times_{\mathscr{T}} \mathscr{T}_2$ is a stack.

Moreover, if $\mathscr{T}_1, \mathscr{T}_2$ and $\mathscr{T}$ are algebraic stacks, then show that $\mathscr{T}_1 \times_{\mathscr{T}} \mathscr{T}_2$ is again an algebraic stack.

(4) Let $f: X_1 \to Y$ and $g: X_2 \to Y$ be two morphisms between schemes. As in Example C.4(b), these induce morphisms $\mathbf{f}: \mathfrak{S}_{X_1} \to \mathfrak{S}_Y$ and $\mathbf{g}: \mathfrak{S}_{X_2} \to \mathfrak{S}_Y$ of stacks. Then show that there is an equivalence of categories over $\mathfrak{S}$ between the stack $\mathfrak{S}_{X_1 \times_Y X_2}$ and the fiber product stack $\mathfrak{S}_{X_1} \times_{\mathfrak{S}_Y} \mathfrak{S}_{X_2}$.

(5) Let $\mathbf{f}_o: \mathfrak{S}_{S_o} \to \mathscr{T}$ be a smooth presentation of an algebraic stack $\mathscr{T}$. Show that a quasi-coherent sheaf $\mathfrak{F}$ on $\mathscr{T}$ corresponds to a quasi-coherent sheaf $\mathfrak{F}_{S_o}$ over S_o together with an isomorphism $g: \pi_1^*(\mathfrak{F}_{S_o}) \simeq \pi_2^*(\mathfrak{F}_{S_o})$ as quasi-coherent sheaves over $(S_o)_{\mathbf{f_o}} \simeq \mathfrak{S}_{S_o} \underset{\mathscr{T}}{\times} \mathfrak{S}_{S_o}$ such that under the two projections

$$\mathfrak{S}_{(S_o)_{\mathbf{f_o}}} \simeq \mathfrak{S}_{S_o} \underset{\mathscr{T}}{\times} \mathfrak{S}_{S_o} \underset{\pi_2}{\overset{\pi_1}{\rightrightarrows}} \mathfrak{S}_{S_o},$$

the isomorphism g satisfies the following cocycle condition on the 3-fold stack-theoretic fiber product $\mathfrak{S}_{S_o} \times_{\mathscr{T}} \mathfrak{S}_{S_o} \times_{\mathscr{T}} \mathfrak{S}_{S_o}$:

$$(\pi_{2,3}^* g) \circ (\pi_{1,2}^* g) = \pi_{1,3}^*(g).$$

(6) Let $S \in \hat{\mathfrak{S}}$ be a scheme and let $\mathfrak{S}_S$ be the corresponding stack (cf. Examples C.4(a), C.11(1) and C.18(a)). Let $\mathbf{Qcoh}_S$ be the category

whose objects are the quasi-coherent sheaves over S and the morphisms are the $\mathcal{O}_S$-module isomorphisms. Show that the two categories $\mathbf{Qcoh}(\mathfrak{S}_S)$ and $\mathbf{Qcoh}_S$ are equivalent under the correspondence of objects:

$$\mathcal{F} \in \mathbf{Qcoh}(\mathfrak{S}_S) \mapsto \mathcal{F}_S, \text{ where } S \text{ lies over } S \text{ via the identity map.}$$

Prove the same result for $\mathbf{Vect}^r(\mathfrak{S}_S)$.

Moreover, for a vector bundle $\mathcal{V}$ over $\mathfrak{S}_S$, the cohomology $H^0(\mathfrak{S}_S, \mathcal{V})$ is canonically isomorphic with $H^0(S, \mathcal{V}_S)$.

(7) Let $E \to B$ be a principal G-bundle over a scheme $B \in \mathfrak{S}$ for an affine algebraic group G. Then show that $[E/G]$ is isomorphic with $\mathfrak{S}_B$.[2]

More generally, if Γ is a $\mathbb{C}$-group functor and $\mathbf{E} \to \mathbf{B}$ is a Γ-torsor (cf. Definition C.16), where $\mathbf{E}$ and $\mathbf{B}$ are $\mathbb{C}$-space functors, then, following Example C.11(1),

$$[\Gamma\backslash\mathbf{E}] \simeq \bar{\mathbf{B}}.$$

Show by an example that $[G\backslash X]$ is in general *not* a scheme for a G-scheme X.

(8) Let Γ be an ind-group scheme acting on an ind-scheme X from the left. Recall the definition of the quotient stack $[\Gamma\backslash X]$ from Example C.18(b). Assume that $[\Gamma \backslash X]$ is an algebraic stack. Show that the category $\mathbf{Vect}^r([\Gamma\backslash X])$ is equivalent with the category of Γ-equivariant rank-r vector bundles $\mathbf{Vect}^r_\Gamma(X)$ over X (cf. Section 1.1) under the functor which takes any object $\mathcal{F} \in \mathbf{Vect}^r([\Gamma\backslash X])$ to the vector bundle $\mathcal{F}_{(\pi,\varphi)}$ over X, where π is the trivial Γ-bundle map $\Gamma \times X \to X$ and φ is the map $\Gamma \times X \to X$, $(g,x) \mapsto g \cdot x$. The morphisms $(\tau_h, \overline{\tau}_h)\colon (\pi,\varphi) \to (\pi,\varphi)$ (for $h \in \Gamma$) given by $\tau_h(g,x) = (gh^{-1}, h\cdot x)$, for $x \in X$, $g \in \Gamma$, and $\overline{\tau}_h(x) = h \cdot x$ (making the following diagram commutative):

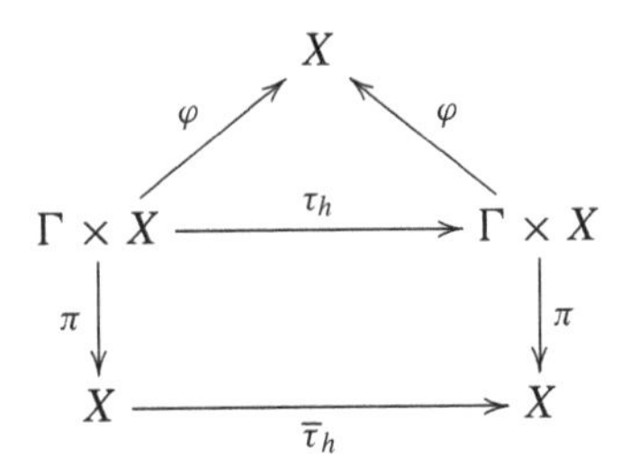

show that $\mathcal{F}_{(\pi,\varphi)}$ is a Γ-equivariant vector bundle over X.

[2] We thank N. Nitsure for correcting this exercise as well as Exercise 5.

In particular, $\mathrm{Pic}([\Gamma\backslash X]) \simeq \mathrm{Pic}_\Gamma(X)$, where $\mathrm{Pic}_\Gamma(X)$ is the group of isomorphism classes of Γ-equivariant line bundles on X.

(9) Let $\mathsf{f}\colon \mathscr{T}_1 \to \mathscr{T}_2$ be a representable morphism of stacks. Then show that if $\mathscr{T}_2$ is a Deligne–Mumford (resp. algebraic) stack, then so is $\mathscr{T}_1$.
Hint: For any schemes S_1 and S_2 and any morphisms of stacks $\mathsf{g}_{S_1} : \mathfrak{S}_{S_1} \to \mathscr{T}_1$ and $\mathsf{h}_{S_2} : \mathfrak{S}_{S_2} \to \mathscr{T}_2$, show that there is an equivalence of categories over $\mathfrak{S}$:

$$\mathfrak{S}_{S_1} \underset{\mathscr{T}_1}{\times} \left(\mathscr{T}_1 \underset{\mathscr{T}_2}{\times} \mathfrak{S}_{S_2} \right) \simeq \mathfrak{S}_{S_1} \underset{\mathscr{T}_2}{\times} \mathfrak{S}_{S_2}, \quad \text{where } \mathfrak{S}_{S_1} \to \mathscr{T}_2 \text{ is the morphism } \mathsf{f} \circ \mathsf{g}_{S_1}.$$

(10) Following the notation in Definition C.24, let $(\mathscr{T}_i)_{i\in I}$ be a family of stacks. Show that the disjoint union $\coprod_{i\in I} \mathscr{T}_i$ is again a stack.

(11) Let Γ be an ind-group variety acting on an ind-variety X from the left. Consider the quotient stack $[\Gamma\backslash X]$ as in Example C.18(b). Let $[\Gamma\backslash X](\mathrm{pt})$ be the set consisting of objects in the fiber of $[\Gamma\backslash X]$ over the one point reduced scheme pt, modulo the isomorphisms. Then, show that there is a natural bijection between $[\Gamma\backslash X](\mathrm{pt})$ with the set-theoretic quotient $\Gamma\backslash X$ (i.e., the set of Γ-orbits in X).

C.C Comments

For a detailed treatment of stacks (original French name *champs*), we refer to the standard book (Laumon and Moret-Bailly, 1999). By now, there are several other sources to learn stacks from, e.g., Simpson (1996); Sorger (1999b); Fantechi (2001); Nitsure (2005a); Vistoli (2005); Kresch (2007); Wang (2011) (among several other sources). We also refer to 'The Stacks Project' with webpage http://stacks.math.columbia.edu

However, our treatment of stacks in this chapter is taken largely from Fantechi (2001).

Proposition C.23 is taken from Beauville and Laszlo (1994, Lemma 7.2).

Appendix D

Rank-Level Duality (A Brief Survey) (by Swarnava Mukhopadhyay)

Introduction. Representation theory of $\mathrm{GL}(r)$ and the intersection theory of Grassmannians $\mathrm{Gr}(r,N)$ are deeply connected. In particular the structure constants of the Grothendieck ring of representations of $\mathrm{GL}(r)$ can be read off the structure constants of the cohomology ring of the Grassmannians. Let $\lambda = (\lambda^1 \geq \cdots \geq \lambda^r \geq 0) \in \mathbb{Z}^r$ parameterize rows of a Young diagram and the corresponding representation of $\mathrm{GL}(r)$ will be denoted by V_λ. Let $\mathcal{Y}_{r,s}$ denote the set of Young diagram, with at most r rows and s columns. For any $\lambda \in \mathcal{Y}_{r,s}$, we obtain a new Young diagram $\lambda^t \in \mathcal{Y}_{s,r}$ by interchanging the rows and the columns of λ. We consider $\lambda, \mu, \nu \in \mathcal{Y}_{r,s}$ such that the total number of boxes $|\lambda|+|\mu|+|\nu| = rs$. Then, using the natural duality $\mathrm{Gr}(r,r+s) \simeq \mathrm{Gr}(s,r+s)$, it follows that

$$\dim_{\mathbb{C}}(V_\lambda \otimes V_\mu \otimes V_\nu)^{\mathrm{SL}(r)} = \dim_{\mathbb{C}}(V_{\lambda^t} \otimes V_{\mu^t} \otimes V_{\nu^t})^{\mathrm{SL}(s)}.$$

The above ‘strange’ dimension equality is not only numerical but it turns out that the vector spaces are canonically dual to each other (see (Belkale, 2004b) and (Belkale, Gibney and Mukhopadhyay, 2015). It is natural to ask for similar results for groups of other types. However, easy computations with Littlewood–Richardson coefficients show that such equalities do not hold in general. Since conformal blocks are refinements of the spaces of invariants of tensor product representations of semisimple Lie algebras, it is natural to consider conformal blocks as the right objects to study such dualities. They were motivated by direct connections in Goodman and Wenzl (1990); Kuniba and Nakanishi (1991) and Naculich and Schnitzer (1990) between the fusion rules of the Wess–Zumino–Witten models of conformal blocks associated to $\mathfrak{sl}(r)$ at level s and $\mathfrak{sl}(s)$ at level r. In this section, we sketch a general approach to formulate rank-level duality questions and recall known rank-level duality results without proof. We mostly focus on the genus zero case due to its direct connection with conformal blocks dealt in the book and only briefly comment on the geometric counterpart of rank-level duality which is known as strange duality. For strange duality questions on surfaces, we refer the reader to Abe (2010, 2015); Marian and Oprea (2009, 2013, 2014) and the references cited there.

D.1 Conformal Embeddings

We will use the notion of conformal embeddings of Lie algebras to formulate rank-level duality questions as a natural map between conformal blocks associated to embedding of Lie algebras and their associated affine branching rules. We refer the reader to Section A.1 of this book for a definition of the Dynkin index of an embedding of simple Lie algebras.

Definition D.1.1 If $(\varphi_1, \varphi_2)\colon \mathfrak{s}_1 \oplus \mathfrak{s}_2 \to \mathfrak{s}$ is an embedding of a semisimple Lie algebra into a simple Lie algebra, we define the Dynkin multi-index to be $(d_{\varphi_1}, d_{\varphi_2})$, where d_{φ_i} is the Dynkin index of the embedding $\mathfrak{s}_i \to \mathfrak{s}$.

Example D.1.2 *Consider the natural embedding of $\varphi\colon \mathfrak{sl}(r) \oplus \mathfrak{sl}(s) \to \mathfrak{sl}(rs)$ given by the tensor product of vector spaces and linear operators on them. The normalized Cartan Killing form on $\mathfrak{sl}(r)$ is given by*

$$(X,Y)_{\mathfrak{sl}(r)} = \mathrm{Trace}(X.Y),$$

where X, Y are $(r \times r)$-matrices with zero trace. The image of X under φ is $rs \times rs$ matrix given by s-diagonal copies of the matrix X. Hence it follows that the Dynkin multi-index of the embedding is (s,r).

Conformal Embeddings and their Classifications

Consider an embedding of Lie algebras $\varphi\colon \mathfrak{s}_1 \oplus \mathfrak{s}_2 \to \mathfrak{s}$ as before and extend it to a map $\widehat{\varphi}\colon \widehat{\mathfrak{s}}_1 \oplus \widehat{\mathfrak{s}}_2 \to \widehat{\mathfrak{s}}$ of affine Lie algebras as follows:

$$\widehat{\varphi}(X \otimes f) = \varphi(X) \otimes f,$$
$$\widehat{\varphi}(c_1) := d_{\varphi_1}.c \text{ and } \widehat{\varphi}(c_2) = d_{\varphi_2}.c.$$

Let ℓ be a non-negative integer and given a weight $\lambda \in D_\ell(\mathfrak{s})$, consider the highest-weight, integrable irreducible $\widehat{\mathfrak{s}}$-module $\mathscr{H}_\lambda(\mathfrak{s},\ell)$ of highest weight λ. The module $\mathscr{H}_\lambda(\mathfrak{s},\ell)$ gets $\widehat{\mathfrak{s}}_1 \oplus \widehat{\mathfrak{s}}_2$-module structure via the map $\widehat{\varphi}$. Since $\mathscr{H}_\lambda(\mathfrak{s},\ell)$ is integrable as a $\widehat{\mathfrak{s}}$-module, it follows that $\mathscr{H}_\lambda(\mathfrak{s},\ell)$ is also integrable as a $\widehat{\mathfrak{s}}_1 \oplus \widehat{\mathfrak{s}}_2$-module at level $(d_{\varphi_1}.\ell, d_{\varphi_2}.\ell)$. By complete reducibility of integrable $\widehat{\mathfrak{s}}_1 \oplus \widehat{\mathfrak{s}}_2$-modules, we get that $\mathscr{H}_\lambda(\mathfrak{s},\ell)$ will decompose as a direct sum of integrable $\widehat{\mathfrak{s}}_1 \oplus \widehat{\mathfrak{s}}_2$-modules at level $(d_{\varphi_1}.\ell, d_{\varphi_2}.\ell)$. However since $\mathscr{H}_\lambda(\mathfrak{s},\ell)$

are infinite dimensional, the number of components in the decomposition may not be finite in general. This motivates us to consider only a special class of embeddings known as *conformal embeddings*.

Remark D.1.3 The notation $\mathscr{H}(\lambda_\ell)$ was used in Chapter 1 to denote the irreducible integrable representation of highest weight λ at level ℓ. However to stress the dependence on the Lie algebra and the level, we use the notation $\mathscr{H}_\lambda(\mathfrak{s},\ell)$ in this section.

Definition D.1.4 An embedding $\varphi\colon \mathfrak{s}_1 \oplus \mathfrak{s}_2 \to \mathfrak{s}$ is conformal at level k if the following holds:

$$\frac{d_{\varphi_1}k.\dim\mathfrak{s}_1}{d_{\varphi_1}k+h^\vee(\mathfrak{s}_1)}+\frac{d_{\varphi_2}k.\dim\mathfrak{s}_2}{d_{\varphi_2}k+h^\vee(\mathfrak{s}_2)}=\frac{k.\dim\mathfrak{s}}{k+h^\vee(\mathfrak{s})}, \tag{1}$$

where $h^\vee(\mathfrak{g})$ is the dual Coxeter number of a simple Lie algebra $\mathfrak{g}$ and d_{φ_i} is the Dynkin-index of the embedding $\mathfrak{s}_i \to \mathfrak{s}$.

It was pointed out in Kac (1990) that (1) is satisfied only when $k = 1$. Conformal embeddings have been classified independently by Bais and Bouwknegt (1987) and Schellekens and Warner (1986). We give some examples below:

- $\mathfrak{sl}(r) \oplus \mathfrak{sl}(s) \to \mathfrak{sl}(rs)$ with Dynkin multi-index (s,r).
- $\mathfrak{sp}(2r) \oplus \mathfrak{sp}(2s) \to \mathfrak{so}(4rs)$ with Dynkin multi-index (s,r).
- $\mathfrak{so}(r) \oplus \mathfrak{so}(s) \to \mathfrak{so}(rs)$ with Dynkin multi-index (s,r), with $r,s \geq 5$.
- $\mathfrak{g}_2 \oplus \mathfrak{f}_4 \to \mathfrak{e}_8$ with Dynkin multi-index $(1,1)$.
- $\mathfrak{so}(r) \to \mathfrak{sl}(r)$ with Dynkin index 2 for $r \geq 4$.

We now list two important properties that make conformal embeddings special. We refer the reader to Kac (1990) for more details:

(i) An embedding $\varphi\colon \mathfrak{s}_1 \oplus \mathfrak{s}_2 \to \mathfrak{s}$ is conformal if and only if any irreducible integrable $\widehat{\mathfrak{s}}$-module $\mathscr{H}_\Lambda(\mathfrak{s},1)$ of level one decomposes into a finite direct sum of $\mathfrak{s}_1 \oplus \mathfrak{s}_2$ modules of level $(d_{\varphi_1}, d_{\varphi_2})$.

(ii) If $\varphi\colon \mathfrak{s}_1 \oplus \mathfrak{s}_2 \to \mathfrak{s}$ is a conformal embedding, then the action of the Virasoro operators are the same, i.e. for any integer N, the following equality holds as operators on $\mathscr{H}_\Lambda(\mathfrak{s},1)$:

$$L_N^{\mathfrak{s}_1} + L_N^{\mathfrak{s}_2} = L_N^{\mathfrak{s}} \in \operatorname{End}(\mathscr{H}_\Lambda(\mathfrak{s},1)),$$

where we refer the reader to Section 3.2 for a definition of Virasoro operators.

Branching Rules of Conformal Embeddings

The theory of conformal embeddings has found very interesting applications in theoretical physics. Given a level-one highest-weight irreducible, integrable $\widehat{\mathfrak{s}}$-module $\mathscr{H}_\Lambda(\mathfrak{s}, 1)$, it is interesting to find the finite list of representations of $\widehat{\mathfrak{s}}_1 \oplus \widehat{\mathfrak{s}}_2$-representations that appear in the decomposition of $\mathscr{H}_\Lambda(\mathfrak{s}, 1)$ along with their multiplicities. In Kac and Peterson (1981), 'string functions' were introduced to study branching rules of conformal embeddings. Branching rules for conformal embeddings were derived by studying asymptotics of these string functions. We recall the following results on the branching rules for some conformal embeddings and we refer the reader to Altschuler, Bauer and Itzykson (1990); Cellini et al. (2006); Hasegawa (1989); Kac and Sanielevici (1988); Kac and Wakimoto (1988); Levstein and Liberati (1995) for more details.

We consider the conformal embedding $\mathfrak{sl}(r) \oplus \mathfrak{sl}(s) \to \mathfrak{sl}(rs)$. The level-one weights of $D_1(\mathfrak{sl}(rs)) = \{\omega_0, \ldots, \omega_{rs-1}\}$. The level-$s$ weights of $\mathfrak{sl}(r)$ are parameterized by the set $\mathcal{Y}_{r-1,s}$. If $\lambda \in \mathcal{Y}_{r-1,s}$, we define the reduced transpose λ^T to be a Young diagram in $\mathcal{Y}_{s-1,r}$ obtained by taking the usual transpose and deleting any column of length s.

Theorem D.1.5 *The module $\mathscr{H}_\lambda(\mathfrak{sl}(r), s) \otimes \mathscr{H}_{\lambda^T}(\mathfrak{sl}(s), r)$ appears with multiplicity one in the branching of $\mathscr{H}_{\omega_{|\lambda|}}(\mathfrak{sl}(rs), 1)$, where $|\lambda|$ denote the number of boxes in the Young diagram of λ. All the of the other components are obtained by the permutations of the weights under the action of the automorphisms of the affine Dynkin diagrams.*

We refer the reader to Altschuler, Bauer and Itzykson (1990); Hasegawa (1989) for complete details of the branching rules for this conformal embedding.

Next we consider the embedding $\mathfrak{sp}(2r) \oplus \mathfrak{sp}(2s) \to \mathfrak{so}(4rs)$. The level-one weights of $\mathfrak{so}(4rs)$ are $\{\omega_0, \omega_1, \omega_+, \omega_-\}$, where $\omega_\pm$ are the spin representations. The level-s representations of the Lie algebra $\mathfrak{sp}(2r)$ are parameterized by the set $\mathcal{Y}_{r,s}$. The branching rules for this conformal embedding can be found in Hasegawa (1989). If Y is in $\mathcal{Y}_{r,s}$, then Y^* denote the Young diagram in $\mathcal{Y}_{s,r}$ obtained by first exchanging the rows and the columns and then taking the complement in a rectangle of size $(s \times r)$.

Theorem D.1.6 *The modules $\mathscr{H}_Y(\mathfrak{sp}(2r), s) \otimes \mathscr{H}_{Y^*}(\mathfrak{sp}(2s), r)$ appears in the branching rules of $\mathscr{H}_{\omega_+}(\mathfrak{so}(4rs), 1)$ (respectively $\mathscr{H}_{\omega_-}(\mathfrak{so}(4rs), 1)$) if and only if the number of boxes of Y is even (respectively odd). If (Y, Y^*) appears in the branching of ω_+ or ω_-, then the multiplicity is always one. Moreover this list if complete.*

Next we consider the case $G_2 \times F_4 \to E_8$. The only level-one representation of $\mathfrak{e}_8$ is ω_0. The level-one representations of $\mathfrak{g}_2$ and $\mathfrak{f}_4$ are $\{\omega_0, \omega_1\}$ and $\{\omega_0, \omega_4\}$, respectively.

Theorem D.1.7 *The module (λ, μ) appears in the branching rule of $\mathscr{H}_{\omega_0}(\mathfrak{e}_8, 1)$ if and only if $(\lambda, \mu) = (\omega_0, \omega_0)$ or (ω_1, ω_4). Moreover if (λ, μ) appears, they always appear with multiplicity one.*

Next we consider the example $\mathfrak{so}(r) \to \mathfrak{sl}(r)$. Let $r \geq 5$, if r is odd and $r \geq 8$, if r is even. With the above assumptions, the following result is due to Kac and Wakimoto (1988, p. 213).

Theorem D.1.8 *The module $\mathscr{H}_{\omega_i}(\mathfrak{sl}(r), 1)$ restricted to $\widehat{\mathfrak{so}}(r)$ decomposes as follows:*

(i) *If $i = 0$, then $\mathscr{H}_{\omega_0}(\mathfrak{sl}(r), 1) \simeq \mathscr{H}_{\omega_0}(\mathfrak{so}(r), 2) \oplus \mathscr{H}_{2\omega_1}(\mathfrak{so}(r), 2)$.*
(ii) *If $1 \leq i \leq \lfloor r/2 \rfloor - 2$, then $\mathscr{H}_{\omega_i}(\mathfrak{sl}(r), 1) \simeq \mathscr{H}_{\omega_i}(\mathfrak{so}(r), 2)$.*
(iii) *If $r = 2m + 1$, then*

  i $\mathscr{H}_{\omega_{m-1}}(\mathfrak{sl}(2m+1), 1) \simeq \mathscr{H}_{\omega_{m-1}}(\mathfrak{so}(2m+1), 2)$,
  ii $\mathscr{H}_{\omega_m}(\mathfrak{sl}(2m+1), 1) \simeq \mathscr{H}_{2\omega_m}(\mathfrak{so}(2m+1), 2)$.

(iv) *If $r = 2m$, then*

  i $\mathscr{H}_{\omega_{m-1}}(\mathfrak{sl}(2m), 1) \simeq \mathscr{H}_{(\omega_{m-1}+\omega_m)}(\mathfrak{so}(2m), 2)$,
  ii $\mathscr{H}_{\omega_m}(\mathfrak{sl}(2m), 1) \simeq \mathscr{H}_{2\omega_{m-1}}(\mathfrak{so}(2m), 2) \oplus \mathscr{H}_{2\omega_m}(\mathfrak{so}(2m), 2)$.

D.2 Rank-Level Duality: General Formulation

Let $\varphi \colon \mathfrak{s}_1 \oplus \mathfrak{s}_2 \to \mathfrak{s}$ be a conformal embedding with Dynkin index (ℓ_1, ℓ_2). For any level-one weight Λ of $\widehat{\mathfrak{s}}$, we denote by I_Λ the set of highest weights of $\widehat{\mathfrak{s}}_1 \oplus \widehat{\mathfrak{s}}_2$ that appear in the decomposition of $\mathscr{H}_\Lambda(\mathfrak{s}, 1)$ as $\widehat{\mathfrak{s}}_1 \oplus \widehat{\mathfrak{s}}_2$-modules via φ. Given an n tuple $\vec{\Lambda} = (\Lambda_1, \ldots, \Lambda_n)$ of level-one weights of $\widehat{\mathfrak{s}}$, we consider two n-tuples $\vec{\lambda} = (\lambda_1, \ldots, \lambda_n)$ and $\vec{\mu} = (\mu_1, \ldots, \mu_n)$ of level ℓ_1 (respectively ℓ_2) of $\widehat{\mathfrak{s}}_1$ (respectively $\widehat{\mathfrak{s}}_2$) such that the following holds:

- For each $1 \leq i \leq n$, we have $(\lambda_i, \mu_i) \in I_{\Lambda_i}$.
- The multiplicity of $\mathscr{H}_{\lambda_i}(\mathfrak{s}_1, \ell_1) \otimes \mathscr{H}_{\mu_i}(\mathfrak{s}_2, \ell_2)$ is one.

Taking tensor product over n chosen factors, we get a map

$$\widetilde{\varphi} \colon \bigotimes_{i=1}^{n} \Big(\mathscr{H}_{\lambda_i}(\mathfrak{s}_1, \ell_1) \otimes \mathscr{H}_{\mu_i}(\mathfrak{s}_2, \ell_2) \Big) \to \bigotimes_{i=1}^{n} \mathscr{H}_{\Lambda_i}(\mathfrak{s}, 1). \tag{1}$$

Let $(\Sigma, \vec{p}, \vec{z})$ be a point of the Deligne–Grothendieck–Knudsen–Mumford moduli stack $\widehat{\overline{\mathcal{M}}}_{g,n}$ of pointed stable curves with n marked points and a choice of formal coordinates at the marked points. The map $\widetilde{\varphi}$ is equivariant with respect to the action of $(\mathfrak{s}_1 \oplus \mathfrak{s}_2) \otimes H^0(\Sigma, \mathcal{O}_\Sigma(*\vec{p}))$ on the left and $\mathfrak{s} \otimes H^0(\Sigma, \mathcal{O}_\Sigma(*\vec{p}))$ action on the right. Taking coinvariants with respect to these, we get a map of the conformal blocks:

$$\widetilde{\varphi}\colon \mathcal{V}_\Sigma(\mathfrak{s}_1, \vec{\lambda}, \ell_1) \otimes \mathcal{V}_\Sigma(\mathfrak{s}_2, \vec{\mu}, \ell_2) \to \mathcal{V}_\Sigma(\mathfrak{s}, \vec{\Lambda}, 1). \tag{2}$$

The above map can be defined as a map of locally free sheaves of covacuas on $\widehat{\overline{\mathcal{M}}}_{g,n}$.

Question D.2.1 One can ask the following natural questions, all of which are broadly known as *rank-level duality questions*:

(i) If $\mathfrak{s}_1$ and $\mathfrak{s}_2$ are both nontrivial and simple and $\dim_{\mathbb{C}} \mathcal{V}_\Sigma(\mathfrak{s}, \vec{\Lambda}, 1)$, then is $\widetilde{\varphi}$ a perfect pairing?
(ii) Is there a natural section in $\mathcal{V}_\Sigma(\mathfrak{s}, \vec{\Lambda}, 1)$ that induces a duality between $\mathcal{V}_\Sigma(, \mathfrak{s}_1, \vec{\lambda}, \ell_1)$ and $\mathcal{V}_\Sigma^\dagger(\mathfrak{s}_2, \vec{\mu}, \ell_2)$?
(iii) If $\mathfrak{s}_2$ is trivial, is $\widetilde{\varphi}$ surjective?

The *propagation of vacua* (Section 2.2) identifies $\mathcal{V}_\Sigma(\mathfrak{g}, \vec{\lambda}, \ell)$ associated to any simple Lie algebra $\mathfrak{g}$ with weights $\vec{\lambda}$ at level ℓ with the sheaf of covacua $\mathcal{V}_\Sigma(\mathfrak{g}, \vec{\lambda}, \omega_0, \ell)$. There is a natural commutative diagram

$$\begin{array}{ccc} \mathcal{V}_\Sigma(\mathfrak{s}_1, \vec{\lambda}, \ell_1) \otimes \mathcal{V}_\Sigma(\mathfrak{s}_2, \vec{\mu}, \ell_2) & \longrightarrow & \mathcal{V}_\Sigma(\mathfrak{s}, \vec{\Lambda}, 1) \\ \downarrow & & \downarrow \\ \mathcal{V}_\Sigma(\mathfrak{s}_1, (\vec{\lambda}, \omega_0), \ell_1) \otimes \mathcal{V}_\Sigma(\mathfrak{s}_2, (\vec{\mu}, \omega_0), \ell_2) & \longrightarrow & \mathcal{V}_\Sigma(\mathfrak{s}, (\vec{\Lambda}, \omega_0), 1). \end{array} \tag{1}$$

Here the horizontal maps are given by branching rules of conformal embeddings and the vertical maps are isomorphisms given by *propagation of vacua*. Using Diagram (1), new rank results on rank-level duality results are obtained.

Remark D.2.2 The rank-level duality map of sheaves of covacua attached to a family of pointed curves with formal coordinates does not descend to a map of locally free sheaves of covacua on $\overline{\mathcal{M}}_{g,n}$. However, using the fact that the embedding is conformal, it was shown in Mukhopadhyay (2016c) that up to some correction factors involving the difference of trace anomaly and Psi-classes, the map $\widetilde{\varphi}$ descends to a map of locally free sheaves on $\overline{\mathcal{M}}_{g,n}$.

Remark D.2.3 In all the known examples, the rank-level duality map fails to be an isomorphism/perfect pairing if Σ is nodal. The behavior of the rank-level duality along the boundary of $\overline{\mathcal{M}}_{g,n}$ has important implications regarding the first Chern class of conformal blocks bundles and their positivity on $\overline{\mathcal{M}}_{0,n}$. We refer the reader to Mukhopadhyay (2016c) for further details.

Applications of Verlinde Formula in Rank-Level Duality

The Verlinde formula has many applications to questions on rank-level duality. The dimensions of level-one conformal blocks have been computed in the works of Fakhruddin (2012); Mukhopadhyay (2016b); Nakanishi and Tsuchiya (1992) using the Verlinde formula formalisms for three points and factorization rules.

Theorem D.2.4 *The following dimensions formula for level-one conformal blocks hold:*

(i) *If $\mathfrak{s} = \mathfrak{sl}(r)$, let $\vec{\Lambda} = (\omega_{i_1}, \dots, \omega_{i_n})$ and Σ be a stable curve of genus g, then the dimension of $\mathcal{V}_\Sigma(\mathfrak{sl}(r), \vec{\Lambda}, 1)$ is r^g if r divides $i_1 + \cdots + i_n$ and is zero otherwise.*

(ii) *If $\mathfrak{s} = \mathfrak{so}(2r+1)$, the level-one weights are ω_0, ω_1 and ω_r. Let $\vec{\Lambda} = (\omega_1, \dots, \omega_1)$, then the dimension of $\mathcal{V}_{\mathbb{P}^1}(\mathfrak{so}(2r+1), \vec{\Lambda}, 1)$ is 1-dimensional if n is even and is zero otherwise. Let $n = m_1 + m_2$ and $\vec{\Lambda}$ consists of m_1-copies of ω_1 and m_2-copies of ω_r. Then the dimension of the space of covacua $\mathcal{V}_{\mathbb{P}^1}(\mathfrak{sl}(2m+1), \vec{\Lambda}, 1)$ is $2^{\frac{m_2}{2}-1}$ if m_2 is even and is zero otherwise.*

(iii) *If $\mathfrak{s} = \mathfrak{so}(2r)$, let $\vec{\Lambda} = (\omega_{i_1}, \dots, \omega_{i_n})$ and Sigma be a stable curve of genus g, then the dimension of $\mathcal{V}_{\mathbb{P}^1}(\mathfrak{so}(2r), \vec{\Lambda}, 1)$ is one if $\omega_{i_1} + \cdots + \omega_{i_n}$ is in the root lattice and is zero otherwise. If $\vec{\Lambda} = (\omega_0, \dots, \omega_0)$, then the dimension of the space of covacua $\mathcal{V}_\Sigma(\mathfrak{so}(2r), \vec{\Lambda}, 1)$ is 4^g, where g is the genus of Σ.*

(iv) *Let $\mathfrak{s}$ be of type G_2 or F_4, and $\vec{\Lambda} = (\omega_1, \dots, \omega_1)$ for $\mathfrak{s} = \mathfrak{g}_2$ or $\vec{\Lambda} = (\omega_4, \dots, \omega_4)$ for $\mathfrak{s} = \mathfrak{f}_4$. Then the dimension of $\mathcal{V}_\Sigma(\mathfrak{g}, \vec{\Lambda}, 1)$ is*

$$\left(\frac{1+\sqrt{5}}{2}\right)^n \left(\frac{5+\sqrt{5}}{2}\right)^{g-1} + \left(\frac{1-\sqrt{5}}{2}\right)^n \left(\frac{5-\sqrt{5}}{2}\right)^{g-1}. \tag{1}$$

Here Σ is any n-pointed stable curve of genus g.

(v) *If $\mathfrak{s} = \mathfrak{e}_6$, the level-one weights are ω_0, ω_1 and ω_6. The representation ω_6 is dual to ω_1. Let $\vec{\Lambda} = (\omega_{i_1}, \dots, \omega_{i_n})$, then the dimension of $\mathcal{V}_{\mathbb{P}^1}(\mathfrak{e}_6, \vec{\Lambda}, 1)$ is one if 3 divides $(i_1 + \cdots + i_n)$ and is zero otherwise.*

(vi) *If $\mathfrak{s} = \mathfrak{e}_7$, the level-one weights are ω_0, ω_7. The representation ω_7 is self dual. Hence the only nontrivial three-pointed conformal blocks on $\mathbb{P}^1$ are associated to the weights $(\omega_0, \omega_0, \omega_0)$ and $(\omega_0, \omega_1, \omega_1)$. The dimension is one in all these cases.*

(viii) *If* $\mathfrak{s} = \mathfrak{e}_8$, *the only level-one weight of* $\mathfrak{e}_8$ *is* ω_0. *The dimension of* $\mathcal{V}_\Sigma(\mathfrak{e}_8, \omega_0, 1)$ *is always one, where* Σ *is any stable curve of genus* g.

We can now use the above calculations to check when the condition that the dimension $\dim_{\mathbb{C}} \mathcal{V}_\Sigma(\mathfrak{s}, \vec{\Lambda}, 1) = 1$ related for Question D.2.1 is satisfied. We can also use the Verlinde formula to compare the dimensions of the $\mathcal{V}_\Sigma(\mathfrak{s}_1, \vec{\lambda}, \ell_1)$ and $\mathcal{V}_\Sigma(\mathfrak{s}_2, \vec{\mu}, \ell_2)$ appearing in the rank-level duality questions. Such comparison involves identities involving the determinant of matrices whole entries are certain cyclotomic polynomials evaluated at the root of unity that appear in the Verlinde formula via the Weyl character formula. We refer the reader to the papers of Altschuler, Bauer and Itzykson (1990); Abe (2008); Donagi and Tu (1994); Mlawer et al. (1991); Mukhopadhyay (2016a,b); Mukhopadhyay and Wentworth (2019); Naculich and Schnitzer (1990); Nakanishi and Tsuchiya (1992); Oxbury and Wilson (1996); Zagier (1996) for detailed calculations.

Results on Rank-Level Duality

In this section, we give a brief survey of the known rank-level duality isomorphisms.

The Case $\mathfrak{sl}(r) \oplus \mathfrak{sl}(s) \to \mathfrak{sl}(rs)$

Let $\vec{\Lambda} = (\omega_{i_1}, \ldots, \omega_{i_n})$ be such that rs divides $i_1 + \cdots + i_n$. Then, the following holds (Nakanishi and Tsuchiya, 1992):

Theorem D.2.5 *For any choice of* $\vec{\lambda}$ *and* $\vec{\mu}$ *such that* $(\lambda_i, \mu_i) \in I_{\Lambda_i}$ *(see Section D.2 for notation), the following rank-level duality map is a perfect pairing:*

$$\widetilde{\varphi}\colon \mathcal{V}_{\mathbb{P}^1}(\mathfrak{sl}(r), \vec{\lambda}, s) \otimes \mathcal{V}_{\mathbb{P}^1}(\mathfrak{sl}(s), \vec{\mu}, r) \to \mathcal{V}_{\mathbb{P}^1}(\mathfrak{sl}(rs), \vec{\Lambda}, 1) \simeq \mathbb{C}.$$

Hence, it induces an isomorphism between $\mathcal{V}^{\dagger}_{\mathbb{P}^1}(\mathfrak{sl}(r), \vec{\lambda}, s) \simeq \mathcal{V}_{\mathbb{P}^1}(\mathfrak{sl}(s), \vec{\mu}, r)$.

Rank-level duality in this case for curves of positive genus has been studied by Belkale (2008a, 2009); Marian and Oprea (2007); Oudompheng (2011) using methods from enumerative geometry. This question was formulated using the language of non-abelian theta functions and is also known as the strange duality conjecture (Donagi and Tu, 1994). The level-one case was proved by Beauville, Narasimhan and Ramanan (1989) and the strange duality conjecture for generic curves was proved by Belkale (2008a). Later, Belkale used the notion of conformal embeddings, uniformization theorems and showed that the strange duality map is flat with respect to the Hitchin connection, hence proving the result for all curves in Belkale (2009). We describe the statement below.

Theorem D.2.6 *Let Σ be a smooth curve of genus $g \geq 2$. Consider the moduli stack $\mathcal{SU}_\Sigma(r)$ (respectively $\mathcal{U}_\Sigma(s, s(g-1))$) of rank r (respectively rank s) vector bundles on curve Σ with trivial determinant (respectively of degree $s(g-1)$). There is a natural duality between $H^0(\mathcal{SU}_\Sigma(r), \mathcal{L}^{\otimes s})$ and $H^0(\mathcal{U}_\Sigma(s, s(g-1)), \Theta_s^{\otimes r})$, where $\mathcal{L}$ (respectively Θ_s) (see Chapter 8) is the generator of the Picard group of $\mathcal{SU}_\Sigma(r)$ (respectively the generalized theta divisor on $\mathcal{U}_\Sigma(s, s(g-1))$).*

R. Oudompheng (2011) proved a parabolic version of the above result. Later, C. Pauly (2014) showed that Theorem D.2.5 in Nakanishi and Tsuchiya (1992) combined with the strange duality result in Beauville, Narasimhan and Ramanan (1989) gives the strange duality stated in Theorem D.2.6.

Remark D.2.7 We could not find a complete reference for $\mathfrak{sl}(r)$ to the proof of the second part of Proposition 1 in Nakanishi and Tsuchiya (1992, Section 5, p. 363). We can use the result of Oudompheng (2011) and results in Mukhopadhyay (2013) to derive Theorem D.2.5. We refer the reader to Section 8.2 of Mukhopadhyay (2013) for further details.

The Case $\mathfrak{sp}(2r) \oplus \mathfrak{sp}(2s) \to \mathfrak{so}(4rs)$

Let n be an even integer and we write $n = 2a + 2b$. Let $\vec{\Lambda} = (\vec{\omega}_+, \vec{\omega}_-)$, where $\vec{\omega}_+$ (respectively $\vec{\omega}_-$) be a $2a$ (respectively $2b$) tuple of the weight ω_+ (respectively ω_-) of $\mathfrak{so}(4rs)$. We state the following theorem (Abe, 2008; Belkale, 2012a):

Theorem D.2.8 *Let Σ be any $2m$-pointed smooth curve of genus g, then there is a natural duality between $\mathcal{V}_\Sigma(\mathfrak{sp}(2r), \vec{Y}, s)$ and $\mathcal{V}_\Sigma(\mathfrak{sp}(2s), \vec{Y}^*, r)$, where $\vec{Y} = (Y_1, \ldots, Y_n)$ is an n-tuple of Young diagrams in $\mathcal{Y}_{r,s}$ and $\sum_{i=1}^{2m} |Y|$ is even.*

In Theorem D.2.8, the natural duality is induced by a canonical projectively flat (Belkale, 2012a) 'Pfaffian' section (Beauville, 2006). We now state a genus-zero version of the rank-level duality result which is due to T. Abe (2008):

Theorem D.2.9 *The rank-level duality map $\widetilde{\varphi}$ induced by the branching rules with $\vec{\Lambda}$ as above is a perfect pairing for any smooth genus-zero curve with n marked points.*

Combining Theorem D.2.9, along with the factorization theorem and a result on projective flatness of Pfaffian sections in Belkale (2012a), Theorem D.2.8 follows directly. Moreover, new rank-level dualities for genus-zero smooth curves with n marked points can be obtained from Theorem D.2.9 by applying diagram automorphisms. We refer the reader to Mukhopadhyay (2013) for further details.

The Case $\mathfrak{so}(2r+1) \oplus \mathfrak{so}(2s+1) \to \mathfrak{so}((2r+1)(2s+1))$

Let d be such that $2d+1 = (2r+1)(2s+1)$ and consider an n-tuple $\vec{\Lambda} = (\Lambda_1, \ldots, \Lambda_n)$ of level-one weights of $\mathfrak{so}(2r+1)$ such that

(i) Each Λ_i is either ω_0 of ω_1.
(ii) The number of ω_1 is even.

The following theorem can be found in Mukhopadhyay (2016c):

Theorem D.2.10 *Let $\vec{\Lambda}$ be as above, then the rank-level duality map $\widetilde{\varphi}$ given by the branching rules is a perfect pairing for any smooth genus-zero curve with n marked points.*

Remark D.2.11 Let $\vec{\Lambda} = (\omega_1, \omega_d, \omega_d)$, then the level-one conformal block on $\mathbb{P}^1$ with three marked points is 1-dimensional. Unlike the previous cases, it was shown in Mukhopadhyay and Wentworth (2019) that the rank-level duality map is not in general an isomorphism. We refer the reader to Mukhopadhyay and Wentworth (2019) to computations with Verlinde formula that show that the source and the target are of different dimensions. However an injectivity result (Mukhopadhyay and Wentworth, 2019) involving the maps of conformal blocks still holds. This shows that the monodromy representations of Knizhnik–Zamolodchikov connections on the space of covacua associated to $\mathfrak{so}(2r+1)$ with spin weights are not in general irreducible.

Remark D.2.12 A conjectural dimensional equality between the source and target of rank-level duality map for conformal blocks involving Lie algebras of type B_r was proposed in Oxbury and Wilson (1996). This conjecture was proved in Mukhopadhyay and Wentworth (2019), however it was shown there that the associated rank-level duality map is not an isomorphism. It remains an open question to find a rank-level duality for conformal blocks of type B_r on curves of positive genus.

The Case $\mathbf{G_2} \times \mathbf{F_4} \to \mathbf{E_8}$

Consider $\vec{\lambda} = (\omega_1, \ldots, \omega_1)$ (respectively $\vec{\mu} = (\omega_4, \ldots, \omega_4)$) to be the n-tuple of the weight ω_1 of $\mathfrak{g}_2$ (respectively the weight ω_4 of $\mathfrak{f}_4$) and $\vec{\Lambda}$ be n-tuple of the weight ω_0 of the vacuum representation of $\mathfrak{e}_8$. The following theorem can be found in Mukhopadhyay (2016b):

Theorem D.2.13 *The map of the space of covacuas induced by the branching rules is a perfect pairing for any n-pointed smooth curve Σ of genus g.*

$$\mathcal{V}_\Sigma(\mathfrak{g}_2,\omega_1,\ldots,\omega_1,1)\otimes\mathcal{V}_\Sigma(\mathfrak{f}_4,\omega_4,\ldots,\omega_4,1)\to\mathcal{V}_\Sigma(\mathfrak{e}_8,\omega_0,\ldots,\omega_0,1).$$

The equality of the dimension of the $\mathfrak{g}_2$ and $\mathfrak{f}_4$ conformal blocks in the statement of Theorem D.2.13 follows from (1).

The Case $\mathfrak{so}(r)\to\mathfrak{sl}(r)$

It follows directly from the branching rules of the conformal embedding $\mathfrak{so}(r)\to\mathfrak{sl}(r)$ that for any stable nodal curve Σ of genus g, the following map is surjective:

$$\mathcal{V}_\Sigma(\mathfrak{so}(r),\omega_0,2)\oplus\mathcal{V}_\Sigma(\mathfrak{so}(r),2\omega_1,2)\to\mathcal{V}_\Sigma(\mathfrak{sl}(r),\omega_0,1).$$

If Σ is smooth, then using the uniformization theorem and the invariance of the rank-level duality map under the action of two torsion points $J_2(\Sigma)$ of the Jacobian, the following was shown in Mukhopadhyay and Zelaci (2020):

Theorem D.2.14 *The natural map between the moduli stack* $\mathrm{Bun}_{\mathrm{SO}(r)}(\Sigma)$ *of* SO(r)*-bundles on a smooth curve Σ and the moduli stack* $\mathrm{Bun}_{\mathrm{SL}(r)}(\Sigma)$ *of* SL(r)*-bundles induces an isomorphism between* $SD\colon H^0(\mathrm{Bun}_{\mathrm{SO}(r)}(\Sigma),\mathcal{D})\simeq H^0(\mathrm{Bun}_{\mathrm{SL}(r)}(\Sigma),\mathcal{D})$ *where $\mathcal{D}$ is the determinant of cohomology line bundles. Moreover the isomorphism SD is flat with respect to the Hitchin connection.*

Other Results

Strange duality results associated to some conformal subalgebra $\mathfrak{p}$ of $\mathfrak{e}_8$ was considered in Boysal and Pauly (2010). Here $\mathfrak{p}$ is the Lie algebra of a simply-connected group P and the list of such P is the following:

- Spin(8) $\times$ Spin(8),
- Spin(16),
- SL(9),
- SL(5) $\times$ SL(5),
- SL(3) $\times$ E_6,
- SL(2) $\times$ E_7.

Their proof is based on flatness of rank-level duality, Verlinde formula and representations of the Heisenberg group associated to the center of the simply-connected group P. Rank-level duality for level-one theta functions for G_2, $SL_2 \times SL_2$ theta functions at level 2 and SL_3 theta functions at level 3 was considered in Grégoire and Pauly (2013). We also refer to the results in Beauville (2006); Pauly and Ramanan (2001); Mukhopadhyay and Zelaci (2020) for level-one rank-level duality results associated to the group $SO(r)$ and theta functions on a Prym variety associated to an étale double cover of a curve.

Remark D.2.15 There are several examples of conformal embeddings for which rank-level duality questions have not been investigated in genus zero. On curves of positive genus, Question D.2.1 usually has a negative answer due to the action of torsion points of the Jacobians of the curves arising from the center of the simply-connected group G. It is not clear how to modify Question D.2.1 for curves of higher genus to accommodate the action of torsion points. Part (1) of Question D.2.1 fails to be a perfect pairing for all curves in $\overline{\mathcal{M}}_{g,n}$; it is interesting to study the stratum of stable nodal curves on which Question D.2.1 holds.

Bibliography

Abe, T. Strange duality for parabolic symplectic bundles on a pointed projective line, *Int. Math. Res. Not.*, Art. ID rnn121 (2008).

Abe, T. Deformation of rank 2 quasi-bundles and some strange dualities for rational surfaces, *Duke Math. J.* **155**, 577–620 (2010).

Abe, T. Strange duality for height zero moduli spaces of sheaves on $\mathbb{P}^2$, *Michigan Math. J.* **64**, 569–586 (2015).

Agnihotri, S. and Woodward, C. Eigenvalues of products of unitary matrices and quantum Schubert calculus, *Math. Res. Lett.* **5**, 817–836 (1998).

Agrebaoui, B. Standard modules and standard modules of level one, *J. Pure Appl. Alg.* **102**, 235–241 (1995).

Alekseev, A., Meinrenken, E. and Woodward, C. Formulas of Verlinde type for non-simply connected groups, ArXiv: math/0005047 (2000).

Altschuler, D., Bauer, M. and Itzykson, C. The branching rules of conformal embeddings, *Comm. Math. Phys.* **132**, 349–364 (1990).

Anchouche, B., Azad, H. and Biswas, I. Harder–Narasimhan reduction for principal bundles over a compact Kähler manifold, *Math. Ann.* **323**, 693–712 (2002).

Arbarello, E., Cornalba, M., Griffiths, P. and Harris, J. *Geometry of Algebraic Curves*, vol. I. Springer-Verlag, Berlin, 1985.

Arbarello, E., Cornalba, M., Griffiths, P. and Harris, J. *Geometry of Algebraic Curves* vol. II. Grundlehren der Mathematischen Wissenschaften, vol. 268. Springer, New York, 2011.

Atiyah, M. F. Vector bundles over an elliptic curve, *Proc. London Math. Soc. (Third Series)* **7**, 412–452 (1957).

Atiyah, M. F. and Bott, R. The Yang–Mills equations over Riemann surfaces, *Phil. Trans. Roy. Soc. London A* **308**, 523–615 (1982).

Atiyah, M.F. and Macdonald, I.G. *Introduction to Commutative Algebra*. Addison–Wesley, Boston, MA, 1969.

Baier, T., Bolognesi, M., Martens, J. and Pauly, C. The Hitchin connection in arbitrary chracteristic, ArXiv: 2002-12288 (Math. AG) (2020).

Bais, F.A. and Bouwknegt, P.G. A classification of subgroup truncations of the Bosonic string, *Nuclear Phys. B* **279**, no. 3–4, 561–570 (1987).

Bakalov, B. and Kirillov, Jr., A. *Lectures on Tensor Categories and Modular Functors.* University Lecture Series, vol. 21. American Mathematical Society, Providence, RI, 2001.

Balaji, V., Biswas, I. and Nagaraj, D.S. Principal bundles over projective manifolds with parabolic structure over a divisor, *Tohoku Math. J.* **53**, 337–367 (2001).

Balaji, V., Biswas, I. and Pandey, Y. Connections on parahoric torsors over curves, *Publ. RIMS* **53**, 551–585 (2017).

Balaji, V. and Parameswaran, A.J. Semistable principal bundles-II (positive characteristics), *Transformation Groups* **8**, 3–36 (2003).

Balaji, V. and Seshadri, C.S. Moduli of parahoric $\mathcal{G}$-torsors on a compact Riemann surface, *J. Alg. Geom.* **24**, 1–49 (2015).

Baldoni, V., Boysal, A. and Vergne, M. Multiple Bernoulli series and volumes of moduli spaces of flat bundles over surfaces, *J. Symbolic Computation* **68**, 27–60 (2015).

Barlet, D. and Magnússon, J. Cycles Analytiques complexes I: Théorèmes de Préparation des Cycles, *Publication Société Mathématique de France, Cours Spécialisés* **22** (2014).

Beauville, A. Conformal blocks, fusion rules and the Verlinde formula, *Israel Math. Conf. Proc.*, vol. **9**, 75–96 (1996).

Beauville, A. The Verlinde formula for PGL_p, *Adv. Ser. Math. Phys.*, vol. **24**, 141–151 (1997).

Beauville, A. Orthogonal bundles on curves and theta functions, *Ann. Inst. Fourier (Grenoble)* **56**, 1405–1418 (2006).

Beauville, A. and Laszlo, Y. Conformal blocks and generalized theta functions, *Commun. Math. Phys.* **164**, 385–419 (1994).

Beauville, A. and Laszlo, Y. Un lemme de descente, *C.R. Acad. Sci. Paris* **320**, 335–340 (1995).

Beauville, A., Laszlo, Y. and Sorger, C. The Picard group of the moduli of G-bundles on a curve, *Compositio Math.* **112**, 183–216 (1998).

Beauville, A, Narasimhan, M.S. and Ramanan, S. Spectral curves and the generalised theta divisor, *J. Reine Angew. Math.* **398**, 169–179 (1989).

Behrend, K. A. *The Lefschetz trace formula for the moduli stack of principal bundles*, PhD thesis, University of California at Berkeley, 1991.

Behrend, K. A. Semi-stability of reductive group schemes over curves, *Math. Ann.* **301**, 281–305 (1995).

Beilinson, A., Bloch, S. and Esnault, H. ϵ-factors for Gauss–Manin determinants, *Moscow Mathematical Journal* **2**, 477–532 (2002).

Beilinson, A. and Drinfeld, V. Quantization of Hitchin's integrable system and Hecke eigensheaves, Preprint (1994).

Beilinson, A., Feigin, B. and Mazur, B. Introduction to algebraic field theory on curves, Preprint (1990).

Beilinson, A. and Ginzburg, V. Infinitesimal structure of moduli spaces of G-bundles, *International Math. Res. Notices*, Issue **4**, 63–74 (1992).

Beilinson, A. and Kazhdan, D. Flat projective connections, Preprint (1990).

Beilinson, A. and Schechtman, V. Determinant bundles and Virasoro algebras, *Commun. Math. Phys.* **118**, 651–701 (1988).

Belkale, P. Local systems on $\mathbb{P}^1 - S$ for S a finite set, *Compositio Math.* **129**, 67–86 (2001).

Belkale, P. Invariant theory of $GL(n)$ and intersection theory of Grassmannians, *Int. Math. Res. Not.* **2004**, 3709–3721 (2004a).

Belkale, P. Transformation formulas in quantum cohomology, *Compos. Math.* **140**, 778–792 (2004b).

Belkale, P. Quantum generalization of the Horn conjecture, *J. Amer. Math. Soc.* **21**, 365–408 (2008a).

Belkale, P. The strange duality conjecture for generic curves, *J. Amer. Math. Soc.* **21**, 235–258 (2008b).

Belkale, P. Strange duality and the Hitchin/WZW connection, *J. Differential Geom.* **82**, 445–465 (2009).

Belkale, P. Orthogonal bundles, theta characteristics and symplectic strange duality, *Contemp. Math.* **564**, 185–193 (2012a).

Belkale, P. Unitarity of the KZ/Hitchin connection on conformal blocks in genus 0 for arbitrary Lie algebras, *J. Math. Pures Appl.* **98**, 367–389 (2012b).

Belkale, P. and Fakhruddin, N. Triviality properties of principal bundles on singular curves, *Algebr. Geom.* **6**, 234–259 (2019).

Belkale, P., Gibney, A. and Mukhopadhyay, S. Vanishing and identities of conformal blocks divisors, *Algebr. Geom.* **2**, 62–90 (2015).

Belkale, P., Gibney, A. and Mukhopadhyay, S. Nonvanishing of conformal blocks divisors on $\bar{M}_{0,n}$, *Transformation Groups* **21**, 329–353 (2016).

Belkale, P. and Kumar, S. The multiplicative eigenvalue problem and deformed quantum cohomology, *Advances in Math.* **288**, 1309–1359 (2016).

Beltrametti, M. and Robbiano, L. Introduction to the theory of weighted projective spaces, *Expo. Math.* **4**, 111–162 (1986).

Berline, N., Getzler, E. and Vergne, M. *Heat Kernels and Dirac Operators*. Grundlehren Text Editions. Springer. New York, 2004.

Bernshtein, I.N. and Shvartsman, O.V. Chevalley's theorem for complex crystallographic Coxeter groups, *Functional Analysis and its Applications* **12**, 308–310 (1978).

Bertram, A. Generalized SU(2) theta functions, *Invent. Math.* **113**, 351-372 (1993).

Bertram, A. and Szenes, A. Hilbert polynomials of moduli spaces of rank 2 vector bundles II, *Topology* **32**, 599-609 (1993).

Bhosle, U. and Ramanathan, A. Moduli of parabolic G-bundles on curves, *Math. Z.* **202**, 161–180 (1989)

Bialynicki-Birula, A. and Swiecicka, J. A reduction theorem for existence of good quotients, *Amer. J. Math.* **113**, 189–201 (1991).

Bismut, J.-M. and Labourie, F. Symplectic geometry and the Verlinde formulas, *Surveys in Differential Geometry* **5**, 97–311 (1999).

Biswas, I. Parabolic bundles as orbifold bundles, *Duke Math. J.* **88**, 305–325 (1997).

Biswas, I. A criterion for the existence of a parabolic stable bundle of rank two over the projective line, *International Journal of Mathematics* **9**, 523–533 (1998).

Biswas, I. and Holla, Y. I. Harder–Narasimhan reduction of a principal bundle, *Nagoya Math. J.* **174**, 201–223 (2004).

Biswas, I. and Holla, Y. I. Principal bundles whose restriction to curves are trivial, *Math. Z.* **251**, 607–614 (2005).

Biswas, I. and Ramanan, S. An infinitesimal study of the moduli of Hitchin pairs, *J. London Math. Soc.* **49**, 219–231 (1994).

Biswas, I. and Raghavendra, N. Determinants of parabolic bundles on Riemann surfaces, *Proc. of the Indian Academy of Sciences - Mathematical Sciences* **103**, 41–71 (1993).

Boden, H. Representations of orbifold groups and parabolic bundles, *Comment. Math. Helvetici* **66**, 389–447 (1991).

Borel, A. *Linear Algebraic Groups*, 2nd enlarged edn. Graduate Texts in Mathematics, vol. 126. Springer, New York, 1991.

Bott, R. An application of the Morse theory to the topology of Lie-groups, *Bull. Soc. Math. France* **84**, 251–281 (1956).

Bott, R. and Samelson, H. Applications of the theory of Morse to symmetric spaces, *Am. J. Math.* **80**, 964–1029 (1958).

Bourbaki, N. *Lie Groups and Lie Algebras*, Chap. 4–6, Springer, New York, 2002.

Bourbaki, N. *Lie Groups and Lie Algebras*, Chap. 7–9, Springer, New York, 2005.

Boutot, J.-F. Singularités rationnelles et quotients par les groupes féductifs, *Invent. Math.* **88**, 65-68 (1987).

Boysal, A. Nonabelian theta functions of positive genus, *Proc. A.M.S.* **136**, 4201–4209 (2008).

Boysal, A. and Kumar, S. Explicit determination of the Picard group of moduli spaces of semistable G-bundles on curves, *Math. Ann.* **332**, 823–842 (2005).

Boysal, A. and Kumar, S. A conjectural presentation of fusion algebras. *Advanced Studies in Pure Math.* vol. **54** (Algebraic Analysis and Around), 95–107 (2009).

Boysal, A. and Pauly, C. Strange duality for Verlinde spaces of exceptional groups at level one, *Int. Math. Res. Not.*, Issue 4, 595–618 (2010).

Boysal, A. and Vergne, M. Multiple Bernoulli series, an Euler–Maclaurin formula, and wall crossings, *Ann. Inst. Fourier, Grenoble* **62**, 821–858 (2010).

Bremner, M. R., Moody, R.V. and Patera, J. *Tables of Dominant Weight Multiplicities for Representations of Simple Lie Algebras*, Marcel Dekker, New York, 1985.

Bröcker, T. and tom Dieck, T. *Representations of Compact Lie Groups*. Graduate Texts in Mathematics, vol. 98. Springer, New York, 1985.

Bruguieres, A. Filtration de Harder-Narasimhan et stratification de Shatz, in: *Module des Fibrés Stables sur les Courbes Algébriques*. Progress in Mathematics, vol. 54. Birkhäuser, Basel, pp. 83–106 (1985).

Cartan, H. Quotient d'un espace analytique par un groupe d'automorphismes, in: *Algebraic Geometry and Topology* (A Symposium in Honor of S. Lefschetz, Princeton, NJ), 90–102 (1957).

Cartan, H. and Eilenberg, S. *Homological Algebra*. Princeton University Press, Princeton, NJ, 1956.

Cellini, P., Kac, V.G., Möseneder, F.P. and Papi, P. Decomposition rules for conformal pairs associated to symmetric spaces and abelian subalgebras of $\mathbb{Z}_2$-graded Lie algebras, *Adv. Math.* **207**, 156–204 (2006).

Chevalley, C. and Eilenberg, S. Cohomology theory of Lie groups and Lie algebras, *Trans. Amer. Math. Soc.* **63**, 85–124 (1948).

Chriss, N. and Ginzburg, V. *Representation Theory and Complex Geometry*. Birkhäuser, Basel, 1997.

Conrad. B., Gabber, O. and Prasad, G. *Pseudo-reductive Groups*, 2nd edn. Cambridge University Press, Cambridge, 2015.

Damiolini, C. Conformal blocks attached to twisted groups, *Math. Z.* **295**, 1643–1681 (2020).

Damiolini, C., Gibney, A. and Tarasca, N. On factorization and vector bundles of conformal blocks from vertex algebras, ArXiv: 1909.04683 v2 (2019).

Damiolini, C., Gibney, A. and Tarasca, N. Vertex algebras of cohFT-type, ArXiv: 1910.01658 v2 (2020).

Daskalopoulos, G. and Wentworth, R. Local degeneration of the moduli space of vector bundles and factorization of rank two theta functions. I, *Math. Ann.* **297**, 417–466 (1993).

Daskalopoulos, G. and Wentworth, R. Factorization of rank two theta functions. II: proof of the Verlinde formula, *Math. Ann.* **304**, 21–51 (1996).

Deligne, P. *Cohomologie Etale, SGA* $4\frac{1}{2}$, Springer Lecture Notes in Mathematics, vol. 569. Springer, New York, 1977.

Deligne, P. and Mumford, D. The irreducibility of the space of curves of a given genus, *Publications Math. IHES* **36**, 75–109 (1969).

Demazure, M. and Gabriel, P. *Introduction to Algebraic Geometry and Algebraic Groups*. North-Holland Mathematics Studies, vol. 39. North-Holland, Amsterdam, 1980.

Demazure, M. and Grothendieck, A. *Schémas en groupes*, SGA III. Lecture Notes in Mathematics, vol. 153. Springer, New York, 1970.

Dolgachev, I. Weighted projective varieties, in: *Group Actions and Vector Fields*. Springer Lecture Notes in Mathematics, vol. 956, 34–71. Springer, New York, 1982.

Donagi, R. and Tu, L. W. Theta functions for $SL(n)$ versus $GL(n)$, *Math. Res. Lett.* **1**, 345–357 (1994).

Donaldson, S.K. A new proof of a theorem of Narasimhan and Seshadri, *J. Diff. Geom.* **18**, 269–277 (1983).

Douady, A. Le problème des modules pour les variétés analytiques complexes, *Séminaire Bourbaki, Exposé* **277**, anneés 1964–65 (1966).

Drezet, J.–M. and Narasimhan, M.S. Groupe de Picard des variétés de modules de fibrés semi-stables sur les courbes algébriques, *Invent. Math.* **97**, 53–94 (1989).

Drinfeld, V. and Simpson, C. B-structures on G-bundles and local triviality, *Math. Res. Letters* **2**, 823–829 (1995).

Dynkin, E.B. Semisimple subalgebras of semisimple Lie algebras, *Am. Math. Soc. Transl. (Ser. II)* **6**, 111–244 (1957).

Edixhoven, B. Néron models and tame ramification, *Compositio Math.* **81**, 291–306 (1992).

Eisenbud, D. *Commutative Algebra with a View Toward Algebraic Geometry*. Graduate Texts in Mathematics, vol. 150. Springer, New York, 1995.

Eisenbud, D. and Harris, J. *The Geometry of Schemes*. Graduate Texts in Mathematics, vol. 197. Springer, New York, 2000.

Fakhruddin, N. Chern classes of conformal blocks, *Contemp. Math.* **564**, 145–176 (2012).

Faltings, G. Stable G-bundles and projective connections, *J. Alg. Geom.* **2**, 507–568 (1993).

Faltings, G. A proof for the Verlinde formula, *J. Alg. Geom.* **3**, 347–374 (1994).

Faltings, G. Algebraic loop groups and moduli spaces of bundles, *J. Eur. Math. Soc.* **5**, 41–68 (2003).

Faltings, G. Theta functions on moduli spaces of G-bundles, *J. Alg. Geom.* **18**, 309–369 (2009).

Fantechi, B. Stacks for everybody, in: *Progress in Mathematics* vol. 201. Birkhäuser, Basel, 2001.

Feigin, B.L., Schechtman, V.V. and Varchenko, A.N. On algebraic equations satisfied by correlators in Wess–Zumino–Witten models, *Letters in Mathematical Physics* **20**, 291–297 (1990).

Fontaine, Jean-Marc. Groupes p-divisible sur les corps locaux, Astérisque Bd. 47/48, Publication Société Mathématique de France, 1977.

Friedman, R., Morgan, J.W. and Witten, E. Principal G-bundles over elliptic curves, *Mathematical Research Letters* **5**, 97–118 (1998).

Fuchs, J. and Schweigert, C. A representation theoretic approach to the WZW Verlinde formula, ArXiv: hep-th/9707069 (1997).

Fulton, W. *Intersection Theory*, 2nd edn. Springer, New York, 1998.

Fulton, W. and Pandharipande, R. Notes on stable maps and quantum cohomology, in: *Algebraic Geometry—Santa Cruz 1995*, Proc. Sympos. Pure Math. **62**, Part 2, 45–96 (1997).

Furuta, M. and Steer, B. Seifert fibred homology 3-spheres and the Yang–Mills equations on Riemann surfaces with marked points, *Adv. Math.* **96**, 38–102 (1992).

Garland, H. The arithmetic theory of loop groups, *Publ. Math. IHES* **52**, 5–136 (1980).

Garland, H. and Lepowsky, J. Lie algebra homology and the Macdonald–Kac formulas, *Invent. Math.* **34** , 37–76 (1976).

Garland, H. and Raghunathan, M.S. A Bruhat decomposition for the loop space of a compact group: A new approach to results of Bott, *Proc. Natl. Acad. Sci. USA* **72**, 4716–4717 (1975).

van Geemen, B. and de Jong, A.J. On Hitchin's connection, *J. Am. Math. Soc.* **11**, 189–228 (1998).

Ginzburg, V. Resolution of diagonals and moduli spaces, in: *The Moduli Space of Curves*. Progress in Mathematics, vol. 129. Birkhäuser, Basel, 231–266 (1995).

Goddard, P., Kent, A. and Olive, D. Virasoro algebras and coset space models, *Physics Letters* **B152**, 88–92 (1985).

Goodman, R. and Wallach, N.R. *Symmetry, Representations, and Invariants*. Graduate Texts in Mathematics, vol. 255. Springer, New York, 2009.

Goodman, F.M. and Wenzl, H. Littlewood–Richardson coefficients for Hecke algebras at roots of unity, *Adv. Math.* **82**, 244–265 (1990).

Grauert, H. and Riemenschneider, O. *Verschwindungssätze für analytische Kohomologiegruppen auf Komplexen Raümen*, Lecture Notes in Mathematics, vol. 155. Springer, New York, 1970.

Grégoire, C. and Pauly, C. The space of generalized G_2-theta functions of level 1, *Michigan Math. J.* **62**, 857–867 (2013).

Griffiths, P. and Harris, J. *Principles of Algebraic Geometry*. Wiley, Chichester, 1978.

Grothendieck, A. *Sur le mémoire de Weil. "généralisation des fonctions abéliennes"*, Séminaire Bourbaki, Exposé **141**, pp. 57–71, 1956–57.

Grothendieck, A. *Techniques de construction et théorèmes d'existence en géometérie algébrique IV: Les schémas de Hilbert*, Séminaire Bourbaki, Exposés **205–222**, talk 221, pp. 249–276, 1960–61.

Grothendieck, A. *Éléments de géométrie algébrique: I*, Publications Math. IHES **4**, pp. 5–228 (1960).

Grothendieck, A. *Éléments de géométrie algébrique II*, Publications Math. IHES **8**, pp. 5–222, (1961a).

Grothendieck, A. *Éléments de géométrie algébrique III (Première Partie)*, Publications Math. IHES **11**, pp. 5–167 (1961b).

Grothendieck, A. *Éléments de géométrie algébrique: IV (Seconde Partie)*, Publications Math. IHES **24**, pp. 5–231(1965).

Grothendieck, A. *Éléments de géométrie algébrique: IV (Quatrième Partie)*, Publications Math. IHES **32**, pp. 5–361 (1967).

Grothendieck, A. *Revêtements Étales et Groupe Fondamental (SGA 1)*. Lecture Notes in Mathematics, vol. 224. Springer, New York, 1971.

Harder, G. Halbeinfache gruppenschemata über Dedekindringen, *Invent. Math.* **4**, 165–191 (1967).

Harder, G. Halbeinfache gruppenschemata über vollständigen kurven, *Invent. Math.* **6**, 107–149 (1968).

Harder, G. and Narasimhan, M.S. On the cohomology groups of moduli spaces of vector bundles on curves, *Math. Ann.* **212**, 215–248 (1975).

Harris, J. and Morrison, I. *Moduli of Curves*. Graduate Texts in Mathematics, vol. 187. Springer, New York, 1998.

Hartshorne, R. *Algebraic Geometry*. Graduate Texts in Mathematics, vol. 52. Springer-Verlag, New York, 1977.

Hasegawa, K. Spin module versions of Weyl's reciprocity theorem for classical Kac–Moody Lie algebras—an application to branching rule duality, *Publ. Res. Inst. Math. Sci.* **25**, 741–828 (1989).

Hatcher, A. *Algebraic Topology*. Cambridge University Press, Cambridge, 2001.

Heinloth, J. Bounds for Behrend's conjecture on the canonical reduction, *Int. Math. Res. Not.*, Issue **9** (2008).

Heinzner, P. and Kutzschebauch, F. An equivariant version of Grauert's Oka principle, *Invent. Math.* **119**, 317–346 (1995).

Helgason, S. *Differential Geometry, Lie Groups, and Symmetric Spaces*. Academic Press, New York, 1978.

Hilton, P.J. and Stammbach, U. *A. Course in Homological Algebra*, 2nd edn. Graduate Texts in Mathematics, vol. 4. Springer, New York, 1997.

Hitchin, N. Flat connections and geometric quantization, *Commun. Math. Phys.* **131**, 347–380 (1990).

Hochschild, G. and Serre, J-P. Cohomology of group extensions, *Trans. Amer. Math. Soc.* **74**, 110–134 (1953).

Hong, J. and Kumar, S. Conformal blocks for Galois covers of algebraic curves, ArXiv: 1807.00118 v3 (2019).

Hotta, R., Takeuchi, K. and Tanisaki, T. *D-Modules, Perverse Sheaves, and Representation Theory*. Progress in Mathematics, vol. 236. Birkhäuser, Basel, 2008.

Huang, Y.-Z. Vertex operator algebras and the Verlinde conjecture, *Comm. Contemp. Math.* **10**, 103–154 (2008).

Humphreys, J.E. *Introduction to Lie Algebras and Representation Theory*. Graduate Texts in Mathematics, vol. 9. Springer, New York, 1972.

Humphreys, J.E. *Conjugacy Classes in Semisimple Algebraic Groups*. American Mathematical Society, Providence, RI, 1995.

Hurtubise, J.C. Holomorphic maps of a Riemann surface into a flag manifold, *J. Diff. Geom.* **43**, 99–118 (1996).

Illusie, L. Grothendieck's existence theorem in formal geometry, in: *Fundamental Algebraic Geometry- Grothendieck's FGA Explained* (edited by B. Fantechi et al.), Mathematical Surveys and Monographs, vol. 123. American Mathematical Society, Providence, RI, 181–233 (2005).

Iwahori, N. and Matsumoto, H. On some Bruhat decomposition and the structure of the Hecke rings of $\mathfrak{p}$–adic Chevalley groups, *Publ. Math. IHES* **25**, 5–48 (1965).

Jantzen, J.C. *Representations of Algebraic Groups*, 2nd edn. American Mathematical Society, Providence, RI, 2003.

Jeffrey, L. and Kirwan, F. Intersection theory on moduli spaces of holomorphic bundles of arbitrary rank on a Riemann surface, *Ann. of Math.* **148**, 101–196 (1998).

Jones, G.A. and Singerman, D. *Complex Functions*. Cambridge University Press, Cambridge, 1987.

Kac, V.G. Highest weight representations of infinite-dimensional Lie algebras, in: *Proceeding of ICM*, Helsinki, 299–304 (1978).

Kac, V.G. *Infinite Dimensional Lie Algebras*, 3rd edn. Cambridge University Press, Cambridge, 1990.

Kac, V.G. and Peterson, D.H. Spin and wedge representations of infinite-dimensional Lie algebras and groups, *Proc. Nat. Acad. Sci. U.S.A.* **78**, 3308–3312 (1981).

Kac, V.G., Raina, A.K. and Rozhkovskaya, N. *Bombay Lectures on Highest Weight Representations of Infinite Dimensional Lie Algebras*, 2nd edn. Advanced Series in Mathematical Physics, vol. 29. World Scientific, Singapore, 2013.

Kac, V.G. and Sanielevici, M.N. Decompositions of representations of exceptional affine algebras with respect to conformal subalgebras, *Phys. Rev. D* **37**, 2231–2237 (1988)

Kac, V.G. and Wakimoto, M. Modular and conformal invariance constraints in representation theory of affine algebras, *Adv. Math.* **70**, 156–236 (1988).

Kazhdan, D. and Lusztig, G. Schubert varieties and Poincaré duality, *Proc. Symp. Pure Math. (A.M.S.)* **36**, 185–203 (1980).

Kazhdan, D. and Lusztig, G. Tensor structures arising from affine Lie algebras. I, *J. Am. Math. Soc.* **6**, 905–947 (1993).

Kempf, G. The Grothendieck–Cousin complex of an induced representation, *Advances in Math.* **29**, 310–396 (1978).

Kirwan, F. The cohomology rings of moduli spaces of bundles over Riemann surfaces, *J. Am. Math. Soc.* **5**, 853–906 (1992).

Knapp, A.W. *Lie Groups Beyond an Introduction*, 2nd edn. Progress in Mathematics, vol. 140, Birkhäuser, Basel, 2002.

Knudsen, F. The projectivity of the moduli space of stable curves, II: The stacks $M_{g,n}$, *Math. Scand.* **52**, 161–199 (1983a).

Knudsen, F. The projectivity of the moduli space of stable curves, III: The line bundles on $M_{g,n}$, and a proof of the projectivity of $\bar{M}_{g,n}$ in charactristic 0, *Math. Scand.* **52**, 200–212 (1983b).

Knudsen, F. and Mumford, D. The projectivity of the moduli space of stable curves I: preliminaries on "det" and "div", *Math. Scand.* **39**, 19–55 (1976).

Kobayashi, S. and Nomizu, K. *Foundations of Differential Geometry*, vol. II. Wiley-Interscience, New York 1969.

Kodaira, K. and Spencer, D.C. On deformations of complex analytic structures, I, *Annals of Math.* **67**, 328–401 (1958a).

Kodaira, K. and Spencer, D.C. A theorem of completeness for complex analytic fiber spaces, *Acta Math.* **100**, 281–294 (1958b).

Kollár, J. *Rational Curves on Algebraic Varieties*. Ergebnisse der Mathematik und ihrer Grenzgebiete, vol. 32. Springer-Verlag, Berlin, 1996.

Kontsevich, M. and Manin, Yu. Gromov–Witten classes, quantum cohomology, and enumerative geometry, *Commun. Math. Phys.* **164**, 525–562 (1994).

Kostant, B. Powers of the Euler product and commutative subalgebras of a complex simple Lie algebra, *Invent Math.* **158**, 181–226 (2004).

Koszul, J.L. *Lectures on Fibre Bundles and Differential Geometry*. Tata Institute of Fundamental Research Lecture Notes, Bombay, 1960.

Kraft, H. Algebraic automorphisms of affine space, in: *Proceedings of the Hyderabad Conference on Algebraic Groups* (edited by S. Ramanan), Manoj Prakashan, 251–274 (1991).

Krepski, D. and Meinrenken, E. On the Verlinde formulas for SO(3)-bundles, *Quarterly J. Math.* **64**, 235–252 (2013).

Kresch, A. *Algebraic Stacks* http://home.kias.re.kr/MKG/h/AlgebraicStacks/ (2007)

Kumar, S. Rational homotopy theory of flag varieties associated to Kac–Moody groups, in: *Infinite-dimensional Groups with Applications*. MSRI Publication 4. Springer-Verlag, Berlin, 233–273 (1985).

Kumar, S. Demazure character formula in arbitrary Kac–Moody setting, *Invent. Math.* **89**, 395–423 (1987).

Kumar, S. Infinite Grassmannians and moduli spaces of G–bundles, in: *Vector Bundles on Curves - New Directions*. Lecture Notes in Mathematics, vol. 1649. Springer-Verlag, Berlin, 1–49, 1997a.

Kumar, S. Fusion product of positive level representations and Lie algebra homology, in: *Geometry and Physics* (edited by J. E. Andersen et al.). Lecture Notes in Pure and Applied Mathematics, vol. 184. Marcel Dekker, New York, 253–259, 1997b.

Kumar, S. *Kac–Moody Groups, their Flag Varieties and Representation Theory*. Progress in Mathematics, vol. 204. Birkhäuser, Basel, 2002.

Kumar, S. and Narasimhan, M.S. Picard group of the moduli spaces of G-bundles, *Math. Ann.* **308**, 155–173 (1997).

Kumar, S., Narasimhan, M.S. and Ramanathan, A. Infinite Grassmannians and moduli spaces of G–bundles, *Math. Ann.* **300**, 41–75 (1994).

Kuniba, A. and Nakanishi, T. Level-rank duality in fusion RSOS models, *Modern quantum field theory, Bombay Quantum Field Theory*, 344–374 (1991).

Kuranishi, M. Two elements generations on semi-simple Lie groups, *Kôdai Math. Sem. Report 5–6*, 9–10 (1949).

Lang, S. *Algebra*. Addison-Wesley, New York, 1965.

Lang, S. *Introduction to Arakelov Theory*. Springer, New York, 1988.

Laszlo, Y. A propos de l'espace des modules de fibrés de rang 2 sur une courbe, *Math. Ann.* **299**, 597–608 (1994).

Laszlo, Y. Local structure of the moduli space of vector bundles over curves, *Comm. Math. Helv.* **71**, 373–401 (1996).

Laszlo, Y. Linearization of group stack actions and the Picard group of the moduli of SL_r/μ_s-bundles on a curve, *Bull. Soc. Math. France* **125**, 529–545 (1997).

Laszlo, Y. Hitchin's and WZW connections are the same, *J. Diff. Geom.* **49**, 547–576 (1998a).

Laszlo, Y. About G-bundles over elliptic curves, *Ann. Inst. Fourier, Grenoble* **48**, 413–424 (1998b).

Laszlo, Y. and Sorger, C. The line bundles on the moduli of parabolic G-bundles over curves and their sections, *Ann. Scient. Éc. Norm. Sup.* **30**, 499–525 (1997).

Laumon, G. and Moret-Bailly, L. *Champs Algébriques*. Springer-Verlag, New York, 1999.

Le Potier, J. *Fibrés Vectoriels sur les Courbes Algébriques*. Cours de DEA, Université Paris 7, 1991.

Le Potier, J. *Lectures on Vector Bundles*. Cambridge University Press, Cambridge, 1997.

Looijenga, E. Root systems and elliptic curves, *Invent. Math.* **38**, 17–32 (1976).

Looijenga, E. Conformal blocks revisited, ArXiv:math/0507086 (2005).

Looijenga, E. The KZ system via polydifferentials, *Adv. Stud. in Pure Math.* **62**, 189–231 (2012).

Looijenga, E. From WZW models to modular functors, in: *Handbook of Moduli*, vol. II. Advanced Lectures in Mathematics, vol. 25. International Press, Boston, MA, pp. 427–466 (2013).

Marian, A. and Oprea, D. The level-rank duality for non-abelian theta functions, *Invent. Math.* **168**, 225-247 (2007).

Marian, A. and Oprea, D. Sheaves on abelian surfaces and strange duality, *Math. Ann.* **343**, 1-33 (2009).

Marian, A. and Oprea, D. On the strange duality conjecture for abelian surfaces, *J. Euro. Math. Soc.* **16**, 1221–1252 (2014).

Marian, A., Oprea, D., Pandharipande, R., Pixton, A. and Zvonkine, D. The Chern character of the Verlinde bundle over $\overline{\mathcal{M}}_{g,n}$, *J. Reine Angew. Math.* **732**, 147–163 (2017).

Mathieu, O. Formules de caractères pour les algèbres de Kac–Moody générales, *Astérisque* **159–160**, 1–267 (1988).

Matsumura, H. *Commutative Ring Theory*. Cambridge University Press, Cambridge, 1989.

Maunder, C.R.F. *Algebraic Topology*. Van Nostrand Reinhold Company, London, 1970.

Mehta, V. B. and Ramadas, T.R. Moduli of vector bundles, Frobenius splitting, and invariant theory, *Ann. of Math.* **144**, 269–313 (1996).

Mehta, V. B. and Ramanathan, A. Restriction of stable sheaves and representations of the fundamental group, *Invent. Math.* **77**, 163–172 (1984).

Mehta, V. B. and Seshadri, C. S. Moduli of vector bundles on curves with parabolic structures, *Math. Ann.* **248**, 205–239 (1980).

Mehta, V.B. and Subramanian, S. On the Harder–Narasimhan filtration of principal bundles, in: *Algebra, Arithmetic and Geometry Part* 1 (edited by R. Parimala). Narosa Publishing House, New Delhi, pp. 405–415 (2002).

Meinrenken, E. and Woodward, C. Hamiltonian loop group actions and Verlinde factorization, *J. Diff. Geom.* **50**, 417–469 (1998).

Milne, J.S. Chap. VII – Jacobian varieties, in: *Arithmetic Geometry* (edited by G. Cornell et al.). Springer-Verlag, Berlin, pp. 167–212 (1986).

Milne, J.S. *Lectures on Étale Cohomology*. Online version 2.21, www.jmilne.org/math/CourseNotes/LEC.pdf (2013).

Milnor, J. *Morse Theory*. Annals of Mathematics Studies, vol. 51. Princeton University Press, Princeton, NJ, 1969.

Milnor, J.W. and Stasheff, J.D. *Characteristic Classes*. Annals of Mathematics Studies, vol. 76. Princeton University Press, Princeton, NJ, 1974.

Mitchell, S.A. Quillen's theorem on buildings and the loops on a symmetric space, *L'Enseignement Math.* **34**, 123–166 (1988).

Mlawer, E.J., Naculich, S.G., Riggs, H.A. and Schnitzer, H.J. Group-level duality of WZW fusion coefficients and Chern–Simons link observables, *Nuclear Phys. B* **352**, 863–896 (1991).

Moore, G. and Seiberg, N. Polynomial equations for rational conformal field theories, *Phys. Lett. B* **212**, 451–460 (1988).

Moore, G. and Seiberg, N. Classical and quantum conformal field theory, *Commun. Math. Phys.* **123**, 177–254 (1989).

Mukhopadhyay, S. Diagram automorphisms and rank-level duality, ArXiv: 1308.1756 (2013).

Mukhopadhyay, S. Strange duality of Verlinde spaces for G_2 and F_4, *Math. Z.* **283**, 387–399 (2016a).

Mukhopadhyay, S. Rank-level duality of conformal blocks for odd orthogonal Lie algebras in genus 0, *Trans. Amer. Math. Soc.* **368**, 6741–6778 (2016b).

Mukhopadhyay, S. Rank-level duality and conformal block divisors, *Adv. Math.* **287**, 389–411 (2016c).

Mukhopadhyay, S. and Wentworth, R. Generalized theta functions, strange duality, and odd orthogonal bundles on curves, *Commun. Math. Phys.* **370**, 325–376 (2019).

Mukhopadhyay, S. and Zelaci, H. Conformal embedding and twisted theta functions at level one, *Proc. Amer. Math. Soc.* **148**, 9–22 (2020).

Mumford, D. *Abelian Varieties*, 2nd edn. Tata Institute of Fundamental Research, Bombay. Oxford University Press, Oxford, 1985.

Mumford, D. *The Red Book of Varieties and Schemes*. Lecture Notes in Mathematics, vol. 1358. Springer-Verlag, Berlin, 1988.

Mumford, D., Fogarty, J. and Kirwan, F. *Geometric Invariant Theory*, 3rd enlarged edn. Ergebnisse der Mathematik und ihrer Grenzgebiete, vol. 34. Springer-Verlag, Berlin, 2002.

Mumford, D. and Oda, T. *Algebraic Geometry II*. Texts and Readings in Mathematics, vol. 73. Hindustan Book Agency, Haryana, 2015.

Naculich, S.G. and Schnitzer, H.J. Duality relations between $SU(N)_k$ and $SU(k)_N$ WZW models and their braid matrices, *Phys. Lett. B*, **244**, 235–240 (1990).

Nakanishi, T. and Tsuchiya, A. Level-rank duality of WZW models in conformal field theory, *Commun. Math. Phys.* **144**, 351–372 (1992).

Narasimhan, M.S. and Ramadas T.R. Factorisation of generalised theta functions. I, *Invent. Math.* **114**, 565–623 (1993).

Narasimhan, M.S. and Ramanan, S. Moduli of vector bundles on a compact Riemann surface, *Ann. of Math.* **89**, 14–51 (1969).

Narasimhan, M.S. and Seshadri, C.S. Holomorphic vector bundles on a compact Riemann surface, *Math. Ann.* **155**, 69–80 (1964).

Narasimhan, M.S. and Seshadri, C.S. Stable and unitary vector bundles on a compact Riemann surface, *Ann. of Math.* **82**, 540–567 (1965).

Newstead, P.E. *Introduction to Moduli Problems and Orbit Spaces*. Narosa Publishing House, New Delhi, 2012.

Nitsure, N. *Theory of Descent and Algebraic Stacks*. KIAS, Seoul, 2005a.

Nitsure, N. Construction of Hilbert and Quot schemes, in: *Fundamental Algebraic Geometry- Grothendieck's FGA Explained* (edited by B. Fantechi et al.), Mathematical Surveys and Monographs, vol. 123. American Mathematical Society, Providence, RI, pp. 107–137 (2005b).

Nitsure, N. Deformation theory for vector bundles, in: *Moduli Spaces and Vector Bundles* (edited by L. Brambila-Paz et al.), London Mathematical Society Lecture Note Series **359**, 128–164 (2009).

Olsson, M. Hom-stacks and restriction of scalars, *Duke Math. J.* **134**, 139–164 (2006).

Oort, F. Algebraic group schemes in characteristic zero are reduced, *Invent. Math.* **2**, 79–80 (1966).

Oprea, D. The Verlinde bundles and the semihomogeneous Wirtinger duality, *J. Reine Angew. Math.* **654**, 181–217 (2011).

Oudompheng, R. Rank-level duality for conformal blocks of the linear group, *J. Alg. Geom.* **20**, 559–597 (2011).

Oxbury, W.M. and Wilson, S.M.J. Reciprocity laws in the Verlinde formulae for the classical groups, *Trans. Amer. Math. Soc.* **348**, 2689–2710 (1996).

Pantev, T. Comparison of generalized theta functions, *Duke Math. J.* **76**, 509–539 (1994).

Pappas, G. and Rapoport, M. Some questions about $\mathscr{G}$-bundles on curves, in: *Algebraic and Arithmetic Structures of Moduli Spaces*. Advanced Studies in Pure Mathematics, vol. 58. Cambridge University Press, Cambridge, pp. 159–171 (2010).

Parthasarathi, P. On parabolic bundles on algebraic surfaces, *J. Ramanujan Math. Soc.* **28**, 379–413 (2013).

Pauly, C. Espaces de modules de fibrés praboliques et blocs conformes, *Duke Math. J.* **84**, 217–235 (1996).

Pauly, C. La dualité étrange [d'après P. Belkale, A. Marian et D. Oprea] *Astérisque* **326**, Exp. No. 994, 363–377 (2009).

Pauly, C. Strange duality revisited, *Math. Res. Lett.* **21**, 1353–1366 (2014).

Pauly, C. and Ramanan, S. A duality for spin Verlinde spaces and Prym theta functions, *J. London. Math. Soc.* **63**, 513–532 (2001).

Popov, V.L. and Vinberg, E.B. *Invariant Theory, Algebraic Geometry IV*. Encyclopaedia of Mathematical Sciences. vol. 55. Springer, New York, pp. 123–278 (1994).

Pressley, A. and Segal, G. *Loop Groups*. Clarendon Press, Oxford, 1988.

Quillen, D. Determinants of Cauchy–Riemann operators over a Riemann surface, *Funct. Anal. Appl.* **19**, 31–34 (1985).

Raghunathan, M. S. Principal bundles on affine space, in: *C.P. Ramanujam– A Tribute*. Springer-Verlag, Berlin, pp. 223–244 (1978).

Ramanan, S. A note on C.P. Ramanujam, in: *C.P. Ramanujam – A Tribute*. Springer-Verlag, Berlin, pp. 13–15 (1978).

Ramanan, S. and Ramanathan, A. Some remarks on the instability flag, *Tôhoku Math. J.* **36**, 269–291 (1984).

Ramanathan, A. Stable principal bundles on a compact Riemann surface, *Math. Ann.* **213**, 129–152 (1975).
Ramanathan, A. *Moduli of principal bundles*, Lecture Notes in Mathematics, vol. 732, 527–533. Springer, Berlin, 1979.
Ramanathan, A. Deformations of principal bundles on the projective line, *Invent. Math.* **71**, 165–191 (1983).
Ramanathan, A. Moduli for principal bundles over algebraic curves: I and II, *Proc. Indian Acad. Soc. (Math. Sci.)* **106**, 301–328 and 421–449 (1996).
Ramanathan, A. and Subramanian, S. Einstein–Hermitian connections on principal bundles and stability, *J. Reine Angew. Math.* **390**, 21–31 (1988).
Remmert, R. Holomorphe und meromorphe abbildungen komplexer räume, *Math. Ann.* **133**, 328–370 (1957).
Rudin, W. *Real and Complex Analysis*. McGraw-Hill, New York, 1966.
Safarevic, I.R. On some infinite–dimensional groups. II, *Math. USSR Izvestija* **18**, 185–194 (1982).
Schellekens, A.N. and Warner, N.P. Conformal subalgebras of Kac–Moody algebras, *Phys. Rev. D* **34**, 3092–3096 (1986).
Schieder, S. The Harder–Narasimhan stratification of the moduli stack of G-bundles via Drinfeld's compactifications, *Selecta Math.* **21**, 763–831 (2015).
Selberg, A. On discontinuous groups in higher-dimensional symmetric spaces, in: *Internat. Colloquium on Function Theory, Bombay, 1960*. Tata Institute of Fundamental Research, Bombay, pp. 147–164 (1960).
Serre, J.-P. Géométrie algébrique et géométrie analytique, *Ann. Inst. Fourier, Grenoble* **6**, 1–42 (1956).
Serre, J.-P. Espaces fibrés algébriques, in: *Anneaux de Chow et applications, Séminaire C. Chevalley*, 1958.
Serre, J.-P. *Algèbres de Lie Semi-simples Complexes*. W.A. Benjamin, Inc., New York, 1966.
Serre, J.-P. *Topics in Galois Theory*. Jones and Bartlett Publishers, Boston, MA, 1992.
Serre, J.-P. *Galois Cohomology*. Springer Monographs in Mathematics, Springer, New York (1997).
Seshadri, C.S. Space of unitary vector bundles on a compact Riemann surface, *Annals of Math.* **85**, 303–336 (1967).
Seshadri, C.S. Moduli of vector bundles on curves with parabolic structures, *Bull. Amer. Math. Soc.* **83**, 124–126 (1977).
Seshadri, C.S. *Vector Bundles on Curves*. Contemporary Mathematics, vol. 153. American Mathematical Society, Providence, RI (1993).
Seshadri, C.S. Moduli of π-vector bundles over an algebraic curve, in: *Collected Papers of C.S. Seshadri*, vol. 250. Hindustan Book Agency, 2011.
Shatz, S. The decomposition and specialization of algebraic families of vector bundles, *Compositio Math.* **35**, 163–187 (1977).
Sheinman, O.K. *Current Algebras on Riemann Surfaces (New results and applications)*. De Gruyter Expositions in Mathematics, vol. 58. De Gruyter, Berlin, 2012.
Simpson, C. Moduli of representations of the fundamental group of a smooth projective variety II, *Publ. Math. I.H.E.S.* **80**, 5–79 (1994).
Simpson, C. *Algebraic (Geometric) n-Stacks*, ArXiv.alg-geom/9609014 (1996).

Slodowy, P. On the geometry of Schubert varieties attached to Kac–Moody Lie algebras, in: *Can. Math. Soc. Conf. Proc. on 'Algebraic Geometry' (Vancouver)*, vol. **6**, 405–442 (1984).

Sorger, C. La formule de Verlinde, *Séminaire Bourbaki*, 47ème année, n^o **794** (1994).

Sorger, C. La semi-caractéristique d'Euler–Poincaré des faisceaux ω-quadratiques sur un schéma de Cohen-Macaulay, *Bull. Soc. Math. France* **122**, 225–233 (1994).

Sorger, C. On moduli of G-bundles on a curve for exceptional G, *Ann. Scient. Éc. Norm. Sup.* **32**, 127–133 (1999).

Sorger, C. Lectures on moduli of principal G-bundles over algebraic curves, School on Algebraic Geometry, Trieste, pp. 1–57 (1999).

Spanier, E.H. *Algebraic Topology*. McGraw-Hill, New York, 1966.

Stacks. *The Stacks project*, Online version, https://stacks.math.columbia.edu/ (2019).

Steenrod, N. *The Topology of Fibre Bundles*. Princeton University Press, Princeton, NJ, 1951.

Steinberg, R. *Lectures on Chevalley Groups*. University Lecture Series, vol. 66. American Mathematical Society, Providence, RI, 2016.

Sun, X. Degeneration of moduli spaces and generalized theta functions, *J. Alg. Geom.* **9**, 459–527 (2000).

Sun, X. Factorization of generalized theta functions in the reducible case, *Ark. Mat.* **41**, 165–202 (2003).

Sun, X. and Tsai, I-H. Hitchin's connection and differential operators with values in the determinant bundle, *J. Diff. Geom.* **66**, 303–343 (2004).

Sun, X. and Zhou, M. *Globally F-regular type of moduli spaces and Verlinde formula*, ArXiv: 1802.08392 (2018).

Suzuki, T. A new proof of the finite-dimensionality of the space of conformal blocks, Preprint (1990).

Szenes, A. Hilbert polynomials of moduli spaces of rank 2 vector bundles I, *Topology* **32**, 587-597 (1993).

Szenes, A. The combinatorics of the Verlinde formula, in: *Vector Bundles in Algebraic Geometry* (edited by N. J. Hitchin et al.). London Mathematical Society Lecture Note Series, vol. 208. Cambridge University Press, Cambridge, pp. 241– 254 (1995).

Teleman, C. Lie algebra cohomology and the fusion rules, *Commun. Math. Phys.* **173**, 265–311 (1995).

Teleman, C. Verlinde factorization and Lie algebra cohomology, *Invent. Math.* **126**, 249–263 (1996).

Teleman, C. Borel–Weil–Bott theory on the moduli stack of G-bundles over a curve, *Invent. Math.* **134**, 1–57 (1998).

Teleman, C. and Woodward, C. Parabolic bundles, products of conjugacy classes and Gromov–Witten invariants, *Ann. Inst. Fourier (Grenoble)* **51**, 713–748 (2001)

Thaddeus, M. Stable pairs, linear systems and the Verlinde formula, *Invent. Math.* **117**, 317-353 (1994).

Tsuchimoto, Y. On the coordinate-free description of the conformal blocks, *J. Math. Kyoto Univ.* **33**, 29–49 (1993).

Tsuchiya, A. and Kanie, Y. Vertex operators in Conformal field theory on $\mathbb{P}^1$ and monodromy representations of braid group, *Adv. Stud. Pure Math.* **16**, 297–372 (1988).

Tsuchiya, A., Ueno, K. and Yamada, Y. Conformal field theory on universal family of stable curves with gauge symmetries, *Adv. Stud. Pure Math.* **19**, 459–566 (1989).

Tu, L.W. Semistable bundles over an elliptic curve, *Adv. in Math.* **98**, 1-26 (1993).

Ueno, K. *Conformal Field Theory with Gauge Symmetry*. Fields Institute Monographs. American Mathematical Society, Providence, RI, 2008.

Verlinde, E. Fusion rules and modular transformations in 2D conformal field theory, *Nucl. Phys. B* **300**, 360–376 (1988).

Vistoli, A. Grothendieck topologies, fibered categories and descent theory, in: *Fundamental Algebraic Geometry– Grothendieck's FGA Explained* (edited by B. Fantechi et al.). Mathematical Surveys and Monographs, vol. 123. American Mathematical Society, Providence, RI, pp. 1–104 (2005).

Wang, J. The moduli stack of G-bundles, arXiv:1104.4828 [math.AG] (2011).

Waterhouse, W.C. *Introduction to Affine Group Schemes*. Graduate Texts in Mathematics, vol. 66. Springer-Verlag, Berlin, 1979.

Weil, A. Généralisation des fonctions abéliennes, *J. Math. Pures Appl.* **17**, 47–87 (1938).

Witten, E. On quantum gauge theories in two dimensions, *Commun. Math. Phys.* **141**, 153–209 (1991).

Zagier, D. On the cohomology of moduli spaces of rank two vector bundles over curves, in: *The Moduli Space of Curves*. Progress in Mathematics, vol. 129. Birkhäuser, Basel, pp. 533–563 (1995).

Zagier, D. Elementary aspects of the Verlinde formula and of the Harder–Narasimhan–Atiyah–Bott formula, *Israel Math. Conf. Proc.*, vol. **9**, Bar-Ilan Univ., pp. 445–462 (1996).

Zhu, X. An introduction to affine Grassmannians and the geometric Satake equivalence, in: *Geometry of Moduli Spaces and Representation Theory*. IAS/Park City Mathematics Series, vol. 24. American Mathematical Society, Providence, RI, pp. 59–154 (2017).

Index

CPSIA information can be obtained
at www.ICGtesting.com
Printed in the USA
LVHW040514301221
707429LV00001B/23